T0185755

To Pili and Eladia, our wives

Electrical Properties of Polymers

Evaristo Riande and
Ricardo Diaz-Calleja

CRC Press
Taylor & Francis Group
Boca Raton London New York

CRC Press is an imprint of the
Taylor & Francis Group, an **informa** business

First published 2004 by Marcel Dekker

Published 2020 by CRC Press
Taylor & Francis Group
6000 Broken Sound Parkway NW, Suite 300
Boca Raton, FL 33487-2742

First issued in paperback 2020

ISBN 13: 978-0-367-57835-0 (pbk)
ISBN 13: 978-0-8247-5346-7 (hbk)

Visit the Taylor & Francis Web site at
http://www.taylorandfrancis.com

and the CRC Press Web site at
http://www.crcpress.com

Library of Congress Cataloging-in-Publication Data
A catalog record for this book is available from the Library of Congress.

Preface

Polymers differ from simple crystalline solids and simple liquids in that they have length and molecular scales larger than atomic, a characteristic that gives them unusual properties. The underlying structure of a polymer is a long chain in which one or more chemical motives repeat along the chain. Most of the skeletal bonds of polymers are of a type that can rotate, giving rise to an unimaginably large number of spatial conformations. As a result, statistical considerations must enter into the description of even the simplest molecular chain. Moreover, the macromolecular nature of polymers vastly broadens the time scale for molecular adjustments to external force fields. The macromolecular size of polymers makes them suitable to develop materials that may combine great elasticity with great toughness, fluidity with a solid-like structure, etc. As a result, polymers are ubiquitous in both nature and industry.

A relevant characteristic of polymers is their ability to withstand high electric fields with negligible conduction due to the large energy differences between the localized valence electronic states and the conduction band. This characteristic coupled with favorable mechanical and processing properties make polymers the obvious choice for insulating applications. However, the versatility of polymers may expand their window of use to include the most unexpected applications. Thus the coupling between the electric properties of some polymers and their mechanical and thermal properties has led to important transducer applications. Moreover, in the last decades increased scientific and technological attention has been given to the development of polymeric electric conductors.

v

This book is mainly concerned with the response of polymers to electric fields. We have approached this subject in such a way that the book meets the requirements of the beginner in the study of the electric properties of polymers as well as those of experienced workers in other type of materials. The book is divided into three parts: Part I deals with the physical fundamentals of dielectrics, Part II with the relation between structure and equilibrium and dynamic dielectric properties, and Part III with the electric response of special polymers to force fields.

An understanding of the response of polymers to electric fields requires knowledge of the basic physical properties of dielectrics, and how these properties are affected by molecular size is offered in Chapters 1 and 2. Chapter 1 deals with the interactions between dipoles and static electric fields and the description of theories that relate the static dielectric permittivity with the polarity of low-molecular-weight amorphous compounds or monomers. These theories are extended to high-molecular-weight chains and inter- and intramolecular dipolar correlations that depend on the molecular structure are considered.

Most processes in nature are stochastic and their description in this discussion requires the use of probability and averages. In Chapter 2, both the Langevin and Schmulokowsky equations that describe the probability distribution in stochastic processes are introduced and further used in the description of dielectric relaxations using the Debye and Onsager models. Attention is paid to the relation between statistical mechanics and linear dielectric responses for systems with nonpolarizable dipoles, and is further extended to polarizable ones. Attempts are also made to relate dielectric and mechanical properties of polymers.

The time rate of change of a polarization vector in a dielectric isotropic system is studied in Chapter 3, using extended irreversible thermodynamics. This is a novel approach not often found in studies of dielectrics. It is shown how conservation equations in conjunction with the entropy production equation make it possible to obtain expressions that in principle could describe dielectric relaxation processes, even in the cases in which the dipoles have one component parallel and the other perpendicular to the chain contour. In contrast with classical irreversible thermodynamics, where the equations are parabolic, the present approach, based on extended irreversible thermodynamics, leads to hyperbolic equations with finite speed for the propagation of electrical signals.

Taking advantage of the fact that the linear responses in the frequency and time domains of systems to a step function field are related through Fourier transforms, experimental devices have been designed that allow determination of the dielectric behavior of polymers over a frequency/time window of about 12 decades. Chapter 4 is focused on the description of these instruments as well as on the underlying physics. Empirical equations that allow the analysis of the dielectric results are described in detail. Special attention is devoted to the behavior of electrets as monitored by thermal stimulated discharge currents.

Owing to the flexibility of molecular chains, including those containing rigid segments in their structure, the square of the dipole moment of a polymer is the average over the square of the dipole moments associated with the large number of conformations of the system. Statistical mechanics methods are described in Chapter 5 that allow the computation of the mean-square dipole moments of polymers by assuming that the skeletal bonds are restricted to a limited number of rotational states. The use of this analysis to obtain both mean-square dipole moments in terms of the chemical structure and the conformational energies associated with rotational states is emphasized.

Among the conformational properties most sensitive to chemical structure, the electric birefringence expressed in terms of the molar Kerr constant stands out. Chapter 6 deals with the experimental measurements of the electric birefringence of polymer solutions and the development of mathematical expressions obtained by statistical mechanical procedures that relate the Kerr constant with the averages of the polarizability tensors associated with the conformations of the chains. The procedure for assigning the polarizability tensor of groups of bonds to each skeletal bond of the chains is illustrated.

Chapter 7 deals with the use of molecular dynamics to compute the trajectory of the dipole moments of molecules in the conformational space. The fundamentals of molecular dynamics techniques are given in detail, emphasizing how the time dipole correlation functions obtained from the trajectories of monomers and low-molecular-weight polymers can be used to compute their mean-square dipole moments and their relaxation spectra in the frequency domain.

Dielectric behavior is an excellent diagnostic property in that it reflects molecular structure and motions. The wide frequency window available in this technique makes it possible to obtain isotherms displaying the glass rubber and the secondary relaxations of polymer melts in the frequency domain. Chapters 8 and 9 discuss how the chemical structure of polymers may affect their relaxation spectra. The relaxation spectra of a few polymers are included and theories interpreting short- and long-range motions are presented in these two chapters.

A step electric field applied to a polymer solution induces a birefringence in the system that increases with time as the molecules rotate, reaching a constant value at equilibrium. Removal of the electric field decreases the birefringence to zero as the Brownian motions randomize the orientations of the molecules. Chapter 10 deals with the study of the buildup and decay functions and how these functions are related to the rotational relaxation times of molecular chains.

Liquid crystals are characterized by the molecules being free to move as in a liquid, although as they do so they tend to spend a little more time pointing along the direction of orientation than along some other direction. Chapter 11 studies the microscopic and macroscopic order parameters of mesophases and their relation with the permittivity. Theories developed for isotropic systems are modified to account for the equilibrium and dielectric relaxation behavior of diverse mesophases. Ferroelectricity in liquid crystals is also discussed.

For certain polymers an intrinsic polarization can be induced by these effects of stress or temperature. These intrinsic piezoelectric and pyroelectric materials frequently obtain their anisotropic polarization through some structural rearrangement involving either crystal packing or dipole alignment of macro-poles. This is the subject of Chapter 12, where the relationships between the polarization vector and the stress tensor in piezoelectrics polymers as well as between temperature and polarization in pyroelectrics are studied. Polymer structures that can develop ferroelectric, pyroelectric, and piezoelectric properties are discussed.

Polymers containing certain chromophore groups in their structure as well as ferroelectric materials are being considered promising candidates for future nonlinear optical (NLO) applications, such as frequency doublers, optical storage devices, electrooptic uses and modulators. Their advantage over traditional inorganic materials such as $LiNbO_3$ basically lies in their high laser damage threshold and their ease of processing and architectural modification. Chapter 13 gives an overview of the physical fundaments of nonlinear optics and second-harmonic generation in polymers, emphasizing the physics underlying the relations between second-order susceptibility and hyperpolarizability, poling decay, etc. Attention is paid to the guidelines that allow the design of polymeric systems containing chromophore groups with good NLO properties.

The synthesis of conductors or semiconductors that retain the desirable polymeric attributes of moldability, flexibility, and toughness is a subject of great importance from a basic and applied point of view. Chapter 14 describes double bond conjugated polymers that conveniently doped could produce good electronic conduction. Semiempirical quantum mechanics methods useful for the computation of the energy gaps between the valence and conduction bands are discussed. The conduction mechanisms in the doped conducting polymers and the nature of the conducting species in the doped polymers are studied. Attention is also paid to the use of these materials in electroluminiscence, batteries, electromagnetic interference shielding, anticorrosion, etc.

This book brings together the coverage of different electrical phenomena in polymers and of how both chemical and the supermolecular structures may affect them. The book is not intended to be an overall review of electric phenomena but rather a description of the fundamentals of these phenomena in relation to the structure of polymers. Some chapters, especially the basic ones, include problem sets that we hope will facilitate the understanding of the subjects discussed in the book, especially for readers who are not familiar with the dielectric behavior of polymeric materials. Some of these problems deal with important aspects of the theory not fully developed in the main text. This book can also be used as a textbook in undergraduate and graduate courses of materials science.

Evaristo Riande
Ricardo Díaz-Calleja

Table of Contents

Part II. Structure Dependence of the Equilibrium and Dynamic Dielectric Properties of Polymers

1

Static Dipoles

1.1. DIPOLES

Owing to the electrical neutrality of matter, dipoles or multipoles are basic elements of the electric structure of many materials. A dipole can be defined as an entity made up by a positive charge q separated a relatively short distance l from an equal negative charge. The dipole moment is a vectorial quantity defined as[1]

$$\mu = q\mathbf{l} \qquad\qquad (1.1.1)$$

The direction of a dipole moment lies along the line connecting the two charges, and, by convection, its positive direction points towards the positively charged end. To find the dipole moment of a molecule containing a few atoms is a relatively simple matter. In more complicated polyatomic molecules it is necessary to take into account not only the geometric structure of the molecule but also the interactions with other surrounding molecules. The determination of the dipole moments of polymers requires a mathematical description of the spatial conformation of the chains, which are mainly random in character. The rotational isomeric state (RIS) model proposed by Volkenstein[2] and Flory[3] proved to be suitable to calculate dipole moments of polymeric systems as a function of the chemical structure. In this case, the mean-square dipole moment is the convenient magnitude to be calculated (see Chapters 2 and 5)

1

1.2. ELECTRIC POTENTIALS ARISING FROM AN ISOLATED DIPOLE

Let us consider a dipole along the z axis with the charges $+q$ and $-q$ placed at distances $\pm d/2$ from the origin (Fig. 1.1). The electrostatic potential $\Phi(r, \theta)$ at a generic point P is given by

$$\Phi(r, \theta) = \frac{q}{r_+} - \frac{q}{r_-} \tag{1.2.1}$$

where r is the radial coordinate, θ is the polar angle, and r_+ and r_- are the distances from the charges to point P (Fig. 1.2). The cosine theorem indicates that

$$\frac{1}{r_\pm} = \frac{1}{r}\left[1 \mp \frac{d}{r}\cos\theta + \left(\frac{d}{2r}\right)^2\right]^{-1/2} \tag{1.2.2}$$

By using the series expansion

$$(1+x)^n = 1 + nx + \frac{n(n-1)}{2!}x^2 + \cdots \tag{1.2.3}$$

the combination of Eqs (1.2.1) and (1.2.2) leads to*

$$\Phi(r, \theta) = \frac{qd}{r^2}\left[P_1(\cos\theta) + \frac{1}{4}\left(\frac{d}{r}\right)^2 P_3(\cos\theta) + \cdots\right] \tag{1.2.4}$$

where P_i are the Legendre polynomials (Appendix A), given by

$$P_0 = 1, \qquad P_1 = \cos\theta, \qquad P_2 = (1/2)(3\cos^2\theta - 1),$$
$$P_3 = (1/2)(5\cos^3\theta - 3\cos\theta), \ldots \tag{1.2.5}$$

Note that $qd\cos\theta = q\mathbf{d} \cdot \mathbf{r}/r$ (where $\mathbf{r}/r = \mathbf{u}$), and the quantity $\boldsymbol{\mu} = q\mathbf{d}$ is called the dipolar moment. The remaining terms in Eq. (1.2.4) are octopolar and successive moments. The high-order terms are of odd order, contrary to the expansion of the electric potential of a single point charge, which includes odd as well as even terms. For a collection of non-interacting dipoles, the total dipole moment can be found by geometric addition of elementary dipole moments.

The unit of dipole moment in the SI system is coulomb $\times$ m (Cm), although the Debye, D, which is the dipole moment corresponding to two electronic charges separated by 0.1 nm, is a commonly used unit.

---

*In agreement with the current literature on dielectrics, the Gaussian unrationalized system of units is used in Chapters 1, 2, and 3. In Appendix 1.D, a glossary of formulae in SI units is given.

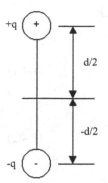

FIG. 1.1   Schematic diagram of a dipole

1.3.   POINT DIPOLE

In the expression for the potential of points close to a dipole, terms depending on high powers of r^{-1} must be included. However, for points located away from the dipole, only the lowest-order terms of the multipole expansion need to be considered. In this way we note that the dipole potential varies as r^{-2}, unlike the potential of a point charge, which varies as r^{-1}. For many purposes it is convenient to introduce a fictitious entity called the point dipole. This is a charge distribution for which the dipole potential is given by the first term of the right-hand side of Eq. (1.2.4) for all the values of r, no matter how close they are to the point dipole. Obviously, this is a contradictory concept in the same way as the point mass, for example.

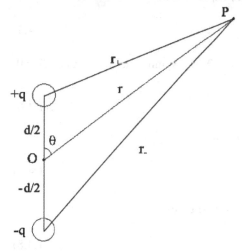

FIG. 1.2   Diagram to calculate the electric field produced by a dipole

1.4. FIELD OF AN ISOLATED POINT DIPOLE

The dipole potential can alternatively be written as

$$\Phi = \frac{\mu \cos \theta}{r^2} = \frac{\boldsymbol{\mu} \cdot \mathbf{u}}{r^2} = \frac{\boldsymbol{\mu} \cdot \mathbf{r}}{r^3} = -\boldsymbol{\mu} \cdot \operatorname{grad} \frac{1}{r} \qquad (1.4.1)$$

The electric field derived from Eq. (1.4.1) is given by

$$\mathbf{E} = -\nabla \Phi = -\nabla \frac{\boldsymbol{\mu} \cdot \mathbf{r}}{r^3} \qquad (1.4.2)$$

After some calculations, one obtains

$$\mathbf{E} = \left(\frac{3(\boldsymbol{\mu} \cdot \mathbf{r})\mathbf{r}}{r^5} - \frac{\boldsymbol{\mu}}{r^3} \right) \qquad (1.4.3)$$

The field can be expressed in matrix form as follows

$$\begin{vmatrix} E_x \\ E_y \\ E_z \end{vmatrix} = \begin{vmatrix} \dfrac{3x^2 - r^2}{r^5} & \dfrac{3xy}{r^5} & \dfrac{3xz}{r^5} \\ \dfrac{3yx}{r^5} & \dfrac{3y^2 - r^2}{r^5} & \dfrac{3yz}{r^5} \\ \dfrac{3zx}{r^5} & \dfrac{3zy}{r^5} & \dfrac{3z^2 - r^2}{r^5} \end{vmatrix} \begin{vmatrix} \mu_x \\ \mu_y \\ \mu_z \end{vmatrix} \qquad (1.4.4)$$

This matrix is the tensor dipole field $\mathbf{T}$, expressed as

$$\mathbf{T} = \frac{1}{r^5} (3\mathbf{rr} - r^2 \mathbf{I}) \qquad (1.4.5)$$

where $\mathbf{I}$ is the unit tensor. Note that $\mathbf{T}$ is symmetric and *trace* $\mathbf{T} = 0$. Consequently

$$\mathbf{E} = \mathbf{T}\boldsymbol{\mu} \qquad (1.4.6)$$

It follows from Eqs (1.4.1) and (1.4.2) that

$$\mathbf{E} = \nabla \left(\boldsymbol{\mu} \cdot \nabla \left(\frac{1}{r} \right) \right) = \left(\nabla \nabla \frac{1}{r} \right) \boldsymbol{\mu} \qquad (1.4.7)$$

This equation in conjunction with Eq. (1.4.6) gives

$$\mathbf{T} = \nabla^2 \left(\frac{1}{r} \right) \qquad (1.4.8)$$

1.5. FORCE EXERTED ON A DIPOLE BY AN EXTERNAL ELECTRIC FIELD

Let us consider a dipole located in an external electric field $\mathbf{E}_0$ where the subscript indicates that the applied field includes no contribution from the dipole itself. The total force acting on the dipole is the sum of the forces acting on each charge, that is

$$\mathbf{F} = q_+ \mathbf{E}_0(\mathbf{r} + \mathbf{d}) + q_- \mathbf{E}_0(\mathbf{r}) = q[\mathbf{E}_0(\mathbf{r} + \mathbf{d}) - \mathbf{E}_0(\mathbf{r})] \tag{1.5.1}$$

If $\mathbf{d}$ as defined in Fig. 1.1 is very small in comparison with $\mathbf{r}$, for example if the external electric field is produced by charges placed far away from the dipole, the first term on the right-hand side of Eq. (1.5.1) can be expanded into a Taylor series around $\mathbf{r}$, giving

$$\mathbf{E}_0(\mathbf{r} + \mathbf{d}) = \mathbf{E}_0(\mathbf{r}) + \mathbf{d}\nabla\mathbf{E}_0(\mathbf{r}) + \cdots \tag{1.5.2}$$

Then, by substituting Eq. (1.5.2) into Eq. (1.5.1) and neglecting terms of order d^2 and higher, we obtain

$$\mathbf{F} = q\mathbf{d}\nabla\mathbf{E}_0 = \boldsymbol{\mu}\nabla\mathbf{E}_0 = \nabla(\boldsymbol{\mu} \cdot \mathbf{E}_0) \tag{1.5.3}$$

The last equality in Eq. (1.5.3) is a consequence of (1) the constancy of $\boldsymbol{\mu}$, (2) the well known electrostatic formula $\nabla \times \mathbf{E}_0 = 0$, and (3) from the vectorial expression

$$\nabla(\boldsymbol{\mu} \cdot \mathbf{E}_0) = (\boldsymbol{\mu} \cdot \nabla)\mathbf{E}_0 + (\mathbf{E}_0 \cdot \nabla)\boldsymbol{\mu} + \boldsymbol{\mu} \times (\nabla \times \mathbf{E}_0)$$
$$+ \mathbf{E}_0 \times (\nabla \times \boldsymbol{\mu}) \tag{1.5.4}$$

The electrical force can also be derived from the potential electrical energy U, noting that

$$\mathbf{F} = -\nabla U \tag{1.5.5}$$

For a single charge q, the ratio of Eq. (1.5.5) to q gives $\mathbf{E}_0 = -\nabla\Phi$. By comparing Eqs (1.5.3) and (1.5.5), the potential for a single dipole is given by

$$U = -(\boldsymbol{\mu} \cdot \mathbf{E}_0) \tag{1.5.6}$$

Equation (1.5.6) suggests that the potential energy reaches a minimum when $\boldsymbol{\mu}$ and $\mathbf{E}_0$ are perpendicular and a maximum when they are parallel.

For an arbitrary external electric field, Eq. (1.5.3) can be written in terms of the components as

$$F_{0x} = \mu_x \frac{\partial E_{0x}}{\partial x} + \mu_y \frac{\partial E_{0x}}{\partial y} + \mu_z \frac{\partial E_{0x}}{\partial z}$$

$$F_{0y} = \mu_x \frac{\partial E_{0y}}{\partial x} + \mu_y \frac{\partial E_{0y}}{\partial y} + \mu_z \frac{\partial E_{0y}}{\partial z} \qquad (1.5.7)$$

$$F_{0z} = \mu_x \frac{\partial E_{0z}}{\partial x} + \mu_y \frac{\partial E_{0z}}{\partial y} + \mu_z \frac{\partial E_{0z}}{\partial z}$$

Obviously, the field forces are null for a uniform electric field and the dipoles do not translate.

1.6. DIPOLE–DIPOLE INTERACTION

The energy of a system formed by two dipoles μ_1 and μ_2 is the work done on one of these dipoles placed in the field of the other dipole. According to Eq. (1.4.3), the energy is

$$U_i = -\mu_2 \cdot \mathbf{E}(\mu_1) = -\mu_1 \cdot \mathbf{E}(\mu_2)$$

$$= \frac{1}{r^3}\left((\mu_1 \cdot \mu_2) - \frac{3}{r^2}(\mu_1 \cdot \mathbf{r})(\mu_2 \cdot \mathbf{r}) \right) \qquad (1.6.1)$$

In tensorial notation

$$U_i = \mu_1 \mathbf{T} \mu_2 \qquad (1.6.2)$$

where $\mathbf{T}$ is given by Eq. (1.4.5). Therefore, the force exerted on one of these dipoles owing to the field produced by the other is

$$
\begin{aligned}
\mathbf{F}_i = -\nabla U_i &= -\left(\nabla \frac{\mu_1 \cdot \mu_2}{r^3} - 3\nabla \frac{(\mu_1 \cdot \mathbf{r})(\mu_2 \cdot \mathbf{r})}{r^5} \right) \\
&= -\left[(\mu_1 \cdot \mu_2)\nabla\left(\frac{1}{r^3}\right) - 3\frac{(\mu_1 \cdot \mathbf{r})}{r^5}\nabla(\mu_2 \cdot \mathbf{r}) \right. \\
&\quad \left. - 3\frac{(\mu_2 \cdot \mathbf{r})}{r^5}\nabla(\mu_1 \cdot \mathbf{r}) - 3(\mu_1 \cdot \mathbf{r})(\mu_2 \cdot \mathbf{r})\nabla\left(\frac{1}{r^5}\right) \right] \\
&= \frac{3}{r^5}\left[(\mu_1 \cdot \mu_2)\mathbf{r} + (\mu_1 \cdot \mathbf{r})\mu_2 + (\mu_2 \cdot \mathbf{r})\mu_1 - \frac{5}{r^2}(\mu_1 \cdot \mathbf{r})(\mu_2 \cdot \mathbf{r})\mathbf{r} \right]
\end{aligned}
$$

$$(1.6.3)$$

Several special cases will be considered.

1. If $\mu_1 = \mu_2 = \mu$, the mutual force will be

$$\mathbf{F}_i = \frac{3}{r^5}\left[\mu^2\mathbf{r} + 2(\mu \cdot \mathbf{r})\mu - \frac{5}{r^2}(\mu \cdot \mathbf{r})^2\mathbf{r}\right] \qquad (1.6.4)$$

2. For parallel dipoles ($\mu_1 = \mu_2$) perpendicular to $\mathbf{r}$

$$\mathbf{F}_i = \frac{3\mu^2}{r^5}\mathbf{r} \qquad (1.6.5)$$

3. For antiparallel dipoles ($\mu_1 = -\mu_2$) perpendicular to $\mathbf{r}$

$$\mathbf{F}_i = -\frac{3\mu^2}{r^5}\mathbf{r} \qquad (1.6.6)$$

4. Finally, for perpendicular dipoles that are both also perpendicular to $\mathbf{r}$

$$\mathbf{F}_i = 0$$

1.7. TORQUES ON DIPOLES

The torque exerted by an electric field on a dipole can be obtained from the contributions of each charge. According to the notations used in Figs 1.1 and 1.2, the total torque is

$$\Gamma = \frac{\mathbf{d}}{2} \times q\mathbf{E}_0(\mathbf{d} + \mathbf{r}) + \left(-\frac{\mathbf{d}}{2}\right) \times (-q)\mathbf{E}_0(\mathbf{r}) \qquad (1.7.1)$$

Neglecting terms of higher order and considering the definition of dipole moment, we obtain

$$\Gamma = \mathbf{d} \times q\mathbf{E}_0 = \mu \times \mathbf{E}_0 \qquad (1.7.2)$$

Obviously, the torque is null when μ and $\mathbf{E}_0$ are parallel. The assumption made on passing from Eq. (1.7.1) to Eq. (1.7.2) is equivalent to considering uniformity in the electric field or a point dipole. In the vicinity of an infinite dipole, the high-order terms in Eq. (1.7.1) must be taken into account.

As a consequence of the torque, the dipole tends to reorient along the field, even in the presence of a uniform field. This is a consequence of the not cancellation of the torque of two opposite noncollinear forces. The work of reorientation from a position perpendicular to the field to another forming an angle θ is given by

$$W = \int_{\pi/2}^{\theta} |\boldsymbol{\Gamma}| \, d\theta' = \int_{\pi/2}^{\theta} |\boldsymbol{\mu}||\mathbf{E}_0| \sin \theta' \, d\theta' = -|\boldsymbol{\mu}||\mathbf{E}_0| \cos \theta \qquad (1.7.3)$$

which coincides with Eq. (1.5.6).

1.8. DIPOLE MOMENT AND DIELECTRIC PERMITTIVITY. MOLECULAR VERSUS MACROSCOPIC PICTURE

In the preceding sections we have only considered isolated dipoles. However, in real situations we deal with a continuous medium having an enormous number of elementary dipoles. Dipolar substances contain permanent dipoles arising from the asymmetrical location of electric charges in the matter. Under an electric field, these dipoles tend to orient as a result of the action of torques, as we have seen in the previous section. The macroscopic result is the orientation polarization of the material. However, the effect of the applied electric field is also to induce a new type of polarization by distortion of the electronic clouds (electronic polarization) and the nucleus (atomic polarization) of the atomic structure of the matter. Owing to the low mass of the entities intervening in the distortion polarization, this type of polarization takes a time of the order of 10^{-10}–10^{-12} s to reach equilibrium; this means that it is practically instantaneous. Under an applied field of frequency 10^{10}–10^{12} Hz, the orientation polarization scarcely contributes to the total polarization. However, this contribution increases when the frequency diminishes (time increases), and this is the essential feature of the dielectric dispersion.

For a static case, that is, for zero-frequency polarizability, the field $\mathbf{E}$ acting on a dipole is, in general, different from the applied field $\mathbf{E}_0$. The average moment for an isotropic material can be written as

$$\mathbf{m} = \alpha \mathbf{E} \qquad (1.8.1)$$

where α is the total polarizability which includes the electronic, α_e, the atomic, α_a, and the orientational, α_o, contributions. Equation (1.8.1) is strictly true only if $\mathbf{E}$ is small. This means that there are no saturation effects and the relationship between the induced moment and the electric field is of the linear type. The total polarization of N_1 dipolar molecules per unit volume, each of them having an average moment $\mathbf{m}$, is

$$\mathbf{P} = N_1 \mathbf{m} = \frac{N}{V} \mathbf{m} \qquad (1.8.2)$$

where N is the total number of molecules. On the other hand, the average field caused by the induced dipoles of the dielectric placed between two parallel plates (see problem 4) is given by

$$\langle E_z \rangle = \frac{1}{V} \int_{\perp} E_z(\mathbf{r}) \, d^3\mathbf{r} = -\frac{4\pi Nm}{V} = -4\pi P \tag{1.8.3}$$

The mean electric field $\langle E_z \rangle$ must be added to the field arising from the charge densities in the plates. This field is given by

$$\mathbf{E}_0 + 4\pi\mathbf{P} = \mathbf{D} \tag{1.8.4}$$

where $\mathbf{D}$ is the electric displacement.

The dielectric susceptibility is defined as

$$\chi = \frac{\varepsilon - 1}{4\pi} = \frac{\mathbf{P}}{\mathbf{E}_0} \tag{1.8.5}$$

As a consequence, the permittivity is given by

$$\varepsilon = 1 + 4\pi\chi = \frac{\mathbf{D}}{\mathbf{E}_0} \tag{1.8.6}$$

This equation can also be written as

$$\mathbf{D} = \varepsilon\mathbf{E}_0 = e_0\varepsilon_r\mathbf{E}_0 \tag{1.8.7a}$$

where

$$\varepsilon_r = \frac{\varepsilon}{e_0} \tag{1.8.7b}$$

is the relative permittivity and e_0 ($= 8.854 \times 10^{-12}\,\mathrm{C^2 kg^{-1}\,m^{-3}\,s^2}$) is the permittivity of the free space. In cgs units, $e_0 = 1$ and $\varepsilon_r = \varepsilon$. Note that, in the present context, D and E_0 are aligned, and thus ε is a scalar. This corresponds to the isotropic case. When there exists anisotropy in the dielectric, ε is a second order tensor. In this case

$$D_i = \varepsilon_{ij}E_{j0} \tag{1.8.8}$$

The work involved in the orientation of dipoles by the effect of an electric field is given by

$$dW = \frac{1}{4\pi}\mathbf{E}_0 \cdot d\mathbf{D} = \frac{1}{4\pi}E_{i0}\,dD_i = \frac{1}{4\pi}\varepsilon_{ij}E_{i0}\,dE_{j0} \tag{1.8.9}$$

Hence

$$4\pi \frac{\partial^2 W}{\partial E_{i0}\, \partial E_{j0}} = \varepsilon_{ij} = 4\pi \frac{\partial^2 W}{\partial E_{j0}\, \partial E_{i0}} = \varepsilon_{ji} \tag{1.8.10}$$

This expression indicates that the permittivity is a symmetric tensor.

1.9. LOCAL FIELD. THE DEBYE STATIC THEORY OF DIELECTRIC PERMITTIVITY

Although in the development of the equations described in the preceding section the dipolar structure of the dielectric was considered, the electric field acting on each dipole that produces the induced moment appearing in Eq. (1.8.3) was not determined. The evaluation of the static dielectric permittivity in terms of the actual dipole moment μ of the molecules of the dielectric requires the determination of the local field acting upon the molecule, and the ratio of the dipole moment μ of the molecule to either the polarizability α_0 or to the average moment m.

The concept of internal field has received considerable attention because it relates macroscopic properties to molecular ones. The problem involved in the determination of the field acting on a single dipole in the dielectric is its dependence on the polarization of the neighboring molecules. The first approach to the analysis of this problem is due to Lorentz. The basic idea is to consider a spherical zone containing the dipole under study immersed in the dielectric. The sphere is small in comparison with the dimensions of the condenser, but large compared with the molecular dimensions. We treat the properties of the sphere at the microscopic level as containing many molecules, but the material outside the sphere is considered a continuum. The field acting at the center of the sphere where the dipole is placed arises from the field due to (1) the charges on the condenser plates, (2) the polarization charges on the spherical surface, and (3) the molecular dipoles inside the spherical region. This latter field cannot be evaluated unless the distribution of charges inside the sphere is known. For the special case with the molecular dipoles on a cubic lattice, this field is zero.

The field due to the polarization charges on the spherical surface, E_{sp}, can be calculated by considering an element of the spherical surface defined by the angles θ and $\theta + d\theta$. The area of this elementary surface is $2\pi r^2 \sin\theta\, d\theta$. The density of charges on this element is given by $P\cos\theta$, and the angle between this polarization and the elementary surface is θ. Integrating over all the values of the angle formed by the direction of the field with the normal vector to the spherical surface at each point, and dividing by the surface of the sphere, we obtain

$$E_{sp} = \frac{1}{r^2} \int_0^\pi 2\pi r^2 P \sin\theta \cos^2\theta\, d\theta = \frac{4\pi P}{3} \tag{1.9.1}$$

From Eq. (1.8.6), P is given by

$$P = \frac{1}{4\pi}(\varepsilon - 1)E_0 \tag{1.9.2}$$

Therefore, the total field will be

$$E_i = E_{sp} + E_0 = E_0 + \frac{\varepsilon - 1}{3}E_0 = \frac{\varepsilon + 2}{3}E_0 \tag{1.9.3}$$

This is the Lorentz local field.

By combining Eqs (1.9.2), (1.9.3), and (1.8.1), the following equation relating the permittivity and the total polarizability is obtained

$$\frac{\varepsilon - 1}{\varepsilon + 2} = \frac{4\pi N_1 \alpha}{3} \tag{1.9.4}$$

Here, the assumption that $\mathbf{m} = \boldsymbol{\mu}$ (dilute systems) is made. If the material has molecular weight M and density ρ, then $N_1 = \rho N_A/M$, where N_A is Avogadro's number. In this case, Eq. (1.9.4) becomes

$$\frac{\varepsilon - 1}{\varepsilon + 2} = \frac{4\pi N_A \rho \alpha}{3M} \tag{1.9.5}$$

This expression is the Clausius–Mossotti equation valid for nonpolar gases at low pressure. As will be shown later, this expression is also valid for the high-frequency limit.

The remaining problem to be solved is the calculation of the dipolar contribution to the polarizability. The first solution to this problem for the static case was proposed by Debye.[4] In the absence of structure, the potential energy of a dipole of moment μ is given by

$$U = -\mu E_i \cos\theta \tag{1.9.6}$$

Note that the internal field $\mathbf{E}_i$ is used instead of the external applied field $\mathbf{E}_0$ because $\mathbf{E}_i$ is the field acting upon the dipole. If a Boltzmann distribution for the orientation of the dipole axis is assumed, the probability to find a dipole within an elementary solid angle $d\Omega$ is given by

$$\frac{\exp(\mu E_i \cos\theta/kT)}{\int \exp(\mu E_i \cos\theta/kT)\,d\Omega}\,d\Omega \tag{1.9.7}$$

The average value of the dipole moment in the direction of the field is

$$\mu\langle\cos\theta\rangle = \frac{\int_0^\pi \mu\cos\theta\exp(\mu E_i \cos\theta/kT)\sin\theta\,d\theta}{\int_0^\pi \exp(\mu E_i \cos\theta/kT)\sin\theta\,d\theta} \tag{1.9.8}$$

After integration by parts

$$\langle \cos \theta \rangle = \coth\left(\frac{\mu E_i}{kT}\right) - \frac{kT}{\mu E_i} = L\left(\frac{\mu E_i}{kT}\right) \tag{1.9.9}$$

where $L(y) = \coth y - y^{-1}$ denotes the Langevin function given by

$$L(y) = \frac{y}{3} - \frac{y^3}{45} + \frac{2y^5}{945} - \frac{y^7}{4725} + \cdots \tag{1.9.10}$$

For $y \ll 1$, the Langevin function is approximately $L(y) \cong y/3$, and the orientational polarization is given by

$$P_o = N_1 \mu \langle \cos \theta \rangle = \frac{N_1 \mu^2 E_i}{3kT} = N_1 \alpha E_i \tag{1.9.11}$$

where N_1 is the number of dipoles per unit volume. If the distortional polarizability α_d is added, the total polarization is

$$P = N_1 E_i \left(\alpha_d + \frac{\mu^2}{3kT} \right) \tag{1.9.12}$$

By substituting Eq. (1.9.3) into Eq. (1.9.12), and taking into account Eq. (1.9.2), one obtains

$$\frac{\varepsilon_0 - 1}{\varepsilon_0 + 2} = \frac{4}{3} \pi N_1 \left(\alpha_d + \frac{\mu^2}{3kT} \right) \tag{1.9.13}$$

This equation can alternatively be written (see Eqs 1.9.4 and 1.9.5) as

$$\frac{\varepsilon_0 - 1}{\varepsilon_0 + 2} \frac{M}{\rho} = \frac{4\pi}{3} N_A \left(\alpha_d + \frac{\mu^2}{3kT} \right) \tag{1.9.14}$$

This expression is the Debye equation for static permittivity and is a generalization of the Clausius–Mossotti equation [Eq. (1.9.5)]. The left-hand side of this equation is called the molar polarization. This latter parameter should be constant for a nonpolar substance, whereas extrapolation of the molar polarization against $1/T$ should give the distortional polarization for a polar substance. From the slope of that representation, the dipole moment can be found.

At very high frequencies, dipoles do not have time for reorientation and the dipolar contribution to permittivity is negligible. Then, according to Eq. (1.9.5)

$$\frac{\varepsilon_\infty - 1}{\varepsilon_\infty + 2} \frac{M}{\rho} = \frac{4\pi N_A \alpha_\infty}{3} \tag{1.9.15}$$

where we consider $\alpha_d = \alpha_\infty$. Since the infinite frequency permittivity is equal to the square of the refraction index, n, Eq. (1.9.15) becomes

$$\frac{n^2 - 1}{n^2 + 2} \frac{M}{\rho} = \frac{4\pi N_A \alpha_\infty}{3} \tag{1.9.16}$$

which is known as the Lorentz–Lorenz equation. Subtraction of Eq. (1.9.15) from Eq. (1.9.14) gives

$$\frac{\varepsilon_0 - 1}{\varepsilon_0 + 2} - \frac{\varepsilon_\infty - 1}{\varepsilon_\infty + 2} = \frac{\rho}{M} \frac{4\pi N_A \mu^2}{3kT} \tag{1.9.17}$$

It is worthy pointing out that Eq. (1.9.16) can be rewritten as

$$\frac{n^2 + 2}{n^2 - 1} C = v \tag{1.9.18}$$

where $v = \rho^{-1}$ is the specific volume and $C = (4\pi N_A \alpha_\infty)/(3M)$. By plotting $C(n^2 + 2)/(n^2 - 1)$ vs. T^{-1}, the isobaric dilatation coefficient can be obtained and, eventually, the glass transition temperatures could be estimated. Note that, from the temperature dependence of the index of refraction of glassy systems, physical aging and related phenomena could be analyzed.

1.10. DRAWBACK OF THE LORENTZ LOCAL FIELD

For highly polar substances, the dipolar contribution to the polarizability is clearly dominant over the distortional one. From Eqs (1.8.6) and (1.9.3), and making the substitution $\alpha_o = \mu^2/3kT$ [see Eq. (1.9.10)], the susceptibility becomes infinite at a temperature given by

$$\Theta = \frac{4\pi N_A \rho \mu^2}{9Mk} \tag{1.10.1}$$

This temperature Θ is called the Curie temperature. This equation suggests that, at temperature Θ, spontaneous polarization should occur and the material should become ferroelectric even in the absence of an electric field. Ferroelectricity is uncommon in nature, and predictions made by Eq. (1.10.1) are not experimentally supported. The failure of the theory arises from considering null the contributions to the local field of the dipoles in the cavity. This fact emphasizes the inadequacy of the Lorentz field [Eq. (1.9.3)] in a dipolar dielectric and leaves open the question of the internal field in the cavity.[5]

1.11. DIPOLE MOMENT OF A DIELECTRIC SPHERE IN A DIELECTRIC MEDIUM

The Clausius–Mossotti equation [Eq. (1.9.5)] represents a first attempt to relate a macroscopic quantity, the dielectric permittivity, to the microscopic polarizability of the substances. The equation obtained is only valid for the displacement polarization. To solve the same problem for the orientational polarizability is much more complicated. As we have seen in the preceding section, the Lorentz field fails in that it predicts spontaneous ferroelectricity.

The starting point to find an alternative solution to the problem of the local field in the context of a theory of the orientational polarization is to consider a spherical cavity of radius a and relative permittivity ε_1 that contains at the center a rigid dipolar molecule with permanent dipole μ. The radius of the cavity is obtained from the relation

$$\frac{4\pi a^3}{3} = \frac{V}{N} \tag{1.11.1}$$

where N is the number of molecules in a volume V. In many cases a can be considered as the "molecular radius." This cavity is assumed to be surrounded by a macroscopic spherical shell of radius $b \gg a$, and relative permittivity ε_2. The ensemble is embedded in a continuum medium of relative permittivity ε_3. In a first stage, the displacement polarization will be ignored. According to Onsager,[6] the internal field in the cavity has two components:

1. The cavity field, $\mathbf{G}$, that is, the field produced in the empty cavity by the external field.
2. The reaction field, $\mathbf{R}$, that is, the field produced in the cavity by the polarization induced by the surrounding dipoles.

A slightly more general case than that considered by Onsager in the determination of the internal field will be analyzed, because of its usefulness (see Problem 8 and the reference therein). In fact, dielectric or conducting shells are technologically important, and biological cells are also examples of layered spherical structures. The spherical concentric dielectric shell to be studied is schematically shown in Fig. 1.3. Following Fröhlich,[7] and in order to simplify calculations, the effects of the external electric field and the dipole into the cavity will be treated together because they lead to the same angular dependence of the potential ϕ.

The general solution of the Laplace equation ($\nabla^2\phi = 0$), in spherical coordinates and for the case of axial symmetry, is

$$\phi_i(r, \theta) = \sum_{n=0}^{\infty} (a_{in}r^n + b_{in}r^{-(n+1)})P_n(\cos \theta) \tag{1.11.2}$$

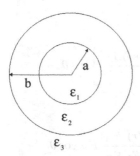

FIG. 1.3 Spherical concentric dielectric shell in a medium of permittivity ε_3 having a shell permittivity ε_2 and a core permittivity ε_1

where subscript i refers to each one of the three zones under consideration. For our particular geometry, the following boundary conditions hold

$$\phi_1 = \phi_2, \qquad \varepsilon_1 \frac{\partial \phi_1}{\partial r} = \varepsilon_2 \frac{\partial \phi_2}{\partial r}, \qquad r = a$$

$$\phi_2 = \phi_3, \qquad \varepsilon_2 \frac{\partial \phi_2}{\partial r} = \varepsilon_3 \frac{\partial \phi_3}{\partial r}, \qquad r = b$$

(1.11.3)

These conditions arise from the fact that the only charges existing in the cavity contributing to ϕ_1 belong to the dipole. Moreover, the contribution of the external field to ϕ_3 for $r \gg b$ is $E_0 r \cos \theta$. Owing to the boundary conditions, only the first term in the Lagrange polynomial (cos θ) appears in Eq. (1.11.2), so that the series of spherical harmonics is greatly simplified. Consequently, the potentials are

$$\phi_1 = \left(-Ar + \frac{\mu}{r^2}\right)\cos \theta, \qquad r < a$$

$$\phi_2 = \left(-Br + \frac{C}{r^2}\right)\cos \theta, \qquad a < r < b$$

(1.11.4)

$$\phi_3 = -\left(E_0 r + \frac{D}{r^2}\right)\cos \theta, \qquad b < r$$

Equations (1.11.4) together with the boundary conditions given by Eqs. (1.11.3), form a linear system of equations with four unknowns A, B, C, and D. After solving the system and substituting the results in Eq. (1.11.4), the resulting potentials are given by

$$\phi_1 = \frac{\mu}{r^2}\left[1 - \left(\frac{r}{a}\right)^3 \frac{R_{21} + (a/b)^3 R_{32}[(R_{21}/2) + (\varepsilon_1/\varepsilon_2)G_{21}]}{1 + (\frac{1}{2})(a/b)^3 R_{21}R_{32}}\right]\cos\theta$$

$$- \frac{G_{32}G_{21}}{1 + (\frac{1}{2})(a/b)^3 R_{21}R_{32}}E_0 r\cos\theta$$

$$\phi_2 = \frac{\mu\varepsilon_1}{\varepsilon_2 r^2}\frac{G_{21}[1 - (r/b)^3 R_{32}]}{1 + \frac{1}{2}(a/b)^3 R_{21}R_{32}}\cos\theta - \frac{G_{32}[1 + \frac{1}{2}(a/r)^3 R_{21}]}{1 + \frac{1}{2}(a/b)^3 R_{21}R_{32}}E_0 r\cos\theta$$

$$\phi_3 = \frac{\mu\varepsilon_1}{\varepsilon_2 r^2}\left[\frac{G_{21}(1 - R_{32})}{1 + \frac{1}{2}(a/b)^3 R_{21}R_{32}}\right]\cos\theta - \left\{1 - \left(\frac{b}{r}\right)^3\right.$$

$$\times \left[1 - \frac{G_{32}}{1 + \frac{1}{2}(a/b)^3 R_{21}R_{32}}\right] + \left(\frac{a}{r}\right)^3 \frac{G_{32}(R_{21}/2)}{1 + \frac{1}{2}(a/b)^3 R_{21}R_{32}}\right\}E_0 r\cos\theta$$

$$(1.11.5)$$

where

$$G_{ij} = \frac{3\varepsilon_i}{2\varepsilon_i + \varepsilon_j}, \qquad R_{ij} = \frac{2(\varepsilon_i - \varepsilon_j)}{2\varepsilon_i + \varepsilon_j}, \qquad G_{ij} - \frac{1}{2}R_{ij} = 1 \qquad (1.11.6)$$

For a $\ll$ b, Eqs (1.11.5) become

$$\phi_1 = \frac{\mu}{r^2}\left[1 - \left(\frac{r}{a}\right)^3 R_{21}\right]\cos\theta - G_{32}G_{21}E_0 r\cos\theta$$

$$\phi_2 = \frac{\varepsilon_1\mu}{\varepsilon_2 r^2}G_{21}\left[1 - \left(\frac{r}{b}\right)^3 R_{32}\right]\cos\theta - G_{32}\left[1 + \frac{1}{2}\left(\frac{a}{r}\right)^3 R_{21}\right]E_0 r\cos\theta$$

$$\phi_3 = \frac{\varepsilon_1\mu}{\varepsilon_2 r^2}G_{21}(1 - R_{32})\cos\theta - \left[1 - \left(\frac{b}{r}\right)^3 (1 - G_{32})\right]E_0 r\cos\theta$$

$$(1.11.7)$$

Next, the following two special cases are considered:

1. Empty cavity.
    In this case, $\varepsilon_1 = 1$, and all the terms in Eq. (1.11.7) containing μ as a factor vanish.

The resulting potentials are given by

$$\phi_1 = -\frac{9\varepsilon_3\varepsilon_2}{(2\varepsilon_3 + \varepsilon_2)(2\varepsilon_2 + 1)} E_0 r \cos\theta$$

$$\phi_2 = -\frac{3\varepsilon_3}{2\varepsilon_3 + \varepsilon_2}\left[1 + \left(\frac{a}{r}\right)^3 \frac{\varepsilon_2 - 1}{2\varepsilon_2 + 1}\right] E_0 r \cos\theta \qquad (1.11.8)$$

$$\phi_3 = -\left[1 + \left(\frac{b}{r}\right)^3 \frac{\varepsilon_3 - \varepsilon_2}{2\varepsilon_3 + \varepsilon_2}\right] E_0 r \cos\theta$$

2. Dipole point in the center of the empty cavity. Absence of external field. In this case, again $\varepsilon_1 = 1$, and the terms containing E_0 vanish. The potentials are given by

$$\phi_1 = \frac{\mu}{r^2}\left[1 - 2\left(\frac{r}{a}\right)^3 \frac{\varepsilon_2 - 1}{2\varepsilon_2 + 1}\right]\cos\theta$$

$$\phi_2 = \frac{\mu}{r^2}\frac{3}{2\varepsilon_2 + 1}\left[1 - 2\left(\frac{r}{b}\right)^3 \frac{\varepsilon_3 - \varepsilon_2}{2\varepsilon_3 + \varepsilon_2}\right]\cos\theta \qquad (1.11.9)$$

$$\phi_3 = \frac{\mu}{r^2}\frac{9\varepsilon_2}{(2\varepsilon_2 + 1)(2\varepsilon_3 + \varepsilon_2)}\cos\theta$$

With regard to these two situations, it is also useful to consider the inner cavity in a continuum, that is, without the external spherical shell. In this case, $\varepsilon_2 = \varepsilon_3$, and consequently $G_{32} = 1$ and $R_{32} = 0$. Thus, Eqs (1.11.8) and (1.11.9) become respectively

$$\phi_1 = -\frac{3\varepsilon_2}{2\varepsilon_2 + 1} E_0 r \cos\theta \qquad (1.11.8a)$$

$$\phi_2 = \phi_3 = -\left[1 + \left(\frac{a}{r}\right)^3 \frac{\varepsilon_2 - 1}{2\varepsilon_2 + 1}\right] E_0 r \cos\theta \qquad (1.11.8b)$$

and

$$\phi_1 = \frac{\mu}{r^2}\left[1 - \left(\frac{r}{a}\right)^3 \frac{2(\varepsilon_2 - 1)}{2\varepsilon_2 + 1}\right]\cos\theta \qquad (1.11.9a)$$

$$\phi_2 = \phi_3 = \frac{\mu}{r^2}\frac{3}{2\varepsilon_2 + 1}\cos\theta \qquad (1.11.9b)$$

These results indicate that the field in an empty cavity embedded in a dielectric medium with permittivity ε_2 is given by

$$\mathbf{G} = \frac{3\varepsilon_2}{2\varepsilon_2 + 1}\mathbf{E}_0 \tag{1.11.10}$$

On the other hand, the dipole induces on the surface of the spherical cavity (even in the absence of an external field) an electric field opposing that of the dipole himself, called the reaction field, given by

$$\mathbf{R} = \frac{1}{a^3}\frac{2(\varepsilon_2 - 1)}{2\varepsilon_2 + 1}\boldsymbol{\mu} = R\boldsymbol{\mu} \tag{1.11.11}$$

3.   Finally, for a concentric shell with $\varepsilon_1 = \varepsilon_3 = 1$, Eq. (1.11.7) leads to

$$\phi_1 = \frac{\mu}{r^2}\left[1 - 2\left(\frac{r}{a}\right)^3\frac{\varepsilon_2 - 1}{2\varepsilon_2 + 1}\right]\cos\theta - \frac{9\varepsilon_2}{(\varepsilon_2 + 2)(2\varepsilon_2 + 1)}E_0 r\cos\theta$$

$$\phi_2 = \frac{\mu}{r^2}\frac{3}{2\varepsilon_2 + 1}\left[1 + 2\left(\frac{r}{b}\right)^3\frac{\varepsilon_2 - 1}{\varepsilon_2 + 2}\right]\cos\theta$$

$$- \frac{3}{\varepsilon_2 + 2}\left[1 + \left(\frac{a}{r}\right)^3\frac{\varepsilon_2 - 1}{2\varepsilon_2 + 1}\right]E_0 r\cos\theta$$

$$\phi_3 = \frac{\mu}{r^2}\frac{9\varepsilon_2}{(2\varepsilon_2 + 1)(\varepsilon_2 + 2)}\cos\theta$$

$$- \left[1 - \left(\frac{b}{r}\right)^3\frac{\varepsilon_2 - 1}{\varepsilon_2 + 2} + \left(\frac{a}{r}\right)^3\frac{3(\varepsilon_2 - 1)}{(\varepsilon_2 + 2)(2\varepsilon_2 + 1)}\right]E_0 r\cos\theta$$

$$\tag{1.11.12}$$

1.12. ACTUAL DIPOLE MOMENTS, DEFINITION AND STATUS

The potential of a rigid nonpolarizable dipole in a medium of relative permittivity ε_1 is given by

$$\phi = \frac{\mu_e\cos\theta}{\varepsilon_1 r^2} \tag{1.12.1}$$

where subscript e denotes external.

Usually, dipoles are associated to molecules having an electronic cloud that shields them. Thus, let us consider a dipole μ in the center of a sphere representing the molecule. According to Eq. (1.9.14), the electronic polarizability of the sphere gives rise, at a macroscopic level, to an instantaneous permittivity ε_∞.

In this case, and according to Eq. (1.11.8a), the potential is given by

$$\phi = \frac{3\varepsilon_1}{2\varepsilon_1 + \varepsilon_\infty} \frac{\mu \cos\theta}{\varepsilon_1 r^2} = \frac{3}{2\varepsilon_1 + \varepsilon_\infty} \frac{\mu \cos\theta}{r^2} \tag{1.12.2}$$

By identifying the potentials in Eqs (1.12.1) and (1.12.2), one finds

$$\mu_e = \frac{3\varepsilon_1}{2\varepsilon_1 + \varepsilon_\infty} \mu \tag{1.12.3}$$

Therefore, μ_e is called the external moment of a molecule in a medium with relative permittivity ε_1. The dipole of the molecule in vacuum will be

$$\mu_v = \frac{3}{\varepsilon_\infty + 2} \mu \tag{1.12.4}$$

which is called the vacuum dipole moment. As a consequence

$$\mu_e = \frac{\varepsilon_1(\varepsilon_\infty + 2)}{2\varepsilon_1 + \varepsilon_\infty} \mu_v \tag{1.12.5}$$

However, the total moment of the molecule, also called the internal moment, is the (vector) sum of its vacuum value and the value induced in it by the reaction field, that is

$$\mu_i = \mu_v + \alpha R = \mu_v + \alpha \frac{1}{a^3} \frac{2(\varepsilon_1 - 1)}{2\varepsilon_1 + 1} \mu_i \tag{1.12.6}$$

which can be written as

$$\mu_i = \frac{\mu_v}{1 - [(2\alpha)/a^3][(\varepsilon_1 - 1)/(2\varepsilon_1 + 1)]} \tag{1.12.7}$$

If the polarizability α is known, Eqs (1.9.15) and (1.11.1) lead to

$$\alpha = a^3 \frac{\varepsilon_\infty - 1}{\varepsilon_\infty + 2} \tag{1.12.8}$$

Then, from Eqs (1.12.6) to (1.12.8) one obtains

$$\mu_i = \mu_v + \alpha R = \mu_v + \frac{2(\varepsilon_1 - 1)(\varepsilon_\infty - 1)}{3(2\varepsilon_1 + \varepsilon_\infty)} \mu_v$$

$$= \frac{(2\varepsilon_1 + 1)(\varepsilon_\infty + 2)}{3(2\varepsilon_1 + \varepsilon_\infty)} \mu_v \tag{1.12.9}$$

which is the expression for the internal dipole moment. An interesting point is that the internal dipole moment is independent of the size of the spherical specimen. Consequently, if a very large sphere containing a dipole with internal moment μ_i in an infinite dielectric medium of relative permittivity ε_1 is considered, the dipole moment of the large sphere is also μ_i. For $\varepsilon_\infty = \varepsilon_1$, the following expression holds

$$\mu_i = \frac{(2\varepsilon_1 + 1)(\varepsilon_1 + 2)}{9\varepsilon_1} \mu_v \tag{1.12.10}$$

Note that the factor appearing in Eq. (1.12.10) also appears in Eq. (1.11.12c). From Eqs (1.12.5) and (1.12.8) the internal moment can be written as

$$\mu_i = \frac{2\varepsilon_1 + 1}{3\varepsilon_1} \mu_e \tag{1.12.11}$$

The internal moment can also be calculated as the geometric sum of the external moment of the dipole and the moment of a sphere with the same permittivity as the medium surrounding the dipole. In these conditions

$$\mu_i = \mu_e + \int_{sphere} \mathbf{P}\, dv \tag{1.12.12}$$

Although this integral can be obtained by a simple electrostatic calculation (see problem 6), it can also be determined by comparing Eqs (1.12.11) and (1.12.12), that is

$$\int_{sphere} \mathbf{P}\, dv = -\frac{\varepsilon_1 - 1}{3\varepsilon_1} \mu_e = -\frac{\varepsilon_1 - 1}{2\varepsilon_1 + 1} \mu_i \tag{1.12.13}$$

The results of this section are summarized in Sch. I.

1.13. DIRECTING FIELD AND THE ONSAGER EQUATION

In the preceding section, the internal dipole moment has been calculated in the absence of an external electric field. When this force field is applied, Eq. (1.12.6) must be modified to take into account this effect. In fact the total field in the cavity is now the superposition of the cavity field $\mathbf{G}$ with the reaction field due to the dipole. Accordingly

$$\mu_i = \mu_v + \alpha\left(\frac{3\varepsilon_1}{2\varepsilon_1 + 1}\mathbf{E}_0 + \frac{1}{a^3}\frac{2(\varepsilon_1 - 1)}{2\varepsilon_1 + 1}\mu_i\right) \tag{1.13.1}$$

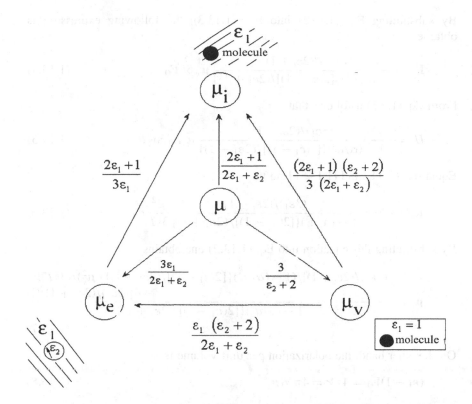

SCHEME 1.1   Relationship between different definitions of dipole moment as given in section 1.12

from which

$$\mu_i = \frac{\mu_v + \alpha[(3\varepsilon_1)/(2\varepsilon_1 + 1)]E_0}{1 - (\alpha/a^3)\{[2(\varepsilon_1 - 1)]/(2\varepsilon_1 + 1)\}} \tag{1.13.2}$$

where α is given by Eq. (1.12.8).

The mean value of the dipole moment parallel to the external field is calculated from the Boltzmann distribution for the orientational polarization given by Eqs (1.9.6) and (1.9.7). However, the energy of the dipole is calculated from the torque acting on the molecule, which according to Eq. (1.13.1) is given by

$$\Gamma = \mu_i \times \left(\frac{3\varepsilon_1}{2\varepsilon_1 + 1}E_0 + \frac{1}{a^3}\frac{2(\varepsilon_1 - 1)}{2\varepsilon_1 + 1}\mu_i \right) \tag{1.13.3}$$

By substituting Eq. (1.13.2) into Eq. (1.13.3), the following expression is obtained

$$\Gamma = \frac{(3\varepsilon_1)/(2\varepsilon_i + 1)}{1 - (\alpha/a^3)\{[2(\varepsilon_1 - 1)]/(2\varepsilon_1 + 1)\}} \boldsymbol{\mu}_v \times \mathbf{E}_0 \qquad (1.13.4)$$

From Eq. (1.7.3) it follows that

$$U = -\frac{(3\varepsilon_1)/(2\varepsilon_i + 1)}{1 - (\alpha/a^3)\{[2(\varepsilon_1 - 1)]/(2\varepsilon_1 + 1)\}} \mu_v E_0 \cos \theta \qquad (1.13.5)$$

Equations (1.9.8) and (1.13.5) lead to

$$\mu_v \langle \cos \theta \rangle = \frac{(3\varepsilon_1)/(2\varepsilon_i + 1)}{1 - (\alpha/a^3)\{[2(\varepsilon_1 - 1)]/(2\varepsilon_1 + 1)\}} \frac{\mu_v^2}{3kT} E_0 \qquad (1.13.6)$$

By substituting this equation into Eq. (1.13.2) one obtains

$$\mu_i = \frac{[(3\varepsilon_1)/(2\varepsilon_i + 1)]/(1 - (\alpha/a^3)\{[2(\varepsilon_1 - 1)]/(2\varepsilon_1 + 1)\})(\mu_v^2)/(3kT)\mathbf{E}_0 \atop +\alpha[(3\varepsilon_1)/(2\varepsilon_1 + 1)]\mathbf{E}_0}{1 - (\alpha/a^3)\{[2(\varepsilon_1 - 1)]/(2\varepsilon_1 + 1)\}}$$

$$(1.13.7)$$

On the other hand, the polarization per unit volume is

$$(\varepsilon_1 - 1)\mathbf{E}_0 = 4\pi\mathbf{P} = 4\pi N_1 \boldsymbol{\mu}_i \qquad (1.13.8)$$

Substitution of Eq. (1.13.7) into Eq. (1.13.8) gives

$$(\varepsilon_1 - 1)\mathbf{E}_0 = 4\pi\mathbf{P} = 4\pi N_1 \boldsymbol{\mu}_i$$

$$= 4\pi N_1 \left[\alpha + \frac{1}{1 - (\alpha/a^3)\{[2(\varepsilon_1 - 1)]/(2\varepsilon_1 + 1)\}} \frac{\mu_v^2}{3kT} \right]$$

$$\times \frac{(3\varepsilon_1)/(2\varepsilon_1 + 1)}{1 - (\alpha/a^3)\{[2(\varepsilon_1 - 1)]/(2\varepsilon_1 + 1)\}} \mathbf{E}_0 \qquad (1.13.9)$$

The quantity

$$\frac{[(3\varepsilon_1)/(2\varepsilon_1 + 1)]}{1 - (\alpha/a^3)\{[2(\varepsilon_1 - 1)]/(2\varepsilon_1 + 1)\}} \mathbf{E}_0 \qquad (1.13.10)$$

is called the directing field $\mathbf{E}_d$. The directing field is calculated as the sum of the cavity field and the reaction field caused by a fictive dipole $\alpha\mathbf{E}_d$.

The directing field must be not confused with the internal field, $\mathbf{E}_i$. The latter field can be calculated as the sum of the directing field and the reaction field caused by the polarizable point dipole averaged over all the possible orientations (see Problem 7). According to these results, Eq. (1.13.9) can be written as

$$(\varepsilon_1 - 1)\mathbf{E}_0 = 4\pi N_1 \left(\alpha \mathbf{E}_i + \frac{\mu_v^2}{3kT} \mathbf{E}_d \right) \tag{1.13.11}$$

Rearrangement of Eq. (1.13.9), taking into account Eq. (1.12.8), gives the Onsager expression

$$\frac{(\varepsilon_1 - \varepsilon_\infty)(2\varepsilon_1 + \varepsilon_\infty)}{\varepsilon_1(\varepsilon_\infty + 2)^2} = \frac{4\pi N_1 \mu_v^2}{9kT} \tag{1.13.12}$$

It is interesting to compare the results obtained from the Debye (Eq. 1.9.11) and Onsager theories. The Lorentz local field [Eq. (1.9.3)] changes Eq. (1.9.11) to

$$(\varepsilon_1 - 1)\mathbf{E}_0 = 4\pi \mathbf{P} = 4\pi N_1 \left(\alpha_d + \frac{\mu_v^2}{3kT} \right) \frac{\varepsilon_1 + 2}{3} \mathbf{E}_0 \tag{1.13.13}$$

Comparison of Eqs (1.13.13) and (1.13.11) leads to the conclusion that the Onsager theory takes for the internal and directing fields more accurate values than older theories in which the Lorentz field is used. Note that, after some rearrangements, the Onsager equation can be written as

$$\varepsilon_1 - 1 = 4\pi N_1 \left(\alpha_d + \frac{(2\varepsilon_1 + 1)(\varepsilon_\infty + 2)}{3(2\varepsilon_1 + \varepsilon_\infty)} \frac{\mu_v^2}{3kT} \right) \frac{\varepsilon_1(\varepsilon_\infty + 2)}{2\varepsilon_1 + \varepsilon_\infty} \tag{1.13.14}$$

In view of the coefficients appearing in Eq. (1.13.14), it is clear that the Onsager equation does not predict ferroelectricity as the Debye equation does. In fact, the cavity field tends to $3E_0/2$ when $\varepsilon_1 \to \infty$. Onsager takes into account the field inside the spherical cavity that is caused by molecular dipoles, which is neglected in the Debye theory. However, following the Boltzmann–Langevin methodology, the Onsager theory neglects dipole–dipole interactions, or, equivalently, local directional forces between molecular dipoles are ignored. The range of applicability of the Onsager equation is wider than that of the Debye equation. For example, it is useful to describe the dielectric behavior of noninteracting dipolar fluids, but in general this is not valid for condensed matter.

1.14. STATISTICAL THEORIES FOR STATIC DIELECTRIC PERMITTIVITY. KIRKWOOD'S THEORY

Onsager treatment of the cavity differs from Lorentz's because the cavity is assumed to be filled with a dielectric material having a macroscopic dielectric permittivity. Also, Onsager studies the dipolar reorientation polarizability on statistical grounds, as Debye does. However, the use of macroscopic arguments to analyze the dielectric problem in the cavity prevents the consideration of local effects which are important in condensed matter. This situation led Kirkwood first[8] and Fröhlich later on[9] to develop a fully statistical argument to determine the short-range dipole–dipole interactions. For methodological reasons, these two approaches will be considered separately.

Kirkwood considers a spherical specimen of volume V containing N molecules with dipole moment μ and subjected to an external field $\mathbf{E}_0$. The potential energy of this specimen is given by

$$U(X, E_0) = U(X, 0) - \mathbf{M}(X) \cdot \mathbf{E}_0 \tag{1.14.1}$$

where the argument X refers to the actual configuration of the specimen (positions and orientations of the dipoles contained in it), and $\mathbf{M}(X)$ is the corresponding moment of this configuration, such that $\mathbf{M} = \sum_{k=1}^{n} \mu_k$. The average moment of the jth molecule in the direction of the applied field is given by

$$\langle \mu_j \cdot \mathbf{e} \rangle = \frac{\int (\mu_j \cdot \mathbf{e}) \exp[-U(X, E_0)/kT] \, dX}{\int \exp[-U(X, E_0)/kT] \, dX} \tag{1.14.2}$$

where $\mathbf{e}$ is the unit vector in the direction of the field.

If the perturbation term in the energy is small, that is, if $\mathbf{M} \cdot \mathbf{E}_0/kT \ll 1$, one can write

$$\exp\left(\frac{\mathbf{M} \cdot \mathbf{E}_0}{kT}\right) \cong 1 + \left(\frac{\mathbf{M} \cdot \mathbf{E}_0}{kT}\right) \tag{1.14.3}$$

In the absence of an electric field, the average moment of the molecule is zero. Neglecting terms of high order in Eq. (1.14.3), the following expression is obtained

$$\langle \mu_j \cdot \mathbf{e} \rangle = \frac{\int (\mu_j \cdot \mathbf{e})[\mathbf{M}(X) \cdot \mathbf{E}_0/kT] \exp[-U(X,0)/kT] \, dX}{\int \exp[-U(X,0)/kT] \, dX}$$

$$= \frac{\langle (\mu_j \cdot \mathbf{e})(\mathbf{M}(X) \cdot \mathbf{E}_0) \rangle}{kT} \tag{1.14.4}$$

where the external field in the weighting factor $\exp[-U(X, 0)/kT]$ is nil. Since for an isotropic medium all the directions have the same probability, Eq. (1.14.4) becomes

$$\langle \mu_j \cdot \mathbf{e} \rangle = \frac{\langle \mu_j \cdot \mathbf{M}(X) \rangle_0}{3kT} E_0 \tag{1.14.5}$$

The average in Eq. (1.14.5) can be carried out in two stages, first over all the configurations of all molecules, except molecule j, and then over all positions and orientations of j. Then

$$\langle \boldsymbol{\mu}_j \cdot \mathbf{M} \rangle_0 = \langle \boldsymbol{\mu}_j \cdot \bar{\mathbf{M}}_j \rangle_0 = \langle \boldsymbol{\mu} \cdot \bar{\mathbf{M}} \rangle_0 \tag{1.14.6}$$

where the subscript has been omitted since the average will be the same for all the molecules. As pointed out by Kirkwood,[8] although the vectors $\boldsymbol{\mu}_j$ and $\bar{\mathbf{M}}_j$ vanish individually when averaged over all orientations of molecule j, this will not be true for their scalar product. On the other hand, it is more convenient to use instead of $\bar{\mathbf{M}}$ a dipole moment more closely related to the properties of a single molecule and its immediate environment. Since $\bar{\mathbf{M}}$ is the total dipole moment induced in the specimen when one molecule is fixed but the others can take all possible configurations, it is easy to show that $\bar{\mathbf{M}}$ is independent of the radius of the specimen (see problem 6). Moreover, $\bar{\mathbf{M}}$ arises from two sources: the induced moment produced by the fixed molecule in the homogeneous specimen and the effects of the molecules surrounding the fixed molecule. This latter contribution takes into account the local dipole–dipole interactions. Kirkwood separates these two contributions by assuming a sphere of radius r surrounding the fixed molecule. This sphere is large enough to include all the local order effects but small enough in comparison with the size of the whole specimen with radius R which will be regarded as a continuum. On this basis, Kirkwood defines

$$\bar{\boldsymbol{\mu}} = \lim_{r \to \infty, R/r \to \infty} \bar{\mathbf{M}}(R, r) \tag{1.14.7}$$

as the dipole moment of the smaller sphere accomplishing the aforementioned conditions; $\bar{\boldsymbol{\mu}}$ is the dipole moment of any finite spherical specimen polarized by a fixed molecule, when embedded in a medium of its own dielectric permittivity, whereas r is the distance beyond which the local permittivity is close to its macroscopic value.

As the sphere is screened by the external vacuum, a reaction field appears and the internal moment of the small sphere is related to $\bar{\mu}$ by [see Eq. (1.B.16) where $\varepsilon_2 = 1$]

$$\mu_i = \frac{2(\varepsilon_1 - 1)}{\varepsilon_1 + 2} \bar{\mu} \tag{1.14.8}$$

Then, the moment of the whole spherical specimen is given by

$$\bar{\mathbf{M}} = \bar{\mu} + \int_{v_0}^{v} \mathbf{P} \, dv \tag{1.14.9}$$

where the integration domain is the concentric shell between the spheres with radii r and R and volumes v_0 and v. This integration has been performed[8] by transforming the volume integration into surface integration using Green's theorem. On account of the relative dimensions of the spheres, the integral over the exterior surface dominates the integral over the interior surface. Thus, according to Eq. (1.12.12) (see also Problem 6), the integral in Eq. (1.14.9) can be written as

$$\int_{\text{outer sphere}} \mathbf{P}\, dv = -\frac{2(\varepsilon_1 - 1)^2}{(\varepsilon_1 + 2)(2\varepsilon_1 + 1)} \bar{\boldsymbol{\mu}} \tag{1.14.10}$$

where the relationship between the internal dipole moment $\boldsymbol{\mu}_i$ and $\bar{\boldsymbol{\mu}}$ given by Eq. (1.14.8) has been used. Carrying Eq. (1.14.10) into Eq. (1.14.9), the following expression is obtained

$$\bar{\mathbf{M}} = \frac{9\varepsilon_1}{(\varepsilon_1 + 2)(2\varepsilon_1 + 1)} \bar{\boldsymbol{\mu}} \tag{1.14.11}$$

Then, Eq. (1.14.5) can be written as

$$\langle \boldsymbol{\mu} \cdot \mathbf{e} \rangle = \frac{9\varepsilon_1}{(\varepsilon_1 + 2)(2\varepsilon_1 + 1)} \frac{\langle \boldsymbol{\mu} \cdot \bar{\boldsymbol{\mu}} \rangle}{3kT} E_0 \tag{1.14.12}$$

Taking m in Eq. (1.8.2) as $\langle \boldsymbol{\mu} \cdot \mathbf{e} \rangle$, and the cavity field as given by Eq. (1.B.14), one obtains

$$\frac{(\varepsilon_1 - 1)(2\varepsilon_1 + 1)}{3\varepsilon_1} = \frac{4\pi N}{V} \frac{\langle \boldsymbol{\mu} \cdot \bar{\boldsymbol{\mu}} \rangle}{3kT} \tag{1.14.13}$$

where Eqs (1.8.5) and (1.14.12) were considered. Equation (1.14.13) is the Kirkwood equation for nonpolarizable dipoles. It is usual to write

$$\langle \boldsymbol{\mu} \cdot \bar{\boldsymbol{\mu}} \rangle = g\mu^2 \tag{1.14.14}$$

where g is the correlation parameter, which is a measure of the local order in the specimen. The correlation parameter is 1 if the average moment of a finite sphere surrounding a fixed molecule in an infinite medium of the material is the same as that of the molecule. In this case there is no dipolar correlation between neighboring molecules, or equivalently a dipole does not influence the positions and orientations of the neighboring ones. If the effect of the dipole is to orient the surrounding dipoles in the same direction, then $g > 1$, but if a fixed dipole tends to orient the neighbors in an antiparallel direction, then $g < 1$. Obviously, g depends on the structure of the material, and for this reason it is a parameter that gives information about the forces of local type.

1.15. FRÖHLICH'S STATISTICAL THEORY

Like Lorentz, Fröhlich considers a macroscopic spherical region within an infinite continuum material. The selected spherical region is big enough to have the same macroscopic permittivity as the medium. This region consists of a great number of elementary charges with positions, relative to their equilibrium positions, given by $X = (\mathbf{r}_i)$. In the absence of an external field, the energy of the spherical region is composed of the interaction of the electric particles inside this region, which depends on X, and the interaction with the surroundings, which depends on both X and the dielectric properties of the infinite medium. Now, the energy is changed by the action of an external field produced by outer sources in the dielectric medium. If, in any state, the spherical region has a dipole moment $\mathbf{M}(X)$, the perturbed energy due to the external field is given by

$$U(X, E_0) = U(X, 0) - \mathbf{M}(X)\frac{3\varepsilon_1}{2\varepsilon_1 + 1}\mathbf{E} \tag{1.15.1}$$

where the Onsager cavity factor has been taken into account. On the other hand, the average dipole moment in the direction of the field over all the states into the spherical cavity is

$$\langle \mathbf{M} \cdot \mathbf{e} \rangle = \frac{\int [\mathbf{M}(X) \cdot \mathbf{e}] \exp[-U(X, \mathbf{E})/kT]\, dX}{\int \exp[-U(X, \mathbf{E})/kT]\, dX} \tag{1.15.2}$$

This expression can be expanded into a power series and after some approximations can be written as

$$\langle \mathbf{M} \cdot \mathbf{e} \rangle = \frac{3\varepsilon_1}{2\varepsilon_1 + 1}\frac{E}{kT}\frac{\int [\mathbf{M}(X) \cdot \mathbf{e}]^2 \exp[-U(X, 0)/kT]\, dX}{\int \exp[-U(X, 0)/kT]\, dX} \tag{1.15.3}$$

Since all the directions have the same probability

$$\langle \mathbf{M} \cdot \mathbf{e} \rangle = \frac{3\varepsilon_1}{2\varepsilon_1 + 1}\frac{E}{3kT}\frac{\int M^2(X) \exp[-U(X, 0)/kT]\, dX}{\int \exp[-U(X, 0)/kT]\, dX} \tag{1.15.4}$$

The third factor on the right-hand side of Eq. (1.15.4) is the mean square dipole moment of the spherical region in the absence of an external field, so that

$$\langle \mathbf{M} \cdot \mathbf{e} \rangle = \frac{3\varepsilon_1}{2\varepsilon_1 + 1}\frac{\langle M^2 \rangle_0}{3kT}E \tag{1.15.5}$$

The remaining part of the Fröhlich analysis is devoted to the calculation of $\langle M^2 \rangle_0$. For this purpose, it is assumed that the spherical region is made up of N similar units (atoms, molecules, or ensembles of such entities), each of which makes the same contribution to $\langle M^2 \rangle_0$. In a liquid, each cell contains a molecule; in a crystal, each cell contains all the atoms corresponding to a crystal cell; in a polymer, the cell contains all the dipoles of a chain and those of the adjacent chains. Thus, if $\mathbf{m}(x_i)$ is the contribution of the ith-cell to $\mathbf{M}(X)$, where x_i represents the configuration of the elementary charges in this cell, the value of $\mathbf{M}(X)$ is given by

$$\mathbf{M}(X) = \sum_{i=1}^{n} \mathbf{m}(x_i) \tag{1.15.6}$$

and

$$M^2(X) = \sum_{i=1}^{n} \mathbf{m}(x_i) \cdot \mathbf{M}(X) = N \mathbf{m}(x) \cdot \mathbf{M}(X) \tag{1.15.7}$$

where subscript i has been dropped because each cell makes the same contribution, and x can represent the generic configuration of any selected unit cell. Consequently

$$\langle M^2 \rangle_0 = N \frac{\int \mathbf{m}(x) \cdot \mathbf{M}(X) \exp[-U(X, 0)/kT] \, dX}{\int \exp[-U(X, 0)/kT] \, dX} \tag{1.15.8}$$

The integration is carried out first over the whole region except for a selected unit cell. The volume element for this integration is denoted by dX', whereas the integration dx corresponds to the selected cell. Then, Eq. (1.15.8) is written in the following way

$$\langle M^2 \rangle_0 = N \frac{\int \mathbf{m}(x) \cdot \{ \int \mathbf{M}(X) \exp[-U(X, 0)/kT] \, dX' \} \, dx}{\int \exp[-U(X, 0)/kT] \, dX'}$$
$$\times \frac{\int \exp[-U(X, 0)/kT] \, dX'}{\int \exp[-U(X, 0)/kT] \, dX} \tag{1.15.9}$$

In this expression, we can write

$$\mathbf{m}^*(x) = \frac{\int \mathbf{M}(X) \exp[-U(X, 0)/kT] \, dX'}{\int \exp[-U(X, 0)/kT] \, dX'}$$

$$\text{and} \quad p(x) = \frac{\int \exp[-U(X, 0)/kT] \, dX'}{\int \exp[-U(X, 0)/kT] \, dX} \tag{1.15.10}$$

In this expression, $\mathbf{m}^*(x)$ is the average dipole moment of the spherical region, including the unit under consideration held in a fixed configuration, the others being allowed all possible configurations. Note that $\mathbf{m}^*(x)$ is the polarization produced in the spherical specimen owing to the charges in the region characterized by the configuration x, and for this reason it includes the short-range dipole–dipole interactions. On the other hand, $p(x)$ is the probability of the cell being in the fixed configuration with no restriction on the configurations of the other cells. Then, one can write

$$\langle M^2(X)\rangle_0 = N \int \mathbf{m}(x) \cdot \mathbf{m}^*(x)p(x)\,\mathrm{d}x = N\langle \mathbf{m}(x) \cdot \mathbf{m}^*(x)\rangle \tag{1.15.11}$$

Introducing this result into Eq. (1.15.5), and taking into account Eq. (1.9.2), the Fröhlich equation for nonpolarizable dipoles is obtained

$$\frac{(\varepsilon_1 - 1)(2\varepsilon_1 + 1)}{3\varepsilon_1} = \frac{4\pi N}{3kTV}\langle \mathbf{m} \cdot \mathbf{m}^*\rangle \tag{1.15.12}$$

Note that if one identifies $\mathbf{m} = \boldsymbol{\mu}$ and $\mathbf{m}^* = \bar{\boldsymbol{\mu}} = g\boldsymbol{\mu}$, then Eq. (1.14.13), or the Kirkwood equation, is recovered. In this case the cell obviously only contains a dipole of moment $\boldsymbol{\mu}$.

1.16. DISTORTION POLARIZATION IN THE KIRKWOOD AND FRÖHLICH THEORIES

Kirkwood deals with distortional polarization by postulating that the polarizability in Eq. (1.9.12) is also affected by a local field given by

$$\frac{3\varepsilon_1}{2\varepsilon_1 + 1}E_i = \frac{3\varepsilon_1}{2\varepsilon_1 + 1} \cdot \frac{3}{\varepsilon_1 + 2}E_0 = \frac{9\varepsilon_1}{(2\varepsilon_1 + 1)(\varepsilon_1 + 2)}E_0 \tag{1.16.1}$$

where the first factor in the second equality is the Onsager cavity field for the dipole [Eq. (1.B.13) with $\varepsilon_1 = \varepsilon_2$] and the second factor is the local field in the dielectric sphere in vacuum [Eq. (1.14.11)] with $\varepsilon_2 = \varepsilon_1$. Accordingly, the Kirkwood equation that includes the distortion polarizability is given by

$$\frac{(\varepsilon_1 - 1)(2\varepsilon_1 + 1)}{3\varepsilon_1} = \frac{4\pi N}{V}\left(\alpha + \frac{g\mu^2}{3kT}\right) \tag{1.16.2}$$

It should be emphasized that the dipolar moment appearing in Eq. (1.16.2) is the internal moment related to the moment in vacuum by Eq. (1.12.10).

On the other hand, Fröhlich takes into account the distortion polarization by assuming the dipoles embedded in a polarizable continuum of permittivity ε_∞.

In this case, the cavity field can be written as

$$G = \frac{3\varepsilon_1}{2\varepsilon_1 + \varepsilon_\infty} E \tag{1.16.3}$$

Then, the increment in permittivity that is due to dipolar reorientation is given by

$$\varepsilon_1 - \varepsilon_\infty = 4\pi \frac{\langle \mathbf{M} \cdot \mathbf{e} \rangle}{VE} \tag{1.16.4}$$

where, according to Eqs (1.15.5) and (1.15.11), the mean square dipole moment of the spherical region in the absence of the field is

$$\langle \mathbf{M} \cdot \mathbf{e} \rangle = \frac{3\varepsilon_1}{2\varepsilon_1 + \varepsilon_\infty} \frac{4\pi N \langle \mathbf{m} \cdot \mathbf{m}^* \rangle}{3kT} E \tag{1.16.5}$$

whence

$$\frac{(\varepsilon_1 - \varepsilon_\infty)(2\varepsilon_1 + \varepsilon_\infty)}{3\varepsilon_1} = \frac{4\pi N \langle \mathbf{m} \cdot \mathbf{m}^* \rangle}{3kTV} \tag{1.16.6}$$

In the Fröhlich theory, the dipoles are not themselves polarizable and the distortional polarizability corresponds to that of the continuum medium surrounding the dipoles. For this reason, when the cells in the Fröhlich theory are single molecules, $\mathbf{m}$ is related to the vacuum moment by

$$\boldsymbol{\mu}_v = \frac{3\mathbf{m}}{\varepsilon_\infty + 2} \tag{1.16.7}$$

By making the identifications carried out in the final sentence of the preceding section, Eqs (1.16.6) and (1.16.7) lead to the following expression

$$\frac{(\varepsilon_1 - \varepsilon_\infty)(2\varepsilon_1 + \varepsilon_\infty)}{\varepsilon_1(\varepsilon_\infty + 2)^2} = \frac{4\pi N g \mu_v^2}{9kTV} \tag{1.16.8}$$

which is a clear generalization of the Onsager equation [Eq. (1.13.9)].

From the preceding analysis it is obvious that Eqs (1.16.2) and (1.16.8) are not equivalent. This is because of the expression used by Kirkwood for the field in the cavity in the evaluation of the distortion polarization. If instead of Eq. (1.16.1) the total local field given by

$$\frac{[(3\varepsilon_1)/(2\varepsilon_1 + 1)]E}{1 - (\alpha/a^3)\{[2(\varepsilon_1 - 1)]/(2\varepsilon_1 + 1)\}} = \frac{[(3\varepsilon_1)/(2\varepsilon_1 + 1)][3/(\varepsilon_1 + 2)]}{1 - (\alpha/a^3)\{[2(\varepsilon_1 - 1)]/(2\varepsilon_1 + 1)\}} E_0$$

$$= \frac{3\varepsilon_1(\varepsilon_\infty + 2)}{(2\varepsilon_1 + \varepsilon_\infty)(\varepsilon_1 + 2)} E_0 \tag{1.16.9}$$

is used in the Kirkwood equation, then

$$\frac{\varepsilon_1 - 1}{\varepsilon_1 + 2} = \frac{4\pi N}{3V}\left[\alpha\frac{3\varepsilon_1(\varepsilon_\infty + 2)}{(2\varepsilon_1 + \varepsilon_\infty)(\varepsilon_1 + 2)} + \frac{9\varepsilon_1}{(2\varepsilon_0 + 1)(\varepsilon_0 + 2)}\frac{g\mu^2}{3kT}\right]$$

(1.16.10)

By writing $\mu \equiv \mu_i$ in terms of μ_v by means of Eq. (1.12.9), and using Eq. (1.12.8), the Fröhlich equation is recovered.

APPENDIX A

Properties of the Legendre Polynomials

Legendre polynomials appear in the theory of the potential as defined by the following expression

$$(1 - 2xh + h^2)^{-1/2}$$

(1.A.1)

By the binomial theorem

$$(1 - 2xh + h^2)^{-1/2} = 1 + xh + \tfrac{1}{2}(3x^2 - 1)h^2 + \tfrac{1}{2}(5x^3 - 3x)h^3 + \cdots$$

(1.A.2)

where

$$P_0(x) = 1, \quad P_1(x) = x, \quad P_2(x) = \tfrac{1}{2}(3x^2 - 1),$$
$$P_3(x) = \tfrac{1}{2}(5x^3 - x),\ldots$$

(1.A.3)

are the Legendre polynomials.

The Legendre polynomials satisfy the Legendre differential equation

$$(1 - x^2)\frac{d^2y}{dx^2} - 2x\frac{dy}{dx} + n(n + 1)y = 0$$

(1.A.4)

if n is an integer.

According to the recurrence Rodrigues formula

$$P_n(x) = \frac{1}{2^n n!}\frac{d^n}{dx^n}(x^2 - 1)^n$$

(1.A.5)

where it is possible to generate the Legendre polynomials. Recurrence formulae for Legendre polynomials are

$$(2n + 1)xP_n(x) = (n + 1)P_{n+1}(x) + nP_{n-1}(x)$$

$$(1 - x^2)\frac{dP_n(x)}{dx} = nP_{n-1}(x) - nxP_n(x)$$

(1.A.6)

The orthogonality condition for the Legendre polynomials is given by

$$\int_{-1}^{1} P_m(x)_n P(x)\, dx = \frac{2}{2n + 1}\, \delta_{mn}$$

(1.A.7)

APPENDIX B
Fröhlich alternative calculations for the polarization of a dielectric sphere in an infinite medium

An alternative to the calculations developed in Section 1.11 is to find the electric field of a dielectric sphere of radius a and permittivity ε_1 in a dielectric infinite medium of permittivity ε_2 under each one of the three following situations:

(a)   A constant external field E_0 taken in the z direction produced by charges at large distances from the sphere, for example by charges placed in the plates of a large condenser.

(b)   A point dipole, also oriented in the z direction, and placed at the center of the sphere.

(c)   A dipole in the z direction, arising from the homogeneously polarized sphere of radius a, such that $\mathbf{M} = (4\pi/3)a^3\mathbf{P}$, where $\mathbf{P}$ is the polarization by unit volume.

As mentioned above, this problem has conveniently been addressed by Fröhlich,[7] and his arguments will also be followed here. Taking the origin of spherical coordinates at the center of the sphere, let $\mathbf{r}$ be the distance from the center of the sphere and θ be the angle between $\mathbf{r}$ and the z axis. In case (a)

$$\phi = -E_0 r \cos\theta \qquad \text{for } r \to \infty$$

(1.B.1)

in case (b)

$$\phi = \frac{\mu \cos\theta}{\varepsilon_1 r^2} \qquad \text{for } r \to 0$$

(1.B.2)

whereas in case (c)

$$\mathbf{D} = \varepsilon_2 \mathbf{E} \qquad \text{for } r > a \tag{1.B.3a}$$

and

$$\mathbf{D} = \varepsilon_1 \mathbf{E} + 4\pi \mathbf{P} \qquad \text{for } r < a \tag{1.B.3b}$$

In the last case, $\mathbf{D}$ is the electric displacement and $\mathbf{P}$ is the polarization. In all cases the normal and tangential components of both the electric displacement and the field must be continuous at $r = a$.

Since there are no free charges inside and outside the sphere, the solution to our problem is equivalent to solving the Laplace equation inside and outside the sphere with adequate continuity conditions. The more general solution of the Laplace equation for the potential $\nabla^2 \phi = 0$ is given by Eq. (1.11.2), here repeated for convenience

$$\phi = \sum_{n=0}^{\infty} \left(a_n r^n + \frac{b_n}{r^{n+1}} \right) P_n(\cos\theta) \tag{1.B.4}$$

Owing to the imposed conditions, only terms proportional to $\cos\theta$ appear, and all the cases can be treated together. Thus, the general solution reduces to

$$\phi = -Ar\cos\theta - \frac{B}{r^2}\cos\theta \tag{1.B.5}$$

The first term on the right-hand side corresponds to the applied uniform field, and the second is the field of a dipole in the center of the sphere. This dipole is produced by the induced polarization charges in the dielectric sphere. The values of the constants are different inside and outside the sphere. Therefore

$$\phi_1 = -A_1 r\cos\theta - \frac{B_1}{r^2}\cos\theta, \qquad r < a$$

$$\phi_2 = -A_2 r\cos\theta - \frac{B_2}{r^2}\cos\theta, \qquad r > a \tag{1.B.6}$$

From Eqs (1.B.1) and (1.B.2), $A_2 = E_e$ and $B_1 = -\mu/\varepsilon_1$ respectively.

The continuity in the tangential components of the field requires continuity at $r = a$ of

$$E_\theta = -\frac{1}{r}\frac{\partial\phi}{\partial\theta} \tag{1.B.7}$$

Consequently

$$\frac{B_2}{a^3} + E_0 = -\frac{\mu}{\varepsilon_1 a^3} + A_1 \tag{1.B.8}$$

The continuity in the normal components of the dielectric displacement ($\varepsilon_1 \partial\phi_1/\partial r = \varepsilon_2 \partial\phi_2/\partial r$ at $r = a$), together with Eq. (1.B.3b), implies that

$$2\frac{\mu_1}{a^3} + \varepsilon_1 A_1 = -\frac{2\varepsilon_2 B_2}{a^3} + \varepsilon_2 E_0 + 4\pi P \tag{1.B.9}$$

From these equations one finds

$$A_1 = \frac{3\varepsilon_2}{2\varepsilon_2 + \varepsilon_1} E_0 - \frac{2}{\varepsilon_1} \frac{\varepsilon_1 - \varepsilon_2}{2\varepsilon_2 + \varepsilon_1} \frac{\mu}{a^3} - \frac{4\pi P}{2\varepsilon_2 + \varepsilon_1}$$

$$\frac{B_2}{a^3} = -\frac{\varepsilon_1 - \varepsilon_2}{2\varepsilon_2 + \varepsilon_1} E_0 - \frac{3}{2\varepsilon_2 + \varepsilon_1} \frac{\mu}{a^3} - \frac{4\pi P}{2\varepsilon_2 + \varepsilon_1}$$

$$\tag{1.B.10}$$

After insertion of Eq. (1.B.10) into Eq. (1.B.6), together with the values found for A_2 and B_1, the general solution of the problem will be

$$\phi_2 = -\frac{a^3}{r^2}\left[\left(\frac{r^3}{a^3} - \frac{\varepsilon_1 - \varepsilon_2}{2\varepsilon_2 + \varepsilon_1}\right)E_0 - \frac{3}{2\varepsilon_2 + \varepsilon_1}\frac{\mu}{a^3} - \frac{4\pi P}{2\varepsilon_2 + \varepsilon_1}\right]\cos\theta, \quad r > a$$

$$\phi_1 = -r\left[\frac{3\varepsilon_2}{2\varepsilon_2 + \varepsilon_1}E_0 - \frac{\mu}{\varepsilon_1 r^3}\left(1 + \frac{2r^3}{a^3}\frac{\varepsilon_1 - \varepsilon_2}{2\varepsilon_2 + \varepsilon_1}\right) - \frac{4\pi P}{2\varepsilon_2 + \varepsilon_1}\right]\cos\theta, \quad r < a$$

$$\tag{1.B.11}$$

Now we can analyze separately the three cases under discussion.

Case (a)

In this case there are no internal sources, that is, $\mu = P = 0$.
The field in the cavity will be

$$\mathbf{G} = \frac{3\varepsilon_2}{2\varepsilon_2 + \varepsilon_1}\mathbf{E}_0 \tag{1.B.12}$$

If the spherical cavity is empty

$$\mathbf{G} = \frac{3\varepsilon_2}{2\varepsilon_2 + 1}\mathbf{E}_0 \tag{1.B.13}$$

If the sphere is in vacuum

$$\mathbf{G_v} = \frac{3}{\varepsilon_1 + 2} \mathbf{E_0} \tag{1.B.14}$$

The field outside the sphere is composed of the external applied field and a dipole field with potential given by

$$\frac{a^3}{r^2} \frac{\varepsilon_1 - \varepsilon_2}{2\varepsilon_2 + \varepsilon_1} E_0 \cos\theta \tag{1.B.15}$$

The dipole vector directed along the z axis is given by

$$\frac{\varepsilon_1 - \varepsilon_2}{2\varepsilon_2 + \varepsilon_1} a^3 E_0 \tag{1.B.16}$$

This is the electric moment of the dielectric sphere. Obviously, this dipole moment is null when $\varepsilon_1 = \varepsilon_2$.

If the sphere is immersed in vacuum, $\varepsilon_2 = 1$ and Eq. (1.B.16) becomes

$$\frac{\varepsilon_1 - 1}{\varepsilon_1 + 2} a^3 E_0 \tag{1.B.17}$$

Note that since the dipole moment per unit volume is given by

$$4\pi\mathbf{P} = (\varepsilon_1 - 1)\mathbf{G_v} \tag{1.B.18}$$

the total polarization can be written as

$$\mathbf{P} = 3\frac{\varepsilon_1 - 1}{\varepsilon_1 + 2} \mathbf{E_0} \tag{1.B.19}$$

where the value of $\mathbf{G_v}$ given by Eq. (1.B.14) was used. Multiplying by the sphere volume $(4\pi a^3/3)$, the total polarization of the sphere given by Eq. (1.B.17) is obtained.

Case (b)

A point dipole is at the center of the sphere without an external field. In this case there are two components of the field inside the sphere. The first is the field created by the central dipole, but the second, called the reaction field, is due to the different permittivities inside and outside the sphere. Its value is given by

$$\mathbf{R} = \frac{2}{a^3} \frac{1}{\varepsilon_1} \frac{\varepsilon_1 - \varepsilon_2}{2\varepsilon_2 + \varepsilon_1} \mu \tag{1.B.20}$$

If the sphere is empty ($\varepsilon_1 = 1$), the reaction field will be

$$\mathbf{R} = -\frac{2}{a^3}\frac{\varepsilon_2 - 1}{2\varepsilon_2 + 1}\boldsymbol{\mu} \tag{1.B.21}$$

The potential outside the sphere is

$$\phi = \frac{3}{2\varepsilon_2 + \varepsilon_1}\frac{\mu\cos\theta}{r^2} = \frac{\mu\cos\theta}{[(2\varepsilon_2 + \varepsilon_1)/3]r^2} \tag{1.B.22}$$

Case (c)

In this case neither the external field nor the point dipoles are inside the sphere. If the sphere is in vacuum ($\varepsilon_1 = \varepsilon_2 = 1$), the field inside the sphere is

$$\mathbf{E}_s = -\frac{4\pi\mathbf{P}}{3} = -\frac{\mathbf{M}}{a^3} \tag{1.B.23}$$

Now, surrounding the empty sphere by a medium with permittivity ε_1, the increase in the field inside the sphere is the reaction field

$$\mathbf{R} = -\frac{\mathbf{P}}{2\varepsilon_2 + 1} + \frac{4\pi\mathbf{P}}{3} = \frac{2(\varepsilon_2 - 1)}{2\varepsilon_2 + 1}\frac{4\pi\mathbf{P}}{3} = \frac{2(\varepsilon_2 - 1)}{2\varepsilon_2 + 1}\frac{\mathbf{M}}{a^3} \tag{1.B.24}$$

in agreement with Eq. (1.B.21).

If the sphere contains a point dipole surrounded by a homogeneous polarized sphere with moment $\mathbf{M}$, the reaction field is obtained from the sum of Eqs (1.B.21) and (1.B.24). Then, the reaction field is

$$\mathbf{R} = \frac{2}{a^3}\frac{\varepsilon_2 - 1}{2\varepsilon_2 + 1}(\mathbf{M} + \boldsymbol{\mu}) \tag{1.B.25}$$

The field outside the sphere is given by

$$\phi = \frac{3}{2\varepsilon_2 + \varepsilon_1}\frac{|\mathbf{M}|\cos\theta}{r^2} \tag{1.B.26}$$

For this reason, outside the sphere, cases (b) and (c) lead to the same result.

APPENDIX C

The Polarization of an Ellipsoid

The problem of the polarization of an ellipsoid is important because, when it is placed in a uniform external field $\mathbf{E}_0$, the field within itself is uniform.[10] Let us assume an ellipsoidal body with dielectric permittivity ε_1 in a dielectric medium of permittivity ε_2 under an electric field. If we choose a coordinate system in such a way that the axes are along the principal axis of the ellipsoid, then the potential inside the ellipsoid is given by[10,11]

$$\mathbf{E}_1 = \frac{E_{0x}}{1 + [(\varepsilon_1 - \varepsilon_2)/\varepsilon_2]A_a}\mathbf{i} + \frac{E_{0y}}{1 + [(\varepsilon_1 - \varepsilon_2)/\varepsilon_2]A_b}\mathbf{j}$$

$$+ \frac{E_{0z}}{1 + [(\varepsilon_1 - \varepsilon_2)/\varepsilon_2]A_c}\mathbf{k} \tag{1.C.1}$$

where A_i, $i = a,b,c$, is the so-called depolarizing factor, given by

$$A_i = \frac{abc}{2}\int_0^\infty \frac{ds}{(s + \alpha^2)R}, \qquad i = a, b, c \tag{1.C.2}$$

with

$$R^2 = (s + a^2)(s + b^2)(s + c^2) \tag{1.C.3}$$

On the other hand, the field $\mathbf{E}_0$ outside the ellipsoid is given by

$$\mathbf{E}_2 = E_{0x}\left[1 - \frac{[(\varepsilon_1 - \varepsilon_2)/\varepsilon_2]A_a(\xi)}{1 + [(\varepsilon_1 - \varepsilon_2)/\varepsilon_2]A_a}\right]\mathbf{i} + E_{0y}\left[1 - \frac{[(\varepsilon_1 - \varepsilon_2)/\varepsilon_2]A_b(\xi)}{1 + [(\varepsilon_1 - \varepsilon_2)/\varepsilon_2]A_b}\right]\mathbf{j}$$

$$+ E_{0z}\left[1 - \frac{[(\varepsilon_1 - \varepsilon_2)/\varepsilon_2]A_c(\xi)}{1 + [(\varepsilon_1 - \varepsilon_2)/\varepsilon_2]A_c}\right]\mathbf{k} \tag{1.C.4}$$

A_i cannot be found in closed form from Eq. (1.C.2), but there exist tables of these elliptic integrals.[12] By adding the integrals given by Eq. (1.C.2), and using as a variable of integration $u = R^2$, we find

$$A_a + A_b + A_c = \frac{abc}{2}\int_{(abc)^2}^\infty \frac{du}{u^{3/2}} = 1 \tag{1.C.5}$$

If the ellipsoid degenerates into a sphere, all the depolarizing factors are equal and $A_a = A_b = A_c = 1/3$. If, in this case, the external field is along the z axis, Eq. (1.C.1) reduces to Eq. (1.16.3) or (1.B.12).

For a cylinder with its axis in the z direction ($c \to \infty$), $A_c = 0$ and $A_a = A_b = 1/2$. For a flat plate, a, $b \to \infty$ and $A_a = A_b = 0$ and $A_c = 1$.

Integrals given by Eq. (1.C.2) can be calculated if the ellipsoid degenerates into a spheroid. For prolate spheroids ($a > b = c$) and eccentricity $e = \sqrt{1 - (b/a)^2}$, we have

$$A_a = \frac{1 - e^2}{2e^3}\left(\log\frac{1 + e}{1 - e} - 2e\right), \qquad A_b = A_c = \tfrac{1}{2}(1 - A_a) \tag{1.C.6}$$

For a nearly spherical prolate spheroid, $e \ll 1$ and

$$A_a = \tfrac{1}{3} - \tfrac{2}{15}e^2, \qquad A_b = A_c = \tfrac{1}{3} + \tfrac{1}{15}e^2 \tag{1.C.7}$$

whereas for an oblate spheroid $(a = b > c)$, with $e = \sqrt{(a/c)^2 - 1}$, the values of A_i are

$$A_c = \frac{1 + e^2}{e^3}(e - \tan^{-1} e), \qquad A_a = A_b = \tfrac{1}{2}(1 - A_c) \tag{1.C.8}$$

If $e \ll 1$, then

$$A_c = \tfrac{1}{3} + \tfrac{2}{15}e^2, \qquad A_a = A_b = \tfrac{1}{3} - \tfrac{1}{15}e^2 \tag{1.C.9}$$

If the external field is in direction z, the fields inside an empty ellipsoidal cavity and for an ellipsoid in vacuum are respectively given by

$$\mathbf{E}_z = \frac{\varepsilon_2}{\varepsilon_2 + (1 - \varepsilon_2)A_c} E_0\mathbf{k}, \qquad \mathbf{E}_z = \frac{1}{1 + (\varepsilon_1 - 1)A_c} E_0\mathbf{k} \tag{1.C.10}$$

which reduce to (1.B.13) and (1.B.14) for a sphere.

APPENDIX D

Important Formulae in SI Units

Electrostatic Potential of a Dipole

$$\Phi(r, \theta) = \frac{1}{4\pi e_0}\left(\frac{q}{r_+} - \frac{q}{r_-}\right) \tag{1.2.1a}$$

where $e_0 = 8.854 \times 10^{-12}\, \mathrm{C^2\,kg^{-1}\,m^{-3}\,s^2}$ is the vacuum permittivity.

Field of an Isolate Point Dipole

$$\Phi = \frac{\mu \cos\theta}{4\pi e_0 r^2} \tag{1.4.1a}$$

Dielectric Susceptibility

$$\chi = \varepsilon - 1 = \frac{\mathbf{P}}{e_0 \mathbf{E}_0} \tag{1.8.5a}$$

where ε is the relative dielectric permittivity.

Relationship Between Electric Displacement and the External Applied Field

$$\mathbf{D} = \varepsilon\mathbf{E} = e_0\varepsilon_r\mathbf{E} \tag{1.8.7a}$$

Clausius–Mossotti Equation

$$\frac{\varepsilon - 1}{\varepsilon + 2} = \frac{N_1\alpha}{3e_0} \tag{1.9.5a}$$

Debye Equation for the Static Permittivity

$$\frac{\varepsilon_0 - 1}{\varepsilon_0 + 2}\frac{M}{\rho} = \frac{N_A}{3e_0}\left(\alpha_d + \frac{\mu^2}{3kT}\right) \tag{1.9.13a}$$

Onsager Formula

$$\frac{(\varepsilon_1 - \varepsilon_\infty)(2\varepsilon_1 + \varepsilon_\infty)}{\varepsilon_1(\varepsilon_\infty + 2)^2} = \frac{N_1\mu_v^2}{9kTe_0} \tag{1.13.12a}$$

Kirkwood Equation for Nonpolarizable Dipoles

$$\frac{(\varepsilon_1 - 1)(2\varepsilon_1 + 1)}{3\varepsilon_1} = \frac{N}{V}\frac{\langle\boldsymbol{\mu}\cdot\bar{\boldsymbol{\mu}}\rangle}{3kTe_0} \tag{1.14.13a}$$

Fröhlich–Kirkwood–Onsager Formula

$$\frac{(\varepsilon_1 - \varepsilon_\infty)(2\varepsilon_1 + \varepsilon_\infty)}{\varepsilon_1(\varepsilon_\infty + 2)^2} = \frac{Ng\mu_v^2}{9kTVe_0} \tag{1.16.8a}$$

PROBLEMS

Problem 1

Show that the quadrupole moment of an isolated dipole is null.

Solution

The quadrupole moment is given by

$$\sum_{i=1}^{2} q_i d_i^2\left[\frac{1}{2}(3\cos^2\theta_i - 1)\right] \tag{P.1.1.1}$$

where the quantity between brackets is the second Legendre polynomial. In this expression all the magnitudes are squared, except for q_i. Since the charges are equal except in the sign, the two terms in the sum mutually cancelout, and thus the quadrupole moment is null.

Problem 2

Find the electric field of an isolated point dipole in polar plane coordinates.

Solution

In polar plane coordinates

$$E_r = -\frac{\partial \Phi}{\partial r}, \qquad E_\theta = -\frac{1}{r}\frac{\partial \Phi}{\partial \theta} \qquad \text{(P.1.2.1)}$$

The values of the components of the electric field obtained by means of Eq. (1.4.1) are

$$E_r = \frac{2\mu \cos \theta}{r^3}, \qquad E_\theta = \frac{\mu \sin \theta}{r^3} \qquad \text{(P.1.2.2)}$$

Therefore, the electric field can be written as

$$\mathbf{E} = \frac{\mu}{r^3}(2 \cos \theta\, \mathbf{u}_r + \sin \theta\, \mathbf{u}_\theta) \qquad \text{(P.1.2.3)}$$

and

$$|\mathbf{E}| = \frac{\mu}{r^3}(1 + 3 \cos^2 \theta)^{1/2} \qquad \text{(P.1.2.4)}$$

The equipotential lines are

$$\Phi = C \qquad \text{(P.1.2.5)}$$

where C is a constant. Consequently

$$r^2 = C_1 \cos \theta \qquad \text{(P.1.2.6)}$$

where $C_1 = \mu/C$.

The field lines are given by

$$\frac{dr}{E_r} = \frac{r\, d\theta}{E_\theta} \qquad \text{(P.1.2.7)}$$

from which

$$r = C_2 \sin^2 \theta \qquad \text{(P.1.2.8)}$$

Problem 3

Consider two point dipoles in such a way that the linejoining the dipoles is the z axis (Fig. 1.4). Find:

(a)   An expression for the angle formed by these dipoles expressed in terms of the precessional and azimutal angles of each dipole.

(b)   The interaction energy in terms of the former result.

(c)   The components of the force in spherical coordinates.

Solution

(a) According to the notation of the figure, the components of each dipole are

$$\boldsymbol{\mu}_1 = |\boldsymbol{\mu}_1|(\sin \theta_1 \cos \varphi_1 \mathbf{i} + \sin \theta_1 \sin \varphi_1 \mathbf{j} + \cos \theta_1 \mathbf{k})$$
$$\boldsymbol{\mu}_2 = |\boldsymbol{\mu}_2|(\sin \theta_2 \cos \varphi_2 \mathbf{i} + \sin \theta_2 \sin \varphi_2 \mathbf{j} + \cos \theta_2 \mathbf{k}) \qquad \text{(P.1.3.1)}$$

The scalar product of the vector representing the dipoles can be expressed as

$$\boldsymbol{\mu}_1 \cdot \boldsymbol{\mu}_2 = |\boldsymbol{\mu}_1||\vec{\boldsymbol{\mu}}_2| \cos \gamma$$
$$= (\sin \theta_1 \sin \theta_2 \cos \varphi_1 \cos \varphi_2 + \sin \theta_1 \sin \theta_2 \sin \varphi_1 \sin \varphi_2$$
$$+ \cos \theta_1 \cos \theta_2)|\boldsymbol{\mu}_1||\boldsymbol{\mu}_2| \qquad \text{(P.1.3.2)}$$

and from Eq. (P.1.3.2) we obtain

$$\cos \gamma = \sin \theta_1 \sin \theta_2 \cos (\varphi_2 - \varphi_1) + \cos \theta_1 \cos \theta_2 \qquad \text{(P.1.3.3)}$$

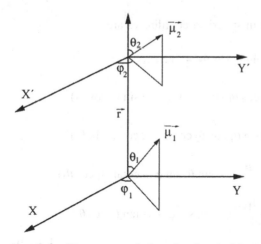

FIG. 1.4   Diagram to calculate the electrical interaction between two point dipoles

(b) The expression for the interaction work is easily obtained as

$$W = \frac{|\boldsymbol{\mu}_1||\boldsymbol{\mu}_2|}{r^3}(\cos \gamma - 3 \cos \theta_1 \cos \theta_2) \tag{P.1.3.4}$$

The work is minimum if $\theta_1 = \theta_2 = 0$ or $\theta_1 = \theta_2 = 0$

$$W_{\min} = -2\frac{|\boldsymbol{\mu}_1||\boldsymbol{\mu}_2|}{r^3} \tag{P.1.3.5}$$

The work is maximum if $\theta_1 = 0$ and $\theta_2 = \pi$ or if $\theta_1 = \pi$ and $\theta_2 = 0$

$$W_{\max} = 2\frac{|\boldsymbol{\mu}_1||\boldsymbol{\mu}_2|}{r^3} \tag{P.1.3.6}$$

Note that, if $\varphi_2 - \varphi_1 = \pi$ and $\theta_1 = \theta_2 = \pi/2$, the dipoles are antiparallel and both perpendicular to $\mathbf{r}$. Then

$$W = -\frac{|\boldsymbol{\mu}_1||\boldsymbol{\mu}_2|}{r^3} \tag{P.1.3.7}$$

The work is null if

$$\tan \theta_1 \tan \theta_2 \cos (\varphi_2 - \varphi_1) = 2 \tag{P.1.3.8}$$

(c) We start from

$$W = \frac{|\boldsymbol{\mu}_1||\boldsymbol{\mu}_2|}{r^3}(\sin \theta_1 \sin \theta_2 \cos \varphi - 3 \cos \theta_1 \cos \theta_2) \tag{P.1.3.9}$$

where $\varphi = \varphi_2 - \varphi_1$.

The components of the force in spherical coordinates are

$$F_r = -\frac{\partial W}{\partial r} = -\frac{3|\boldsymbol{\mu}_1||\boldsymbol{\mu}_2|}{r^4}(\sin \theta_1 \sin \theta_2 - 3 \cos \theta_1 \cos \theta_2)$$

$$F_\theta = -\frac{1}{r}\frac{\partial W}{\partial \theta_1} = -\frac{|\boldsymbol{\mu}_1||\boldsymbol{\mu}_2|}{r^4}(\cos \theta_1 \sin \theta_2 \cos \varphi + 2 \sin \theta_1 \cos \theta_2)$$

$$F_{\theta_2} = -\frac{1}{r}\frac{\partial W}{\partial \theta_2} = -\frac{|\boldsymbol{\mu}_1||\boldsymbol{\mu}_2|}{r^4}(\sin \theta_1 \cos \theta_2 \cos \varphi + 2 \cos \theta_1 \sin \theta_2)$$

$$F_{\varphi_1} = -\frac{1}{r \sin \theta_1}\frac{\partial W}{\partial \varphi_1} = -\frac{|\boldsymbol{\mu}_1||\boldsymbol{\mu}_2|}{r^4}(- \sin \theta_1 \sin \varphi + \cotan \theta_1 \cos \theta_2)$$

$$F_{\varphi_2} = -\frac{1}{r \sin \theta_2}\frac{\partial W}{\partial \varphi_2} = -\frac{|\boldsymbol{\mu}_1||\boldsymbol{\mu}_2|}{r^4}(\sin \theta_1 \sin \varphi + \cotan \theta_2 \cos \theta_1)$$

$$\tag{P.1.3.10}$$

Problem 4

Let us consider two equal and opposite charges, $+q$ and $-q$, located at points $\pm a$ on the z axis. Calculate the volume integral of the z component of the electric field extended over a sphere centered on $z = 0$ and radius $R \gg a$[13].

Solution

The electric potential ψ at a point P, at distantce **r** from the origin, is given by

$$\psi(\mathbf{r}) = q\left(\frac{1}{r_+} - \frac{1}{r_-}\right) \tag{P.1.4.1}$$

where cylindrical coordinates (ρ, φ, z) have been used and

$$r_\pm = [\rho^2 + (z \mp a)^2]^{1/2} \tag{P.1.4.2}$$

The component of the electric field along z is

$$E_z = -q\frac{\partial}{\partial z}\{[\rho^2 + (z - a)^2]^{-1/2} - [\rho^2 + (z + a)^2]^{-1/2}\} \tag{P.1.4.3}$$

The volume integral requested is

$$\int_{r<R} E_z(\mathbf{r})\,d^3\mathbf{r} = \int_0^{2\pi} d\varphi \int_0^R \rho\,d\rho \int_{-(R^2-\rho^2)^{1/2}}^{(R^2-\rho^2)^{1/2}} E_z(\rho, \varphi, z)\,dz \tag{P.1.4.4}$$

Integrating first with respect z, and changing from the variable ρ to $u = (R^2 - \rho^2)^{1/2}$, the final result becomes

$$\int_{r<R} E_z(\mathbf{r})\,d^3\mathbf{r} = -\frac{8\pi aq}{3} = -\frac{4\pi m}{3} \tag{P.1.4.5}$$

where m is the dipole moment of the pair charges. It can be demonstrated that the components of the field along x and y are zero.

This result has been extended by Scaife,[14] taking the integral over infinite slabs between parallel plates (a) perpendicular and (b) parallel to the axis of the dipole. Splitting the infinite slabs into two regions, one over a sphere enclosing the pair of charges and the other over the region between the sphere and the bounding plates, respectively yields

$$\int E_z(\mathbf{r})\,d^3\mathbf{r} = -4\pi m, \qquad \int E_z(\mathbf{r})\,d^3\mathbf{r} = 0 \tag{P.1.4.6}$$

We notice that in all the cases the result does not depend upon the dimensions of the region of integration.

Problem 5

In a cubic lattice, the lattice constant is 2 Å and the refraction index is $n = 2.07$. The material, which is assumed to be incompressible (Poisson's ratio = 0.5), is uniaxially strained 2% along one of the cube edges. Make an estimation of the refractive index of the strained material when an external electric field $\mathbf{E}$ is applied (a) parallel and (b) perpendicular to the strained axis. Assume the atomic polarizability α to be a scalar and constant under small deformations.

Solution

As mentioned in Section 1.9, the field acting at the center of the cavity where the dipole is placed, that is, the internal field, arises from three sources: (1) the external field, (2) the polarization charges on the inner part of the cavity, and (3) the molecular dipoles inside the cavity. For a cubic lattice, the contribution to the internal field of the dipoles inside the cavity is null owing to symmetry factors (see, for example, Classical Electrodynamics, by J.D. Jackson, 2nd Ed., 152–154). However, for a distorted lattice, the medium becomes anisotropic and this contribution is not zero. In fact, it can be calculated from the dipole–dipole tensor field (Eq. 1.4.5), which expresses the electric field due to the nearest neighbors located at $\mathbf{r}$ with the dipole moment given by $\mathbf{p} = \alpha\mathbf{E_i}$, where $\mathbf{E_i}$ is the internal field. Thus, by considering a cubic cavity enclosing only the six nearest atoms to the one under consideration, the cavity field will be given by

$$\mathbf{E_c} = \sum_{i=1}^{6} \frac{3(\mathbf{p}_i \cdot \mathbf{r}_i)\mathbf{r}_i - \mathbf{p}_i r^2}{r_i^5} \tag{P.1.5.1}$$

where subscript c denotes cavity, $[3(\mathbf{p}_i \cdot \mathbf{r}_i)\mathbf{r}_i - \mathbf{p}_i r^2]/r_i^5$ is the contribution from the dipole induced in the ith atom within the cavity, and $\mathbf{p}_i = \alpha\mathbf{E_i}$.

It is worth noting that, for an incompressible material, a 2% strain elongation along one edge implies a 1% strain contraction along the other two edges. If x_l ($\cong x = 2$ Å) and x_s ($\cong x = 2$ Å) denote, respectively, the elongated and constrained interatomic distances, then, in case (a), two atoms have x_l and the other four have x_s. Moreover, in this case, all the dipoles have the same magnitude and direction along $\mathbf{E}$. Accordingly, the following expression is obtained

$$2\frac{(3\mathbf{p} \cdot \mathbf{x}_l)\mathbf{x}_l - \mathbf{p}x_l^2}{x_l^5} + 4\frac{(3\mathbf{p} \cdot \mathbf{x}_s)\mathbf{x}_s - \mathbf{p}x_s^2}{x_s^5} \tag{P.1.5.2}$$

Since $\mathbf{p} \parallel \mathbf{x}_l$ and $\mathbf{p} \perp \mathbf{x}_s$, the preceding expression gives

$$4\mathbf{p}\left(\frac{1}{x_l^3} - \frac{1}{x_s^3}\right) \cong -12\mathbf{p}\frac{x_l - x_s}{x^4} \tag{P.1.5.3}$$

and consequently

$$\mathbf{E}_c = -12\alpha\frac{x_l - x_s}{x^4}\mathbf{E}_i \tag{P.1.5.4}$$

where the approximation $x_l^2 + x_l x_s + x_s^2 \cong 3x^2$ has been made.

Proceeding in a similar way, one obtains for case (b)

$$\mathbf{E}_c = 6\alpha\frac{x_l - x_s}{x^4}\mathbf{E}_i \tag{P.1.5.5}$$

which is one-half of the magnitude and opposite in sign with respect to case (a).

Let us first calculate the atomic polarizability α. From Eqs (1.8.1), (1.8.2) and (1.9.2) the following equation is obtained

$$\mathbf{P} = N_1\alpha\mathbf{E}_i = \frac{\varepsilon - 1}{4\pi}\mathbf{E} \tag{P.1.5.6}$$

For the given lattice constant, the volume of the unit cell is $8 \times 10^{-30}\,\mathrm{m}^3$, thus finding $N_1 = 1.25 \times 10^{29}/\mathrm{m}^3$. On the other hand, from the electromagnetic theory, $\varepsilon = n^2 = 4.2849$, where n is the refractive index. For the undistorted lattice, $\mathbf{E}_i$ is given by Eq. (1.9.3), and from Eq. (1.9.4) the polarizability of the atoms of the lattice is obtained as

$$\alpha = \frac{3}{4\pi N_1}\frac{\varepsilon - 1}{\varepsilon + 2} \cong 1 \times 10^{-30}\,\mathrm{m}^3 \tag{P.1.5.7}$$

For the strained lattice

$$\mathbf{E}_i = \mathbf{E} + \frac{4\pi}{3}N_1\alpha\mathbf{E}_i + \mathbf{E}_c \tag{P.1.5.8}$$

and taking into account Eq. (P.1.5.4), one obtains

$$\mathbf{E}_i = \mathbf{E}\left(1 - \frac{4\pi}{3}N_1\alpha + 12\alpha\frac{x_l - x_s}{x^4}\right)^{-1} \tag{P.1.5.9}$$

where $x = 2 \times 10^{-10}\,\mathrm{m}$, and $x_l - x_s = [2.04 - 1.98] \times 10^{-10}\,\mathrm{m} = 0.06 \times 10^{-10}\,\mathrm{m}$.

Carrying this result into Eq. (P.1.5.6), the following expression holds for case (a)

$$N_1\alpha = \frac{\varepsilon - 1}{4\pi}\left(1 - \frac{4\pi}{3}N_1\alpha + 12\alpha\frac{x_l - x_s}{x^4}\right) \tag{P.1.5.10}$$

from which $\varepsilon = 4.0127$ and $n = 2.003$, which are lower than the undistorted values.

In a similar way, for case (b) one obtains $\varepsilon = 4.4607$ and $n = 2.112$.

Problem 6

Calculate the dipolar moment of a sphere with macroscopic permittivity ε_1 surrounding a rigid dipole with external moment μ_e [7, p. 167].

Solution

If μ_e points in the z direction, then

$$P_z = \frac{1}{4\pi}(\varepsilon_1 - 1)E_z = -\frac{\varepsilon_1 - 1}{4\pi}\frac{\partial\phi}{\partial z} \tag{P.6.1}$$

Note that

$$\phi = \frac{\mu_e \cos\theta}{\varepsilon_1 r^2} \quad \text{and} \quad \frac{\cos\theta}{r^2} = -\frac{\partial}{\partial z}\left(\frac{1}{r}\right) \tag{P.6.2}$$

Since the three axes are equivalent

$$\int_{\text{sphere}} P_z\, dv = (\varepsilon_1 - 1)\frac{\mu_e}{4\pi\varepsilon_1}\frac{4\pi}{3}\int_{\text{sphere}}\Delta\left(\frac{1}{r}\right)dv = -\frac{\varepsilon_1 - 1}{3\varepsilon_1}\mu_e \tag{P.6.3}$$

which is the same result as Eq. (1.12.13).

Problem 7

Find expressions for the directing and internal fields in terms of the applied field.

Solution

According to section (1.13), the directing field can be found by addition of the cavity field and the reaction field produced by a fictive dipole αE_d caused by the action of the directing field upon the matter inside the cavity. This is calculated without taking into account the dipole itself because the reaction field does not influence the direction of the permanent dipole moment of the molecule; in fact they have the same direction. Thus, E_d will be given by

$$E_d = G + \frac{\alpha}{a^3} \frac{2(\varepsilon - 1)}{2\varepsilon + 1} E_d \qquad \text{(P.1.7.1)}$$

Taking into account the value of the cavity field given by Eq. (1.11.10), E_d can be written as

$$E_d = \frac{(3\varepsilon)/(2\varepsilon + 1)}{1 - (\alpha/a^3)\{[2(\varepsilon - 1)]/(2\varepsilon + 1)\}} E_0 \qquad \text{(P.1.7.2)}$$

The internal field is given by [11, p. 176]

$$E_i = E_d + \frac{R}{1 - \alpha R} \bar{\mu} \qquad \text{(P.1.7.3)}$$

where R is given by Eq. (1.11.11) and $\bar{\mu}$ is the value of μ averaged over all orientations. As a consequence

$$E_i = E_d + \frac{(1/a^3)\{[2(\varepsilon - 1)]/(2\varepsilon + 1)\}}{1 - (\alpha/a^3)\{[2(\varepsilon - 1)]/(2\varepsilon + 1)\}} \frac{\mu_v^2}{3kT} E_d \qquad \text{(P.1.7.4)}$$

The substitution of Eq. (P.1.7.2) into Eq. (P.1.7.4) leads to

$$E_i = \left(1 + \frac{(1/a^3)\{[2(\varepsilon - 1)]/(2\varepsilon + 1)\}}{1 - (\alpha/a^3)\{[2(\varepsilon - 1)]/(2\varepsilon + 1)\}} \frac{\mu_v^2}{3kT}\right)$$
$$\times \frac{1}{1 - (\alpha/a^3)\{[2(\varepsilon - 1)]/(2\varepsilon + 1)\}} \frac{3\varepsilon}{2\varepsilon + 1} E_0 \qquad \text{(P.1.7.5)}$$

The internal field can also be obtained by adding the cavity field and the reaction field produced by μ_i, the internal dipole moment, that is

$$E_i = G + \frac{1}{a^3} \frac{2(\varepsilon - 1)}{2\varepsilon + 1} \mu_i \qquad \text{(P.1.7.6)}$$

Substitution of Eq. (1.11.10) and the value of μ_i implicit in Eq. (1.13.9) lead once more to Eq. (P.1.7.5).

Problem 8

Find the field far away from a dielectric ellipsoid with permittivity ε_1 in a medium of permittivity ε_2 under an external applied field. Find the alignment torque.

Solution

According to Eq. (1.C.5), when $\xi \gg a, b, c$ and $\xi \cong r^2$ we obtain

$$A(\xi) \cong \frac{abc}{3r^3} \tag{P.1.8.1}$$

By substituting this expression into Eq. (1.C.4) and taking the external applied field along the x axis as

$$\varphi_0 = -E_{x0} r \cos \theta \tag{P.1.8.2}$$

the potential outside the ellipsoid is given in polar coordinates by

$$\varphi_2 = \varphi_0 - \frac{E_{x0} \cos \theta}{r^2} \frac{abc(\varepsilon_1 - \varepsilon_2)}{3[\varepsilon_2 + (\varepsilon_1 - \varepsilon_2)A]} \tag{P.1.8.3}$$

and here the depolarizing factor is given by

$$A = \frac{abc}{2} \int_0^\infty \frac{ds}{(s + a^2)R_s} \tag{P.1.8.4}$$

Equation (P.1.8.3) means that the external applied field "sees" the ellipsoid as a dipole whose effective dipole moment is given by

$$\mu_x = \varepsilon_2 \frac{abc(\varepsilon_1 - \varepsilon_2)}{3[\varepsilon_2 + (\varepsilon_1 - \varepsilon_2)A_a]} E_{0x} = \frac{1}{4\pi} V(\varepsilon_1 - \varepsilon_2) E_x \tag{P.1.8.5}$$

where the second equality is a consequence of Eq. (1.C.1) and V is the volume of the spheroid.

According to Eq. (1.7.2), the x component of the torque will be given by

$$\Gamma_x = \frac{1}{4\pi} V \frac{(\varepsilon_1 - \varepsilon_2)^2}{\varepsilon_2} \frac{A_c - A_b}{[1 + ((\varepsilon_1 - \varepsilon_2)/\varepsilon_2)A_b][1 + ((\varepsilon_1 - \varepsilon_2)/\varepsilon_2)A_c]} E_{0y} E_{0z} \tag{P.1.8.6}$$

with similar expressions for Γ_y and Γ_z.

The sign of each of the three components of the torque is determined by the relative values of the depolarizing factors.

More specific results can be found for prolate and oblate spheroids.[15]

Problem 9

An empty ellipsoidal cavity in a dielectric medium with permittivity ε has a nonpolarizable dipole μ oriented along the principal axis of the ellipsoid. Determine the reaction field in the cavity that is due to the dipole.

Solution

The expressions for the potential are

$$\varphi = \frac{\mu \cos \theta}{r^2} - Rr \cos \theta, \qquad r < a$$

$$\varphi = \frac{\mu * \cos \theta}{\varepsilon r^2} = \frac{\mu \cos \theta}{[\varepsilon + (1 - \varepsilon)A]r^2}$$

$$(P.1.9.1)$$

where R denotes the reaction field.

Over the ellipsoidal surface the two expressions for the potentials are equal and consequently

$$R = \frac{\mu}{abc} \frac{(\varepsilon - 1)(1 - A)}{\varepsilon - (\varepsilon - 1)A} \qquad\qquad (P.1.9.2)$$

which reduces to Eq. (1.11.11) when $A = 1/3$ and $a = b = c$, that is, for a spherical cavity.

Problem 10

An empty ellipsoidal cavity is in a dielectric medium with permittivity ε under an external electric field E_0. A molecule without a permanent dipolar moment and polarizability α is placed in the center of the cavity. Determine the dipole moment acquired by the molecule.

Solution

The dipole moment of the molecule will be the sum of the polarizations produced by (a) the internal field and (b) the reaction field, that is

$$\boldsymbol{\mu} = \alpha(\mathbf{E}_i + \mathbf{R}) \qquad\qquad (P.1.10.1)$$

where according to Eqs (1.C.9a) and (P.1.9.2) the following expressions hold

$$\mathbf{E}_i = \frac{\mathbf{E}_0}{1 + ((1 - \varepsilon)/\varepsilon)A}$$

$$\mathbf{R} = \frac{\boldsymbol{\mu}}{abc} \cdot \frac{(\varepsilon - 1)(1 - A)}{\varepsilon - (\varepsilon - 1)A} \qquad\qquad (P.1.10.2)$$

Substitution of Eq. (P.1.10.2) into Eq. (P.1.10.1) and further rearrangements of the expression obtained gives

$$\mu = \frac{(\alpha\varepsilon)/[\varepsilon - (\varepsilon - 1)A]}{1 - (\alpha/(abc))\{[(\varepsilon - 1)(1 - A)]/[\varepsilon - (\varepsilon - 1)A]\}} E_0 \qquad (P.1.10.3)$$

REFERENCES

1. Riande, E.; Saiz, E. *Dipole Moments and Birefringence of Polymers*; Prentice-Hall, Englewood Cliffs: NJ, 1992.
2. Volkenstein, M.V. *Configurational Statistics of Polymeric Chains*; translated from the Russian; Wiley Interscience: New York, NY, 1963.
3. Flory, P.J. *Statistical Mechanics of Chain Molecules*; Wiley Interscience, 1969.
4. Debye, P. *Polar Molecules*; Dover: New York, NY, 1936.
5. Powles, J.G. J. Chem. Phys. **1953**, *21*, 633.
6. Onsager, L. J. Chem. Phys. **1936**, *58*, 1486.
7. Fröhlich, H. *Theory of Dielectrics*, 2nd Ed.; Oxford, 1958; Appendix A.2, 163 pp
8. Kirkwood, J.G. J. Chem. Phys. **1939**, *7*, 911.
9. Fröhlich, H. Trans. Far. Soc. **1948**, *44*, 238.
10. Landau, L.; Lifshitz, E.M.; Pitaevskii, L.P. *Electrodynamics of Continuous Media*, 2nd Ed.; Butterworth-Heinemann: Oxford, 1984; 40 pp.
11. Böttcher, C.J.F. *Theory of Electric Polarization*, 2nd Ed.; Elsevier: Amsterdam, 1973; Vol. 1, 79 pp.
12. Osborn, J.A. Phys. Rev. **1945**, *67*, 351.
13. Lundblad, P. Ann. Physik. **1918**, *57*, 183.
14. Scaife, B.K.P. *Principles of Dielectrics*; Oxford University Press, 1989; 324 pp.
15. Jones, T.B. *Electromechanics of Particles*; Cambridge University Press, Cambridge, 1995.

2

Quasi-static Dipoles

2.1. BROWNIAN MOTION

Many processes in nature are stochastic. A stochastic process is a set of random time-dependent variables. The physical description of a stochastic process requires the formalization of the concepts of "probability" and "average". When the time average is equal to the ensemble average, the process is said to be ergodic. Ergodicity is a property of stationary processes. A Markov process is a stochastic process whose future behavior is only determined by the present state, not the earlier states. The best-known example of a Markov process in physics is the Brownian motion.[1,2] The first account of Brownian motion was given by the English botanist R. Brown, who was concerned with the transport of pollen grains into the ovule of a plant.

Brownian motion refers to the trajectory of a heavy particle immersed in a fluid of light molecules that collide randomly with it. The velocity of the particle varies by a number of uncorrelated jumps. After a number of collisions, the velocity of the particle has a certain value v. The probability for a certain change Δv in the velocity depends on the present value of v but not on preceding values of the velocity. An ensemble of dipoles can be considered a Brownian system.

Einstein[3] made conclusive mathematical predictions of the diffusive effect arising from the random motions of Brownian particles bombarded by other particles of the surrounding medium. Einstein's idea was to combine the Maxwell–Boltzmann distribution of velocities with the elementary Markov process known as random walk. The first clear explanation of random walk was made by Pearson,[4] who stated: "A man starts from a point O and walks l yards in a straight line; he then turns through any angle whatever and walks another l yards in a second straight line. He repeats this process n times. I require the probability that after these n stretches he is at a distance between r and $r + dr$ from his starting point."

The random walk model is very useful in many branches of physics and in particular can be used to describe the molecular chains of amorphous polymers. It can be demonstrated[5] that the probability distribution function for the end-to-end distance R of freely jointed chains can be obtained by solving the following partial differential equation

$$\frac{\partial P}{\partial n} = \frac{b^2}{6} \frac{\partial^2 P}{\partial R^2} \tag{2.1.1}$$

where P is the probability distribution, b is the length of each segment, and n is the number of bonds. It is also assumed that $n \gg 1$ and $R \gg b$ in order to consider P as a continuous function of n and R. Solving Eq. (2.1.1) under the condition that R is at the origin when $n = 0$ gives

$$P(R, n) = \left(\frac{3}{2\pi nb^2}\right)^{3/2} \exp\left(-\frac{3R^2}{2nb^2}\right) \tag{2.1.2}$$

Note that this parabolic differential equation is the same as that obtained for both the diffusion and the heat conduction and other random walk phenomena. The only difference is the coefficient $b^2/6$ appearing in it. Note also that Eq. (2.1.1) is the simplest case of a probability density diffusion equation known as the Fokker–Planck equation.

2.2. BRIEF ACCOUNT OF EINSTEIN'S THEORY OF BROWNIAN MOTION

If a particle in a fluid without friction collides with a molecule of the fluid, its velocity changes. However, if the fluid is very viscous, the change in the velocity is quickly dissipated and the net result of the impact is a change in the position of the particle. Thus, what it is generally observed at intervals of time in Brownian motion is the displacement of the particle after many variations in the velocity. As a result, random jumps in the position of the particle are observed. This is a consequence of the fact that the time interval between observations is larger than the time between collisions. Accordingly, the kinetic energy of translation of a Brownian particle behaves as a noninteractive molecule of gas, as required in the kinetic elementary theory of gases. Assuming very small jumps, Einstein obtained for the probability distribution of the displacement of particles the following partial differential equation

$$\frac{\partial f}{\partial t} = D\frac{\partial^2 f}{\partial x^2} \tag{2.2.1}$$

where

$$D = \frac{\langle \Delta x^2 \rangle}{2\tau} \tag{2.2.2}$$

is the diffusion coefficient, $\langle \Delta x^2 \rangle$ is the mean-square displacement, and τ is a time interval such that the motion of the particle at time t is independent of its motion at time $t \pm \tau$. The time interval τ is small in comparison with the experimental time. Note the close similarity between Eqs (2.1.1) and (2.2.1). Equation (2.2.2) suggests that the mean-square displacement of a Brownian particle is a linear function of the reciprocal of the jumping rate of such a particle. Finally, by using the Maxwell distribution of velocities, the diffusion coefficient is obtained as

$$D = \frac{kT}{\zeta} \tag{2.2.3}$$

where T is the temperature, k is the Boltzmann constant, and ζ is the friction coefficient.

2.3. LANGEVIN TREATMENT OF BROWNIAN MOTION

In 1908, Langevin[6] introduced the concept of the equation of motion of a random variable and subsequently initiated a new dynamic theory of Brownian motion in the context of stochastic differential equations. Langevin's approach is very useful in finding the effect of fluctuations in macroscopic systems.

Langevin developed the equation of motion of a Brownian particle following Newton's second law and using two assumptions: (1) the Brownian particle experiences a viscous force that represents a dynamic friction given by $-\zeta dv/dt = -\zeta \dot{x}$; (2) a fluctuating force $F(t)$ due to the impacts of the molecules of the surrounding fluid on the particle in question appears. The force fluctuates rapidly and is called white noise. Accordingly, the equation of motion of the particle is

$$m\ddot{x}(t) = -\zeta\dot{x}(t) + F(t) \tag{2.3.1}$$

The friction force is assumed in our context to be governed by Stokes's law. The pertinent expression for the friction coefficient of a spherical particle of radius a

and mass m, moving in a medium of viscosity η, is

$$\zeta = 6\pi\eta a \tag{2.3.2}$$

The force $F(t)$ is unpredictable, but its averaged effects over a certain ensemble are relatively simple. This ensemble may simply consist of a series of successive observations of the same particle in such a way that these observations do not influence each other. This is attained by leaving sufficient time intervals between observations. This is why $F(t)$ may be treated as a stochastic variable. In fact its average vanishes

$$\langle F(t) \rangle = 0 \tag{2.3.3}$$

2.4. CORRELATION FUNCTIONS

In general, a correlation is an interdependence between measurable random variables. In the present context we consider processes in which the variables evaluated at different times are such that their stochastic properties do not change with time. These processes are said to be stationary and the correlation between two variables in these processes is expressed by

$$C_{BA}(t) = \langle A(t_0)B(t_0 + t) \rangle = \langle A(0)B(t) \rangle \tag{2.4.1}$$

An autocorrelation function is a correlation function of the same variable at different times, that is

$$C(t)_{AA} = \langle A(0)A(t) \rangle \tag{2.4.2}$$

The normalized autocorrelation function is expressed as

$$C^0(t)_{AA} = \frac{\langle A(0)A(t) \rangle}{\langle (A(0))^2 \rangle} \tag{2.4.3}$$

where the denominator in Eq. (2.4.3) is the zero time, or static autocorrelation function.

Owing to the fact that the random force $F(t)$ in Eq. (2.3.1) is caused by the collisions of the molecules of the surrounding fluid on the Brownian particle, we can write

$$\langle \mathbf{F}(0) \cdot \mathbf{F}(t) \rangle = \Gamma\delta(t) \tag{2.4.4}$$

where Γ is a constant and $\delta(t)$ is the Dirac delta function. Equation (2.4.4) expresses the idea that the collisions are practically instantaneous and uncorrelated. It also indicates that $F(t)$ has a white spectrum. In more formal terms, Eq. (2.4.4) can be written as

$$\langle \mathbf{F}(0) \cdot \mathbf{F}(t) \rangle = \lim_{T \to \infty} \frac{1}{T} \int_0^T \mathbf{F}(0) \cdot \mathbf{F}(t') \, dt' \tag{2.4.5}$$

2.5. MEAN-SQUARE DISPLACEMENT OF A BROWNIAN PARTICLE

By multiplying both sides of Eq. (2.3.1) by $x(t)$ and taking into account that

$$\dot{x}x = \frac{1}{2} \frac{d(x^2)}{dt} \quad \text{and} \quad \ddot{x}x = \frac{1}{2} \frac{d}{dt} \left[\frac{d(x^2)}{dt} \right] - \dot{x}^2 \tag{2.5.1}$$

we obtain

$$\frac{1}{2} m \frac{d}{dt} \left[\frac{d(x^2)}{dt} \right] - m\dot{x}^2 = -\frac{1}{2} \zeta \frac{d(x^2)}{dt} + Fx \tag{2.5.2}$$

Averaging over all the particles, and assuming that the Maxwell distribution of velocities holds, that is

$$\tfrac{1}{2} m \langle \dot{x}^2 \rangle = \tfrac{1}{2} kT \tag{2.5.3}$$

the following expression is obtained

$$m \frac{d}{dt} \left[\frac{d\langle x^2 \rangle}{dt} \right] + \zeta \frac{d\langle x^2 \rangle}{dt} = 2kT \tag{2.5.4}$$

Note that $\langle Fx \rangle = 0$ since the random force is uncorrelated with the displacement.
Let us write

$$\frac{d\langle x^2 \rangle}{dt} = u \tag{2.5.5}$$

Then Eq. (2.5.4) may be expressed as

$$m\dot{u} + \zeta u = 2kT \tag{2.5.6}$$

whose solution is

$$u = \frac{2kT}{\zeta} + K \exp\left(-\frac{\zeta}{m} t \right) \tag{2.5.7}$$

where K is a constant of integration. For long times or friction constants larger than the mass, the exponential term has no influence on the motion of the particle after a small time interval. This is equivalent to excluding inertial effects. In these conditions, we have

$$u = \frac{d\langle x^2 \rangle}{dt} = \frac{2kT}{\zeta} \tag{2.5.8}$$

Integrating Eq. (2.5.8) from $t = 0$ to $t = \tau$, and assuming that $x = 0$ for $t = 0$, we obtain

$$\langle \Delta x^2 \rangle = \frac{2kT}{\zeta} \tau \tag{2.5.9}$$

where Δx is written instead of x because of its small value. This is the same result as that obtained by Einstein [see text between Eqs (2.2.1) and (2.2.3)]. This result indicates that we only observe the mean-square displacement of the elementary particle, neither its velocity nor its detailed path.

2.6. FLUCTUATION–DISSIPATION THEOREM

The Langevin equation can also be solved for the velocity, taking into account that $\dot{x} = v$ in Eq. (2.3.1). The resulting expression is

$$v(t) = \exp\left(-\frac{\zeta}{m} t\right)\left(v_0 + \frac{1}{m}\int_0^t \exp\left(\frac{\zeta}{m} t'\right) F(t') \, dt'\right) \tag{2.6.1}$$

where $v(0) = v_0$. By averaging this equation for an ensemble of particles all having the same initial velocity v_0, and noting that the noise term $F(t')$ is null in average and uncorrelated with velocity, we obtain

$$\langle v(t) \rangle = v_0 \exp\left(-\frac{\zeta}{m} t\right) \tag{2.6.2}$$

Squaring Eq. (2.6.1) and further integration of the resulting expression leads to

$$\langle v^2(t) \rangle = v_0^2 \exp\left(-2\frac{\zeta}{m} t\right) + \frac{\Gamma}{2\zeta}\left(1 - \exp\left(-2\frac{\zeta}{m} t\right)\right) \tag{2.6.3}$$

where use of Eq. (2.6.2) was made, Γ is given by Eq. (2.4.4), and it was taken into account that $\langle F(t) \cdot F(t') \rangle = \Gamma \delta(t - t')$.

In order to identify the constant Γ, we note that, for $t \to \infty$, Eq. (2.6.3) becomes

$$\langle v^2(t) \rangle = \frac{\Gamma}{2\zeta} = \frac{kT}{m} \tag{2.6.4}$$

where the Maxwell expression for $\langle v^2 \rangle$ [Eq. (2.5.3)] was used (see reference [7] p. 39 for details).

Equation (2.6.4) relates the size of the fluctuating term Γ to the damping constant ζ. In other words, fluctuations induce damping. This equation, the seminal form of which is Eq. (2.2.3), is in fact a first version of the fluctuation–dissipation theorem whose main relevant aspect is that it relates the mean microscopic noise to the macroscopic friction. As Van Kampen said (Ref. [1] p. 221): "The physical picture is that the random kicks tend to spread out v, while the damping term tries to bring v back to zero. The balance between these two opposite tendencies is the equilibrium distribution." The solution of the Langevin equation subject to a Maxwell–Boltzmann distribution of velocities is called the stationary solution.

2.7. SMOLUCHOWSKI EQUATION

A further step in the approach to Brownian motion is to formulate a general master equation that models more accurately the properties of the particles in question. In this respect, Eq. (2.2.1) can be written as a continuity equation for the probability density

$$\frac{\partial f}{\partial t} = -\text{div } \mathbf{j} \tag{2.7.1}$$

where $\mathbf{j}$ is the flux of Brownian grains or, in general, the flux of events suffered by the random variable whose probability density has been designed as f. Part of this flux is diffusive in origin and, according to a first constitutive assumption, can be written as

$$\mathbf{j}_{\text{diff}} = -D \text{ grad } f \tag{2.7.2}$$

where D is the diffusion coefficient given in this instance by kT/ζ. In this way, thermal agitation of Brownian particles is considered. The combination of Eqs (2.7.1) and (2.7.2) together with the assumption of the uniformity of D gives Eq. (2.2.1), according to Einstein's methodology.

Now, we introduce a second constitutive assumption. Let us suppose that the particles are subject to an external force that derives of a potential function, so that

$$\mathbf{F} = -\text{grad } U \tag{2.7.3}$$

Then, the current density of the particles that is due to this external force can be expressed as

$$\mathbf{j}_{\text{d}} = f\mathbf{v} \tag{2.7.4}$$

where $\mathbf{v}$ is the drift velocity of the particle. If the external force is related to the velocity through the equation

$$\mathbf{F} = \zeta \mathbf{v} \tag{2.7.5}$$

then the drift current due to this effect will be

$$\mathbf{j}_d = -\frac{f}{\zeta} \, \text{grad } U \tag{2.7.6}$$

The sum of the two current fluxes ($\mathbf{j} = \mathbf{j}_{\text{diff}} + \mathbf{j}_d$) and subsequent substitution into Eq. (2.7.1) yields

$$\frac{\partial f}{\partial t} = \text{div}\left(D \, \text{grad} f + \frac{f}{\zeta} \, \text{grad } U \right) \tag{2.7.7}$$

This equation, which describes the evolution of f, is known as the Smoluchowski equation. This is in fact a special type of Fokker–Planck master equation. The first term on the right-hand side of Eq. (2.7.7) is diffusive, and the second is convective. This second term is also called the transport or drift term, and it is represented by a hydrodynamic derivative. The Smoluchowski equation assumes that the velocity distribution has reached its equilibrium value given by the Maxwell distribution.

2.8. ROTATIONAL BROWNIAN MOTION

The arguments used to establish the Smoluchowski equation can be applied to the rotational motion of a dipole in suspension. In this case the fluctuating quantity is the angle or angles of rotation. In fact the Debye theory of dielectric relaxation has as its starting point a Smoluchowski equation for the rotational Brownian motion of a collection of homogeneous spheres each containing a rigid electric dipole $\boldsymbol{\mu}$, where the inertia of the spheres is neglected. The motion is due entirely to random couples with no preferential direction. We take at the center of the sphere a unit vector $\mathbf{u}(t)$ in the direction of $\boldsymbol{\mu}$. The orientation of this unitary vector is described only in terms of the polar and azimuthal angles, θ and φ. Having the system spherical symmetry, we must take the divergence of Eq. (2.7.1) in spherical coordinates, that is

$$\text{div } \mathbf{j} = \left(\frac{\partial}{\partial \theta} + \cot \theta \right) \mathbf{j}_\theta + \frac{1}{\sin \theta} \frac{\partial \mathbf{j}_\varphi}{\partial \varphi} \tag{2.8.1}$$

where the contribution to the divergence in the r coordinate axis is missing because there is no flux in this direction. The current density contains two terms, a diffusion term defined in Eq. (2.7.2) given in spherical coordinates by

$$\mathbf{j}_{\text{diff}} = -D \text{ grad } f = -D\left(\frac{\partial f}{\partial \theta}\mathbf{e}_\theta + \frac{1}{\sin\theta}\frac{\partial f}{\partial\varphi}\mathbf{e}_\varphi\right) \tag{2.8.2}$$

where $\mathbf{e}_\theta$ and $\mathbf{e}_\varphi$ are the unitary vectors corresponding to the θ and φ coordinates, and the term $f(\theta, \varphi, t)$ represents the density of dipole moment orientations on a sphere of unit radius.

On the other hand, the convective contribution to the current is due to the electric force field acting upon the dipole. The corresponding equation for such a current is similar to Eq. (2.7.6), but now U is the potential for the force produced by the electric field. This force can be calculated by combining the kinematic equation for the rate of change in the unity vector $\mathbf{u}$, given by

$$\dot{\mathbf{u}}(t) = \dot{\boldsymbol{\theta}}(t) \times \mathbf{u}(t) \tag{2.8.3}$$

with the noninertial Langevin equation for the rotating dipole, that is

$$\zeta\dot{\boldsymbol{\theta}}(t) - \mathbf{F}(t) - \boldsymbol{\mu} \times \mathbf{E} = 0 \tag{2.8.4}$$

where $\mathbf{F}$ is the white noise driving torque and $\boldsymbol{\mu} \times \mathbf{E}$ is the torque $\boldsymbol{\Gamma}$ due to an external field. For this reason, between two impacts, Eq. (2.8.4) can be written as

$$\dot{\boldsymbol{\theta}} = \boldsymbol{\Gamma}/\zeta \tag{2.8.5}$$

This is, in fact, the angular velocity of the dipoles under the effect of the applied field. Equation (2.8.4) is the differential equation for the rotational Brownian motion of a molecular sphere enclosing the dipole $\boldsymbol{\mu}$. Introduction of Eq. (2.8.4) into Eq. (2.8.3) yields

$$\zeta\dot{\boldsymbol{\mu}}(t) = \mathbf{F}(t) \times \boldsymbol{\mu} + (\boldsymbol{\mu} \times \mathbf{E}) \times \boldsymbol{\mu} \tag{2.8.6}$$

Equation (2.8.6) is the Langevin equation for the dipole in the noninertial limit. It is important to note that it is also valid for a linear molecule rotating in space.

The applied electric field is the negative gradient of a scalar potential U, which, in spherical coordinates, and omitting the r coordinate as before, is given by

$$\mathbf{E}(t) = -\text{ grad } U = -\frac{\partial U}{\partial\theta}\mathbf{e}_\theta - \frac{1}{\sin\theta}\frac{\partial U}{\partial\varphi}\mathbf{e}_\varphi = E_\theta\mathbf{e}_\theta + E_\varphi\mathbf{e}_\varphi \tag{2.8.7}$$

According to Eq. (2.8.7), and neglecting the noise term, we obtain

$$(\mathbf{u} \times \mathbf{E}) \times \mathbf{u} = E_\theta\mathbf{e}_\theta + E_\varphi\mathbf{e}_\varphi \tag{2.8.8}$$

As a consequence, the drift current density given by Eq. (2.7.6) is

$$\mathbf{j}_d = -\frac{f(\theta, \varphi, t)}{\zeta}\left(\frac{\partial U}{\partial \theta}\mathbf{e}_\theta + \frac{1}{\sin\theta}\frac{\partial U}{\partial \varphi}\mathbf{e}_\varphi\right) \tag{2.8.9}$$

where ζ is the drag coefficient for a sphere of radius a rotating about a fixed axis in a viscous fluid, given by

$$\zeta = 8\pi a^3 \eta \tag{2.8.10}$$

By combining Eqs (2.8.2) and (2.8.9) and using the continuity Eq. (2.7.1), the Smoluchowski equation for Brownian rotational motion is obtained

$$\frac{\partial f}{\partial t} = D\left(\frac{1}{\sin\theta}\frac{\partial}{\partial\theta}\left(\sin\theta\frac{\partial f}{\partial\theta}\right) + \frac{1}{\sin^2\theta}\frac{\partial^2 f}{\partial\varphi^2}\right)$$
$$+ \frac{1}{\zeta}\left[\frac{1}{\sin\theta}\frac{\partial}{\partial\theta}\left(\sin\theta f\frac{\partial U}{\partial\theta}\right) + \frac{1}{\sin^2\theta}\frac{\partial}{\partial\varphi}\left(f\frac{\partial U}{\partial\varphi}\right)\right] \tag{2.8.11}$$

At equilibrium, the classic Maxwell–Boltzmann distribution must hold and f is given by

$$f = A\exp\left(-\frac{U}{kT}\right) \tag{2.8.12}$$

Substitution of Eq. (2.8.12) into Eq. (2.8.11) leads to the Einstein relationship

$$D = \frac{kT}{\zeta} \tag{2.8.13}$$

The constant A in Eq. (2.8.12) can be determined from the normalization condition for the distribution function. Actually

$$\int_0^{2\pi} d\varphi \int_0^\pi f\sin\theta\, d\theta\, d\varphi = 1 \tag{2.8.14}$$

where f and U are given by Eqs (2.8.12) and (1.9.6) respectively. After integrating Eq. (2.8.14) we obtain

$$A = \frac{\mu E}{4\pi kT}\left(\sinh\frac{\mu E}{kT}\right)^{-1} \tag{2.8.15}$$

where sinh is the hyperbolic sine function. In general, $\mu E \ll kT$, so that

$$A \cong \frac{1}{4\pi} \tag{2.8.16}$$

If we define the Debye relaxation time as

$$\tau_D = \frac{\zeta}{2kT} \qquad (2.8.17)$$

then Eq. (2.8.11) becomes

$$2\tau_D \frac{\partial f}{\partial t} = \frac{1}{\sin\theta} \frac{\partial}{\partial\theta}\left(\sin\theta \frac{\partial f}{\partial\theta}\right) + \frac{1}{\sin^2\theta} \frac{\partial^2 f}{\partial\varphi^2}$$
$$+ \frac{1}{kT}\left[\frac{1}{\sin\theta} \frac{\partial}{\partial\theta}\left(\sin\theta f \frac{\partial U}{\partial\theta}\right) + \frac{1}{\sin^2\theta} \frac{\partial}{\partial\varphi}\left(f \frac{\partial U}{\partial\varphi}\right)\right] \qquad (2.8.18)$$

2.9. DEBYE THEORY OF RELAXATION PROCESSES

When the applied field is constant in direction but variable in time, and the selected direction is the z axis, we have

$$U = -\mu E \cos\theta \qquad (2.9.1)$$

where use was made of Eq. (1.9.6). In this case, we recover the equation obtained by Debye in his detailed original derivation[8] based on Einstein's ideas. According to this approach

$$\frac{\partial f}{\partial t} = \frac{1}{\sin\theta} \frac{\partial}{\partial\theta}\left[\sin\theta\left(\frac{kT}{\zeta} \frac{\partial f}{\partial\theta} + \frac{\mu E}{\zeta} \sin\theta f\right)\right] \qquad (2.9.2)$$

To calculate the transition probability, we assume that $U = 0$. Then Eq. (2.9.2) becomes

$$\sin\theta \frac{\partial f}{\partial t} = D \frac{\partial}{\partial\theta}\left(\sin\theta \frac{\partial f}{\partial\theta}\right) \qquad (2.9.3)$$

It is possible to calculate the mean-square value of $\sin\theta$ without solving the preceding equation. First, we note that

$$\langle \sin^2\theta \rangle = 2\pi \int_0^\pi f \sin^3\theta \, d\theta \qquad (2.9.4)$$

Multiplying Eq. (2.9.3) by $2\pi \sin^2\theta$ and further integration over θ yields

$$2\pi \int_0^\pi \sin^3\theta \frac{\partial f}{\partial t} d\theta = \frac{d}{dt}\langle \sin^2\theta \rangle = 2\pi D \int_0^\pi \sin^2\theta \frac{\partial}{\partial\theta}\left(\sin\theta \frac{\partial f}{\partial\theta}\right) d\theta \qquad (2.9.5)$$

After integrating by parts, 2 times, the second integral in Eq. (2.9.5), and taking into account the normalization condition, we obtain

$$\frac{d}{dt}\langle \sin^2 \theta \rangle = 4D - 6D\langle \sin^2 \theta \rangle \tag{2.9.6}$$

Integration of Eq. (2.9.6) with the condition $\langle \sin^2 \theta \rangle = 0$ for $t = 0$ yields

$$\langle \sin^2 \theta \rangle = \tfrac{2}{3}[1 - \exp(-6Dt)] \tag{2.9.7}$$

For small values of θ, $\langle \sin^2 \theta \rangle \cong \langle \theta^2 \rangle$. By taking the first term in the development of the exponential in Eq. (2.9.7), the following expression is obtained

$$\langle \theta^2 \rangle = 4Dt \tag{2.9.8}$$

which is equivalent to Eq. (2.2.2) for translational Brownian motion. When $t \to \infty$ in Eq. (2.9.7), the mean-square value for the sinus tends to $2/3$, which corresponds to the condition of equiprobability in all directions.

The non-negative solution of Eq. (2.9.2) can be expressed as

$$f = \frac{1}{4\pi}\sum_{n=0}^{\infty}(2n + 1)P_n(\cos \theta)\exp(-[n(n + 1)Dt]) \tag{2.9.9}$$

where P_n are Legendre polynomials. However, as far as dielectric relaxation is concerned, we are generally only interested in a linear approximation to the solution. In this case, we assume that

$$f(\theta, t) = \frac{1}{4\pi}\left(1 + \frac{\mu E}{kT}\psi\cos \theta\right) \tag{2.9.10}$$

where ψ is a time-dependent function to be determined.

Once the distribution function has been found, the mean-square dipole moment can be calculated from

$$\mu\langle \cos \theta \rangle = \frac{\int_0^{2\pi}\int_0^{\pi}\mu f \cos \theta \sin \theta \, d\theta \, d\varphi}{\int_0^{2\pi}\int_0^{\pi} f \sin \theta \, d\theta \, d\varphi} \tag{2.9.11}$$

Several cases could be considered. For a static field, $\mathbf{E} = \mathbf{E}_0$, substitution of Eq. (2.9.10) into Eq. (2.9.2) gives

$$f(\theta, t) = \frac{1}{4\pi}\left[1 + \frac{\mu E_0}{kT}\exp\left(-\frac{t}{\tau_D}\right)\cos\theta\right] \tag{2.9.12}$$

Therefore, according to Eq. (2.9.10)

$$\psi(t) = \exp\left(\frac{-t}{\tau_D}\right) \tag{2.9.13}$$

and

$$\mu\langle\cos\theta\rangle = \frac{\mu^2 E_0}{3kT}\exp\left(-\frac{t}{\tau_D}\right) \tag{2.9.14}$$

For an alternating field, $E = E_m\exp(i\omega t)$, substitution of Eq. (2.9.10) into Eq. (2.9.2) gives

$$\psi(\omega) = \frac{1}{1 + i\omega\tau_D} \tag{2.9.15}$$

Note that, in this equation, $\psi(\omega)$ is $L[-\dot{\psi}(t)]$, where L is the Laplace transform and $\psi(t)$ is given by Eq. (2.9.13). The mean dipole can be written as

$$\mu\langle\cos\theta\rangle = \frac{\mu^2 E}{3kT}\frac{1}{1 + i\omega\tau_D} \tag{2.9.16}$$

or

$$\mu\langle\cos\theta\rangle = \frac{\mu^2 E_m}{3kT}\frac{1}{1 + \omega^2\tau_D^2}[(\cos\omega t + \omega\tau_D\sin\omega t)$$
$$+ i(\sin\omega t - \omega\tau_D\cos\omega t)] \tag{2.9.17}$$

Note that the difference in phase between $\mu\langle\cos\theta\rangle$ and $\mathbf{E}$ persists if the real or imaginary parts of E are taken. On the basis of Eq. (2.8.10), Debye[8] estimated τ_D for several polar liquids, finding values of the order of $\tau_D = 10^{-11}$ s. Thus the maximum absorption should occur at the microwave region.

2.10. DEBYE EQUATIONS FOR THE DIELECTRIC PERMITTIVITY

Debye used the Lorentz field and replaced the static permittivity with the dynamic permittivity. From Eq. (1.8.5), the polarization in an alternating field can be written as

$$\mathbf{P} = \frac{\varepsilon^*(\omega) - 1}{4\pi}\mathbf{E} = N_1\left[\alpha_\infty + \frac{\mu^2}{3kT(1 + i\omega\tau_D)}\right]\frac{\varepsilon^*(\omega) + 2}{3}\mathbf{E} \qquad (2.10.1)$$

from which

$$\frac{\varepsilon^*(\omega) - 1}{\varepsilon^*(\omega) + 2} = \frac{4\pi}{3}N_1\left(\alpha_\infty + \frac{\mu^2}{3kT(1 + i\omega\tau_D)}\right) \qquad (2.10.2)$$

At low and high frequencies ($\omega \to 0$, $\omega \to \infty$), the two following limiting equations are obtained

$$\frac{\varepsilon_0 - 1}{\varepsilon_0 + 2} = \frac{4\pi}{3}N_1\left(\alpha_\infty + \frac{\mu^2}{3kT}\right)$$

$$\frac{\varepsilon_\infty - 1}{\varepsilon_\infty + 2} = \frac{4\pi}{3}N_1\alpha_\infty \qquad (2.10.3)$$

where ε_0 and ε_∞ are, respectively, the unrelaxed ($\omega = \infty$) and relaxed ($\omega = 0$) dielectric permittivities. Rearrangement of Eq. (2.10.2) using Eq. (2.10.3) gives

$$\varepsilon^*(\omega) = \frac{\varepsilon_0/(\varepsilon_0 + 2) + i\omega\tau_D[\varepsilon_\infty/(\varepsilon_\infty + 2)]}{1/(\varepsilon_0 + 2) + i\omega\tau_D[1/(\varepsilon_\infty + 2)]} \qquad (2.10.4)$$

or more conveniently

$$\frac{\varepsilon^*(\omega) - 1}{\varepsilon^*(\omega) + 2} = \frac{\varepsilon_\infty - 1}{\varepsilon_\infty + 2} + \left(\frac{\varepsilon_0 - 1}{\varepsilon_0 + 2} - \frac{\varepsilon_\infty - 1}{\varepsilon_\infty + 2}\right)\frac{1}{1 + i\omega\tau_D} \qquad (2.10.5)$$

If we define the reduced polarizability as

$$\alpha = \frac{\varepsilon - 1}{\varepsilon + 2} \qquad (2.10.6)$$

we can write

$$\alpha^*(\omega) = \alpha_\infty + (\alpha_0 - \alpha_\infty)\frac{1}{1 + i\omega\tau_D} = \alpha_\infty + (\alpha_0 - \alpha_\infty)L[-\dot{\psi}(t)] \qquad (2.10.7)$$

More physical insight into the reduced polarizability will be developed in sections 2.12 and 2.24.

Equation (2.10.5) can alternatively be written as

$$\frac{\varepsilon^*(\omega) - \varepsilon_\infty}{\varepsilon_0 - \varepsilon_\infty} = \frac{1}{1 + i\omega\tau} \tag{2.10.8}$$

where $\tau = \tau_D[(\varepsilon_0 + 2)/(\varepsilon_\infty + 2)]$ is the Debye macroscopic relaxation time. Splitting Eq. (2.10.8) into the real and imaginary parts, the following expressions are obtained for the permittivity and loss

$$\varepsilon'(\omega) = \varepsilon_\infty + \frac{\varepsilon_0 - \varepsilon_\infty}{1 + \omega^2\tau^2}$$

$$\varepsilon''(\omega) = \frac{(\varepsilon_0 - \varepsilon_\infty)\omega\tau}{1 + \omega^2\tau^2} \tag{2.10.9}$$

If we define the dielectric loss angle as $\delta = \tan^{-1}(\varepsilon''/\varepsilon')$, the following expression for $\tan\delta$ is obtained

$$\tan\delta(\omega) = \frac{(\varepsilon_0 - \varepsilon_\infty)\omega\tau}{\varepsilon_0 + \varepsilon_\infty\omega^2\tau^2} \tag{2.10.10}$$

2.11. MACROSCOPIC THEORY OF THE DIELECTRIC DISPERSION

Debye equations can also be obtained by considering a first-order kinetics for the rate of rise or decay of the dipolar polarization. When an electric field E is applied to a dielectric, the distortion polarization, P_∞, is very quickly established (nearly instantaneously). However, the dipolar part of the polarization, P_d, takes a time to reach its equilibrium value. Assuming that the increase rate of the polarization is proportional to the departure from its equilibrium value, we have*

$$\frac{dP_d}{dt} = \frac{P - P_d - P_\infty}{\tau} \tag{2.11.1}$$

where τ is the macroscopic relaxation time, and P is the equilibrium value of the total polarization which is related to the applied electric field E through an equation similar to Eq. (1.8.5), namely

$$P = \chi_0 E = \frac{\varepsilon_0 - 1}{4\pi} E \tag{2.11.2}$$

Consequently, Eq. (2.11.1) can alternatively be written as

$$\frac{dP_d}{dt} = \frac{\chi_0 E - P_d - P_\infty}{\tau} \tag{2.11.3}$$

---

*P_d and P_∞ are related to the polarizabilities α_d and α_∞ by $P_i = (1/3)\pi N_A \alpha_i$, i = d, ∞, where N_A is Asogadro's number.

where $P_\infty = \chi_\infty E = [(\varepsilon_\infty - 1)/(4\pi)]E$. Equation (2.11.3) can also be written as

$$P_d + \tau \frac{dP_d}{dt} = (\chi_0 - \chi_\infty)E \tag{2.11.4}$$

Integrating Eq. (2.11.4) under a steady electric field, using the initial condition $P = \chi_\infty E$ for $t = 0$, gives

$$P_d = \left\{ \chi_\infty + (\chi_0 - \chi_\infty)\left[1 - \exp\left(-\frac{t}{\tau}\right)\right] \right\} E \tag{2.11.5}$$

where

$$\chi(t) = \chi_\infty + (\chi_0 - \chi_\infty)\left[1 - \exp\left(-\frac{t}{\tau}\right)\right] \tag{2.11.6}$$

The first term of Eq. (2.11.6) is the time-dependent dielectric susceptibility. The complex susceptibility is defined as

$$\chi^* = s\chi(s) \tag{2.11.7}$$

where, according to the usual definition, $\chi(s)$ is the Laplace transform of the susceptibility. By taking the Laplace transform on both sides of Eq. (2.11.6) and making

$$\chi^* = \frac{\varepsilon^* - 1}{4\pi}, \qquad \chi_0 = \frac{\varepsilon_0 - 1}{4\pi}, \qquad \chi_\infty = \frac{\varepsilon_\infty - 1}{4\pi} \tag{2.11.8}$$

the Debye equation [Eq. (2.10.8)] is recovered.

2.12. DIELECTRIC BEHAVIOR IN TIME-DEPENDENT ELECTRIC FIELDS

The analysis of the dielectric response to dynamic fields can be performed in terms of the polarizability or, alternatively, in terms of the permittivity. In the first case we choose the geometry of the sample material in order to ensure a uniform polarization. The simplest geometry accomplishing such a condition is spherical. The advantage of this approach is that the macroscopic polarizability is directly related to the total dipole moment of the sphere, that is, the sum of the dipole moments of the individual dipoles contained in the sphere. Obviously such dipoles are microscopic in character. However, analysis in terms of permittivities only requires the consideration of a parallel-plate capacitor with a thickness smaller than the largest wavelength involved in the measurements to ensure spatial uniformity in the electric field. Moreover, the sphere must be considered macroscopic because it still contains thousands of polar molecules.

The relationship between polarizability and permittivity in the case of spherical geometry is given by

$$\alpha = \frac{3V}{4\pi} \frac{\varepsilon - 1}{\varepsilon + 2} \tag{2.12.1}$$

where the term $3/(\varepsilon + 2)$ arises from the local field factor. Accordingly, dielectric analysis can be made in terms of the polarizability instead of the experimentally accessible permittivity if we assume the material to be a macroscopic spherical specimen of radius large enough to contain all the dipoles under study.[1] Additional advantages of the polarizability representation are: (1) from a microscopic point of view, long-range dipole–dipole coupling is reduced to a minimum in a sphere; (2) the effects of the high permittivity values at low frequencies are minimized in polarizability plots. We will return to this point in section 2.24.

Since superposition holds in linear systems[9], the linear relationship between electric displacement and electric field is given by

$$D(t) = \int_{-\infty}^{t} \varepsilon(t - \theta) \, dE(\theta) \tag{2.12.2}$$

where ε is a tensor-valued function which is reduced to a scalar for isotropic materials. Through a simple variables change $t - \theta = \tau$ and, after integration by parts, one obtains the more convenient expression

$$D(t) = \varepsilon_\infty E(t) + \int_{0}^{\infty} E(t - \tau) \dot{\varepsilon}(\tau) \, d\tau \tag{2.12.3}$$

where ε_∞ is $\varepsilon(t = 0)$ accounts for the instantaneous response of the electrons and nuclei to the electric field, thus corresponding to the instantaneous or distortional polarization in the material.

For a linear relaxation system,[10] the function $\dot{\varepsilon}$ is a monotone decreasing function of its argument, that is

$$(-1)^n \varepsilon^{(n)}(t) < 0 \tag{2.12.4}$$

For sinusoidal fields, that is, taking $E = E_0 \exp(i\omega t)$, and for times large enough to make the displacement also sinusoidal, a decay function $\Phi(t)$ for the present geometrical configuration can be defined as

$$\frac{d\Phi}{dt} = -\frac{\dot{\varepsilon}}{\varepsilon_0 - \varepsilon_\infty} \tag{2.12.5}$$

Then, Eq. (2.12.3) can be written in terms of the decay function as

$$D(t) = E_0 \exp(i\omega t)\left[\varepsilon_\infty + (\varepsilon_0 - \varepsilon_\infty)\int_0^\infty \exp(-i\omega\tau)\left(-\frac{d\Phi}{dt}\right)dt\right] \quad (2.12.6)$$

If we define the complex dielectric permittivity as the ratio between the displacement $D(t)$ and the sinusoidal applied field $E_0 \exp(i\omega t)$, the following expression for the permittivity is obtained

$$\varepsilon^*(\omega) = \varepsilon_\infty + (\varepsilon_0 - \varepsilon_\infty)L\left(-\frac{d\Phi}{dt}\right) \quad (2.12.7)$$

where L is the Laplace transform. The function Φ accounts for the decay of the orientational polarization after the removal of a previously applied constant field. In general, a field factor must be introduced into Eq. (2.12.7) in order to describe more complex geometries. Note the close relationship between Eq. (2.12.7) and the ψ function defined by Eq. (2.9.14). If one takes for Φ the simplest expression, namely a decreasing exponential of the form $\Phi = \exp(-t/\tau)$, the Debye equation [Eq. (2.10.8)] is recovered. As we will see later, Φ can be identified with the normalized and macroscopic correlation function.

Williams and Watts proposed[11] to extend the applicability of Eq. (2.10.7) by using in Eq. (2.12.6) a stretched exponential function for Φ of the form

$$\Phi = \exp\left(-\left(\frac{t}{\tau}\right)^\beta\right) \quad (2.12.8)$$

where $0 < \beta \leqslant 1$. A similar equation had been proposed many years earlier by Kohlrausch to describe electrical discharges in Leyden jars. For this reason, Eq. (2.12.8) is known as the Kohlrausch–Williams–Watts (KWW) equation. Calculation of the dynamic permittivity after insertion of Eq. (2.12.7) into Eq. (2.12.6) leads to an asymptotic development except for the case where $\beta = 1/2$. The final result is

$$\varepsilon^*(\omega) = \varepsilon_\infty + (\varepsilon_0 - \varepsilon_\infty)\sum_{n=1}^\infty \frac{(-1)^{n-1}}{(\omega\tau_0)^{\beta n}}\frac{\Gamma(n\beta + 1)}{\Gamma(n + 1)}\exp\left(\frac{in\pi\beta}{2}\right) \quad (2.12.9)$$

Note that, for some ranges of frequencies and for some values of β, a bad convergency of the series is observed.

For a periodic field given by

$$E = E_0 \cos \omega t \quad (2.12.10)$$

direct insertion of the electric field into Eq. (2.12.3) leads to

$$D(t) = E_0[\varepsilon'(\omega)\cos \omega t + \varepsilon''(\omega)\sin \omega t] \quad (2.12.11)$$

where

$$\varepsilon'(\omega) = \varepsilon_\infty + \int_0^\infty \dot{\varepsilon}(\tau) \cos \omega\tau \, d\tau$$

(2.12.12)

$$\varepsilon''(\omega) = \int_0^\infty \dot{\varepsilon}(\tau) \sin \omega\tau \, d\tau$$

These expressions are closely related to the cosine and sine transforms of the decay function. Moreover, the dielectric loss tangent is defined as

$$\tan \delta = \frac{\varepsilon''}{\varepsilon'}$$

(2.12.13)

Note that ε' and ε'' are respectively even and odd functions of the frequency, that is

$$\varepsilon'(\omega) = \varepsilon'(-\omega)$$
$$\varepsilon''(\omega) = -\varepsilon''(-\omega)$$

(2.12.14)

Equivalently

$$\varepsilon^*(\omega) = \overline{\varepsilon^*}(-\omega)$$

(2.12.15)

where the overbar denotes the complex conjugate.

2.13. DISSIPATED ENERGY IN POLARIZATION

It is well known from thermodynamics that the power spent during a polarization experiment is given by

$$\dot{W} = \frac{1}{4\pi} E(t) \frac{dD(t)}{dt}$$

(2.13.1)

By substituting Eqs (2.12.10) and (2.12.11) into Eq. (2.13.1), and with further integration of the resulting expression, the following equation is obtained for the work of polarization per cycle and volume unit

$$W = \int_0^{2\pi/\omega} \dot{W} \, dt = \frac{1}{4\pi} \omega E_0^2 \left[- \int_0^{2\pi/\omega} \varepsilon' \sin \omega t \cos \omega t \, dt + \int_0^{2\pi/\omega} \varepsilon'' \cos^2 \omega t \, dt \right]$$

(2.13.2)

The first integral on the right-hand side is zero, because the dielectric work done on part of the cycle is recovered during the remaining part of it. On the other hand, the second integral is related to the dielectric dissipation, and the total work in the complete cycle corresponds to the dissipated energy. This value is

$$W = \left(\frac{\varepsilon'' E_0^2}{4}\right) \tag{2.13.3}$$

The rate of loss of energy will be given by

$$\frac{\omega \varepsilon'' E_0^2}{8\pi} \tag{2.13.4}$$

2.14. DISPERSION RELATIONS

The formal structure of Eqs (2.12.12) indicates that the real and imaginary parts of the complex permittivity are, respectively, the cosine and sine Fourier transforms of the same function, that is, $\dot{\varepsilon}(\tau)$. As a consequence, ε' and ε'' are not independent. The inverse Fourier transform of Eqs (2.12.12) leads to

$$\dot{\varepsilon}(u) = \frac{2}{\pi} \int_0^\infty [\varepsilon'(x) - \varepsilon_\infty] \cos xu \, dx$$

$$\dot{\varepsilon}(u) = \frac{2}{\pi} \int_0^\infty \varepsilon''(x) \sin xu \, dx \tag{2.14.1}$$

After insertion of Eq. (2.14.1b) into Eq. (2.12.12a) and Eq. (2.14.1a) into Eq. (2.12.12b), the following equations are obtained

$$\varepsilon'(\omega) = \frac{2}{\pi} \int_0^\infty \left[\int_0^\infty \varepsilon''(x) \sin xu \, dx \right] \cos \omega u \, du$$

$$= \frac{2}{\pi} \int_0^\infty \left[\int_0^\infty \sin xu \cos \omega u \, du \right] \varepsilon''(x) \, dx$$

$$\tag{2.14.2}$$

$$\varepsilon''(\omega) = \frac{2}{\pi} \int_0^\infty \left[\int_0^\infty [\varepsilon'(x) - \varepsilon_\infty] \cos xu \, dx \right] \sin \omega u \, du$$

$$= \frac{2}{\pi} \int_0^\infty \left[\int_0^\infty \cos xu \sin \omega u \, du \right] [\varepsilon'(x) - \varepsilon_\infty] \, dx$$

Taking into account that

$$\int_0^\infty \sin xu \cos \omega u \, du = \frac{x}{x^2 - \omega^2} \qquad \text{and}$$

$$\tag{2.14.3}$$

$$\int_0^\infty \sin \omega u \cos xu \, du = \frac{\omega}{\omega^2 - x^2}$$

Eqs (2.14.2) become

$$\varepsilon'(\omega) - \varepsilon_\infty = \frac{2}{\pi} \int_0^\infty \frac{x\varepsilon''(x)}{x^2 - \omega^2} \, dx$$

$$\varepsilon''(\omega) = \frac{2}{\pi} \int_0^\omega \frac{\omega[\varepsilon'(x) - \varepsilon_\infty]}{\omega^2 - x^2} \, dx$$

(2.14.4)

where the integrals are taken in the sense of their principal values. These formulae relating dispersion and absorption are known as the Kramers–Kronig relationship.[12] They are a consequence of the linearity and causality of the systems under study and make it possible to calculate the real and imaginary parts of the dielectric permittivity one from the other.

The limit $\omega \to 0$ of Eq. (2.14.4a) leads to

$$\varepsilon_0 - \varepsilon_\infty = \Delta\varepsilon = \frac{2}{\pi} \int_\infty^\infty \varepsilon''(x) \, d\ln x$$

(2.14.5)

where $\varepsilon_0 = \varepsilon(\omega = 0)$. Equation (2.14.5) can be used to calculate the relaxation strength from the dielectric loss spectrum.

2.15. ENERGY DISSIPATION AND THE DEBYE PLATEAU

It was shown in section (2.13) that the power loss is proportional to $\omega\varepsilon''(\omega)$. Assuming that the Debye equations [Eq. (2.10.7)] are valid over the entire range of frequencies, the total energy loss diverges

$$\int_{-\infty}^\infty \omega\varepsilon''(\omega) \, d\omega \to \infty$$

(2.15.1)

The divergence of Eq. (2.15.1) arises from the constancy of the integrand in this equation at large frequencies. This constancy is known as the Debye plateau as it can be seen in the absorption coefficient (reference [13] p.33 or reference [14] p.72). Since the total energy loss for a finite specimen should be finite, the results show that Debye equations become inadequate at the high frequency limit. The physical reason is that new mechanisms at atomic or electronic level appear at very high frequencies. These mechanisms are roughly included in the context of the Debye theory into the relaxed permittivity ε_∞ but become relevant at these frequencies, also making relevant the inertial term in the Langevin equation. In fact, dipoles cannot follow high-frequency fields because of the molecular inertia.

2.16. INERTIAL EFFECTS

In previous analysis, inertial effects have been neglected. However, at micro-wave frequencies the rotational kinetic energy of the molecules, $\frac{1}{2}I\dot{\theta}^2$, may be comparable with its thermal energy. The importance of this effect was first considered by Rocard[15] and will be extended here following Powles' approach.[16] The Powles treatment is a modification of the arguments given in section 2.8. The starting point is the Langevin equation for a dipole under an electric field where the inertial term is included but the terms corresponding to the white noise driving force are neglected. The corresponding scalar equation is

$$I\ddot{\theta} + \zeta\dot{\theta} = \Gamma = -\mu E \sin \theta \tag{2.16.1}$$

The solution of this equation is

$$\dot{\theta} = \frac{\Gamma}{\zeta + i\omega I} = \frac{\Gamma}{\zeta}\frac{1}{1 + i\omega I/\zeta} \tag{2.16.2}$$

The effect of the inertia is to modify Eq. (2.8.5) by the factor $(1 + i\omega I/\zeta)^{-1}$, where I/ζ is called the friction time τ_F. Also, the free rotation time is defined as

$$\tau_R = \left(\frac{I}{kT}\right)^{1/2} \tag{2.16.3}$$

This time is the inverse of the angular velocity of a rotating molecule with rotational energy given by $\frac{1}{2}kT$, the mean kinetic energy in statistical equilibrium. The relationship between the three relaxation times is given by

$$\tau_D = \frac{\tau_R^2}{2\tau_F} \tag{2.16.4}$$

Then, the corresponding Debye–Smoluchowsky equation for the polar inertial molecules is, instead of Eq. (2.9.3), the following expression

$$\frac{\partial f}{\partial t} = \frac{1}{\sin \theta}\frac{\partial}{\partial \theta}\left[\sin \theta\left(D\frac{\partial f}{\partial \theta} - \frac{\Gamma}{\zeta}\frac{1}{1 + i\omega I/\zeta}f\right)\right] \tag{2.16.5}$$

Following Debye, a new diffusion coefficient could be derived when the steady-state condition $(\partial f/\partial t = 0)$ is applied to Eq. (2.6.5) taking into account

Eqs (2.8.11) and (2.8.15). The resulting value for a frequency-dependent D is

$$D = \frac{kT}{\zeta} \frac{1}{1 + i\omega I/\zeta} \tag{2.16.6}$$

which reduces to Eq. (2.2.3), the classical expression at $\omega = 0$.

In view of the uncertainty of D at high frequencies, one can consider a more general expression for D by making

$$D = \frac{kT}{\zeta} \frac{1}{1 + i\alpha\omega I/\zeta} \tag{2.16.7}$$

where α is an empirical parameter lying in the interval $0 \leqslant \alpha \leqslant 1$. Two extreme cases will be considered for $\alpha = 0$ and $\alpha = 1$. Assuming that the distribution function is described by Eq. (2.9.10), then, on account of Eqs (2.16.5) and (2.16.7), we obtain

$$\frac{d\psi}{dt} = -\frac{1}{\tau_D} \frac{\psi}{1 + i\alpha\omega\tau_F} \tag{2.16.8}$$

Here, high-order terms were neglected and Eqs (2.8.17), (2.16.3), and (2.16.4) were taken into account. Note that, when $\tau_F = 0$, that is $I = 0$, we recover the exponential function for $\psi(t)$. For alternating fields, the following expression for ψ is obtained

$$\psi = \frac{1}{(1 + i\omega\tau_F)[i\omega\tau_D + 1/(1 + i\alpha\omega\tau_F)]} \tag{2.16.9}$$

which reduces to Eq. (2.9.5) when $\tau_F = 0$. Since $\omega\tau_F \ll 1$, it is possible to make the approximation $(1 + i\alpha\omega\tau_F)^{-1} \cong 1 - i\alpha\omega\tau_F$, and Eq. (2.16.9) becomes

$$\psi = \frac{1}{1 + i\omega\tau_D + i(1 - \alpha)\omega\tau_F - \omega^2\tau_D\tau_F} \tag{2.16.10}$$

By assuming $\alpha = 0$ in this equation, the following expression, known as the Rocard equation, is obtained

$$\psi = \frac{1}{(1 + i\omega\tau_D)(1 + i\omega\tau_F)} \tag{2.16.11}$$

By writing $\alpha = 1$ in Eq. (2.16.10), then

$$\psi = \frac{1}{1 + i\omega\tau_D - \omega^2\tau_D\tau_F} \tag{2.16.12}$$

which, on account of Eqs (2.16.3) and (2.16.4), can be written as

$$\psi = \frac{1}{1 + i\omega\tau - (I\omega^2)/(2kT)} \tag{2.16.13}$$

Owing to the small values of the inertia moments of the rotating dipoles, inertial effects are only relevant at high frequencies ($\geqslant 10^{12}$ Hz) (see section 2.28).

The lack of the fluctuating force that appears in the equation of dipole motions [Eq. (2.16.1)] implies the elimination of an essential ingredient of Brownian motion. This is the force due to the impacts of the molecules of the surrounding fluid on the dipole. In fact, as pointed out by McConnell,[13] the angle θ changes continuously in the actual rotational Brownian motion, but the impacts produce discontinuous changes in $\dot{\theta}$. Since $\dot{\theta}$ is discontinuous, Eq. (2.16.1), which includes $\ddot{\theta}$, has meaningless consequences, and therefore the analysis of Powles is inadequate. However, the Rocard–Powles equation [Eq. (2.16.10)] can be obtained as a limit of a more elaborate theory (see reference [13]) corresponding to inertial motions of rotating discs or spherical models.

2.17. LANGEVIN EQUATION FOR THE DIPOLE VECTOR

As we have seen in the preceding section, the Langevin equation for Brownian rotational motion is a good starting point to investigate the relaxation problems in condensed matter. In order to write down this equation for the more general case, we include in the total torque term a white noise torque, $\Lambda(t)$, in such a way that

$$\mathbf{\Gamma}(t) = \mathbf{\Lambda}(t) + \mathbf{N}(t) \tag{2.17.1}$$

where $\mathbf{N}(t)$ is now the external torque due to the applied field, that is

$$\mathbf{N}(t) = \boldsymbol{\mu} \times \mathbf{E} \tag{2.17.2}$$

Accordingly, the Langevin equation is given by

$$I\ddot{\boldsymbol{\theta}} + \zeta\dot{\boldsymbol{\theta}} = \mathbf{\Lambda}(t) + \mathbf{N}(t) \tag{2.17.3}$$

where the angle of libration–rotation as well as its derivatives now have vectorial character.

In this section, the Langevin equation will be obtained in terms of the dipole moment and its time derivatives.[17] For simplicity, only linear molecules will be considered. The kinematic equation of the dipole moment can be written as

$$\dot{\boldsymbol{\mu}}(t) = \dot{\boldsymbol{\theta}}(t) \times \boldsymbol{\mu}(t) = \boldsymbol{\omega}(t) \times \boldsymbol{\mu}(t) \tag{2.17.4}$$

The angular velocity vector $\boldsymbol{\omega}(t)$ in the case of linear molecules can be decomposed into the sum of two components according to the usual procedure

$$\boldsymbol{\omega} = \boldsymbol{\omega}_z + \boldsymbol{\omega}_p \tag{2.17.5}$$

where $\boldsymbol{\omega}_z$ and $\boldsymbol{\omega}_p$ are respectively the components along the dipole moment vector and precesional of the angular velocity. Consequently

$$\dot{\boldsymbol{\mu}}(t) = \boldsymbol{\omega}_p(t) \times \boldsymbol{\mu}(t) \tag{2.17.6}$$

The time derivative of Eq. (2.17.5) gives

$$\ddot{\boldsymbol{\mu}} = \dot{\boldsymbol{\omega}}_p \times \boldsymbol{\mu} + \boldsymbol{\omega}_p \times \dot{\boldsymbol{\mu}} = \dot{\boldsymbol{\omega}}_p \times \boldsymbol{\mu} + \boldsymbol{\omega}_p \times (\boldsymbol{\omega}_p \times \boldsymbol{\mu})$$

$$= \boldsymbol{\omega}_p(\boldsymbol{\omega}_p \cdot \boldsymbol{\mu}) - \omega_p^2 \boldsymbol{\mu} + \dot{\boldsymbol{\omega}}_p \times \boldsymbol{\mu} \tag{2.17.7}$$

By using the vectorial equality

$$\mathbf{a} \times (\mathbf{b} \times \mathbf{c}) = \mathbf{b}(\mathbf{a} \cdot \mathbf{c}) - \mathbf{c}(\mathbf{a} \cdot \mathbf{b}) \tag{2.17.8}$$

the Langevin equation becomes

$$\ddot{\boldsymbol{\mu}} = (-\zeta \dot{\boldsymbol{\mu}} + \boldsymbol{\Lambda}(t) + \boldsymbol{\mu} \times \mathbf{E})I^{-1} + \boldsymbol{\omega}_p(\boldsymbol{\omega}_p \cdot \boldsymbol{\mu}) - \omega_p^2 \boldsymbol{\mu} \tag{2.17.9}$$

This equation, in conjunction with Eqs (2.17.8) and (2.17.6), gives

$$\ddot{\boldsymbol{\mu}} + I^{-1}\zeta \dot{\boldsymbol{\mu}} + \omega_p^2 \boldsymbol{\mu} - \boldsymbol{\omega}_p(\boldsymbol{\omega}_p \cdot \boldsymbol{\mu})$$

$$= I^{-1}(\boldsymbol{\Lambda}(t) \times \boldsymbol{\mu}) + I^{-1}(\mu^2 \mathbf{E} - \boldsymbol{\mu}(\boldsymbol{\mu} \cdot \mathbf{E})) \tag{2.17.10}$$

Note that μ^2 is constant because the dipole is rigid with constant modulus.

If $\boldsymbol{\omega}$ is normal to $\boldsymbol{\mu}$, Eq. (2.17.10) reduces to

$$\ddot{\boldsymbol{\mu}} + I^{-1}\zeta \dot{\boldsymbol{\mu}} + \omega_p^2 \boldsymbol{\mu} = I^{-1}(\boldsymbol{\Lambda}(t) \times \boldsymbol{\mu}) + I^{-1}(\mu^2 \mathbf{E} - \boldsymbol{\mu}(\boldsymbol{\mu} \cdot \mathbf{E})) \tag{2.17.11}$$

On the other hand, if the external field is applied along the Z axis of the laboratory reference frame and $\boldsymbol{\mu}$ is projected onto that axis

$$\boldsymbol{\mu}|_Z = p = \mu \cos \theta \tag{2.17.12}$$

then Eq. (2.17.11) is given by

$$\ddot{p} + \zeta I^{-1} \dot{p} + \omega_p^2 p = I^{-1} E(t)(\mu^2 - p^2) - I^{-1}(\boldsymbol{\mu} \times \boldsymbol{\Lambda}(t))_Z \qquad (2.17.13)$$

which can alternatively be written as

$$\ddot{p} + \zeta I^{-1} \dot{p} + [\omega_p^2 + I^{-1} E(t)p]p = I^{-1} \mu^2 E(t) - I^{-1} \mu \sin \theta \Lambda(t) \qquad (2.17.14)$$

From this equation, the Debye equation is obtained. In order to linearize this equation, we average over all the positions and velocities of all the dipoles, taking, for simplicity, a constant electric field

$$\langle \ddot{p} \rangle + I^{-1} \zeta \langle \dot{p} \rangle + \langle (\omega_p^2 + pE_0)p \rangle = \mu^2 E_0 - \langle \mu \sin \theta \Lambda(t) \rangle \qquad (2.17.15)$$

The last term on the right-hand side vanishes because the random torque varies in an arbitrary way. From trigonometry, we have

$$\langle p^2 \rangle E_0 = \frac{\mu^2 E_0}{2} \langle 1 + \cos 2\theta \rangle \qquad (2.17.16)$$

where $\langle \cos 2\theta \rangle = 0$. Thus, for a linear response

$$\langle p^2 \rangle E_0 = \frac{\mu^2 E_0}{2} \qquad (2.17.17)$$

Assuming a Maxwell distribution for the angular velocity

$$\langle \ddot{p} \rangle = 0 \qquad (2.17.18)$$

and if the angular velocity and the dipole orientation of the aforementioned distribution are decoupled, then

$$\langle \omega^2 p \rangle = \frac{kT \langle p \rangle}{I} \qquad (2.17.19)$$

As a consequence, the linearized version of Eq. (2.17.15) can be written as

$$\zeta \langle \dot{p} \rangle + kT \langle p \rangle = \frac{\mu^2 E_0}{2} \qquad (2.17.20)$$

After the application of a step electric field, the solution of Eq. (2.17.20) is the Debye response

$$\langle p \rangle = \frac{\mu^2 E_0}{2kT} \left[1 - \exp\left(-\frac{kT}{\zeta} t \right) \right] \qquad (2.17.21)$$

Note the factor 2 in the expression for $\langle p \rangle$, in contrast to the factor 3 appearing in Eq. (2.9.14). This is due to the fact that we are dealing with a plane rotor in contrast to the three-dimensional rotation studied in section 2.9. Having in mind the preceding consideration, a similar equation to (2.9.17) could be obtained for an ac field.

2.18.   DIFFUSIVE THEORY OF DEBYE AND THE ONSAGER MODEL

In section (1.10) we emphasized the limitations imposed by the assumption of the Lorentz field on the static theory of permittivity. The Onsager model [Eq. (1.13.9)], though with limitations, gives a better representation of the static dielectric properties of liquids. In these conditions, it is interesting to extend the Onsager theory to the case of frequency-dependent fields. This problem was first addressed by Collie, Hasted and Ritson.[18] The calculation will be carried out in two steps. The reason for this is that, while the cavity field varies with frequency in the same way as the applied field does, the reaction field also varies because the thermal motion of the molecules is the same at high and low frequencies.

When an alternating field is applied to a dielectric medium, the field in an empty cavity is given by

$$E_i = \frac{3\varepsilon^*}{2\varepsilon^* + 1} E + \frac{2(\varepsilon^* - 1)\alpha}{2\varepsilon^* + 1} \frac{\alpha}{a^3} E_i \qquad (2.18.1)$$

where $\alpha/a^3 = (\varepsilon_\infty - 1)/(\varepsilon_\infty + 2)$. Equation (2.18.1) leads to

$$E_i = \frac{\varepsilon^*(\varepsilon_\infty + 2)}{2\varepsilon^* + \varepsilon_\infty} E \qquad (2.18.2)$$

The reaction field in the cavity containing a molecule with dipole moment $\mathbf{m}$ can be written as

$$R = \frac{2(\varepsilon_0 - 1)}{2\varepsilon_0 + 1} \frac{\mathbf{m}}{\varepsilon_0 a^3} \qquad (2.18.3)$$

where $\mathbf{m} = \boldsymbol{\mu} + [[2(\varepsilon_0 - 1)]/(2\varepsilon_0 + 1)(\alpha/a^3)]\mathbf{m}$. Consequently

$$\mathbf{m} = \frac{(\varepsilon_\infty + 2)(2\varepsilon_0 + 1)}{3(2\varepsilon_0 + \varepsilon_\infty)} \boldsymbol{\mu} \qquad (2.18.4)$$

and

$$R = \frac{2(\varepsilon_\infty + 2)(\varepsilon_0 - 1)}{3(2\varepsilon_0 + \varepsilon_\infty)} \frac{\boldsymbol{\mu}}{a^3} \qquad (2.18.5)$$

The total field acting on the molecule is obtained from the sum of Eqs (2.18.2) and (2.18.5)

$$\frac{\varepsilon^*(\varepsilon_\infty + 2)}{2\varepsilon^* + \varepsilon_\infty} \mathbf{E} + \frac{2(\varepsilon_\infty + 2)(\varepsilon_0 - 1)}{3(2\varepsilon_0 + \varepsilon_\infty)} \frac{\boldsymbol{\mu}}{a^3} \tag{2.18.6}$$

and the total moment is

$$\frac{(\varepsilon_\infty + 2)(2\varepsilon_0 + 1)}{3(2\varepsilon_0 + \varepsilon_\infty)} \boldsymbol{\mu} + \alpha \frac{\varepsilon^*(\varepsilon_\infty + 2)}{2\varepsilon^* + \varepsilon_\infty} \mathbf{E} \tag{2.18.7}$$

The torque will be

$$\boldsymbol{\Gamma} = \mathbf{m} \times (\mathbf{E_i} + \mathbf{R}) = \frac{\varepsilon^*(\varepsilon_\infty + 2)}{2\varepsilon^* + \varepsilon_\infty} \boldsymbol{\mu} \times \mathbf{E} \tag{2.18.8}$$

which, according to Debye, determines the mean orientation of the molecules as

$$\langle \cos \theta \rangle = \frac{\varepsilon^*(\varepsilon_\infty + 2)}{2\varepsilon^* + \varepsilon_\infty} \frac{\mu E}{3kT} \frac{1}{1 + i\omega\tau_D} \tag{2.18.9}$$

Then, the average dipole moment in the direction of the field is given by

$$\langle \boldsymbol{\mu}^* \rangle = \left[\frac{\varepsilon^*(2\varepsilon_0 + 1)(\varepsilon_\infty + 2)^2}{3(2\varepsilon_0 + \varepsilon_\infty)(2\varepsilon^* + \varepsilon_\infty)} \frac{\mu^2 E}{3kT} \frac{1}{1 + i\omega\tau_D} + \frac{\alpha\varepsilon^*(\varepsilon_\infty + 2)}{2\varepsilon^* + \varepsilon_\infty} \right] \mathbf{E} \tag{2.18.10}$$

The polarization by unit volume can be written as

$$\varepsilon^* - 1 = \frac{4\pi \mathbf{P}}{\mathbf{E}}$$

$$= 4\pi N_1 \left[\frac{\varepsilon^*(2\varepsilon_0 + 1)(\varepsilon_\infty + 2)^2}{3(2\varepsilon_0 + \varepsilon_\infty)(2\varepsilon^* + \varepsilon_\infty)} \frac{\mu^2 E}{3kT} \frac{1}{1 + i\omega\tau_D} + \frac{\alpha\varepsilon^*(\varepsilon_\infty + 2)}{2\varepsilon^* + \varepsilon_\infty} \right] \mathbf{E} \tag{2.18.11}$$

Since $(4\pi/3)N_1\alpha = (\varepsilon_\infty - 1)/(\varepsilon_\infty + 2)$, Eq. (2.18.11) becomes

$$\frac{(\varepsilon^* - \varepsilon_\infty)(2\varepsilon^* + 1)(2\varepsilon_0 + \varepsilon_\infty)}{\varepsilon^*(\varepsilon_\infty + 2)^2(2\varepsilon_0 + 1)} = \frac{4\pi N_1 \mu^2}{9kT} \frac{1}{1 + i\omega\tau_D} \tag{2.18.12}$$

This expression may be written as

$$\varepsilon^* - \varepsilon_\infty = \frac{A}{1 + i\omega\tau_D} \frac{\varepsilon^*}{\varepsilon^* + \frac{1}{2}} \tag{2.18.13}$$

where

$$A = \frac{2\pi N_1 \mu^2}{9kT} \frac{(2\varepsilon_0 + 1)(\varepsilon_\infty + 2)^2}{2\varepsilon_0 + \varepsilon_\infty} \qquad (2.18.14)$$

is frequency independent.

2.19. RELATIONSHIP BETWEEN MACROSCOPIC DIELECTRIC AND MECHANICAL PROPERTIES

In the development of the Debye theory for dielectric relaxation via the Smoluchowski equation, the friction coefficient ζ is the drag force of a rotating sphere in a viscous fluid [see Eq. (2.8.10)]. The torque is due to the external field acting on the dipole embedded in the sphere. It has been suggested by DiMarzio and Bishop[19] that a constant friction coefficient is a too crude description of how a viscous medium would react to an applied dynamic force field. Instead, a dynamic viscosity is required. This implies a generalization of Eq. (2.8.5). Accordingly

$$\dot{\theta}(\omega) = \frac{\Gamma(\omega)}{\zeta(\omega)} \qquad (2.19.1)$$

where ω is the angular frequency of the applied field. In the time domain, Eq. (2.19.1) can be written as a linear "memory" equation with all the technical qualifications of these integral representations. That is

$$\dot{\theta}(t) = \int_{-\infty}^{t} V(t - \tau)\Gamma(\tau)\,d\tau \qquad (2.19.2)$$

where

$$\zeta^{-1}(\omega) = V(\omega) \qquad (2.19.3)$$

is the Fourier transform of $V(t)$. In this case, the Smoluchowski equation becomes

$$\frac{\partial f}{\partial t} = \frac{1}{\sin\theta}\frac{\partial}{\partial\theta}\left\{\sin\theta\left[\frac{\partial}{\partial\theta}\int_{-\infty}^{t} D(t - \tau)f(\theta, \tau)\,d\tau - f\int_{-\infty}^{t} V(t - \tau)\Gamma(\tau)\,d\tau\right]\right\}$$

$$(2.19.4)$$

with $D(\omega) = kT/\zeta(\omega)$ and $D(t) = kTV(t)$. Following the same procedure as in sections 2.9 and 2.10 for an alternating field, one obtains

$$\frac{\varepsilon^* - \varepsilon_\infty}{\varepsilon_s - \varepsilon_\infty} = \left[1 + i\omega\tau_D(\omega)\frac{\varepsilon_s + 2}{\varepsilon_\infty + 2}\right] \tag{2.19.5}$$

This equation with the dynamic counterpart of Eq. (2.8.10) leads to

$$\tau_D(\omega) = \frac{\zeta(\omega)}{2kT} = \frac{8\pi R^3 \eta^*(\omega)}{2kT} \tag{2.19.6}$$

where R is the radius of the rotating unit considered as a hard sphere.

The second equality in Eq. (2.19.6) relates the dynamic friction to the dynamic viscosity $\eta^*(\omega)$. This latter parameter is related to the dynamic shear modulus $G^*(\omega)$ by {reference,[9] Eq. (6.10)}

$$\eta^*(\omega) = -i\frac{G^*(\omega)}{\omega} \tag{2.19.7}$$

This equation allows to write Eq. (2.19.5) in the following way

$$\frac{\varepsilon^*(\omega) - \varepsilon_\infty}{\varepsilon_s - \varepsilon_\infty} = \frac{1}{1 + KG^*(\omega)} \tag{2.19.8}$$

where $K = (4\pi R^3/kT)(\varepsilon_0 + 2)/(\varepsilon_\infty + 2)$.

As a consequence, from the knowledge of the dynamic modulus and permittivity of the material, the radius R of the rotating unit can be estimated. Some results can be found in reference.[20] Within the validity of the Debye theory, the agreement between theory and experiments is reasonably good.

2.20. STATISTICAL MECHANICS AND LINEAR RESPONSE

One of the most important objectives in dielectric theory is to relate macroscopic properties to molecular properties. For the static case this task has been accomplished by the Onsager–Kirkwood–Fröhlich theories, as described in the previous chapter. Equations (1.14.13), (1.15.12), and (1.16.8) represent the main results of these theories. To solve the same problem in the case of time-dependent fields requires the use of statistical mechanics time-dependent processes. The introduction of time as a variable carries the system out of equilibrium and implies the use of time correlation functions as defined in section 2.4.

On the other hand, Eq. (2.12.3) can be considered a linear pheno-menological relationship between macroscopic quantities. The input is the external force field, and the output, also called the flux, is the induced polarization or the electrical induction. In the first case, the relationship between these quantities is the dielectric susceptibility; in the second case it is the dielectric permittivity. Linear relations between forces and fluxes are common in physics and they are particular cases of constitutive equations. In classical irreversible thermodynamics, the relations between forces and fluxes are established through the so-called kinetic coefficients. Very simple examples of linear phenomenological equations are Ohm's law of electric conduction, the Fick's laws of diffusion, and Fourier's law of thermal conduction. In thermostatics these types of equation are simple and are called equations of state. This name is based on the assumption of the existence of a space of phase upon which these equations are defined. Instantaneous response is a characteristic of the equations of state in classical thermodynamics. In fact, equilibrium thermodynamics is the tool for handling this type of phenomenology. However, when one considers non-uniformity of temperature or concentration, that is, gradients of these magnitudes, the scenario changes and one deals with irreversible thermodynamics. The Fourier and Fick laws, although linear, are representative examples of irreversible processes. In general, interactions between different force fields are involved and no clear distinction can be made between their effects. In our case, the electric polarization is the combined effect of changes in both the spatial distribution of electric charges and the orientational distribution of permanent dipoles under the simultaneous effects of temperature and electric field gradients. Linear relationships such as Eq. (2.12.3) are also representative of irreversible processes. They are also called relaxation or retardation laws. In this case, irreversibility appears because the instantaneous value of the flux response function depends on the past history of the applied force. This is a consequence of the causality principle. The material has memory of the past history of forces in such a way that their effects are superimposed, and the Markovian character of the phenomenological laws is lost.

It is our purpose to discuss the central problem of nonequilibrium statistical mechanics for a dielectric, that is, the relaxation from nonequilibrium to equilibrium polarization and the response of a dielectric system to a dynamic electrical field. In this respect, we can advance that the main finding of the theory is that, for relatively weak interactions, the dynamic permittivity can be expressed in terms of the correlation functions of fluctuating dipole moments at equilibrium.

As is well known in statistical mechanics, the behavior of a complex system is described in terms of the generalized coordinates q_i and moments p_i of all the particles of the system. The Hamiltonian of the system $H(p, q)$ gives the rate of change in the corresponding coordinates and moments for each particle according to

$$\dot{q}_i = \frac{\partial H}{\partial p_i}$$

$$\dot{p}_i = -\frac{\partial H}{\partial q_i}$$

(2.20.1)

Let $\mathbf{M}(p, q)$ be the dipole moment of the system which depends on time through changes in p and q. Equations (2.20.1) lead to the following expression for $\mathbf{M}$

$$\frac{d\mathbf{M}}{dt} = \sum_i \left(\frac{\partial \mathbf{M}}{\partial q_i} \frac{\partial H}{\partial p_i} - \frac{\partial \mathbf{M}}{\partial p_i} \frac{\partial H}{\partial q_i} \right) \tag{2.20.2}$$

which is called in classical mechanics the Poisson bracket $[\mathbf{M}, H]$. The right-hand side of Eq. (2.20.2) defines a linear operator, the Liouville operator denoted by iL, which operates on $\mathbf{M}$, according to

$$\dot{\mathbf{M}} = -iL\mathbf{M} \tag{2.20.3}$$

Equation (2.20.3) can explicitly be solved by expanding $\mathbf{M}(p(t), q(t))$ into a Taylor series. Noting that

$$\exp(-itL) = 1 - itL - \frac{1}{2!}t^2L^2 + \cdots \tag{2.20.4}$$

the pertinent expression is

$$\mathbf{M}(p(t), q(t)) = \exp(-itL)\mathbf{M}(p(0), q(0)) \tag{2.20.5}$$

Equation (2.20.5) expresses the motion of the system along a trajectory in the phase space. In the present case one is dealing with linear dissipative systems, upon which an external force acts. The dynamic variable under consideration is $\mathbf{M}$ and the coupled dynamic external force is the electric field $\mathbf{E}_0$. In these conditions

$$\mathbf{M} = -\frac{\partial H(p, q, t)}{\partial \mathbf{E}_0(t)} \tag{2.20.6}$$

If the interaction of the electrical field is weak, the perturbed Hamiltonian can be assumed to be

$$H(p, q, t) = H_0(p, q) - \mathbf{M}(p, q) \cdot \mathbf{E}_0(t) \tag{2.20.7}$$

In classical statistical mechanics, a statistical ensemble is represented by a distribution function in the phase space of the canonical variables describing the system. This distribution function is associated with the Hamiltonian. As usual, for a dipole system in equilibrium the Maxwell–Boltzmann distribution holds

$$f_0(p, q) = \frac{\exp(-H_0/kT)}{\int \exp(-H_0/kT)\,d\Gamma} \tag{2.20.8}$$

where subscript 0 denotes equilibrium distribution and $d\Gamma = dp\, dq$ is an element of the phase space. In our case the associated distribution function can be written as

$$f(p, q, t) = f_0(p, q) + \delta f(p, q, t) \qquad (2.20.9)$$

where δf can be assumed to be linear with $\mathbf{E}_0$.

As a dynamic variable, the distribution function can also be expressed explicitly as a function of time, so that the total derivative can be written as

$$\frac{df}{dt} = \frac{\partial f}{\partial t} + [f, H] \qquad (2.20.10)$$

where the second term on the right-hand side is the Poisson bracket of the pair f, H. The time derivative at equilibrium is zero, and consequently

$$\frac{\partial f_0}{\partial t} = iL_0 f_0 = -[f_0, H_0] \qquad (2.20.11)$$

where L_0 is the Liouville operator corresponding to the equilibrium Hamiltonian. On the other hand, the evolution of the distribution function with time is given by

$$\frac{\partial f}{\partial t} + [f, H] = 0 \qquad (2.20.12)$$

By substituting Eqs (2.20.8) and (2.20.9) into Eq. (2.20.12), and bearing in mind the assumption made after Eq. (2.20.9), we obtain

$$\frac{\partial}{\partial t}\delta f + [\delta f, H_0] - [f_0, \mathbf{M} \cdot \mathbf{E}_0] = 0 \qquad (2.20.13)$$

From Eq. (2.20.6)

$$\frac{\partial f_0}{\partial p_i} = -\frac{f_0}{kT}\frac{\partial H_0}{\partial p_i} \qquad (2.20.14)$$

Moreover, since the third term on the left-hand side of Eq. (2.20.13) can be written as

$$[f_0, \mathbf{M} \cdot \mathbf{E}_0] = -\frac{1}{kT}f[H_0, \mathbf{M} \cdot \mathbf{E}_0] = \frac{1}{kT}\, f[\mathbf{M} \cdot \mathbf{E}_0, H_0] \qquad (2.20.15)$$

Eq. (2.20.13) is finally given by

$$\frac{\partial}{\partial t}\delta f + [\delta f, H_0] - \frac{1}{kT}f_0[\mathbf{M} \cdot \mathbf{E}_0, H_0] = 0 \qquad (2.20.16)$$

or in terms of the Liouville operator

$$\frac{\partial}{\partial t}\delta f - iL_0\delta f + \frac{1}{kT}f_0 iL_0 \mathbf{M} \cdot \mathbf{E}_0 = 0 \tag{2.20.17}$$

The solution of Eq. (2.20.17) is equivalent to the following integral equation

$$\delta f(p, q, t) = -\frac{1}{kT}f_0(p, q) \int_{-\infty}^{t} \exp\left[i(t - \tau)L_0\right] iL_0 \mathbf{M}(p, q)$$

$$\cdot \mathbf{E}_0(\tau)\,d\tau \tag{2.20.18}$$

as it is easily found after differentiating both sides of Eq. (2.20.17) with respect to time. Substitution of Eq. (2.20.18) into Eq. (2.20.9) yields

$$f(p, q, t) = f_0(p, q) - \frac{1}{kT}f_0(p, q) \int_{-\infty}^{t} \exp\left[i(t - \tau)L_0\right] iL_0 \mathbf{M}(p, q) \cdot \mathbf{E}_0(\tau)\,d\tau \tag{2.20.19}$$

According to the distribution function $f(p, q, t)$, the phase average of the dynamic variable $\mathbf{M}(p, q)$ is given by

$$\langle \mathbf{M} \rangle = \int \mathbf{M}(p, q)f(p, q, t)\,d\Gamma \tag{2.20.20}$$

and at equilibrium

$$\langle \mathbf{M} \rangle_0 = \int \mathbf{M}(p, q)f_0(p, q)\,d\Gamma \tag{2.20.21}$$

where the distribution function is assumed to be normalized in both cases. Substituting Eq. (2.20.19) into Eq. (2.20.20), and bearing in mind that the dipolar moment in the absence of an electric field is zero, we obtain

$$\langle \mathbf{M} \rangle = -\frac{1}{kT} \int_{-\infty}^{t} \mathbf{E}_0(\tau) \cdot \langle \mathbf{M} \exp\left[i(t - \tau)L_0\right] iL_0 \mathbf{M} \rangle\,d\tau \tag{2.20.22}$$

From Eq. (2.20.3), and considering that, according to Eq. (2.20.5), the effect of the operator $\exp[i(t - \tau)L_0]$ is to shift the function backwards in the phase trajectory, the average of $\mathbf{M}(t)$ is given by

$$\langle \mathbf{M}(t) \rangle = \frac{1}{kT} \int_{-\infty}^{t} \mathbf{E}_0(\tau) \cdot \langle \mathbf{M}(t)\dot{\mathbf{M}}(\tau) \rangle\,d\tau \tag{2.20.23}$$

which is the dielectric version of a formula proposed by Kubo[21] and in dielectric literature is recognized as a second version of the fluctuation–dissipation theorem.[22]

The average $\langle \mathbf{M}(t)\dot{\mathbf{M}}(\tau)\rangle$ in Eq. (2.20.23) is the correlation between the dipolar moment and the derivative of the dipolar moment in a generic time τ. By writing $t - \tau = \theta$, Eq. (2.20.23) becomes

$$\langle \mathbf{M}(t)\rangle = \frac{1}{kT}\int_0^\infty \mathbf{E}_0(t-\theta)\cdot\langle \mathbf{M}(t)\dot{\mathbf{M}}(t-\theta)\rangle\,\mathrm{d}\theta \qquad (2.20.24)$$

According to the properties of stationary processes

$$\langle \mathbf{M}(t)\rangle = \frac{1}{kT}\int_0^\infty \mathbf{E}_0(t-\theta)\cdot\langle \mathbf{M}(0)\dot{\mathbf{M}}(-\theta)\rangle\,\mathrm{d}\theta$$

$$= -\frac{1}{kT}\int_0^\infty \mathbf{E}_0(t-\theta)\cdot\langle \mathbf{M}(0)\dot{\mathbf{M}}(\theta)\rangle\,\mathrm{d}\theta \qquad (2.20.25)$$

Note that, for a homogeneous and isotropic dielectric, the elements off the diagonal in the average $\langle \mathbf{M}(t)\mathbf{M}(\tau)\rangle$ are zero, while each diagonal element is $\frac{1}{3}\langle \mathbf{M}(t)\cdot\dot{\mathbf{M}}(\tau)\rangle$. Consequently

$$\langle \mathbf{M}(t)\rangle = \frac{1}{3kT}\int_{-\infty}^{t} \mathbf{E}_0(\tau)\langle \mathbf{M}(t)\cdot\dot{\mathbf{M}}(\tau)\rangle\,\mathrm{d}\tau$$

$$= -\frac{1}{3kT}\int_0^\infty \mathbf{E}_0(t-\theta)\langle \mathbf{M}(0)\cdot\dot{\mathbf{M}}(\theta)\rangle\,\mathrm{d}\theta \qquad (2.20.26)$$

2.21. RELATIONSHIP BETWEEN THE FREQUENCY-DEPENDENT PERMITTIVITY AND THE AUTOCORRELATION FUNCTION FOR DIPOLE MOMENTS

In order to proceed further in the preceding analysis, more specific material systems need to be analysed. Let us consider an isotropic molecular sample of spherical shape and radius a with frequency-dependent permittivity $\varepsilon_2(\omega)$, embedded in an infinite medium of permittivity $\varepsilon_1(\omega)$[22]. It is assumed further that an external field along the z axis having a constant value $E_0(t)$ is present at large distances from the sphere. According to Eq. (2.20.26), the dipole moment of the sphere in the absence of the field, expressed in terms of the autocorrelation function of the dipole moment, is given by

$$\langle \mathbf{M}(t)\rangle = -\frac{1}{3kT}\int_0^\infty \mathbf{E}(t-\tau)\langle \mathbf{M}(0)\cdot\dot{\mathbf{M}}(\tau)\rangle\,\mathrm{d}\tau \qquad (2.21.1)$$

where $E(t)$ is the field in the empty cavity when the molecular sample is removed, that is

$$E(t) = \frac{3\varepsilon_1^*(\omega)}{2\varepsilon_1^*(\omega) + 1} E_0(t) \tag{2.21.2}$$

with $E_0(t) = E_0 \exp(i\omega t)$. On the other hand, the total dipole moment per unit volume of a specimen located inside the cavity of dielectric permittivity $\varepsilon_2^*(\omega)$, under the external electric field $\mathbf{E}_0(t)$, is given by

$$\langle \mathbf{M}(t) \rangle = \frac{\varepsilon_2^*(\omega) - 1}{4\pi} \frac{4}{3} \pi a^3 \frac{3\varepsilon_1^*(\omega)}{2\varepsilon_1^*(\omega) + \varepsilon_2^*(\omega)} \mathbf{E}_0(t) \tag{2.21.3}$$

where the field inside the spherical sample has been considered.

Equalizing Eqs (2.21.1) and (2.21.3) and taking into account Eq. (2.21.2) gives

$$\frac{[\varepsilon*_2(\omega) - 1][2\varepsilon_1^*(\omega) + 1]}{3[2\varepsilon_1^*(\omega) + \varepsilon_2^*(\omega)]} a^3 = -\frac{1}{3kT} \int_0^\infty \exp(-i\omega\tau)$$
$$\langle \mathbf{M}(0) \cdot \dot{\mathbf{M}}(\tau) \rangle \, d\tau \tag{2.21.4}$$

For a static field, $E_0(t) = E_0$, and Eq. (2.21.4) yields

$$\frac{(\varepsilon_{20} - 1)(2\varepsilon_{10} + 1)}{3(2\varepsilon_{10} + \varepsilon_{20})} a^3 = \frac{1}{3kT} \langle \mathbf{M}(0) \cdot \mathbf{M}(0) \rangle = \frac{1}{3kT} \langle M^2(0) \rangle \tag{2.21.5}$$

Note that, if $\varepsilon_1^*(\omega) = \varepsilon_2^*(\omega) = \varepsilon(0) = \varepsilon_0$, and taking into account that $4\pi a^3 / 3 = V$, an expression similar to the Kirkwood equation for the case of a single dipole [Eq. (1.14.13)] is recovered.

For a sphere in vacuum, $\varepsilon_1^*(\omega) = 1$, and Eq. (2.21.4) reduces to

$$\frac{\varepsilon^*(\omega) - 1}{\varepsilon^*(\omega) + 2} a^3 = -\frac{1}{3kT} \int_0^\infty \exp(-i\omega\tau) \langle \mathbf{M}(0) \cdot \mathbf{M}(\tau) \rangle \, d\tau \tag{2.21.5a}$$

If $\omega = 0$, the following expression is obtained

$$\frac{\varepsilon_0 - 1}{\varepsilon_0 + 2}a^3 = \frac{1}{3kT}\langle M^2(0)\rangle \tag{2.21.5b}$$

which is sometimes called the Kirkwood–Fröhlich susceptibility theorem.[23]
    Let us define the normalized autocorrelation moment function as

$$\phi(t) = \frac{\langle \mathbf{M}(0) \cdot \mathbf{M}(t)\rangle}{\langle \mathbf{M}(0) \cdot \mathbf{M}(0)\rangle} \tag{2.21.6}$$

Then, Eqs (2.21.4) to (2.21.6) lead to

$$\frac{[\varepsilon_2^*(\omega) - 1][2\varepsilon_1^*(\omega) + 1](2\varepsilon_{10} + \varepsilon_{20})}{(\varepsilon_{20} - 1)(2\varepsilon_{10} + 1)(2\varepsilon_1^*(\omega) + \varepsilon_2^*(\omega))} = L[-\dot{\phi}(t)] \tag{2.21.7}$$

where the right-hand side of Eq. (2.21.7) is the Laplace transform of the normalized autocorrelation function. Equation (2.21.7) relates the macroscopic properties of the system, expressed by $\varepsilon_1^*(\omega)$ and $\varepsilon_2^*(\omega)$, to the fluctuations of the dipole moment into the cavity *in the absence of an electric field*. In order to clarify the consequences of (2.21.7), let us consider three different cases:

(a)   The molecular sample has the same permittivity as the medium in which it is embedded. Then $\varepsilon_1^*(\omega) = \varepsilon_2^*(\omega) = \varepsilon^*(\omega)$, and

$$\frac{[\varepsilon^*(\omega) - 1][2\varepsilon^*(\omega) + 1]\varepsilon_0}{(\varepsilon_0 - 1)(2\varepsilon_0 + 1)\varepsilon^*(\omega)} = L[-\dot{\phi}(t)] \tag{2.21.8}$$

a result obtained by Fatuzzo and Mason[24] for nonpolarizable dipoles.

(b)   The molecular sample has a dielectric permittivity given by $\varepsilon_2^*(\omega) = \varepsilon^*(\omega)$ and the surrounding medium responds instantaneously to fields of any frequency as a static dielectric, that is, $\varepsilon_{20} = \varepsilon_{10} = \varepsilon_0$. In this case the expression resulting from Eq. (2.21.7) is

$$\frac{3[\varepsilon^*(\omega) - 1]\varepsilon_0}{[\varepsilon^*(\omega) + 2\varepsilon_0](\varepsilon_0 - 1)} = L[-\dot{\phi}(t)] \tag{2.21.9}$$

as obtained by Glarum.[25]

(c)   Finally, if the external region is vacuum, then $\varepsilon_1 = 1$ and $\varepsilon_2^*(\omega) = \varepsilon^*(\omega)$. In this case, Eq. (2.21.7) becomes

$$\frac{[\varepsilon^*(\omega) - 1](\varepsilon_0 + 2)}{[\varepsilon^*(\omega) + 2](\varepsilon_0 - 1)} = L[-\dot{\phi}(t)] \tag{2.21.10}$$

and this result corresponds to a sphere in vacuum, as was found by Cole.[26]

2.22. EXTENSION TO POLARIZABLE DIPOLES

Equation (2.21.10) can easily be extended to polarizable dipoles, as Cole[26] did. As in case (c) of the previous section, the system is a sphere containing N molecules, each with permanent dipole and polarizability, μ and α respectively. For a time-dependent field, $\mathbf{M}(t)$ can be written as

$$\langle \mathbf{M}(t) \rangle = \frac{\varepsilon_\infty - 1}{\varepsilon_\infty + 2} \frac{3}{4\pi} \mathbf{E}_0(t) - \frac{1}{3kT} \int_0^\infty \mathbf{E}_0(t - \theta) \langle \mathbf{M}(0) \cdot \dot{\mathbf{M}}(\theta) \rangle \, d\theta \qquad (2.22.1)$$

where the first term on the right-hand side describes the instantaneous polarization of the sphere. In this case, the corresponding correlation function is given by

$$\phi(t) = \frac{1}{\alpha} \left(\frac{3V}{4\pi} \frac{\varepsilon_\infty - 1}{\varepsilon_\infty + 2} H(t) - \frac{1}{3kT} \langle \mathbf{M}(0) \cdot \mathbf{M}(t) \rangle \right) \qquad (2.22.2)$$

where $H(t)$ is the Heaviside step function. On this basis and taking into account that

$$\langle \mathbf{M}(t) \rangle = \frac{\varepsilon^*(\omega) - 1}{4\pi} \frac{3}{\varepsilon^*(\omega) + 2} V \mathbf{E}(t) \qquad (2.22.3)$$

where Eq. (1.B.14) for the sphere in vacuum has been used, we obtain

$$\frac{\varepsilon^*(\omega) - 1}{\varepsilon^*(\omega) + 2} = \frac{\varepsilon_\infty - 1}{\varepsilon_\infty + 2} - \frac{1}{3kT} \int_0^\infty \exp(-i\omega\tau) \langle \mathbf{M}(0) \cdot \dot{\mathbf{M}}(\tau) \rangle \, d\tau \qquad (2.22.4)$$

Therefore, for the static case

$$\frac{\varepsilon_0 - 1}{\varepsilon_0 + 2} = \frac{\varepsilon_\infty - 1}{\varepsilon_\infty + 2} - \frac{1}{3kT} \langle M^2(0) \rangle \qquad (2.22.5)$$

From Eqs (2.22.4) and (2.22.5), the following expression is obtained

$$\frac{[\varepsilon^*(\omega) - \varepsilon_\infty](\varepsilon_0 + 2)}{(\varepsilon_0 - \varepsilon_\infty)[\varepsilon^*(\omega) + 2]} = L[-\dot{\phi}(t)] \qquad (2.22.6)$$

This result represents the extension of Eq. (2.21.10) for polarizable dipoles and is formally equivalent to Eq. (2.10.6)

$$\frac{\varepsilon^*(\omega) - 1}{\varepsilon^*(\omega) + 2} = \frac{\varepsilon_\infty - 1}{\varepsilon_\infty + 2} + \left[\frac{\varepsilon_0 - 1}{\varepsilon_0 + 2} - \frac{\varepsilon_\infty - 1}{\varepsilon_\infty + 2}\right] L[-\dot{\phi}(t)] \qquad (2.22.7)$$

which becomes the Debye equation when $\phi = \exp(-t/\tau_D)$. For very dilute gaseous polarizable molecules, $\varepsilon_0 \cong \varepsilon^*(\omega)$ and Eq. (2.22.6) becomes

$$\frac{\varepsilon^*(\omega) - \varepsilon_\infty}{\varepsilon_0 - \varepsilon_\infty} = L[-\dot{\phi}(t)] \qquad (2.22.8)$$

In a similar way, the Fatuzzo–Mason[24] and Glarum[25] equations indicated above can be extended to polarizable dipoles. The final results are respectively

$$\frac{[\varepsilon^*(\omega) - \varepsilon_\infty][2\varepsilon^*(\omega) + 1]\varepsilon_0}{[\varepsilon_0 - \varepsilon_\infty](2\varepsilon_0 + 1)\varepsilon^*(\omega)} = L[-\dot{\phi}(t)] \qquad (2.22.9)$$

and

$$\frac{3[\varepsilon^*(\omega) - \varepsilon_\infty]\varepsilon_0}{[\varepsilon^*(\omega) + 2\varepsilon_0](\varepsilon_0 - \varepsilon_\infty)} = L[-\dot{\phi}(t)] \qquad (2.22.10)$$

2.23. MACROSCOPIC AND MICROSCOPIC CORRELATION FUNCTIONS

The correlation function ϕ considered until now for condensed matter is macroscopic, in the sense that it incorporates long-range dipole–dipole interactions. To obtain a microscopic correlation function, these interactions should be eliminated, as Kirkwood [Eq. (1.16.6)] and Fröhlich [Eq. (1.16.8)] did for the static field case. Note that in the Onsager theory the macroscopic correlation function is identical to the microscopic one because dipole–dipole interactions are not considered. For time-dependent fields, the solution to this problem has been addressed independently by Klug et al. [27] and Rivail[28] on the basis of the work by Fatuzzo–Mason.[24]

As in the Kirkwood model, let us consider the spherical region under study having a time-dependent dipole moment $\mathbf{M}(t)$ and a dipole $\boldsymbol{\mu}(t)$ in the centre of this region. The spherical region is small enough to contain only the central dipole. In the case of time-dependent fields, these moments differ according to the following expression

$$
\mathbf{M}(t) = \boldsymbol{\mu}(t) - \alpha \int_{-\infty}^{t} \mathbf{R}(\tau) \dot{\phi}'(t - \tau) \, d\tau
$$

$$
= \boldsymbol{\mu}(t) + \alpha \left(\mathbf{R}(0)\phi'(t) + \int_{0}^{t} \mathbf{R}(t - \theta) \dot{\phi}'(\theta) \, d\theta \right) \tag{2.23.1}
$$

where the reaction field $\mathbf{R}$ at instant t is the superposition of the aftereffect of the polarization of the sphere, assuming the central dipole to be frozen. Here, ϕ' refers to the external region outside the cavity, in a shell with a very small inner radius. The second equality of Eq. (2.23.1) has been obtained by breaking the integral in Eq. (2.23.1) into two integrals, one from $-\infty$ to 0, and the other from 0 to ∞, changing the variables $t - \tau = \theta$, integrating by parts, and taking into account that $\dot{\boldsymbol{\mu}} = 0$ for $-\infty < t < 0$. Note that the nonconvolutive term in the second equality of Eq. (2.23.1) was introduced through the integration of Dirac delta functions in order to account for an initial discontinuity in the reaction field history. Then, according to Eq. (1.12.9) we can write

$$
\mathbf{M}(t) = \boldsymbol{\mu}(t) + \frac{2(\varepsilon_0 - 1)(\varepsilon_\infty - 1)}{3(2\varepsilon_0 + \varepsilon_\infty)} \left(\boldsymbol{\mu}(0)\phi'(t) + \int_{0}^{t} \boldsymbol{\mu}(t - \tau) \dot{\phi}'(\tau) \, d\tau \right)
$$

$$
= \boldsymbol{\mu}(t) + B(\varepsilon_0) \left(\boldsymbol{\mu}(0)\phi'(t) + \int_{0}^{t} \boldsymbol{\mu}(t - \tau) \dot{\phi}'(\tau) \, d\tau \right) \tag{2.23.2}
$$

Multiplying both sides of Eq. (2.23.2) by $\boldsymbol{\mu}(0)$, and dividing by $\langle \mu^2(0) \rangle$, the following expression is obtained

$$
\frac{\langle \boldsymbol{\mu}(0) \cdot \mathbf{M}(t) \rangle}{\langle \mu^2(0) \rangle} = \frac{\langle \boldsymbol{\mu}(0) \cdot \boldsymbol{\mu}(t) \rangle}{\langle \mu^2(0) \rangle} + B(\varepsilon_0) \left(\phi'(t) + \int_{0}^{t} \frac{\langle \boldsymbol{\mu}(0) \cdot \boldsymbol{\mu}(t - \theta) \rangle}{\langle \mu^2(0) \rangle} \dot{\phi}'(\theta) \, d\theta \right)
$$

$$
= \frac{\langle \boldsymbol{\mu}(0) \cdot \boldsymbol{\mu}(t) \rangle}{\langle \mu^2(0) \rangle} + B(\varepsilon_0) \tag{2.23.3}
$$

Since $\phi'(0) = 1$, Eq. (2.23.3) becomes

$$
\frac{\langle \boldsymbol{\mu}(0) \cdot \mathbf{M}(0) \rangle}{\langle \mu^2(0) \rangle} = \frac{\langle \boldsymbol{\mu}(0) \cdot \boldsymbol{\mu}(0) \rangle}{\langle \mu^2(0) \rangle} + B(\varepsilon_0)
$$

$$
= 1 + B(\varepsilon_0) = \frac{(2\varepsilon_0 + 1)(\varepsilon_\infty + 2)}{3(2\varepsilon_0 + \varepsilon_\infty)} = C(\varepsilon_0) \tag{2.23.4}
$$

Also, the left-hand side of Eq. (2.23.4) can be written as

$$
\frac{\langle \boldsymbol{\mu}(0) \cdot \mathbf{M}(t) \rangle}{\langle \boldsymbol{\mu}(0) \cdot \mathbf{M}(0) \rangle} \frac{\langle \boldsymbol{\mu}(0) \cdot \mathbf{M}(0) \rangle}{\langle \mu^2(0) \rangle} \tag{2.23.5}
$$

and Eq. (2.23.3) becomes

$$C(\varepsilon_0)\phi(t) = \varphi(t) + B(\varepsilon_0)\left(\phi'(t) + \int_0^\infty \frac{\langle \boldsymbol{\mu}(0) \cdot \boldsymbol{\mu}(t-\theta) \rangle}{\langle \mu^2(0) \rangle} \dot{\phi}'(\theta) \, d\theta\right)$$

(2.23.6)

where φ is the microscopic correlation function in terms of which the permittivity will be expressed. Taking the Laplace transform of both sides of Eq. (2.23.6), the following expression is obtained

$$C(\varepsilon_0)\phi(s) = \varphi(s) + B(\varepsilon_0)[\phi'(s) + \varphi(s)\dot{\phi}'(s)]$$

(2.23.7)

Now, noting that $L(\dot{f}) = sL(f) - 1$, where L means the Laplace transform, we obtain

$$C(\varepsilon_0)L(-\dot{\phi}) = L(-\dot{\varphi})[1 + B(\varepsilon_0)L(-\dot{\phi}')]$$

(2.23.8)

The explicit value of $L(-\dot{\phi}')$ can be obtained from $B[\varepsilon^*(\omega)]$ as follows. Multiplying the Laplace transform of $-\dot{\phi}'$ by $B(\varepsilon_0)\mu(0)\exp i\omega t$ leads to

$$B(\varepsilon_0)\mu(0)L(-\dot{\phi}')\exp i\omega t = -B(\varepsilon_0)\int_0^\infty \mu(0)\exp\left[i\omega(t-\theta)\right]\dot{\phi}'(\theta)\,d\theta$$

(2.23.9)

In the case of a sinusoidal field, this is the steady-state response of the contribution to the dipole moment inside the cavity induced by the reaction field. For this reason, the left-hand side of Eq. (2.23.9) is $B[\varepsilon(\omega)]\mu(t)$, where

$$B[\varepsilon^*(\omega)] = \frac{2[\varepsilon^*(\omega) - 1](\varepsilon_\infty - 1)}{3[2\varepsilon^*(\omega) + \varepsilon_\infty]}$$

(2.23.10)

(in the original paper by Klug et al. [27] there is a typographical error here).

The complex quantity $B[\varepsilon^*(\omega)]$ determines the amplitude and phase difference between the reaction field and the oscillating permanent moment. Consequently

$$L(-\dot{\phi}') = \frac{B[\varepsilon^*(\omega)]}{B(\varepsilon_0)} = \frac{[\varepsilon^*(\omega) - 1](2\varepsilon_0 + \varepsilon_\infty)}{(\varepsilon_0 - 1)[2\varepsilon^*(\omega) + \varepsilon_\infty]}$$

(2.23.11)

from which

$$\frac{L(-\dot{\phi})}{L(-\dot{\varphi})} = \frac{C[\varepsilon^*(\omega)]}{C(\varepsilon_0)} = \frac{[2\varepsilon^*(\omega) + \varepsilon_\infty](2\varepsilon_0 + 1)}{(2\varepsilon_0 + \varepsilon_\infty)[2\varepsilon(\omega) + 1]}$$

(2.23.12)

Therefore, the desired expression is

$$\frac{\varepsilon_0[\varepsilon^*(\omega) - \varepsilon_\infty][2\varepsilon^*(\omega) + \varepsilon_\infty]}{\varepsilon^*(\omega)(\varepsilon_0 - \varepsilon_\infty)(2\varepsilon_0 + \varepsilon_\infty)} = L(-\dot{\varphi}) \tag{2.23.13}$$

where Eq. (2.22.8) has been used. Note that, if $\phi = \phi'$ in Eq. (2.23.8), then equation (2.22.9) is recovered. The field factor $f(\omega)$ in Eq. (2.23.13) is defined as

$$f(\omega) = \frac{[2\varepsilon^*(\omega) + \varepsilon_\infty]\varepsilon_0}{(2\varepsilon_0 + \varepsilon_\infty)\varepsilon^*(\omega)} \tag{2.23.14}$$

In many cases $f(\omega) \cong 1$, and Eq. (2.12.7) is recovered, meaning that macroscopic and microscopic correlation functions coincide.

It is noteworthy that, if the microscopic sphere under consideration only contains a molecular dipole, then the microscopic correlation function is the dipole–dipole correlation function. In the case of gases or dilute solutions of polar molecules, this identification is generally correct. In the static case this corresponds to the situation where $g = 1$ in Eq. (1.14.14). However, when one is dealing with condensed matter, the microscopic correlation function also incorporates dipole–dipole interactions. A comparison between the macroscopic and microscopic relaxation behavior has been carried out by Böttcher and Bordewijk.[29]

2.24. COMPLEX POLARIZABILITY

If $\varepsilon_1^*(\omega)$ is set equal to 1 in Eq. (2.21.3), the following expression is obtained

$$\mathbf{M}(t) = \frac{\varepsilon^*(\omega) - 1}{\varepsilon^*(\omega) + 2} a^3 \mathbf{E}_0 \tag{2.24.1}$$

which corresponds to a spherical specimen in vacuum. This equation is similar to Eq. (2.12.1). In fact, to consider the dielectric body under study as spherical in shape avoids unnecessary calculations and has been the classical way to start the analysis of dispersion and fluctuations in dielectrics.[13,19] Equation (2.24.1), together with Eq. (2.22.7), led Scaife[30] to consider that the reduced complex polarizability $\alpha(\omega)$ of a spherical specimen of radius $a = 1$, defined as

$$\alpha^*(\omega) = \frac{\varepsilon^*(\omega) - 1}{\varepsilon^*(\omega) + 2} \tag{2.24.2}$$

is a more convenient way to represent the macroscopic dielectric properties of a material system. Note that, for convenience, the factor $3V/4\pi$ has been omitted.

A better separation of the relaxation mechanisms is obtained in comparison with the permittivity representation. Moreover, owing to the normalizing character of Eq. (2.24.1), large values of the permittivity arising, for example, from electrode effects or any other phenomena may appear, whereas the real and imaginary components of $\alpha^*(\omega)$ maintain reasonable values. Indeed, the semicircular shape of the Cole–Cole representation is preserved and at high frequencies the use of polarizability instead of permittivity avoids spurious effects sometimes detected in the ε'' vs. ε' representations.[31] From a molecular point of view, the use of Eq. (2.24.2) is equivalent to assuming the whole system of arbitrary shape to be a dielectric sphere surrounded by the evacuated space. In this way, the long-range, intermolecular, dipole–dipole coupling is strongly minimized.

2.25. DISPERSION RELATIONS CORRESPONDING TO THE POLARIZABILITY. A NEW VERSION OF THE FLUCTUATION–DISSIPATION THEOREM

As a consequence of the preceding considerations, the dipole moment of a spherically shaped body in response to an arbitrary electric force field is given by an expression similar to Eq. (2.12.2)

$$\mathbf{M}(t) = \int_{-\infty}^{t} \alpha(t - \theta)\, d\mathbf{E}(\theta) \tag{2.25.1}$$

or after integration by parts and a change of variable

$$\mathbf{M}(t) = \alpha_{\infty}\mathbf{E}(t) + \int_{0}^{\infty} \mathbf{E}(t - \tau)\dot{\alpha}(\tau)\, d\tau \tag{2.25.2}$$

where α_{∞} means $\alpha|_{\omega=\infty} = \alpha|_{t=0}$. Equation (2.25.2) is similar in structure to Eq. (2.12.3). For the periodic field described in Eq. (2.12.10), Eq. (2.25.2) leads to

$$M(t) = E_0[\alpha'(\omega)\cos \omega t + \alpha''(\omega)\sin \omega t] \tag{2.25.3}$$

where

$$\alpha'(\omega) = \alpha_{\infty} + \int_{0}^{\infty} \alpha(\tau)\cos \omega\tau\, d\tau$$

$$\alpha''(\omega) = \int_{0}^{\infty} \alpha(\tau)\sin \omega\tau\, d\tau \tag{2.25.4}$$

Taking the inverse Fourier transform of Eqs (2.25.4) and following the procedure described in section (2.14) for dielectric dispersions, the following dispersion equations are obtained

$$\alpha'(\omega) - \alpha_\infty = \frac{2}{\pi} \int_0^\infty \frac{x\alpha''(x)}{x^2 - \omega^2} \, dx$$

$$\alpha''(\omega) = \frac{2}{\pi} \int_0^\infty \frac{\omega[\alpha'(x) - \alpha_\infty]}{\omega^2 - x^2} \, dx$$

(2.25.5)

These integrals are taken in the sense of their principal value and represent a new version of the Krönig–Kramers relationships (2.14.4). Note that, in these dispersion relations, similar conditions to those given by Eqs (2.12.14) and (2.12.15) must be fulfilled. A particular case of these relations is obtained when $\omega \to 0$. By assuming that $\alpha_\infty \to 0$, Eq. (2.25.5a) becomes

$$\alpha_0 = \frac{2}{\pi} \int_{-\infty}^\infty \alpha''(x) \, d\ln x$$

(2.25.6)

By combining Eqs (2.25.6) and (2.21.5a), changing the argument in the integral, and assuming that the fluctuations are ergodic, that is, the time average is equivalent to the ensemble average, we obtain

$$\langle M^2(0) \rangle = \frac{6kT}{\pi} \int_{-\infty}^\infty \frac{\alpha''(\omega)}{\omega} \, d\omega$$

(2.25.7)

The meaning of this result is that an equilibrium property of the system is related to a dynamic property, the loss polarizability. On the other hand, from the definition of the non-normalized autocorrelation function

$$\psi(t) = \langle \mathbf{M}(0) \cdot \mathbf{M}(t) \rangle$$

(2.25.8)

the Fourier transform of $\psi(t)$ is defined as

$$\psi(\omega) = \int_{-\infty}^{+\infty} \langle \mathbf{M}(0) \cdot \mathbf{M}(\tau) \rangle \exp(-i\omega\tau) \, d\tau$$

(2.25.9)

which is called the spectral density of the function $\psi(t)$. In fact, $\psi(\omega)$ is a generalized susceptibility. From Eq. (2.25.9) it is clear that $\psi(-\omega) = \bar{\psi}(\omega)$, where the overbar indicates the complex conjugate. This implies that the real and imaginary parts of $\psi(\omega)$ are even and odd respectively.

Owing to the odd character of the function $\alpha''(x)$, that is $\alpha''(x) = -\alpha''(-x)$, the following equality holds

$$\pi\alpha'(\omega) = 2\int_0^\infty \frac{x\alpha''(x)}{x^2 - \omega^2}\,dx = \int_0^\infty \frac{\alpha''(x)}{x - \omega}\,dx + \int_0^\infty \frac{\alpha''(x)}{x + \omega}\,dx$$

$$= \int_{-\infty}^\infty \frac{\alpha''(x)}{x - \omega}\,dx \qquad (2.25.10)$$

As a consequence, Eq. (2.25.5a) can also be written as

$$\alpha'(\omega) - \alpha_\infty = \frac{1}{\pi}\int_{-\infty}^\infty \frac{\alpha''(x)}{x - \omega}\,dx \qquad (2.25.11)$$

For the special case $\omega = 0$

$$\psi(0) = \int_{-\infty}^{+\infty} \langle \mathbf{M}(0) \cdot \mathbf{M}(\tau) \rangle\,d\tau \qquad (2.25.12)$$

The inverse of the Fourier transform of Eq. (2.25.9) is given by

$$\psi(t) = \langle \mathbf{M}(0) \cdot \mathbf{M}(t) \rangle = \frac{1}{2\pi}\int_{-\infty}^{+\infty} \psi(\omega)\exp\,(i\omega t)\,d\omega \qquad (2.25.13)$$

which is called the Wiener–Kintchine theorem [21b]. For $t = 0$

$$\langle M^2(0) \rangle = \frac{1}{2\pi}\int_{-\infty}^{+\infty} \psi(\omega)\,d\omega = \frac{1}{\pi}\int_0^\infty \psi(\omega)\,d\omega \qquad (2.25.14)$$

where the second equality is a consequence of the even character of $\psi(\omega)$. From Eqs (2.25.7) and (2.25.14) the following expression is obtained

$$\psi(\omega) = \frac{6kT}{\omega}\alpha''(\omega) \qquad (2.25.15)$$

This equation represents a new version of the fluctuation–dissipation theorem. Note that the present microscopic fluctuations are related to a state of equilibrium specified by the absence of an electric field. The fluctuation–dissipation theorem asserts that, on average, fluctuations follow the same law as macroscopic dissipation. The theorem links the static and dynamic properties of a system consisting of an ensemble of dipoles.

2.26. FLUCTUATIONS IN A SPHERICAL SHELL

The results of the preceding section can be applied to evaluate the correlation functions for the spontaneous fluctuations in the net dipole moment of a dielectric sphere embedded in a spherical shell. This type of analysis was initiated by Scaife[23] and followed by Titulaer and Deutch.[22] The basic idea is to replace the infinite medium considered in section 2.21 with a spherical shell of external radius R and permittivity $\varepsilon_2^*(\omega)$ surrounding an inner sphere of permittivity $\varepsilon_1^*(\omega)$ and radius r. As usual, the magnitude of the respective radius is small, in comparison with the atomic distances, but large compared with the wavelength of the corresponding frequency.[14] The dipole moments of the embedded sphere and the shell are respectively $\mathbf{m}_1$ and $\mathbf{m}_2$.

An essential ingredient of the present development is that these two distinct dielectric specimens are respectively subject to different external fields $\mathbf{E}_1(t)$ and $\mathbf{E}_2(t)$. These fields are in the z direction, and only these z components of the dipole moment will be considered. The response of the dipole moments in terms of the external fields may be expressed by the following system of equations

$$m_i(\omega) = \sum_{j=1,2} \alpha_{ij}^*(\omega) E_j(\omega) \tag{2.26.1}$$

For a spherical specimen in vacuum the preceding system is reduced to a single equation and the value of α^* is given by Eq. (2.24.2) times the radius of the specimen {see reference,[23] Eq. (20)}.

According to Eqs (2.25.8) and (2.25.13), the cross-correlation function of the dipole moments of the shell and inner sphere, both exhibiting the same oscillatory time frequency, is given by

$$\psi_{ij}(t) = \langle \mathbf{m}_i(0) \cdot \mathbf{m}_j(t) \rangle = \frac{1}{2\pi} \int_{-\infty}^{+\infty} \psi_{ij}(\omega) \exp(i\omega t)\, d\omega \tag{2.26.2}$$

where, according to Eq. (2.25.9), $\psi_{ij}(\omega)$ is the Fourier transform of $\psi_{ij}(t)$. The corresponding time derivative of Eq. (2.26.2) is

$$\dot{\psi}_{ij}(t) = \langle \mathbf{m}_i(0) \cdot \dot{\mathbf{m}}_j(t) \rangle = \frac{i}{2\pi} \int_{-\infty}^{+\infty} \omega \psi_{ij}(\omega) \exp(i\omega t)\, d\omega \tag{2.26.3}$$

Then, the Laplace transform of this derivative can be written as

$$L\langle \mathbf{m}_i(0) \cdot \dot{\mathbf{m}}_j(t) \rangle = \int_0^\infty \exp(-i\omega\tau) \left[\frac{i}{2\pi} \int_{-\infty}^{+\infty} \omega' \psi_{ij}(\omega') \exp(i\omega' t)\, d\omega' \right] d\tau \tag{2.26.4}$$

Substitution of Eq. (2.25.15) into Eq. (2.26.4) gives

$$L\langle \mathbf{m}_i(0) \cdot \dot{\mathbf{m}}_j(t)\rangle = \frac{3ikT}{\pi} \int_0^\infty \exp(-i\omega\tau)\left[\int_{-\infty}^{+\infty} \alpha_{ij}''(\omega') \exp(i\omega't) \, d\omega'\right] d\tau$$

(2.26.5)

Changing the order of integration, and after some rearrangements, this equation becomes

$$L\langle \mathbf{m}_i(0) \cdot \dot{\mathbf{m}}_j(t)\rangle = \frac{3ikT}{\pi} \int_{-\infty}^{+\infty} \alpha_{ij}''(\omega') \, d\omega' \int_0^\infty \exp[-i(\omega - \omega')\tau] \, d\tau \quad (2.26.6)$$

By using the following mathematical expression

$$\int_0^\infty \exp[-i(\omega - \omega')\tau] \, d\tau = -iP\left(\frac{1}{\omega - \omega'}\right) + \pi\delta(\omega' - \omega)$$

(2.26.7)

where P is the principal value of the integral and δ is the Dirac function, we obtain

$$L\langle \mathbf{m}_i(0) \cdot \dot{\mathbf{m}}_j(t)\rangle = \frac{3kT}{\pi}\left[\int_{-\infty}^{+\infty} \frac{\alpha_{ij}''(\omega')}{\omega - \omega'} \, d\omega' + i\pi \int_{-\infty}^{+\infty} \delta(\omega' - \omega)\alpha_{ij}''(\omega') \, d\omega'\right]$$

(2.26.8)

From the properties of the Dirac function and after application of the Krönig–Kramers relationship [Eq. (2.25.5) or Eq.(2.25.11)], the following expression is finally obtained

$$-L\langle \mathbf{m}_i(0) \cdot \dot{\mathbf{m}}_j(t)\rangle = 3kT[\alpha'(\omega) + i\alpha''(\omega)] = 3kT\alpha_{ij}^*(\omega)$$

$$= L[-\dot{\psi}_{ij}(t)]$$

(2.26.9)

For $t = 0$

$$\langle \mathbf{m}_i(0) \cdot \dot{\mathbf{m}}_j(0)\rangle = 3kT\alpha_0$$

(2.26.10)

By taking the ratio of Eq. (2.26.9) to Eq. (2.26.10), we obtain

$$L\left[-\frac{\langle \mathbf{m}_i(0) \cdot \dot{\mathbf{m}}_j(t)\rangle}{\langle \mathbf{m}_i(0) \cdot \dot{\mathbf{m}}_j(0)\rangle}\right] = L[-\dot{\varphi}(t)] = \frac{\alpha_{ij}^*(\omega)}{\alpha_{ij}(0)}$$

(2.26.11)

where φ is the normalized cross-correlation function.

The calculation of $\alpha_{ij}^*(\omega)$ in Eq. (2.26.1) can be made macroscopically. Let us consider first the spherical shell under the external field E_2 with the dipole m_1 in the centre. The potentials are like those in Eqs (1.11.4), but Eq. (1.11.4c) is written in the following way

$$\phi_3 = \left(-E_2 r + \frac{m_1 + m_2}{r^2}\right) \cos \theta, \qquad r > R \tag{2.26.12}$$

The boundary conditions are now

$$\phi_1 = \phi_2, \qquad \frac{\partial \phi_1}{\partial r} = \varepsilon_2 \frac{\partial \phi_2}{\partial r}, \qquad r = a$$

$$\phi_2 = \phi_3, \qquad \varepsilon_2 \frac{\partial \phi_2}{\partial r} = \frac{\partial \phi_3}{\partial r}, \qquad r = R \tag{2.26.13}$$

From Eq. (2.26.13) and by using Eq. (2.26.12) we obtain

$$
m_2 = \frac{(\varepsilon_2^* - 1)(\varepsilon_2^* + 1)R^3(R^3 - a^3)}{(\varepsilon_2^* + 2)(2\varepsilon_2^* + 1)R^3 - 2(\varepsilon_2^* - 1)^2 a^3} E_2
$$
$$
- \frac{2(\varepsilon_2^* - 1)^2(R^3 - a^3)}{(\varepsilon_2^* + 2)(2\varepsilon_2^* + 1)R^3 - 2(\varepsilon_2^* - 1)^2 a^3} m_1 \tag{2.26.14}
$$

In this and in the following equations, the frequency dependence is omitted for the sake of clarity.

The corresponding results for A, B, and C of Eqs (1.11.4), expressed in terms of E_2 and m_1, are

$$
A = \frac{9\varepsilon_2^* R^3}{(\varepsilon_2^* + 2)(2\varepsilon_2^* + 1)R^3 - 2(\varepsilon_2^* - 1)^2 a^3} E_2
$$
$$
- \frac{2(\varepsilon_2^* - 1)(\varepsilon_2^* + 2)(R^3 - a^3)}{(\varepsilon_2^* + 2)(2\varepsilon_2^* + 1)R^3 - 2(\varepsilon_2^* - 1)^2 a^3} \cdot \frac{m_1}{a^3}
$$
$$
B = \frac{3(2\varepsilon_2^* + 1)R^3}{(2\varepsilon_2^* + 1)(\varepsilon_2^* + 2)R^3 - 2(\varepsilon_2^* - 1)^2 a^3} E_2
$$
$$
- \frac{6(\varepsilon_2^* - 1)}{(2\varepsilon_2^* + 1)(\varepsilon_2^* + 2)R^3 - 2(\varepsilon_2^* - 1)^2 a^3} m_1
$$
$$
C = - \frac{3(\varepsilon_2^* - 1)R^3 a^3}{(2\varepsilon_2^* + 1)(\varepsilon_2^* + 2)R^3 - 2(\varepsilon_2^* - 1)^2 a^3} E_2
$$
$$
+ \frac{3(\varepsilon_2^* + 2)R^3}{(2\varepsilon_2^* + 1)(\varepsilon_2^* + 2)R^3 - 2(\varepsilon_2^* - 1)^2 a^3} m_1
$$

$$\tag{2.26.15}$$

Now, the inner sphere is considered. In addition to the external field, E_1, the field due to the surrounding shell, $A - E_2$, needs to be considered. Therefore

$$m_1 = \frac{\varepsilon_1^* - 1}{\varepsilon_1^* + 2} a^3 (E_1 + A - E_2) \qquad (2.26.16)$$

where

$$A = E_2 \left(\frac{R}{a}\right)^3 - \frac{\varepsilon_2^* + 2}{\varepsilon_2^* - 1} \frac{m_2}{a^3} \qquad (2.26.17)$$

From Eqs (2.26.16) and (2.26.17), and after some rearrangements, the following expression is obtained

$$(\varepsilon_1^* - 1)(\varepsilon_2^* + 2)m_2 + (\varepsilon_2^* - 1)(\varepsilon_1^* + 2)m_1$$
$$= (\varepsilon_1^* - 1)(\varepsilon_2^* - 1)a^3 E_1 + (\varepsilon_1^* - 1)(\varepsilon_2^* - 1)(R^3 - a^3)E_2 \qquad (2.26.18)$$

The system of equations formed by Eqs (2.26.14) and (2.26.18) can be solved, giving the following expressions

$$\alpha_{11}^* = \frac{(\varepsilon_1^* - 1)\left[(2\varepsilon_2^* + 1)(\varepsilon_2^* + 2)R^3 - 2(\varepsilon_2^* - 1)^2 a^3\right]a^3}{3(\varepsilon_2^* + 2)(2\varepsilon_2^* + \varepsilon_1^*)R^3 + 6(\varepsilon_2^* - 1)(\varepsilon_1^* - \varepsilon_2^*)a^3}$$

$$\alpha_{22}^* = \frac{(\varepsilon_2^* - 1)\left[3(2\varepsilon_2^* + \varepsilon_1^*)R^3 - 2(\varepsilon_2^* - 1)(\varepsilon_1^* - 1)a^3\right](R^3 - a^3)}{3(\varepsilon_2^* + 2)(2\varepsilon_2^* + \varepsilon_1^*)R^3 + 6(\varepsilon_2^* - 1)(\varepsilon_1^* - \varepsilon_2^*)a^3}$$

$$\alpha_{12}^* = \alpha_{21}^* = -\frac{2(\varepsilon_2^* - 1)^2(\varepsilon_1^* - 1)a^3(R^3 - a^3)}{3(\varepsilon_2^* + 2)(2\varepsilon_2^* + \varepsilon_1^*)R^3 + 6(\varepsilon_2^* - 1)(\varepsilon_1^* - \varepsilon_2^*)a^3}$$

$$(2.26.19)$$

If $a \ll R$, these equations can be simplified to give

$$\alpha_{11}^* = \frac{(\varepsilon_1^* - 1)(2\varepsilon_2^* + 1)}{3(2\varepsilon_2^* + \varepsilon_1^*)} a^3$$

$$\alpha_{22}^* = \frac{\varepsilon_2^* - 1}{\varepsilon_2^* + 2} R^3 \qquad (2.26.20)$$

$$\alpha_{12}^* = \alpha_{21}^* = -\frac{2(\varepsilon_2^* - 1)^2(\varepsilon_1^* - 1)}{3(\varepsilon_2^* + 2)(2\varepsilon_2^* + \varepsilon_1^*)} a^3$$

For $a = 0$, Eqs (2.26.20) lead to

$$m_2 = \frac{\varepsilon_2^* - 1}{\varepsilon_2^* + 2} R^3 E_2 \tag{2.26.21}$$

In the same way, when $\varepsilon_2^* = 1$

$$m_1 = \frac{\varepsilon_1^* - 1}{\varepsilon_1^* + 2} a^3 E_1 \tag{2.26.22}$$

Obviously, these two situations correspond to the case of a sphere in vacuum.

For $\varepsilon_1^* = \varepsilon_2^* = \varepsilon^*$, a situation that corresponds to a sphere embedded in an infinite medium with the same permittivity, the term α_{11} adopts the following form

$$\alpha_{11}^* = \frac{(\varepsilon^* - 1)(2\varepsilon^* + 1)}{9\varepsilon^*} a^3 \tag{2.26.23}$$

If $\varepsilon_1^* = \varepsilon^*$ and $\varepsilon_2^* = \varepsilon_1(0) = \varepsilon_0$, the sphere embedded in a medium is characterized for all the frequencies by the static dielectric constant of the inner sphere. Finally, the cross-term $\alpha_{12}^* = \alpha_{21}^*$ corresponds to the polarization induced by the inner sphere (surrounding shell) in the surrounding shell (inner sphere).

By substituting each of the preceding cases into Eq. (2.26.11), we obtain:

(a)   Sphere in vacuum

$$L[-\dot{\psi}_{22}(t)] = \frac{(\varepsilon^* - 1)(\varepsilon_0 + 2)}{(\varepsilon^* + 2)(\varepsilon_0 - 1)} \tag{2.26.24}$$

which is the same as Eq. (2.21.10)

(b)   Sphere embedded in a medium with the same permittivity

$$L[-\dot{\psi}_{11}(t)] = \frac{\varepsilon_0(\varepsilon^* - 1)(2\varepsilon^* + 1)}{\varepsilon^*(\varepsilon_0 - 1)(2\varepsilon_0 + 1)} \tag{2.26.25}$$

which is equivalent to Eq. (2.21.8).

(c)   Sphere embedded in a medium with a dielectric constant equal to the static value of the sphere

$$L[-\dot{\psi}_{11}(t)] = \frac{3\varepsilon_0(\varepsilon^* - 1)}{(\varepsilon_0 - 1)(\varepsilon^* + 2\varepsilon_0)} \tag{2.26.26}$$

the same result as Eq. (2.21.9). Accordingly, there is agreement between the results obtained from the linear theory and those obtained by means of the fluctuation – dissipation theorem.

2.27. DIELECTRIC FRICTION

In Onsager static theory (see section 1.13), the treatment of the long-range dipolar forces is taken into account by using the cavity [Eq. (1.11.10)] and the reaction fields [Eq. (1.11.11)]. However, in the time-dependent case, that is, under alternating electric fields, rotations of dipoles produce a time-varying electric field outside the cavity, and energy is dissipated in the surroundings because of dielectric loss. This energy dissipation has the effect of slowing down rotations of dipoles. We refer to this effect as "dielectric friction". It should be noted that dielectric friction is part of the total friction on the dipole, to be added to any local effects due, for example, to the viscosity. The molecular picture implies a situation in which the dipoles librate around their equilibrium positions. A lag of the reaction field arises from this model because the reaction field at each moment is not in phase with the orientation of the dipole which librates into the cavity. As pointed out by Fatuzzo and Mason,[32] the reaction field lag has two consequences: it produces a retarding torque on the molecules, and it induces on average a dipole moment in the molecule, forming an angle with the permanent dipole. This suggests that simple extensions of the Onsager equation might not be the more convenient starting point to solve the time-dependent problem of the relationship between the dipole moment and dielectric permittivity. The pertinent calculations have been addressed independently by Fatuzzo and Mason,[32] Scaife,[33] and Nee and Zwanzig.[34] The final expression given in references[32] and[34] is

$$\frac{\varepsilon_0(\varepsilon^* - \varepsilon_\infty)(2\varepsilon^* + \varepsilon_\infty)}{\varepsilon^*(\varepsilon_0 - \varepsilon_\infty)(2\varepsilon_0 + \varepsilon_\infty)} = \left[1 + i\omega\tau + \frac{(\varepsilon_0 - \varepsilon_\infty)(\varepsilon_0 - \varepsilon^*)}{\varepsilon_0(2\varepsilon^* + \varepsilon_\infty)} \right]^{-1} \qquad (2.27.1)$$

where the last term between brackets on the right-hand side is due to the long-range frictional forces arising from the reaction field lag. Here, $\tau = \tau_D$ is the Debye relaxation time given by Eq. (2.8.17).

In particular, Nee and Zwanzig[34] obtained the following explicit expression for the dielectric friction coefficient

$$\zeta_D(\omega) = \frac{2kT}{i\omega} \frac{(\varepsilon_0 - \varepsilon_\infty)(\varepsilon_0 - \varepsilon^*)}{\varepsilon_0(2\varepsilon^* + \varepsilon_\infty)} \qquad (2.27.2)$$

The relationship between the dielectric friction and the relaxation time due to dielectric friction is a generalization of Eq. (2.8.17), namely

$$\tau_{DF}(\omega) = \frac{\zeta_D(\omega)}{2kT} \qquad (2.27.3)$$

and consequently

$$\tau_{DF} = \frac{1}{i\omega} \frac{(\varepsilon_0 - \varepsilon_\infty)(\varepsilon_0 - \varepsilon^*)}{\varepsilon_0(2\varepsilon^* + \varepsilon_\infty)} \tag{2.27.4}$$

The total friction coefficient is given by

$$\frac{\zeta(\omega)}{2kT} = \tau_D + \frac{1}{i\omega} \frac{(\varepsilon_0 - \varepsilon_\infty)(\varepsilon_0 - \varepsilon^*)}{\varepsilon_0(2\varepsilon^* + \varepsilon_\infty)} \tag{2.27.5}$$

On the other hand, Scaife[33] gives a different expression for the microscopic time due to the contribution of all frictional forces, which can be written as

$$\tau_{total} = \tau_D + \frac{(\varepsilon_0 - \varepsilon_\infty)^2}{\varepsilon_0(2\varepsilon_0 + \varepsilon_\infty)} \bar{\tau}_M \tag{2.27.6}$$

and the relaxation time due to dielectric friction is given by

$$\tau_{DF} = \frac{(\varepsilon_0 - \varepsilon_\infty)^2}{\varepsilon_0(2\varepsilon_0 + \varepsilon_\infty)} \bar{\tau}_M \tag{2.27.7}$$

In Eqs (2.27.6) and (2.27.7), $\bar{\tau}_M$ is a mean macroscopic relaxation time defined by

$$\left.\frac{\partial \varepsilon''}{\partial \omega}\right|_{\omega \to 0} = (\varepsilon_0 - \varepsilon_\infty)\bar{\tau}_M \tag{2.27.8}$$

The parameter τ_D in Eq. (2.27.6) is the contribution of short-range forces to the relaxation time which we can identify as the relaxation time in the Debye theory. The second term on the right-hand side of Eq. (2.27.6) is the contribution to long-range forces from the lag of the reaction field. It can be demonstrated that Eq. (2.27.4) coincides with Eq. (2.27.7) at low frequencies if $\bar{\tau}_M = -(i\omega)^{-1}$ and a Debye equation is assumed.

After taking into account the inertial effects together with the dielectric friction, we can write the following expression for the dielectric permittivity

$$\frac{\varepsilon_0(\varepsilon^* - \varepsilon_\infty)(2\varepsilon^* + \varepsilon_\infty)}{\varepsilon^*(\varepsilon_0 - \varepsilon_\infty)(2\varepsilon_0 + \varepsilon_\infty)} = \left[1 - \frac{I\omega^2}{2kT} + i\omega\tau_D + \frac{(\varepsilon_0 - \varepsilon_\infty)(\varepsilon_0 - \varepsilon^*)}{\varepsilon_0(2\varepsilon^* + \varepsilon_\infty)}\right]^{-1} \tag{2.27.9}$$

which, according to Eq. (2.27.5), can alternatively be written as

$$\frac{\varepsilon_0(\varepsilon^* - \varepsilon_\infty)(2\varepsilon^* + \varepsilon_\infty)}{\varepsilon^*(\varepsilon_0 - \varepsilon_\infty)(2\varepsilon_0 + \varepsilon_\infty)} = \left(1 - \frac{I\omega^2}{2kT} + \frac{i\omega I\zeta'(\omega)}{2kT}\right)^{-1} \tag{2.27.10}$$

where $\zeta(\omega) = I\zeta'(\omega)$. Equation (2.27.10) will be discussed later in the context of the memory function.

An expression similar to Eq. (2.27.10) has been found by Lobo et al. [35] in the study of the high-frequency dielectric response of a condensed system of molecules with permanent dipole moments. Coffey and Kalmykov[36] obtained from the Langevin equation the linear response of a system of dipoles, in terms of the dielectric permittivity. The calculation involves the cavity as well as the reaction field in the Onsager cavity. The result can be compared with Eq. (2.27.1).

Let us write the averaged Eq. (2.17.11) as

$$\tau_D \langle \dot{\mu} \rangle + \langle \mu \rangle = \frac{1}{2kT} \langle \mu^2 \langle F \rangle - \langle \mu(\mu \cdot F) \rangle \rangle \qquad (2.27.11)$$

where the term $I\ddot{\mu}$ is deleted. Moreover, it is assumed that $\langle \Lambda(t) \times \mu(t) \rangle = 0$ because the random torque varies arbitrarily, $\omega_p^2 = (2kT)/I$ and $\tau_D = \zeta/(2kT)$. In this equation, F is the electric field acting upon the dipole, that is, the sum of the cavity field $G(t)$ and the reaction field $R(t)$. Note that the dipole moment is a random variable and for this reason is averaged over the ensemble of dipoles. The Z component of Eq. (2.27.11) can be written as

$$\tau_D \langle \dot{p} \rangle + \langle p \rangle + \frac{G(t)}{2kT} \langle p^2 - \mu^2 \rangle + \frac{1}{2kT} (\langle p(\mu \cdot R) \rangle - \mu^2 \langle R_Z \rangle) = 0 \qquad (2.27.12)$$

In the same way as μ and p, R and R_Z also are random variables and the pair μ, R is not statistically independent because of the dielectric friction.

For an alternating field, the instantaneous dipole moment in the linear limit will be

$$\mu = \mu_0 + u\mu_{1Z} \qquad (2.27.13)$$

where μ_0 denotes the dipole moment in the absence of an external field, u is the unit vector along the Z axis, and $\mu_Z^1(t) = \mu_Z^1 \exp(i\omega t)$ is the component of the dipole moment that is linear in the applied field. In the same way, the reaction field is time dependent and the corresponding Fourier transform will be

$$r(\omega)\mu(\omega) = [r_0(\omega) + r_1(\omega)]\mu(\omega) \qquad (2.27.14)$$

where

$$r(\omega) = \frac{1}{a^3} \frac{1}{\varepsilon_\infty} \frac{2[\varepsilon^*(\omega) - \varepsilon_\infty]}{2\varepsilon^*(\omega) + \varepsilon_\infty} \qquad \text{and}$$

$$r_0(\omega) = \frac{1}{a^3} \frac{1}{\varepsilon_\infty} \frac{2(\varepsilon_0 - \varepsilon_\infty)}{2\varepsilon_0 + \varepsilon_\infty} \qquad (2.27.15)$$

Obviously

$$
\begin{aligned}
r_1(\omega) &= \frac{1}{a^3} \frac{1}{\varepsilon_\infty} \left\{ \frac{2[\varepsilon^*(\omega) - \varepsilon_\infty]}{2\varepsilon^*(\omega) + \varepsilon_\infty} - \frac{2(\varepsilon_0 - \varepsilon_\infty)}{2\varepsilon_0 + \varepsilon_\infty} \right\} \\
&= \frac{4\pi N_1}{3\varepsilon_\infty} \left(\frac{2[\varepsilon^*(\omega) - \varepsilon_\infty]}{2\varepsilon^*(\omega) + \varepsilon_\infty} - \frac{2(\varepsilon_0 - \varepsilon_\infty)}{2\varepsilon_0 + \varepsilon_\infty} \right) \\
&= \frac{8\pi N_1}{3\varepsilon_\infty} \frac{3\varepsilon_\infty[\varepsilon^*(\omega) - \varepsilon_0]}{[2\varepsilon^*(\omega) + \varepsilon_\infty](2\varepsilon_0 + \varepsilon_\infty)}
\end{aligned}
\tag{2.27.16}
$$

This is the only component of the reaction field that acts on the dipole.

The linear response disregards all non-linear terms in G, the cavity field, and in p_1. Therefore, Eq. (2.27.11) can be written as

$$
\begin{aligned}
\tau_D \langle \dot{p}_1 \rangle + \langle p_1 \rangle &= \frac{G(t)}{2kT}(\mu_0^2 - p_0^2) + \frac{\mu_0^2}{2kT} \int_{-\infty}^{t} r_1(t - \tau)\langle p_1(\tau) \rangle \, d\tau \\
&\quad + \frac{p_0 p_1}{2kT} \int_{-\infty}^{t} r_1(t - \tau)\langle p_0(\tau) \rangle \, d\tau \\
&\quad - \frac{1}{2kT} \int_{-\infty}^{t} r_1(t - \tau)\langle [p_1 \boldsymbol{\mu}_0 \cdot \boldsymbol{\mu}_0(\tau) \\
&\quad + p_0 p_0(\tau) + p_0^2 p_1(\tau)] \rangle \, d\tau
\end{aligned}
\tag{2.27.17}
$$

We can write

$$
\begin{aligned}
G(t) &= \int_{-\infty}^{t} g(t - \tau) E \exp(i\omega\tau) \, d\tau = E \exp(i\omega t) \int_{0}^{\infty} \exp -(i\omega\tau) g(\tau) \, d\tau \\
&= E \exp(i\omega t) g(\omega)
\end{aligned}
\tag{2.27.18}
$$

where $g(\omega) = [3\varepsilon^*(\omega)]/[2\varepsilon^*(\omega) + \varepsilon_\infty]$.

In a similar way

$$
\int_{-\infty}^{t} r_1(t - \tau)\langle p_1(\tau) \rangle \, d\tau = \langle p_1 \rangle r_1(\omega) \exp(i\omega t)
\tag{2.27.19}
$$

Then, on account of Eqs (2.27.12), (2.27.17) and (2.27.19), Eq. (2.27.16) leads to

$$
\left[1 + i\omega\tau_D - \frac{\mu_0^2}{2kT} r_1(\omega) \right] \langle p_1 \rangle
$$

$$
= \frac{\mu_0^2}{3kT} g(\omega) E + \frac{p_0 p_1}{2kT} \int_{-\infty}^{t} r_1(t-\tau) \langle p_0(\tau) \rangle \, d\tau
$$

$$
- \frac{r_1(\omega)}{2kT} \langle p_1 p_0^2 \rangle - \frac{1}{2kT} \int_{-\infty}^{t} r_1(t-\tau) \langle p_1 [p_0 p_0(\tau)] + \boldsymbol{\mu}_0 \cdot \boldsymbol{\mu}_0(\tau) \rangle \, d\tau
$$

$$
(2.27.20)
$$

This equation gives the frequency dependence of the Z component of the dipolar moment. Note that the last three nonlinear terms on the right-hand side of Eq. (2.17.20) arise from the reaction field. If we omit these effects, then Eq. (2.27.20) reduces to

$$
\left[1 + i\omega\tau_D - \frac{\mu_0^2}{2kT} r_1(\omega) \right] \langle p_1 \rangle = \frac{\mu_0^2}{3kT} g(\omega) E \qquad (2.27.21)
$$

Equations (2.27.15), (2.27.18), (1.13.9) and (1.12.4), in conjunction with the expression

$$
\varepsilon^*(\omega) - \varepsilon_\infty = 4\pi N_1 \frac{\langle p_1 \rangle}{E} \qquad (2.27.22)
$$

which is the dynamic counterpart of Eq. (1.8.5), lead to Eq. (2.21.7).

An expression that partially includes the effect of the nonlinear terms in Eq. (2.27.20) can be obtained by assuming that p_0^2 and p_1 are uncorrelated, that is

$$
\langle p_1 p_0^2 \rangle = \langle p_1 \rangle \langle p_0^2 \rangle = \frac{\langle p_1 \rangle \mu_0^2}{3} \qquad (2.27.23)
$$

Then, Eq. (2.27.21) is modified to give

$$
\left[1 + i\omega\tau_D - \frac{\mu_0^2}{3kT} r_1(\omega) \right] \langle p_1 \rangle = \frac{\mu_0^2}{3kT} g(\omega) E \qquad (2.27.24)
$$

and, following the same procedure as above, we obtain

$$
\frac{\varepsilon_0 (\varepsilon^* - \varepsilon_\infty)(2\varepsilon^* + \varepsilon_\infty)}{\varepsilon^* (\varepsilon_0 - \varepsilon_\infty)(2\varepsilon_0 + \varepsilon_\infty)} = \left[1 + i\omega\tau_D + \frac{2}{3} \frac{(\varepsilon_0 - \varepsilon_\infty)(\varepsilon_0 - \varepsilon^*)}{\varepsilon_0 (2\varepsilon^* + \varepsilon_\infty)} \right]^{-1}
$$

$$
(2.27.25)
$$

Equation (2.27.25) differs from the Fatuzzo–Mason equation only in the factor 2/3.

2.28. RESONANCE ABSORPTION

As mentioned in the section 1.8 electric fields provoke a distortion of the electronic clouds and nucleus positions, giving rise to electronic and atomic polarizations. These polarization processes, which take place at frequencies of the order of 10^{12} Hz, are physically very different from those corresponding to the orientational polarization. Molecules can be assumed to be elastic systems of electric charges that can oscillate around their equilibrium positions. In this sense, the induced polarization is basically an elastic process governed by a second-order differential equation. These systems of particles are characterized by a natural frequency of vibration. When the frequency of the applied electric field coincides with the natural frequency, a resonance phenomenon appears that causes an absorption of energy together with a dispersion of the dielectric constant. Consequently, the imaginary part of the permittivity shows a maximum at the resonance frequency.

Under an external periodic force field $E_0 \exp(i\omega t)$, the equation of motion of a molecular oscillator is described by the equation

$$m\ddot{x}(t) + 2\beta'\dot{x}(t) + kx(t) = qE_i \tag{2.28.1}$$

where the first term on the left-hand side is the inertial term containing the mass of the system under oscillation, the second refers to the attenuation, and the third corresponds to the elastic restoring force. On the right-hand side of Eq. (2.28.1), q is the electric charge and E_i is the local field acting upon the oscillating system.

Owing to the fact that a charge q displaced from its equilibrium position by a distance x produces a dipole moment given by

$$P = N_1 qx \tag{2.28.2}$$

Eq. (2.28.1) can be written as

$$\ddot{P} + 2\beta\dot{P} + \omega_0^2 P = aE_i \tag{2.28.3}$$

where ω_0 is the natural frequency of the oscillator and $a = N_1 q^2/m$. Assuming the Lorentz field as the local field [Eq. (1.9.3)], we have

$$E_i = E + \frac{4\pi P}{3} \tag{2.28.4}$$

Substitution of Eq. (2.28.4) into Eq. (2.28.3) gives

$$\ddot{P} + 2\beta\dot{P} + \omega_0'^2 P = aE \tag{2.28.5}$$

where

$$\omega_0'^2 = \omega_0^2 - \frac{4\pi a}{3} \tag{2.28.6}$$

Assuming for the polarization an equation of the form

$$P = P_0 \exp(i(\omega t + \delta)) \tag{2.28.7}$$

the complex permittivity is obtained as

$$\chi^* = \frac{P_0}{E_0} \exp(i\delta) = \frac{a}{\omega_0^2 - \omega^2 + 2\beta i\omega} \tag{2.28.8}$$

Accordingly, if the attenuation β is nonzero, the polarization vector is not in phase with the electric field. By writing

$$\chi^* = \frac{\varepsilon^* - \varepsilon_\infty}{4\pi}, \qquad \varepsilon_0 - \varepsilon_\infty = \frac{4\pi a}{\omega_0'^2}, \qquad \gamma = \frac{2\beta}{\omega_0'^2} \tag{2.28.9}$$

the dielectric properties of a material on the high-frequency side of the spectrum (optical frequencies) can be described by means of an equation of the form

$$\varepsilon^* = \varepsilon_\infty + \frac{\varepsilon_0 - \varepsilon_\infty}{1 + i\omega\gamma - (\omega/\omega_0')^2} \tag{2.28.10}$$

the real and imaginary parts of which are

$$\varepsilon' = \varepsilon_\infty + \frac{(\varepsilon_0 - \varepsilon_\infty)[1 - (\omega/\omega_0')^2]}{\left[1 - (\omega/\omega_0')^2\right]^2 + \omega^2\gamma^2}$$

$$\varepsilon'' = \frac{\omega\gamma(\varepsilon_0 - \varepsilon_\infty)}{\left[1 - (\omega/\omega_0')^2\right]^2 + \omega^2\gamma^2}$$

$$\tag{2.28.11}$$

The parameter ω_0' of Eqs (2.28.11) can be interpreted in terms of the inertia moment of the oscillator. In fact, the comparison of Eqs (2.28.10) and (2.16.13) leads to

$$\omega_0'^2 = \frac{2kT}{I} \tag{2.28.12}$$

and Eq. (2.28.9) can be written as

$$\frac{\varepsilon^* - \varepsilon_\infty}{\varepsilon_0 - \varepsilon_\infty} = \frac{1}{1 + i\omega\gamma - [(I\omega^2)/(2kT)]} \tag{2.28.13}$$

On the other hand, $\varepsilon' - \varepsilon_\infty$ vanishes at the resonant frequency, and ε'' attains its maximum value given by

$$\varepsilon''_{max} = \frac{\varepsilon_0 - \varepsilon_\infty}{\omega'_0 \gamma} \tag{2.28.14}$$

The smaller the value of γ, the sharper is the resonance. Owing to inertial effects, $\varepsilon' - \varepsilon_\infty$ is negative above the resonant frequency. There is a range of frequencies around ω'_0 for which $d\varepsilon'/d\omega$ is negative, so, in this range of frequencies, normal and anomalous dispersion occur.

Another type of resonance absorption is described by the Van Vleck–Weisskopf–Fröhlich equation.[37]

$$\Phi = \exp\left(-\frac{t}{\tau}\right) \cos \omega_0 t \tag{2.28.15}$$

According to Eq. (2.12.6), the complex dielectric permittivity is given by

$$\varepsilon^*(\omega) = \varepsilon_\infty + \frac{\varepsilon_0 - \varepsilon_\infty}{2} \left[\frac{1 - i\omega_0\tau}{1 - i(\omega + \omega_0)\tau} + \frac{1 + i\omega_0\tau}{1 - i(\omega - \omega_0)\tau} \right] \tag{2.28.16}$$

The corresponding real and imaginary parts of the complex permittivity are

$$\varepsilon' = \varepsilon_\infty + \frac{\varepsilon_0 - \varepsilon_\infty}{2} \left[\frac{1 + \omega_0(\omega + \omega_0)\tau^2}{1 + (\omega + \omega_0)^2\tau^2} + \frac{1 - \omega_0(\omega - \omega_0)\tau^2}{1 + (\omega - \omega_0)^2\tau^2} \right]$$

$$\varepsilon'' = \frac{\varepsilon_0 - \varepsilon_\infty}{2} \left[\frac{\omega\tau}{1 + (\omega + \omega_0)^2\tau^2} + \frac{\omega\tau}{1 + (\omega - \omega_0)^2\tau^2} \right] \tag{2.28.17}$$

The imaginary part has a maximum given by

$$\varepsilon''_{max} = \frac{\varepsilon_0 - \varepsilon_\infty}{2}(1 + \omega_0^2\tau^2)^{1/2} \tag{2.28.18}$$

located at frequency

$$\omega = \tau^{-1}(1 + \omega_0^2\tau^2)^{1/2} \tag{2.28.19}$$

The corresponding electric circuit is presented in Fig. 1.8 of reference.[14]

2.29. MEMORY FUNCTIONS

Correlation functions, as defined in section 2.4, are efficient tools to describe the dynamic response of a multibody system. The corresponding application to dipolar fluids was made in sections 2.20 to 2.26. A similar treatment can be done for the viscoelastic relaxation. In both cases, the description of the relaxation behavior arising from molecules between equilibrium sites requires, from a formal point of view, the solution of the Liouville equation (2.20.3) for each molecule. Although this problem can be solved with the aid of a computer by using molecular mechanics, there is an alternative way to treat it, based on the memory function formalism.[38-41] The calculation of memory functions is, at present, an important step in the theoretical analysis of the dynamics of fluids. In fact, correlation and memory functions are closely related, and each one can be obtained from the other. Moreover, important properties of the time correlation functions are preserved in the context of the memory functions.

Let us start with the phenomenological description of a Brownian particle with velocity v. The evolution of a fluctuating quantity, such as the velocity of the particle, can be described by the following stochastic differential equation of the Langevin type (see Eq. (2.3.1))

$$m\dot{\mathbf{v}} + \zeta \mathbf{v} = -\mathbf{F}(t) \tag{2.29.1}$$

The damping term is linear in v, ζ being the dissipation coefficient. The right-hand side of Eq. (2.29.1) is the random force representing the interaction with other particles, and its average vanishes according to Eq. (2.3.3). Equation (2.29.1) can be solved by using the methodology outlined in section 2.6. It is also possible to obtain an equation for the correlation function by multiplying Eq. (2.29.1) by $v(0)$ and subsequent averaging, that is

$$\frac{d}{dt}\langle \mathbf{v}(t) \cdot \mathbf{v}(0) \rangle + \frac{\zeta}{m}\langle \mathbf{v}(t) \cdot \mathbf{v}(0) \rangle = 0 \qquad \text{or}$$

$$\dot{C}(t) + \frac{\zeta}{m} C(t) = 0 \tag{2.29.2}$$

In this equation, C accounts for the correlation function of the observed variable. However, in order to describe the dynamics of a system without restrictions on the timescale, a generalization of Eq. (2.29.2) for a non-Markoffian situation can be written as

$$\dot{C}(t) = -\int_0^t K(\tau)C(t-\tau)\,d\tau \tag{2.29.3}$$

which is called the memory function equation (ME). A detailed proof of Eq. (2.29.3) can be found in reference.[41] The technical tool that is needed for the aforementioned proof is the projection operator which has the effect of separating the time-dependent part of the phase-space function.[38] The projector operator is a mathematical formulation for the contraction of the information about the system. The parameter $C(t)$ represents the velocity correlation function in hydrodynamics, the correlation function for the end-to-end distance $\mathbf{r}$ in viscoelasticity, and the microscopic dipole–dipole correlation function as defined in section 2.23 for a dipolar system. In Eq. (2.29.3) the fluctuation term is deleted. In the same way, any propagation process associated with the time evolution of the stochastic variable is ruled out.

The kernel of Eq. (2.29.3) is a memory function. This equation is one of the major results in the statistical mechanical study of multibody complex systems. In fact, it is a key tool in the analysis of the dynamic behavior of the system. In spite of the complexity of the memory function for complex condensate matter, and in contrast to the conventional expressions for the dielectric permittivity in terms of the dipole–dipole correlation function (see section 2.26), ME does not depend on the shape of the sample.

The physical meaning of the memory function can easily be understood if we take the long time limit in Eq. (2.29.3)

$$\lim_{t \to \infty} \int_0^t K(\tau)C(t - \tau)\mathrm{d}\tau = C(t) \int_0^\infty K(\tau)\,\mathrm{d}\tau \qquad (2.29.4)$$

By comparing Eqs (2.29.4) and (2.29.2b), the following expression for the friction coefficient is obtained

$$\zeta = m \int_0^\infty K(\tau)\,\mathrm{d}\tau \qquad (2.29.5)$$

According to Eqs (2.4.4) and (2.6.4)

$$\int_0^\infty K(s)\,\mathrm{d}s = \frac{\Gamma}{2kT} = \frac{1}{2kT} \int_0^\infty \langle \mathbf{F}(0) \cdot \mathbf{F}(t) \rangle \,\mathrm{d}t \qquad (2.29.6)$$

The Laplace transform of Eq. (2.29.3) makes it possible to obtain the normalized correlation function $[C(0) = 1]$ from the memory function, that is

$$C(s) = \frac{1}{s + K(s)} \qquad (2.29.7)$$

For a single relaxation time, the memory function is

$$K(t) = \frac{\delta(t)}{\tau} \qquad (2.29.8)$$

where $\delta(t)$ is the Dirac function. Then, according to Eq. (2.12.7), and neglecting field factor effects, we obtain

$$\frac{\varepsilon^* - \varepsilon_\infty}{\varepsilon_0 - \varepsilon_\infty} = \frac{1}{1 + s\tau} = \frac{1}{1 + i\omega\tau} \qquad (2.29.9)$$

which is the classical Debye result. The correlation function is obviously a single exponential, $C(t) = \Phi(t) = \exp(-(t/\tau))$.

Equation (2.29.7) represents the Laplace transform of the correlation function in terms of a first-order moment given by the memory function. For this reason it is called the first-order memory function. Therefore, $K_1(t)$ represents the first-order memory function. Note that the same treatment can be repeated for subsequent memory functions in such a way that

$$\dot{K}_{i-1}(t) = -\int_0^t K_i(t-\tau)K_{i-1}(\tau)\,d\tau \qquad (2.29.10)$$

or in terms of the Laplace transform

$$K_{i-1}(s) = \frac{K_{i-1}(0)}{s + K_i(s)} \qquad (2.29.11)$$

This "cascade" process constitutes the basis of the continued fraction representation (CFR) of Mori.[38] The virtue of CFR is that it provides an analytical contact with models initially introduced as ansatz for the memory function. The continued fraction takes the following form

$$C(s) = \cfrac{K_0(0)}{s + \cfrac{K_1(0)}{s + \cfrac{K_2(0)}{s + \cfrac{K_3(0)}{s + \cdots}}}} \qquad (2.29.12)$$

If we have a model that prescribes a particular $K(s)$, then the continued fraction is closed at that stage and it is possible to calculate $C(s)$ and consequently the dielectric permittivity. In this way, Douglas and Hubbard[42] considered particular forms of $K(t)$ and recovered several empirical models for representing the experimental data. Recently, Williams and Fournier[43] have described the first-order memory function (FOMF) in terms of the so-called apparent memory. They found the corresponding FOMF for the Cole–Cole and KWW representations of dielectric permittivity for polymers and glass-forming materials. They also emphasized that, if the dynamic heterogeneity of Schmidt-Rohr and Spiess[44] is considered in the α-relaxation zone, the physical meaning of the memory may change. Dynamic heterogeneity explains the relaxation function as an average of a series of parallel and independent decaying elementary processes with a broad distribution of relaxation times. From this point of view, the existence of memory functions would be only apparent.

Conversely, the knowledge of R' and R'' from the experimental data or from a model function enables K_1' and K_1'' to be determined. Thus, Eq. (2.23.13) can be

written as

$$R^*(\omega) = \frac{\varepsilon^*(\omega) - \varepsilon_\infty}{\varepsilon_0 - \varepsilon_\infty} f(\omega) = L\left[-\frac{dC(t)}{dt}\right] \qquad (2.29.13)$$

where

$$f(\omega) = \frac{\varepsilon_0(2\varepsilon^* + \varepsilon_\infty)}{\varepsilon^*(2\varepsilon_0 + \varepsilon_\infty)} \qquad (2.29.14)$$

is the field factor. Then, the real and imaginary parts of the FOMF can easily be obtained from Eq. (2.29.7), yielding

$$K_1'(\omega) = \frac{\omega R''}{(1 - R')^2 + R''^2}$$

$$K_2''(\omega) = \frac{\omega\left[(1 - R')R' - R''^2\right]}{(1 - R')^2 + R''^2} \qquad (2.29.15)$$

In a similar way, we can calculate the second-order memory function (SOMF).[45,46] According to Eq. (2.29.11)

$$K_1(s) = \frac{K_1(0)}{s + K_2(s)} \qquad (2.29.16)$$

with $s = i\omega$. Hence, the following expression is obtained

$$R^* = \left[1 + i\omega\frac{i\omega + K_2^*(\omega)}{K_1(0)}\right]^{-1} = \frac{K_1(0)}{K_1(0) + i\omega K_2^* + (i\omega)^2} \qquad (2.29.17)$$

and the value of the SOMF is given by

$$K_2^*(\omega) = K_2' - iK_2'' = \frac{K_1(0)[1 - R^*(\omega)]}{i\omega R^*(\omega)} - i\omega \qquad (2.29.18)$$

From Eq. (2.28.18), the real and imaginary parts of the SOMF are easily obtained as

$$K_2' = \frac{K_1(0)}{\omega}\left[\frac{R''}{R'^2 + R''^2}\right]$$

$$K_2'' = -\frac{K_1(0)}{\omega}\left(1 - \frac{R'}{R'^2 + R''^2}\right) + \omega \qquad (2.29.19)$$

By comparing Eqs (2.27.9) and (2.29.16) it is found that

$$K_1(0) = \frac{2kT}{I} \tag{2.29.20}$$

This expression represents the square of the resonance frequency of a free rotor, and therefore the SOMF may be identified with the generalized friction coefficient as follows

$$K_2^*(\omega) = \zeta^*(\omega) \tag{2.29.21}$$

Note that, for small-size molecules, the following inequality is fulfilled

$$\omega^2 \ll K_1(0) \tag{2.29.22}$$

which is consistent with the fact that the contribution to the memory of any resonance process is nil at moderate frequencies ($\omega \leqslant 10^8$). Then, Eq. (2.29.18) reduces to

$$R^*(\omega) = \left[1 + \frac{i\omega\zeta'(\omega)}{K_1(0)}\right]^{-1} \tag{2.29.23}$$

and consequently

$$K_2'' = -\frac{K_1(0)}{\omega}\left(1 - \frac{R'}{R'^2 + R''^2}\right) \tag{2.29.24}$$

An alternative interpretation of the SOMF is obtained by analyzing the frequency-dependent diffusion coefficient $D(\omega)$, namely

$$\frac{\zeta(\omega)}{2kT} = \tau(\omega) = \frac{1}{2D(\omega)} \tag{2.29.25}$$

Furthermore, according to Eq. (2.19.6), the SOMF can be related to the complex shear viscosity, $\eta(\omega)$, which is also related to the complex shear modulus [Eq. (2.19.7)]. In this way, the storage and loss parts of the dynamic shear modulus are given by

$$G'(\omega) = \frac{\omega K_2''(\omega)}{AK_1(0)}$$

$$G''(\omega) = \frac{\omega K_2'(\omega)}{AK_1(0)} \tag{2.29.26}$$

where $A = (8\pi R^3)/(2kT)$, R being the radius of the rotating unit. Therefore, Eqs (2.29.21), (2.21.25) and (2.21.26) show that the experimental data can be explained in any of the three following ways:

(a)   the frictional interpretation, which assumes that the energy is dissipated by the internal mechanisms;

(b)   a diffusive interpretation according to which the molecular dipoles undergo diffusion processes by means of rotational Brownian motion;

(c)   the description of the evolution of the shear viscosity by generalized hydrodynamics.

2.30. FIRST-ORDER MEMORY FUNCTION AND MACROSCOPIC RELAXATION TIME

As we have seen in section 2.23, the relationship between the dielectric permittivity and the fluctuations of the dipolar moment in a spherical region embedded in its own medium is given by Eq. (2.23.13), that is

$$\frac{\varepsilon_0[\varepsilon^*(\omega) - \varepsilon_\infty][2\varepsilon^*(\omega) + \varepsilon_\infty]}{\varepsilon^*(\omega)(\varepsilon_0 - \varepsilon_\infty)(2\varepsilon_0 + \varepsilon_\infty)} = L(-\dot{\varphi}_{FM}) \tag{2.30.1}$$

where subscript FM indicates Fatuzzo–Mason, to whom Eq. (2.30.1) is due.

When the spherical specimen is embedded in a medium of permittivity ε_0, the Glarum–Cole equation holds

$$\frac{3\varepsilon_0[\varepsilon^*(\omega) - \varepsilon_\infty]}{(\varepsilon_0 - \varepsilon_\infty)[\varepsilon^*(\omega) + 2\varepsilon_0]} = L(-\dot{\varphi}_{GC}) \tag{2.30.2}$$

where subscript GC indicates that the correlation function refers to the Glarum–Cole model.

On the other hand, according to Eq. (2.29.7), the relation between the Laplace transform of the dipolar correlation function φ and the first-order memory function (FOMF) K_1 is given by

$$\varphi(s) = \frac{1}{s + K_1(s)} \tag{2.30.3}$$

Substitution of Eq. (2.30.3) into Eqs (2.30.1) and (2.30.2) gives respectively

$$\frac{\varepsilon^*(\omega) - \varepsilon_\infty}{\varepsilon_0 - \varepsilon_\infty} = \frac{1}{1 + \{(i\omega)/[K_1(\omega)]\}} \frac{\varepsilon^*(\omega)(2\varepsilon_0 + \varepsilon_\infty)}{\varepsilon_0[2\varepsilon^*(\omega) + \varepsilon_\infty]}$$

$$\frac{\varepsilon^*(\omega) - \varepsilon_\infty}{\varepsilon_0 - \varepsilon_\infty} = \frac{1}{1 + \{(i\omega)/[K_1(\omega)]\}} \frac{\varepsilon^*(\omega) + 2\varepsilon_0}{3\varepsilon_0}$$

(2.30.4)

According to the definition of the dynamic permittivity [Eq. (2.20.1)], the following equation holds in a polarizable medium

$$\varepsilon^*(\omega) - \varepsilon_\infty = 4\pi \frac{P(\omega)}{E(\omega)} = f(\omega) \tag{2.30.5}$$

$P(\omega)$ and $E(\omega)$ in this equation can be interpreted as Laplace or Fourier transforms so that

$$\varepsilon^*(\omega) - \varepsilon_\infty = \int_0^\infty f(t) \exp(-i\omega t)\, dt \tag{2.30.6}$$

Equation (2.30.6) can be expanded as follows

$$\varepsilon^*(\omega) - \varepsilon_\infty = \int_0^\infty f(t)(1 - i\omega t + \cdots)\, dt \tag{2.30.7}$$

and, for relatively low frequencies, Eq. (2.30.7) is approximately given by

$$\varepsilon^*(\omega) - \varepsilon_\infty \cong \int_0^\infty f(t)\, dt - i\omega \int_0^\infty tf(t)\, dt \tag{2.30.8}$$

Let

$$\varepsilon_0 - \varepsilon_\infty = \int_0^\infty f(t)\, dt \tag{2.30.9}$$

then

$$\varepsilon^*(\omega) \cong \varepsilon_0 - i\omega \int_0^\infty tf(t)\, dt \tag{2.30.10}$$

If we define the macroscopic relaxation time as

$$\tau_\mu = \frac{\int_0^\infty tf(t)\, dt}{\int_0^\infty f(t)\, dt} \tag{2.30.11}$$

then the following result holds

$$\varepsilon^*(\omega) \cong \varepsilon_0 - i\omega\tau_\mu \int_0^\infty f(t)\,dt = \varepsilon_0 - i\omega\tau_\mu(\varepsilon_0 - \varepsilon_\infty) \tag{2.30.12}$$

from which

$$\tau_\mu = \lim_{\omega \to 0} \frac{\varepsilon''(\omega)}{(\varepsilon_0 - \varepsilon_\infty)\omega} \tag{2.30.13}$$

Carrying Eq. (2.30.12b) to the right-hand side of Eq. (2.30.4), and after some simplifications, we obtain

$$\frac{\varepsilon^*(\omega) - \varepsilon_\infty}{\varepsilon_0 - \varepsilon_\infty} \cong \frac{1}{1 + \{(i\omega)/[K_1(\omega)]\}} \frac{1}{1 + i\omega\tau_\mu(\varepsilon_0 - \varepsilon_\infty)/(3\varepsilon_0)}$$

$$\cong \frac{1}{1 + i\omega\{1/[K_1(\omega)] + \tau_\mu(\varepsilon_0 - \varepsilon_\infty)/(3\varepsilon_0)\}}$$

$$\frac{\varepsilon^*(\omega) - \varepsilon_\infty}{\varepsilon_0 - \varepsilon_\infty} \cong \frac{1}{1 + \{i\omega/[K_1(\omega)]\}} \frac{1}{1 + i\omega\tau_\mu[\varepsilon_\infty(\varepsilon_0 - \varepsilon_\infty)]/[\varepsilon_0(2\varepsilon_0 + \varepsilon_\infty)]}$$

$$\cong \frac{1}{1 + i\omega\{1/[K_1(\omega)] + \tau_\mu[\varepsilon_\infty(\varepsilon_0 - \varepsilon_\infty)]/[\varepsilon_0(2\varepsilon_0 + \varepsilon_\infty)]\}} \tag{2.30.14}$$

Equations (2.30.14) can be expressed in the classical form

$$\frac{\varepsilon^*(\omega) - \varepsilon_\infty}{\varepsilon_0 - \varepsilon_\infty} = \frac{1}{1 + i\omega\tau'_\mu} \tag{2.30.15}$$

only if

$$\tau_{\mu FM} = \frac{\varepsilon_0(2\varepsilon_0 + \varepsilon_\infty)}{\varepsilon_\infty(\varepsilon_0 + 2\varepsilon_\infty)} \frac{1}{K_1(\omega)}$$

$$\tau_{\mu GC} = \frac{3\varepsilon_0}{(2\varepsilon_0 + \varepsilon_\infty)} \frac{1}{K_1(\omega)} \tag{2.30.16}$$

Neglecting the short-range correlations

$$K_1(\omega) = \tau_D^{-1} \tag{2.30.17}$$

transforms Eq. (2.30.15) into

$$\frac{\varepsilon^*(\omega) - \varepsilon_\infty}{\varepsilon_0 - \varepsilon_\infty} = \frac{1}{1 + i\omega\tau_D(3\varepsilon_0)/(2\varepsilon_0 + \varepsilon_\infty)} \quad \text{(GC)}$$

$$\frac{\varepsilon^*(\omega) - \varepsilon_\infty}{\varepsilon_0 - \varepsilon_\infty} = \frac{1}{1 + i\omega\tau_D(2\varepsilon_0 + \varepsilon_\infty)/(\varepsilon_0 + 2\varepsilon_\infty)} \quad \text{(FM)}$$

(2.30.18)

Equations (2.30.18) were respectively obtained in different ways by Powles[47] and Fatuzzo and Mason.[33]

Although the difference between Eqs (2.30.18) is of little experimental relevance, these results express two different approaches to the relation between the macroscopic and the molecular relaxation times. Thus, when $\varepsilon_0 \to \infty$.

$$\tau'_{\mu GC} \to \frac{3}{2}\tau_D \quad \text{and} \quad \tau'_{\mu FM} \to 2\tau_D \tag{2.30.19}$$

In view of Eqs. (2.30.19), both the Glarum–Cole and the Fatuzzo–Mason theories rule out the prediction of ferroelectricity.

2.31. MODE COUPLING THEORIES

The calculations of the memory functions carried out in the two previous sections have been largely based on empirical models. A more general scheme to derive expressions for time correlation functions (called correlators in the present context) and their associated memory kernels could be developed within the framework of the mode coupling theories (MCTs). As a consequence, an MCT can be considered to be mathematical model for the dynamics of a multiparticle system as it is a collective of interacting dipoles. In MCT theories one deals with a microscopic approach where correlators can generally be expressed as sums of products of the correlation functions of conserved single-particle or collective variables after a decoupling or factorization process in the time domain.

In the simplest form, the model is defined by the equations of motion of a set of n correlators, Φ_q, $q = 1, \ldots, n$, in such a way that $\Phi_q(0) = 1$, and where Φ_q is the normalized density correlation function defined as

$$\Phi_q(t) = \frac{\langle \Delta\rho_q(t)\Delta\rho_q(0)\rangle}{\langle \Delta\rho_q^2\rangle} \tag{2.31.1}$$

These equations relate the inertial term with the elastic restoring force and the frictional forces as follows

$$\ddot{\Phi}_q(t) + \Omega_q^2 \Phi_q(t) + \gamma_q \dot{\Phi}_q(t) + \int_0^t m_q(t - \tau)\dot{\Phi}_q(\tau)\,d\tau = 0 \qquad (2.31.2)$$

where Ω_q is the characteristic frequency of the liquid dynamics, γ_q is a Newtonian friction, and m_q is the kernel which describes the correlated processes. This kernel represents the so-called cage effect, arising from repeated correlated collisions. In this way, a particle is trapped by the surrounding molecules for some time. The cage effect introduces time-dependent potential barriers. When the liquid is cooled, the trapping becomes more and more effective. This mechanism is the origin of the glass transition according to the MCT.

Equation (2.31.2) can be solved by taking the Laplace transforms to give

$$\Phi_q(s) = \frac{s + M_q(s)}{s^2 + \Omega_q^2 + sM_q(s)} = \frac{1}{s + \{\Omega_q^2/[s + M_q(s)]\}} \qquad (2.31.3)$$

where $M_q(s) = i\gamma_q + m_q(s)$. It is noticeable that Eq. (2.31.3) has the same form as Eq. (2.29.7) with $K(s) = \Omega_q^2/[s + M_q(s)]$.

The more relevant part of the formalism is the hypothesis according to which m_q can be expressed in terms of the correlators Φ_i as

$$m_q(t) = \sum_{k+p=q} V(q, kp)\Phi_k(t)\Phi_p(t) \qquad (2.31.4)$$

where $V \geqslant 0$ are the coupling constants of the model.

Equation (2.31.3) describes collective effects arising from cooperative motions of any particle and its surroundings. In this way the fluctuations of a dynamic variable decay predominantly into pairs of hydrodynamic modes associated with conserved single particle systems. The observation that motivates Eq. (2.31.3) is that, if a density fluctuation ρ_k relaxes at rate r_k, product modes such as $\rho_k\rho_p$ relax at rate $r_k + r_p$ in the first approximation. Therefore, all product modes relax within the same frequency window as the single variable ρ_q. Treating the dynamics of the density pairs with a factorization approximation yields closed equations of motion which provide a formally correct approach of the cage effect for particle motion in liquids.

The theory has been used to obtain predictions about the relaxation behavior of supercooled liquids near the glass transition giving rise to the α-relaxation.[48–52] In fact, by assuming

$$m_q(t) = v_1 \Phi_q(t) + v_2 \Phi_q^2(t) \qquad (2.31.5)$$

the model shows a two-step decrease in $\Phi_q(t)$: the faster contribution is associated with the β-relaxation, whereas the slow one is related to the dynamic glass transition, that is, the α-relaxation. As a consequence, for $T > T_c$ (where T_c is a critical temperature), the α-relaxation time scales as

$$\tau_\alpha \propto \left(\frac{T_c}{T - T_c}\right)^\gamma \tag{2.31.6}$$

where γ is a constant. Moreover, a KWW law, as defined empirically in Eq. (2.12.8), has been proposed as an approximate solution for Eq. (2.31.5). Close to this critical temperature (where the system experiences an ergodic to nonergodic transition), the minimum of the permittivity between α- and β-relaxations shows the following power law

$$\varepsilon''_{min} \propto \left|\frac{T - T_c}{T_c}\right|^{1/2} \tag{2.31.7}$$

Also, in the range between α- and β-relaxations, the following expression holds

$$\varepsilon'' = \frac{\varepsilon''_{min}}{a + b}\left[b\left(\frac{\omega}{\omega_{min}}\right)^a + a\left(\frac{\omega_{min}}{\omega}\right)^b\right] \tag{2.31.8}$$

which is a Jonscher-type equation [see Eq. 4.4.6.1], where

$$\frac{\Gamma^2(1 + b)}{\Gamma(1 + 2b)} = \frac{\Gamma^2(1 - a)}{\Gamma(1 - 2a)} \tag{2.31.9}$$

where Γ represents the gamma function and a and b are empirical parameters.

The theory also predicts the appearance of the β-relaxation as a "universal" subglass phenomenon. More specifically, the Cole–Cole formula [Eq. (4.4.2.1)] has been found as an exact asymptotic implication of the MCT.[53]

PROBLEMS

Problem 1

Show that for a Debye relaxation the following expression holds

$$\varepsilon'' = [(\varepsilon_0 - \varepsilon')(\varepsilon' - \varepsilon_\infty)]^{1/2}$$

Solution

Applying the height theorem for a right-angle triangle inscribed in a Debye semicircle, it is obvious that the square of the height corresponding to the right angle is the product of the length of the two segments in which the height divides the hypotenuse.

The proposed expression could serve as a test to check if the relaxation phenomenon under study is Debye or not.

Problem 2

Solve the macroscopic equation for the polarization [Eq. (2.11.4)] in the case of an alternating field with angular frequency given by $\omega = 2\pi f$.

Solution

Let

$$E^* = E_0 \exp(i\omega t) \tag{P.2.2.1}$$

be the intensity of the applied field. In the steady state we shall expect, for the polarization, a solution of the form

$$P_d^* = A \exp(i\omega t) \tag{P.2.2.2}$$

Substitution of this type of solution in Eq. (2.11.5) gives

$$A = \frac{\chi - \chi_\infty}{1 + i\omega\tau} E_0 \tag{P.2.2.3}$$

and

$$P_d^* = \frac{\chi - \chi_\infty}{1 + i\omega\tau} E_0 \exp(i\omega t) = \frac{\chi - \chi_\infty}{1 + i\omega\tau} E^* \tag{P.2.2.4}$$

This result indicates that the dipolar part of the polarization is out of phase with the field. The total polarization can be written as

$$P^* = P_\infty + P_d^* = \left(\chi_\infty + \frac{\chi - \chi_\infty}{1 + i\omega\tau} \right) E^* \tag{P.2.2.5}$$

from which, according to the definition of the susceptibility [Eq. (1.8.5)], we can recover the Debye equation

$$\varepsilon^* = \varepsilon_\infty + \frac{\varepsilon - \varepsilon_\infty}{1 + i\omega\tau} \tag{P.2.2.6}$$

Problem 3

Calculate the decay function for a dielectric system consisting of very diluted polarizable dipoles in a matrix that responds instantaneously only at very high frequencies.

Solution

According to Eqs (2.26.11) and (2.26.20a), making $\varepsilon_1^* \to \varepsilon^*$ and taking into account the problem conditions $\varepsilon_2^* \to \varepsilon_\infty$, we obtain

$$L[-\dot{\psi}_{11}(t)] = \frac{(\varepsilon^* - 1)(\varepsilon_0 + 2\varepsilon_\infty)}{(\varepsilon_0 - 1)(\varepsilon_1 + 2\varepsilon_\infty)} \tag{P.2.3.1}$$

In the case of polarizable dipoles, a procedure analogous to that developed in section 2.23 leads to

$$L[-\dot{\psi}_{11}(t)] = \frac{(\varepsilon^* - \varepsilon_\infty)(\varepsilon_0 + 2\varepsilon_\infty)}{(\varepsilon_0 - \varepsilon_\infty)(\varepsilon^* + 2\varepsilon_\infty)} \tag{P.2.3.2}$$

Problem 4

Let us consider a dielectric sphere of large radius compared with the molecular distances in an alternating unidirectional field. The sphere is embedded in a dielectric shell with the same dielectric permittivity as the sphere. The outer radius of the shell is very large in comparison with the radius of the sphere. First, calculate the induced moment in the shell by the dipole in the spherical cavity. Second, by using the techniques outlined in section 2.26, find the Laplace transform for the derivative of the decay function corresponding to the polarization induced by the inner sphere in the surrounding shell.

Solution

The problem of the polarization of any finite macroscopic spherical specimen polarized by a fixed molecule when immersed in a medium of its own dielectric permittivity has been conveniently addressed by Kirkwood (see reference[8] of Chapter 1). The result is given by Eq. (1.14.11)

$$\mathbf{M} = \frac{9\varepsilon^*}{(\varepsilon^* + 2)(2\varepsilon^* + 1)} \boldsymbol{\mu} \tag{P.2.4.1}$$

If the contribution to the polarization due to the central dipole is subtracted from Eq (P.2.4.1), the following expression [similar to Eq. (1.14.10)] is obtained

$$m_2 = -\frac{2(\varepsilon^* - 1)^2}{(\varepsilon^* + 2)(2\varepsilon^* + 1)} \boldsymbol{\mu} \tag{P.2.4.2}$$

where $\varepsilon^* = \varepsilon_0$ is the static permittivity of the medium according to Kirkwood's theory.

This result has been extended by Glarum[21a] and Fatuzzo and Mason[20] for the case of equilibrium fluctuations in a sphere that contains more than a unique molecular dipole. In such a case it is taken that $\mu = m_1$. They write

$$\mathbf{M}_\infty(t) = -\int_{-\infty}^{t} \mathbf{M}_\infty^0(\tau) \frac{d\psi(t-\tau)}{d\tau} \, d\tau \tag{P.2.4.3}$$

where $\mathbf{M}_\infty(t)$ is the moment induced in the shell considered as a continuum, $\mathbf{M}_\infty^0$ is the induced polarization in the outer shell by the dipole in the sphere if it is held stationary, and ψ is the decay function corresponding to the shell, that is, corresponding to the complete specimen where the inner sphere has been removed. At this point, the Fatuzzo and Mason's calculation differs from Glarum's calculation in which the decay function for the polarization induced in the complete spherical specimen is considered. In order to unify the notation, one writes $\mathbf{M}_\infty(t) = \mathbf{m}_2(t)$ and, on account of Eq. (P.2.4.2), we have

$$\mathbf{m}_2(t) = \frac{2(\varepsilon_0 - 1)^2}{(\varepsilon_0 + 2)(2\varepsilon_0 + 1)} \int_0^\infty \mathbf{m}_1(\tau)\dot{\psi}'(t-\tau) \, d\tau \tag{P.2.4.4}$$

where the time is taken from $t = 0$.

Multiplying by $\mathbf{m}_1(0)$ and averaging

$$\langle \mathbf{m}_1(0) \cdot \mathbf{m}_2(t) \rangle = \frac{2(\varepsilon_0 - 1)^2}{(\varepsilon_0 + 2)(2\varepsilon_0 + 1)} \int_0^\infty \langle \mathbf{m}_1(0) \cdot \mathbf{m}_1(\tau) \rangle \dot{\psi}'(t-\tau) \, d\tau$$

$$\tag{P.2.4.5}$$

On account of Eq. (2.26.9), and taking the Laplace transform of Eq. (P.2.4.5), we find

$$\begin{aligned}
L[-\dot{\psi}'(t)] &= -\frac{\alpha_{12}^*(\omega)}{\alpha_{22}^*} \frac{2(\varepsilon_0 - 1)^2}{(\varepsilon_0 + 2)(2\varepsilon_0 + 1)} \\
&= \frac{2(\varepsilon_1^* - 1)^2}{(\varepsilon_1^* + 2)(2\varepsilon_1^* + 1)} \frac{(\varepsilon_0 + 2)(2\varepsilon_0 + 1)}{2(\varepsilon_0 - 1)^2} \\
&= \frac{\alpha_{12}^* \alpha_{22}^*(0)}{\alpha_{12}(0)\alpha_{22}^*}
\end{aligned} \tag{P.2.4.6}$$

which gives the decay of the polarization induced by the inner sphere in the surrounding shell.

When the surrounding shell has a constant dielectric permittivity, then $\psi = 1$, which is the case where the medium responds.

REFERENCES

1. van Kampen, N.G. *Stochastic Processes in Physics and Chemistry*; North-Holland, 1992.
2. Coffey, W. Development and application of the theory of Brownian motion. In *Dynamical Processes in Condensed Matter* (Adv. Chem. Phys. **1985**, *63*, 69).
3. Einstein, A. *Investigations on the Theory of the Brownian Movement*; Fürth, R.H., Ed.; Dover: New York, 1956.
4. Pearson, K. In *Selected Papers on Noise and Stochastic Processes*; Wax, N., Ed.; Dover: New York, 1954.
5. Doi, M. *Introduction to Polymer Physics*; Oxford Sci. Publ.: 1995; 4 pp.
6. Langevin, P. Comptes Rendus Acad. Sci. (Paris) **1908**, *146*, 530; See also Uhlenbeck, G.E.; Ornstein, L.S. Phys. Rev. **1930**, *34*, 823.
7. Coffey, W.T.; Kalmykov, Yu.P.; Waldron, J.T. *The Langevin Equation. World Scientific Series in Contemporary Chemical Physics*; Vol. 10.
8. Debye, P. *Polar Molecules*; (reprinted by Dover: New York), 1929; Chap. V.
9. Riande, E.; Díaz-Calleja, R.; Masegosa, R.; Gonzalez Prolongo, M.; Salom, C. *Polymer Viscoelasticity: Stress and Strain in Practice*; Marcel Dekker: New York, 1999.
10. Christensen, R.M. *The Theory of Viscoelasticity: An Introduction*; Academic Press: New York, 1971.
11. Kohlrausch, R. Am. Phys. Chem. **1954**, *91*, 179; Williams, G.; Watts, D.C. Trans. Far. Soc. **1970**, *66*, 80.
12. Krönig, R.I.. J. Opt. Soc. Am. **1926**, *12*, 547; Kramer, H.A. Atti Cong. Fisici Como, 1927; 545 pp.
13. McConnell, J. *Rotational Brownian Motion and Dielectric Theory*; Academic Press: London, 1980; 37 pp.
14. Scaife, B.K.P. *Principles of Dielectrics*; Oxford University Press: New York, 1989.
15. Rocard, J. J. Phys. Rad. **1933**, *4*, 247.
16. Powles, J.G. Trans. Far. Soc. **1948**, *44*, 802.
17. Coffey, W.T.; Kalmykov, Yu.P. J. Molecular Liquids **1991**, *49*, 79.
18. Collie, C.H.; Hasted, J.B.; Ritson, D.M. Proc. Phys. Soc. **1948**, *60*, 145.
19. Dimarzio, E.A.; Bishop, M. J. Chem. Phys. **1974**, *60*, 3802.
20. Díaz-Calleja, R.; Riande, E.; San Román, J. Polymer **1991**, *32*, 2995.
21. Kubo, R. J. Phys. Soc. **1957**, *12*, 570; Kubo, R.; Toda, M.; Hatshitsume, N. *Statistical Physics*, 2nd Ed.; Springer-Verlag: Berlin, Heidelberg, 1991.
22. Titulaer, U.M.; Deutch, J.M. J. Chem. Phys. **1974**, *60*, 1502.
23. Scaife, B.K.P. In *Complex Permittivity*; The English University Press, 1971.
24. Fatuzzo, E.; Mason, P.R. Proc. Phys. Soc. Lond. **1967**, *90*, 741.
25. Glarum, S.H.; J. Chem. Phys. **1960**, *33*, 1371; Mol. Phys. **1972**, *24*, 1327.
26. Cole, R.H. J. Chem. Phys. **1965**, *42*, 637.
27. Klug, D.D.; Kranbuehl, D.E.; Vaughan, W.E. J. Chem. Phys. **1969**, *50*, 3904.
28. Rivail, J.L. J. Chim. Phys. **1969**, *66*, 981.
29. Böttcher, C.J.F.; Bordewijk, P. *Theory of Electric Polarization*; Elsevier: Amsterdam, 1978; Vol. Ii, 161 pp.
30. Scaife, B.K.P. Proc. Phys. Soc. Lond. **1963**, *81*, 124.

31. Havriliak, S.; Havriliak, S.J. J. Non-Crys. Solids **1994**, *172–174*, 297.
32. Fatuzzo, E.; Mason, P.R. Proc. Phys. Soc. **1967**, *90*, 729.
33. Scaife, B.K.P. Proc. Phys. Soc. **1964**, *84*, 616.
34. Nee, T.; Zwanzig, R. J. Chem. Phys. **1970**, *52*, 6353.
35. Lobo, R.; Robinson, J.E.; Rodríguez, S. J. Chem. Phys. **1973**, *59*, 5992.
36. Coffey, W.T.; Kalmykov, Yu.P. Chem. Phys. **1993**, *169*, 165.
37. Van Vleck, J.H.; Weisskopf, V.F. Rev. Mod. Phys. **1945**, *17*, 227.
38. Mori, H. Prog. Theor. Phys. **1965**, *33*, 423; **1965**, *34*, 399.
39. Zwanzig, R. Ann. Rev. Phys. Chem. **1965**, *16*, 67.
40. Berne, B.J. *Physical Chemistry: An Advanced Treatise*; Academic Press, 1971; Vol. VIIIB, 539 pp.
41. Boon, J.P.; Yip, S. *Molecular Hydrodynamics*; Dover, 1991.
42. Douglas, J.F.; Hubbard, J.B. Macromolecules **1991**, *24*, 3163.
43. Williams, G.; Fournier, J. J. Chem. Phys. **1996**, *104*, 5690.
44. Schmidt-Rohr, K.; Spiess, H.W. J. Chem. Phys. **1991**, *66*, 3020.
45. Díaz-Calleja, R.; Sanchis, M.J.; del Castillo, L.F. J. Chem. Phys. **1998**, *109*, 9057.
46. Díaz-Calleja, R.; García-Bernabé, A.; Sanchis, M.J.; del Castillo, L.F. J. Chem. Phys. **2000**, *113*, 11258.
47. Powles, J.G. J. Chem. Phys. **1953**, *21*, 633.
48. Götze, W.; Sjörgren, L. J. Phys. C: Solid State Phys. **1987**, *20*, 879.
49. Götze, W.; Sjörgren, L. J. Phys. C: Solid State Phys. **1988**, *21*, 3407.
50. Sjörgren, L. J. Phys.: Condens. Matter **1991**, *3*, 5023.
51. Fuchs, M.; Götze, W.; Hofacker, I.; Latz, A. J. Phys.: Condens. Matter **1991**, *3*, 5047.
52. Götze, W. Ferroelectrics **1992**, *128*, 307.
53. Götze, W.; Sjörgren, L. J. Phys.: Condens. Matter **1989**, *1*, 4183.

3

Thermodynamics of Dielectric Relaxations in Complex Systems

3.1. THERMODYNAMICS OF IRREVERSIBLE PROCESSES

Despite its claims of universality, thermodynamics is usually concerned with very specific systems at equilibrium. However, processes in nature are mainly irreversible and their description requires going beyond equilibrium. The first approaches to extending the concepts of equilibrium thermodynamics to non-equilibrium situations were made by Prigogine,[1] Glansdorff and Prigogine,[2] Onsager,[3] Meixner and Reik,[4] and others about seventy years ago. The four main postulates of the theory are:

1. The local and instantaneous relations between thermal and mechanical properties of a physical system are the same as for a uniform system at equilibrium. This is the so-called local equilibrium hypothesis.
2. The internal entropy arising from irreversible phenomena inside a volume element is always a non-negative quantity. This is a local formulation of the second law of thermodynamics.
3. The internal entropy has a very simple character. It is a sum of terms, each being the product of a flux and a thermodynamic force.

4. The theory is completed with a set of phenomenological or constitutive
    equations relating irreversible fluxes to thermodynamic forces. As a
    first approach, these relations are assumed to be linear.

The theory is a continuum theory in the sense that the balance equations can be
formulated as field equations. The corresponding formalism is usually referred to
as classical or linear irreversible thermodynamics (LIT). Although LIT gives a
good account of many processes in biophysics, chemistry, and engineering
science, the inherent theory suffers from serious limitations. In particular, the
linear character assumed for the constitutive equations defining irreversible
processes restricts the applicability to states close to equilibrium.

Efforts have been made to enlarge the range of applicability of the LIT from a
macroscopic point of view. Two lines of thought have been developed. The first is
the so-called rational thermodynamics developed by Noll,[5] Coleman,[6] and
Truesdell and Toupin,[7] among others. In this approach, the absolute temperature
and entropy are considered as primitive concepts. The basic idea of rational
thermodynamics is to check the restrictions on the form of the constitutive
equations that a version of the second law of thermodynamics (Clausius–Duhem
equation) implies. A second way to go beyond the scope of the LIT is to generalize
the Gibbs equation for the entropy by using, in addition to the classical conserved
variables, new space variables such as the dissipative fluxes appearing in the
balance equations. The independent character of the fluxes is made evident in high-
frequency phenomena. The fluxes are "fast" variables in the sense that they decay
to their equilibrium values after a short relaxation time. The idea is to describe
phenomena at frequencies comparable with the inverse of the relaxation time of the
fluxes. Consequently, for such timescales, fluxes must be included as independent
variables. This new approach is called extended irreversible thermodynamics
(EIT).[8,9] The description of a system of n particles in terms of this set of variables
is called "mesoscopic" description, because it is intermediate between the microscopic description, in terms of $6n$ variables, and the macroscopic one, in terms of
only two conserved quantities. It should be remarked that the mesoscopic approach
involved in EIT remains macroscopic in character. It should also be noted that, in
contrast to the classical LIT which is concerned with parabolic equations, EIT
leads to hyperbolic equations with finite speed for the propagation of thermal and
viscous signals. The aim of the present chapter is to study the time rate of change in
the polarization vector in a dielectric isotropic material on the basis of EIT.[10]

3.2. DIELECTRIC RELAXATION IN THE
        FRAMEWORK OF LIT

In the simplest case, and under some simplifying assumptions concerning the
field effect, the macroscopic theory of dielectric relaxations can be expressed by

Eq. (2.11.4). It should be noted that De Groot and Mazur developed, within the framework of LIT, an expression {reference, [11b], Eq. (130) of Chapter XIV} similar to Eq. (2.11.4). In this case the time evolution of the polarizability is exponentially governed by a single relaxation time. Moreover, an instantaneous response of the system is predicted by these equations, corresponding to an infinite velocity of propagation of an external signal in the material. The frequency dependence of the permittivity can subsequently be obtained by taking Laplace transforms or by considering the step response of a simple electrical circuit.

In practice, deviations of the responses from exponential behavior have largely been observed, leading to a nonexponential decay behavior. An extra parameter can artificially be introduced in the exponential decay function, giving rise to a stretched exponential function [or KWW equation, that is, Eq. (2.12.8)]. In the present approach a generalization of the Debye equation is obtained within the framework of EIT. The equation obtained is not only consistent with the Maxwell equations but also corresponds to a situation in which the disturbance produced by the field propagates through the isotropic medium with a finite velocity.

3.3. MAXWELL EQUATIONS

Let us write the Maxwell and the conservation equations for a polarizable fluid, specifically

$$\operatorname{rot} \mathbf{E} + \frac{1}{c} \frac{\partial \mathbf{B}}{\partial t} = 0$$

$$\operatorname{rot} \mathbf{B} = \frac{4\pi}{c} \mathbf{J} + \frac{1}{c} \frac{\partial \mathbf{E}}{\partial t} \tag{3.3.1a}$$

$$\operatorname{div} \mathbf{E} = 4\pi\rho$$

$$\operatorname{div} \mathbf{B} = 0$$

where c is the velocity of light. The total charge density, ρ, and the total current density, $\mathbf{J}$ are taken as the sources of the field. Moreover, if the magnetization is assumed to be zero

$$\mathbf{J} = \mathbf{J}_f + \frac{\partial \mathbf{P}}{\partial t} \tag{3.3.2}$$

and

$$\rho = \rho_f + \rho_p \tag{3.3.3}$$

where $\mathbf{J}$ is the total current, $\mathbf{J_f}$ is the electric current of free charges, ρ_f is the density of free charges, ρ_p is the density of polarization charges, and $\partial\mathbf{P}/\partial t$ is the polarization current. Moreover

$$\mathbf{D} = \mathbf{E} + 4\pi\mathbf{P}, \qquad \mathbf{H} = \mathbf{B} \tag{3.3.4}$$

or, equivalently,

$$\mathbf{D} = \boldsymbol{\varepsilon}\mathbf{E} \tag{3.3.5}$$

where $\boldsymbol{\varepsilon}$ is the dielectric permittivity tensor as defined in Section 1.8 and $\boldsymbol{\mu}$ is considered to be 1. On the other hand, if ρ_f and $\mathbf{J_f}$, are taken as the sources, then a different version of the Maxwell equation can be written as

$$\text{div } \mathbf{D} = 4\pi\rho_f$$

$$\text{rot } \mathbf{E} + \frac{1}{c}\frac{\partial\mathbf{B}}{\partial t} = 0$$

$$\frac{4\pi}{c}\mathbf{J_f} = \text{rot } \mathbf{H} - \frac{1}{c}\frac{\partial\mathbf{D}}{\partial t} \tag{3.3.1b}$$

$$\text{div } \mathbf{B} = 0$$

Equations (3.3.1a) and (3.3.1b) allow two alternative ways to select conserved and nonconserved independent variables. From a physical point of view, the solution of Eqs (3.3.1b) requires knowledge of appropriate constitutive equations, such as Eq. (3.3.4).

3.4. CONSERVATION EQUATIONS

3.4.1. Conservation of Mass

In the absence of chemical reactions, the rate of change in mass within a volume V can be written as the flux through the surface $d\mathbf{S}$ according to

$$\int_V \frac{\partial\rho_m}{\partial t}\, dV = -\int_{\partial V} \rho_m \mathbf{v} \cdot d\mathbf{S} \tag{3.4.1.1}$$

where ρ_m and $\mathbf{v}$ are, respectively, the mass density and velocity. By using Gauss' theorem

$$\frac{\partial\rho_m}{\partial t} = -\text{div}(\rho_m\mathbf{v}) \tag{3.4.1.2}$$

and the definition of the baricentric derivative

$$\frac{d}{dt} = \frac{\partial}{\partial t} + \mathbf{v}\, \text{grad} \tag{3.4.1.3}$$

the alternative equation for the mass balance is obtained as

$$\frac{d\rho_m}{dt} + \rho_m\, \text{div}\, \mathbf{v} = 0 \tag{3.4.1.4}$$

3.4.2. Conservation of Charge

The flux or current equation for free charges is given by

$$\frac{\partial \rho_f}{\partial t} + \text{div}\, \mathbf{J}_f = 0 \tag{3.4.2.1}$$

while this expression for polarization charges can be written as

$$\frac{\partial \rho_p}{\partial t} + \text{div}\, \frac{\partial \mathbf{P}}{\partial t} = 0 \tag{3.4.2.2}$$

The sum of Eqs (3.4.2.1) and (3.4.2.2) yields the balance for the total density charge as

$$\frac{\partial \rho_t}{\partial t} + \text{div}\, \mathbf{J} = 0 \tag{3.4.2.3}$$

3.4.3. Conservation of Linear Momentum

The equation of the conservation of linear momentum in the absence of electric fields can be written as (see Problem 2)

$$\rho_m \dot{\mathbf{v}} = \rho_m \mathbf{b} - \text{div}\, \mathbf{T} \tag{3.4.3.1}$$

where $\mathbf{T}$ is the mechanical stress tensor and $\mathbf{b}$ represents the body force. In this case, the forces $\rho_m \mathbf{b}$ can be determined from the momentum balance equation for the electromagnetic field. Let us start with the vector $(\text{rot}\, \mathbf{E}) \times \mathbf{D}$. The i component of this vector can be written as

$$(\text{rot}\, \mathbf{E}) \times \mathbf{D}|_i = \left(\frac{\partial E_i}{\partial x^k} - \frac{\partial E_k}{\partial x^i}\right) D_k - \left(\frac{\partial E_j}{\partial x^i} - \frac{\partial E_i}{\partial x^j}\right) D_j \tag{3.4.3.2}$$

This equation can alternatively be written as

$$(\text{rot } \mathbf{E}) \times \mathbf{D}|_i = \left(\frac{\partial E_i}{\partial x^k} - \frac{\partial E_k}{\partial x^i}\right)D_k - \left(\frac{\partial E_j}{\partial x^i} - \frac{\partial E_i}{\partial x^j}\right)D_j$$

$$= \frac{\partial}{\partial x^k}(E_i D_k - \delta_{ik}\mathbf{E} \cdot \mathbf{D}) - E_i \operatorname{div} \mathbf{D} + \delta_{ik}\mathbf{E}\frac{\partial \mathbf{D}}{\partial x^k} \qquad (3.4.3.3)$$

By using the fact that both permittivity and permeability are symmetric tensors, it is easy to show that

$$\mathbf{E} \cdot \frac{\partial \mathbf{D}}{\partial x^k} = \frac{1}{2}\frac{\partial}{\partial x^k}(\mathbf{E} \cdot \mathbf{D}) \qquad (3.4.3.4)$$

Equations (3.4.3.3), (3.4.3.4), and (3.3.1b1) lead to

$$(\text{rot } \mathbf{E}) \times \mathbf{D} = \operatorname{div}(\mathbf{ED} - \tfrac{1}{2}(\mathbf{E} \cdot \mathbf{D})\mathbf{I}) - 4\pi\rho_f\mathbf{E} \qquad (3.4.3.5)$$

where $\mathbf{I}$ is the unity tensor. In a similar way, and taking into account Eq. (3.3.1b4), one can write

$$(\text{rot } \mathbf{H}) \times \mathbf{B} = \operatorname{div}(\mathbf{HB} - \tfrac{1}{2}(\mathbf{H} \cdot \mathbf{B})\mathbf{I}) \longrightarrow (\text{rot } \mathbf{H}) \times \mathbf{H}$$

$$= \operatorname{div}(\mathbf{HH} - \tfrac{1}{2}(\mathbf{H} \cdot \mathbf{H})\mathbf{I}) \qquad (3.4.3.6)$$

The sum of Eqs (3.4.3.5) and (3.4.3.6) gives

$$\left(\rho_f\mathbf{E} + \frac{1}{c}\mathbf{J}_f \times \mathbf{H}\right) + \frac{1}{4\pi c}\frac{\partial}{\partial t}(\mathbf{D} \times \mathbf{H})$$

$$= \frac{1}{4\pi}\operatorname{div}\left(\mathbf{ED} + \mathbf{HH} - \frac{1}{2}(\mathbf{E} \cdot \mathbf{D} + \mathbf{H} \cdot \mathbf{H})\mathbf{I}\right) \qquad (3.4.3.7)$$

where use was made of Eqs (3.3.1b2) and (3.3.1b3). Integration of Eq. (3.4.3.7) over the volume and application of the divergence theorem gives

$$\int_V \left(\rho_f\mathbf{E} + \frac{1}{c}\mathbf{J}_f \times \mathbf{H}\right)dV + \frac{1}{4\pi c}\frac{\partial}{\partial t}\int_V (\mathbf{D} \times \mathbf{H})\,dV$$

$$= \frac{1}{4\pi}\int_{\partial V} \boldsymbol{\tau}\,d\mathbf{S} \qquad (3.4.3.8)$$

where $4\pi(\rho\mathbf{E} + (1/c)\mathbf{J}_f \times \mathbf{H})$ is the density of the Lorentz force identified with the body forces in Eq. (3.4.3.1), $\mathbf{D} \times \mathbf{B}$ is the momentum density, and $\boldsymbol{\tau} = \mathbf{ED} + \mathbf{HB} - \frac{1}{2}(\mathbf{E} \cdot \mathbf{D} + \mathbf{H} \cdot \mathbf{B})\mathbf{I}$ is the Maxwell stress tensor according to the usual definition (reference [12] p.181); $\boldsymbol{\tau}$ can be interpreted as the moment flux density.

Equation (3.4.3.8) indicates that the force exerted by the electromagnetic field on the material within the volume V is equal to the rate of decrease in electromagnetic momentum within V plus the rate at which electromagnetic momentum is transferred into V across the surface ∂V. The combination of Eqs (3.4.3.1) and (3.4.3.7) leads to

$$\rho_m \frac{d\mathbf{v}}{dt} = \text{div}(-\mathbf{T} + \boldsymbol{\tau}) - \frac{1}{c}\frac{\partial}{\partial t}(\mathbf{D} \times \mathbf{H}) \tag{3.4.3.9}$$

or alternatively

$$\rho_m \frac{d\mathbf{v}}{dt} = -\text{div}\,\mathbf{T} + 4\pi\left(\rho\mathbf{E} + \frac{1}{c}\mathbf{J}_f \times \mathbf{H}\right) \tag{3.4.3.10}$$

This expression in conjunction with Eq. (P.3.1.5) gives

$$\frac{\partial}{\partial t}\left(\rho_m \mathbf{v} + \frac{1}{c}\mathbf{D} \times \mathbf{H}\right) = \text{div}(-\mathbf{T} + \boldsymbol{\tau} - \rho_m \mathbf{vv}) \tag{3.4.3.11}$$

3.4.4. Conservation of Energy

Equations (3.3.1b2) and (3.3.1b3), and (3.4.3.4), in combination with the vectorial identity

$$\text{div}(\mathbf{H} \times \mathbf{E}) = \mathbf{E} \cdot \text{rot}\,\mathbf{H} - \mathbf{H} \cdot \text{rot}\,\mathbf{E} \tag{3.4.4.1}$$

lead to

$$\frac{c}{4\pi}\,\text{div}(\mathbf{H} \times \mathbf{E}) = \mathbf{J}_f \cdot \mathbf{E} + \frac{1}{8\pi}\frac{\partial}{\partial t}(E^2 + H^2) \tag{3.4.4.2}$$

Integrating, taking into account the definitions given by Eqs (3.3.5), and applying the divergence gives

$$\int_V \mathbf{J}_f \cdot \mathbf{E}\,dV = -\frac{c}{4\pi}\int_{\partial V}(\mathbf{E} \times \mathbf{H}) \cdot d\mathbf{S} - \frac{1}{8\pi}\frac{\partial}{\partial t}\int_V (E^2 + H^2)\,dV \tag{3.4.4.3}$$

where $c/4\pi(\mathbf{E} \times \mathbf{H})$ is called the Poynting vector. The surface integral expresses the electromagnetic energy flux through the surface. The Poynting vector determines the density and direction of this flux at each point of the surface. The integral on the left-hand side of Eq. (3.4.4.3) represents the work per time unit spent in production of conduction currents. The last term on the right hand side of this equation can be interpreted as the time rate of change in the field energy within the region.

Taking Eq. (3.3.4) into account, Eq. (3.4.4.3) can alternatively be written in the absence of magnetization ($\mathbf{M} = 0$) as

$$\mathbf{J} \cdot \mathbf{E} = -\operatorname{div}(\mathbf{E} \times \mathbf{H}) - \frac{1}{2} \frac{\partial}{\partial t} \left(\varepsilon_0 E^2 + \frac{B^2}{\mu_0} \right) \tag{3.4.4.4}$$

Equations (3.4.4.3) and (3.4.4.4) express the law of conservation of energy owing to the electromagnetic field.

In order to obtain a more general expression of the conservation law for the energy, the mechanical effects must be taken into account. Let us start with the local form of Eq. (3.4.3.8)

$$\mathbf{f} = -\operatorname{div} \boldsymbol{\tau} - \frac{1}{c} \frac{\partial}{\partial t} (\mathbf{D} \times \mathbf{H}) \tag{3.4.4.5}$$

where $\mathbf{f} = 4\pi(\rho_f \mathbf{E} + (1/c)\mathbf{J}_f \times \mathbf{H})$ is the local Lorentz force. Multiplying this equation by $\mathbf{v}$ and using the vectorial result

$$\operatorname{div}(\boldsymbol{\tau} \mathbf{v}) = \mathbf{v} \cdot \operatorname{div} \boldsymbol{\tau} + \boldsymbol{\tau} \cdot \operatorname{grad} \mathbf{v} \tag{3.4.4.6}$$

the following expression is obtained

$$\mathbf{v} \cdot \mathbf{f} = -\operatorname{div}(\boldsymbol{\tau} \mathbf{v}) + \boldsymbol{\tau} \cdot \operatorname{grad} \mathbf{v} - \frac{\mathbf{v}}{c} \cdot \frac{\partial}{\partial t} (\mathbf{D} \times \mathbf{H}) \tag{3.4.4.7}$$

Multiplying Eq. (3.4.3.1) by $\mathbf{v}$, where the body forces are identified with the Lorentz force, $\rho_m \mathbf{b} = \mathbf{f}$, making use of Eq. (3.4.4.6), and utilizing the following vectorial result

$$\mathbf{v} \cdot \frac{d\mathbf{v}}{dt} = \frac{\partial}{\partial t} \left(\frac{1}{2} v^2 \right) + \operatorname{div} \left(\frac{1}{2} v^2 \mathbf{v} \right) \tag{3.4.4.8}$$

we obtain

$$\frac{\partial}{\partial t} \left(\frac{1}{2} \rho_m v^2 \right) = \mathbf{v} \cdot \mathbf{f} - \frac{1}{2} \operatorname{div}(\rho_m v^2 \mathbf{v} + \mathbf{T}\mathbf{v}) + \mathbf{T} \cdot \operatorname{grad} \mathbf{v} \tag{3.4.4.9}$$

The substitution of Eq. (3.4.4.7) into Eq. (3.4.4.9) and the use of the expression

$$\frac{\partial}{\partial t} [\mathbf{v} \cdot (\mathbf{D} \times \mathbf{H})] = \mathbf{v} \cdot \frac{\partial}{\partial t} (\mathbf{D} \times \mathbf{H}) + (\mathbf{D} \times \mathbf{H}) \cdot \frac{\partial \mathbf{v}}{\partial t} \tag{3.4.4.10}$$

leads to

$$\frac{\partial}{\partial t}\left(\frac{1}{2}\rho_m v^2 + \mathbf{v}\cdot(\mathbf{D}\times\mathbf{H})\right) = -\text{div}\left(\frac{1}{2}\rho_m v^2\mathbf{v} + (\mathbf{T}+\boldsymbol{\tau})\mathbf{v}\right)$$
$$+ (\mathbf{T}+\boldsymbol{\tau})\cdot\text{grad}\,\mathbf{v} + \frac{1}{c}(\mathbf{D}\times\mathbf{H})\cdot\frac{\partial\mathbf{v}}{\partial t}$$

$$(3.4.4.11)$$

The sum of Eqs (3.4.4.3) and (3.4.4.11) gives

$$\frac{\partial}{\partial t}\left(\frac{1}{2}\rho_m v^2 + \mathbf{v}\cdot(\mathbf{D}\times\mathbf{H}) + \frac{1}{8\pi}(E^2+H^2)\right)$$
$$= -\text{div}\left(\frac{1}{2}\rho_m v^2\mathbf{v} + (\mathbf{T}+\boldsymbol{\tau})\mathbf{v} + \frac{c}{4\pi}(\mathbf{E}\times\mathbf{H})\right)$$
$$+ (\mathbf{T}+\boldsymbol{\tau})\cdot\text{grad}\,\mathbf{v} + \frac{1}{c}(\mathbf{D}\times\mathbf{H})\cdot\frac{\partial\mathbf{v}}{\partial t} - \mathbf{J}\cdot\mathbf{E} \qquad (3.4.4.12)$$

This expression is the balance equation for energy.

3.4.5. Internal Energy Equation

Let us assume the following balance equation for the conservation of the total energy

$$\frac{\partial e_u}{\partial t} = -\text{div}\,\mathbf{J}_u \qquad (3.4.5.1)$$

where e_u is the total specific energy, which includes all the forms of energy of the system, and $\mathbf{J}_u$ is the energy flux. Let us consider further Eq. (P.3.2.16), corresponding to the first law of thermodynamics, in which the heat radiation term is disregarded, that is, $r = 0$. By using Eq. (P.3.1.5) with $a = u$, Eq. (P.3.2.16) can alternatively be written as

$$\frac{\partial}{\partial t}(\rho_m u) = -\text{div}(\mathbf{q}+\rho_m u\mathbf{v}) - \mathbf{T}\cdot\text{grad}\,\mathbf{v} \qquad (3.4.5.2)$$

According to Eqs (3.4.5.2) and (3.4.4.12), the total energy can be expressed as

$$e_u = \rho_m u + \frac{1}{2}\rho_m v^2 + \frac{\mathbf{v}}{c}(\mathbf{D}\times\mathbf{H}) + \frac{1}{8\pi}(E^2+H^2) \qquad (3.4.5.3)$$

In the same way, the total flux of energy, $\mathbf{J}_u$, is given by

$$\mathbf{J}_u = \mathbf{q} + \frac{1}{2}\rho_m v^2\mathbf{v} + (\mathbf{T}+\boldsymbol{\tau})\mathbf{v} + \frac{c}{4\pi}\mathbf{E}\times\mathbf{H} + \rho_m u\mathbf{v} \qquad (3.4.5.4)$$

Combining Eqs (3.4.5.2) and (3.4.5.3) with Eq. (3.4.4.12), the following balance equation for internal energy is obtained

$$\frac{\partial}{\partial t}(\rho_m u) = -\text{div}(\mathbf{q} + \rho_m u \mathbf{v}) - (\mathbf{T} + \boldsymbol{\tau}) \cdot \text{grad } \mathbf{v} + \mathbf{J} \cdot \mathbf{E} - \frac{1}{c}(\mathbf{D} \times \mathbf{H}) \cdot \frac{\partial \mathbf{v}}{\partial t}$$

(3.4.5.5)

or alternatively

$$\rho_m \frac{du}{dt} = -\text{div } \mathbf{q} - (\mathbf{T} + \boldsymbol{\tau}) \cdot \text{grad } \mathbf{v} + \mathbf{J} \cdot \mathbf{E} - \frac{1}{c}(\mathbf{D} \times \mathbf{H}) \cdot \frac{\partial \mathbf{v}}{\partial t} \qquad (3.4.5.6)$$

In the absence of free charges flow, $\mathbf{v} = 0$ and Eq. (3.4.5.5) reduces to

$$\rho_m \frac{du}{dt} = -\text{div } \mathbf{q} + \mathbf{J} \cdot \mathbf{E} \qquad (3.4.5.7)$$

which, on account of

$$\rho_m \frac{dq}{dt} = -\text{div } \mathbf{q} \qquad (3.4.5.8)$$

can be written as

$$\rho_m \frac{du}{dt} = \rho_m \frac{dq}{dt} + \mathbf{E} \cdot \mathbf{J} \qquad (3.4.5.9)$$

In this case

$$\mathbf{J} = \frac{\partial \mathbf{P}}{\partial t} \qquad (3.4.5.10)$$

3.5. ENTROPY EQUATION

Classical irreversible thermodynamics (CIT) extends the results of equilibrium thermodynamics using the hypothesis of local equilibrium (LE). This implies that the entropy depends on the conserved variables of the system. For a single-component system

$$s = \bar{s}(u, \rho_m) \qquad (3.5.1)$$

and the corresponding Gibbs equation is

$$ds = \frac{du}{T} - \frac{R}{\rho_m} d\rho_m \longrightarrow \rho_m \, ds = \frac{\rho_m du}{T} - R \, d\rho_m \qquad (3.5.2)$$

where R is the gas constant. In polarized systems, ρ_p must be included in the entropy function as a new variable, since it is a conserved variable according to Eq. (3.4.2.2). At the same time, and for the sake of simplicity, the changes in the mass density are considered to be negligible. As a result of being in this case the electric potential, ϕ, the conjugate pair of the charge density, the Gibbs equation is given by

$$\rho_m \, ds = \frac{\rho_m}{T} \, du + \frac{\phi}{T} \, d\rho_p \tag{3.5.3}$$

Note that, in the absence of free charges, $\rho = \rho_p$.

Extended irreversible thermodynamics proposes to enlarge the space of basic independent variables through the introduction of nonequilibrium variables. These new variables are the dissipative fluxes appearing in the balance equations. The new entropy function, η, is additive, convex in the whole set of variables, and its production rate is locally positive. The formal expression for the generalized entropy is

$$\eta = \bar{\eta}(u, \rho_p, \mathbf{q}, \mathbf{J}) \tag{3.5.4}$$

This expression leads to the following generalized Gibbs equation

$$\rho_m \, d\eta = \frac{\rho_m}{T} \, du + \frac{\phi}{T} \, d\rho_p + \frac{\alpha_1}{T} \cdot d\mathbf{q} + \frac{\alpha_2}{T} \cdot d\mathbf{J} \tag{3.5.5}$$

where α_1 and α_2 are vectors to be identified.

The balance equation postulated for η is

$$\rho_m \frac{d\eta}{dt} = -\text{div} \, \mathbf{J}_\eta + \sigma_\eta \tag{3.5.6}$$

where $\mathbf{J}_\eta$ and σ are, respectively, the flux and production of entropy. Morcover

$$\mathbf{J}_\eta = \beta_1 \mathbf{q} + \beta_2 \mathbf{J} \tag{3.5.7}$$

To compare the last two equations, Eq. (3.5.5) can be written as

$$\rho_m \frac{d\eta}{dt} = \frac{\rho_m}{T} \frac{du}{dt} + \frac{\phi}{T} \frac{d\rho_p}{dt} + \frac{\alpha_1}{T} \cdot \frac{d\mathbf{q}}{dt} + \frac{\alpha_2}{T} \cdot \frac{d\mathbf{J}}{dt} \tag{3.5.8}$$

It is worth noting that the state four-dimensional space can be split into two subspaces, the first containing the two conserved variables defining the local equilibrium state, the second containing the two nonconserved variables. The latter can be considered as a deviation from equilibrium. For this reason the scalar coefficients appearing on the right-hand side of Eq. (3.5.8) can be expanded into a series as follows

$$T^{-1} = T_0^{-1} + \frac{1}{2}\left(\frac{\partial^2 T^{-1}}{\partial q^2}\right)q^2 + \frac{1}{2}\left(\frac{\partial^2 T^{-1}}{\partial J^2}\right)J^2 + \cdots$$

$$\frac{\phi}{T} = \frac{\phi_0}{T_0} + \frac{1}{2}\left[\frac{\partial^2(\phi/T)}{\partial q^2}\right]q^2 + \frac{1}{2}\left[\frac{\partial^2(\phi/T)}{\partial J^2}\right]J^2 + \cdots$$

(3.5.9)

where T_0 and ϕ_0 are respectively the local temperature and potential at equilibrium. Linear terms are absent in Eq. (3.5.9) because the temperature and the electrical potential are scalar magnitudes.

According to the usual methodology of CIT, the vectorial forces appearing in Eq. (3.5.8) can be expressed in terms of the corresponding fluxes as

$$\alpha_1 = \alpha_{11}\mathbf{q} + \alpha_{12}\mathbf{J}$$
$$\alpha_2 = \alpha_{21}\mathbf{q} + \alpha_{22}\mathbf{J}$$

(3.5.10)

where α_{ij} are coefficients that may be expanded into a series such as Eqs (3.5.9), that is

$$\alpha_{ij} = \alpha_{ij0} + \alpha_{ij1}q^2 + \alpha_{ij2}J^2 + \cdots$$

(3.5.11)

In the same way, β_i in Eq. (3.5.7) can also be expanded as

$$\beta_i = \beta_{i0} + \beta_{i1}q^2 + \beta_{i2}J^2 + \cdots$$

(3.5.12)

From Eqs (3.4.5.8), (3.5.6), (3.5.7), (3.5.12), and (3.5.8) an expression for the entropy production can be derived. By neglecting third-order terms or higher in the products involving fast variables, and taking into account Eqs (3.4.2.2) and (3.4.5.9), the following expression is obtained

$$\sigma_\eta = \mathbf{q} \cdot \left(\text{grad }\beta_{10} + \frac{\alpha_{110}}{T_0}\frac{d\mathbf{q}}{dt} + \frac{\alpha_{210}}{T_0}\frac{d\mathbf{J}}{dt}\right)$$
$$+ \mathbf{J} \cdot \left(\frac{\mathbf{E}}{T_0} + \text{grad }\beta_{20} + \frac{\alpha_{120}}{T_0}\frac{d\mathbf{q}}{dt} + \frac{\alpha_{220}}{T_0}\frac{d\mathbf{J}}{dt}\right)$$
$$+ \text{div }\mathbf{q}\left(\beta_{10} - \frac{1}{T_0}\right) + \text{div }\mathbf{J}\left(\beta_{20} - \frac{\phi_0}{T_0}\right)$$

(3.5.13)

Under the assumption of local equilibrium

$$\frac{d\mathbf{q}}{dt} = \frac{d\mathbf{J}}{dt} = 0$$

(3.5.14)

Eq. (4.5.13) reduces to

$$\sigma_\eta = \mathbf{q} \cdot \text{grad}\, \beta_{10} + \mathbf{J} \cdot \left(\frac{\mathbf{E}}{T_0} + \text{grad}\, \beta_{20} \right) + \text{div}\, \mathbf{q} \left(\beta_{10} - \frac{1}{T_0} \right)$$

$$+ \text{div}\, \mathbf{J} \left(\beta_{20} - \frac{\phi_0}{T_0} \right) \tag{3.5.15}$$

Equation (3.5.15) must be equal to that obtained in the context of LIT. By comparing with the expression obtained for the entropy production in LIT[11]

$$\sigma_{\eta_{(LIT)}} = \mathbf{q} \cdot \text{grad}\, \beta_{10} + \mathbf{J} \cdot \left(\frac{\mathbf{E}}{T_0} + \text{grad}\, \beta_{20} \right) \tag{3.5.16}$$

the following equalities are obtained

$$\beta_{10} = \frac{1}{T_0}, \quad \beta_{20} = \frac{\phi_0}{T_0} \rightarrow \phi_0 = \frac{\beta_{20}}{\beta_{10}} \tag{3.5.17}$$

As a result

$$\sigma_{\eta_{(EIT)}} = \mathbf{q} \cdot \left(\text{grad}\, \beta_{10} + \frac{\alpha_{110}}{T_0} \frac{d\mathbf{q}}{dt} + \frac{\alpha_{210}}{T_0} \frac{d\mathbf{J}}{dt} \right)$$

$$+ \mathbf{J} \cdot \left(\frac{\mathbf{E}}{T_0} + \text{grad}\, \beta_{20} + \frac{\alpha_{120}}{T_0} \frac{d\mathbf{q}}{dt} + \frac{\alpha_{220}}{T_0} \frac{d\mathbf{J}}{dt} \right) \tag{3.5.18}$$

The entropy production σ_η is a scalar that can be expressed in terms of the scalar invariants as follows

$$\sigma_\eta = \sigma_0 + \mu_{11}\mathbf{q} \cdot \mathbf{q} + \mu_{12}\mathbf{J} \cdot \mathbf{J} \tag{3.5.19}$$

Equations (3.5.18) and (3.5.19) lead to $\sigma_0 = 0$. Therefore, Eq. (3.5.18) contains quantities, such as $\text{grad}\, T_0^{-1}$ and $(\mathbf{E}/T_0) + \text{grad}\,(\phi_0/T_0)$, not belonging to the space of state variables. For this reason, σ_η is a scalar that depends on other quantities whose presence arises from the conservation equations. Then, Eq. (3.5.19) can be rewritten as

$$\sigma_\eta = \mathbf{X_1} \cdot \mathbf{q} + \mathbf{X_2} \cdot \mathbf{J} \tag{3.5.20}$$

where

$$\mathbf{X_i} = \mu_{i1}\mathbf{q} + \mu_{i2}\mathbf{J} + \mu_{i3}\left(\frac{\mathbf{E}}{T_0} + \text{grad}\, \frac{\phi_0}{T_0} \right) \tag{3.5.21}$$

Here, the coefficients μ_{ij} only depend on the conserved variables. This means that Eq. (3.5.21) is linear. Carrying Eq. (3.5.21) into Eq. (3.5.20) gives

$$\sigma_\eta = \left[\mu_{11}\mathbf{q} + \mu_{12}\mathbf{J} + \mu_{13}\left(\frac{\mathbf{E}}{T_0} + \text{grad }\frac{\phi_0}{T_0}\right) \right] \cdot \mathbf{q}$$

$$+ \left[\mu_{21}\mathbf{q} + \mu_{22}\mathbf{J} + \mu_{23}\left(\frac{\mathbf{E}}{T_0} + \text{grad }\frac{\phi_0}{T_0}\right) \right] \cdot \mathbf{J}$$

$$= \mu_{11}q^2 + (\mu_{12} + \mu_{21})\mathbf{q} \cdot \mathbf{J} + \mu_{22}\mathbf{J}^2 + \left(\frac{\mathbf{E}}{T_0} + \text{grad }\frac{\phi_0}{T_0}\right) \cdot (\mu_{13}\mathbf{q} + \mu_{23}\mathbf{J})$$

$$(3.5.22)$$

By comparing Eqs (3.5.18) and (3.5.22), the following expressions are obtained

$$\mu_{11}\mathbf{q} + \mu_{12}\mathbf{J} + \mu_{13}\left(\frac{\mathbf{E}}{T_0} + \text{grad }\frac{\phi_0}{T_0}\right) = \text{grad }\beta_{10} + \frac{\alpha_{110}}{T_0}\frac{d\mathbf{q}}{dt} + \frac{\alpha_{210}}{T_0}\frac{d\mathbf{J}}{dt}$$

$$\mu_{21}\mathbf{q} + \mu_{22}\mathbf{J} + \mu_{23}\left(\frac{\mathbf{E}}{T_0} + \text{grad }\frac{\phi_0}{T_0}\right) = \frac{\mathbf{E}}{T_0} + \text{grad }\beta_{20} + \frac{\alpha_{120}}{T_0}\frac{d\mathbf{q}}{dt} + \frac{\alpha_{220}}{T_0}\frac{d\mathbf{J}}{dt}$$

$$(3.5.23)$$

By using

$$\text{grad }\frac{\phi_0}{T_0} = \phi_0 \text{ grad }\frac{1}{T_0} + \frac{1}{T_0}\text{grad }\phi_0 \qquad (3.5.24)$$

Eq. (3.5.23) becomes

$$\mu_{11}\mathbf{q} - \frac{\alpha_{110}}{T_0}\frac{d\mathbf{q}}{dt} - (1 - \mu_{13}\phi_0)\text{grad }\frac{1}{T_0} = \frac{\alpha_{210}}{T}\frac{d\mathbf{J}}{dt} - \mu_{12}\mathbf{J} - \frac{\mu_{13}}{T_0}(\mathbf{E} + \text{grad }\phi_0)$$

$$\mu_{21}\mathbf{q} - \frac{\alpha_{120}}{T_0}\frac{d\mathbf{q}}{dt} - \phi_0(1 - \mu_{23})\text{grad }\frac{1}{T_0} = \frac{\alpha_{220}}{T}\frac{d\mathbf{J}}{dt} - \mu_{22}\mathbf{J}$$

$$+ \frac{1}{T_0}(1 - \mu_{23})(\mathbf{E} + \text{grad }\phi_0)$$

$$(3.5.25)$$

These equations form a system of differential equations that can be solved for $\mathbf{q}$ and $\mathbf{J}$ taking into account Eq. (3.4.5.9). Note that α_{ij0} and μ_{ij} are only dependent on the equilibrium variables.

3.6. RELAXATION EQUATION

Eliminating $d\mathbf{q}/dt$ in Eqs (3.5.25) and making use of both Eq. (3.4.5.9) and grad $\phi_0 = -\mathbf{P}/\chi$, where $\chi\{=(\varepsilon - 1)/(4\pi)\}$ is the dielectric susceptibility, yields

$$\tau_1 \frac{\partial^2 \mathbf{P}}{\partial t^2} + \tau_2 \frac{\partial \mathbf{P}}{\partial t} + (\mathbf{P} - \chi\mathbf{E}) = \gamma_1 \, \text{grad} \, \frac{1}{T_0} + \gamma_2 \mathbf{q} \qquad (3.6.1)$$

where

$$\tau_1 = \chi \frac{\alpha_{210} - (\alpha_{110}\alpha_{220})/\alpha_{120}}{\mu_{13} + [1 - \mu_{23}(\alpha_{110}/\alpha_{120})]}$$

$$\tau_2 = \chi T_0 \frac{\mu_{22}(\alpha_{110}/\alpha_{120}) - \mu_{12}}{\mu_{13} + (\alpha_{110}/\alpha_{120})(1 - \mu_{23})}$$

$$\gamma_1 = \chi T_0 \left[\phi_0 - \frac{1}{\mu_{13} + (\alpha_{110}/\alpha_{120})(1 - \mu_{23})} \right] \qquad (3.6.2)$$

$$\gamma_2 = \chi T_0 \frac{\mu_{11} - (\alpha_{110}/\alpha_{120})\mu_{21}}{\mu_{13} + (\alpha_{110}/\alpha_{120})(1 - \mu_{23})}$$

By solving Eq. (3.6.1), the time dependence of $\mathbf{P}$ is obtained if both grad$(1/T_0)$ and $\mathbf{q}$ are known. Equation (3.6.1) predicts a noninstantaneous propagation of any perturbation of $\mathbf{P}$ in the system.

It is worth noting that, if T_0 is constant and $\mathbf{q} = d\mathbf{J}/dt = 0$, Eq. (3.6.1) reduces to the well-known Debye equation [11]

$$\tau \frac{\partial \mathbf{P}}{\partial t} = \chi\mathbf{E} - \mathbf{P} \qquad (3.6.3)$$

which predicts instantaneous propagation of the perturbations in the system. In this case, not all the μ_{ij} are independent because

$$\mu_{12}\mathbf{J} + \mu_{13} \left(\frac{\mathbf{E}}{T_0} + \text{grad} \, \frac{\phi_0}{T_0} \right) = 0$$

$$\mu_{22}\mathbf{J} + (\mu_{23} - 1) \left(\frac{\mathbf{E}}{T_0} + \text{grad} \, \frac{\phi_0}{T_0} \right) = 0 \qquad (3.6.4)$$

which implies

$$\frac{\mu_{12}}{\mu_{22}} = \frac{\mu_{13}}{\mu_{23} - 1} \qquad (3.6.5)$$

As a consequence, Eq. (3.6.3) can be written in two different ways

$$\chi T_0 \frac{\mu_{12}}{\mu_{13}} \frac{\partial \mathbf{P}}{\partial t} = \mathbf{P} - \chi \mathbf{E}$$

$$\chi T_0 \frac{\mu_{22}}{\mu_{23} - 1} \frac{\partial \mathbf{P}}{\partial t} = \mathbf{P} - \chi \mathbf{E} \tag{3.6.6}$$

In steady-state conditions $\partial \mathbf{P}/\partial t = 0$ and $\mathbf{P} = \chi \mathbf{E}$.

As is well known, the Debye equations derived from Eq. (3.6.3) do not adequately represent the experimental behavior of polymers. Instead of a symmetric semicircular arc, an asymmetric and skewed arc is observed. To represent in a more accurate way the actual behavior, some modifications to the former theory must be made. To start with, one is forced to consider a more general relationship between forces and fluxes [Eq. (3.5.21)] as follows

$$\mathbf{X}_i = \mu_{i1}\mathbf{q} + \sum_{n=1}^{\infty} \mu_{i2n} D_t^{n\alpha} \mathbf{P} + \mu_{i3} \left(\frac{\mathbf{E}}{T_0} + \mathrm{grad}\, \frac{\phi_0}{T_0} \right) \tag{3.6.7}$$

where the operator D_t^{α} represents the fractional derivatives of order $\alpha (0 < \alpha < 1)$. Fractional derivatives were introduced in the theory of viscoelastic relaxations by Philippoff[13] and more recently have been used by Scott-Blair[14] and Slonimsky[15] to give account of the deviations of the experimental data from those predicted by classical linear models, such as Maxwell and Kelvin–Voigt models, which are combinations of springs and dashpots. A summary concerning definitions and some properties of fractional operators is given in the Appendix of this chapter.

If Eq. (3.6.7) is used instead of Eq. (3.5.21), Eqs (3.5.25) lead to

$$\mu_{11}\mathbf{q} - \frac{\alpha_{110}}{T_0} \frac{d\mathbf{q}}{dt} - (1 - \mu_{13}\phi_0)\,\mathrm{grad}\,\frac{1}{T_0} = \frac{\alpha_{210}}{T} \frac{d^2\mathbf{P}}{dt^2} - \sum_{n=1}^{\infty} \mu_{i2n} D_t^{n\alpha}\mathbf{P}$$

$$- \frac{\mu_{13}}{T_0}(\mathbf{E} + \mathrm{grad}\,\phi_0)$$

$$\mu_{21}\mathbf{q} - \frac{\alpha_{120}}{T_0} \frac{d\mathbf{q}}{dt} - \phi_0(1 - \mu_{23})\,\mathrm{grad}\,\frac{1}{T_0} = \frac{\alpha_{220}}{T} \frac{d^2\mathbf{P}}{dt^2} - \sum_{n=1}^{\infty} \mu_{i2n} D_t^{n\alpha}\mathbf{P}$$

$$+ \frac{1}{T_0}(1 - \mu_{23})(\mathbf{E} + \mathrm{grad}\,\phi_0) \tag{3.6.8}$$

where Eq. (3.5.26) has been used. By eliminating $\dot{\mathbf{q}}$ in Eq. (3.6.8) and taking $\mathrm{grad}\,\phi_0 = -\mathbf{P}/\chi$, where $\chi = (\varepsilon - 1)/(4\pi)$, the following expression is obtained instead of Eq. (3.6.1)

$$\tau_1 D^2 \mathbf{P} + \sum_{n=1}^{\infty} \lambda_n (\tau D_t)_t^\alpha \mathbf{P} + (\mathbf{P} - \chi \mathbf{E}) = \gamma_1 \operatorname{grad} \frac{1}{T_0} + \gamma_2 \mathbf{q} \tag{3.6.9}$$

where

$$\lambda_n \tau^{n\alpha} = \chi T_0 \frac{\mu_{22n}(\alpha_{110}/\alpha_{120}) - \mu_{12n}}{\mu_{13} + (\alpha_{110}/\alpha_{120})(1 - \mu_{23})}, \qquad \lambda_n = \binom{\beta}{n} \tag{3.6.10}$$

The symbols used on the right-hand side of Eq. (3.6.10a) have the same meaning as in Eq. (3.6.2b), and β is a shape parameter. If, as before, T_0 is assumed to be a constant and $\mathbf{q} = D_t^2 \mathbf{P} = 0$, Eq. (3.6.9), with $\mu_{11} = \mu_{21} = \alpha_{210} = \alpha_{220}$, reduces to

$$\chi \mathbf{E} = \mathbf{P} + \sum_{n=1}^{\infty} \lambda_n (\tau D_t)_t^\alpha \mathbf{P} = \sum_{n=0}^{\infty} \binom{\beta}{n} (\tau D)_t^{n\alpha} \mathbf{P}$$

$$= \left[(1 + (\tau D_t)^\alpha)^\beta \right] \mathbf{P} \tag{3.6.11}$$

If the Laplace transform (with $s = i\omega$) of Eq. (3.6.11) is taken, the term corresponding to the instantaneous polarization ($\mathbf{P}_\infty = \chi_\infty \mathbf{E}$) is added, and, noting that $\mathbf{P} = \chi^* \mathbf{E}$ and $\mathbf{P}_0 = (\chi_0 - \chi_\infty)\mathbf{E}$, one obtains

$$\chi^*(\omega) = \chi_\infty + \frac{\chi_0 - \chi_\infty}{[1 + (i\omega\tau)^\alpha]^\beta} \tag{3.6.12}$$

or in terms of permittivities

$$\varepsilon^*(\omega) = \varepsilon_\infty + \frac{\varepsilon_0 - \varepsilon_\infty}{[1 + (i\omega\tau)^\alpha]^\beta} \tag{3.6.13}$$

which is the Havriliak–Negami equation (see Section 4.4.5). This equation, contrary to the Debye equation, adequately predicts the shape of the actual dielectric data in the relaxation zones.

3.7. CORRELATION AND MEMORY FUNCTIONS

To establish a connection with the theories described in Chapter 2, it is interesting to calculate both the correlation and memory functions of the dynamic description of the dielectric relaxation processes according to EIT. By making $\gamma_1 = \gamma_2 = 0$, Eq. (3.6.1) becomes

$$\tau_1 \frac{\partial^2 \mathbf{P}}{\partial t^2} + \tau_2 \frac{\partial \mathbf{P}}{\partial t} + (\mathbf{P} - \chi \mathbf{E}) = 0 \tag{3.7.1}$$

Multiplication by $\mathbf{P}(0)$ and averaging gives

$$\tau_1 \ddot{\Phi} + \tau_2 \dot{\Phi} + \Phi = \chi \langle \mathbf{E}(t) \cdot \mathbf{P}(0) \rangle \tag{3.7.2}$$

where $\Phi(t) = \langle \mathbf{P}(t) \cdot \mathbf{P}(0) \rangle$ is the time autocorrelation function for the polarization. If only the time dependence of the electric field is considered, then

$$\langle \mathbf{E}(t) \cdot \mathbf{P}(0) \rangle = \mathbf{E}(t) \cdot \langle \mathbf{P}(0) \rangle = 0 \tag{3.7.3}$$

In this case, the right-hand side of Eq. (3.7.2) vanishes so that the solution of the resulting equation is given by

$$\Phi(t) = C_1 \exp(-k_1 t) + C_2 \exp(-k_2 t) \tag{3.7.4}$$

where

$$k_1, k_2 = \frac{1}{2\tau_1} [\tau_2 \pm (\tau_2^2 - 4\tau_1)^{1/2}] \tag{3.7.5}$$

By using the boundary conditions

$$\Phi(0) = \Phi_0 \quad \text{and} \quad \dot{\Phi}(0) = 0 \tag{3.7.6}$$

the integration constants in Eq. (3.7.4) are

$$C_1 = \Phi_0 \frac{k_2}{k_2 - k_1}$$

$$C_2 = -\Phi_0 \frac{k_1}{k_2 - k_1} \tag{3.7.7}$$

where

$$k_2 - k_1 = -2 \left(\left[\frac{\tau_2}{2\tau_1} \right]^2 - \frac{1}{\tau_1} \right)^{1/2} \tag{3.7.8}$$

If $k_1 = k_2 = \tau_r^{-1}$, where τ_r is the rotational diffusion time, the correlation function can be written as

$$\Phi(t) = \Phi(0) \exp\left(\frac{-t}{\tau_r}\right) \tag{3.7.9}$$

where

$$\tau_r = \tau_1^{1/2} = \frac{\tau_2}{2} \tag{3.7.10}$$

Equation (3.7.9) is a Debye-type decay function.

The memory function can be evaluated by taking the Laplace transform of the following equation

$$\tau_1 \ddot{\Phi} + \tau_2 \dot{\Phi} + \Phi = 0 \tag{3.7.11}$$

obtained from Eq. (3.7.2), taking into account Eqs (3.7.3) and (3.7.6). After regrouping terms, the pertinent expression for $\Phi(s)$ is

$$\Phi(s) = \frac{\tau_1 s + \tau_2}{\tau_1 s^2 + \tau_2 s + 1} \Phi_0 \tag{3.7.12}$$

which, according to the continuous fraction Mori–Zwanzig[13] formalism, can be further written as

$$\Phi(s) = \frac{1}{s + \tau_1^1/(s + \tau_2/\tau_1)} \Phi_0 \tag{3.7.13}$$

The corresponding memory function equation is

$$\dot{\Phi}(t) = -\int_0^t K(t - t')\Phi(t')\,dt' \tag{3.7.14}$$

The Laplace transform of Eq. (3.7.14) is given by

$$s\Phi(s) - \Phi(0) = -K(s)\Phi(s) \tag{3.7.15}$$

from which

$$\Phi(s) = \frac{\Phi_0}{s + K(s)} \tag{3.7.16}$$

Then the memory function is

$$K(s) = \frac{1}{\tau_1} \frac{1}{s + \tau_2/\tau_1} \tag{3.7.17}$$

The inverse of the Laplace transform of Eq. (3.7.17) is given by

$$K(t) = \frac{1}{\tau_1} \exp\left(-\frac{\tau_2}{\tau_1} t\right) \tag{3.7.18}$$

which may alternatively be written as

$$K(t) = k_1 k_2 \exp(-k_1 + k_2)t \qquad (3.7.19)$$

where the two exponentials appearing in Eq. (4.7.19) are the same as those appearing in the correlation function. This indicates that the memory function is also the superposition of two relaxation processes, each governed by one characteristic relaxation time (k_1^{-1}, k_2^{-1}).

3.8. DIELECTRIC RELAXATIONS IN POLAR FLUIDS

3.8.1. Introduction

This section deals with the coupling between viscoelastic and dielectric effects using the EIT methodology. The basic idea is that relaxation moduli and permittivities are related to the complex friction coefficient through relaxation and retardation times respectively. On the other hand, the rotational motion of a Brownian particle is affected by the force field.[17–19]

3.8.2. Balance Equations

Using the results of Problem 1, and taking into account Eq. (3.4.2.2), the balance of the density of polarization charges is given by

$$\frac{d\rho_p}{dt} = -\text{div}\,\frac{\partial P}{\partial t} + v\,\text{grad}\,\rho_p = -\text{div}\,J + v \cdot \text{grad}\,\rho_p \qquad (3.8.2.1)$$

Let us consider now dipoles rotating with angular velocity $\boldsymbol{\omega}$ in a fluid moving with velocity v. Taking into account Eq. (3.4.5.5) the pertinent expression for the balance of energy can be written as

$$\rho_m \frac{du}{dt} = J \cdot E_l - (\tau + T) \cdot (\text{grad}\,v + \varepsilon w) \qquad (3.8.2.2)$$

where E_l is the local field.

In what follows, Q denotes the total stress tensor which includes the mechanical effects (viscous or viscoelastic) T and the Maxwell stress tensor τ,[12] that is

$$Q = -(\tau + T) \qquad (3.8.2.3)$$

The tensor Q can be split into the symmetric and antisymmetric parts

$$Q = Q^s + Q^a \qquad (3.8.2.4)$$

where

$$\mathbf{Q}^s = \tfrac{1}{2}(\mathbf{Q} + \mathbf{Q}^T), \qquad \mathbf{Q}^a = \tfrac{1}{2}(\mathbf{Q} - \mathbf{Q}^T) \tag{3.8.2.5}$$

The symmetric part of the stress tensor, $\mathbf{Q}^s$, gives account of the translational hydrodynamic friction, whereas the antisymmetric stress tensor, $\mathbf{Q}^a$, plays the role of a local force mainly associated with the rotational friction.

The velocity gradient admits a similar decomposition as follows

$$\operatorname{grad} \mathbf{v} = \mathbf{L}, \qquad \mathbf{L} = \mathbf{D} + \mathbf{W} \tag{3.8.2.6}$$

where $\mathbf{D}$ is the deformation velocity and $\mathbf{W}$ is the vorticity or vector spin. Since the inner product of a symmetric tensor and another antisymmetric tensor is zero, we have

$$\mathbf{Q} \operatorname{grad} \mathbf{v} = \mathbf{Q}^s \mathbf{D} + \mathbf{Q}^a \mathbf{W} \tag{3.8.2.7}$$

It is worth noting that the tensor $\mathbf{W}$ is associated with an angular velocity vector $\mathbf{w}$, in such a way that

$$\mathbf{W} = \varepsilon \mathbf{w} \tag{3.8.2.8}$$

where ε (or ε_{ijk}) is the alternating tensor (permutation symbol). The total axial vector $\mathbf{w}'$ must include not only the angular velocity of the fluid but also the angular velocity of the dipoles into the fluid $\boldsymbol{\omega}$, that is

$$\mathbf{w} = \tfrac{1}{2}\operatorname{rot} \mathbf{v} + \boldsymbol{\omega} = \tfrac{1}{2}\operatorname{rot}(\mathbf{w} \times \mathbf{r}) + \boldsymbol{\omega} \tag{3.8.2.9}$$

We are mainly interested in the antisymmetric part of the tensor $\mathbf{Q}$. Assuming that $\mathbf{v} = 0$, the balance equation for the internal energy is given by

$$\rho_m \frac{du}{dt} = \mathbf{J} \cdot \mathbf{E}_l + \mathbf{Q}^a \varepsilon \boldsymbol{\omega} \tag{3.8.2.10}$$

On the basis of the preceding hypothesis, the balance equation for the linear momentum can be written as

$$\rho_m \frac{d\mathbf{v}}{dt} = \operatorname{div} \mathbf{Q}^a - \operatorname{grad} p \tag{3.8.2.11}$$

where p is the hydrostatic pressure.

Finally, the balance equation for the angular momentum is given by

$$I \frac{d\boldsymbol{\omega}}{dt} = \varepsilon \mathbf{Q}^a \tag{3.8.2.12}$$

where I is the mean moment of inertia of the ensemble of dipoles.

When the antisymmetric stress tensor is zero, there is no inertial effect and the Debye equation for the polarization evolution is obtained. On the other hand, when $\mathbf{Q}^a$ interacts, an additional effect on the polarization appears.

3.8.3. Entropy Equation

It will be assumed that there exists a sufficiently regular and continuous function η, the generalized entropy, defined over the complete set of independent variables as

$$\eta = \bar{\eta}(u, \rho_p, \mathbf{J}, \mathbf{Q}^a) \tag{3.8.3.1}$$

This equation differs from Eq. (3.5.4) in that the heat flux has been replaced by the antisymmetric part of the total stress tensor. The basic idea in the EIT is to generalize the Gibbs equation of equilibrium thermodynamics. Accordingly, by restricting the analysis to lowest order in the fast variables, the extended entropy takes the form

$$\frac{d\eta}{dt} = \frac{1}{T}\frac{du}{dT} + \frac{\phi}{\rho_m T}\frac{d\rho_p}{dt} + \frac{a_1 \mathbf{J}}{\rho_m}\cdot\frac{d\mathbf{J}}{dt} + \frac{a_2 \mathbf{Q}^a}{\rho_m}\cdot\frac{d\mathbf{Q}^a}{dt} \tag{3.8.3.2}$$

where a_1 and a_2 are coefficients that depend linearly upon the equilibrium variables, that is, T and $\mathbf{E}$.

The entropy balance is represented by Eq. (3.5.6), that is

$$\sigma_\eta = \rho_m \frac{d\eta}{dt} + \operatorname{div}\mathbf{J}_\eta \tag{3.8.3.3}$$

where

$$\mathbf{J}_\eta = \beta_1 \mathbf{J} + \beta_2 \mathbf{J}\mathbf{Q}^a \tag{3.8.3.4}$$

Substituting Eqs (3.8.3.2) and (3.8.3.4) into Eq. (3.8.3.3), and taking into account Eq. (3.4.2.3), yields

$$\sigma_\eta = \frac{1}{T}(\mathbf{E}_l \cdot \mathbf{J} + \mathbf{Q}^a \boldsymbol{\varepsilon}\boldsymbol{\omega}) + \frac{\phi}{T}(-\operatorname{div}\mathbf{J}) + a_1 \mathbf{J}\cdot\frac{d\mathbf{J}}{dt} + a_2 \mathbf{Q}^a \cdot\frac{d\mathbf{Q}^a}{dt} + \beta_1 \operatorname{div}\mathbf{J}$$

$$+ \mathbf{J}\cdot\operatorname{grad}\frac{\phi}{T} + \beta_2 \mathbf{J}\cdot\operatorname{div}\mathbf{Q}^a + \beta_2 \mathbf{Q}^a \cdot(\operatorname{grad}\mathbf{J})^a + \mathbf{J}\cdot\mathbf{Q}^a\operatorname{grad}\beta_2$$

$$\tag{3.8.3.5}$$

Making $\beta_1 = \phi/T$, as in Eq. (3.5.17), gives

$$\sigma_\eta = \frac{1}{T}(\mathbf{E}_l \cdot \mathbf{J} + \mathbf{Q}^a \boldsymbol{\varepsilon}\boldsymbol{\omega}) + a_1 \mathbf{J} \cdot \frac{\mathrm{d}\mathbf{J}}{\mathrm{d}t} + a_2 \mathbf{Q}^a \cdot \frac{\mathrm{d}\mathbf{Q}^a}{\mathrm{d}t}$$

$$+ \mathbf{J} \cdot \operatorname{grad}\frac{\phi}{T} + \beta_2 \mathbf{J} \cdot \operatorname{div}\mathbf{Q}^a + \beta_2 \mathbf{Q}^a(\operatorname{grad}\mathbf{J})^a + \mathbf{J} \cdot \mathbf{Q}^a \operatorname{grad}\beta_2$$

$$(3.8.3.6)$$

where it was considered that $\mathbf{Q}^a(\operatorname{grad}\mathbf{J})^s = 0$. In what follows, and for the sake of simplicity, it is assumed that $\operatorname{grad}\beta_2 = 0$. By regrouping terms, Eq. (3.8.3.6) becomes

$$\sigma_\eta = \mathbf{J} \cdot \left(\frac{\mathbf{E}_l}{T} + a_1 \frac{\mathrm{d}\mathbf{J}}{\mathrm{d}t} + \operatorname{grad}\frac{\phi}{T} + \beta_2 \operatorname{div}\mathbf{Q}^a\right)$$

$$+ \mathbf{Q}^a \cdot \left(\frac{\boldsymbol{\varepsilon}\boldsymbol{\omega}}{T} + a_2 \frac{\mathrm{d}\mathbf{Q}^a}{\mathrm{d}t} + \beta_2(\operatorname{grad}\mathbf{J})^a\right) \qquad (3.8.3.7)$$

This entropy production must equalize to

$$\sigma_\eta = \mathbf{J} \cdot \left(\mu_1 \mathbf{J} + \mu_2 \operatorname{div}(\operatorname{grad}\mathbf{P}) + \mu_3 \operatorname{div}\left(\operatorname{grad}\frac{\mathrm{d}\mathbf{P}}{\mathrm{d}t}\right)\right)$$

$$+ \mathbf{Q}^a \cdot (\mu_4 \mathbf{Q}^a + \mu_5(\operatorname{grad}\mathbf{P}) + \mu_6(\operatorname{grad}\dot{\mathbf{P}})) \qquad (3.8.3.8)$$

where the terms between brackets represent the generalized forces which preserve the tensorial order and the spatial character of the variation in polarization. The identification of the coefficients of Eqs (3.8.3.7) and (3.8.3.8) leads to

$$\frac{\mathbf{E}_l}{T} + a_1 \frac{\mathrm{d}\mathbf{J}}{\mathrm{d}t} + \operatorname{grad}\frac{\phi}{T} + \beta_2 \operatorname{div}\mathbf{Q}^a = \mu_1 \mathbf{J} + \mu_2 \operatorname{div}(\operatorname{grad}\mathbf{P})^a$$

$$+ \mu_3 \operatorname{div}(\operatorname{grad}\dot{\mathbf{P}})^a$$

$$\frac{\boldsymbol{\varepsilon}\boldsymbol{\omega}}{T} + a_2 \frac{\mathrm{d}\mathbf{Q}^a}{\mathrm{d}t} + \beta_2(\operatorname{grad}\mathbf{J})^a = \mu_4 \mathbf{Q}^a + \mu_5(\operatorname{grad}\mathbf{P})^a + \mu_6(\operatorname{grad}\dot{\mathbf{P}})^a$$

$$(3.8.3.9)$$

Assuming isothermal conditions,

$$\operatorname{grad}\frac{\phi}{T} = \frac{1}{T}\operatorname{grad}\phi = -\frac{\mathbf{E}_l}{T} = -\frac{4\pi\mathbf{P}}{(\varepsilon - 1)T} = -\frac{\mathbf{P}}{\chi T} \qquad (3.8.3.10)$$

On the other hand, taking Eq. (3.4.5.10), into account Eq. (3.8.3.9) can be written as

$$a_1 \ddot{\mathbf{P}} - \mu_1 \dot{\mathbf{P}} - \frac{1}{\chi T}(\mathbf{P} - \chi \mathbf{E}_l) + \beta_2 \operatorname{div} \mathbf{Q}^a - \mu_2 \operatorname{div}(\operatorname{grad} \mathbf{P})^a$$
$$- \mu_3 \operatorname{div}(\operatorname{grad} \dot{\mathbf{P}})^a = 0$$
$$a_2 \dot{\mathbf{Q}}^a - \mu_4 \mathbf{Q}^a + \frac{\varepsilon \omega}{T} + \beta_2(\operatorname{grad} \mathbf{J})^a - \mu_5(\operatorname{grad} \mathbf{P})^a - \mu_6(\operatorname{grad} \dot{\mathbf{P}})^a = 0$$

$$(3.8.3.11)$$

By defining

$$\lambda_1 = -\frac{a_1}{\mu_1}, \qquad \tau_D = \mu_1 \chi T, \qquad D_2 = -\frac{\beta_2}{\mu_1}, \qquad D_3 = -\frac{\mu_2}{\mu_1}, \qquad \gamma_1 = \frac{\mu_3}{\mu_2}$$

$$\lambda_2 = -\frac{a_2}{\mu_4}, \qquad \zeta_0 = -\frac{1}{\mu_4 T}, \qquad \gamma_2 = -\frac{\beta_2 - \mu_6}{\mu_4}, \qquad D_1 = -\frac{\mu_5}{\mu_4}$$

$$(3.8.3.12)$$

Eqs (3.8.3.11) become

$$\left(1 + \lambda_1 \frac{d}{dt}\right)\dot{\mathbf{P}} + \frac{1}{\tau_D}(\mathbf{P} - \chi \mathbf{E}_l) + D_2 \operatorname{div} \mathbf{Q}^a - D_3\left(1 + \gamma_1 \frac{d}{dt}\right)\operatorname{div}(\operatorname{grad} \mathbf{P})^a$$
$$= 0$$
$$\left(1 + \lambda_2 \frac{d}{dt}\right)\mathbf{Q}^a - D_1\left(1 + \gamma_2 \frac{d}{dt}\right)(\operatorname{grad} \mathbf{P})^a + \zeta_0 \varepsilon \omega = 0$$

$$(3.8.3.13)$$

Equations (3.8.3.13) describe the interdependent dynamics between the polarization vector $\mathbf{P}$ and the antisymmetric stress tensor $\mathbf{Q}^a$. The coupling between these two equations arises when the polarization effect, induced by the external electric field, is coupled with spatial inhomogeneities in the polarization through the gradients of $\mathbf{P}$. The parameter γ_1 in Eq. (3.8.3.13a) is identified with a relaxation time. It should be stressed that the viscoelastic effects lie in the parameters λ_2, γ_1, and γ_2. They are relaxation or retardation times. Parameter D_2 is a stress coupling effect coefficient that enhances the dielectric dispersion and friction due to the shear flow. Making $D_2 = D_3 = 0$ in Eq. (3.8.3.13a), a generalized Debye equation [Eq. (3.7.1)] that includes inertial effects is obtained. If $D_2 = \gamma_1 = 0$ but $D_3 \neq 0$, then the uncoupled inhomogeneous case is obtained. In this case, the use of the vectorial identity

$$\operatorname{div}(\operatorname{grad} \mathbf{P})^a = \tfrac{1}{2}(\nabla^2 \mathbf{P} - \operatorname{grad}(\operatorname{div} \mathbf{P})) \qquad (3.8.3.14)$$

together with

$$\text{div}\, \mathbf{P} = -\rho_p \qquad (3.8.3.15)$$

transforms Eq. (3.3.8.13a) into

$$\lambda_1 \ddot{\mathbf{P}} + \dot{\mathbf{P}} + \frac{1}{\tau_D}\mathbf{P} = \frac{\chi}{\tau_D}\mathbf{E}_l + \frac{D_3}{2}(\nabla^2 \mathbf{P} + \text{grad}\,\rho_p) \qquad (3.8.3.16)$$

Equation (3.8.3.16) is a particular type of telegrapher's equation for the propagation of polarization. The perturbation is propagated with a velocity given by

$$c = \left(\frac{D_3}{2\lambda_1}\right)^{1/2} \qquad (3.8.3.17)$$

In the literature,[17–19] other cases have been analyzed.

3.9. DIELECTRIC SUSCEPTIBILITIES AND PERMITTIVITIES

In this Section, use is made of the Fourier–Laplace transform to study the frequency dependence of the polarization $\mathbf{P}$ and to obtain the corresponding dielectric characteristics of the system. For this purpose, let us start with Eq. (3.8.3.13) and the balance of the angular momentum [Eq. (3.8.2.11)]. Let us also assume a harmonic variation for the angular velocity

$$w = \exp i(\omega t - \mathbf{k} \cdot \mathbf{r}) \qquad (3.9.1)$$

where $\mathbf{k}$ is the wave vector defining the propagation of the wave. Taking Laplace transforms in Eqs (3.8.2.11) and (3.8.3.13) with $s = i\omega$, the following expressions are obtained

$$s(1 + \lambda_1 s)\mathbf{P} + \frac{1}{\tau_D}(\mathbf{P} - \chi\mathbf{E}_l) + D_2 i\mathbf{k} \cdot \mathbf{Q}^a + D_3(1 + \gamma_1 s)\mathbf{k} \cdot (\mathbf{kP})^a = 0$$

$$(1 + \lambda_2 s)\mathbf{Q}^a - D_1(1 + \gamma_2 s)(i\mathbf{kP})^a + \zeta_0 \varepsilon \omega = 0 \qquad (3.9.2)$$

$$sI\omega = \varepsilon \mathbf{Q}^a$$

Premultiplying Eq. (4.9.2c) by ε, the following result holds

$$\varepsilon\varepsilon\mathbf{Q}^a = \varepsilon_{ijk}\varepsilon_{khl}Q^a_{hl} = (\delta_{ih}\delta_{jl} - \delta_{il}\delta_{hj})Q^a_{hl} = Q^a_{ij} - Q^a_{ji} = 2Q^a_{ij} \qquad (3.9.3)$$

From here

$$sI\varepsilon\omega = 2Q^a \longrightarrow \varepsilon\omega = \frac{2Q^a}{sI} \tag{3.9.4}$$

Carrying Eq. (3.9.4) into Eq. (3.9.2b) gives

$$(1 + \lambda_2 s)Q^a - D_1(1 + \gamma_2 s)(ik\mathbf{P})^a + \zeta_0 \frac{2Q^a}{sI} = 0 \longrightarrow$$

$$\left[(1 + \lambda_2 s) + \frac{2\zeta_0}{sI}\right]Q^a = D_1(1 + \gamma_2 s)(ik\mathbf{P})^a \tag{3.9.5}$$

Multiplied by $\mathbf{k}$, this expression becomes

$$\mathbf{k} \cdot Q^a = \frac{D_1(1 + \gamma_2 s)i\mathbf{k} \cdot (\mathbf{kP})^a}{(1 + \lambda_2 s) + (2\zeta_0)/(sI)} \tag{3.9.6}$$

Carrying Eq. (3.9.6) into Eq. (3.9.2a), yields

$$s(1 + \lambda_1 s)\mathbf{P} + \frac{1}{\tau_D}(\mathbf{P} - \chi\mathbf{E}_l)$$

$$- \frac{D_1 D_2(1 + \gamma_2 s)\mathbf{k} \cdot (\mathbf{kP})^a}{(1 + \lambda_2 s) + (2\zeta_0)/(sI)} + D_3(1 + \gamma_1 s)\mathbf{k} \cdot (\mathbf{kP})^a = 0 \tag{3.9.7}$$

or equivalently

$$\mathbf{P}\left[s(1 + \lambda_1 s) + \frac{1}{\tau_D}\right]$$

$$+ \left[D_3(1 + \gamma_1 s) - \frac{D_1 D_2(1 + \gamma_2 s)}{(1 + \lambda_2 s) + (2\zeta_0)/(sI)}\right]\mathbf{k} \cdot (\mathbf{kP})^a = \frac{\chi}{\tau_D}\mathbf{E}_l$$

$$\tag{3.9.8}$$

According to the expression

$$\mathbf{k} \cdot (\mathbf{kP})^a = \tfrac{1}{2}(k^2\mathbf{P} - \mathbf{k}(\mathbf{k} \cdot \mathbf{P})) \tag{3.9.9}$$

obtained from the Laplace transform of Eq. (3.8.3.14), and using index notation, Eq. (3.9.8) becomes

$$
P_j \left\{ s(1 + \lambda_1 s) + \frac{1}{\tau_D} + \left[D_3(1 + \gamma_1 s) - \frac{1}{2} \frac{D_1 D_2(1 + \gamma_2 s)}{(1 + \lambda_2 s) + (2\zeta_0)/(sI)} \right] k^2 \right\} \delta_{ji}
$$
$$
- \left[D_3(1 + \gamma_1 s) - \frac{1}{2} \frac{D_1 D_2(1 + \gamma_2 s)}{(1 + \lambda_2 s) + (2\zeta_0)/(sI)} \right] k_i k_j P_j = \frac{\chi}{\tau_D} E_{li}
$$

(3.9.10)

where $P_j = \chi_{ji}(\mathbf{k}, \omega) E_{li}$.

Let $\mathbf{k} = (k, 0, 0)$ and $E_l = (E_1, E_2, E_3)$. In this case $\chi_{ji} = 0$ for $(i \neq j)$ and $k_i k_j = 0$. Then, the only non-null components of the susceptibility are $\chi_{11} = \chi_P$ and $\chi_{22} = \chi_{33} = \chi_\perp$, that is, the longitudinal and transversal components of the dielectric susceptibility. Accordingly, it is finally found that

$$
\frac{\chi_P}{\chi} = \left\{ 1 + i\omega\tau_D^L \left[1 + i\omega\lambda_1 + \frac{D_3 k^2(1 + i\omega\gamma_1)}{i\omega} \right] \right\}^{-1}
$$
$$
\frac{\chi_\perp}{\chi} = \left(1 + i\omega\tau_D^T \left\{ 1 + i\omega\lambda_1 \right. \right.
$$
$$
\left. \left. + \left[\frac{D_3(1 + i\omega\gamma_1)}{i\omega} - \frac{1}{2} \frac{D_1 D_2(1 + i\omega\gamma_2)}{i\omega(1 + i\omega\lambda_2) + (2\zeta_0)/I} \right] k^2 \right\} \right)^{-1}
$$

(3.9.11)

By taking the limit $\omega \to 0$, Eq. (3.9.11b) becomes

$$
\frac{\chi_\perp}{\chi} = (1 + D_3 \tau_D^T k^2)^{-1}
$$

(3.9.12)

The dielectric susceptibility is $\mathbf{k}$ dependent for two reasons. First, if the dielectric susceptibility is related to the external field instead of the local field, a dependence on $\mathbf{k}$ is introduced for uniform and isotropic systems. The longitudinal and transversal components of the dielectric permittivity have traditionally been expressed as[20]

$$
\frac{\chi_P}{\chi} = \frac{\varepsilon^*(\omega) - \varepsilon_\infty}{\varepsilon_0 - \varepsilon_\infty} \frac{\varepsilon_0}{\varepsilon_\infty}
$$
$$
\frac{\chi_\perp}{\chi} = \frac{\varepsilon^*(\omega) - \varepsilon_\infty}{\varepsilon_0 - \varepsilon_\infty}
$$

(3.9.13)

The second reason for a $\mathbf{k}$-dependent dielectric susceptibility is the spatial nonuniformity induced in the interaction between dipoles. Therefore, there are diffusion contributions to the transversal mode and this mode, is different from the longitudinal one.

3.10. GENERALIZATION AND SPECIAL CASES

The results of the preceding section can easily be generalized by considering more general expressions for the balance equations. In particular, the balance equation for the angular momentum can be generalized to

$$I\frac{d\boldsymbol{\omega}}{dt} = D_s\nabla^2\boldsymbol{\omega} + \varepsilon \mathbf{Q}^a \tag{3.10.1}$$

where D_s is a diffusion coefficient for the spin transport, which accounts for the inhomogeneity of the system. In this case, taking into account that the Laplace transform of Eq. (3.10.1) is given by

$$(sI + D_s k^2)\boldsymbol{\omega} = \varepsilon \mathbf{Q}^a(s) \tag{3.10.2}$$

and following the methodology outlined in the preceding section, the transversal susceptibility is obtained as

$$\frac{\chi_\perp}{\chi} = \left[1 + i\omega\tau_D\left(1 + i\omega\lambda_1 + \left\{\frac{D_3(1 + i\omega\gamma_1)}{i\omega}\right.\right.\right.$$
$$\left.\left.\left. - \frac{D_1 D_2(1 + i\omega\gamma_2)(i\omega I + D_s k^2)}{2i\omega[(1 + i\omega\lambda_2)(i\omega I + D_s k^2) + 2\zeta_0]}\right\}k^2\right)\right]^{-1} \tag{3.10.3}$$

Let us consider now the noninertial case. Then, Eq. (3.10.3) leads to the following expression for the transversal susceptibility

$$\frac{\chi_\perp}{\chi} = \left[1 + i\omega\tau_D\left(1 + i\omega\lambda_1 + \left\{\frac{D_3(1 + i\omega\gamma_1)}{i\omega}\right.\right.\right.$$
$$\left.\left.\left. + \frac{k^2 D_1 D_2 D_s(1 + i\omega\gamma_2)}{2i\omega[2\zeta_0 + D_s k^2(1 + i\omega\lambda_2)]}\right\}k^2\right)\right]^{-1} \tag{3.10.4}$$

Moreover, if $\lambda_1 = D_3 = 0$, then

$$\frac{\chi_\perp}{\chi} = \left(1 + i\omega\tau_D - \frac{k^4 \tau_D D_1 D_2 D_s(1 + i\omega\gamma_2)}{2(D_s k^2(1 + i\omega\lambda_2) + 2\zeta_0)}\right)^{-1} \tag{3.10.5}$$

By defining

$$\tau_2^{-1} = k^2 D_1 D_2 \tag{3.10.6}$$

as the polarization diffusion characteristic time, Eq. (3.10.5) becomes

$$\frac{\chi_{\perp}}{\chi} = \left\{1 + i\omega\tau_D - \frac{\tau_D k^2 D_s(1 + i\omega\gamma_2)}{2\tau_2[D_s k^2(1 + i\omega\lambda_2) + 2\zeta_0]}\right\}^{-1} \tag{3.10.7}$$

Note that the inclusion of the coefficient for the spin transport D_s is necessary to avoid reduction to the Debye equation in the noninertial case.

Equations (3.10.5) and (3.10.7) correspond to the usual expression for the dielectric segmental mode relaxation. By making $I \neq 0$, and $\lambda_1 = D_3 = D_s = 0$, Eq. (3.10.3) can be written as

$$\frac{\chi_{\perp}}{\chi} = \left\{1 + i\omega\tau_D - \frac{k^2 \tau_D D_1 D_2(1 + i\omega\gamma_2)}{2([(1 + i\omega\lambda_2) + (2\zeta_0)/(i\omega I)])}\right\}^{-1} \tag{3.10.8}$$

If we further define

$$\tau_F' = \tfrac{1}{2}\zeta_0 \tag{3.10.9}$$

as a rotational relaxation time [note that τ_F' is not the same as that defined by Eq. (2.16.4)], then Eq. (3.10.8) can also be written as

$$\frac{\chi_{\perp}}{\chi} = \left\{1 + i\omega\tau_D - \frac{\tau_D(1 + i\omega\gamma_2)}{2\tau_2([(1 + i\omega\lambda_2) + 1/(i\omega\tau_F')])}\right\}^{-1} \tag{3.10.10}$$

where the definitions (3.10.6) and (3.10.9) have been considered. Hence, the condition $D_s = 0$ does not lead to the Debye equation in the inertial case.

3.11. MEMORY FUNCTION

As discussed in Chapter 2 (Section 2.29), the response of a system to an external field can formally be described in terms of the memory functions. In fact, within the framework of the linear response theory (Section 2.20), the dielectric susceptibility is given in terms of the memory by

$$\frac{\chi^*(\omega,\mathbf{k})}{\chi} = L(-\dot{C}) \tag{3.11.1}$$

where C is the correlation function and L denotes the Laplace transform. Making use of Eq. (2.29.7), Eq. (3.11.1) becomes

$$\frac{\chi^*(\omega,\mathbf{k})}{\chi} = (1 + i\omega K(\omega,\mathbf{k}))^{-1} \tag{3.11.2}$$

where $K(\omega, \mathbf{k})$ is the memory function. Comparison of Eqs (3.11.2) and (3.10.3) leads to the determination of all the contributions to the first-order memory function, that is[19]

$$
K^{-1}(\omega, \mathbf{k}) = \tau_D + i\omega\lambda_1\tau_D + \frac{D_3\tau_D k^2(1 + i\omega\gamma_1)}{i\omega}
$$
$$
- \frac{\tau_D k^2 D_1 D_2 (1 + i\omega\gamma_2)(i\omega I + D_s k^2)}{2i\omega[(1 + i\omega\lambda_2)(i\omega I + D_s k^2) + 2\zeta_0]} \tag{3.11.3}
$$

The first term on the right-hand side of Eq. (3.11.3) is the Debye time, the second is the inertial term for the polarization, the third gives the polarization diffusion contribution through the diffusion coefficient D_3, and the fourth term represents a dielectric memory produced by the coupling of the antisymmetric stress tensor and the polarization diffusion. The actual contribution of this last term to the dielectric memory is equivalent to a polarization diffusion process, coupled with the stress via a diffusion coefficient given by the inverse of

$$
- \frac{2i\omega[(1 + i\omega\lambda_2)(i\omega I + D_s k^2) + 2\zeta_0]}{\tau_D k^2 D_1 D_2 (1 + i\omega\gamma_2)(i\omega I + D_s k^2)} \tag{3.11.4}
$$

For $D_s = 0$, this term can be written as

$$
- \frac{2\tau_2[1 + i\omega\tau_F'(1 + i\omega\lambda_2)]}{\tau_D \tau_F'(1 + i\omega\gamma_2)} \tag{3.11.5}
$$

where use was made of Eqs. (3.10.6) and (3.10.9). This contribution can produce important modifications in the Debye absorption band.

3.12. NORMAL MODE ABSORPTION

As is well known,[21] polymers containing dipoles with a component along the chain contour present a characteristic low-frequency relaxation process (see Chapter 8). This process, called normal mode relaxation, is strongly dependent on chain length. Because the total dipole moment along the chain contour is proportional to the end-to-end distance. In this case, new relaxation band appears, associated with long-range dynamics. The results of the preceding sections can easily be extended to the case where the normal hydrodynamic friction is present.[22] Therefore, the symmetric part of the total stress tensor must be taken into account. This strategy allows description of the molecular motions involving the whole chain.

Let us start by including in the entropy equation [Eq. (4.8.3.2)] the symmetric part of the stress tensor $\mathbf{Q}$. In this case

$$\frac{d\eta}{dt} = \frac{1}{T}\frac{du}{dT} + \frac{\phi}{\rho_m t}\frac{d\rho_p}{dt} + \frac{a_1\mathbf{J}}{\rho_m}\cdot\frac{d\mathbf{J}}{dt} + \frac{a_2\mathbf{Q}^a}{\rho_m}\cdot\frac{d\mathbf{Q}^a}{dt} + \frac{a_3\mathbf{Q}^s}{\rho_m}\cdot\frac{d\mathbf{Q}^s}{dt} \qquad (3.12.1)$$

with

$$\mathbf{J}_\eta = \beta_1\mathbf{J} + \beta_2\mathbf{J}\mathbf{Q}^a + \beta_3\mathbf{J}\mathbf{Q}^s \qquad (3.12.2)$$

Then, according to Eq. (3.8.3.3), the following expression is obtained for the entropy production

$$\sigma_\eta = \frac{1}{T}(\mathbf{E}_l\cdot\mathbf{J} + \mathbf{Q}^a\cdot\boldsymbol{\varepsilon}\boldsymbol{\omega} + \mathbf{Q}^s\cdot\mathbf{D}) + \frac{\phi}{T}(-\operatorname{div}\mathbf{J}) + a_1\mathbf{J}\cdot\frac{d\mathbf{J}}{dt} + a_2\mathbf{Q}^a\cdot\frac{d\mathbf{Q}^a}{dt}$$

$$+ a_3\mathbf{Q}^s\cdot\frac{d\mathbf{Q}^s}{dt} + \beta_1\operatorname{div}\mathbf{J} + \mathbf{J}\cdot\operatorname{grad}\frac{\phi}{T} + \beta_2\mathbf{J}\cdot\operatorname{div}\mathbf{Q}^a$$

$$+ \beta_2\mathbf{Q}^a\cdot(\operatorname{grad}\mathbf{J})^a + \mathbf{J}\cdot\mathbf{Q}^a\operatorname{grad}\beta_2 + \beta_3\mathbf{J}\cdot\operatorname{div}\mathbf{Q}^s$$

$$+ \beta^s\mathbf{Q}^s\cdot(\operatorname{grad}\mathbf{J})^s(3 + \mathbf{J}\cdot\mathbf{Q}^s(\operatorname{grad}\beta_3) \qquad (3.12.3)$$

where, as above, $\mathbf{D} = (\operatorname{grad}\mathbf{v})^s$ and $\beta_1 = \phi/T$.

Let us assume, for simplicity, that $\operatorname{grad}\beta_2 = \operatorname{grad}\beta_3 = 0$. After regrouping terms, the production of entropy is given by

$$\sigma_\eta = \mathbf{J}\cdot\left(\frac{\mathbf{E}_l}{T} + a_1\frac{d\mathbf{J}}{dt} + \operatorname{grad}\frac{\phi}{T} + \beta_2\operatorname{div}\mathbf{Q}^a + \beta_3\operatorname{div}\mathbf{Q}^s\right)$$

$$+ \mathbf{Q}^a\cdot\left(\frac{\boldsymbol{\varepsilon}\boldsymbol{\omega}}{T} + a_2\frac{d\mathbf{Q}^a}{dt} + \beta_2(\operatorname{grad}\mathbf{J})^a\right)$$

$$+ \mathbf{Q}^s\cdot\left(\frac{\mathbf{D}}{T} + a_3\frac{d\mathbf{Q}^s}{dt} + \beta_3(\operatorname{grad}\mathbf{J})^s\right) \qquad (3.12.4)$$

This equation must be equal to the following expression

$$\sigma_\eta = \mathbf{J}\cdot(\mu_1\mathbf{J} + \mu_2\operatorname{div}(\operatorname{grad}\mathbf{P}) + \mu_3\operatorname{div}(\operatorname{grad}\dot{\mathbf{P}})) + \mathbf{Q}^a\cdot(\mu_4\mathbf{Q}^a$$

$$+ \mu_5(\operatorname{grad}\mathbf{P})^a + \mu_6(\operatorname{grad}\dot{\mathbf{P}})^a) + \mathbf{Q}^s\cdot(\mu_7\mathbf{Q}^s + \mu_8(\operatorname{grad}\mathbf{P})^s$$

$$+ \mu_9(\operatorname{grad}\dot{\mathbf{P}})^s) \qquad (3.12.5)$$

Identifying forces in Eqs (3.12.4) and (3.12.5), and neglecting terms containing both $(\text{grad } \mathbf{u})^s = \mathbf{D}$ and derivatives of $\mathbf{w}$, the following equations are obtained

$$
\frac{\mathbf{E}_l}{T} + a_1 \frac{d\mathbf{J}}{dt} + \text{grad } \frac{\phi}{T} + \beta_2 \text{ div } \mathbf{Q}^a + \beta_3 \text{ div } \mathbf{Q}^s
$$

$$
= \mu_1 \mathbf{J} + \mu_2 \text{ div}(\text{grad } \mathbf{P}) + \mu_3 \text{ div}(\text{grad } \dot{\mathbf{P}})
$$

$$
\frac{\varepsilon \omega}{T} + a_2 \frac{d\mathbf{Q}^a}{dt} + \beta_2 (\text{grad } \mathbf{J})^a = \mu_4 \mathbf{Q}^a + \mu_5 (\text{grad } \mathbf{P})^a + \mu_6 (\text{grad } \dot{\mathbf{P}})^a
$$

$$
a_3 \frac{d\mathbf{Q}^s}{dt} + \beta_3 (\text{grad } \mathbf{J})^s = \mu_7 \mathbf{Q}^s + \mu_8 (\text{grad } \mathbf{P})^s + \mu_9 (\text{grad } \dot{\mathbf{P}})^s
$$

$$(3.12.6)$$

Equation (3.12.6a), in conjunction with Eqs (3.4.5.9) and (3.8.3.10), together with the definitions of (3.8.3.12), gives

$$
\left(1 + \lambda_1 \frac{d}{dt}\right)\dot{\mathbf{P}} + \frac{1}{\tau_D}(\mathbf{P} - \chi \mathbf{E}_l) + D_2 \text{ div } \mathbf{Q}^a + D_5 \text{ div } \mathbf{Q}^s
$$

$$
- D_3\left(1 + \gamma_1 \frac{d}{dt}\right) \text{div}(\text{grad } \mathbf{P}) = 0 \qquad (3.12.7a)
$$

where $D_5 = \beta_3/\mu_1$. Equation (3.12.7a) is obviously a generalization of Eq. (3.8.3.13a).

Equation (3.12.6b) gives Eq. (3.8.3.13b) and it will be renamed Eq. (3.12.7b). On the other hand, Eq. (3.12.6c) can be written as

$$
\left(1 + \lambda_5 \frac{d}{dt}\right)\mathbf{Q}^s - \lambda_7\left(1 + \lambda_8 \frac{d}{dt}\right)(\text{grad } \mathbf{P})^s = 0 \qquad (3.12.7c)
$$

where the coefficients are given by

$$
\lambda_5 = -\frac{a_3}{\mu_7}, \qquad \lambda_7 = -\frac{\mu_8}{\mu_7}, \qquad \lambda_8 = \frac{\lambda_6}{\lambda_7} \qquad \text{with} \qquad \lambda_6 = \frac{\beta_3 - \mu_9}{\mu_7}
$$

$$(3.12.8)$$

Equations (3.12.7) describe the interdependent dynamics between the polarization vector $\mathbf{P}$ and the symmetric and antisymmetric stress tensors $\mathbf{Q}^s$ and $\mathbf{Q}^a$. The coupling between these three equations arises from the fact that the polarization effect induced by the external electric field is coupled with spatial inhomogeneities in the polarization through the gradients of $\mathbf{P}$. The parameter in Eq. (3.12.7a) is identified with a relaxation time. It should be stressed that the viscoelastic effects lie in the parameters λ_2, λ_5, λ_7, λ_8, γ_1 and γ_2. They are relaxation or retardation times. Parameters D_2 and D_5 are stress coupling effect coefficients that enhance the dielectric dispersion and friction due to the shear flow.

The system formed by Eqs (3.12.7a,b,c) can be closed by means of the balance of the angular momentum that will be performed, as in Eq. (3.10.1). The solution of the systems by means of the Laplace transform is outlined in section 3.9. The final result (see Problem 6) is

$$
\frac{\chi_\perp}{\chi} = \left(1 + i\omega\tau_D\left\{1 + i\omega\lambda_1 + \frac{1}{i\omega}(D_3(1 + i\omega\gamma_1)\right.\right.
$$
$$
\left.\left. - \frac{1}{2}\left[\frac{D_1 D_2(1 + i\omega\gamma_2)}{1 + i\omega\lambda_2 + (2\zeta_0)/(i\omega I + D_s k^2)} + \frac{D_5\lambda_7(1 + i\omega\lambda_8)}{1 + i\omega\lambda_5}\right]k^2\right\}\right)^{-1}
$$

$$(3.12.9)$$

For $D_5 = 0$ or $\lambda_7 = 0$, the equation for the segmental mode (3.9.11) is recovered.

Taking into account the equations relating the susceptibility to the permittivity

$$
\chi_\perp = \frac{\varepsilon(\omega) - \varepsilon_\infty}{4\pi}, \qquad \chi = \frac{\varepsilon_0 - \varepsilon_\infty}{4\pi} \tag{3.12.10}
$$

where the distortional contribution to the permittivity has been subtracted, Eq. (3.12.9) can be written as

$$
\frac{\varepsilon^*(\omega) - \varepsilon_\infty}{\varepsilon_0 - \varepsilon_\infty} = \left[1 + i\omega\tau_D\left(1 + i\omega\lambda_1 + \frac{1}{i\omega}\left\{D_3(1 + i\omega\gamma_1)\right.\right.\right.
$$
$$
- \frac{1}{2}\left[\frac{D_1 D_2(1 + i\omega\gamma_2)}{1 + i\omega\lambda_2 + (2\zeta_0)/(i\omega I + D_s k^2)}\right.
$$
$$
\left.\left.\left.\left. + \frac{D_5\lambda_7(1 + i\omega\lambda_8)}{1 + i\omega\lambda_5}\right]\right\}k^2\right)\right]^{-1}
$$

$$(3.12.11)$$

In order to simplify calculations, let us make $\lambda_1 = D_3 = 0$. Then, Eq. (3.12.11) becomes

$$
\frac{\varepsilon^*(\omega) - \varepsilon_\infty}{\varepsilon_0 - \varepsilon_\infty} = \left\{1 + i\omega\tau_D - \frac{1}{2}\tau_D\left[\frac{D_1 D_2(1 + i\omega\gamma_2)}{1 + i\omega\lambda_2 + (2\zeta_0)/(i\omega I + D_s k^2)}\right.\right.
$$
$$
\left.\left. + \frac{D_5\lambda_7(1 + i\omega\lambda_8)}{1 + i\omega\lambda_5}\right]k^2\right\}^{-1}
$$

$$(3.12.12)$$

The normal mode is obtained when $D_2 = 0$ or decoupling between Eqs (3.12.7a) and (4.12.7b) occurs by taking $D_1 = 0$. In this situation

$$
\frac{\varepsilon^*(\omega) - \varepsilon_\infty}{\varepsilon_0 - \varepsilon_\infty} = \left[1 + i\omega\tau_D - \frac{D_5\lambda_7\tau_D k^2(1 + i\omega\lambda_8)}{2(1 + i\omega\lambda_5)}\right]^{-1} \tag{3.12.13}
$$

If $\lambda_7 = 0$ or $D_5 = 0$, there is no diffusion polarization effect and the Debye equation for the Rouse mode is recovered.

If there is no diffusion for the spin transport, $D_s = 0$, and, assuming once more that $\lambda_1 = D_3 = 0$, the following expression for the complex permittivity is obtained

$$
\frac{\varepsilon^*(\omega) - \varepsilon_\infty}{\varepsilon_0 - \varepsilon_\infty} = \left\{ 1 + i\omega\tau_D - \frac{1}{2}\tau_D \left[\frac{D_1 D_2 (1 + i\omega\gamma_2)}{1 + i\omega\lambda_2 + (2\zeta_0)/(i\omega I)} \right. \right.
$$
$$
\left. \left. + \frac{D_5\lambda_7(1 + i\omega\lambda_8)}{1 + i\omega\lambda_5} \right] k^2 \right\}^{-1}
\tag{3.12.14}
$$

By using Eqs (3.10.6) and (3.10.9), Eq. (3.12.14) may be written as

$$
\frac{\varepsilon^*(\omega) - \varepsilon_\infty}{\varepsilon_0 - \varepsilon_\infty} = \left\{ 1 + i\omega\tau_D - \frac{1}{2} \left[\frac{1}{\tau_2} \frac{\tau_D(1 + i\omega\gamma_2)}{1 + i\omega\lambda_2 + 1/(i\omega\tau_F')} \right. \right.
$$
$$
\left. \left. + \frac{D_5\lambda_7\tau_D k^2(1 + i\omega\lambda_8)}{1 + i\omega\lambda_5} \right] \right\}^{-1}
\tag{3.12.15}
$$

A further simplification of Eq. (3.12.15) can be made by taking $\lambda_2 = \gamma_2 = \lambda_8 = 0$. In these conditions, the following equation is obtained

$$
\frac{\chi_\perp}{\chi} = \left\{ 1 + i\omega\tau_D - \frac{\tau_D}{2} \left[\frac{\tau_2^{-1}}{1 + 1(i\omega\tau_F')} + \frac{D_5\lambda_7 k^2}{1 + i\omega\lambda_5} \right] \right\}^{-1}
\tag{3.12.16}
$$

The effect of the molecular weight on the normal mode can be envisaged by modifying the parameter λ_5 (Fig. 3.1)

Comparison with the experimental data suggests (see, for example, reference[23]) that the absorption bands corresponding to the normal as well as the segmental processes are considerably broader than predicted by the simple Debye-like terms appearing in Eqs (3.12.9) and (3.12.11) to (3.12.16) of this section. A better agreement with the actual data could be obtained by modifying Eqs (3.12.6a) and (3.12.6c) according to the same premises as Eq. (3.6.7) to give

$$
\frac{\mathbf{E}_t}{T} + a_1 \frac{d\mathbf{J}}{dt} + \operatorname{grad} \frac{\phi}{T} + \beta_2 \operatorname{div} \mathbf{Q}^a + \beta_3 \operatorname{div} \mathbf{Q}^s
$$
$$
= \sum_{n=1}^{\infty} \mu_{1n} D_t^{n\alpha} \mathbf{P} + \mu_2 \operatorname{div}(\operatorname{grad}\mathbf{P}) + \mu_2 \operatorname{div}(\operatorname{grad}\dot{\mathbf{P}})
$$
$$
a_3 \frac{d\mathbf{Q}^s}{dt} + \beta_2 (\operatorname{grad}\mathbf{J})^s = \sum_{n=0}^{\infty} \mu_{1n} D_t^{n\alpha} \mathbf{Q}^s + \mu_8 (\operatorname{grad}\mathbf{P})^s + \mu_9 (\operatorname{grad}\mathbf{P})^s
$$

$$
\tag{3.12.17}
$$

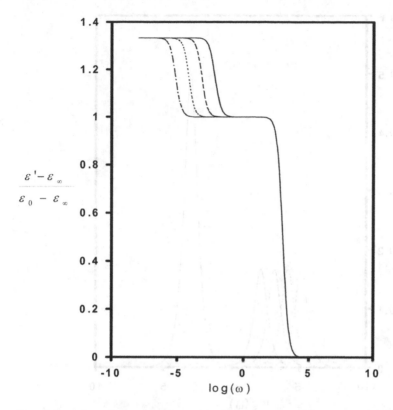

FIG. 3.1 Effect of the molecular weight envisaged by modifying λ_5, the coefficient of the symmetric part of the stress tensor responsible for the translational hydrodynamic friction. Parameters: $\tau_D = 10^{-3}$ s; $D_5\lambda_7 k^2 = 5 \times 10^2$ s^{-1}; $D_1 D_2 k^2 = 10^3$ s^{-1}; $I = 10^{-40}$ g cm^{-1}; $\zeta_0 = 10^{-36}$ g cm^{-1} s^{-1}; $\lambda_5 = (——)10^2, (— —)10^3, (\cdots\cdots)10^4, (—\cdot—)10^5$ s

It is noticeable that, for a first approach, Eq. (3.12.6b) should be maintained. After the pertinent calculations, similar to those made in section (3.9), the following equation is obtained

$$\frac{\varepsilon^*(\omega) - \varepsilon_\infty}{\varepsilon_0 - \varepsilon_\infty} = \left([1 + (i\omega\tau)^{\alpha_1}]^{\beta_1} - \frac{k^2}{2}\left\{\frac{D_1 D_2(1 + i\omega\gamma_2)}{1 + (2\zeta_0)/(i\omega I + D_2 k^2)}\right.\right.$$
$$\left.\left.+ \frac{D_5\lambda_2(1 + i\omega\lambda_s)}{[1 + (i\omega\lambda_5)^{\alpha_2}]^{\beta_2}}\right\}\right)^{-1}$$

(3.12.18)

where, in order to simplify calculations, $\lambda_1 = D_3 = 0$ and $Q^s = Q^a = 0$

The first and third terms on the right-hand side of Eq. (3.12.17) corresponds respectively to the segmental and normal modes, the corresponding relaxation

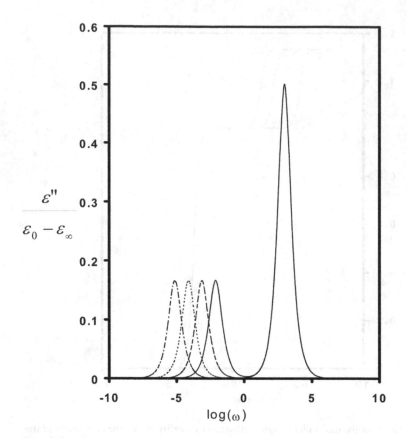

FIG. 3.1 Continued

times being respectively τ and λ_5. The appearance of the parameters α_i and $\beta_i(i = 1, 2)$ in Eq. (3.12.17) lead to a much more accurate Haviliak–Negami terms for the representation of the segmental and normal modes of actual data. Figure 3.2 shows the fit of Eq. (3.12.17) to the experimental data reported in reference.[23] The agreement is fairly good.

APPENDIX

Consider the differential equation

$$y^{(n)}(t) = F(t), \qquad n = 1, 2, 3, \ldots, y^{(n)} = \frac{d^n y}{dx^n} \tag{3.A.1}$$

where $F(t)$ is a continuous function in $[0, t]$.

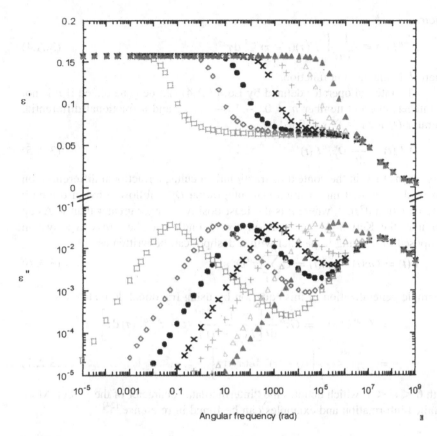

FIG. 3.2 Effect of the molecular weight envisaged by modifying λ_5, the coefficient of the symmetric part of the stress tensor responsible for the translational hydrodynamic friction, on the normal mode relaxation corresponding to the date of reference [23] as a function of the angular frequency. Parameters: $\tau = 5 \times 10^8$ s; $D_5\lambda_7 k^2 = 1.38$; $D_1 D_2 k^2 = 10^2$; $I = 10^{40}\,\mathrm{g\,cm^{-1}}$; $\zeta_0 = 10^{36}\,\mathrm{g\,cm^{-1}s^{-1}}$, $\alpha_1 = 0.95$, $\beta_1 = 0.5$, $\alpha_2 = 0.8$, $\beta_2 = 0.5$, $\lambda_8 = \gamma_2 = 0$, λ_5 in $s = 1.2$, 0.14, 0.012, 0.0008, 0.000012, 0.0000012, 0.00000014, in order of increasing frequencies for the peak corresponding to the normal mode

Let the initial values be

$$y^{(i)}(t = 0) = y_i, \qquad 0 \leqslant i \leqslant n - 1 \tag{3.A.2}$$

Then, the solution (unique) of the preceding problem is given by

$$y(t) = \sum_{i=0}^{n-1} \frac{y_i t^i}{i!} + D_t^{-n} F(t) \tag{3.A.3}$$

where

$$D_t^{-n}F(t) \equiv \frac{1}{\Gamma(n)} \int_0^t F(\tau)(t-\tau)^{n-1}\,d\tau \qquad (3.A.4)$$

where Γ is the gamma function.

The integral operator defined by Eq. (3.A.4) can be generalized if n is not an integer, but any number $n \neq 0, -1, -2, \ldots$, and a fractional differential operator D_t^v for $v > 0$ is defined by

$$D_t^v F(t) \equiv \frac{d^n}{dt^n} D_t^{v-n} F(t) \qquad (3.A.5)$$

This means that, in the context of fractional calculus, a fractional differentiation can be decomposed into a fractional integration D_t^{v-n} followed by an ordinary differentiation d^n/dt^n, where n is the least positive integer greater than v. As an example, the Kelvin–Voigt equation corresponding to the stress in a system composed of a spring in parallel with a dashpot can be written as

$$\sigma(t) = G\varepsilon(t) + \eta\dot{\varepsilon}(t) = G\sum_{0,1} \tau^\alpha D_t^\alpha \varepsilon(t) \qquad (3.A.6)$$

A simple generalization of this equation by using fractional derivatives is

$$\sigma_\alpha(t) = G\tau^\alpha D_t^\alpha \varepsilon(t) = G\tau^\alpha \frac{d}{dt}\left[\frac{1}{\Gamma(1-\alpha)} \int_0^t (t-\tau)^{-\alpha}\varepsilon(\tau)\,d\tau \right]$$

$$= \frac{G\tau^\alpha}{\Gamma(1-\alpha)}\left[\int_0^t (t-\tau)^{-\alpha}\dot{\varepsilon}(\tau)\,d\tau \right] \qquad (3.A.7)$$

with $0 < \alpha < 1$, which denotes the "intermediate" character of the model. More detailed information and examples can be found in reference.[24]

PROBLEMS

Problem 1

Let a be a quantity (scalar, vector, or tensor valued). Show that

$$\rho\frac{da}{dt} = \frac{\partial(\rho a)}{\partial t} + \operatorname{div}(\rho a\mathbf{v})$$

Particularize for $a = \mathbf{v}$. The quantity $\rho a\mathbf{v}$ is called the convective flux of a.

Solution

The application of the baricentric derivative to the product ρa gives

$$\frac{d(\rho a)}{dt} = \frac{\partial(\rho a)}{\partial t} + \mathbf{v}\operatorname{grad}(\rho a) \qquad (P.3.1.1)$$

Developing the preceding expression

$$\rho \frac{da}{dt} = -a \frac{d\rho}{dt} + \frac{\partial(\rho a)}{\partial t} + \mathbf{v} \operatorname{grad}(\rho a) \qquad \text{(P.3.1.2)}$$

and using Eq. (4.4.1.3), one obtains

$$\rho \frac{da}{dt} = -a \frac{\partial \rho}{\partial t} - a\mathbf{v} \operatorname{grad} \rho + \frac{\partial(\rho a)}{\partial t} + \mathbf{v} \operatorname{grad}(\rho a) \qquad \text{(P.3.1.3)}$$

The continuity equation [Eq. (3.4.1.2)] transforms Eq. (P.3.1.3) into

$$\rho \frac{da}{dt} = a \operatorname{div}(\rho \mathbf{v}) - a\mathbf{v} \operatorname{grad} \rho + \frac{\partial(\rho a)}{\partial t} + \mathbf{v} \operatorname{grad}(\rho a)$$

$$= \rho a \operatorname{div} \mathbf{v} + \frac{\partial(\rho a)}{\partial t} + \mathbf{v}\operatorname{grad}(\rho a) \qquad \text{(P.3.1.4)}$$

This expression can finally be written as

$$\rho \frac{da}{dt} = \frac{\partial(\rho a)}{\partial t} + \operatorname{div}(\rho a \mathbf{v}) \qquad \text{(P.3.1.5)}$$

When $a = \mathbf{v}$, the following equation is obtained

$$\rho \frac{d\mathbf{v}}{dt} = \frac{\partial(\rho \mathbf{v})}{\partial t} + \operatorname{div}(\rho \mathbf{v} \cdot \mathbf{v}) \qquad \text{(P.3.1.6)}$$

Problem 2

Obtain the equations of the conservation of mass, linear momentum, angular momentum, and energy, in local form, from the global equation of the conservation of the energy, taking into account the invariance of that equation under translations and rotations.

Solution

The power rate can be expressed as

$$P = \dot{K} + \dot{U} \qquad \text{(P.3.2.1)}$$

where the kinetic energy, K, and the internal energy, U, are given by

$$K = \int_B \frac{1}{2} \mathbf{v} \cdot \mathbf{v} \, dm$$

$$U = \int_B u \, dm \qquad \text{(P.3.2.2)}$$

On the other hand, the first law of thermodynamics expresses that

$$\dot{K} + \dot{U} = \dot{W} + Q \qquad \text{(P.3.2.3)}$$

where the working (mechanical power) and the heating rates are respectively

$$W = \int_{\partial B} \mathbf{t} \mathbf{v}\, dS + \int_B \mathbf{b} \cdot \mathbf{v}\, dm$$

$$Q = -\int_{\partial B} \mathbf{q} \cdot \mathbf{n}\, dS + \int_B r\, dm$$

(P.3.2.4)

In these equations, $\mathbf{t}$ is the surface force, $\mathbf{b}$ represents the body forces, $\mathbf{q}$ is the heat flux across the surface, and r is the heat per unit of mass. By taking a control volume using Eq. (P.3.1.5), and applying the divergence theorem, Eq. (P.3.2.3) becomes

$$\int_V \rho(r + \mathbf{b} \cdot \mathbf{v})\, dV + \int_{\partial V} (\mathbf{t}\mathbf{v} - \mathbf{q} \cdot \mathbf{n})\, dS$$

$$= \frac{d}{dt} \int_V \left(\frac{1}{2}\mathbf{v} \cdot \mathbf{v} + u \right) \rho\, dV = \int_V \frac{\partial}{\partial t} \left(\frac{1}{2}\mathbf{v} \cdot \mathbf{v} + u \right) dV$$

$$+ \int_V \mathrm{div}\left[\rho \left(\frac{1}{2}\mathbf{v} \cdot \mathbf{v} + u \right) \mathbf{v} \right] dV$$

(P.3.2.5)

where ρ is the mass density.

By using the definition of the baricentric time derivative

$$\frac{df}{dt} = \frac{\partial f}{\partial t} + (\mathrm{grad} f) \cdot \mathbf{v}$$

(P.3.2.6)

Eq. (P.3.2.5) can be written as

$$\frac{d}{dt} \int_V \left(\frac{1}{2}\mathbf{v} \cdot \mathbf{v} + u \right) \rho\, dV$$

$$= \int_V \frac{\partial}{\partial t} \left(\frac{1}{2}\mathbf{v} \cdot \mathbf{v} + u \right) dV + \int_V \mathrm{div}\left[\rho \left(\frac{1}{2}\mathbf{v} \cdot \mathbf{v} + u \right) \mathbf{v} \right] dV$$

$$= \int_V \frac{\partial}{\partial t} \left(\frac{1}{2}\mathbf{v} \cdot \mathbf{v} + u \right) dV + \int_V \mathrm{grad}\left[\rho \left(\frac{1}{2}\mathbf{v} \cdot \mathbf{v} + u \right) \right] \mathbf{v}\, dV$$

$$+ \int_V \rho \left(\frac{1}{2}\mathbf{v} \cdot \mathbf{v} + u \right) \mathrm{div} \mathbf{v}\, dV$$

$$= \int_V \frac{d}{dt} \rho \left(\frac{1}{2}\mathbf{v} \cdot \mathbf{v} + u \right) dV + \int_V \rho \left(\frac{1}{2}\mathbf{v} \cdot \mathbf{v} + u \right) \mathrm{div} \mathbf{v}\, dV$$

$$= \int_V \rho(\dot{\mathbf{v}} \cdot \mathbf{v} + \dot{u})\, dV + \int_V (\dot{\rho} + \rho\, \mathrm{div}\, \mathbf{v}) \left(\frac{1}{2}\mathbf{v} \cdot \mathbf{v} + u \right) dV$$

(P.3.2.7)

Consequently

$$\int_V \rho(\dot{\mathbf{v}} \cdot \mathbf{v} + \dot{u}) \, dV + \int_V (\dot{\rho} + \rho \operatorname{div} \mathbf{v})\left(\frac{1}{2}\mathbf{v} \cdot \mathbf{v} + u\right) dV$$

$$= \int_V \rho(r + \mathbf{b} \cdot \mathbf{v}) \, dV - \int_{\partial V} (t\mathbf{v} + \mathbf{q} \cdot \mathbf{n}) \, dS \qquad \text{(P.3.2.8)}$$

Let us consider a translation given by

$$\mathbf{v}^* = \mathbf{v} + \mathbf{c} \qquad \text{(P.3.2.9)}$$

The invariance under translations requires fulfillment of Eq. (P.3.2.8) for both $\mathbf{v}^*$ and $\mathbf{v}$. Then, substitution of Eq. (P.3.2.9) into Eq. (P.3.2.8) and further subtraction of Eq. (P.3.2.8), gives

$$\int_V \rho \mathbf{b} \, dV + \int_{\partial V} \mathbf{t} \, dS - \int_V \rho \dot{\mathbf{v}} \, dV - \int_V (\dot{\rho} + \rho \operatorname{div} \mathbf{v})\mathbf{v} \, dV$$

$$= \tfrac{1}{2}\dot{\mathbf{c}} \int_V (\dot{\rho} + \rho \operatorname{div} \mathbf{v}) \, dV \qquad \text{(P.3.2.10)}$$

which is valid for all $\dot{\mathbf{c}}$ and all V. For smooth enough situations, two conditions result from Eq. (P.3.2.10) after application of the divergence theorem

$$\dot{\rho} + \rho \operatorname{div} = 0$$
$$- \operatorname{div} \mathbf{T} + \rho \mathbf{b} = \rho \mathbf{v}$$

$$\text{(P.3.2.11)}$$

where the following definition of the mechanical (Cauchy) stress tensor $\mathbf{T}$ has been used

$$\mathbf{t} = -\mathbf{Tn} \qquad \text{(P.3.2.12)}$$

The resulting global balance energy equation will be

$$\int_V \rho(r + \mathbf{b} \cdot \mathbf{v}) \, dV + \int_{\partial V} (t\mathbf{v} - \mathbf{q} \cdot \mathbf{n}) \, dS = \int_V \rho(\mathbf{v} \cdot \dot{\mathbf{v}} + \dot{u}) \, dV \qquad \text{(P.3.2.13)}$$

Application of the divergence theorem transforms Eq. (P.3.2.13) into

$$\int_V \rho(r + \mathbf{b} \cdot \mathbf{v}) \, dV - \int_V \operatorname{div}(\mathbf{Tv} + \mathbf{q}) \, dV = \int_V \rho(\dot{\mathbf{v}} \cdot \mathbf{v} + \dot{u}) \, dV \qquad \text{(P.3.2.14)}$$

By using the vector equality and Eq. (P.3.2.11b)

$$\operatorname{div}(\mathbf{Tv}) = \mathbf{v} \cdot \operatorname{div} \mathbf{T} + \mathbf{T} \operatorname{grad} \mathbf{v} \qquad \text{(P.3.2.15)}$$

the following local expression for the conservation of energy is obtained

$$\rho r - \mathbf{T} \cdot \operatorname{grad} \mathbf{v} - \operatorname{div} \mathbf{q} = \rho \dot{u} \qquad (P.3.2.16)$$

which is equivalent to the first law of thermodynamics.

Let us consider now a rotation $\mathbf{Q}$ such that

$$\mathbf{F}^* = \mathbf{Q}\mathbf{F} \qquad (P.3.2.17)$$

where $\mathbf{F}$ is called the displacement gradient tensor and $\mathbf{Q}$ is a rotation, that is $\mathbf{Q} = \mathbf{Q}^t$, where the subscript denotes transposed. It should be noted also that

$$\dot{\mathbf{F}} = \mathbf{G}\mathbf{F} \qquad (P.3.2.18)$$

where $\mathbf{G}$ is the tensor gradient of velocity. The time derivative of Eq. (P.3.2.17) together with Eq. (P.3.2.18) and the properties of $\mathbf{Q}$ gives

$$\dot{\mathbf{F}}^* = \mathbf{G}^*\mathbf{F}^* = \mathbf{Q}\dot{F} + \dot{\mathbf{Q}}F = \mathbf{Q}\mathbf{G}F + \dot{\mathbf{Q}}F$$

$$= \mathbf{Q}\mathbf{G}\mathbf{Q}^t\mathbf{F}^* + \dot{\mathbf{Q}}\mathbf{Q}^t\mathbf{F}^* \qquad (P.3.2.19)$$

Since $\mathbf{F}^*$ is no singular

$$\mathbf{G}^* = \mathbf{Q}\mathbf{G}\mathbf{Q}^t + \dot{\mathbf{Q}}\mathbf{Q}^t \qquad (P.3.2.20)$$

where $\mathbf{A} = \dot{\mathbf{Q}}\mathbf{Q}^t$ is an antisymmetric matrix called spin. Let us assume for simplicity that $\mathbf{Q} = \delta_{ij}$, then

$$\mathbf{G}^* = \mathbf{G} + \mathbf{A} \qquad (P.3.2.21)$$

and Eq. (P.3.2.16) can be written as

$$\rho r - Tr(\mathbf{T}\mathbf{G}^t) - \operatorname{div} \mathbf{q} = \rho \dot{u} \qquad (P.3.2.22)$$

Carrying Eq. (P.3.2.21) into Eq. (P.3.2.22) and subtracting Eq. (P.3.2.22) from the resulting expression yields

$$Tr(\mathbf{T}\mathbf{A}^t) = 0 \qquad (P.3.2.23)$$

where $A_{ij} = -A_{ji}$ from which

$$(T_{12} - T_{21})A_{12} + (T_{13} - T_{31})A_{13} + (T_{23} - T_{32})A_{23} = 0 \qquad (P.3.2.24)$$

This expression is valid for all A_{ij}, and consequently

$$T_{ij} = T_{ji} \tag{P.3.2.25}$$

that is

$$\mathbf{T} = \mathbf{T}^t \tag{P.3.2.26}$$

The symmetry of the stress tensor is equivalent to the law of conservation of angular momentum (second Cauchy law).

Problem 3

Find in the frequency domain the real and imaginary parts of the dielectric permittivity corresponding to the correlation function given by Eq. (3.7.4).

Solution

According to Eq. (2.12.7), the complex permittivity can be written as

$$\frac{\varepsilon^*(\omega) - \varepsilon_\infty}{\varepsilon_0 - \varepsilon_\infty} = L\left(-\frac{\dot{\Phi}(t)}{\Phi(0)}\right) \tag{P.3.3.1}$$

where L denotes the Laplace transform and $\Phi(t)$ is the relation decay function.
   On the other hand, according to Eq. (3.7.4)

$$-\dot{\Phi}(t) = C_1 k_1 \exp(-k_1 t) + C_2 k_2 \exp(-k_2 t) \tag{P.3.3.2}$$

Therefore

$$L\left(-\frac{\dot{\Phi}(t)}{\Phi(0)}\right) = \frac{C_1 k_1}{C_1 + C_2}\frac{1}{s + k_1} + \frac{C_2 k_2}{C_1 + C_2}\frac{1}{s + k_2} \tag{P.3.3.3}$$

where $s = i\omega$. From Eqs (P.3.3.3) and (3.7.5), the following result holds

$$\frac{\varepsilon^*(\omega) - \varepsilon_\infty}{\varepsilon_0 - \varepsilon_\infty} = \frac{1}{1 - \omega^2 \tau_1 + i\omega\tau_2} \tag{P.3.3.4}$$

The real and imaginary parts of the complex permittivity are

$$
\mathrm{Re}\frac{\varepsilon^*(\omega) - \varepsilon_\infty}{\varepsilon_0 - \varepsilon_\infty} = \frac{1 - \omega^2 \tau_1}{(1 - \omega^2 \tau_1)^2 + \omega^2 \tau_2^2}
$$

$$
\mathrm{Im}\frac{\varepsilon^*(\omega) - \varepsilon_\infty}{\varepsilon_0 - \varepsilon_\infty} = \frac{-\omega \tau_2}{(1 - \omega^2 \tau_1)^2 + \omega^2 \tau_2^2}
$$

(P.3.3.5)

Note that, if $\tau_1 = 0$, the Debye equations are recovered. This means that the discrepancies with the Debye model can be expressed only in terms of an additional parameter. On the other hand, the Cole–Cole arc is given by

$$
\left[\varepsilon' - \frac{1}{2(1 - \omega^2 \tau_1)}\right]^2 + \varepsilon''^2 \left[\frac{1}{2(1 - \omega^2 \tau_1)}\right]^2 = 0
$$

(P.3.3.6)

Problem 4 (see reference [10b])

Find the conditions under which the KWW relaxation function is consistent with EIT.

Solution

Let us consider the KWW equation [Eq. (2.12.8)]

$$
\Phi(t) = \exp\left[-\left(\frac{t}{\tau}\right)^\beta\right]
$$

(P.3.4.1)

Equation (P.3.4.1) is the solution of the following differential equation

$$
\tau_{\mathrm{KWW}}\,\dot\Phi(t) = -\Phi(t)
$$

(P.3.4.2)

with

$$
\tau_{\mathrm{KWW}} = \frac{\tau}{\beta}\left(\frac{t}{\tau}\right)^{1-\beta}
$$

(P.3.4.3)

Equation (P.3.4.3) is only compatible with Eq. (3.6.1) if $\tau_1 = \mathrm{grad}\,T_0^{-1} = \mathbf{q} = 0$, that is, in the absence of heat flux, for nil temperature gradients, and for $\tau_1 = 0$. According to Eqs (3.6.2b) and (P.3.4.3), the following time dependence for τ_2 is required

$$
\tau_2 = \chi T_0 \frac{\mu_{22}(\alpha_{110}/\alpha_{120}) - \mu_{12}}{\mu_{13} + \alpha_{110}/\alpha_{120}(1 - \mu_{23})} = \frac{\tau}{\beta}\left(\frac{t}{\tau}\right)^{1-\beta}
$$

(P.3.4.4)

In conclusion, as long as another variable different from the polarization is present, Eq. (P.3.4.1) is unable to represent the polarization correlation function.

Problem 5

Repeat the calculations of the Section 3.9 for dielectric susceptibility in the case where the friction ζ is time dependent.

Solution

If the rotational friction is considered to be time dependent, then Eq. (3.8.3.13b) can be written as

$$\left(1 + \lambda_2 \frac{d}{dt}\right)\mathbf{Q}^a - D_1\left(1 + \gamma_2 \frac{d}{dt}\right)(\text{grad }\mathbf{P})^a + \int_0^\infty \zeta(\tau)\boldsymbol{\varepsilon}\boldsymbol{\omega}(t - \tau)\,d\tau = 0$$

(P.3.5.1)

The Laplace transform of this expression yields

$$(1 + s\lambda_2)\mathbf{Q}^a(s) - D_1(1 + s\gamma_2)ik\mathbf{P}^a + \zeta(s)\boldsymbol{\varepsilon}\boldsymbol{\omega}(s) = 0 \qquad (P.3.5.2)$$

Following the same strategy as in Section 3.9, the transversal susceptibility is given by

$$\frac{\chi_\perp}{\chi} = \left[1 + i\omega\tau_D\left(1 + i\omega\lambda_1 + \left\{\frac{D_3(1 + i\omega\gamma_1)}{i\omega}\right.\right.\right.$$
$$\left.\left.\left. - \frac{D_1D_2(1 + i\omega\gamma_2)(i\omega I + D_sk^2)}{2i\omega[(1 + i\omega\lambda_2)(i\omega I + D_sk^2) + 2\zeta(\omega)]}\right\}k^2\right)\right]^{-1}$$

(P.3.5.3)

If $\lambda_1 = D_3 = I = 0$, then

$$\frac{\chi_\perp}{\chi} = \left\{1 + i\omega\tau_D - \frac{\tau_D D_s k^4(1 + i\omega\gamma_2)}{2\tau_2[D_sk^2(1 + i\omega\lambda_2) + 2\zeta(\omega)]}\right\}^{-1} \qquad (P.3.5.4)$$

where the definition given by Eq. (3.10.6) was considered.

Problem 6

Deduce Eq. (3.12.9).

Solution

As in section 3.9, use is made of the Fourier–Laplace transform to study the frequency dependence of the polarization $\mathbf{P}$ from which the corresponding dielectric characteristics of the system are obtained. For this purpose, we start with Eqs (3.12.7) and the balance of the angular momentum [Eq. (3.10.1)]. A harmonic variation is also assumed for the angular velocity

$$\boldsymbol{\omega} = \exp\mathrm{i}(\omega t - \mathbf{k} \cdot \mathbf{r}) \tag{P.3.6.1}$$

where $\mathbf{k}$ is the wave vector defining the propagation of the wave. The system to be solved is

$$\left(1 + \lambda_1 \frac{\mathrm{d}}{\mathrm{d}t}\right)\dot{\mathbf{P}} + \frac{1}{\tau_\mathrm{D}}(\mathbf{P} - \chi\mathbf{E}_l) + D_2 \operatorname{div} \mathbf{Q}^\mathrm{a} + D_5 \operatorname{div} \mathbf{Q}^\mathrm{s}$$

$$- D_3\left(1 + \gamma_1 \frac{\mathrm{d}}{\mathrm{d}t}\right)\operatorname{div}(\operatorname{grad} \mathbf{P}) = 0$$

$$\left(1 + \lambda_2 \frac{\mathrm{d}}{\mathrm{d}t}\right)\mathbf{Q}^\mathrm{a} - \varepsilon\omega D_1\left(1 + \gamma_2 \frac{\mathrm{d}}{\mathrm{d}t}\right)(\operatorname{grad} \mathbf{P})^\mathrm{a} + \zeta_0 \varepsilon\boldsymbol{\omega} = 0 \tag{P.3.6.2}$$

$$\left(1 + \lambda_5 \frac{\mathrm{d}}{\mathrm{d}t}\right)\mathbf{Q}^\mathrm{s} - \lambda_7\left(1 + \lambda_8 \frac{\mathrm{d}}{\mathrm{d}t}\right)(\operatorname{grad} \mathbf{P})^\mathrm{s} = 0$$

$$I \frac{\mathrm{d}\boldsymbol{\omega}}{\mathrm{d}t} = D_\mathrm{s}\nabla\boldsymbol{\omega} + \varepsilon\mathbf{Q}^\mathrm{a}$$

Taking Laplace transforms in Eq. (P.3.6.2), with $s = \mathrm{i}\omega$, gives

$$s(1 + \lambda_1 s)\mathbf{P} + \frac{1}{\tau_\mathrm{D}}(\mathbf{P} - \chi\mathbf{E}_l) + D_2 \mathrm{i}(\mathbf{k} \cdot \mathbf{Q})^\mathrm{a} + D_5 \mathrm{i}(\mathbf{k} \cdot \mathbf{Q})^\mathrm{s}$$

$$+ D_3(1 + \gamma_1 s)\mathbf{k} \cdot (\mathbf{k}\mathbf{P}) = 0$$

$$(1 + \lambda_2 s)\mathbf{Q}^\mathrm{a} - D_1(1 + \gamma_2 s)\mathrm{i}(\mathbf{k}\mathbf{P})^\mathrm{a} + \zeta_0 \varepsilon\boldsymbol{\omega} = 0$$

$$(1 + \lambda_5 s)\mathbf{Q}^\mathrm{s} - \lambda_7(1 + \lambda_8 s)\mathrm{i}(\mathbf{k}\mathbf{P})^\mathrm{s} = 0$$

$$(sI + D_\mathrm{s}k^2)\boldsymbol{\omega} = \varepsilon\mathbf{Q}^\mathrm{a} \tag{P.3.6.3}$$

Premultiplying Eq. (P.3.6.3d) by $\boldsymbol{\varepsilon}$, Eq. (3.9.3) holds

$$\boldsymbol{\varepsilon}\boldsymbol{\varepsilon}\mathbf{Q}^{\mathrm{a}} = \varepsilon_{ijk}\varepsilon_{khl}Q_{hl}^{\mathrm{a}} = (\delta_{ih}\delta_{jl} - \delta_{il}\delta_{hj})Q_{hl}^{\mathrm{a}} = Q_{ij}^{\mathrm{a}} - Q_{ji}^{\mathrm{a}} = 2Q_{ij}^{\mathrm{a}} \qquad \text{(P.3.6.4)}$$

From here

$$(sI + D_s k^2)\boldsymbol{\varepsilon}\boldsymbol{\omega} = 2\mathbf{Q}^{\mathrm{a}} \rightarrow \boldsymbol{\varepsilon}\boldsymbol{\omega} = \frac{2\mathbf{Q}^{\mathrm{a}}}{sI + D_s k^2} \qquad \text{(P.3.6.5)}$$

Carrying Eq. (P.3.6.5) into (P.3.6.3b), the following expression is obtained

$$(1 + \lambda_2 s)\mathbf{Q}^{\mathrm{a}} - D_1(1 + \gamma_2 s)\mathrm{i}(\mathbf{kP})^{\mathrm{a}}$$

$$+ \zeta_0 \frac{2\mathbf{Q}^{\mathrm{a}}}{sI + D_s k^2} = 0 \rightarrow \left[(1 + \lambda_2 s) + \frac{2\zeta_0}{sI + D_s k^2}\right]\mathbf{Q}^{\mathrm{a}}$$

$$= D_1(1 + \gamma_2 s)\mathrm{i}(\mathbf{kP})^{\mathrm{a}} \qquad \text{(P.3.6.6)}$$

Multiplying this equation by $\mathrm{i}\mathbf{k}$ yields

$$\mathrm{i}\mathbf{k} \cdot \mathbf{Q}^{\mathrm{a}} = -\frac{D_1(1 + \gamma_2 s)\mathbf{k} \cdot (\mathbf{kP})^{\mathrm{a}}}{(1 + \lambda_2 s) + (2\zeta_0)/(sI + D_s k^2)} \qquad \text{(P.3.6.7)}$$

Similarly, multiplying Eq. (P.3.6.3c) by $\mathrm{i}\mathbf{k}$ gives

$$\mathrm{i}\mathbf{k} \cdot \mathbf{Q}^{\mathrm{s}} = -\frac{\lambda_7(1 + \lambda_8 s)\mathbf{k} \cdot (\mathbf{kP})^{\mathrm{s}}}{1 + \lambda_5 s} \qquad \text{(P.3.6.8)}$$

Substitution of Eq. (P.3.6.7) and (P.3.6.8) into Eq. (P.3.6.3a) leads to

$$s(1 + \lambda_1 s)\mathbf{P} + \frac{1}{\tau_{\mathrm{D}}}(\mathbf{P} - \chi\mathbf{E}_l) - \frac{D_1 D_2(1 + \gamma_2 s)\mathbf{k} \cdot (\mathbf{kP})^{\mathrm{a}}}{(1 + \lambda_2 s) + (2\zeta_0)/(sI + D_s k^2)}$$

$$- \frac{D_5 \lambda_7(1 + \lambda_8 s)\mathbf{k} \cdot (\mathbf{kP})^{\mathrm{s}}}{1 + \lambda_5 s} + D_3(1 + \gamma_1 s)\mathbf{k} \cdot (\mathbf{kP}) = 0 \qquad \text{(P.3.6.9)}$$

Equation (P.3.6.9) can be more conveniently written as

$$
\mathbf{P}\left[s(1 + \lambda_1 s) + \frac{1}{\tau_D} \right] + \left[D_3(1 + \gamma_1 s) - \frac{D_1 D_2(1 + \gamma_2 s)}{(1 + \lambda_2 s) + 2\zeta_0/sI + D_s k^2} \right]
$$

$$
\times \tfrac{1}{2}(k^2 \mathbf{P} - \mathbf{k}(\mathbf{k} \cdot \mathbf{P})) + \left[D_3(1 + \gamma_1 s) - \frac{D_5 \lambda_7(1 + \lambda_8 s)}{1 + \lambda_5 s} \right]
$$

$$
\times \tfrac{1}{2}(k^2 \mathbf{P} + \mathbf{k}(\mathbf{k} \cdot \mathbf{P})) = \frac{\chi}{\tau_D} \mathbf{E}_l \tag{P.3.6.10}
$$

where the following expressions are used

$$
\mathbf{k} \cdot (\mathbf{kP})^a = \tfrac{1}{2}(k^2 \mathbf{P} - \mathbf{k}(\mathbf{k} \cdot \mathbf{P}))
$$

$$
\mathbf{k} \cdot (\mathbf{kP})^s = \tfrac{1}{2}(k^2 \mathbf{P} + \mathbf{k}(\mathbf{k} \cdot \mathbf{P})) \tag{P.3.6.11}
$$

$$
\mathbf{k} \cdot (\mathbf{kP})^s + \mathbf{k} \cdot (\mathbf{kP})^a = \mathbf{k} \cdot (\mathbf{kP}) = k^2 \mathbf{P}
$$

Note that the first of these expressions is simply the Laplace transform of Eq. (3.8.3.14). Using Eqs (P.3.6.11) in vectorial notation, Eq. (P.3.6.10) can be written as

$$
P_i\left[s(1 + \lambda_1 s) + \frac{1}{\tau_D} + D_3(1 + \gamma_1 s) \right.
$$

$$
- \tfrac{1}{2}\left(\frac{D_1 D_2(1 + \gamma_2 s)}{(1 + \lambda_2 s) + (2\zeta_0)/(sI + D_s k^2)} + \frac{D_5 \lambda_7(1 + \lambda_8 s)}{1 + \lambda_5 s} \right) k^2 \bigg] \delta_{ij}
$$

$$
- \tfrac{1}{2}\left(\frac{D_5 \lambda_7(1 + \lambda_8 s)}{1 + \lambda_5 s} - \frac{D_1 D_2(1 + \gamma_2 s)}{(1 + \lambda_2 s) + (2\zeta_0)/(sI)} \right) k_j k_i P_i
$$

$$
= \frac{\chi}{\tau_D} E_{lj} \tag{P.3.6.12}
$$

Let us consider that the wave vector lies in the x direction, that is, $\mathbf{k} = (k, 0, 0)$ and $E_l = (0, E_2, E_3)$. In this case, the susceptibility matrix is diagonal, so that $\chi_{ij} = 0$ for $(i \neq j)$ and $k_i k_j = 0$. The only non-null components of the susceptibility are $\chi_{22} = \chi_{33} = \chi_\perp$, that is, the transversal components of the dielectric susceptibility. From

$$
P_i = \chi_{ij}(\mathbf{k}, \omega) E_{l_j} \tag{P.3.6.13}
$$

and taking $s = i\omega$, the following expression is found

$$\frac{\chi_\perp}{\chi} = \left[1 + i\omega\tau_D\left(1 + i\omega\lambda_1 + \frac{1}{i\omega}\left\{D_3(1 + i\omega\gamma_1) - \frac{1}{2}\right.\right.\right.$$

$$\times \left[\frac{D_1 D_2(1 + i\omega\lambda_3)}{1 + i\omega\lambda_2 + (2\zeta_0)/(i\omega I + D_s k^2)}\right.$$

$$\left.\left.\left.\left. + \frac{D_5\lambda_7(1 + i\omega\lambda_8)}{1 + i\omega\lambda_5}\right]\right\}k^2\right)\right]^{-1} \tag{P.3.6.14}$$

In the limit $\omega \to 0$, Eq. (P.3.6.14) is given by

$$\frac{\chi_\perp}{\chi} = \left[1 + \left(D_3 - \frac{1}{2}D_5\lambda_7\right)\tau_D k^2\right]^{-1} \tag{P.3.6.15}$$

This means that the dielectric susceptibility is **k** dependent, as previously reported.[16]

REFERENCES

1. Prigogine, I. *Introduction to Thermodynamics of Irreversible Processes*; Interscience: New York, 1961.
2. Glansdorff, P.; Prigogine, I. *Thermodynamic Theory of Structure, Stability and Fluctuations*; J. Wiley & Sons: New York, 1971.
3. Onsager, L. Phys. Rev. **1931**, *37*, 405; **1931**, *38*, 2265.
4. Meixner, J.; Reik, H. *Thermodynamik der Irreversiblen Prozesen (Handbuch de Physik)*; Flugge, S., Ed.; Springer: Berlin, 1959.
5. Noll, W. *The Foundations of Thermodynamics*; Springer: Berlin, 1974.
6. Coleman, B.D. Arch. Rat. Mech. Anal. **1964**, *17*, 1.
7. Truesdell, C.; Toupin, R.A. *The Classical Field Theories (Handbuch der Physik, III/1)*; Springer: Berlin, 1960; The *Non-Linear Field Theories (Handbuch der Physik, III/3)*; Springer: Berlin, 1965.
8. Jou, D.; Casas-Vazquez, J.; Lebon, G. *Extended Irreversible Thermodynamics*; Springer: Berlin, 1993; see also Rep. Prog. Phys. **1988**, *51*, 1105; Rep. Prog. Phys. **1999**, *62*, 1035.
9. García-Colín, L. Revista Mexicana de Física **1988**, *34*, 344.
10. del Castillo, L.F.; García-Colín, L. Phys. Rev. B **1986**, *33*, 4944; **1988**, *37*, 448.
11. De Groot, S.R.; Mazur, P. *Non-Equilibrium Thermodynamics*; North-Holland: Amsterdam, 1962. Reprinted by Dover in 1984 [11b].
12. Panofsky, W.; Phillips, M. *Classical Electricity and Magnetism*; Addison-Wesley: Reading, 1962.
13. Philippoff, G. Phys. Z. **1934**, *35*, 884.

14.   Scott-Blair, P. Proc. Roy. Soc. (Lond.) **1947**, *189A*, 69.
15.   Slonimsky, H. J. Pol. Sci. C **1967**, *16*, 1667.
16.   Mori, H. Prog. Theor. Phys. **1965**, *33*, 423; **1965**, *34*, 399.
17.   del Castillo, L.F.; Dávalos-Orozco, L.A. J. Chem. Phys. **1990**, *93*, 5147.
18.   Dávalos-Orozco, L.A.; del Castillo, L.F. J. Chem. Phys. **1992**, *96*, 9102.
19.   del Castillo, L.F.; Díaz-Calleja, R. Physica A **1999**, *268*, 469.
20.   Fulton, R.L. Mol. Phys. **1975**, *29*, 405; J. Chem. Phys. **1975**, *62*, 4355.
21.   Stockmayer, W.H. Pure and Appl. Chem. **1967**, *15*, 539.
22.   del Castillo, L.F.; Hernández, S.I.; García-Zavala, A.; Díaz-Calleja, R. J. Non Crystalline Solids **1998**, *235–237*, 677.
23.   Adachi, K.; Kotaka, T. Prog. Polym. Sci. **1993**, *18*, 585.
24.   Nonenmacher, T.F. In *Lecture Notes in Physics 381*; Casas-Vázquez, J., Jou, D., Eds.; Springer, 1991; 309 pp.

4

Experimental Techniques

The interactions between electromagnetic waves and dielectric materials can be determined by broadband measurement techniques. Dielectric relaxation spectrosocopy is nowadays a well-developed method that allows the study of molecular structure through the orientation of dipoles under the action of an electric field. The experimental devices to perform this task may cover the frequency range $10^{-4}-10^{11}$ Hz. Obviously, this frequency window cannot be covered with a single apparatus. Several methods are available for a given frequency band, as indicated in Scheme 4.1.

4.1. MEASUREMENT SYSTEMS IN THE TIME DOMAIN

It is a fundamental fact of linear systems that the time-dependent response to a step function field and the frequency-dependent response to a sinusoidal electric field are related through Fourier transforms. For this reason, from a mathematical point of view, there is no essential difference between these two types
of measurement. However, over a long period of time the equipment for measurements in the time domain has been far less developed than that used in the frequency domain. As a result, available experimental data in the time domain

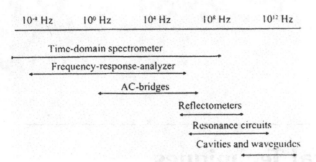

SCHEME 4.1   Dielectric measurement techniques as a function of the frequency

are much less abundant than those in the frequency domain. To cover the lowest
frequency range (from 10^{-4} to 10^1 Hz), time domain spectrometers have recently
been developed.[1-3] In these devices, a voltage step V_0 is applied to the sample
placed between the plates of a plane parallel capacitor, and the current $I(t)$ is
recorded. The total current traversing the condenser is given by

$$I^* = (G + i\omega C)V_0 \tag{4.1.1}$$

Equivalently, one can write

$$\frac{I^*}{V_0} = i\omega C_0 \varepsilon^* = Y^* \tag{4.1.2}$$

where Y^* is the admittance and C_0 is the vacuum capacitance which in SI units
for a plane capacitor is given by

$$C_0 = e_0 \frac{S}{d} \tag{4.1.3}$$

In Eq. (4.1.3), $e_0 = (10^{-10}/3.6\pi)\,\mathrm{F\,m^{-1}}$ is the permittivity of the evacuated
space, S is the surface area of the electrodes, and d is the distance between the
plates of the condenser. For circular disc-shaped and cylindrical electrodes C_0
can be written as

$$C_0 = \frac{10^{-2}r^2}{3.6d} \tag{4.1.4a}$$

$$C_0 = \frac{10^{-2}l}{1.8\ln(R/r)} \tag{4.1.4b}$$

where r in Eq. (4.1.4a) is the radius of the electrodes. In Eq. (4.1.4b), l is the
length of the cylinder wheras R abd r are respectively the radii of the external and
internal cylinders. If the values of r, d, l, and R are given in m, C_0 is obtained in

pF. By writing Eq. (4.1.2) in the time domain, one finds

$$\frac{I(t)}{V_0} = C_0 \frac{d\varepsilon}{dt} \qquad (4.1.5)$$

This equation can be written as

$$\frac{d\varepsilon}{dt} = \frac{I(t)}{C_0 V_0} \qquad (4.1.6)$$

and according to the plane parallel geometry

$$\frac{d\varepsilon}{dt} = \frac{I(t)}{e_0 S E_0} \qquad (4.1.7)$$

where E_0 is the applied electric field. Integration of Eq. (4.1.7) leads to

$$\varepsilon(t) = \frac{1}{C_0 V_0} \int_0^t I(t') \, dt' = \frac{C(t)}{C_0} \qquad (4.1.8)$$

where

$$C(t) = V_0^{-1} \int_0^t I(t') \, dt' \qquad (4.1.9)$$

The main item in the equipment is the electrometer, which must be able to measure currents as low as 10^{-16} A. In many cases the applied voltage can be taken from the internal voltage source of the electrometer. Also low-noise cables with high insulation resistance must be used. For the depolarization step, a relay separates the voltage source from the circuit that connects the sample to the electrometer. Technical details and pertinent figures can be found in the literature.[4,5]

Recently [6], an alternative way to measure dielectric properties in the time domain has been proposed. The method consists in the application of an electrical displacement D and further recording of the time-dependent field $E(t)$ to keep D constant. The measured property is now the electric modulus, $M(t)$. In the time domain the relationship between permittivity and electric modulus is similar to the relation between mechanical compliance and modulus,[7] that is

$$\int_0^\infty M(\tau)\varepsilon(t - \tau) \, d\tau = t \qquad (4.1.10)$$

After taking the Laplace transform of Eq. (4.1.10), the following relationship between M and ε is found

$$M^*(\omega) = [\varepsilon^*(\omega)]^{-1} \qquad (4.1.11)$$

The advantage of measuring the electric modulus is that the timescale of the experiment is more convenient than that of the permittivity. This is more clearly seen when conductive contributions to the dipolar relaxations are important, as is usual in measurements performed at relatively low frequencies well above T_g.

4.2. MEASUREMENT SYSTEMS IN THE FREQUENCY DOMAIN

In the intermediate frequency range $10^{-1} - 10^6$ Hz, capacitance bridges have been the common tools used to measure dielectric permittivities. The devices are based on the Wheatstone bridge principle where the arms are capacitance–resistance networks. In the earlier types of capacitance bridge (Schering or Scheiber bridges[8]) the generator was not coupled via a transformer to the bridge. Subsequent developments incorporate transformer ratio arms. In this way, the residual impedance and the guard circuit impedance are eliminated from the balance conditions, giving the permittivity and loss with a high level of accuracy. The principle of measurement of capacitance bridges is based on the balance of the bridge placing the test sample in one of the arms. The sample is represented by an RC network in parallel or series. When the null detector of the bridge is at its minimum value (as close as possible to zero), the equations of the balanced bridge provide the values of the capacitance and loss factor (or conductivity) for the test sample (see Problem 1).

Frequency response analyzers have proved to be very useful in measuring dielectric permittivities in the frequency range $10^{-2} - 10^6$ Hz, and their use has recently exceeded that of bridges. An a.c. voltage V_1 is applied to the sample, and then a resistor R, or alternatively a current-to-voltage converter for low frequencies, converts the sample current I_s into a voltage V_2 (Fig. 4.1). By comparing the amplitude and the phase angle between these two voltages, the complex impedance of the sample Z_s can be calculated as

$$Z_s = \frac{V_1 - V_2}{I_s} = \frac{V_1 - V_2}{V_2} R \tag{4.2.1}$$

For frequencies higher than 10^5 Hz, a buffer amplifier decouples the sample current from the analyzer and the current-to-voltage converter by means of an impedance variable. Owing to parasitic inductances, the high-frequency limit is about 1 MHz. For more details, see reference [4], where impedance bridges and their advantages are also reviewed. For a sample of disclike geometry, 10 mm in diameter and 100 μm thick, a capacitance of about 42 pF is obtained for $\varepsilon' = 3$. It is necessary to be very careful with the temperature control, and for this purpose it is advisable to measure the temperature as close as possible to the sample.

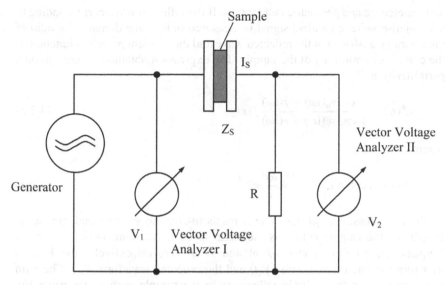

FIG. 4.1 Frequency response analyzer

At frequencies ranging from 1 MHz to 10 GHz, the inductance of the connecting cables contributes to the measured impedance. In fact, standing waves preclude the measurements at frequencies higher than 10^7 Hz. Resonance circuits have been used in the past from 10^5 to 10^8 Hz. However, the measurements performed in a broad range of temperatures require the lengths of the electric wires connecting the sample and the resonance circuit to be as short as possible. This problem can be avoided by putting the sample capacitor as the termination of a high-precision coaxial line. That is, a line with shape-defined and temperature-independent propagation parameters.

At frequencies above 1 GHz we move from interference optics to geometrical optics. Network analysis is used and the dielectric properties of materials from infrared to ultraviolet are studied in terms of the amplitude and phase of the reflected as well as the transmitted waves.

The dielectric relaxation response of polymer solutions to perturbation fields of frequency above 10 MHz often reflects conformational changes and internal local motions of molecular chains. Therefore, a precise analysis of the dielectric spectra obtained in the high-frequency region is fundamental to understanding the dynamics of polymer chains. One of the techniques often used to obtain dielectric spectra at high frequencies is time domain reflectometry. The technique is based on the reflection of an electric wave, transported through a coaxial line, in a dielectric sample cell attached at the end of the line. In this case, the reflective coefficient is a function of the complex permittivity of the sample, and the electric and geometric cell lengths. If the reflected wave corresponding to a step-pulse voltage incident signal is measured in the time domain, the ratio of the Fourier transforms of the reflected wave and the incident pulse is a function of the complex permittivity of the sample. The expression obtained for the complex permittivity is[9]

$$\varepsilon^*(\omega) = \frac{c}{i\omega\gamma d} \frac{v_0(i\omega) - r(i\omega)}{v_0(i\omega) + r(i\omega)} f(x) \tag{4.2.2}$$

where

$$f(x) = x\cot x, \qquad x = \frac{\omega d}{c} \varepsilon^{*1/2} \tag{4.2.3}$$

In these expressions, γd and d are, respectively, the electric and geometric lengths of the sample cell, c is the light speed in vacuum, ω is the angular frequency, $i = (-1)^{1/2}$, and $v_0(i\omega)$ and $r(i\omega)$ are, respectively, the Fourier transforms of the incident pulse $V(t)$ and the reflected waveform $R(t)$. The term $f(x)$ accounts for the multiple reflections in the sample section. By using Eq. (4.2.2), the complex dielectric permittivity of a sample x, ε_x^*, can be expressed in terms of the known permittivity ε_s^* of a sample by means of the following equation

$$\varepsilon_x^* = \varepsilon_s^* \frac{1 + \{(cf_s)/[i\omega(\gamma d)\varepsilon_s^*]\}\rho f_x}{1 + \{[i\omega(\gamma d)]\varepsilon_s^*/cf_s\}\rho \ f_s} \tag{4.2.4}$$

where

$$\rho = \frac{\int_0^\infty [R_s(t) - R_x(t)] e^{-i\omega t} \, dt}{[2/(i\omega)] \int_0^\infty [(d/dt)R_s(t)] e^{-i\omega t} \, dt - \int_0^\infty [R_s(t) - R_x(t)] e^{-i\omega t} \, dt} \tag{4.2.5}$$

Note that $R_s(t)$ and $R_x(t)$ are, respectively, the reflected waveforms from the known and unknown samples. A time window within which $R_s - R_x$ reaches zero is employed to obtain the waveform of $R_s - R_x$. By choosing a time t_1 as that at which $R_s - R_x = 0$, Eq. (4.2.5) becomes

$$\rho = \frac{\int_0^{t_1} [R_s(t) - R_x(t)] e^{-i\omega t} \, dt}{2 \int_0^{t_1} [R_s(t)] e^{-i\omega t} \, dt - \int_0^{t_1} [R_s(t) - R_x(t)] e^{-i\omega t} \, dt} \tag{4.2.6}$$

The limit of the highest frequency obtained from the Fourier transform is determined by the time interval Δt as $f_m = 1/2\Delta t$. Therefore, the measurements of R_s and R_t in shorter ranges are required to obtain precise values in the high-frequency region (see references [9] for more details).

4.3. IMMITTANCE ANALYSIS

4.3.1. Basic Immittance Functions

In many cases, it is possible to reproduce the electric properties of a dipolar system by means of passive elements such as resistors, capacitors or combined elements. One of the advantages of the models is that they often easily describe the response of a system to polarization processes. However, it is necessary to stress that the models in general only provide an approximate way to represent the actual behavior of the system. Although the analysis of dielectric materials is commonly made in terms of the complex permittivity function ε^* or its inverse, the electric modulus M^*, it can also be considered as a part of the science called electrochemistry of solids. In this context, both electrical impedance and admittance are the appropriate functions to represent the response of the corresponding circuits. As a consequence, the four basic immittance functions are permittivity, electric modulus, impedance and admittance. They are related by the following formulae

$$M^* = (\varepsilon^*)^{-1}$$

$$Y^* = (Z^*)^{-1}$$

$$Y^* = i\omega C_0 \varepsilon^*$$

$$M^* = i\omega C_0 Z^*$$

(4.3.1.1)

where C_0 is the capacitance of the empty capacitor. It is a matter of convenience, dictated by the type of phenomenon to be made relevant through the chosen function, to use one of these functions. For this reason it is very useful to express the electrical analogy of Debye as well as other empirical equations in terms of passive elements such as resistors and capacitors. Conversely, it is more realistic to find an empirical equation giving account of a set of experimental results. In this respect a further potential problem is that there are many circuits, some of them equivalent, that can describe the problem. In this context, the simplest circuit, compatible with the phenomenality of the problem under study, is the best.

On the basis of the analogies between the phenomenological treatment of dielectric and viscoelastic materials, it is also easy to build the viscoelastic analogy for the electric circuit. First, note that, in the electromechanical analogy, elements in series (parallel) or in parallel (series) appear. A spring corresponds to a capacitor and a damping element to a resistor. Consequently, the mechanical analog system to the Debye material is the standard solid (or Zener material) consisting of a spring in series with a Kelvin–Voigt element (spring and dashpot in parallel).[7]

4.3.2. Series and Parallel RC Networks

As mentioned in section 4.2, an RC assembly, in series or parallel, represents the test sample in many bridges or lumped circuits. It is interesting to obtain the relationship between the parameters corresponding to these simple networks. The admittance of an RC assembly is given by

$$Y^* = R_p^{-1} + i\omega C_p \tag{4.3.2.1}$$

while the impedance is

$$Z^* = R_s + \frac{1}{i\omega C_s} \tag{4.3.2.2}$$

By noting that, for the equivalency of both assemblies, the impedance must be equal to the reciprocal of the admittance, the following expressions after some calculations are obtained

$$R_p^{-1} = G = \frac{R_s}{R_s^2 + (\omega^2 C_s^2)^{-1}}, \qquad R_p = R_s\left(1 + \frac{1}{\omega^2 R_s^2 C_s^2}\right)$$

$$C_p = \frac{C_s}{1 + \omega^2 R_s^2 C_s^2}$$

$$R_s = \frac{R_p}{1 + \omega^2 R_p^2 C_p^2} = \frac{G}{G^2 + \omega^2 C_p^2} \tag{4.3.2.3}$$

$$C_s = C_p\left(1 + \frac{G^2}{\omega^2 C_p^2}\right)$$

Since

$$\tan \delta = \omega C_s R_s = \frac{1}{\omega C_p R_p} \tag{4.3.2.4}$$

parameters G, R_s, C_p, and C_2 can be written in terms of $\tan \delta$ as

$$G = \omega C_s \frac{\tan \delta}{1 + \tan^2 \delta}$$

$$R_s = \frac{1}{G} \frac{1}{1 + \tan^{-2} \delta}$$

$$C_p = \frac{C_s}{1 + \tan^2 \delta}$$

$$C_s = C_p(1 + \tan^2 \delta)$$

(4.3.2.5)

4.3.3. Mixed Circuit. Debye Equations

By adding a capacitor of capacitance C_1 in series to a parallel RC_2 assembly, a mixed circuit as in Fig. 4.2 is obtained. The corresponding admittance is given by

$$Y^* = i\omega C_1 + \frac{i\omega C_2}{1 + i\omega C_2 R}$$

(4.3.3.1)

This expression in conjunction with Eq. (4.3.1.1c) leads to a Debye equation if we make $\tau = RC_2$, $C_1 = \varepsilon_\infty C_0$, and $C_2 = (\varepsilon_0 - \varepsilon_\infty)C_0$. It is easy to show that the polarization of this circuit under a steady field gives Eq. (2.11.4), from which the Debye equations are obtained. As shown in Chapter 2, these equations can be obtained in three different ways: (1) on the grounds of some simplifying assumptions concerning rotational Brownian motion (Section 2.10), (2) assuming time-dependent orientational depolarization of a material governed by first-order kinetics (Section 2.11), and (3) from the linear response theory assuming the time dipole correlation function described by a simple decreasing exponential (Section 2.22). The actual expressions are given by

$$\varepsilon^*(\omega) = \varepsilon_\infty + \frac{\varepsilon_0 - \varepsilon_\infty}{1 + i\omega\tau}$$

$$\varepsilon'(\omega) = \varepsilon_\infty + \frac{\varepsilon_0 - \varepsilon_\infty}{1 + \omega^2\tau^2}$$

(4.3.3.2)

$$\varepsilon''(\omega) = \frac{(\varepsilon_0 - \varepsilon_\infty)\omega\tau}{1 + \omega^2\tau^2}$$

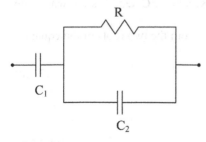

FIG. 4.2 Mixed circuit to represent the single relaxation time response

where the positive sign is taken in the imaginary part. Plots corresponding to the frequency dependence of the permittivity, loss, and loss tangent are shown in reference [4].

It is easy to prove that the loss permittivity attains a maximum given by

$$\varepsilon''_{max} = \frac{\varepsilon_0 - \varepsilon_\infty}{2} = \frac{\Delta\varepsilon}{2} \qquad (4.3.3.3)$$

at

$$\omega = \tau^{-1} \qquad (4.3.3.4)$$

At this frequency, $\varepsilon' = (\varepsilon_0 + \varepsilon_\infty)/2$.

On the other hand, the loss tangent has a maximum given by

$$\tan\delta = \frac{\varepsilon_0 - \varepsilon_\infty}{2(\varepsilon_0\varepsilon_\infty)^{1/2}} \qquad (4.3.3.5)$$

at

$$\omega = \left(\frac{\varepsilon_0}{\varepsilon_\infty}\right)^{1/2} \qquad (4.3.3.6)$$

After eliminating $\omega\tau$ in Eqs (4.4.1.1b) and (4.4.1.1c), the following expression is obtained

$$\left(\varepsilon' - \frac{\varepsilon_0 + \varepsilon_\infty}{2}\right)^2 + \varepsilon''^2 = \left(\frac{\varepsilon_0 - \varepsilon_\infty}{2}\right)^2 \qquad (4.3.3.7)$$

which represents the analytical equation of a circumference of radius $(\varepsilon_0 - \varepsilon_\infty)/2$ and center at $((\varepsilon_0 + \varepsilon_\infty)/2, 0)$. In fact, if an Argand diagram is constructed by plotting $|\varepsilon''|$ against ε', a semicircle of radius $(\varepsilon_0 - \varepsilon_\infty)/2$ with its center located at $((\varepsilon_0 + \varepsilon_\infty)/2, 0)$ is obtained. This circle is called the Debye circle (Fig. 4.3). This type of representation, introduced by K. and R. Cole,[10] is known as the Cole–Cole plot.

The half-width of the peak is calculated from the two roots of the equation

$$\frac{1}{2}\varepsilon''_{max} = \frac{\Delta\varepsilon}{4} = \frac{\Delta\varepsilon}{1 + \omega^2\tau^2}\omega\tau \qquad (4.3.3.8)$$

That is

$$\omega\tau = 2 \pm \sqrt{3} \qquad (4.3.3.9)$$

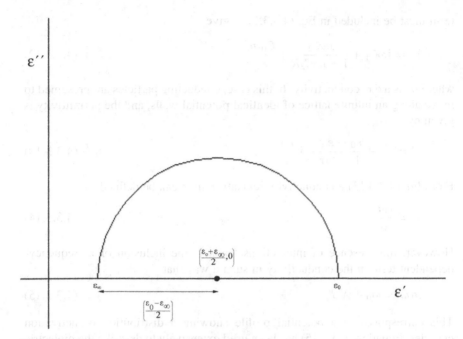

FIG. 4.3   Cole–Cole plot for the Debye dispersion relations

from which the half-width is given by

$$\log\frac{2+\sqrt{3}}{2-\sqrt{3}} \cong 1.144 \qquad (4.3.3.10)$$

Some polar liquids such as chloroform and chlorobenzene show a semicircular behavior described by Eq. (4.3.3.10). From the semicircle, the values of ε_0, ε_∞, and τ can be estimated. Note that

$$\varepsilon' = \varepsilon_\infty + \frac{\varepsilon''}{\omega\tau} \qquad (4.3.3.11)$$

$$\varepsilon_0 - \varepsilon' = \varepsilon''\omega\tau$$

From these linear plots it is also possible to estimate ε_0, ε_∞, and τ.

Under certain circumstances, the admittance is increased on account of hopping conductivity processes. Then, a frequency-independent conductivity term must be included in Eq. (4.3.3.1) to give

$$Y^* = i\omega C_1 + \frac{i\omega C_2}{1 + i\omega C_2 R} + \frac{C_0 \sigma_0}{e_0} \tag{4.3.3.12}$$

where σ_0 is a d.c. conductivity. In this case, conducting particles are presumed to move along an infinite lattice of identical potential wells, and the permittivity is given by

$$\varepsilon^* = \varepsilon_\infty + \frac{\varepsilon_0 - \varepsilon_\infty}{1 + i\omega\tau} - i\frac{\sigma_0}{\omega e_0} \tag{4.3.3.13}$$

From Eq. (4.3.3.13) a conductivity relaxation time can be defined as

$$\tau_\sigma = \frac{e_0 \varepsilon_0}{\sigma_0} \tag{4.3.3.14}$$

However, the presence of interactions leads to the inclusion of a frequency-dependent term in the conductivity in such a way that

$$\sigma(\omega) = \sigma_0 + A\omega^n \tag{4.3.3.15}$$

This corresponds to a potential profile showing a distribution of activation energies. Equation (4.3.3.15) has been used extensively to describe the dielectric properties of ionically conducting glasses.

4.4. EMPIRICAL MODELS TO REPRESENT DIELECTRIC DATA

This section includes a description of several analytic methods together with the extension of the single-relaxation Debye theory to account for the broad relaxations normally observed in bulk polymers.

4.4.1. Retardation Time Spectra

The assumptions upon which the Debye equations are based imply, in practice, that very few systems display Debye behavior. In fact, relaxations in complex and disordered systems deviate from this simple behavior. An important body of literature has consequently merged to give a more or less empirical account of the experimental results. An alternative way to extend the scope of the Debye dispersion relations is to include more than one relaxation time in the physical description of relaxation phenomena. The idea can be formalized by generalizing

Eq. (4.4.1.1a) as follows

$$\varepsilon^* = \varepsilon_\infty + (\varepsilon_0 - \varepsilon_\infty) \int_0^\infty \frac{N(\tau)\,d\tau}{1 + i\omega\tau} \qquad (4.4.1.2)$$

where

$$\int_0^\infty N(\tau)\,d\tau = 1 \qquad (4.4.1.3)$$

is the normalization condition. The term $N(\tau)$ represents the distribution of relaxation (or better retardation) times representing the fraction of the total dispersion that has a retardation time between τ and $\tau + d\tau$. The real and imaginary parts of the complex permittivity are given in terms of the retardation times by (see reference [7], Chapter 9)

$$\varepsilon' = \varepsilon_\infty + (\varepsilon_0 - \varepsilon_\infty) \int_0^\infty \frac{N(\tau)\,d\tau}{1 + \omega^2\tau^2}$$
$$\qquad (4.4.1.4)$$
$$\varepsilon'' = (\varepsilon_0 - \varepsilon_\infty) \int_0^\infty \frac{N(\tau)\omega\tau\,d\tau}{1 + \omega^2\tau^2}$$

Alternatively, the retardation spectrum can be defined as

$$L(\tau) = \tau N(\tau) \qquad (4.4.1.5)$$

In this case, Eqs (4.4.1.2) to (4.4.1.4) can be written as

$$\varepsilon^* = \varepsilon_\infty + (\varepsilon_0 - \varepsilon_\infty) \int_{-\infty}^\infty \frac{L(\tau)\,d\ln\tau}{1 + i\omega\tau}$$

$$\int_{-\infty}^\infty L(\tau)\,d\ln\tau = 1$$
$$\qquad (4.4.1.6)$$
$$\varepsilon' = \varepsilon_\infty + (\varepsilon_0 - \varepsilon_\infty) \int_{-\infty}^\infty \frac{L(\tau)\,d\ln\tau}{1 + \omega^2\tau^2}$$

$$\varepsilon'' = (\varepsilon_0 - \varepsilon_\infty) \int_{-\infty}^\infty \frac{L(\tau)\omega\tau\,d\ln\tau}{1 + \omega^2\tau^2}$$

When the retardation spectrum reduces to a delta function, Eqs (4.4.1.6) become Eqs (4.4.1.1). The integral reduces to a discrete sum for a discrete set of retardation times. Equations (4.4.1.4b) and (4.4.1.6d) can be inverted to express the functions $N(\tau)$ or $L(\tau)$ in terms of the loss permittivity. The result[7] is given by

$$N(\tau) = \frac{L(\tau)}{\tau} = \frac{1}{2\pi(\varepsilon_0 - \varepsilon_\infty)}\left[\varepsilon''\left(\frac{i}{\tau}\right) + \varepsilon''\left(\frac{-i}{\tau}\right)\right] \tag{4.4.1.7}$$

For thermally activated processes, as dipolar relaxations in polymers, the Arrhenius equation leads to

$$\tau = \tau_0 \exp\left(\frac{E}{kT}\right) \tag{4.4.1.8}$$

where E is the activation energy. Therefore, we may define a distribution of activation energies $K(E)$ such that

$$K(E)\,dE = N(\tau)\,d\tau \tag{4.4.1.9}$$

and, from Eq. (4.4.1.8), the following relationship between both types of spectrum holds

$$K(E) = \frac{\tau N(\tau)}{kT} = \frac{L(\tau)}{kT} \tag{4.4.1.10}$$

Equation (4.4.1.10) indicates that, if $K(E)$ is independent of T, then $L(\tau)$ must be dependent on this magnitude. In general, both the maximum and width of the distribution of retardation times change with temperature. The calculation of the retardation time spectra and the distribution of activation energies can, in principle, be made numerically, starting from empirical data, as well as analytically by fitting the experimental data to some empirical equation as described in the following sections. Macdonald and Brachman[11] have described a general treatment.

A general methodology has been outlined in viscoelasticity theory to obtain relaxation or retardation time distributions from the transient or dynamic functions and their derivatives.[12] Following the same strategy, it is also possible to obtain similar relationships for the case of dielectric functions such as the permittivity.

4.4.2. Cole–Cole Equation

As mentioned above, experimental data rarely fit to a Debye semicircle. Studying several organic crystalline compounds, Cole and Cole[10] found that the centers of the experimental arcs were displaced below the real axis, the experimental data thus having the shape of a depressed arc. Consequently, the simple Debye semicircle is not adequate to represent the actual data. To overcome this difficulty, Cole and Cole proposed to use the following equation

$$\varepsilon^* = \varepsilon_\infty + \frac{\varepsilon_0 - \varepsilon_\infty}{1 + (i\omega\tau_0)^{1-\alpha}}, \qquad 0 \leqslant \alpha < 1 \qquad (4.4.2.1)$$

The corresponding electrical circuit in terms of the passive elements is represented in Fig. 4.4. The admittance is given by

$$Y^* = j\omega\left[C_\infty + \frac{C_0 - C_\infty}{1 + (i\omega\tau_0)^{1-\alpha}}\right] \qquad (4.4.2.2)$$

Note that the circuit contains a new element, namely a constant phase element (CPE), the admittance of which is given by

$$Y^*_{CPE} = R^{-1}(i\omega\tau_0)^{\alpha} \qquad (4.4.2.3)$$

The admittance reduces to R^{-1} when $\alpha = 0$. The physical origin of a CPE is the function

$$f(t) = At^{\alpha} \qquad \text{with } 0 \leqslant \alpha < 1 \qquad (4.4.2.4)$$

which is called the parabolic creep function and corresponds to a parabolic shaped function for a creep experiment. In Eq. (4.4.2.4), $A = \alpha!/\tau_0^{\alpha}$. It is easy to show that the Laplace transform of Eq. (4.4.2.4) gives $(i\omega\tau_0)^{-\alpha}$, which is proportional to the impedance of the CPE.

It has been assumed that the presence of a CPE induces a distribution of relaxation times.[13] This is right from a mathematical point of view; however, it is an open question whether we are facing a true distribution of relaxation times

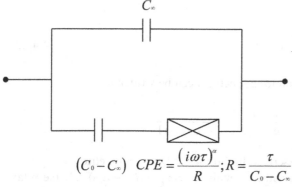

FIG. 4.4 Electrical circuit corresponding to the Cole–Cole equation

or instead a distribution of environmental sites of the molecules, the motions of which cause the relaxation. This problem has recently been addressed by several authors.[14-16]

Splitting Eq. (4.4.2.1) into real and imaginary parts, one obtains

$$\frac{\varepsilon' - \varepsilon_\infty}{\varepsilon_0 - \varepsilon_\infty} = \frac{1 + (\omega\tau_0)^{1-\alpha} \sin \pi\alpha/2}{1 + (\omega\tau_0)^{2(1-\alpha)} + 2(\omega\tau_0)^{1-\alpha} \sin \pi\alpha/2}$$

$$\frac{\varepsilon''}{\varepsilon_0 - \varepsilon_\infty} = \frac{(\omega\tau_0)^{1-\alpha} \cos \pi\alpha/2}{1 + (\omega\tau_0)^{2(1-\alpha)} + 2(\omega\tau_0)^{1-\alpha} \sin \pi\alpha/2}$$

(4.4.2.5)

Eliminating $(\omega\tau_0)^{1-\alpha}$ between these equations, the following equation is obtained

$$\left[\frac{1}{2}(\varepsilon_0 + \varepsilon_\infty) - \varepsilon'\right]^2 + \left[\varepsilon'' + \frac{1}{2}(\varepsilon_0 - \varepsilon_\infty) \tan \frac{\pi\alpha}{2}\right]^2$$

$$= \frac{1}{4}(\varepsilon_0 - \varepsilon_\infty)^2 \sec^2 \frac{\pi\alpha}{2}$$

(4.4.2.6)

which corresponds to a circle with its center at

$$\left[\frac{\varepsilon_0 + \varepsilon_\infty}{2}, -\frac{\varepsilon_0 - \varepsilon_\infty}{2} \tan \frac{\pi\alpha}{2}\right]$$

(4.4.2.7)

and with radius

$$\frac{\varepsilon_0 - \varepsilon_\infty}{2} \sec \frac{\pi\alpha}{2}$$

(4.4.2.8)

The values of ε_0 and ε_∞ can be found from the intercept of the arc with the abscissa axis. The maximum of the loss is located at $\omega\tau_0 = 1$, and its value is given by

$$\varepsilon''_{max} = \frac{\varepsilon_0 - \varepsilon_\infty}{2} \tan \frac{\pi}{4}(1 - \alpha)$$

(4.4.2.9)

The corresponding relaxation time spectrum can be written as[13]

$$G(\tau) = \frac{\sin \alpha\pi}{2\pi\tau\{\cosh[(1 - \alpha) \ln (\tau/\tau_0)] - \cos \pi\alpha\}}$$

(4.4.2.10)

Obviously, the spectrum described by Eq. (4.4.2.10) is symmetric.

According to the Arrhenius formalism, $\tau/\tau_0 = (E_a - E_{a0})/kT$, the relaxation time distribution [Eq. (4.4.2.10)] can be transformed into an activation energy distribution according to the following expression

$$K(E_a) = \frac{\sin \pi \alpha}{2\pi k T \{\cosh[(1-\alpha)(E_a - E_{a0})/(kT)] - \cos \pi \alpha\}} \tag{4.4.2.11}$$

where E_a is the activation energy, T is the absolute temperature, and k is the Boltzmann constant.

It should be noted that CPE can also be used to describe more complex conductivity phenomena. For example, in order to take into account interfacial effects close to the electrodes, a frequency-dependent impedance can be added to the dipolar impedance. To simplify calculations, this dipolar impedance is represented by a parallel RC arrangement in such a way that the total impedance is given by

$$
\begin{aligned}
Z^* = Z_e + Z_{R\|C} &= Z_0(i\omega\tau)^{-\alpha} + \frac{R(1 - i\omega RC)}{1 + (\omega RC)^2} \\
&= \left[Z_0(\omega\tau)^{-\alpha} \cos\frac{\pi}{2}\alpha + \frac{R}{1 + (\omega RC)^2} \right] \\
&\quad - i\left[Z_0(\omega\tau)^{-\alpha} \sin\frac{\pi}{2}\alpha + \frac{\omega C R^2}{1 + (\omega RC)^2} \right]
\end{aligned} \tag{4.4.2.12}
$$

Taking into account the following expression relating the complex permittivity and the impedance

$$\varepsilon^* = (i\omega C_0 Z)^{-1} \tag{4.4.2.13}$$

the permittivity and loss corresponding to Eq. (4.4.2.12) are given by

$$
\varepsilon' = \frac{1}{\omega C_0} \frac{[1 + (\omega CR)^2]\operatorname{Im} Z_e + \omega CR^2 |Z_e|^2}{1 + (\omega CR)^2 + R^2|Z_e|^2 + 2(R\operatorname{Re} Z_e + \omega CR^2 \operatorname{Im} Z_e)}
$$

$$
\varepsilon'' = \frac{1}{\omega C_0} \frac{[1 + (\omega CR)^2]\operatorname{Re} Z_e + R|Z_e|^2}{1 + (\omega CR)^2 + R^2|Z_e|^2 + 2(R\operatorname{Re} Z_e + \omega CR^2 \operatorname{Im} Z_e)}
\tag{4.4.2.14}
$$

where Re and Im denote respectively the real and imaginary parts. It is interesting to note that the asymptotic behavior of Eqs (4.4.2.14)

$$
\begin{aligned}
\omega \to 0, &\quad \varepsilon' \to \omega^{-(1+\alpha)} \quad \varepsilon'' \to \omega^{-1} \\
\omega \to \infty, &\quad \varepsilon', \varepsilon'' \to 0
\end{aligned} \tag{4.4.2.15}
$$

agrees fairly well with the current experimental data. More complex methods to fit the actual data to empirical or semiempirical equations have been proposed.[17,18]

4.4.3. Fuoss–Kirkwood Equation

By noting that the imaginary part of the permittivity in the Debye model can be written as

$$\frac{\varepsilon''(\omega)}{\varepsilon''_{max}} = \text{sech} \ln\left(\frac{\omega}{\omega_0}\right) \tag{4.4.3.1}$$

Fuoss and Kirkwood[19] proposed the following extended equation to represent symmetric relaxations

$$\frac{\varepsilon''(\omega)}{\varepsilon''_{max}} = \text{sech} m \ln\left(\frac{\omega}{\omega_0}\right) \tag{4.4.3.2}$$

or equivalently

$$\varepsilon''(\omega) = 2\varepsilon''_{max} \frac{(\omega\tau_0)^m}{1 + (\omega\tau_0)^{2m}} \tag{4.4.3.3}$$

from which the maximum loss can be obtained as

$$\varepsilon''_{max} = \frac{(\varepsilon_0 - \varepsilon_\infty)m}{2} \tag{4.4.3.4}$$

The Fuoss–Kirkwood equation can be expressed in terms of the temperature by using a simple Arrhenius dependence for the relaxation time, that is, $\tau/\tau_0 = \exp E_a/RT$, as follows

$$\varepsilon''(T) = \varepsilon''_{max} \text{sech} \, m \frac{E_a}{R}\left(\frac{1}{T} - \frac{1}{T_{max}}\right) \tag{4.4.3.5}$$

By comparing Eqs (5.4.2.9) and (5.4.3.4), the parameters of the Cole–Cole and Fuoss–Kirkwood equations are related by

$$m = \tan\frac{(1 - \alpha)\pi}{4} \tag{4.4.3.6}$$

From an experimental point of view, m can be found from a plot of $\cosh^{-1}(\varepsilon''_{max}/\varepsilon'')$ vs. $\ln \omega$.

The distribution of relaxation times for the Fuoss–Kirkwood equation is given by[13]

$$G(\tau) = \frac{2}{\pi} \frac{\cos(\pi\alpha/2)\cosh(\alpha\tau/\tau_0)}{\cos^2(\pi\alpha/2) + \sinh^2(\alpha\tau/\tau_0)} \tag{4.4.3.7}$$

The real part of the complex permittivity corresponding to the Fuoss–Kirkwood equation could in principle be found from the Krönig–Kramers equations,[20] but the equation obtained for ε' is not a closed expression.

4.4.4. Davidson–Cole Equation

The Cole–Cole and Fuoss–Kirkwood equations are very useful for symmetric relaxations. However, actual curves obtained from ε'' vs. ε' plots show "skewness" on the high-frequency side. For this reason, Davidson and Cole proposed to fit the experimental data to the following empirical equation[21]

$$\varepsilon^* = \varepsilon_\infty + \frac{\varepsilon_0 - \varepsilon_\infty}{(1 + i\omega\tau_0)^\beta} \tag{4.4.4.1}$$

The real and imaginary parts are

$$\varepsilon' = \varepsilon_\infty + (\varepsilon_0 - \varepsilon_\infty)\cos\beta y \cos^\beta y$$

$$\varepsilon'' = (\varepsilon_0 - \varepsilon_\infty)\sin\beta y \cos^\beta y \tag{4.4.4.2}$$

where $y = \tan^{-1}\omega\tau_0$. At high frequencies

$$\lim_{\omega\to\infty} \frac{\varepsilon^* - \varepsilon_\infty}{\varepsilon_0 - \varepsilon_\infty} \cong (i\omega\tau_0)^{-\beta} \tag{4.4.4.3}$$

Therefore, β can be determined from the slope of the angle of the arc with the abscissa at high frequencies. In fact, differentiation of Eqs (4.4.4.2) with respect to y and further division of the expressions obtained yields

$$\frac{d\varepsilon''}{d\varepsilon'} = -\cotan(1 + \beta)y \tag{4.4.4.4}$$

In the limit $\omega \to \infty$, that is, $y \to \infty$, the following result holds

$$\lim_{y\to\infty} \frac{d\varepsilon''}{d\varepsilon'} = \tan\frac{\pi}{2}\beta \tag{4.4.4.5}$$

On the other hand (see Fig. 4.5)

$$\lim_{\omega \to \infty} \frac{\varepsilon''}{\varepsilon' - \varepsilon_\infty} = \tan \frac{\pi}{2} \beta \qquad (4.4.4.6)$$

and

$$\frac{\varepsilon''(\omega = \tau_0^{-1})}{\varepsilon'(\omega = \tau_0^{-1}) - \varepsilon_\infty} = \tan \frac{\pi}{4} \beta \qquad (4.4.4.7)$$

The product of the frequency at which the maximum loss is located and the mean relaxation time is given by

$$\omega_{max} \tau_0 = \tan \frac{\pi}{2(1 + \beta)} \qquad (4.4.4.8)$$

and the value of the loss at the peak maximum can be written as

$$\varepsilon''_{max} = (\varepsilon_0 - \varepsilon_\infty) \sin \frac{\pi \beta}{2(1 + \beta)} \cos^\beta \frac{\pi}{2(1 + \beta)} \qquad (4.4.4.9)$$

Finally, the distribution of relaxation times is given by[4]

$$G(\tau) = \frac{\sin \beta \pi}{\pi} \left(\frac{\tau}{\tau_0 - \tau} \right)^\beta \qquad (4.4.4.10)$$

4.4.5. Havriliak–Negami Equation

The generalization of the Cole–Cole and Davidson–Cole equations was proposed by Havriliak and Negami (HN).[22] The flexibility of the HN five-parameter equation makes it one of the most widely used methods of representing

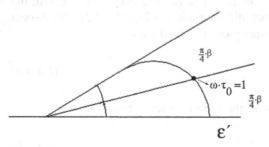

FIG. 4.5 Cole–Cole arc for the Davidson–Cole equation

dielectric relaxation data. The formal expression is

$$\varepsilon^* = \varepsilon_\infty + \frac{\varepsilon_0 - \varepsilon_\infty}{[1 + (i\omega\tau_0)^{1-\alpha}]^\beta} \qquad (4.4.5.1)$$

After splitting into real and imaginary parts, we obtain

$$\varepsilon' = \varepsilon_\infty + \frac{(\varepsilon_0 - \varepsilon_\infty)\cos\beta\phi}{1 + 2(\omega\tau_0)^{1-\alpha}\sin\left(\frac{\pi}{2}\right)\alpha + (\omega\tau_0)^{2(1-\alpha)}}$$

$$\varepsilon'' = \frac{(\varepsilon_0 - \varepsilon_\infty)\sin\beta\phi}{1 + 2(\omega\tau_0)^{1-\alpha}\sin\left(\frac{\pi}{2}\right)\alpha + (\omega\tau_0)^{2(1-\alpha)}}$$

$$(4.4.5.2)$$

with

$$\phi = \tan^{-1}\frac{(\omega\tau_0)^{1-\alpha}\cos\left(\frac{\pi}{2}\right)\alpha}{1 + (\omega\tau_0)^{1-\alpha}\sin\left(\frac{\pi}{2}\right)\alpha} \qquad (4.4.5.3)$$

Note that

$$\lim_{\omega\to\infty}\frac{\varepsilon''}{\varepsilon' - \varepsilon_\infty} = \lim_{\omega\to\infty}\tan\beta\phi = \tan\beta\cot^{-1}\frac{\pi}{2}\alpha$$

$$= \tan\frac{\beta(1-\alpha)\pi}{2} \qquad (4.4.5.4)$$

On the other hand

$$\tan^{-1}\frac{\varepsilon''(\omega\tau_0 = 1)}{\varepsilon'(\omega\tau_0 = 1) - \varepsilon_\infty} = \frac{\beta(1-\alpha)\pi}{2} \qquad (4.4.5.5)$$

On the low-frequency side, the HN equation reduces to

$$\varepsilon^* \cong \varepsilon_0 - (\varepsilon_0 - \varepsilon_\infty)\beta(i\omega\tau_0)^{1-\alpha} + O(\omega)^{1-\alpha} \qquad (4.4.5.6)$$

where ε' and $\varepsilon'' \approx \omega^{1-\alpha}$, O denotes high orders, and $\approx$ expresses approximation. At high frequencies

$$\varepsilon^* \cong \varepsilon_\infty + (\varepsilon_0 - \varepsilon_\infty)(i\omega\tau_0)^{-\beta(1-\alpha)} + O(\omega)^{-\beta(1-\alpha)} \qquad (4.4.5.7)$$

The frequency at which the loss permittivity attains a maximum can be obtained from $d\varepsilon''/d\omega = 0$. This condition applied to Eq. (4.4.5.2b) gives

$$\omega_{max}\tau_0 = \left[\frac{\sin\{[\pi(1-\alpha)]/[2(1+\beta)]\}}{\sin\{[\pi(1-\alpha)\beta]/[2(1+\beta)]\}}\right]^{1/\alpha} \qquad (4.4.5.8)$$

which reduces to the values obtained for Cole–Cole and Davidson–Cole Eq. (4.4.4.8), taking, respectively, $\alpha = 0$ and $\beta = 1$ in Eq. (4.4.5.8). The value of the maximum loss at the frequency given by Eq. (4.4.5.8) is

$$\varepsilon_{max}'' = (\varepsilon_0 - \varepsilon_\infty) \frac{\left(\sin\left\{ [\pi(1 - \alpha)\beta]/[2(1 + \beta)] \right\} \right)^{1+\beta}}{\left(\sin\left\{ [\pi(1 - \alpha)]/2 \right\} \right)^{\beta}} \tag{4.4.5.9}$$

The corresponding distribution of relaxation times is given by[23]

$$G(\ln \tau) = \frac{1}{\pi} \frac{(\tau/\tau_0)^{\beta(1-\alpha)} \sin \beta\phi}{[1 + 2(\tau/\tau_0)^{1-\alpha} \cos \pi(1 - \alpha) + (\tau/\tau_0)^{2(1-\alpha)}]^{\beta/2}} \tag{4.4.5.10}$$

4.4.6. Jonscher Model

In the context of the "universal dielectric response" (UDR) theory, Jonscher[24] proposed the following expression for the loss permittivity

$$\varepsilon'' = \frac{a}{(\omega/\omega_1)^{-m} + (\omega/\omega_2)^{1-n}} \tag{4.4.6.1}$$

where $m, n < 1$, and ω_1 and ω_2 are characteristic frequencies. Note that this equation becomes the Fuoss–Kirkwood equation for $\omega_1 = \omega_2$ and $m + n = 1$. For this reason, independent of the physical meaning of the frequency power dependence of the loss proposed by the UDR, Eq. (4.4.6.1) can be envisaged as a generalization of the Fuoss–Kirkwood equation. The frequency at which the loss attains its maximum value is given by

$$\omega_{max} = \left(\frac{m}{1 - n} \omega_1^m \omega_2^{1-n} \right)^{1/(m-n+1)} \tag{4.4.6.2}$$

which reduces to

$$\omega_{max} = \omega_0 \left(\frac{m}{1 - n} \right)^{1/(m+1-n)} \tag{4.4.6.3}$$

if $\omega_1 = \omega_2 = \omega_0$, and to $\omega_{max} = \omega_0$ if $m + n = 1$ in Eq. (4.4.6.3). As a consequence, the value of parameter a appearing in Eq. (4.4.6.1) is given by

$$a = \varepsilon_{max}'' \left(\frac{\omega_1}{\omega_2} \right)^{m(1-n)/(m+1-n)}$$
$$\times \left[\left(\frac{m}{1 - n} \right)^{-[m/(m-n+1)]} + \left(\frac{m}{1 - n} \right)^{(1-n)/(m-n+1)} \right] \tag{4.4.6.3}$$

which, for $\omega_1 = \omega_2$, reduces to

$$a = \varepsilon''_{max} \left[\left(\frac{m}{1-n} \right)^{-[m/(m+1-n)]} + \left(\frac{m}{1-n} \right)^{(1-n)/(m+1-n)} \right] \qquad (4.4.6.4)$$

If $m + n = 1$, Eq. (4.4.6.4) gives $a = 2\varepsilon''_{max}$.

The limiting frequency behavior of the Jonscher model is given by

$$\lim_{\omega \to 0} \frac{d\varepsilon''}{d\ln(\omega/\omega_1)} = m$$

$$(4.4.6.5)$$

$$\lim_{\omega \to \infty} \frac{d\varepsilon''}{d\ln(\omega/\omega_2)} = n - 1$$

A formally equivalent equation to (4.4.6.1) for $\omega_1 = \omega_2 = \tau_0^{-1}$ has been recently proposed by Schneider.[25]

4.4.7. Hill Model

Hill[26] proposed the following equation to represent the dielectric loss

$$\frac{\varepsilon''}{\varepsilon_0 - \varepsilon_\infty} = \frac{\omega^m}{(\omega_p^{2s} + \omega^{2s})^{(1-n+m)/(2s)}} \qquad (4.4.7.1)$$

where $0 < n, m, s < 1$. The necessity for the three parameters m, n and s is a consequence of the shape of the loss curves which are characterized by two power laws and an intermediate curved region. Note that Eq. (4.4.7.1) becomes the Debye equation for $m = 1 - n = s$, the Fuoss–Kirkwood or Cole–Cole equations for $m = 1 - n$, and the Davidson–Cole equation for $m = 1$ and $\beta_{DC} = 1 - n$.

The frequency of the maximum loss is given by

$$\omega_{max} = \omega_p \left(\frac{m}{1-n} \right)^{1/(2s)} \qquad (4.4.7.2)$$

from which it is clear that $\omega_{max} = \omega_p$ only when $m = 1 - n$. The corresponding value for the maximum loss is given by

$$\frac{\varepsilon''_{max}}{\varepsilon_0 - \varepsilon_\infty} = \frac{\omega_{max}^{n-1}}{[(1-n+m)/m]^{(1-n+m)/(2s)}} \qquad (4.4.7.3)$$

which reduces to

$$\frac{\varepsilon''_{\max}}{\varepsilon_0 - \varepsilon_\infty} = (2)^{-1/s} \omega_{\max}^{n-1} \qquad (4.4.7.4)$$

if $1 - n = m$. Equation (4.4.7.1) can alternatively be written as a function of the maximum frequency by

$$\frac{\varepsilon''_{\max}}{\varepsilon_0 - \varepsilon_\infty} = \frac{\omega^m}{(\omega_{\max}^{2s}(1 - n)/m + \omega^{2s})^{(1-n+m)/(2s)}} \qquad (4.4.7.5)$$

The limiting frequency behavior is given by

$$\lim_{\omega \to 0} \frac{d \ln \varepsilon''}{d \ln \omega} = m$$

$$\lim_{\omega \to \infty} \frac{d \ln \varepsilon''}{d \ln \omega} = n - 1 \qquad (4.4.7.6)$$

4.4.8. KWW Model

As mentioned in Chapter 2, Williams and Watts[27] proposed to use a stretched exponential for the decay function $\phi(t)$, in a similar way to Kohlrausch many years ago. In this way, the normalized dielectric permittivity can be written as

$$\frac{\varepsilon^* - \varepsilon_\infty}{\varepsilon_0 - \varepsilon_\infty} = 1 - i\omega L[\phi(t)] \qquad (4.4.8.1)$$

where L denotes the Laplace transformation. The resulting expression does not have a closed form but can be expressed as a series expansion[28]

$$\frac{\varepsilon^* - \varepsilon_\infty}{\varepsilon_0 - \varepsilon_\infty} = \sum_{n=1}^{\infty} \frac{(-1)^{n-1}}{(\omega\tau)^{\beta n}} \frac{\Gamma(n\beta + 1)}{\Gamma(n+1)} \exp\left(in\beta \frac{\pi}{2}\right) \qquad (4.4.8.2)$$

where Γ is the gamma function. For $\beta = 1$ the Debye equations are recovered. From Eq. (4.4.8.2) it follows that the angles of intersection in a Cole–Cole plot are $\pi/2$ and $\beta\pi/2$ at low and high frequencies respectively.

For low values of $\omega\tau$ and $\beta > 0.25$, the convergence of the series in Eq. (4.4.8.2) is slow, and the following equation is proposed[28]

$$\frac{\varepsilon^* - \varepsilon_\infty}{\varepsilon_0 - \varepsilon_\infty} = \sum_{n=1}^{\infty} (-1)^{n-1} \frac{(\omega\tau_0)^{n-1}}{\Gamma(n)} \Gamma\left(\frac{n + \beta - 1}{\beta}\right) \exp\left(\frac{i\pi(n-1)}{2}\right)$$

$$(4.4.8.3)$$

The KWW equations are nonsymmetrical in shape and for this reason it is particularly useful to describe the nonsymmetrical α-relaxations. The corresponding distribution function has been given by Montroll and Bendler.[29] A simple expression for $\beta = 1/2$ can be obtained, given by

$$G(\tau) = \frac{1}{2\pi^{1/2}} \left(\frac{\tau}{\tau_0}\right)^{1/2} \exp\left(-\frac{\tau}{4\tau_0}\right) \tag{4.4.8.4}$$

4.4.9. Dissado–Hill Model

Based on the regression properties of the response function of a linear system, Dissado and Hill[30] proposed the following equation for the susceptibility

$$\varepsilon^* \cong \frac{\omega_p^{1-n}}{(\omega_p + i\omega)^{1-n}} \, {}_2F_1\left(1 - n, \, 1 - m; \, 2 - n; \, \frac{\omega_p}{(\omega_p + i\omega)}\right) F_0^{-1} \tag{4.4.9.1}$$

where ${}_2F_1(\ldots, \, \ldots; \, \ldots; \ldots)$ is the Gaussian hypergeometric function and

$$F_0 = \frac{\Gamma(2 - n)\Gamma(m)}{\Gamma(1 + m - n)} \tag{4.4.9.2}$$

This equation is a consequence of the presence of power-law behavior in the frequency response of the dielectric permittivity of solids as a general phenomenon, as proposed by Jonscher [see Eq. (4.4.6)]. It can also be considered a generalization of the Hill equation. The limiting behavior is given by

$$\lim_{\omega \to 0} \frac{\varepsilon^* - \varepsilon_\infty}{\varepsilon_0 - \varepsilon_\infty} = \omega^m$$

$$\lim_{\omega \to \infty} \frac{\varepsilon^* - \varepsilon_\infty}{\varepsilon_0 - \varepsilon_\infty} = \omega^{n-1} \tag{4.4.9.3}$$

In a recent paper, Dissado et al.[31] have compared the present model to the KWW and Havriliak–Negami empirical equations.

Van Turnhout (unpublished results, 1984) derived the relaxation time distribution function from the Dissado–Hill function. The results are

$$G\left(\ln\frac{\tau}{\tau_0}\right) = \frac{\sin(1-n)\pi}{\pi F_0}\left(\frac{\tau}{\tau_0}\right)^{1-n}$$

$$\times\ _2F_1\left(1-n,\ 1+m-n;\ 2-n;\ \frac{\tau}{\tau_0}\right) \qquad \text{for } \left(\frac{\tau}{\tau_0}\right) < 1$$

$$G\left(\ln\frac{\tau}{\tau_0}\right) = \frac{\sin m\pi}{\pi F_0}\left(\frac{1-n}{m}\right)\left(\frac{\tau}{\tau_0}\right)^{-m}$$

$$\times\ _2F_1\left(m,\ 1+m-n;\ 1+m;\ \frac{\tau_0}{\tau}\right) \qquad \text{for } \left(\frac{\tau}{\tau_0}\right) > 1$$

$$(4.4.9.4)$$

A nasty problem in these distributions is the appearance of a cusp or infinity value for $\tau = \tau_0$. It seems clear that the non-Debye shape is related to a specific type of broadening of the delta function appropriate for the Debye process, revealed via the mentioned cusp, rather than to an arbitrary set of relaxation times.

4.4.10. Friedrich Model

Among the most recent models, we note that proposed by Friedrich[32] in which the dynamic modulus is expressed by

$$M^* = M_0 + (M_\infty - M_0)\frac{(i\omega\tau)^d}{1+(i\omega\tau)^c} \qquad (4.4.10.1)$$

where $c > d$. Note that if $c = d$ we recover a Cole–Cole type equation. The corresponding compliance or permittivity function is given by

$$\varepsilon^* = \frac{\varepsilon_0\varepsilon_\infty(1+x^c)}{\varepsilon_\infty(1+x^c)+\Delta\varepsilon x^d} \qquad (4.4.10.2)$$

where $x = iy = i\omega\tau$, $\Delta\varepsilon = \varepsilon_0 - \varepsilon_\infty$, $\varepsilon_0 = M_0^{-1}$, and $\varepsilon_\infty = M_\infty^{-1}$. Splitting the complex permittivity into the real and imaginary parts gives

$$\varepsilon' = \varepsilon_0\varepsilon_\infty \frac{\varepsilon_\infty(1 + 2y^c \cos c(\pi/2) + y^{2c}) + \Delta\varepsilon y^d[\cos d(\pi/2) \atop +y^c \cos(d-c)(\pi/2)]}{\varepsilon_\infty^2(1 + 2y^c \cos c(\pi/2) + y^{2c}) + 2\varepsilon_\infty\Delta\varepsilon y^d[\cos d(\pi/2) \atop +y^c \cos(d-c)(\pi/2)] + \Delta\varepsilon^2 y^{2d}}$$

$$\varepsilon'' = \Delta\varepsilon \frac{\varepsilon_0\varepsilon_\infty[y^c \sin(d-c)(\pi/2) + \sin d(\pi/2)]}{\varepsilon_\infty^2(1 + 2y^c \cos c(\pi/2) + y^{2c}) + 2\varepsilon_\infty\Delta\varepsilon y^d(\cos d(\pi/2) \atop +y^c \cos(d-c)(\pi/2)) + \Delta\varepsilon^2 y^{2d}}$$

$$(4.4.10.3)$$

The limiting behavior of these functions is as follows

$$
\begin{aligned}
&\omega \to 0, &&\varepsilon' \to \varepsilon_0, &&\varepsilon'' \to 0 \\
&\omega \to \infty, &&\varepsilon' \to 0 &&\text{if } d > c, &&\varepsilon' \to \varepsilon_\infty &&\text{if } d = c \\
&&&\varepsilon'' \to 0
\end{aligned}
$$

$$(4.4.10.4)$$

Note that, when $\omega \to \infty, \varepsilon' \to \varepsilon_0$ if $d < c$, a conclusion without physical meaning. On the other hand, it is clear from the high-frequency behavior that the model predicts a zero value for the real part of the complex permittivity for $d > c$. Obviously, this is not the actual case. In the same way, Eq. (4.4.10.3) predict a constant and finite value for $\tan \delta[\tan \delta = \tan(d-c)\pi/2]$ when $d > c$. For these reasons, the present model is only useful when Eq. (4.10.2) reduces to

$$\varepsilon^* = \frac{\varepsilon_0\varepsilon_\infty}{\varepsilon_\infty + \Delta\varepsilon(i\omega\tau)^c/[1 + (i\omega\tau)^c]} \qquad (4.4.10.5)$$

when $c = d$. If we define $\tau' = \tau(\varepsilon_0/\varepsilon_\infty)^{1/c}$, Eq. (4.4.10.5) becomes the Cole–Cole equation

$$\varepsilon^* = \varepsilon_\infty + \frac{\varepsilon_0 - \varepsilon_\infty}{1 + (i\omega\tau')^c} \qquad (4.4.10.6)$$

By using the modulus formalism, Eq. (4.4.10.1) can give other interesting alternatives consistent with mechanical measurements for $d > c$.

4.4.11. Model of Metzler, Schick, Kilian, and Nonnenmacher

A new model has recently been proposed by Metzler, Schick, Kilian and Nonnenmacher[33] in which dynamic mechanical data are given by

$$
M^* = M_0 + \frac{M_m + M_e(i\omega\tau)^{-\mu}}{1 + (i\omega\tau)^{-q}}
\tag{4.4.11.1}
$$

where $q \geq \mu$. Note that this model includes a new parameter and is slightly more general than the Friedrich model.[32] The corresponding compliance or permittivity function can be written as

$$
\varepsilon^* = \varepsilon_0 \frac{1 + (i\omega\tau)^q}{1 + (1 + (\varepsilon_0/\varepsilon_m))(i\omega\tau)^q + \varepsilon_0/\varepsilon_e(i\omega\tau)^{q-\mu}}
\tag{4.4.11.2}
$$

The real and imaginary parts of the complex permittivity function are

$$
\varepsilon' = \varepsilon_0 \frac{
\begin{bmatrix}
1 + \left(2 + \dfrac{\varepsilon_0}{\varepsilon_m}\right) y^q \cos q\dfrac{\pi}{2} + \dfrac{\varepsilon_0}{\varepsilon_e} y^{q-\mu} \cos(q-\mu)\dfrac{\pi}{2} \\[2mm]
+ \left(1 + \dfrac{\varepsilon_0}{\varepsilon_m}\right) y^{2q} + \dfrac{\varepsilon_0}{\varepsilon_e} y^{2q-\mu} \cos \mu\dfrac{\pi}{2}
\end{bmatrix}
}{
\begin{bmatrix}
1 + \left(1 + \dfrac{\varepsilon_0}{\varepsilon_m}\right)^2 y^{2q} + \left(\dfrac{\varepsilon_0}{\varepsilon_e}\right)^2 y^{2(q-\mu)} + 2\left(1 + \dfrac{\varepsilon_0}{\varepsilon_m}\right) y^q \cos q\dfrac{\pi}{2} \\[2mm]
+ 2\dfrac{\varepsilon_0}{\varepsilon_e} y^{q-\mu} \cos (q-\mu)\dfrac{\pi}{2} + 2\left(1 + \dfrac{\varepsilon_0}{\varepsilon_m}\right)\dfrac{\varepsilon_0}{\varepsilon_e} y^{2q-\mu} \cos \mu\dfrac{\pi}{2}
\end{bmatrix}
}
$$

$$
\varepsilon'' = \varepsilon_0 \frac{
\dfrac{\varepsilon_0}{\varepsilon_e} y^{2q-\mu} \sin \mu\dfrac{\pi}{2} - \dfrac{\varepsilon_0}{\varepsilon_m} y^q \sin q\dfrac{\pi}{2} - \dfrac{\varepsilon_0}{\varepsilon_e} y^{q-\mu} \sin (q-\mu)\dfrac{\pi}{2}
}{
\begin{bmatrix}
1 + \left(1 + \dfrac{\varepsilon_0}{\varepsilon_m}\right)^2 y^{2q} + \left(\dfrac{\varepsilon_0}{\varepsilon_e}\right)^2 y^{2(q-\mu)} + 2\left(1 + \dfrac{\varepsilon_0}{\varepsilon_m}\right) y^q \cos q\dfrac{\pi}{2} \\[2mm]
+ 2\dfrac{\varepsilon_0}{\varepsilon_e} y^{q-\mu} \cos (q-\mu)\dfrac{\pi}{2} + 2\left(1 + \dfrac{\varepsilon_0}{\varepsilon_m}\right)\dfrac{\varepsilon_0}{\varepsilon_e} y^{2q-\mu} \cos \mu\dfrac{\pi}{2}
\end{bmatrix}
}
\tag{4.4.11.3}
$$

where we have used the same notation as in the previous equation. The limiting behavior of these equations is as follows

$$
\begin{aligned}
\omega \to 0, \qquad & \varepsilon' \to \varepsilon_0, \qquad & \varepsilon'' \to 0 \\[2mm]
\omega \to \infty, \qquad & \varepsilon' \to \frac{\varepsilon_0 \varepsilon_m}{\varepsilon_0 + \varepsilon_m}, \qquad & \varepsilon'' \to 0
\end{aligned}
\tag{4.4.11.4}
$$

From these results it is clear that $\varepsilon_m < \varepsilon_0$ and

$$\varepsilon_m = \frac{\varepsilon_0 \varepsilon_\infty}{\varepsilon_0 - \varepsilon_\infty} \tag{4.4.11.5}$$

When $q = \mu$, Eq. (4.4.11.2) becomes

$$\varepsilon^* = \varepsilon_0 \frac{1 + (i\omega\tau)^q}{1 + \varepsilon_0/\varepsilon_e + (1 + \varepsilon_0/\varepsilon_m)(i\omega\tau)^q} \tag{4.4.11.6}$$

the limiting behavior of which is given by

$$\omega \to 0, \qquad \varepsilon' \to \frac{\varepsilon_0 \varepsilon_e}{\varepsilon_0 + \varepsilon_e}, \qquad \varepsilon'' \to 0$$

$$\omega \to \infty, \qquad \varepsilon' \to \frac{\varepsilon_0 \varepsilon_m}{\varepsilon_0 + \varepsilon_m}, \qquad \varepsilon'' \to 0 \tag{4.4.11.7}$$

These expressions indicate that, if $\varepsilon_{relaxed} \neq \varepsilon_0$, Eq. (4.4.11.5) holds, and $\varepsilon_e > \varepsilon_m$.

4.4.12. Biparabolic Model

Huet[34] proposed a representation of the dynamic viscoelastic properties of some bituminous asphalts by means of a biparabolic model

$$E^* = E_0 + \frac{E_\infty - E_0}{1 + \delta(i\omega\tau_0)^{-k} + (i\omega\tau_0)^{-h}} \tag{4.4.12.1}$$

that includes one parameter more than the HN equation. In this equation, $0 < k < h < 1$. Equation (4.4.12.1) formally coincides with the limiting behavior of a viscoelastic material whose response has been modelled by using the annihilation of defects in the matrix of that material.[35] The corresponding compliance or permittivity is given by

$$\varepsilon^* = \frac{\varepsilon_0 \varepsilon_\infty \left[1 + \delta(i\omega\tau_0)^{-k} + (i\omega\tau_0)^{-h} \right]}{\varepsilon_0 + \varepsilon_\infty \left[\delta(i\omega\tau_0)^{-k} + (i\omega\tau_0)^{-h} \right]} \tag{4.4.12.2}$$

To describe the properties of the model, it is convenient to make some rearrangements in Eq. (4.4.12.2), giving

$$\varepsilon^* = \varepsilon_\infty + \frac{(\varepsilon_0 - \varepsilon_\infty)\left[1 + \delta'(i\omega\tau_0')^{h-k} \right]}{1 + \delta'(i\omega\tau_0')^{h-k} + (i\omega\tau_0')^{h}} \tag{4.4.12.3}$$

or in an even more convenient way

$$\varepsilon^* = \varepsilon_0 - \frac{\varepsilon_0 - \varepsilon_\infty}{1 + \delta'(i\omega\tau_0')^{-k} + (i\omega\tau_0')^{-h}} \qquad (4.4.12.4)$$

where

$$\delta' = \delta\left(\frac{\varepsilon_0}{\varepsilon_\infty}\right)^{(k/h)-1} \qquad \text{and} \qquad \tau_0' = \tau_0\left(\frac{\varepsilon_0}{\varepsilon_\infty}\right)^{1/h} \qquad (4.4.12.5)$$

Splitting ε^* into its real and imaginary parts yields

$$\varepsilon' = \varepsilon_\infty + (\varepsilon_0 - \varepsilon_\infty)$$

$$\times \frac{\left[\begin{array}{c} 1 + \delta' y^{2h-k}\cos k\frac{\pi}{2} + y^h\cos h\frac{\pi}{2} \\ + 2\delta' y^{h-k}\cos(h-k)\frac{\pi}{2} + \delta'^2 y^{2(h-k)} \end{array}\right]}{1 + 2\left[\delta' y^{h-k}\cos(h-k)\frac{\pi}{2} + y^h\cos h\frac{\pi}{2} + \delta' y^{2h-k}\cos k\frac{\pi}{2}\right]} \\ + \delta'^2 y^{2(h-k)} + y^{2h}$$

$$\varepsilon'' = (\varepsilon_0 - \varepsilon_\infty)$$

$$\times \frac{\delta' y^{2h-k}\sin k\frac{\pi}{2} + y^h\sinh\frac{\pi}{2}}{1 + 2\left[\delta' y^{h-k}\cos(h-k)\frac{\pi}{2} + y^h\cosh\frac{\pi}{2} + \delta' y^{2h-k}\cos k\frac{\pi}{2}\right]} \\ + \delta'^2 y^{2(h-k)} + y^{2h}$$

$$(4.4.12.6)$$

where $y = \omega\tau_0'$.

The relaxation time spectrum is given by

$$G(\tau) = \frac{1}{\pi}\left(\frac{\tau}{\tau_0'}\right)^k$$

$$\times \frac{\sin h\pi\left(\frac{\tau}{\tau_0'}\right)^{h-k} + \delta'\sin k\pi}{\left(\frac{\tau}{\tau_0'}\right)^{2h} + \delta'^2\left(\frac{\tau}{\tau_0'}\right)^{2k} + 2\delta'\cos(h-k)\pi\left(\frac{\tau}{\tau_0'}\right)^{h+k}} \qquad (4.4.12.7) \\ + 2\cos h\pi\left(\frac{\tau}{\tau_0'}\right)^h + 2\delta'\cos k\pi\left(\frac{\tau}{\tau_0'}\right)^k + 1$$

The arc intersects the abscissa axis with angles $k\pi/2$ and $-h\pi/2$ on the high- and the low-frequency sides respectively. In this way, a comparison with the parameters corresponding to the Havriliak–Negami equation can be made, obtaining

$$\begin{aligned} \alpha &= h \\ \alpha\beta &= k \end{aligned} \qquad (4.4.12.8)$$

More details about the characteristics and properties together with the corresponding electrical circuit of this type of representation of the dielectric data can be found in reference [36].

4.5. THERMOSTIMULATED CURRENTS

4.5.1. Electrets

An electret is a piece of dielectric material having quasi-permanent electrical charges. As a consequence, electrets have intrinsic theoretical interest as well as important technical applications. Electret charges can be dipolar in origin or space charges that with time form layers of surface charges or are trapped in the material in bulk. Electrets can be viewed as the electrostatic counterparts of magnets, the only difference between them being that magnetic properties only arise from magnetic dipoles. The classical electrets were made of carnauba wax or similar substances. However, polymers are good candidates to form electrets. These materials can contain permanent dipoles, and, under electric fields, induced dipoles can appear in the polymeric matrix. At the same time the presence of free ions as well as electrons cannot in principle be ruled out. Artificial polymer electrets can be formed by heating the material at temperatures well above T_g and further applying an electric field that polarizes the electric charges and reorients the dipoles. Then, the material is cooled under the applied electric field. In this way, free charges are trapped by structural defects, the dipoles remain frozen, and charges are stored via a nearly permanent polarization of the material. In the present context the words "nearly permanent" indicate that the time constant for the decay of these charges is much longer than the experimental time. The specimen thus formed is called a thermoelectret. Although it was Faraday[37] who first theorized about electrets (the name was coined by Heaviside[38]), Eguchi[39] was the first to begin systematic research on electrets in the first decades of the 20th century.

4.5.2. Thermostimulated Depolarization and Polarization

As mentioned above, the charges are virtually immobile at low temperatures. To stimulate the electrical discharge, electrets must be heated in order that the charges recover their mobility. This process is called thermal stimulation. In fact, when the electret is heated, a weak electric current is generated that, plotted as a function of the temperature, gives a curve with peaks. The size, shape, and location of these peaks are a characteristic of both the material and the parameters of the thermostimulated process. They are also strongly dependent on the way the electrets are prepared. Analysis of the peaks yields information on the dipolar relaxation and other charge carriers. Figure 46a shows the field, temperature and current during the process of electret formation and during the discharge process.

The thermostimulated depolarization technique was introduced by Gross[40] but was first applied by Bucci et al.[41] under the name of ionic thermo-currents. Thermostimulated depolarization (TSD) currents as a method of research is similar to other thermostimulated processes used in solid-state physics[42] such as thermoluminescence, thermogravimetry, and differential thermal analysis. It has in common with these methods the same advantages and drawbacks. A less common procedure used that involves polarization of the sample under both heating and an electrical field is called the thermostimulated polarization process (see Problem 5).

4.5.3. Microscopic Mechanisms and Applications of TSD Currents

Polarization under high electric fields may arise from different types of process:

1. Displacement of electrons and nuclei, an instantaneous process.
2. Migration of electrons, ions, and ion vacancies, and their trapping close to the electrodes.
3. Charge injection from electrodes to the sample.
4. Dipolar polarization.

Only the first type of polarization takes place at any temperature immediately after the application of the electric field. All of the other processes approach equilibrium with a characteristic relaxation time, which is temperature dependent.

Ideally, a separate current peak in the depolarization spectrum is produced for each relaxation process. In practice, these peaks may overlap each other, forming a broad peak. In fact, the similarity between dipole and space charge peaks makes it difficult to distinguish between the two. In principle, only space charge peaks are present in nonpolar materials. To observe peaks due to ionic space charge, the polarizing field must be applied at temperatures where the ion mobility is important. For this reason, space charges near the electrodes are only observed at high temperatures. The peaks occur at higher temperatures than the dipolar ones.

If the material is also electronic conducting, the application of an electric field may cause injection of electronic charge carriers into the sample. After cooling and short-circuiting the sample, thermodepolarization current can also be observed. However, since the processes involved are complex, the interpretation of these currents must be done in terms of the trapping of these carriers in the band structure near the electrodes. In this respect, hopping conduction across potential barriers can explain strong retrapping of charge carriers. As a general rule, space charge peaks are strongly dependent on the type of electrodes, sample thickness, preparation of the sample (doping, etc.), and external radiation. Moreover, whereas the intensity of the dipole peaks is linearly dependent on the electric field, this dependence is clearly nonlinear for the space charge peaks. Finally, note that the integral along the time of the current spectrum yields the total stored charge retained by the electret.

In the present context we are mainly interested in depolarization currents arising from dipolar disorientation while sample heating is in progress. In this respect a comparison between a.c. and TSDC data can be useful to elucidate the origin of the mechanisms. As mentioned above, the depolarization spectrum often includes several broad peaks, each of them corresponding to different molecular groups associated with the corresponding dipole. These are called β-peaks. The disorientation of the dipoles of the chains by the effect of cooperative motions of the main chain backbone causes α-relaxation. In fact, disorientation of dipoles requires a certain amount of energy which can be interpreted as the height of the barrier energy opposing reorientation. This picture is quite similar to the molecular interpretation of conventional a.c. dielectric measurements. In fact, the information provided by these two types of experimental technique is also quite similar. As will be shown later, it is possible to interconvert TSDC data into permittivity data. Accordingly, the TSDC technique not only is a useful tool to study the decay of the charge of the electrets but also may supply information concerning the storage and transport of charge carriers in insulators or semiconductors. The TSDC technique also provides information on interfacial barriers near the electrode contacts. It has been applied to the study of impurities in ionic crystals and to the analysis of the frequency relaxation spectra of solids and glass-forming materials such as polymers.

4.5.4. Basic Equations for Dipolar Depolarization

The equation for the thermally stimulated current technique caused by dipolar disorientation was first developed by Bucci et al.,[41] who called it ionic thermocurrents. They applied it to the current spectra of crystals with dipolar defects that can be oriented by external electric fields. Thermally stimulated currents also appear at low temperatures in polymers as a result of disorientation of dipoles either rigidly attached to the backbone or located in flexible side chains. This disorientation is produced when polymers are heated at temperatures well above the glass transition or softening temperature. If only a single type of non-interacting dipole is assumed, a first-order kinetic equation governs the process, that is

$$\frac{\mathrm{d}P(t)}{\mathrm{d}t} = -\frac{P(t)}{\tau[T(t)]} \tag{4.5.4.1}$$

where $\tau[T(t)]$ is the temperature-dependent relaxation time. The time derivative of P on the left-hand side of Eq. (4.5.4.1) is total because the temperature and the time are mutually dependent through the heating rate. Thus, $\tau[T(t)]$ in this equation can directly be written as $\tau(t)$. On the other hand, the density current generated by the decaying polarization is given by

$$i(t) = -\frac{\mathrm{d}P(t)}{\mathrm{d}t} \tag{4.5.4.2}$$

Integration of Eq. (4.5.4.1) yields

$$P(t) = P_0 \exp\left\{-\int_0^t \frac{\mathrm{d}t}{\tau[T(t)]}\right\} \tag{4.5.4.3}$$

where P_0 is the total or equilibrium dipole polarization which in the case of freely rotating dipoles is given by Eq. (1.9.10), that is

$$P_0 = e_0 \Delta \varepsilon E_\mathrm{p} = \frac{N_1 E_\mathrm{p} \mu^2}{3kT} = \int_{T_0}^\infty i(T')\,\mathrm{d}T' \tag{4.5.4.4}$$

In this equation, e_0 is the vacuum permittivity, $\Delta \varepsilon$ is the relaxation strength of dipolar relaxation, and E_p is the applied polarization field. The total polarization is also equivalent to the area of the TSC peak, as expressed in the last equality of Eq. (4.5.4.4). According to Eqs. (4.5.4.3), the depolarization current is given by

$$i(T) = -\tau^{-1}(T)P_0 \exp\left[-s^{-1}\int_{T_0}^T \frac{\mathrm{d}T}{\tau(T)}\right] \tag{4.5.4.5}$$

where $s(= \mathrm{d}T/\mathrm{d}t)$ is the heating rate.

For systems obeying the Arrhenius behavior (that is, for $T < T_g$), the temperature dependency of the relaxation time is given by

$$\tau(T) = \tau_0 \exp\left(\frac{H}{RT}\right)$$ (4.5.4.6)

where H is the activation energy. Only secondary or subglass relaxations are thermally activated processes described by Eq. (4.5.4.6). The glass rubber or α-relaxation is governed by the free volume, and the temperature dependence of the relaxation times is described by the Williams–Landel–Ferry equation

$$\tau(T) = \tau_g \exp\frac{C_1(T - T_g)}{C_2 + T - T_g}$$ (4.5.4.7)

The maximum of the current can be obtained by differentiating Eq. (4.5.4.5), from which one obtains

$$\frac{d\tau(T)}{dT} = -s^{-1}$$ (4.5.4.8)

This equation, in conjunction with Eq. (4.5.4.6), gives

$$\frac{RT_{\max}^2}{\tau(T)sH} = 1$$ (4.5.4.9)

from which the equivalent frequency of the maximum is easily obtained as

$$f_m = \frac{sH}{2\pi RT_{\max}^2}$$ (4.5.4.10)

By substituting Eq. (4.5.4.6) into Eq. (4.5.4.5), the following approximate expression is obtained

$$\int_0^T \frac{dT}{\tau(T)} = \tau(T)RT^2 \frac{1 + 0.682RT/H}{H/RT + 2.663}$$ (4.5.4.11)

4.5.5. Isothermal Measurements

For isothermal depolarizations, Eq. (4.5.4.3) can easily be integrated, giving

$$P(t) = P_0\left[1 - \exp\left(-\frac{t}{\tau}\right)\right]$$ (4.5.5.1)

Therefore, the depolarization current ($i = dP(T)/dt$) will be

$$i(t) = \frac{P_0}{\tau} \exp\left(-\frac{t}{\tau}\right) \tag{4.5.5.2}$$

This equation corresponds to a "first-order" approximation and rarely fits actual data. Therefore, alternative models have been proposed. The first empirical equation to represent the isothermal depolarization current was proposed by Hamon.[43] This equation has the following form

$$i(t) = \frac{I(t)}{S} = Ae_0 E_p t^{-n} + \sigma E_p \tag{4.5.5.3}$$

where σ is the bulk conductivity and e_0 is the permittivity of the evacuated space.

The importance of the measurements of isothermal depolarization currents lies in their allowing the calculation of the dielectric permittivity at very low frequencies, where the relaxations are widely separated. The basis of this calculation is the relation between the dielectric permittivities in the time and frequency domains. According to Eq. (2.12.12), the pertinent expressions for the real and imaginary components of the complex dielectric permittivity, including conductivity, are

$$\varepsilon'(\omega) = \varepsilon_\infty + \int_0^\infty \dot{\varepsilon}(t) \cos \omega t \, dt$$

$$\varepsilon''(\omega) = \frac{\sigma}{\omega\varepsilon} + \int_0^\infty \dot{\varepsilon}(t) \sin \omega t \, dt \tag{4.5.5.4}$$

Note that the bulk conductivity of the sample does not contribute to the real part of the permittivity. The term corresponding to bulk conductivity is missing in a depolarization experiment, whereas it must be retained in an experiment carried out with polarization of the sample under an electric field.

Under an electric field E, the current crossing a sample placed between the plates of a plane parallel capacitor of active surface area S, is given by

$$I(t) = SE(e_0\dot{\varepsilon} + \sigma) \tag{4.5.5.5}$$

from which

$$\dot{\varepsilon} = \frac{I(t)}{e_0 ES} - \frac{\sigma}{e_0} \tag{4.5.5.6}$$

Equations (4.5.5.4) can be transformed into the following expressions

$$\varepsilon'(\omega) = \varepsilon_\infty + \int_0^\infty \left(\frac{I(t)}{e_0 ES} - \frac{\sigma}{e_0} \right) \cos \omega t \, dt$$

$$\varepsilon''(\omega) = \frac{\sigma}{\varepsilon \omega} + \int_0^\infty \left(\frac{I(t)}{e_0 ES} - \frac{\sigma}{e_0} \right) \sin \omega t \, dt$$

(4.5.5.7)

Since the experimental measurements are restricted to relatively short intervals of time (no more than 10^4 s), it is necessary to extend the values of $i(t)$ beyond the experimental window to obtain reliable values of the permittivity. Truncation errors in evaluating integrals of Eqs. (4.5.5.7) are a function of the frequency at which the components of the permittivity are calculated. In general, reduction in the experimental time interval of measurements leads to a more pronounced reduction in the frequency interval in which admissible values of the permittivity are obtained. Insertion of Eq. (4.5.5.3) into Eqs (4.5.5.7) gives

$$\varepsilon'(\omega) = \varepsilon_\infty + \omega^{n-1} A \Gamma(1 - n) \sin \frac{n\pi}{2}, \qquad 0 < n < 1$$

$$\varepsilon''(\omega) = \frac{\sigma}{\omega e_0} + \omega^{n-1} A \Gamma(1 - n) \cos \frac{n\pi}{2}, \qquad 0 < n < 2$$

(4.5.5.8)

where Γ is the gamma function.

From the knowledge of A and n, the complex dielectric permittivity can in principle be found. However, it is clear that no loss maximum in the frequency domain is predicted by Eq. (4.5.5.8b). To overcome this difficulty, Hamon wrote Eq. (4.5.5.8) neglecting conductivity as

$$\varepsilon''(\omega) = \frac{i(t)}{e_0 E \omega} \left[\frac{\Gamma(1 - n) \cos n\pi/2}{(\omega t)^{-n}} \right]$$

(4.5.5.9)

By noting that

$$\omega t = \left[\Gamma(1 - n) \cos \frac{n\pi}{2} \right]^{-1/n}$$

(4.5.5.10)

is close to $\pi/5$ for $0.1 < n < 1.2$, (within a 3% error), the loss permittivity can be obtained from the expression

$$\varepsilon''(\omega) = \frac{i(t)}{e_0 E \omega}, \qquad \omega t = \frac{\pi}{5}$$

(4.5.5.11)

Note that, since $\omega = 2\pi f$, where f is in Hz, $f = 1/10t$ for $\omega t = \pi/5$. The real part of the permittivity can easily be obtained as

$$\varepsilon'(\omega) = \varepsilon_\infty + \varepsilon''(\omega)\tan\frac{n\pi}{2}, \qquad \omega t = \frac{\pi}{5}, \qquad 0.1 < n < 1 \qquad (4.5.5.12)$$

This expression is of limited use because it includes n and ε_∞, two parameters difficult to find. Hamon-type equations can be derived by assuming a distribution of relaxation times that is consistent with some empirical equation representing a.c. dielectric data. As a general criterion, Hamon equations can be used when a double logarithmic current vs. time curve has a slope less than unity.

Following a similar procedure to that reported by Schwarzl and Struik[44] for viscoelastic functions, Brather[45] proposed several approximate expressions for the calculation of the real and imaginary parts of the complex dielectric permittivity from the current intensity. This new procedure is far more accurate than the Hamon method. In practice, these expressions include terms accounting for the current at times $t = t_0 2^{l-1}$, with $l = 1, 2, 3, \ldots$, or the loss and permittivity at frequencies $2^{l-1}\omega_0$, $l = 1, 2, 3, \ldots$ The knowledge of values of the current or loss at higher/lower times/frequencies is necessary to span an entire loss curve. For example, for the loss factor, Brather proposes

$$\varepsilon''(\omega) = \frac{1}{e_0 \omega E}\left[\sum_{j=-3}^{l-1} a_j^{(l)} i\left(\frac{1}{2^j \omega}\right) + \sum_{j=1}^{l+2} b_j^{(l)} \varepsilon''(2^j \omega)\right] \pm \text{error term}$$

$$(4.5.5.13)$$

where i is the density current.

In particular, for $l = 7$, the real part component is given by

$$\varepsilon'(\omega) = \frac{1}{e_0 \omega E}\left[\begin{array}{c} 0.4248i\left(\dfrac{4}{\omega}\right) - 1.9127i\left(\dfrac{2}{\omega}\right) + 1.0066i\left(\dfrac{1}{\omega}\right) + 0.1134i\left(\dfrac{1}{2\omega}\right) \\[2mm] + 0.1927i\left(\dfrac{1}{4\omega}\right) + 0.7429\sum_{j=3}^{l-4}\dfrac{1}{2^j}i\left(\dfrac{1}{2^j\omega}\right) \\[2mm] + \dfrac{1}{2^l}\left\{3.5864i\left(\dfrac{1}{2^{l-3}\omega}\right) + 4.9800i\left(\dfrac{1}{2^{l-2}\omega}\right)\right. \\[2mm] \left. + 1.9778i\left(\dfrac{1}{2^{l-1}\omega}\right)\right\} \end{array}\right]$$

$$+ 1.4579\varepsilon'(2^l\omega) - 0.4890\varepsilon'(2^{l+1}\omega) - 0.4890\varepsilon'(2^{l+2}\omega)$$

$$+ \text{error term} \qquad\qquad (4.5.5.14)$$

whereas for $l \geqslant 4$ the loss can be written as

$$
\varepsilon''(\omega) = \frac{1}{\omega e_0 E} \left[\begin{array}{l} 0.240i\left(\frac{8}{\omega}\right) - 1.167i\left(\frac{4}{\omega}\right) + 1.083i\left(\frac{2}{\omega}\right) \\[2mm] + \left(\frac{1.3338}{4^l} + 0.61073\right)i\left(\frac{1}{\omega}\right) \\[2mm] + 0.6998 \sum_{j=1}^{l-3} \frac{1}{4^j} i\left(\frac{1}{2^j\omega}\right) + \frac{1}{4^l}\left\{ 8.4752i\left(\frac{1}{2^{l-2}\omega}\right) \right. \\[2mm] \left. + 5.12i\left(\frac{1}{2^{l-1}\omega}\right)\right\} \end{array} \right]
$$
$$
+ \frac{1}{2^l}\left[-0.5407\varepsilon''(2^l\omega) + 1.185\varepsilon''(2^{l+1}\omega) - 0.2047\varepsilon''(2^{l+2}\omega)\right]
$$
$$
+ \text{error term} \qquad\qquad (4.5.5.15)
$$

Starting from the Krönig–Kramers relations, $\varepsilon'(\omega)$ and $\varepsilon''(\omega)$ can mutually be interconverted. For example, a pair of such equations are (see reference [45])

$$
\varepsilon''(\omega) - \frac{\sigma}{e_0\omega} = 0.576\left\{\varepsilon'\left(\frac{\omega}{2}\right) - \varepsilon'(2\omega)\right\} + 0.0125\left\{\varepsilon'\left(\frac{\omega}{4}\right) - \varepsilon'(4\omega)\right\}
$$
$$
+ 0.852 \sum_{j=2}^{\infty} \frac{1}{2^{2j-1}}\left\{\varepsilon'\left(\frac{\omega}{2^{2j-1}}\right) - \varepsilon'(2^{2j-1}\omega)\right\}
$$
$$
+ 0.031 \sum_{j=2}^{\infty} \frac{1}{4^j}\left\{\varepsilon'\left(\frac{\omega}{4^j}\right) - \varepsilon'(4^j\omega)\right\} + \text{error term}
$$

$$
(4.5.5.16)
$$

$$
\varepsilon'\left(\frac{\omega}{2}\right) - \varepsilon'(2\omega) = 1.5267\varepsilon''(\omega) - 0.047625\left\{\varepsilon''(2\omega) - \varepsilon''\left(\frac{\omega}{2}\right)\right\}
$$
$$
- 0.25063\left\{\varepsilon''(4\omega) - \varepsilon''\left(\frac{\omega}{4}\right)\right\}
$$
$$
- 0.013716\left\{\varepsilon''(8\omega) - \varepsilon''\left(\frac{\omega}{8}\right)\right\}
$$
$$
- 0.008116\left\{\varepsilon''(16\omega) - \varepsilon''\left(\frac{\omega}{16}\right)\right\}
$$
$$
- 0.001528\left\{\varepsilon''(32\omega) - \varepsilon''\left(\frac{\omega}{32}\right)\right\}
$$
$$
- 1.6551 \sum_{j=6}^{\infty} \frac{1}{4^j}\left\{\varepsilon''(2^j\omega) - \varepsilon''\left(\frac{\omega}{2^j}\right)\right\} + \text{error term}
$$

$$
(4.5.5.17)
$$

Truncation errors together with an interactive program to apply the Brather method are given in reference [46]. The so-called "Brather method" is able to obtain permittivities in a broad range of frequencies, from ultralow frequencies to 10^6 Hz (or higher).

4.5.6. Thermal Windowing

The analysis indicated above [cf. Eq. (4.5.4.1)] is based on the assumption that the orientation polarization decreasing rate is proportional to the orientation polarization. This simple kinetic assumption is called the Debye hypothesis. Equation (4.5.4.5), in conjunction with Eqs (4.5.4.4) and (4.5.4.6), leads to

$$i(T) = \frac{NE_p\mu^2}{3\tau_0 kT} \exp\left(-\frac{H}{kT}\right) \exp\left(-\frac{1}{\tau_0 s} \int_{T_0}^{T} \exp\left(-\frac{H}{kT'}\right) dT'\right) \qquad (4.5.6.1)$$

This equation describes a Debye-type relaxation. For ionic crystals, in general, a single-relaxation model fits quite well to experimental data. However, when dipole–dipole interactions occur, as happens in polar polymers, the processes are broader than those predicted by the single-relaxation model. This indicates that distributed relaxation processes are most likely in complex systems. As in conventional dielectric relaxation, this behavior is caused by a distribution of activation energies, a distribution of pre-exponential factors, or both. Distributions such as the Cole–Cole, Davidson–Cole, Fuoss–Kirkwood, and Havriliak–Negami distributions can be useful in representing the empirical curves. However, a more convenient way to study the fine structure of the relaxations consists in performing partial depolarizations of the sample in small temperature windows covering the experimental range of temperatures where the relaxation peak is present. The technique is known as thermal windowing. Obviously, the elementary peaks are much lower in intensity than the global peak, and consequently this imposes a restriction on the type of electrometer to be used to detect the depolarization currents. A better resolution in the elementary relaxation modes is obtained for narrower windows. However, the current peak is also lower, and for this reason it is necessary to reach a compromise between sensitivity and resolution.

The mean relaxation time associated with each partial depolarization curve is given by[47]

$$\tau_i(T) = \frac{\int_0^t I(t') \, dt'}{I(t)} = \frac{\int_{T_i}^{T} I(T') \, dT'}{sI(T)} \qquad (4.5.6.2)$$

where T_i is the initial temperature of each depolarization curve. The temperature dependence of the relaxation times calculated from Eq. (4.5.6.2) follows Arrhenius behavior up to at least the half-width temperature of the peak. This allows determination of the pre-exponential factors, τ_0, and the activation energies, H, associated with each peak. In general, the values of the activation energies show a tendency to increase with increasing temperatures in each one of the relaxational zones. The use of the Eyring equation instead of the Arrhenius equation allows determination of the activation enthalpy and activation entropy, and the Gibbs free energy shows indeed a linear tendency with temperature. Surprisingly, the activation entropy shows, in many cases, values close to zero but negative.

4.5.7. Thermostimulated Depolarization Currents Versus Conventional Dielectric Measurements

The use of thermostimulated depolarization currents is a complementary way of analyzing dielectric loss phenomena. The results obtained with a.c. and TSDC techniques are comparable, and it is the purpose of this section to establish the quantitative basis for such a comparison. The first fact to bear in mind is that TSDC is a non-isothermal technique, in contrast to dielectric measurements which are typically isochronal or isothermal techniques. As a consequence, the relaxation peaks obtained with these two techniques may differ in both position and intensity. Dielectric a.c. measurements are carried out in the frequency–temperature domain, whereas TSDC measurements are made in the time–temperature domain. The results obtained by the two techniques are related by Fourier transforms. In practice, approximate methods such as those described in Section 4.5.5 or in the monograph by Van Turnhout[47] are very useful. The general method is based on the mutual adjustment of the kernels of the integral equations representing the permittivity and the charge released during the depolarization process. For example, a first-order formula for this interconversion can easily be obtained from the Hamon approximation which, after a simple transformation in Eq. (4.5.5.12), gives

$$\varepsilon''(\omega, T_0) = \frac{1.59 I(t) k T_0^2}{e_0 s E S}, \qquad \omega = \frac{\pi}{5t} \qquad (4.5.7.1)$$

Note that $5/\pi = 1.59$.

Hino[48] has described the conversion of TSDC into ε'' data. Another convenient way to obtain the dielectric permittivity from the TSDC data can be found in the literature (see references [49] and [50]). According to these approaches, the real and imaginary parts of the complex permittivity can be obtained from the following expressions

$$\varepsilon'(\omega) = \varepsilon_\infty + \sum_{i=1}^{n} \frac{\Delta \varepsilon_i}{1 + \omega^2 \tau_i^2}$$

$$\varepsilon''(\omega) = \sum_{i=1}^{n} \frac{\Delta \varepsilon_i (\omega \tau_i)}{1 + \omega^2 \tau_i^2}$$

$$(4.5.7.2)$$

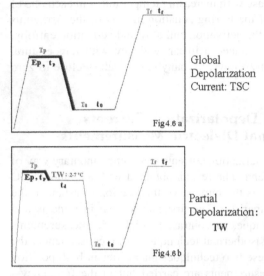

Global
Depolarization
Current: TSC

Partial
Depolarization:
TW

FIG. 4.6 Electric field and temperature profiles for (a) global and (b) single-window depolarization experiments, (c) global TSDC curves for some polymers defined in Table 1 of Chapter 9, and (d) partial depolarization curves for PAMED in the glassy state (see Table 1 of Chapter 9)[51]

where τ_i are given by Eq. (4.5.7.2) and the partial relaxation strength, $\Delta\varepsilon_i$, is obtained from

$$\Delta\varepsilon_i(T_p) = \frac{1}{e_0 sEA} \int_{T_0}^{T_f} I(T') \, dT' \tag{4.5.7.3}$$

In Fig. 4.6 the electric field and temperature profiles for global (Fig. 4.6a) and for single-window (Fig. 4.6b) depolarization experiments are schematized. In Figs 4.6c and d, the global TSDC curves for some polymers defined in Table 1 of Chapter 9, and the partial depolarization curves for poly(5-acryloxmethyl-5-ethyl-1,3-dioxacyclohexane) (PAMED) in the glassy state are shown.[51]

APPENDIX A

Edge Corrections

The dielectric permittivity of a sample can accurately be measured using guarded electrodes when good contact between electrodes and sample is warranted. However, in some cases, the use of a guarded electrode is precluded, for example when surface effects are studied. In this case, fringing fields close to the edges of the capacitor plates are produced. The problem of the corrections for a vacuum

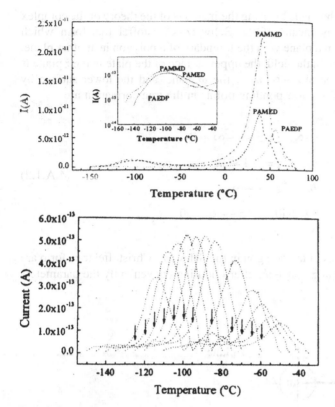

FIG. 4.6 Continued

capacitor was addressed by Kirchhoff many years ago.[52] In this appendix, edge corrections for the most important parallel plate geometries are analyzed.

A.1. Long Rectangular Strip Capacitor

Let us consider a capacitor consisting of two parallel thin plates of width $2a$, separated by a distance $2h$. The aspect ratio h/a is such that the two edge corrections do not interfere with one another. The geometric capacitance by unit length of such a system is

$$C_g = e_0 \frac{a}{h} \qquad (4.A.1.1)$$

where ε_0 is the permittivity of the evacuated space. Owing to the distortion of the electric field near the electrodes, the true capacitance is larger than that given by Eq. (4.A.1.1).

Corrections can be made by using the methods of the theory of the complex variable function[53] by means of the Schwarz–Christoffel transform which connects the real axis in a plane with the boundary of a polygon in another plane. By assuming symmetry in the field, the upper surface of the plate in the z plane is represented by the interval $(-\infty, a)$ in the ω plane, and the lower surface by (a, b) (see Fig. 4.7). The corresponding points in the transformation are

$$
\begin{aligned}
A \to A', \quad & \omega \to -\infty, \quad & z \to -\infty + ih \\
B \to B', \quad & \omega = a, \quad & z = -1 + ih \\
C \to C', \quad & \omega = b, \quad & z \to -\infty \\
D \to D', \quad & \omega \to \infty + i0, \quad & z \to \infty + i0
\end{aligned}
\tag{4.A.1.2}
$$

Since the interior angles of the polygon in the Schwarz–Christoffel transform are 0 at $z \to -\infty$ and 2π at $z = 0 + ih$, after choosing conveniently the parameters

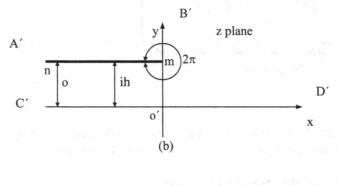

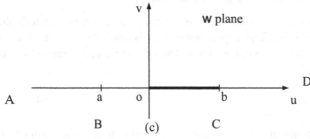

FIG. 4.7 Complex variable transformation for the study of the edge effects in a rectangular strip capacitor

$a = -1$ and $b = 0$, we have

$$\frac{dz}{d\omega} = C\frac{\omega + 1}{\omega} \tag{4.A.1.3}$$

and after integration

$$z = C(\omega + \ln\omega + D) \tag{4.A.1.4}$$

Since $z = -1 + ih$ and $\omega = -1$ are corresponding points in the transformation, and since C is real, we obtain

$$z = \frac{h}{\pi}(\omega + \ln\omega) \tag{4.A.1.5}$$

According to the theory of the conformal transformation for an electric field in the plane, the potential in the ω-plane is

$$w = \frac{V}{\pi}\ln\omega \tag{4.A.1.6}$$

where $+V$ and $-V$ are the potentials of the plates. Substitution of Eq. (4.A.1.6) into Eq. (4.A.1.5) gives

$$z = \frac{h}{\pi}\left(\frac{\pi w}{V} + \exp\left(\frac{\pi w}{V}\right)\right) \tag{4.A.1.7}$$

According to the conformal transform for a plane electric field, the modulus of the electric field is given by

$$E = \frac{dw}{dz} = \frac{1}{dz/dw} = \frac{V}{h}\frac{1}{|1 + \exp(\pi/V)w|} \tag{4.A.1.8}$$

Note that inside the capacitor (point C) $\omega = 0$, that is, $w \to -\infty$ and $E \cong V/h$. According to Eq. (4.A.1.8), the charge density at any point inside the capacitor will be

$$\rho = \rho_0 \frac{1}{|1 + \exp(\pi/V)w|} \tag{4.A.1.9}$$

For this reason, the additional charge in the inner surfaces of the plates on account of the fringing effects can be written as

$$
\int_{-\infty+ih}^{ih} (\rho - \rho_0)\, dz = \int_{-\infty}^{iV} \rho_0 \left[\frac{1}{1 + \exp(\pi w/V)} - 1 \right]
$$

$$
\times \frac{h}{V}\left(1 + \exp\frac{\pi w}{V} \right) dw = \rho_0 \frac{h}{\pi}
\tag{4.A.1.10}
$$

This additional charge is equivalent to enlarging *each* plate a distance given by Eq. (4.A.1.10).

The effect of the charges on the outer surfaces will be calculated according to the same integral, but now from $(ih, -\infty)$. However this integral diverges. This is a consequence of the fact that, close to the edges, $w = \pm iV$, and the intensity of the field given by Eq. (4.A.1.8) tends to infinity. However, since the density of charge decreases rapidly away from the edges, we can integrate between $(ih, -a + ih)$, and in this case

$$
\int_{ih}^{-a+ih} \rho\, dz = \rho_0 \frac{h}{V} \ln \frac{\pi a}{h}
\tag{4.A.1.11}
$$

Taking into account the effects on the two edges, we obtain

$$
C \simeq C_g + 2C_{in} + 2C_{out} = e_0 \frac{a}{h}\left[1 + \frac{h}{a\pi}\left(1 + \ln\left(\frac{\pi a}{h}\right) \right) \right]
\tag{4.A.1.12}
$$

For an aspect ratio of 0.1, Eq. (4.A.1.12) predicts a relative error of about 14%.

Equation (4.A.1.12) has also been obtained by Wintle and Kurylowicz.[54] It represents a simplified version of a more rigorous calculation in terms of elliptic functions by Love[55] who, omitting high-order terms, obtains

$$
C \cong e_0 \frac{a}{h}\left[1 + \frac{h}{a\pi}\left(1 + \ln\left(\frac{2\pi a}{h}\right) \right) \right]
\tag{4.A.1.13}
$$

This equation differs from Eq. (5.A.1.12) in a factor of 2 in the last term on the right-hand side.

A.2　Disc Electrodes

A pair of plane parallel coaxial discs is the more usual arrangement to measure permittivities of solid materials. The edge corrections in this case also date back to more than a century and are originally due to Kirchhoff[52] who, for very thin electrodes, proposed the following expression for the capacity

$$C = e_0 a \left(\frac{\pi a}{h} + \ln\left(\frac{16\pi a}{h}\right) - 1 \right) \tag{4.A.2.1}$$

where a is the radius of the capacitor and h is the distance between electrodes. In modern treatments,[56] a pair of dual integral equations[57,58] are recast in the form of a single integral equation from which analytical approximations can be made[59] and numerical results can be obtained.[60] The results are in agreement in a first approach with Eq. (4.A.2.1). An improved computational method has recently been proposed by Wintle and Kurylowicz[54] for the aspect ratio ($\kappa = h/a$) over the range $1 \geq \kappa \geq 10^{-4}$, where the accuracy of the former approximations is made clear.

More important are the corrections when a sheet of dielectric is placed between the electrodes. In this case, the field due to charges placed outside or inside a dielectric slab can be conveniently computed by the method of images.[61] Following this procedure and a methodology outlined by Tranter,[62] Wintle and Goad[63] found the following expression for the capacitance

$$C = 2e_0 a (\varepsilon_1 + \varepsilon_2) \int_0^1 y(x)\, dx \tag{4.A.2.2}$$

where ε_1 and ε_2 are respectively the relative permittivities outside and inside the electrodes, and $y(x)$ is the solution of the following integral equation

$$y(x) = 1 + \frac{1}{\pi} \int_{-1}^{1} y(s) \sum_{n=1}^{\infty} (1 - K) K^{n-1} \left(\frac{nh/a}{(nh/a)^2 + (s - x)^2} \right) ds \tag{4.A.2.3}$$

where $K = (\varepsilon_1 - \varepsilon_2)/(\varepsilon_1 + \varepsilon_2)$.

When Eq. (4.A.2.3) is solved numerically, the normalized excess capacitance can be represented as a function of the ratio a/h for the case where $\varepsilon_1 = 1$ to give

$$\frac{C_{ex}}{\pi e_0 \varepsilon_2 a} = \frac{C}{\pi e_0 \varepsilon_2 a} - \frac{a}{h} \tag{4.A.2.4}$$

The geometric capacitance is obviously given by

$$C_g = \frac{\pi e_0 \varepsilon_2 a^2}{h} \tag{4.A.2.5}$$

In reference,[63] tabulated edge corrections are given. The results obtained lead to the conclusion that the excess capacitance for $\varepsilon_2 = 1$ approaches the Kirchhoff limit for values of $a/h \geq 10$.

A.3 Guard Gaps

More accurate dielectric measurements are made when guard rings are used. For a geometry like that shown in Fig. 4.8, where g is the gap width, r is the radius of the electrodes, and h is the thickness of the sample, the usual correction to be made is to add to the radius one-half of the gap width. Thus, for a sample with relative permittivity ε, the value for the capacitance will be

$$C = \frac{\pi e_0 \varepsilon}{h}\left(a + \frac{g}{2}\right)^2 = C_g + C_{ex} \qquad (4.A.3.1)$$

where the subscripts g and ex denote geometric and excess respectively. Consequently

$$\frac{C_{ex}}{P} \simeq \frac{\pi e_0 \varepsilon g}{2h} \qquad (4.A.3.2)$$

where $P = 2\pi a$ is the perimeter of the disc-shaped electrodes.

A more elaborate correction[64] that uses the conformal transformation yields

$$\frac{C_{ex}}{P} \cong \frac{\pi e_0 \varepsilon B g}{2h} \qquad (4.A.3.3)$$

where, for the case $\varepsilon \gg 1$, the following expression holds

$$B = 1 - \frac{4h}{\pi g}\ln\cosh\frac{\pi g}{4h} \cong 1 - \frac{1}{4}\left(\frac{\pi g}{2h}\right) + \frac{1}{96}\left(\frac{\pi g}{2h}\right)^3 + \cdots \qquad (4.A.3.4)$$

Here, the approximation is valid for $g/2h \ll 1$, as is the usual case.

A general discussion, including the effect of the thickness of the electrodes, has been carried out by Goad and Wintle.[65]

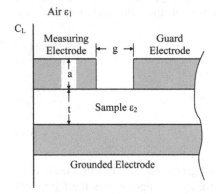

FIG. 4.8 Schematic representation of a guard ring capacitor

APPENDIX B
Single-Surface Interdigital Electrode

Single-surface electrodes are widely used in dielectric characterization of polymers and other materials.[66] These capacitors consist of a periodic pattern of interdigitized metallic lines deposited on a flat substrate of ceramic or fused silica. These planar comblike electrodes are commonly used in applications where the cure of thermosetting resins is monitored, and in general as a part of the techniques associated with microdielectrometry.[67] The potential use of these sensors has recently been recognized[68] in the study of normal-mode peaks in type A chains (according to Stockmayer's nomenclature[69]), in liquid crystal flow analysis,[70] and in general when the dynamics of a flowing polymer near a solid surface is under study.

Figure 4.9 shows schematically the sensor layout. The two-dimensional and periodic geometry of this type of electrode implies that the electric field on the sensor is itself two-dimensional and periodic, as schematically illustrated in

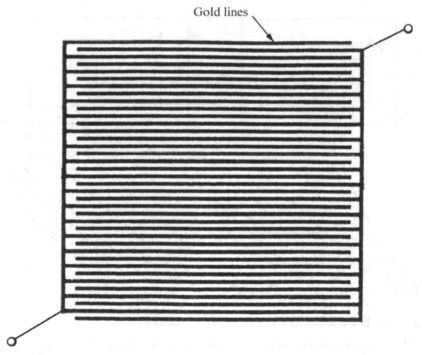

Gold lines

FIG. 4.9   Scheme of sensor layout

Fig. 4.10. The periodic line pattern imparts a finite "penetration depth to the field" comparable in magnitude with the spacing between lines. This electric field can be calculated from electrostatics[71] by assuming that the metallic lines are alternatively sources and sinks and using an analogy between flow and electrostatic potentials.[72]

The complex potential can be modeled by generalizing the electrostatic potential for a system of infinite parallel lines carrying uniformly distributed charge densities (reference[71] p. 209), according to

$$\Phi = -2\lambda \left(\ln z + \sum_{n=1}^{\infty} \ln (z^2 + n^2 a^2) \right) = -2\lambda \ln \sinh \frac{nz}{a} \qquad (4.B.1)$$

where λ is the charge density, $z = x + iy$, and a is the distance between two lines.
According to the theory[70]

$$\Phi = \phi + i\psi \qquad (4.B.2)$$

where ϕ and ψ are respectively the potential and the streamlines. For a configuration of alternative positive and negative lines now spaced $a/2$, the

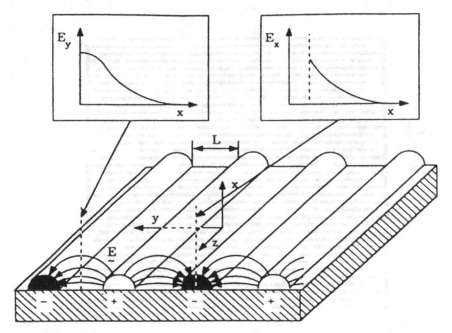

FIG. 4.10   Scheme of an interdigital electrode. Top figures represent the components of the electric field

potential will be

$$\Phi = -2\lambda\left[\ln\sinh\frac{\pi}{a}\left(z - i\frac{a}{4}\right) - \ln sh\frac{\pi}{a}\left(z + i\frac{a}{4}\right)\right]$$

$$= -2\lambda\ln\frac{\sinh(\pi/a)(z - i(a/4))}{\sinh(\pi/a)(z + i(a/4))} \tag{4.B.3}$$

From Eq. (4.B.3) we obtain

$$\phi = -\lambda\ln\frac{\begin{array}{l}\sinh^2((\pi x)/a)\cos^2(\pi/a)(y - (a/4))\\+\cosh^2((\pi x)/a)\sin^2(\pi/a)(y - (a/4))\end{array}}{\begin{array}{l}\sinh^2((\pi x)/a)\cos^2(\pi/a)(y + (a/4))\\+\cosh^2((\pi x)/a)\sin^2(\pi/a)(y + (a/4))\end{array}}$$

$$\tag{4.B.4}$$

$$\psi = -2\lambda\left[\tan^{-1}\frac{\tan(\pi/a)(y - (a/4))}{\tanh((\pi x)/a)}\right.$$

$$\left. - \tan^{-1}[\tan(\pi/a)(y + (a/4))]/\tanh(\pi x)/a\right]$$

From the well-known trigonometric expressions

$$\sin^2 x + \cos^2 x = 1$$

$$\cosh^2 x - \sinh^2 x = 1$$

$$2\cosh^2\frac{x}{2} = \cosh x + 1 \tag{4.B.5}$$

$$2\cos^2\frac{x}{2} = \cos x + 1$$

$$\tan^{-1} x - \tan^{-1} y = \tan^{-1}\frac{x - y}{1 + xy}$$

and, after some rearrangements, Eqs (4.B.4) can be rewritten as

$$\phi = -\lambda\ln\frac{\cosh(2\pi x)/a - \sin(2\pi y)/a}{\cosh(2\pi x)/a + \sin(2\pi y)/a}$$

$$\psi = -2\lambda\tan^{-1}\frac{\sinh(2\pi x)/a}{\cos(2\pi y)/a} \tag{4.B.6}$$

The components of the electric field are given by

$$E_x = -\frac{\partial \phi}{\partial x} = \frac{8\pi\lambda}{a} \left[\frac{\sinh((2\pi x)/a)\sin((2\pi y)/a)}{\cosh((4\pi x)/a) + \cos((4\pi y)/a)} \right]$$

$$\text{(4.B.7)}$$

$$E_y = -\frac{\partial \phi}{\partial y} = -\frac{8\pi\lambda}{a} \left[\frac{\cosh((2\pi x)/a)\cos((2\pi y)/a)}{\cosh((4\pi x)/a) + \cos((4\pi y)/a)} \right]$$

Noting that $a = 2L$, and the fact that the components of the field attain their respective maxima at $y = L/2 = a/4$ and $y = 0$, the corresponding values are

$$E_x(x, y = L/2) = \frac{2\pi\lambda}{L} \frac{1}{\sinh(\pi x)/L}$$

$$\text{(4.B.8)}$$

$$E_y(x, y = 0) = -\frac{2\pi\lambda}{L} \frac{1}{\cosh(\pi x)/L}$$

Note that, for $x \to \infty$, both E_x and E_y approach zero, whereas, for $x \to 0$, $E_x(x, L/2) \to \infty$ and $E_y(x, 0) = -2\pi\lambda/L = E_{y,\max}$. The divergent behavior in E_x for $y = L/2$, at $x \to 0$, can be avoided assuming that Eq. (4.B.8a) is applicable everywhere except in the inner part of ths small cylinders surrounding the lines where the singularity is present.

The effective penetration depth for the electric field can be estimated assuming that the local permittivity is a linear function of the dipole density. Accordingly, the changes in the measured permittivity are a consequence of the energy-density-averaged concentration of dipoles in the specimen. In formal terms this can be expressed as

$$\langle \varepsilon \rangle = \int_v \varepsilon \, \Psi(\mathbf{r}) \, dv \tag{4.B.9}$$

where

$$\Psi(\mathbf{r}) = \frac{E^2(\mathbf{r})}{\int_v E^2(\mathbf{r}) \, dv} \tag{4.B.10}$$

the energy being proportional to the squared field.[73] If we assume that

$$E^2(r) \simeq E_y^2(x, 0) \tag{4.B.11}$$

then the denominator in Eq. (4.B.10) can easily be calculated as

$$\int_0^\infty E^2(\mathbf{r}) \, dv \simeq \int_0^\infty E_{y,\max}^2 \frac{1}{\cosh^2(\pi x)/L} \, dx = E_{y,\max}^2 \frac{L}{\pi} \tag{4.B.12}$$

where $\delta = L/\pi$ is the penetration depth. Then

$$\Psi(x) = \frac{\pi}{L} \frac{1}{\cosh^2(\pi x)/L} \tag{4.B.13}$$

This equation is used in connection with Eq. (4.B.9).

An improved calculation can be done by taking into account the finite line width.[70] This calculation shows that the field penetration depth is about 15% smaller than that estimated by the previous procedure.

PROBLEMS

Problem 1

The schematic diagram of a Schering bridge is shown in Fig. 4.11. Express the values of C_x and $\tan \delta_x$ in terms of the remaining parameters of the circuit.

Solution

For sample *in* arm whose passive elements have subscript 1, balance of the bridge requires

$$Z_1 Z_4 |_{in} = Z_2 Z_3 |_{in} \tag{P.4.1.1}$$

For sample *out* only the capacitors C_1 and C_4 need to be adjusted, then

$$Z_1 Z_4 |_{out} = Z_2 Z_3 |_{out} \tag{P.4.1.2}$$

Since arms 2 and 4 are not adjusted

$$Z_2 Z_3 |_{in} = Z_2 Z_3 |_{out} \tag{P.4.1.3}$$

and consequently

$$Z_1 Z_4 |_{in} = Z_1 Z_4 |_{out} \quad \rightarrow \quad Y_1 Y_4 |_{in} = Y_1 Y_4 |_{out} \tag{P.4.1.4}$$

Developing Eq. (P.4.1.4b), one obtains

$$\left[\frac{1}{R_x} + i\omega(C_{1i} + C_x) \right] \left[\frac{1}{R_4} + i\omega C_{4i} \right] = \left[\frac{1}{R_4} + i\omega C_{4o} \right] i\omega C_{1o} \tag{P.4.1.5}$$

where subscript o refers to sample *out* and subscript i refers to sample *in*.

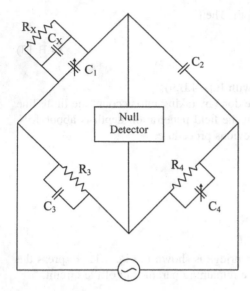

FIG. 4.11 Scheme of a Schering bridge

After splitting into real and imaginary parts, we obtain

$$C_x = \frac{(C_{1o} - C_{1i}) + \omega^2 C_{4i} R_4^2 (C_{4o} C_{1o} - C_{4i} C_{1i})}{1 + \omega^2 C_{4i}^2 R_4^2}$$

(P.4.1.6)

$$R_x = \frac{1 + \omega^2 C_{4i}^2 R_4^2}{\omega^2 R_4 C_{1o} (C_{4i} - C_{4o})}$$

From the definition

$$\tan \delta = \frac{1}{\omega C_x R_x}$$

(P.4.1.7)

and, after substitution of Eq. (P.4.1.6) into Eq. (P.4.1.7), we obtain

$$\tan \delta = \frac{\omega R_4 C_{1o} (C_{4i} - C_{4o})(1 + \omega^2 R_4^2 C_{4i}^2)}{(C_{1o} - C_{1i}) + \omega^2 R_4^2 C_{4i} (C_{4o} C_{1o} - C_{4i} C_{1i})}$$

(P.4.1.8)

For small loss tangents, the following approximate formulae hold

$$C_x = C_{1o} - C_{1i}$$

$$\tan \delta_x = \frac{\omega R_4 C_{1o}(C_{4i} - C_{4o})}{C_{1o} - C_{1i}} \tag{P.4.1.9}$$

Problem 2

Calculate the impedance and the admittance of the RC circuit of Fig. 4.12. Represent both in the complex plot (Argand diagram). Calculate the complex permittivity.

Solution

Simple calculations give the following expression for the admittance

$$Y^* = i\omega C_1 + \frac{1 + \omega R_2 C_2}{R_1 + R_2 + \omega C_2 R_1 R_2}$$

$$= \frac{R_1 + R_2 + \omega^2 C_2^2 R_2^2 R_1}{(R_1 + R_2)^2 + \omega^2 C_2^2 R_2^2 R_1^2} + i\omega \left[C_1 + \frac{R_2^2 C_2}{(R_1 + R_2)^2 + \omega^2 C_2^2 R_2^2 R_1^2} \right]$$

$$\tag{P.4.2.1}$$

The impedance can be calculated from

$$Z^* = (Y^*)^{-1} \tag{P.4.2.2}$$

to give

$$Z^* = \frac{R_1 + R_2 + \omega^2 C_2^2 R_2^2 R_1}{(1 - \omega^2 R_1 R_2 C_1 C_2)^2 + \omega^2 [C_1(R_1 + R_2) + R_2 C_2]^2}$$

$$- i\omega \frac{R_2^2 C_2 + C_1(R_1 + R_2)^2 + \omega^2 C_2^2 R_2^2 R_1^2 C_1}{(1 - \omega^2 R_1 R_2 C_1 C_2)^2 + \omega^2 [C_1(R_1 + R_2) + R_2 C_2]^2} \tag{P.4.2.3}$$

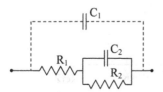

FIG. 4.12 Complex RC circuit

The complex plots are represented schematically in Fig. 4.13, where the arrows indicate increasing frequencies.

From the definition (4.3.1.1c), the frequency-dependent permittivity can be calculated as

$$\varepsilon^* = \frac{C_1}{C_0} + \frac{\omega\tau R_2 - i\left[(R_1 + R_2) + \omega^2\tau^2 R_1\right]}{C_0\omega\left[\omega^2\tau^2 R_1^2 + (R_1 + R_2)^2\right]}$$

(P.4.2.4)

where C_0 is the capacitance of the empty cell and $\tau = C_2 R_2$.

Problem 3

In the plane parallel configuration of the electrodes of a dielectric cell, a layer of air between specimen and electrodes cannot easily be avoided. For this reason, a layer of known thickness has been currently used, and the measured results can be corrected for its effects. This is known as the air gap method. The procedure is reliable when the diameter of the sample is large in comparison with the thickness. Let us consider a disc-shaped sample of radius r and thickness d, placed between circular electrodes with an air gap e. Denoting the capacitance of the empty cell by C_0, calculate the effective permittivity and loss tangent of a dielectric specimen represented by a parallel RC circuit.

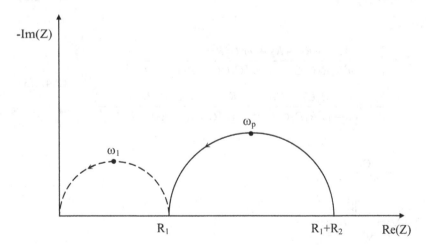

FIG. 4.13   Impedance plane plot corresponding to the circuit of Fig. 4.12

Solution

The capacitance of the empty cell is given by

$$C_0 = e_0 \frac{S}{d+e} \tag{P.4.3.1}$$

where ε is the vacuum permittivity and $S = \pi r^2$.

Assuming that the air gap is in series with the sample, the capacitance will be given by

$$C = \frac{C_s C_a}{C_s + C_a} = \frac{e_0 \varepsilon S}{\varepsilon e + d} \tag{P.4.3.2}$$

where C_s and C_a are, respectively, the capacitances of the sample and the air.

From Eqs (P.4.3.1) and (P.4.3.2) the following expression is obtained

$$\varepsilon = \frac{1}{1 - \{((C - C_0)/C)(d + e)/d\}} \tag{P.4.3.3}$$

For the loss tangent calculation, a capacitor C_a in series with an $R_p C_s$ parallel circuit can be used. According to Eqs (4.3.2.3) to (4.3.2.5), this scheme can be transformed into that of Fig. 4.14 to give an RC series circuit characterized by the following values

$$R = R_p \frac{\tan^2 \delta}{1 + \tan^2 \delta} \quad \text{and} \quad C = \frac{C_s + C_a}{C_s C_a} \tag{P.4.3.4}$$

where $\tan \delta = (\omega R_p C_s)^{-1}$ according to the notation of the figure.

According to Eq. (4.3.2.4), the loss $\tan \delta_m$ is given by

$$\tan \delta_m = \omega \frac{C_s C_a}{C_s + C_a} R_p \frac{\tan^2 \delta}{1 + \tan^2 \delta} \tag{P.4.3.5}$$

Since

$$\tan \delta = \frac{1}{\omega C_p R_p} \tag{P.4.3.6}$$

Eq. (P.4.3.5) can be written as

$$\tan \delta_m = \tan \delta \frac{C_a}{C_s + C_a} \tag{P.4.3.7}$$

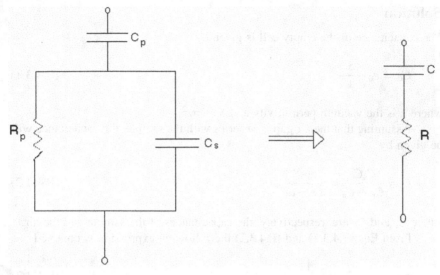

FIG. 4.14 Scheme for the conversion of a complex circuit into a single-series RC arrangement

from which

$$\tan \delta = \tan \delta_m \left(1 + \frac{C_s}{C_a}\right) = \tan \delta_m \left(1 + \frac{\varepsilon e}{d}\right) \tag{P.4.3.8}$$

where ε is given by Eq. (P.4.3.3).

Hence, two measurements must be made, with and without inserting the sample between the electrodes. However, the accuracy of the results depends upon the accuracy with which the two thickness measurements have been made.

It is interesting to estimate the relative error, ε_r, in the permittivity when an undesired air gap is present. Let us denote by ε_m the measured permittivity, that is

$$\varepsilon_m = \frac{C}{C_0} \tag{P.4.3.9}$$

The relative error will be given by

$$\varepsilon_r = \left|\frac{\varepsilon_m - \varepsilon}{\varepsilon}\right| \tag{P.4.3.10}$$

Substitution of Eqs (P.4.3.3) and (P.4.3.9) into Eq. (P.4.3.10) and further rearrangement gives

$$\varepsilon_\varepsilon = \left| \left(\frac{C}{C_0} - 1 \right) \frac{e}{d} \right| \qquad\qquad (P.4.3.11)$$

For a relative permittivity $\varepsilon \approx C/C_0 = 3$, a sample thickness of 1 mm, and an air gap of 0.1 mm, the relative error amounts to 20%. For the same values of the parameters involved in the problem, the relative error in $\tan \delta$ given by the following expression

$$\varepsilon_{r\tan\delta} = \left| \frac{\tan \delta_m - \tan \delta}{\tan \delta} \right| = \left| \frac{\varepsilon e/d}{1 + \varepsilon e/d} \right| \qquad\qquad (P.4.3.12)$$

is even larger (23%). For this reason, when possible, vacuum metallized samples are strongly recommended.

Problem 4

To model the effect of small air bubbles or free spaces between samples and electrodes, it is convenient to assume that these air spaces are circular in shape, as indicated in Fig. 4.15. Estimate the relative error in the permittivity and loss tangent when these circular air free spaces are present between the electrodes and a sample of relative permittivity ε.

Solution

The capacitance and the loss tangent of the system can be evaluated, decomposing the capacitance of the scheme given in Fig. 4.15, according to Fig. 4.16, where the relevant parameters are indicated. The final result after the

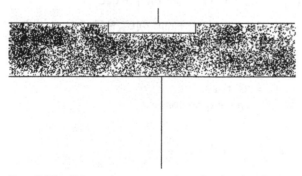

FIG. 4.15   Schematic representation of a circular plane parallel capacitor with a circular air gap

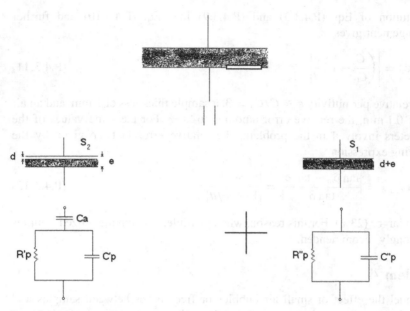

FIG. 4.16 Scheme for the calculation of the permittivity and loss tangent corresponding to the scheme represented in Fig. 4.15

pertinent algebra is given by

$$C_p = C_p'' + \frac{C_a C_p'(1 + \tan^2 \delta)[C_a + C_p'(1 + \tan^2 \delta)]}{[C_a + C_p'(1 + \tan^2 \delta)]^2 + C_a^2 \tan^2 \delta}$$

$$\tan D = \frac{C_p''(R_p'' C_a'(1 + \tan^2 \delta) + R_p' \tan^2 \delta\{C_a'^2 + [C_a + C_p'(1 + \tan^2 \delta)]^2\})}{R_p' \tan \delta(C_p''\{C_a'^2 + [C_a + C_p'(1 + \tan^2 \delta)]^2\} + C_a C_p'(1}$$
$$+ \tan^2 \delta)[C_a + C_p'(1 + \tan^2 \delta)])$$

$$(P.4.4.1)$$

where $\tan \delta = (\omega R_p' C_p')^{-1}$ and $C_a' = C_a \tan \delta$.
   Taking into account the following expressions

$$C_p' = e_0 \varepsilon \frac{S_2}{d}, \qquad C_a = e_0 \frac{S_2}{e}, \qquad C_p'' = e_0 \varepsilon \frac{S - S_2}{d + e},$$

$$R_p' = \rho \frac{d}{S_2}, \qquad R_p'' = \rho \frac{d + e}{S - S_2}$$

$$\tan \delta = (\omega e_0 \varepsilon \rho)^{-1}, \qquad R'_p = \omega e_0 \varepsilon \tan \delta \frac{d}{S_2}, \qquad R''_p = \omega e_0 \varepsilon \tan \delta \frac{d+e}{S-S_2}$$

$$(P.4.4.2)$$

Eq (P.4.4.1) may be written respectively as

$$C_p = e_0 \varepsilon \left\{ \frac{S-S_2}{d+e} + \frac{[S_2/(de)](1+\tan^2 \delta)[1/e + \varepsilon/d(1+\tan^2 \delta)]}{(\tan^2 \delta/e^2) + [1/e + \varepsilon/d(1+\tan^2 \delta)]^2} \right\} \quad (P.4.4.3)$$

$$\tan D = \tan \delta \frac{((1+\tan^2 \delta) + ((S-S_2)/S_2)(d/(d+e))\{\tan^2 \delta \\ + [1 + ((\varepsilon e)/d)(1+\tan^2 \delta)]^2\})}{(1+\tan^2 \delta)[1 + ((\varepsilon e)/d(1+\tan^2 \delta)] + ((S-S_2)/S_2)(d/(d \\ +e))\{\tan^2 \delta + [1 + ((\varepsilon e)/d)(1+\tan^2 \delta)]^2\}}$$

$$(P.4.4.4)$$

Let us denote by ε_m the permittivity measured assuming that the sample occupies all the space between the electrodes. Then

$$\varepsilon_m = C_p \frac{d+e}{\varepsilon S} \qquad (P.4.4.5)$$

and from Eq. (P.4.4.3) we obtain

$$\varepsilon_m = \varepsilon \left(1 - \frac{S_2}{S} \left\{ 1 - \left(1 + \frac{e}{d} \right) \frac{(1+\tan^2 \delta)[1 + ((\varepsilon e)/d)(1+\tan^2 \delta)]}{\tan^2 \delta + [1 + ((\varepsilon e)/d)(1+\tan^2 \delta)]^2} \right\} \right)$$

$$(P.4.4.6)$$

For the cases where the loss tangent is low, typically $\tan \delta < 0.1$, Eqs (P.4.4.6) and (P.4.4.4)can be simplified to give

$$\varepsilon_m \cong \varepsilon \left[1 - \frac{S_2}{S} \left(1 - \frac{d+e}{d+\varepsilon e} \right) \right] = \varepsilon \left[1 - \frac{S_2}{S} \frac{e(\varepsilon-1)}{d+\varepsilon e} \right]$$

$$\tan D \cong \tan \delta \frac{1}{1 + \dfrac{\varepsilon e/d}{1 + \dfrac{S-S_2}{S_2} \dfrac{d}{d+e} \left(1 + \dfrac{\varepsilon e}{d} \right)^2}} \qquad (P.4.4.7)$$

For the set of values $\tan \delta = 0.1$, $d = 0.09 \,\mathrm{cm}$, $e = 0.01 \,\mathrm{cm}$, $\varepsilon = 3$, $S_2 = 1 \,\mathrm{cm}^2$, $S = 3.14 \,\mathrm{cm}^2$, the relative errors in the permittivity and loss tangent are respectively close to 5 and 7%.

Note that, when $S_2 = S$, the following formulae for low loss tangent hold

$$\varepsilon_m \cong \varepsilon \frac{d+e}{d+\varepsilon e} \simeq \varepsilon \frac{d+e}{d}$$

$$\tan D \cong \tan \delta \frac{1}{1+(\varepsilon e)/d}$$

(P.4.4.8)

where the second approximation in (P.4.4.8) holds if $d \gg \varepsilon e$. In this case, Eqs (P.4.4.8) are in good agreement with Eqs (P.4.3.3) and (P.4.3.8).

Problem 5

Develop and solve the equation for the thermally stimulated charging of a polar sample.

Solution

Equation (4.5.4.1) must be modified in order to introduce the effect of the presence of an electric field E. Then, the pertinent equation is

$$\frac{dP(t)}{dt} + \frac{P(t)}{\tau[T(t)]} = \frac{1}{\tau[T(t)]} P_0(T)$$

(P.4.5.1)

where $P_0(T)$ is the equilibrium value of P given by Eq. (4.5.4.4)

$$P_0(T) = \varepsilon \Delta \varepsilon E_p = \varepsilon \Delta \varepsilon E_p = \frac{N\mu^2}{3kT} E_p$$

(P.4.5.2)

For a small temperature dependence of the relaxation strength and an initially uncharged sample, the following approximate solution for Eq. (P.4.5.1) holds

$$P(t) \simeq P_0(t)\left(1 - \exp\left\{-\int_0^t \frac{dt}{\tau[T(t)]}\right\}\right)$$

(P.4.5.3)

Equation (4.5.4.2) must be modified in order to include the conduction current through an ohmic conductivity $\sigma(T)$, which gives a total current

$$i(t) = \frac{P_0(T)}{\tau[T(t)]} \exp\left\{-\int_0^t \frac{dt}{\tau[T(t)]}\right\} + \sigma(T)E_p$$

(P.4.5.4)

or equivalently

$$i(T) = \frac{P_0(T)}{\tau(T)} \exp\left[-s \int_{T_0}^{T} \frac{dT}{\tau(T)}\right] + \sigma(T)E_p \tag{P.4.5.5}$$

The two contributions to the current appearing in Eq. (P.4.5.5) correspond, respectively, to the transient process giving rise to a depolarization peak and to a conduction current which increases with temperature. When the sample is newly reheated, no more dipoles reorientate and only the conductive contribution is observed. In this way, conductive contributions can be conveniently separated. Thermally stimulated charge currents reveal how the polarization proceeds.

Problem 6

In the context of a many-body approach for the "universal dielectric response", the following equation was proposed by Jonscher to describe an isothermal depolarization experiment

$$i(t) = \frac{dP}{dt} = \frac{aP_0}{\tau(T)} \frac{1}{(t/\tau)^n + (t/\tau)^{m+1}} \tag{P.4.6.1}$$

where a is a normalization factor given by

$$a = \frac{m+1-n}{\pi} \sin\left[\frac{\pi(1-n)}{m+1-n}\right]$$

and $0 < n, m < 1$.

It is requested to find the total polarization for the case where $m = 1 - n$.

Solution

The total polarization can in principle be found by integration of Eq. (P.4.6.1). However, in most cases it is not possible to find a closed expression for the polarization. Only for $m = 1 - n$ is a closed solution to our present problem feasible. In this case and by noting that

$$\int \frac{dx}{x^{1+m} + x^{1-m}} = \frac{1}{m} \arctan x^m \tag{P.4.6.2}$$

the time dependence of the polarization for isothermal conditions is given by

$$P(t) = \frac{aP_0}{m} \arctan\left(\frac{t}{\tau}\right)^m \tag{P.4.6.3}$$

In principle, the parameters m and τ could be found from long or very short time data.

Problem 7

Obtain the equipotentials in the $x-y$ plane for the rectangular strip capacitor analyzed in the section A.1 of Appendix A.

Solution

By setting $z = x + iy$ and $w = u + iv$ in Eq. (4.A.1.7) and splitting into real and imaginary parts, we obtain

$$x = \frac{h}{\pi}\left[\frac{\pi}{V}u + \exp\left(\frac{\pi}{V}u\right)\cos\left(\frac{\pi}{V}v\right)\right]$$

$$y = \frac{h}{\pi}\left[\frac{\pi}{V}v + \exp\left(\frac{\pi}{V}u\right)\sin\left(\frac{\pi}{V}v\right)\right]$$

(P.4.7.1)

Then, for $v = V$, the former transformations yield

$$x = \frac{h}{V}u - \frac{h}{\pi}\exp\left(\frac{\pi}{V}u\right)$$

$$y = h$$

(P.4.7.2)

where the function x attains its maximum value $x = -h/\pi$ for $u = 0$.

Let us consider now the intensity of the field along the equipotentials. Since the magnitude of the derivative of an analytic function does not depend on the sense along which this derivative is calculated, we can perform such a calculation along the force field lines ($u = $ constant). Then $|dw| = |dv|$ and $|dz| = ds$, where

$$ds = [(dx)^2 + (dy)^2]^{1/2}$$

$$= \frac{h}{V}\left[1 + \exp\left(\frac{2\pi}{V}u\right) + 2\exp\left(\frac{\pi}{V}u\right)\cos\left(\frac{\pi}{V}v\right)\right]^{1/2} dv$$

(P.4.7.3)

Accordingly

$$E = \frac{dv}{ds} = \frac{V}{h}\left[1 + \exp\left(\frac{2\pi}{V}u\right) + 2\exp\left(\frac{\pi}{V}u\right)\cos\left(\frac{\pi}{V}v\right)\right]^{-1/2}$$

(P.4.7.4)

The maximum value of the field along the equipotential will be the minimum of the brackets in Eq. (P.4.7.4). The extreme condition gives

$$\exp\left(\frac{\pi}{V}u\right) + \cos\left(\frac{\pi}{V}v\right) = 0$$

(P.4.7.5)

For $v = \pm V/2$, the maximum is obtained for $u = -\infty$, that is, on the edge of the capacitor. The field is

$$E = \frac{V}{h}\left(1 + \exp\left(\frac{2\pi}{V}u\right)\right)^{-1/2} \tag{P.4.7.6}$$

Moreover, in this case Eqs (P.4.7.1) lead to

$$x = \frac{h}{V}u$$

$$y = \pm h\left(\frac{1}{2} + \frac{1}{\pi}\exp\left(\frac{\pi}{V}u\right)\right) \tag{P.4.7.7}$$

from which

$$y = \pm h\left(\frac{1}{2} + \frac{1}{\pi}\exp\left(\frac{\pi}{h}x\right)\right) \tag{P.4.7.8}$$

This is the equation of the equipotentials in the x–y plane. If the plates of the capacitor adopt the form of the equipotentials (Rogovsky capacitor), the field decreases as the edges are approached. In Fig. 4.17 the corresponding force field lines and the equipotentials are shown.

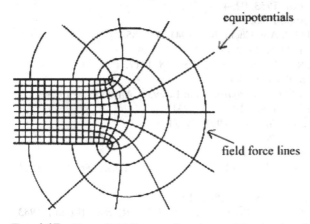

equipotentials

field force lines

FIG. 4.17 Force field lines and equipotentials showing the shape to the plates of a Rogovsky capacitor

REFERENCES

1. Mopsik, F.I. Rev. Sci. Insrum. **1984**, *55*, 79.
2. Garwe, F.; Schönnhals, A.; Lockwenz, H.; Beiner, M.; Schröter, K.; Donth, E. Macromolecules **1996**, *290*, 247.
3. Feldman, Y. Dielectric Newsletter of Novocontrol; Novocontrol Co.: Hundsangen, Germany, March 1995.
4. McCrum, N.G.; Read, B.E.; Williams, G. *Anelastic and Dielectric Effects in Polymeric Solids*; Dover: New York, 1991.
5. Kremer, F.; Arndt, M. In *Dielectric Spectroscopy of Polymeric Materials*; In Dielectric Spectroscopy of Polymeric Materials; Runt, J.P., Fitzgeral, J.J., Eds.; ACS: 1997; 67 pp.
6. Wagner, H.; Richert, R. Polymer **1997**, *38*, 255 and 5801.
7. Riande, E.; Díaz-Calleja, R.; Prolongo, M.G.; Masegosa, R.M.; Salom, C. *Polymer Viscoelasticity: Stress and Strain in Practice*; M. Dekker: 2000; 219 pp.
8. Scheiber, D.J. J. Res. Nat. Bur. Stds. Washington **1961**, *65C*, 23.
9. Cole, R.H. J. Phys. Chem. **1975**, *79*, 1459, 1469; Cole, R.H. Ann. Phys. Rev. Chem. **1977**, *28*, 283; Cole, R.H.; Mashimo, S.; Winsor, P. J. Phys. Chem. **1980**, *84*, 786; Mashimo, S. In *Dielectric Spectroscopy of Polymeric Materials*; Runt, J.P., Fitzgerald, J.J., Eds.; ACS: 1997; 201 pp.
10. Cole, K.; Cole, R.H. J. Chem. Phys. **1941**, *9*, 341.
11. Macdonald, J.R.; Brachman, M.K. Rev. Mod. Phys. **1956**, *28*, 393.
12. Tschoegl, N.W. *The Phenomenological Theory of the Linear Viscoelastic Behavior*; Springer-Verlag: Heidelberg, 1989.
13. Macdonald, J.R. *Impedance Spectroscopy*; Wiley: New York, 1987; 39 pp.
14. Jonscher, A.K. *Dielectric Relaxation in Solids*; Chelsea Dielectric: London, 1983.
15. Williams, G. In *Keynote Lectures in Selected Topics of Polymer Science*; Riande, E., Ed.; ICTP, CSIC: Madrid, 1995.
16. Douglas, J.F.; Hubbard, B. Macromolecules **1991**, *24*, 3163.
17. MacDonald, J.R. Phys. Rev. **1953**, *92*, 4.
18. Coelho, R. Rev. Phys. Appl. **1983**, *18*, 137.
19. Fuoss, R.; Kirkwood, J.G. J. Am. Chem. Soc. **1941**, *63*, 385.
20. Macdonald, J.R. J. Chem. Phys. **1952**, *20*, 1107.
21. Davidson, D.W.; Cole, R.H. J. Chem. Phys. **1950**, *18*, 1417.
22. Havriliak, S.; Negami, S. Polymer **1967**, *8*, 161.
23. Böttcher, C.J.F. *Theory of Electric Polarization*; Elsevier: Amsterdam, 1973.
24. Jonscher, A.K. Colloid Polym. Sci. **1975**, *253*, 231.
25. Schneider, A. J. Polym. Sci. Phys. Ed. **1991**, *29*, 1171.
26. Hill, R.M. Nature **1978**, *275*, 96.
27. Williams, G.; Watts, D.C. Trans. Far. Soc. **1970**, *66*, 80.
28. Williams, G.; Watts, D.C.; Dev, S.B.; North, A.M. Trans. Far. Soc. **1971**, *67*, 1323.
29. Montroll, E.W.; Bendler, J.T. J. Statist. Phys. **1984**, *34*, 129.
30. Dissado, L.; Hill, R.M. Nature **1979**, *279*, 685; Proc. R. Soc. (Lond.) **1983**, *A390*, 131.
31. Dissado, L.; Nigmatullin, R.; Hill, R.M. Adv. Chem. Phys. **1985**, *63*, 253.

32. Friedrich, C. Rheol. Acta **1992**, *31*, 309.
33. Meltzer, R.; Schick, W.; Kilian, H.G.; Nonnenmacher, T.F. J. Chem. Phys. **1995**, *103*, 7180.
34. Huet, C. Ann. Ponts Chaussees **1965**, *VI*, 5.
35. Pérez, J. Acta Metall. **1964**, *12*, 2163.
36. Díaz-Calleja, R.; Motos, J.; Sanchis, M.J. Polymer **1996**, *37*, 4003.
37. Faraday, M. *Experimental Researches in Electricity*; Richard and John Edward Taylor: London, 1839.
38. Heaviside, O. *Electrical Papers*; Chelsea: New York, 1892; 488–493.
39. Eguchi, M. Proc. Phys. Math. Soc Jap. **1919**, *1*, 326; Philos. Mag. **1925**, *49*, 178.
40. Gross, B. An. Acad. Bras. **1945**, *17*, 219; Phys. Rev. **1944**, *66*, 26; J. Chem. Phys. **1949**, *17*, 866.
41. Bucci, C.; Fieschi, R. Phys. Rev. Lett. **1964**, *12*, 16; Bucci, C.; Fieschi, R.; Guidi, G. Phys. Rev. **1966**, *148*, 816.
42. Chen, R.; Kirsh, Y. *Analysis of Thermostimulated Processes (International Series on the Science of the Solid State)*; Pergamon Press: Oxford, 1981; Vol. 15.
43. Hamon, B.V. Proc. Inst. Elect. Engrs. Part 4 **1952**, *27*, 99.
44. Schwarzl, F.R.; Struik, L.C.E. Adv. Mol. Relaxation Processes **1967–68**, *1*, 201.
45. Brather, A. Colloid Polym. Sci. **1979**, *467*, 785.
46. Simon, G.P.; Williams, G. Polymer **1993**, *34*, 2038.
47. Van Turnhout, J. Thermally stimulated discharge of polymer electrets. Ph.D. thesis, Leiden, 1972.
48. Hino, T. J. Appl. Phys. **1975**, *46*, 1956.
49. Perlman, M.M.; Unger, S. J. Appl. Phys. **1974**, *45*, 2389.
50. Shimizu, H.; Nakayama, K. J. Appl. Phys. **1991**, *74*, 1597.
51. Díaz-Calleja, R.; Sanchis, M.J.; Riande, E.; Guzmán, J. J. Appl. Phys. **1998**, *84*, 4436.
52. Kirchhoff, G. Monat. Deutsch. Akad. Wiss. Berlin **1877**, 144.
53. Binns, K.J.; Lawrenson, P.J.; Trowbridge, C.W. *The Analytical and Numerical Solution of Electric and Magnetic Fields*; Wiley: Chichester, UK, 1992.
54. Wintle, H.J.; Kurylowicz, S. IEEE Trans. Instrum. Measmt. **1985**, *34*, 41.
55. Love, A.E.H. Proc. Lond. Math. Soc. **1924**, *22*, 337.
56. Hutson, V. Proc. Camb. Phil. Soc. **1963**, *59*, 211.
57. Love, E.R. Can. J. Math **1963**, *15*, 631.
58. Cooke, J.C. Angew. Z. Math. Mech **1958**, *38*, 349.
59. Leppington, F.; Levine, H. Proc. Camb. Phil. Soc. **1970**, *68*, 235.
60. Sneddon, I.N. *Mixed Boundary Value Problems in Potential Theory*; North-Holland, 1966; Ch. 8.
61. Ianovici, M.; Morf, J.J. IEEE Trans. Elect. Instrum. **1977**, *12*, 165.
62. Tranter, C.J. *Integral Transforms in Mathematical Physics*, 2nd Ed.; Methuen: London, 1956; 117 pp.
63. Wintle, H.J.; Goad, D.G.W. J. Phys. D Appl. Phys. **1989**, *22*, 1620.
64. Amey, W.G.; Hamburger, F. ASTM Proc. **1949**, *49*, 1079.
65. Goad, D.G.W.; Wintle, H.J. Measmt. Sci. and Technol. **1990**, *1*, 965.
66. Senturia, S.D.; Garverick, S.L. US Patent 4,423,371.

67. Sheppard, Jr. N.F.; Garverick, S.L.; Day, D.R.; Senturia, S.D. Proc. 26th SAMPE Symposium, 1981; 65 pp.
68. Fodor, J.; Hill, D. Macromolecules **1992**, *25*, 3511; **1995**, *28*, 1271.
69. Stockmayer, W.H. Pure Appl. Chem. **1967**, *15*, 247.
70. Fodor, J.S.; Hill, D. J. Rheol. **1994**, *38*, 1071.
71. Durand, E. *Electrostatique*; Masson et Cie: 1964; Vol. 1.
72. Lamb, H. *Hydrodynamics*; Cambridge University Press: Cambridge, 1932.
73. Landau, L.D.; Lifshitz, E.M.; Pitaevskii, L.P. *Electrodynamics of Continuous Media*, 2nd Ed.; Butterworth: Oxford, 1995.

5

Mean-Square Dipole Moments of Molecular Chains

5.1. INTRODUCTION

After the concept of the macromolecule has been established in the mid-1920s,[1] it became clear that some properties of polymers, such as their anomalous viscoelastic behavior, were dependent on the internal degrees of freedom of the molecular chains. Kuhn[2] and Guth and Mark[3] made the first attempts at a mathematical description of the spatial conformations of flexible chains. The skeletal bonds were considered steps in a random walk of three dimensions, the steps being uncorrelated one to another. A more realistic approach to the description of the conformation-dependent properties of molecular chains, resting on the rotational isomeric states (RIS) model, was developed in large measure by Volkenstein[4] and others[5] in Leningrad in the late 1950s and early 1960s. The model, which takes into account skeletal bond lengths and angles, rotational angles associated with each skeletal bond, and their probabilities, as well as the contribution of each skeletal bond, to the property to be measured, was rationalized by Flory and coworkers[6] in the 1960s. The model has proved to be suitable for calculation of conformation-dependent properties at equilibrium, such as the mean-square end-to-end distances, the mean-square dipole moments, the molar Kerr constants, optical configuration parameters, etc., as a function of the chemical structure.

5.2. DIPOLE MOMENTS OF GASES

As indicated in section 1.9 of Chapter 1 [see Eq. (1.9.14)], the molar polarization, P, of a gas has two components: the orientation polarization, P_o, and the induced polarization, P_d, given by[7,8]

$$P = P_o + P_d = \frac{\varepsilon - 1}{\varepsilon + 2} \frac{M}{\rho} = \frac{4}{3} \pi N_A \alpha = \frac{4}{3} \pi N_A \left(\frac{\mu_p^2}{3k_B T} + \alpha_d \right) \qquad (5.2.1)$$

Here $\alpha = \alpha_0 + \alpha_d$, where α_0 and α_d are respectively the permanent and the induced polarizabilities of the molecule, μ_p is its permanent dipole moment, N_A and k_B are respectively Avogadro's number and the Boltzmann constant, and ρ and M are respectively the density and molecular weight of the gas.

The induced polarizability is governed by the strength with which the nuclear charges prevent the distortion of the electronic cloud by the applied field. This parameter increases with the atomic number, atomic size, and low ionic potential of the atoms comprising the molecules of the gas. Therefore, the induced polarizability results from the electronic polarizability, α_e, arising from the distortion of the electronic cloud by the action of the electric field, and the atomic polarizability, α_a, is caused by small displacements of atoms and groups of atoms in the molecule by the effect of the electric field. The magnitude of α_e can directly be obtained from Eq. (5.2.1) by making $\mu_p = 0$ and considering the Maxwell relationship $\varepsilon(\lambda) = n^2(\lambda)$. Because α_d corresponds to a static electric field, the index of refraction should be obtained at different wavelengths and its value extrapolated to $1/\lambda \to 0$. Hence, [Eq. (1.9.5)]

$$\frac{n^2 - 1}{n^2 + 2} = \frac{4\pi \rho N_A}{3M} \alpha_e \qquad (5.2.2)$$

This expression is called the Clausius–Mossotti equation [9,10]. The atomic polarizability cannot be determined directly, but its value is small and often negligible. Combining Eqs. (5.2.1) and (5.2.2) gives the Debye equation [Eq. (1.9.18)]

$$\frac{\varepsilon - 1}{\varepsilon + 2} - \frac{n^2 - 1}{n^2 + 2} = \frac{4\pi \rho N_A}{3k_B M T} \mu_p^2 \qquad (5.2.3)$$

This equation was found to hold for a variety of gases and vapors at ordinary pressures.

5.3. DIPOLE MOMENTS OF LIQUIDS AND POLYMERS

Since the Debye equation can only be used to determine the dipole moments of gases, its extension to measurements of the polarity of liquids requires measuring in conditions such that these substances may be considered to behave like gases. This situation can be achieved if the molecules of liquids are sufficiently separated from one another by nonpolar molecules, thus reducing the interactions between their permanent dipole moments.

In a solution containing n_1 molecules of nonpolar solvent and n_2 molecules of solute of molecular weights M_1 and M_2 respectively, the total molar polarization can be written as

$$P = \frac{4}{3}\pi N_A \bar{\alpha} = \frac{\varepsilon - 1}{\varepsilon + 2}\frac{\bar{M}}{\rho} \tag{5.3.1}$$

where $\bar{\alpha}$ is the polarizability of the solution, which in principle could be calculated from the partial molar polarizabilities of the solute and solvent, and the average molecular weight of the solution is considered to be

$$\bar{M} = x_1 M_1 + x_2 M_2 \tag{5.3.2}$$

where x_1 and x_2 are respectively the molar fraction of solvent and solute. For very dilute solutions ($x_2 \rightarrow 0$), intermolecular interactions between the molecules of solute will be negligible and $\bar{\alpha}$ in Eq. (5.3.1) will be the average of the polarizabilities of the solute α_2 and solvent α_1

$$\bar{\alpha} = x_1 \alpha_1 + x_2 \alpha_2 \tag{5.3.3}$$

Since the molar polarization of the solvent is given by

$$\frac{4}{3}\pi N_A \alpha_1 = \frac{\varepsilon_1 - 1}{\varepsilon_1 + 2}\frac{M_1}{\rho_1} \tag{5.3.4}$$

Eqs (5.3.1) to (5.3.4) lead to the following expression for the molar polarization of the solute

$$\begin{aligned}
P_2 &= \frac{4}{3}\pi N_A \alpha_2 = \frac{4}{3}\pi N_A \bar{\alpha}\left(\frac{\bar{\alpha}}{x_2} - \frac{x_1}{x_2}\alpha_1\right) \\
&= \frac{\varepsilon - 1}{\varepsilon + 2}\left(\frac{x_1 M_1}{\rho x_2} + \frac{M_2}{\rho}\right) - \frac{x_1(\varepsilon - 1)M_1}{x_2(\varepsilon + 2)\rho_1}
\end{aligned} \tag{5.3.5}$$

At very low concentrations, both the density and the dielectric permittivity of the solution can be expanded into a series, giving

$$\rho = \rho_1 + \frac{\partial \rho}{\partial w_2} w_2 = \rho_1 + \beta' w_2$$

$$\varepsilon = \varepsilon_1 + \frac{\partial \varepsilon}{\partial w_2} w_2 = \varepsilon_1 + \beta w_2$$

(5.3.6)

where w_2 is the weight fraction of solute, and ε_1 and ρ_1 are respectively the dielectric permittivity and density of the solvent. By substituting Eqs (5.3.6) into Eq. (5.3.5), and taking into account that

$$\frac{x_1}{x_2} = \frac{w_1}{w_2} \frac{M_2}{M_1}$$

(5.3.7)

the following expression for the total molar polarization of the solute is obtained

$$P_2 = \frac{\varepsilon_1 - 2}{\varepsilon_1 + 2} \frac{M_2}{\rho_1} + \frac{M_2}{\rho_1} \frac{3\beta - (\varepsilon_1 + 2)(\varepsilon_1 - 1)(\beta'/\rho_1)}{(\varepsilon_1 + 2)^2}$$

(5.3.8)

where the quadratic terms in w_2 were neglected. Therefore, it is assumed that $\rho \to \rho_1$, $\varepsilon \to \varepsilon_1$ when $w_2 \to 0$. After regrouping terms in Eq. (5.3.8), and considering that $\beta' = (-1/v^2)\partial v/\partial w_2$, the so-called Halverstadt and Kumler equation[11] is obtained

$$P_2 = M_2 \frac{3v_1}{(\varepsilon_1 + 2)^2} \frac{\partial \varepsilon}{\partial w_2} + M_2 \left(v_1 + \frac{\partial v}{\partial w_2} \right) \frac{\varepsilon_1 - 1}{\varepsilon_1 + 2}$$

(5.3.9)

where v and v_1 represent the specific volume of the solution and solvent, respectively.

The molar electronic polarization of the solute, P_{e2}, can be obtained from Eq. (5.3.9) by taking into account that, at very high frequencies, $\varepsilon_1 = n_1^2$ and $\varepsilon = n^2$, where n and n_1 are the index of refraction of the solution and the solvent respectively. Accordingly

$$P_{e2} = M_2 \frac{6v_1 n_1}{(n_1^2 + 2)^2} \frac{\partial n}{\partial w_2} + M_2 \left(v_1 + \frac{\partial v}{\partial w_2} \right) \frac{n_1^2 - 1}{n_1^2 + 2}$$

(5.3.10)

The molar orientation polarization P_{o2} of the solute is given by

$$P_{o2} = P_2 - P_{e2} - P_{a2}$$

(5.3.11)

In most systems, the molar atomic polymerization P_{a2} amounts to only 5–10% of the molar electronic polarization. Therefore, this contribution is often neglected in the calculations of dipole moments. It follows from Eq. (5.2.1) that

$$\mu_p^2 = \frac{9k_BT}{4\pi N_A}P_{o2} \tag{5.3.12}$$

This expression is often used for the experimental determination of the dipole moments of molecules without internal degrees of freedom.

Flexible molecules are continuously changing their spatial conformations, and, because the dipole moment associated with each conformation is generally different, the dipole moments that are measured are average values. Then, Eq. (5.3.12) should be written as

$$\langle\mu_p^2\rangle = \frac{9k_BT}{4\pi N_A}P_{o2} \tag{5.3.13}$$

where the angular brackets denote average.

By defining a fictitious atomic polarizability for the solute as

$$\alpha_{a2}' = \alpha_{a1}\frac{V_2}{V_1} \tag{5.3.14}$$

where α_{a1} is the polarizability of the solvent and V_2 and V_1 are the molar volume of the solute and the solvent respectively, the application of the Debye equation to solutions leads to

$$\frac{4}{3}\pi N_A\left(\alpha_{a2} - \alpha_{a2}' + \frac{\langle\mu_p^2\rangle}{3k_BT}\right) = 3M_2v_1\left[\frac{1}{(\varepsilon_1 + 2)^2}\frac{\partial\varepsilon}{\partial w_2} - \frac{2n}{(n_1^2 + 2)^2}\frac{\partial n}{\partial w_2}\right] \tag{5.3.15}$$

In principle, the atomic polarizability of nonpolar solvents ($\mu_p = 0$) can be obtained by means of the Debye equation. Actually

$$\frac{4}{3}\pi N_A(\alpha_{a1} + \alpha_{e1}) = \frac{\varepsilon_1 - 1}{\varepsilon_1 + 2}\frac{M_1}{\rho_1}$$

$$\tag{5.3.16}$$

$$\frac{4}{3}\pi N_A\alpha_{e1} = \frac{n_1^2 - 1}{n_1^2 + 2}\frac{M_1}{\rho_1}$$

Experiments involving the determination of the dielectric permittivity and the index of refraction of nonpolar solvents by means of Eq. (5.3.16) show that P_{a1} is roughly one-tenth and even less of P_{e1}. There is no reason to believe that P_a for polar substances is larger than the molar polarization for nonpolar ones. In view of these circumstances, the assumption that $P_{a2} = P_{a2}'$ is usually made. Hence,

Eq. (5.3.15) is customarily written as

$$\langle \mu_p^2 \rangle = \frac{27 k_B T M}{4 \pi \rho_1 N_A (\varepsilon_1 + 2)^2} \left(\frac{\partial \varepsilon}{\partial w_2} - 2n \frac{\partial n}{\partial w_2} \right) \tag{5.3.17}$$

where the approximation $\varepsilon_1 = n_1^2$ is made. Equation (5.3.17) is known as the Guggenheim–Smith equation.[12,13]

A shortcoming of the Debye-based equations is that although progressive dilution eliminates intermolecular dipole–dipole interactions, intramolecular dipole–dipole correlations are not considered. Models such as those developed by Kirkwood[14] and Fröhlich[15] (see Sections 1.14 and 1.15 of Chapter 1), which allow for the interaction of surrounding dipoles by correlation function treatment, would then be more appropriate, though their application introduces difficult computations that are often rather arbitrary. However, many dipole moments obtained for oligomers and polymers using Debye-type equations such as the Halverstadt–Kumler and Guggenheim–Smith equations show consistency among them, presumably because intramolecular dipole–dipole interactions in flexible chains fade away for dipoles separated by four or more flexible skeletal bonds. Therefore, these equations are among the most used to evaluate the dipole moments of isolated flexible chains.

5.4. EFFECT OF THE ELECTRIC FIELD ON THE MEAN-SQUARE DIPOLE MOMENT

Let us consider a macromolecular system under an external electric field acting along the x axis. The energy V_F associated with a given conformation of a molecular chain is the result of the energy of that conformation in the absence of an electric field, which will be called V, plus the interactions of the permanent and induced dipole moments of the conformation with the electric field, that is

$$V_F = V - \mu_x F - \tfrac{1}{2} \alpha'_{xx} F^2 \tag{5.4.1}$$

where F is the effective electric field. The component of the polarizability tensor in the direction of the field, α'_{xx}, can be neglected for polar systems so that Eq. (5.4.1) becomes

$$V_F = V - \mu_x F \tag{5.4.1a}$$

The rotational partition function of the chains in the presence of a field can be written as

$$Z_F = \frac{\int_{ext} \cdots \int_{int} \exp[-V/(kT)] \exp[(\mu_x F)/(kT)] \, d\{\phi_{int}\} \, d\Omega_{ext}}{\int_{ext} d\Omega_{ext}} \tag{5.4.1b}$$

where $d\Omega_{ext} = \sin \xi \, d\varphi \, d\zeta$, ξ, φ and ζ being the Euler angles. Note that $d\{\phi_{int}\}$ means that the integral in Eq. (5.4.2) must be performed over all the conformations of the chains. By assuming that $\mu_x F < kT$, Eq. (5.4.1b) can be expanded into a series, giving[16]

$$Z_F = \frac{1}{8\pi^2} \int_{ext} \cdots \int_{int} \exp\left(-\frac{V}{kT}\right)$$

$$\times \left[1 + \frac{F}{kT}\mu_x + \frac{1}{2}\left(\frac{F}{kT}\right)^2 \mu_x^2 + \frac{1}{3!}\left(\frac{F}{kT}\right)^3 \mu_x^3 + \frac{1}{4!}\left(\frac{F}{kT}\right)^4 \mu_x^4 + \cdots\right]$$

$$\times d\{\phi_{int}\} \, d\Omega_{ext} \tag{5.4.2}$$

Terms higher than $(F/kT)^4$ were neglected in the series expansion of the second exponential. In the absence of perturbation fields, the conformational partition function Z of the isolated molecular chains can be written as

$$Z = \frac{1}{8\pi^2} \int_{ext} \cdots \int_{int} \exp\left(-\frac{V}{kT}\right) d\{\phi_{int}\} \, d\Omega_{ext}$$

$$= \int_{int} \exp\left(-\frac{V}{kT}\right) d\{\phi_{int}\} \tag{5.4.3}$$

Note that the exponent in this integral does not depend on the external coordinates. On the other hand (see Chapter 6, Section 6.7)

$$\int_{int} \mu_x \exp\left(-\frac{V}{kT}\right) d\phi_{int} = Z\frac{F}{kT}\langle \mu_x \rangle \tag{5.4.4}$$

where the angular brackets denote averages.

$$\int_{int} \mu_x^2 \exp\left(-\frac{V}{kT}\right) d\{\phi_{int}\} = Z\langle \mu_x^2 \rangle \tag{5.4.5}$$

$$\int_{int} \mu_x^4 \exp\left(-\frac{V}{kT}\right) d\{\phi_{int}\} = Z\langle \mu_x^4 \rangle \tag{5.4.6}$$

Note that odd terms vanish because, in the absence of a field, the three directions of space are equivalent and for each conformation will be another one with the same energy having opposite sign. Therefore

$$\langle \mu_x \rangle = \langle \mu_x \rangle = \langle \mu_x \rangle = \langle \mu \rangle = \langle \mu^3 \rangle = 0 \tag{5.4.7}$$

On the other hand, symmetry dictates

$$\langle \mu_x^2 \rangle = \langle \mu_y^2 \rangle = \langle \mu_z^2 \rangle = \tfrac{1}{3} \langle \mu^2 \rangle$$

$$\langle \mu_x^4 \rangle = \frac{1}{9} \langle \mu^4 \rangle \tag{5.4.8}$$

Hence, Eq. (5.4.2) can be written as

$$Z_F = Z \left[1 + \frac{1}{6} \left(\frac{F}{kT} \right)^2 \langle \mu^2 \rangle + \frac{1}{216} \left(\frac{F}{kT} \right)^4 \langle \mu^4 \rangle + \cdots \right] \tag{5.4.9}$$

which suggests that $Z_F = Z$ if $F \ll kT$.

Let us now consider the relation between the mean-square dipole moment in the presence of a field F, denoted as $\langle \mu^2 \rangle_F$, and the mean square dipole moment in the absence of a field, $\langle \mu^2 \rangle$. The value of the former quantity can be written as

$$\langle \mu^2 \rangle_F = \frac{1/8\pi^2 \int_{\text{ext}} \cdots \int_{\text{int}} \mu^2 \exp[-V/(kT)] \exp[(\mu_x F)/(kT)] \, d\{\phi_{\text{int}}\} \, d\Omega_{\text{ext}}}{Z_F}$$

$$= \left(\frac{Z}{Z_F} \right) \left[\langle \mu^2 \rangle + \frac{1}{2} \left(\frac{F}{kT} \right)^2 \langle \mu^2 \mu_x^2 \rangle + \frac{1}{4!} \left(\frac{F}{kT} \right)^4 \langle \mu^2 \mu_x^4 \rangle + \cdots \right] \tag{5.4.10}$$

The symmetry of the even components transforms Eq. (5.4.10) into

$$\langle \mu^2 \rangle_F = \left(\frac{Z}{Z_F} \right) \left[\langle \mu^2 \rangle + \frac{1}{6} \left(\frac{F}{kT} \right)^2 \langle \mu^4 \rangle + \frac{1}{216} \left(\frac{F}{kT} \right)^4 \langle \mu^6 \rangle + \cdots \right] \tag{5.4.11}$$

In most cases the deviations of $\langle \mu_F^2 \rangle$ from $\langle \mu^2 \rangle$ can adequately be described by means of Eq. (5.4.11). Equation (5.4.11) suggests that, if $F \ll kT$, then $\langle \mu^2 \rangle_F = \langle \mu^2 \rangle$. The higher moments for short chains can directly be computed from the dipoles associated with the conformations and statistical weights. On the other hand, expressions have been derived that make it possible to obtain the exact values of even moments higher than 2 for polymer chains.[6] The value of the mean-square dipole moment of 1,4-dibromoethane at 25°C, in the absence of an electric field, is $5.78D^2$. The values of this quantity for $F/kT = 0.03 \, D^{-1}$ and $0.09 \, D^{-1}$ are $5.78 \, D^2$ and $5.81 \, D^2$ respectively. It can be seen that, for low electric fields, the magnitude of the field has a negligible influence on the value of the mean-square dipole moment of 1,4-dibromoethane.

The upper bound values for the even moments corresponding to Gaussian distributions of infinite chain length are[16]

$$\langle \mu^{2j} \rangle = \frac{(2j+1)!}{j!} \left(\frac{\langle \mu^2 \rangle}{6} \right)^j \tag{5.4.12}$$

Therefore, the effect of the electric field on the mean-square dipole moment of molecular chains can directly be calculated from the mean-square dipole moment in the absence of an external electric field. Actually, substitution of Eq. (5.4.12) into Eqs (5.4.11) and (5.4.10) yields the maximum value of Z_F and the mean-square dipole moment for the normal range of T, F, and $\langle \mu^2 \rangle$ as[16]

$$Z_F = Z \left[1 + \frac{1}{6} \left(\frac{F}{kT} \right)^2 \langle \mu^2 \rangle + \frac{5}{648} \left(\frac{F}{kT} \right)^4 \langle \mu^2 \rangle^2 + \cdots \right] \tag{5.4.13}$$

$$\langle \mu^2 \rangle_F = \left(\frac{Z}{Z_F} \right) \left[\langle \mu^2 \rangle + \frac{5}{18} \left(\frac{F}{kT} \right)^2 \langle \mu^2 \rangle^2 + \frac{35}{1944} \left(\frac{F}{kT} \right)^4 \langle \mu^2 \rangle^3 + \cdots \right] \tag{5.4.14}$$

Substitution of Eq. (5.4.13) into Eq. (5.4.14) yields

$$\langle \mu^2 \rangle_F = \langle \mu^2 \rangle + \frac{1}{9} \left(\frac{F}{kT} \right)^2 \langle \mu^2 \rangle^2 - \frac{2}{243} \left(\frac{F}{kT} \right)^4 \langle \mu^2 \rangle^3 + \cdots \tag{5.4.15}$$

Since the interactions between the electric field and the dipole moments decrease the energy of the system, those conformations with higher energy are favored by the field effect.

5.5. EXCLUDED VOLUME EFFECTS

Dipole moments can rigidly be attached to the skeletal bonds or associated with flexible side groups. In the former case, dipoles can be parallel or perpendicular to the chain contour, and, according to Stockmayer's notation,[17] these dipoles are of type A and B respectively. Dipoles located in flexible side groups are of type C. Some polar polymers, such as poly(propylene oxide), characterized for not having the repeat unit appropriate symmetry elements,[18] display dipole moments with components parallel and perpendicular to the chain contour, and these chains are of type AB. As indicated in Chapter 2, dipole moments and end-to-end distance are uncorrelated for chains with dipoles of types B and C, and therefore the mean-square dipole moment of these chains should not exhibit excluded volume effects. However, the dipoles of type A and AB are correlated with the end-to-end distance of the chains, $\mathbf{r}$, and therefore present an excluded volume effect.

Following Fixman's[19] application of the multivariate Gaussian distribution for $\mathbf{r}$ and $\mathbf{r}_{ij}$ for the purpose of evaluating the expansion coefficient α_r^2, Nagai and Ishikawa[18] employed the multivariate Gaussian distribution for $\boldsymbol{\mu}$ and $\mathbf{r}_{ij}$ to calculate α_μ^2. As usual, $\boldsymbol{\mu}$ and $\mathbf{r}$ represent the dipole and the end-to-end distance vectors for a specific conformation respectively, and $\mathbf{r}_{ij}$ is the distance vector from segment i to segment j in that conformation. Moreover, $\alpha_\mu^2 = \langle \mu^2 \rangle / \langle \mu^2 \rangle_0$ and $\alpha_r^2 = \langle r^2 \rangle / \langle r^2 \rangle_0$, where the averages with subscript zero refer to the mean-square dipole moment and the mean-square end-to-end distance in the absence of excluded volume effects. The expansion factor α_μ^2 for a chain of infinite length up to the term z^3 is given by[19]

$$\alpha_\mu^2 - 1 = \left(\alpha_r^2 - 1 \right) \frac{\langle \mathbf{r} \cdot \boldsymbol{\mu} \rangle_0^2}{\langle r^2 \rangle_0 \langle \mu^2 \rangle_0} \tag{5.5.1}$$

where $\alpha_r^2 = 1 + C_1 z + C_2 z^2 + C_3 z^3$. In these expressions $z = (3/2\pi C_{rr})^{3/2} \beta N^{1/2}$, the parameter β being the binary cluster integral for units or segments,[20] N the number of segments, and $C_{rr} = \lim \langle r^2 \rangle_0 / N$ for $N \to \infty$. The values of C_1, C_2, and C_3 are $4/3$, -2.075, and 6.459 respectively.[18] The fact that

$$0 \leqslant \frac{(\mathbf{r} \cdot \boldsymbol{\mu})^2}{r^2 \mu^2} \leqslant 1 \tag{5.5.2}$$

leads to

$$0 \leqslant \frac{\langle \mathbf{r} \cdot \boldsymbol{\mu} \rangle_0^2}{\langle r^2 \rangle_0 \langle \mu^2 \rangle_0} \leqslant 1 \tag{5.5.3}$$

Hence, Eq. (5.6.1) suggests that

$$\alpha_\mu^2 \leqslant \alpha_r^2 \tag{5.5.4}$$

For most conventional polymers $\langle \mathbf{r} \cdot \mu \rangle = 0$. Among the polymers with $\langle \mathbf{r} \cdot \mu \rangle \neq 0$ are polyesters and polyamides (and polypeptides) derived from α,ω-hydroxy acids and α,ω-amino acids respectively, poly(propylene oxide), poly(propylene sulfide, cis-poly(isoprene), etc. A more detailed list of these polymers is given in the Table 1 of Chapter 8.

5.6. DIPOLE MOMENTS FOR FIXED CONFORMATIONS

The dipole moment associated with a given conformation can be written as

$$\boldsymbol{\mu} = \sum_i \mathbf{m}_i \tag{5.6.1}$$

where $\mathbf{m}_i$ is the contribution to $\boldsymbol{\mu}$ of bond i, and the sum expands over all bonds of the molecular chain. Equation (5.6.1) rests upon the assumption of additivity of the contribution $\boldsymbol{\mu}_i$ from each bond of the polymer. If additivity holds, this means that the contribution to the dipole moment has to be the same regardless of the environment. Thus, it would be invariant with the type of molecule and with the conformation it may adopt. For this reason, the contributions of groups of bonds, rather than the contribution of a single bond, are used in the evaluation of dipole moments of molecular chains.

The computation of $\boldsymbol{\mu}$ is customarily made by assigning a reference frame to each skeletal bond of the chain, and defining an operator matrix that makes it possible to transform the reference frame defined for bond i into coincidence with that of bond $i - 2$, repeating the transformation from $i = n$ to $i = 2$. Of course, the contribution of each bond to $\boldsymbol{\mu}$ is given on its own reference frame. As indicated in Fig. 5.1, the skeletal bond i corresponds to the x axis of the reference frame, the y axis is located in the plane defined by the ith and the $i - 1$th skeletal bonds, whereas the z axis is perpendicular to this plane to complete a right-handed coordinate system. The matrix that performs the transformation of coordinates from skeletal bond $i + 1$ to i is given by

$$\mathbf{T}_i = \begin{pmatrix} \cos\theta & \sin\theta & 0 \\ \sin\theta\cos\phi & -\cos\theta\cos\phi & \sin\phi \\ \sin\theta\sin\phi & -\cos\theta\sin\phi & -\cos\phi \end{pmatrix}_i \tag{5.6.2}$$

where θ_i is the supplement of the valence angle between bonds i and $i + 1$, and ϕ_i is the rotational angle of bond i with $\phi = 0$ for the *trans* conformation.

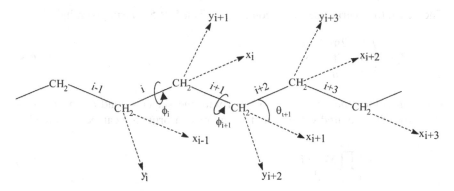

FIG. 5.1 Segment of a polymer chain in all-*trans* ($\phi = 0$) conformation. Coordinate systems affixed to each skeletal bond are indicated

Equation (5.6.1) can be written in a more explicit form as

$$\mu = m_1 + T_1\{m_2 + T_2[m_3 + T_3(m_4 + \cdots)]\} \tag{5.6.3}$$

The use of Eq. (5.6.3) to evaluate μ^2 gives expressions that are too cumbersome even for small chains. A more elegant and easy way to calculate this magnitude is provided by the matrix multiplication scheme developed by Flory and coworkers.[21] It uses generator matrices constructed in such a way that their product generates all the terms required to compute μ. The generator matrix associated with skeletal bond i is a 4×4 matrix given by

$$A_i = \begin{pmatrix} T & m \\ 0 & 1 \end{pmatrix}_i \tag{5.6.4}$$

where T and μ are respectively the transformation matrix and dipole moment associated with bond i. The value of μ is then given by

$$\mu = A_1 \left[\prod_{i=2}^{n-1} A_i \right] A_n \tag{5.6.5}$$

where the generator matrices associated with the first and the last bonds of the chains are 3×4 and 4×1 matrices given by

$$A_1 = (T_1 \quad m_1) \qquad \text{and} \qquad A_n = \text{column}(m_n \quad 1) \tag{5.6.6}$$

Usually, the magnitude to be calculated is μ^2. This quantity for a given conformation can be written as

$$\mu^2 = \sum_{i=1}^{n} m_i \cdot \sum_{j=1}^{n} m_j = \sum_{i=1}^{n} m_i^2 + \sum_{i \neq j} \sum_{j} m_i \cdot m_j \tag{5.6.7}$$

The generator matrix needed to compute μ^2 is a 5×5 matrix given by[21]

$$M_i = \begin{pmatrix} 1 & 2m^T & \mu^2 \\ 0 & T & m \\ 0 & 0 & 1 \end{pmatrix}_i \tag{5.6.8}$$

where m is the dipole vector associated with the skeletal bond i and m^T is its transpose. The required value of μ^2 for a given conformation can be computed as

$$\mu^2 = M_1 \left[\prod_{i=2}^{n-1} M_i \right] M_n \tag{5.6.9}$$

Here M_1 is a 1×5 matrix appearing as the first row of M for the first bond of the chains, whereas M_n is a 5×1 matrix corresponding to the last column of M for the last bond. The construction of generator matrices is described in detail elsewhere.[6,21] However, it is easy to check that Eqs. (5.6.5) and (5.6.9) generate the same products as Eqs (5.6.3) and (5.6.7) respectively, with the advantage of being easily programmable for computer calculations.

5.7. AVERAGE VALUES OF μ^2

The computation of conformation-dependent properties is often carried out by means of the rotational isomeric states (RIS) model. The model assumes that each skeletal bond can adopt a discrete number of rotational states $\alpha, \beta, \ldots, \nu$. Intramolecular interactions are accounted for by assigning statistical weights to conformations about two or more consecutive skeletal bonds. In most cases, conformations defined by the rotations about only two consecutive skeletal bonds are considered, and the statistical weights or Boltzmann factor associated with skeletal bond i can be written as

$$\sigma_i = \exp\left(-\frac{E_{\alpha\beta}}{RT}\right) \tag{5.7.1}$$

where $E_{\alpha\beta}$ represents the conformational energy of the rotational state β of bond i provided that bond $i - 1$ is the α-state. In this case, the statistical of each allowed conformation for this pair of bonds is written as a matrix of dimensions $\nu \times \nu$ defined as[6]

$$\mathbf{U}_i = \begin{pmatrix} \sigma_{\alpha\alpha} & \sigma_{\alpha\beta} & \cdot & \cdot & \cdot & \sigma_{\alpha\nu} \\ \sigma_{\beta\alpha} & \sigma_{\beta\beta} & \cdot & \cdot & \cdot & \sigma_{\beta\nu} \\ \cdot & & & & & \cdot \\ \cdot & & & & & \cdot \\ \cdot & & & & & \cdot \\ \sigma_{\nu\alpha} & \sigma_{\nu\beta} & \cdot & \cdot & \cdot & \sigma_{\nu\nu} \end{pmatrix}_i \tag{5.7.2}$$

The rotational partition function, Z, of a chain of n skeletal bonds can be calculated as

$$Z = \mathbf{U}_1 \left(\prod_{i=2}^{n-1} \mathbf{U}_i \right) \mathbf{U}_n \tag{5.7.3}$$

where the initial and final statistical weight matrices are, respectively, a row of one unity followed by $\nu - 1$ zeros and a column of ν unities, that is

$$\mathbf{U}_1 = \text{row}(1 \quad 0 \quad 0 \quad \ldots \quad 0)$$
$$\mathbf{U}_n = \text{col}(1 \quad 1 \quad 1 \quad \ldots \quad 1) \tag{5.7.4}$$

The computation of the mean-square dipole moment $\langle \mu^2 \rangle$ involves the generation of the product of statistical weight of each conformation by the square of the dipole moment associated with it, normalization with the rotational partition function, and further summation of the values obtained. This can be accomplished by means of a serial product of supermatrices formed by combination of the statistical weight matrices $\mathbf{U}$ and the generator matrices $\mathbf{M}$. By defining a pseudodiagonal matrix as[6,21]

$$\|\mathbf{M}\|_i = \begin{pmatrix} \mathbf{M}(\alpha) & \mathbf{0} & \cdot & \cdot & \mathbf{0} \\ \mathbf{0} & \mathbf{M}(\beta) & \cdot & \cdot & \mathbf{0} \\ \cdot & \cdot & \cdot & \cdot & \cdot \\ \cdot & \cdot & \cdot & \cdot & \cdot \\ \mathbf{0} & \mathbf{0} & \cdot & \cdot & \mathbf{M}(\nu) \end{pmatrix}_i \qquad (5.7.5)$$

the required supermatrices are

$$\mathbf{G}_i = (\mathbf{U}_i \otimes \mathbf{E}_5)\|\mathbf{M}\|_i \qquad (5.7.6)$$

where $\mathbf{E}_5$ is the 5×5 identity matrix and $\mathbf{G}_i$ is a $5\nu \times 5\nu$ matrix, where ν is the number of rotational states. The supermatrices corresponding to the first and last bonds are

$$\begin{aligned} \mathbf{G}_1 &= \mathbf{U}_1 \otimes \mathbf{M}_1 \\ \mathbf{G}_n &= \mathbf{U}_n \otimes \mathbf{M}_n \end{aligned} \qquad (5.7.7)$$

In this expression, $\mathbf{M}_1$ and $\mathbf{M}_n$ represent respectively the first row and the last column of $\mathbf{M}$ corresponding to the first and last skeletal bond of the chains, whereas $\mathbf{U}_1$ and $\mathbf{U}_n$ are given by Eq. (5.7.4).

Once all the $\mathbf{G}_i$ matrices have been formed, the statistical average of μ^2 over all the conformations can be finally computed as

$$\langle \mu^2 \rangle = Z^{-1} \mathbf{G}_1 \left[\prod_{i=2}^{n-1} \mathbf{G}_i \right] \mathbf{G}_n \qquad (5.7.8)$$

where Z is the partition function evaluated according to Eq. (5.7.3).

5.8. ROTATIONAL STATES AND CONFORMATIONAL ENERGIES

The location of the rotational states and the establishment of their relative energies are obtained, whenever possible, by direct measurements, either spectroscopic or thermodynamic, on small molecules having structural characteristics similar to the chain under investigation. When this information is not available, precise values of conformational energies can be obtained using a complete *ab initio* quantum mechanical approach. However, the size of the associated secular equation may be so large that the time involved in the solution becomes prohibitively large even for the faster computer. Simplified energy calculations were developed to account for the interaction between nonbonded atoms, trying to combine the sophistication of the potential functions with the minimization of the time required to solve them.[22]

Even molecules having no permanent dipoles have instantaneous dipoles arising from the continuous fluctuations of their electronic clouds. A molecule with an instantaneous dipole $\boldsymbol{\mu}_1^*$ will polarize another neighboring molecule, inducing upon it an instantaneous dipole $\boldsymbol{\mu}_2^*$ giving rise to an attractive interaction between them that does not average to zero. Since the values of $\boldsymbol{\mu}_1^*$ and $\boldsymbol{\mu}_2^*$ depend on the looseness of the control of the nuclear potential on the outer electrons in the two molecules, the interaction is governed by the polarizabilities of the interacting molecules. The potential associated with this interaction has been rigorously obtained for a pair of identical, spherically symmetrical, and chemically saturated molecules, and has been found to vary as R^{-6}, where R is the distance between the two molecules.[23] By analogy, the attractive term for the potential of two nonbonded atoms or atomic groups is taken as $-A_{ij}/r_{ij}^{-6}$, where r_{ij} is the distance between them. The term A_{ij} can be estimated by means of the Slater–Kirkwood equation[24] as modified by Scott and Scheraga[25]

$$A_{ij} = \frac{3e\hbar\,\alpha_i\alpha_j}{2m^{1/2}C_{ij}} \tag{5.8.1}$$

where

$$C_{ij} = \left(\frac{\alpha}{N_{\text{eff}}}\right)_i^{1/2} + \left(\frac{\alpha}{N_{\text{eff}}}\right)_j^{1/2} \tag{5.8.2}$$

In these equations, e is the electronic charge, $\hbar = h/2\pi$, where h is Planck's constant, m is the electronic mass, α is the atomic polarizability, and N_{eff} is a correcting factor that takes into account the nature, strength, and orientation of the chemical bond; N_{eff} can be considered as the number of effective electrons surrounding the atoms in the chemical bond.

As the separation between two molecules or atomic groups decreases, a distance is reached at which the electronic and nuclear repulsions begin to dominate the attractive forces, and the repulsive interactions rise very steeply with decreasing distance. The potential curve as a function of the distance that accounts for the van der Waals interactions is reasonably well approximated by the equation

$$V(r) = \frac{B_{ij}}{r_{ij}^n} - \frac{A_{ij}}{r_{ij}^6} \tag{5.8.3}$$

When $n = 12$, Eq. (5.8.3) is called the Lennard–Jones potential. The value of B_{ij} is currently obtained by assuming that V_{ij} reaches a minimum at a distance R_0 approximately equal to the sum of the van der Waals radii of the interacting atoms. Hence

$$B_{ij} = \frac{A_{ij}}{2} R_0^6 \tag{5.8.4}$$

Other van der Waals potential functions are given elsewhere.[22] Orbital–orbital interactions for rotations about perfect single bonds are zero owing to the cylindrical symmetry of these bonds. However, most of the the skeletal bonds are not pure single bonds and, as a consequence, a torsional potential appears whose value increases as the deviation from pure single bonds increases. For near double and triple bonds, orbital–orbital interactions clearly dominate and rotations about these bonds are forbidden. The torsional potential can be written as

$$V(\phi) = \frac{V^*}{2}(1 - \cos n\phi) \tag{5.8.5}$$

where V^* is the barrier height, n is the periodicity of the function, and ϕ is the rotational angle. The value of V^* is usually obtained from microwave spectroscopy and also from the temperature dependence of NMR spectra of model compounds of the bonds of the chains under consideration.

Finally, coulombic contributions arise from permanent dipole interactions. Residual charges $+\delta$ and $-\delta$ associated with the atoms of the chains can be determined either from permanent dipoles or by using semi-empirical quantum mechanics methods. Coulombic contributions can be expressed by

$$V_C = \sum \frac{\delta_i \delta_j}{\varepsilon r_{ij}} \tag{5.8.6}$$

where ε is the dielectric constant of the medium.

By considering the atoms as spheres, rotational angles and conformational energies can be obtained by means of the standard equation

$$V(\phi_i, \phi_j) = \frac{V_i^*}{2}(1 - \cos n\phi_i) + \frac{V_j^*}{2}(1 - \cos n\phi_j)$$

$$+ \sum\sum\left(\frac{B_{ij}}{r_{ij}^{12}} - \frac{A_{ij}}{r_{ij}^6}\right) + V_C \tag{5.8.7}$$

This expression includes torsional, van der Waals, and Coulombic contributions to the potential. In this equation, ϕ_i and ϕ_j are the rotational angles of the consecutive i,j skeletal bonds, taking as trans state $\phi_i = \phi_j = 0$. More sophisticated potential functions that include bond length distortion, bond angle bending, etc., are described in Chapter 7.

Average values of the rotational partition function, Z, conformational energies, $\langle E \rangle$, and rotational angles associated with the potential wells are given by

$$\langle E \rangle = Z^{-1}\sum\sum V(\phi_i, \phi_j)\exp\left[-\frac{V(\phi_i, \phi_j)}{RT}\right]$$

$$\tag{5.8.8}$$

$$\langle\phi\rangle_j = Z^{-1}\sum\sum \phi_j\exp\left[-\frac{V(\phi_i, \phi_j)}{RT}\right]$$

where Z is the rotational potential function given by

$$Z = \sum\sum\exp\left[-\frac{V(\phi_i,\phi_j)}{RT}\right] \tag{5.8.9}$$

5.9 DIPOLE AUTOCORRELATION COEFFICIENT OF POLYMERS

As occurs with the molecular dimensions, the mean-square dipole moment of polymers increases with molecular weight. It is convenient to express this quantity as the dimensionless parameter g, also called the dipolar autocorrelation coefficient, given by

$$g = \frac{\langle\mu^2\rangle_0}{\langle\mu^2\rangle_{fj}} \tag{5.9.1}$$

where $\langle \mu^2 \rangle_{fj}$ is the mean-square dipole moment for a freely jointed chain. According to Eq. (5.6.7)

$$\langle \mu^2 \rangle = \left\langle \sum_{i=1}^{n} \mathbf{m}_i \cdot \sum_{j=1}^{n} \mathbf{m}_j \right\rangle = \sum_{i=1}^{n} m_i^2 + \left\langle \sum_{i \neq j} \sum_{j} \mathbf{m}_i \cdot \mathbf{m}_j \right\rangle \qquad (5.9.2)$$

For a freely jointed chain the dipoles associated with the i and j bonds are uncorrelated, that is, any angle among them between 0 and 2π has the same probability of occurrence, and the average of its cosine vanishes, giving the well-known result for this model

$$\langle \mu^2 \rangle_{fj} = \sum_{i=1}^{n} m_i^2 = nm^2 \qquad (5.9.3)$$

where m^2 is the average of the dipole moments of the skeletal bonds of the chains. Therefore, the dipole autocorrelation coefficient is customarily written as[6]

$$g = \frac{\langle \mu^2 \rangle}{nm^2} \qquad (5.9.4)$$

The dipole autocorrelation coefficient, like the dimension autocorrelation coefficient or characteristic ratio C $(=\langle r^2 \rangle_0 / nl^2$, where $\langle r^2 \rangle_0$ and l^2 are respectively the mean-square end-to-end distance and the average of the squares of the skeletal bond lengths) of short chains is molecular weight dependent. However, the values of these quantities remain nearly constant, independent of molecular weight, for long chains.

With respect to other more traditional properties, such as the unperturbed dimensions, dipole moments present some advantages for the study of conformation-dependent properties of polymer chains. These include the following:

1. Dipole moments can be measured for chains of any length, whereas the unperturbed dimensions can only be experimentally obtained for long chains.
2. Dipole moments of most polymer chains do no present excluded volume effects.
3. Since skeletal bonds change more in polarity than they do in length, dipole moments are usually more sensitive to structure than unperturbed dimensions.

5.10. DETERMINATION OF CONFORMATIONAL ENERGIES

The usefulness of dipole moments for the determination of conformational energies is illustrated in polyformals whose chemical structure is $O-CH_2-O-(CH_2)_y$.[26]

These polymers can be considered alternating copolymers of methylene oxide and an alkane oxide. The first member of the series ($y = 1$) is polymethylene oxide, the second ($y = 2$) is poly(1,3-dioxolane), a copolymer of methylene oxide and ethylene oxide (see Fig. 5.2), and so on.

Polymethylene oxide (PMO) is a crystalline polymer with a high melting point ($\approx 210°C$) and is insoluble in most organic solvents, but information on the conformational properties of this polymer is scarce. Poly(1,3-dioxolane) (PXL) has a relatively low melting point ($\approx 55°C$) and it is soluble in common organic solvents. Moreover, the acetalic a, e bonds of this polymer are similar to the bonds of PMO, and therefore critical interpretation of the dipole moments of PXL leads to the determination of the conformational energies of PMO. Study of the conformational properties of poly(ethylene oxide) shows that the rotational states about CH_2-O and CH_2-CH_2 skeletal bonds are located at 0 (*trans*), $+110°$ (gauche positive), $-110°$ (*gauche* negative), and 0, $\pm 120°$ respectively. The dipole moment associated with $O-CH_2$ bonds, $m_{O-C} = -m_{C-O} = 1.07\,D$, whereas that corresponding to CH_2-CH_2 bonds, m_{C-C}, is zero.

The statistical weight matrices corresponding to the skeletal bonds of PXL indicated in Fig. 5.2 are[26]

$$
\mathbf{U}_a = \begin{pmatrix} 1 & \sigma_\eta & \sigma_\eta \\ 1 & \sigma_\eta & 0 \\ 1 & 0 & \sigma_\eta \end{pmatrix}; \quad
\mathbf{U}_b = \begin{pmatrix} 1 & \sigma'' & \sigma'' \\ 1 & \sigma'' & \omega\sigma'' \\ 1 & \omega\sigma'' & \sigma'' \end{pmatrix}
$$

$$
\mathbf{U}_c = \begin{pmatrix} 1 & \sigma' & \sigma' \\ 1 & \sigma' & \omega\sigma' \\ 1 & \omega\sigma' & \sigma' \end{pmatrix}
$$

$$
\mathbf{U}_d = \begin{pmatrix} 1 & \sigma'' & \sigma'' \\ 1 & \sigma'' & \omega\sigma'' \\ 1 & \omega\sigma'' & \sigma'' \end{pmatrix}; \quad
\mathbf{U}_e = \begin{pmatrix} 1 & \sigma_\eta & \sigma_\eta \\ 1 & \sigma_\eta & \omega'\sigma_\eta \\ 1 & \omega'\sigma_\eta & \sigma_\eta \end{pmatrix}
$$

$$(5.10.1)$$

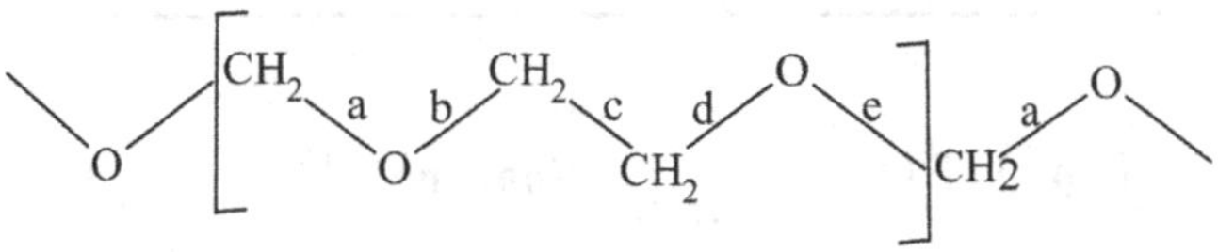

FIG. 5.2 Repeating unit of poly(1,3-dioxolane) (alternating copolymer of methylene oxide and ethylene oxide) in all-*trans* conformation

Values of g for PXL were calculated at 30°C by using the values of -0.5, 0.9, and 0.56 kcal mol^{-1} for the conformational energies $E_{\sigma'}$, $E_{\sigma''}$ and E_{ω} respectively, obtained from the conformational analysis of poly(ethylene oxide). Calculations using semiempirical potential functions suggest that second order interactions involving rotations of different sign about de bonds in Fig. 5.2 have an energy $E_{\omega'}$ nearly 0.8 kcal mol^{-1} above that of the tt conformation. The computation of g was performed as a function of $E_{\sigma\eta}$, the energy associated with $gauche$ states about the acetalic O–CH$_2$–O bonds. The results, presented in Fig. 5.3, indicate that g decreases markedly with increase in the preference for $gauche$ states about the acetalic bonds, because the $g^{\pm}g^{\pm}$ states in the bond pairs ea place the dipoles in an antiparallel direction, as shown in Figure 5.4. Agreement between calculated and experimental values of g is reached for $E_{\sigma\eta} = -1.2$ kcal mol^{-1}. This analysis leads to the conclusion that the $gauche$ states about OCH$_2$–CH$_2$O bonds have an energy 1.2 kcal mol^{-1} below that of the alternative $trans$ state.

As shown in Fig. 5.4, an increase in temperature gives rise to an increase in the fraction of all $trans$ conformations in PXL that place the dipoles in a near-parallel direction. The temperature dependence of the dipole moments of molecular chains is usually expressed in terms of d ln $\langle\mu^2\rangle$/dT. The value of this

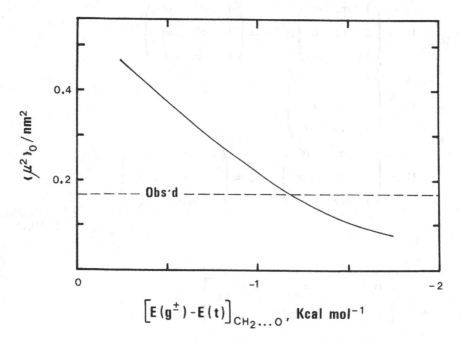

FIG. 5.3 Dependence of the dipolar correlation coefficient on the energy of $gauche$ states about acetalic CH$_2$–O bonds with respect to the alternative $trans$ states [26], [27]

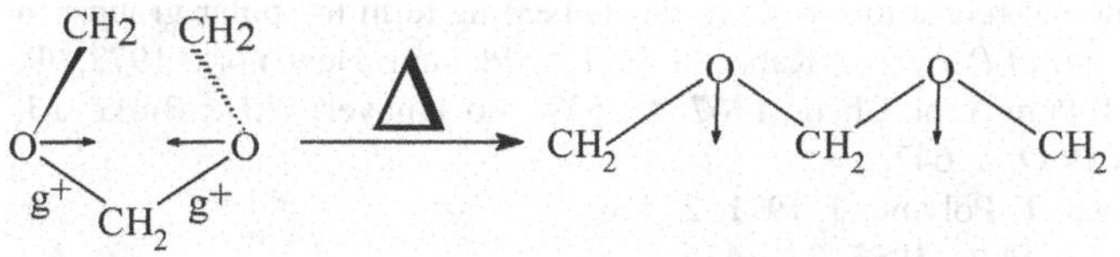

FIG. 5.4   Scheme representing the orientation of dipoles for poly(1,3-dioxolane) when the two acetalic bonds are in tt and $g^{\pm}g^{\pm}$ conformations

quantity computed by the RIS model is $6.0 \times 10^{-3} \, \mathrm{K}^{-1}$, in very good agreement with the experimental result, $6.1 \times 10^{-3} \, \mathrm{K}^{-1}$.

A detailed description of the relationships between polarity and structure as well as the evaluation of conformational energies from the critical interpretation of the mean-square dipole moments of polymer chains is given in reference.[27]

REFERENCES

1. Staudinger, H. Ber. **1920**, *53*, 1073; **1924**, *57*, 1203; Staudinger, H.; Fritschi, J. Helv. Chim. Acta. **1922**, *5*, 785.

2. Kuhn, W. Kolloid-Z. **1934**, *68*, 2.

3. Guth, E.; Mark, H. Monatsh. Chem. **1934**, *65*, 93.

4. Volkenstein, M.V. *Configurational Statistics of Polymeric Chains*, translation from 1959 Russian edition by Timasheff, N.S., Timasheff, M.J.; Wiley-Interscience: New York, 1963.

5. Birsthein, T.M.; Ptitsyn, O.B. *Conformations of Macromolecules*, translation from 1959 Russian edition by Timasheff, N.S., Timasheff, M.J.; Wiley-Interscience: New York, 1964.

6. Flory, P.J. *Statistical Mechanics of Chain Molecules*; Wiley-Interscience: New York, 1969.

7. Debye, P. Phys. Z. **1912**, *13*, 97; *Collected Papers, Polar Molecules*; Wiley Interscience: New York, 1954; 173 pp.

8. Debye, P. *Handbuch der Radiologie*; Akademische Verlagsgesellschaft mb H: deipzig, 1925; VI, 597–653; Chemical Catalog: New York, 1929.

9. Massotti, P.F. Bibl. Univ. Modena **1847**, *6*, 193.

10. Clausius, R. *Die Mechanische Wärmetheory*; Braunschweich, 1879; Vol II.

11. Halverstadt, I.F.; Kumler, W.D. J. Am. Chem. Soc. **1942**, *64*, 2988.

12. Guggenheim, E.A. Trans. Far. Soc. **1949**, *45*, 714 ; **1951**, *47*, 573.

13. Smith, J.W. Trans. Far. Soc. **1950**, *46*, 394.

14. Kirkwood, J.G. Ann. NY Acad. Sci. **1949**, *40*, 315; Trans. Far. Soc. **1946**, *42*, 7.

15. Fröhlich, H. Trans. Faraday Soc. **1948**, *44*, 238; *Theory of Dielectrics*; Oxford University Press: London, 1958.

16. Jeringan, R.L. Internal relaxation in short chains bearing terminal polar groups. In *Dielectric Properties of Polymers*; Karasz, F.E., Ed.; Plenum: New York, 1972; 99.

17. Stockmayer, W.H. Pure Appl. Chem. **1967**, *15*, 539; Stockmayer, W.H.; Burke, J.J. Macromolelcules **1969**, *2*, 647.

18. Nagai, K.; Ishikawa, T. Polymer J. **1971**, *2*, 416.

19. Fixman, M. J. Chem. Phys. **1955**, *23*, 1656.

20. Yamakawa, H.; Tanaka, G. J. Chem. Phys. **1967**, *47*, 3991.

21. Flory, P.J. Macromolecules **1974**, *7*, 381.

22. Hopfinger, A.J. *Conformational Properties of Macromolecules*; Academic Press: New York, 1973.

23. Atkins, P.W. *Physical Chemistry*; Oxford University Press: New York, 1978.

24. Slater, J.C.; Kirkwood, J.G. Phys. Rev. **1931**, *37*, 682.

25. Scott, R.A.; Scheraga, H.A. J. Chem. Phys. **1966**, *45*, 2091.

26. Riande, E.; Mark, H. Macromolecules **1978**, *11*, 956.

27. Riande, E.; Saiz, E. *Dipole Moments and Birefringence of Polymers*; Prentice Hall: Englewood Cliffs, NJ, 1992.

6

Electric Birefringence of Polymers under Static Fields

6.1. INTRODUCTION

By action of the thermal energy, the spatial conformations of amorphous flexible polymers above the glass transition temperature are continuously changing. For molecular weights large enough, the average distribution of the segments of the chains around their centres of gravity is Gaussian, having spherical symmetry.[1] Even amorphous rigid polymers are subject to Brownian motions that randomize both orientation and position of the molecules. A beam of light traveling through isotropic materials has the same interactions regardless of its direction or plane of polarization. Since the index of refraction is the ratio of the speed of the light in vacuum, c, to that in the medium, v ($n = c/v$), the value of this quantity for amorphous materials above T_g is independent of the direction. However, if the chains are preferentially oriented in a direction, a beam of light traveling with a plane of polarization that contains the direction in which the chains are oriented will contemplate a different situation to an identical beam with the plane of polarization perpendicular to the orientation. Owing to the different interactions of the two beams with the polymer, they may travel with different speed, change their planes of polarization to a different degree, be absorbed by a different amount, etc. The two incident beams, differing only in their planes of polarization, may emerge from the oriented material different in many other properties depending on the structure of the matter through which they are traveling.

265

6.2.   BIREFRINGENCE: BASIC PRINCIPLES

Let us assume a coordinate reference frame with the x axis in the direction of an electric or mechanical force field, the z axis in the direction of propagation of the beam light used as a probe in the measurements, and the y axis completing a right-handed coordinate system. Let us assume further that an incident radiation is polarized by means of a suitable oriented polarizer in the plane bisecting the first quadrant in such a way that the plane of polarization makes an angle of $45°$ with the xz and yz planes. In this way, the radiation may be taken as the sum of two components of the same amplitude, polarized, respectively, on the xz and the yz axis.[2] The first and second components are polarized, respectively, in parallel and perpendicular directions to the electric field. Since both incident components have the same phase, the radiation emerging from the isotropic material is given by

$$A_x = A_y = A_0 \cos \omega t \qquad\qquad (6.2.1)$$

If, as a result of the interactions between the traveling beams and the oriented sample, the component polarized along the x axis travels, for example, at higher speed than that polarized along the y axis, then the sample exhibits birefringence. In this case, $\Delta n = n_{yz} - n_{xz} > 0$. If the opposite occurs, $\Delta n < 0$ and the oriented sample exhibits negative birefringence. If l is the thickness of the sample, the difference in the optical path between the two components can be measured in units of length or retardation, R, number of waves, v, or phase difference, δ. These quantities are related to the birefringence by

$$R = l\,\Delta n \quad (m) \qquad\qquad (6.2.2)$$

$$v = \frac{R}{\lambda} = l\frac{\Delta n}{\lambda} \qquad\qquad (6.2.3)$$

$$\delta = 2\pi v = 2\pi l\frac{\Delta n}{\lambda} \qquad\qquad (6.2.4)$$

where λ is the wavelength of the radiation. The emerging radiation has two components given by

$$A_x = A_0 \cos(\omega t), \qquad A_y = A_0 \cos(\omega t \pm \delta) \qquad\qquad (6.2.5)$$

Therefore, the oriented substance transforms plane-polarized light into elliptically polarized radiation.

Chain orientation may be accomplished by electric, mechanical, magnetic, and shear flow fields, respectively, which give rise to mechanical, optical, and shear flow birefringence respectively.[2] In this chapter we shall focus on the electric birefringence, emphasizing its usefulness in studying the conformation-dependent properties of polymer chains. Among the various types of birefringence arising from different perturbation fields, only the mechanical birefringence is measured in solid amorphous polymers, specifically in crosslinked polymers. The other types of birefringence are usually measured in solution, or in sheared melts.

6.3. ELECTRIC BIREFRINGENCE

At the end of the nineteenth century, John Kerr[3] carried out a great number of experiments on the induced birefringence of solids and liquids under electric force fields, finding that the following empirical relationship holds

$$\delta = 2\pi l B E^2 \qquad (6.3.1)$$

where δ is given by Eq. (6.2.4), l is the thickness of the sample traversed by the light, E is the electric field, and B, also called the Kerr constant, is a parameter characteristic of the substance. The units of B in the International System are $m^{-1} V^{-2}$.

Both the solute and solvent, represented by subscripts 2 and 1 respectively, contribute to B in solutions so that, representing the volume fraction by ϕ, B can be expressed as[4]

$$B = B_1 \phi_1 + B_2 \phi_2 \qquad (6.3.2)$$

Since $\phi_1 + \phi_2 = 1$, the value of B_2 is given by

$$B_2 = B_1 + \frac{\Delta B}{\phi_2} \qquad (6.3.3)$$

where ΔB represents the difference between the Kerr constant for the solution and the solvent.

A schematic representation of the experimental device used to measure electric birefringence is shown in Fig. 6.1. A thorough description of the devices used in electric birefringence is given elsewhere.[2] Briefly, the experimental device is made up of:

1. A light source, for example, an He–Ne laser ($\lambda = 632.8$ nm) with a power of about 5 mW.
2. A polarizer and analyzer with extinction ratios of $10^5/1$ are required. Glan-Thompson type polarizers are suitable for birefringence measurements.
3. A detection system to measure the intensity of light emerging from the analyzer. A photomultiplier with a spectral response adequate to the wavelength of the laser is recommended.

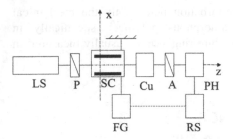

FIG. 6.1 Experimental device for the measurement of electric birefringence: LS = light source; P = polarizer; SC = sample cell with the electrodes for the application of an electric field; CU = compensation unit; A = analyzer; PH = photomultiplier; FG = electric field generator; RS = recording system

4. Sample cell. This component, not commercially available, is a glass cavity to hold the liquid, provided with two parallel conducting plates in contact with the sample. Air bubbles, reflections and strained and birefringence glass should be avoided.

5. Compensating unit. The elliptically polarized light emerging from the sample cell can be transformed back into plane-polarized light by introducing a phase shift of $\lambda/4$ between its components. The planes of polarization of the initial (before the sample) and final (after the $\lambda/4$ shift) beam make an angle α that is proportional to the birefringence produced by the sample (i.e., $\alpha = \delta/2$). The measurement of bire-fringence is converted into a simple polarimetry in which the angle α is determined by rotating the analyzer until it gives a complete extinction. Details of other compensating units are given elsewhere.[2]

6.4. INDUCED DIPOLE MOMENTS AND POLARIZABILITY

By action of an electric field, the electric charges of molecules are distorted, giving rise to an induced dipole moment, **m**. The higher the electric field, **E**, the larger is the induced dipole moment, and for low enough values of E, $\mathbf{m} = \alpha\mathbf{E}$, where α is the polarizability tensor. This parameter is related to the looseness of the electric charges (mostly electrons) in the molecules. The directions of the induced dipole moment and the electric field may not coincide owing to the tensorial character of polarizability. Therefore, the relation between **m** and **E** is expressed by

$$m_i = \alpha_{ij}E_j \tag{6.4.1}$$

which is a generalization of Eq. (1.8.1) for anisotropic media. The units of α_{ij} in the International System are $C\,m^2\,V^{-1}$.

In the laboratory reference frame xyz, the α_{ij} tensor has nine components

$$\alpha = \begin{pmatrix} \alpha_{xx} & \alpha_{xy} & \alpha_{xz} \\ \alpha_{yx} & \alpha_{yy} & \alpha_{yz} \\ \alpha_{zx} & \alpha_{zy} & \alpha_{zz} \end{pmatrix} \tag{6.4.2}$$

The polarizability tensor can be diagonalized by rotating the xyz reference frame until coordinate axes XYZ are found in which the off-diagonal components of the tensor are zero. Therefore, XYZ are the main axes of the molecule, and α_{XX}, α_{YY}, and α_{ZZ} are the main polarizabilities. In this reference frame, the electric field and the induced dipole moment have the same direction.

As has been shown in Chapter 1, the Lorentz–Lorenz equation relates the index of refraction of a molecule to its polarizability, and therefore it may be used to obtain the birefringence of nonpolar systems as follows.[5–7] The refractive indices in directions parallel ($\parallel$) and perpendicular ($\perp$) to the direction of the electric field are related to the respective polarizabilities, $\alpha_\parallel$ and $\alpha_\perp$, by (see also Chapter 11 concerning liquid crystals)

$$\frac{n_\parallel^2 - 1}{n_\parallel^2 + 2} = \frac{4\pi}{3} \sum_{i=1}^{N} \alpha_{\parallel i} = \frac{4\pi}{3} \alpha_\parallel \tag{6.4.3}$$

$$\frac{n_\perp^2 - 1}{n_\perp^2 + 2} = \frac{4\pi}{3} \sum_{i=1}^{N} \alpha_{\perp i} = \frac{4\pi}{3} \alpha_\perp \tag{6.4.4}$$

where N is the number of molecules per unit volume. In these equations, all the molecules are assumed to have the same polarizability. By subtracting the second equation from the first, we obtain

$$\Delta n = n_\parallel - n_\perp = \frac{2\pi N}{9n} (n^2 + 2)^2 (\alpha_\parallel - \alpha_\perp) \tag{6.4.5}$$

Since $n_\parallel - n_\perp$ is very small, the following approximations are made

$$(n_\parallel^2 + 2)(n_\perp^2 + 2) \cong (n^2 + 2)^2$$

$$n_\parallel + n_\perp \cong 2n \tag{6.4.6}$$

Equations (6.2.4), (6.3.1), and (6.4.5) lead to

$$B = \frac{2\pi N(n^2 + 2)^2}{9n\lambda} \frac{\Delta\alpha}{E^2} \tag{6.4.7}$$

where $\Delta\alpha = \alpha_\parallel - \alpha_\perp$.

Therefore, the determination of $\Delta\alpha$ in terms of the electric field permits determination of the Kerr constant B.

The molar Kerr constant is defined in terms of the molar refraction and the electric field by

$$_mK = \frac{R_\parallel - R_\perp}{F^2} \tag{6.4.8}$$

where F is the effective or local electric field, while R is given by Eq. (6.2.2). By writing in Eq. (6.4.5) the number of molecules per unit volume as $N = (\rho/M)N_A$, where ρ, M and N_A are respectively the density, molecular weight and Avogadro's number, the following expression is obtained

$$R_\parallel - R_\perp = \left(\frac{n_\parallel^2 - 1}{n_\parallel^2 + 2} - \frac{n_\perp^2 - 1}{n_\perp^2 + 2} \right) \frac{M}{\rho} = \frac{4\pi}{3} N_A(\alpha_\parallel - \alpha_\perp)$$

$$= \frac{6Mn}{(n^2 + 2)^2 \rho} \Delta n \tag{6.4.9}$$

where use was made of Eq. (6.4.6).

Taking the effective electric field [see Eq. (1.9.3), Chapter 1] as

$$F = \frac{\varepsilon + 2}{3} E \tag{6.4.10}$$

the molar Kerr constant can be written as

$$_mK = \frac{54\lambda Mn}{(n^2 + 2)^2(\varepsilon + 2)^2 \rho} B = \varphi \frac{M}{\rho} B \tag{6.4.11}$$

In the development of this equation, it was considered that

$$E^2 = \frac{\Delta n}{\lambda B} \tag{6.4.12}$$

obtained by combining Eqs. (6.2.4) and (6.3.1).

The SI units of $_mK$ are $m^5 \, V^{-2} \, mol^{-1} \equiv 8.988 \times 10^{14} \, cm^5 \, statV^{-2} \, mol^{-1}$ in cgs units.

6.5. ORIENTATION FUNCTION OF RIGID RODS

The model for determination of the molar Kerr constant of polymers firstly used assumes that the molecular chains are formed by a number of segments, each having cylindrical symmetry. Let us define a laboratory reference frame (see Fig. 6.2) in which the x axis coincides with the direction of the electric field, the

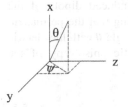

FIG. 6.2   Orientation of a rigid rod in a laboratory reference frame with the X and Z axes in the directions of the electric field and the beam light respectively

z axis has the same direction as the traveling beam light, and the y axis completes a right-handed coordinate system. For a rigid rod, symmetry reasons dictate that α is a diagonal tensor whose nonzero elements are $\alpha_{\parallel}, \alpha_{\perp}$, and $\alpha_{\perp}$, where the symbols $\parallel$ and $\perp$ denote the elements of the diagonal parallel and perpendicular to the main axis of the rod. By assuming that the main axis of the rod forms an angle θ with the x axis and its projection forms an angle ψ with the axis y, the necessary rotation to bring the axis of the rod in coincidence with the laboratory reference frame can be accomplished by using the transformation matrix $\mathbf{T}$ defined in Eq. (26) (see dipole moments)

$$\alpha = \mathbf{T}\,\mathrm{diag}(\alpha_{\parallel}, \alpha_{\perp}, \alpha_{\perp})\mathbf{T}^{\mathrm{T}} \tag{6.5.1}$$

where superscript T denotes the transpose of $\mathbf{T}$. Equation (6.5.1) leads to

$$\alpha_{xx} = \alpha_{\parallel} \cos^2 \theta + \alpha_{\perp} \sin^2 \theta$$
$$\alpha_{yy} = \alpha_{\perp} \sin^2 \psi + \alpha_{\perp} \cos^2 \theta \cos^2 \psi + \alpha_{\parallel} \sin^2 \theta \cos^2 \psi \tag{6.5.2}$$

In the absence of a force field, all the orientations are equivalent and, as a result, $\langle \Delta\alpha \rangle = 0$. However, in the presence of an electric field, the equivalence of the orientations is destroyed and the energy of the orientations forming an angle θ with the x axis is independent of ψ. Since the average of $\sin^2 \psi$ and $\cos^2 \psi$ is $\frac{1}{2}$ (see Problem 2), the average of $\Delta\alpha$ becomes

$$\langle \Delta\alpha \rangle = (\alpha_{\parallel} - \alpha_{\perp})\left\langle \frac{(3\cos^2 \theta - 1)}{2} \right\rangle = \Lambda \langle P_2(\cos \theta) \rangle = \Lambda \Phi \tag{6.5.3}$$

where $P_2\langle \cos \theta \rangle$ is the Legendre second-order polynomial, $\Lambda = \alpha_{\parallel} - \alpha_{\perp}$ is the optical anisotropy, and Φ is the orientation factor. To find Φ requires knowledge of the interaction energy between the particle and the electric field, $V(\theta)$, given by

$$V(\theta) = -\mathbf{F} \cdot (\boldsymbol{\mu} + \mathbf{m}) \tag{6.5.4}$$

where μ and $\mathbf{m}$ are respectively the permanent and induced dipole of the molecule, and $\mathbf{F}$ is the effective electric field. By assuming that the permanent dipole moment lies along the axis of the rod, forming an angle θ with the electric field and represented by $\Lambda' = \alpha'_\parallel - \alpha'_\perp$, the electrostatic anisotropy of the molecule, $V(\theta)$, is given by

$$-V(\theta) = F\mu\cos\theta + \frac{1}{2}F^2\Lambda'\cos^2\theta \tag{6.5.5}$$

We should note that α and α' represent the electrooptic and electrostatic polarizability arising from electron contributions in the former case and both electron and atoms in the latter case.

The orientation factor can be obtained from

$$\Phi = \frac{3\int_0^\pi \cos^2\theta\exp(-V(\theta)/kT)\sin\theta\,d\theta}{2\int_0^\pi \exp(-V(\theta)/kT)\sin\theta\,d\theta} - \frac{1}{2} \tag{6.5.6}$$

where $V(\theta)$ is given by Eq. (6.5.5). Equation (6.5.6) can be simplified by writing $u = \cos\theta$, $du = -\sin\theta\,d\theta$, $F\mu/kT = \beta$, and $\gamma = \Lambda'F^2/2kT$. The analytical solution of this equation is quite complicated but it leads to important results in some limiting situations, as indicated below.[8-10]

For very low electric fields ($\beta \to 0$ and $\gamma \to 0$), the orientation factor becomes

$$\Phi = \frac{\beta^2 + 2\gamma}{15} = \frac{1}{15kT}\left(\frac{\mu^2}{kT} + \Lambda'\right)F^2 \tag{6.5.7}$$

where the expansion into a series of the exponential terms of order higher than F^2 has been neglected. In this situation, the anisotropy $\langle\Delta\alpha\rangle$ and the birefringence Δn, both proportional to the orientation function, increase with the square of the electric field, as indicated by Kerr's law. Therefore, $_m K$ is a constant independent of the field.

If the orientation mainly arises from permanent dipole moments so that $\beta \gg \gamma \equiv 0$, the integration of Eq. (6.5.6) gives

$$\Phi = 1 - \frac{3(\coth\beta - 1/\beta)}{\beta} = 1 - \frac{3\mathcal{L}(\beta)}{\beta} \tag{6.5.8}$$

where $\mathcal{L}(\beta)$ is the Langevin function [Eq. (1.9.10)].

Finally, if the contribution of the permanent dipoles is negligible in comparison with that of the induced dipoles ($\gamma \gg \beta \cong 0$), the orientation function is given by

$$\Phi = \frac{3}{4}\left[\frac{\exp(g^2)}{g\int_0^g \exp(x^2)\,dx} - \frac{1}{g^2}\right] - \frac{1}{2} \tag{6.5.9}$$

with $g = \gamma^{1/2}$, and where the integral in Eq. (6.5.9) is the Dawson integral which has been evaluated numerically.

A detailed analysis of Φ as a function of both the field strength and the contribution of permanent vs. induced dipole moment is given elsewhere.[9] [See also problem 7 of Chapter 11].

6.6. EVALUATION OF $_mK$ FOR POLYMER CHAINS

The contribution of each rod of the molecular chain to the molar Kerr constant can be obtained by combining Eqs (6.4.8) and (6.4.9) which leads to

$$_mK_i = \frac{4\pi}{3}\frac{N(\alpha_\parallel - \alpha_\perp)_i}{F^2} = \frac{4\pi N}{3F^2}\Delta\alpha \tag{6.6.1}$$

This expression, together with Eqs (6.5.3) and (6.5.7), gives

$$_mK_i = \frac{4\pi N}{45kT}\left(\frac{\mu_i^2}{kT} + \Lambda_i'\right)\Lambda_i \tag{6.6.2}$$

where μ_i, Λ_i and Λ_i' are respectively the permanent dipole moment and the electrooptic and electrostatic anisotropies of molecule i.

The first approach to the evaluation of the molar Kerr constant for polymers was firstly carried out for chains with freely jointed segments, each of the segments having cylindrical symmetry.[11] The contribution of segment i to the total molar Kerr constant of the chains is given by Eq. (6.6.2). For polymer chains made up of N_s freely jointed segments, average quantities such as molecular dimensions, dipole moments, optical anisotropies, etc., scale with N_s. The average molar Kerr constant is given by

$$\langle _mK\rangle = \frac{4\pi N_A N_s}{45kT}\left(\frac{\mu_i^2}{kT} + \Lambda_i'\right)\Lambda_i \tag{6.6.3}$$

For rigid molecules, the conformational properties indicated scale with N_s^2, and the average molar Kerr constant can be written as

$$\langle _mK\rangle = \frac{4\pi N_A N_s^2}{45kT}\left(\frac{\mu_i^2}{kT} + \Lambda_i'\right)\Lambda_i \tag{6.6.4}$$

6.7. REALISTIC MODEL FOR THE EVALUATION OF $_mK$ IN FLEXIBLE POLYMERS

The physical properties of polymers are straightforwardly related to the spatial conformations of the chains. The realistic determination of a conformation-dependent property requires the evaluation of the property for each conformation and further averaging of the property over all the conformations to which the chains have accessibility. In the model first developed by Nagai and Ishikawa[12] and later improved by Flory and coworkers,[13-15] the optical anisotropy $\Delta\alpha$ ($\Delta\alpha = \alpha_{xx} - \alpha_{yy}$ in the laboratory reference frame) is averaged over all the possible orientations of the chain in the field and over all possible conformations of the chain. Thus, averaging of $\Delta\alpha$ can be accomplished by integration over all rotational angles ϕ_{int} defining the spatial conformations, called the internal coordinates, and a fixed orientation with respect to the electric field that is determined by a set of Euler angles Ω_{ext}, called the external coordinates.

The energy V_F of the chain in a given conformation is the result of the contribution of the energy of that conformation in the absence of an electric field, V, plus the interactions of the permanent and induced dipole moments with the electric field. Accordingly

$$V_F = V - \mu_x F - \frac{1}{2}\alpha'_{xx}F^2 \tag{6.7.1}$$

where F is the effective electric field applied in the direction of the x axis. The partition function can be written as an integral of the Boltzmann function, that is

$$Z = \frac{\int_{ext} \cdots \int_{int} \exp(-(V_F)/(kT))\,d\phi_{int}\,d\Omega_{ext}}{\int_0^\pi \int_0^{2\pi} \int_0^{2\pi} \sin\xi\,d\varphi\,d\xi\,d\varsigma}$$

$$= \frac{1}{8\pi^2}\int_{ext}\cdots\int_{int}\exp\left(-\frac{V_F}{kT}\right)d\phi_{int}\sin\xi\,d\xi\,d\varphi\,d\varsigma \tag{6.7.2}$$

where ξ, φ and ζ are the Euler angles. By assuming that the interaction energy between the permanent and induced dipoles with the electric field is much lower than the thermal energy kT, Eq. (6.7.2) can be expanded into a series so that

$$Z_F = \frac{1}{8\pi^2}\int_{ext}\cdots\int_{int}\exp\left(-\frac{V}{kT}\right)\exp\left(\frac{\mu_x F + (1/2)\alpha'_{xx}F^2}{kT}\right)d\phi_{int}\,d\Omega_{ext}$$

$$= \int_{ext}\cdots\int_{int}\exp\left(-\frac{V}{kT}\right)$$

$$\times\left[1 + \frac{F}{kT}\mu_x + \frac{1}{2}\frac{F^2}{kT}\alpha'_{xx} + \frac{1}{2}\left(\frac{F}{kT}\right)^2\mu_x^2\right]d\phi_{int}\,d\Omega_{ext} \tag{6.7.3}$$

where terms higher than $(F/kT)^2$ were neglected in the series expansion of the second exponential. In the absence of perturbation fields, the conformational partition function Z can be written as

$$Z = \frac{1}{8\pi^2} \int_{\text{ext}} \cdots \int_{\text{int}} \exp\left(-\frac{V}{kT}\right) d\phi_{\text{int}} \, d\Omega_{\text{ext}}$$

$$= \int_{\text{int}} \exp\left(-\frac{V}{kT}\right) d\phi_{\text{int}} \tag{6.7.4}$$

because the exponent of the integral does not depend on the external coordinates. Moreover

$$\int_{\text{int}} \mu_x \exp\left(-\frac{V}{kT}\right) d\phi_{\text{int}} = Z\langle\mu_x\rangle = 0 \tag{6.7.5}$$

where the angular brackets denote the average. On the other hand

$$\int_{\text{int}} \mu_x^2 \exp\left(-\frac{V}{kT}\right) d\phi_{\text{int}} = Z\langle\mu_x^2\rangle = Z\frac{\langle\mu^2\rangle}{3} \tag{6.7.6}$$

$$\int_{\text{int}} \alpha'_{xx} \exp\left(-\frac{V}{kT}\right) d\phi_{\text{int}} = Z\langle\alpha'_{xx}\rangle = Z\frac{\langle\text{trace}(\alpha')\rangle}{3} \tag{6.7.7}$$

The term $\langle\mu_x\rangle$ in Eq. (6.7.5) vanishes because, in the absence of a field, the three directions of space are equivalent, and for each conformation having a given value μ_x there will be another one with the same energy having $-\mu_x$. In Eqs (6.7.6) and (6.7.7), advantage was taken of the symmetry of space which leads to

$$\langle\mu_x^2\rangle = \langle\mu_y^2\rangle = \langle\mu_z^2\rangle = \frac{1}{3}\langle\mu^2\rangle$$

$$\langle\alpha'_{xx}\rangle = \langle\alpha'_{yy}\rangle = \langle\alpha'_{zz}\rangle = \frac{1}{3}\langle\text{trace}(\alpha')\rangle \tag{6.7.8}$$

Then, Eq. (6.7.3) can be written as

$$Z_F = Z\left[1 + \frac{1}{6}\left(\frac{F}{kT}\right)^2 \langle\mu^2\rangle + \frac{F^2}{6kT}\langle\text{trace}(\alpha')\rangle\right] \tag{6.7.9}$$

where it has been taken into account that the integrals over the conformations do not depend on the external coordinates. Obviously, $Z = Z_F$ when $F = 0$ for a nonpolar ($\mu = 0$) and nonpolarizable ($\alpha' = 0$) molecule. Let us now consider the average $\langle \Delta \alpha \rangle = \langle \alpha_{xx} - \alpha_{yy} \rangle_F$. In this case

$$(\Delta \alpha)_F = \langle \alpha_{xx} - \alpha_{yy} \rangle = \frac{1}{Z_F} \int_{\text{ext}} \cdots \int_{\text{int}} (\alpha_{xx} - \alpha_{yy}) \exp\left(-\frac{V_F}{kT}\right) d\phi_{\text{int}}$$

$$= \frac{1}{2Z_F} \int_{\text{ext}} \cdots \int_{\text{int}} (\alpha_{xx} - \alpha_{yy}) \left[1 + \frac{F}{kT}\mu_x + \left(\frac{F}{kT}\right)^2 \mu_x^2 \right.$$

$$\left. + \frac{1}{2}\alpha'_{xx}\frac{F^2}{kT} \right] \exp\left(-\frac{V}{kT}\right) d\phi_{\text{int}} \tag{6.7.10}$$

Because of the symmetry of the space

$$\int_{\text{int}} (\alpha_{xx} - \alpha_{yy}) \exp\left(-\frac{V}{kT}\right) d\phi_{\text{int}} = Z\langle \alpha_{xx} - \alpha_{yy} \rangle = 0 \tag{6.7.11}$$

and

$$\int_{\text{int}} (\alpha_{xx} - \alpha_{yy})\mu_x \exp\left(-\frac{V}{kT}\right) d\phi_{\text{int}} = Z\langle (\alpha_{xx} - \alpha_{yy})\mu_x \rangle = 0 \tag{6.7.12}$$

The component of the dipole moment in the direction of the electric field (the x axis of the laboratory reference frame) can be written as $\mu_x = \mu \cos \theta$. On the other hand, the α_{xx} and α_{yy} components of the tensor $\boldsymbol{\alpha}$ in the xyz laboratory coordinates are related to those in which the tensor is diagonalized (XYZ) by

$$\alpha_{xx} = \frac{\partial x}{\partial X}\frac{\partial x}{\partial X}\alpha_{XX} + \frac{\partial x}{\partial Y}\frac{\partial x}{\partial Y}\alpha_{YY} + \frac{\partial x}{\partial Z}\frac{\partial x}{\partial Z}\alpha_{ZZ}$$

$$= \alpha_X \cos^2 \tau_X^x + \alpha_Y \cos^2 \tau_Y^x + \alpha_Z \cos^2 \tau_Z^x \tag{6.7.13}$$

where the elements of the diagonal of the polarizability tensor in the XYZ reference frame are represented by α_X, α_Y, and α_Z, and τ are the angles formed by the axis of the xyz and XYZ coordinate axes. In the same way

$$\alpha_{yy} = \alpha_X \cos^2 \tau_X^y + \alpha_Y \cos^2 \tau_Y^y + \alpha_Z \cos^2 \tau_Z^y \tag{6.7.14}$$

Hence, $\mu_x^2(\alpha_{xx} - \alpha_{yy})$ can be written as

$$(\alpha_{xx} - \alpha_{yy})\mu_x^2 = \mu^2 \sum_i \alpha_i \cos^2 \tau_x(\cos^2 \tau_i^x - \cos^2 \tau_i^y) \tag{6.7.15}$$

where $i = X, Y, Z$. With averaging of $\cos^2 \tau_x \cos^2 \tau_i^x$ and $\cos^2 \tau_x \cos^2 \tau_i^y$ for a fixed conformation ϕ over the external coordinates Ω_{ext} (i.e., over all possible orientations of $\boldsymbol{\mu}$ and XYZ relative to the laboratory coordinate axes xyz)

Eq. (6.7.15) becomes (see Problem 3)

$$\langle(\alpha_{xx} - \alpha_{yy})\mu_x^2\rangle_\phi = \frac{\mu^2}{15}\sum_i \alpha_i[(2\cos^2\tau_i^\mu + 1) - (2 - \cos^2\tau_i^\mu)]$$

$$= \frac{\mu^2}{15}\sum_i (3\alpha_i\cos^2\tau_i^\mu - \alpha_i) \tag{6.7.16}$$

where τ_i^μ represents the angle between μ and the main axes (XYZ) of α. If μ is given in the XYZ coordinate reference frame, $\tau_i^\mu = 0$. In this case

$$\mu^2 \sum_i \alpha_i\cos^2\tau_i^\mu = \mu^T\alpha\mu \tag{6.7.17}$$

where superscript T denotes the transpose of vector μ. On the other hand

$$\mu^2 \sum_i \alpha_i = \mu^2(\alpha_X + \alpha_Y + \alpha_Z) = \mu^2\,\text{trace}(\alpha) \tag{6.7.18}$$

From Eqs (6.7.16) to (6.7.18)

$$\langle(\alpha_{xx} - \alpha_{yy})\mu_x^2\rangle_\phi = \frac{1}{15}[(3\mu^T\alpha\mu) - \mu^2\,\text{trace}(\alpha)] \tag{6.7.19}$$

Following the procedure outlined above, the average of the term $(\alpha_{xx} - \alpha_{yy})\alpha'_{xx}$ in Eq. (6.7.10) over the external coordinates, for a fixed conformation ϕ, is given by

$$\langle(\alpha_{xx} - \alpha_{yy})\alpha'_{xx}\rangle_\phi = \frac{1}{15}\sum_i \alpha_i\alpha'_j(3\cos^2\tau_j^i - 1) \tag{6.7.20}$$

where τ_j^i is the angle between axis i of the tensor α and axis j of the tensor α' in such a way that $\tau_j^i = 0°$ for $i = j$ and $90°$ for $i \neq j$. Therefore, Eq. (6.7.20) becomes

$$\langle(\alpha_{xx} - \alpha_{yy})\alpha'_{xx}\rangle_\phi = \frac{1}{15}[3(\alpha_X\alpha'_X + \alpha_Y\alpha'_Y + \alpha_Z\alpha'_Z)$$

$$- (\alpha_X + \alpha_Y + \alpha_Z)(\alpha'_X + \alpha'_Y + \alpha'_Z)]$$

$$= \frac{1}{15}[3\,\text{trace}(\alpha\alpha') - \text{trace}(\alpha)\text{trace}(\alpha')] \tag{6.7.21}$$

Averaging Eqs (6.7.19) and (6.7.21) over the internal coordinates gives

$$\frac{1}{15}\int_{\text{int}} [(3\boldsymbol{\mu}^T\boldsymbol{\alpha}\boldsymbol{\mu}) - \mu^2 \text{trace}(\boldsymbol{\alpha})]\exp\left(-\frac{V(\phi)}{kT}\right) d\phi$$

$$= \frac{Z}{15}[3\langle\boldsymbol{\mu}^T\boldsymbol{\alpha}\boldsymbol{\mu}\rangle - \langle\mu^2 \text{trace}(\boldsymbol{\alpha})\rangle] \tag{6.7.22}$$

and

$$\frac{1}{15}\int_{\text{int}} [3\,\text{trace}(\boldsymbol{\alpha}\boldsymbol{\alpha}') - \text{trace}(\boldsymbol{\alpha})\text{trace}(\boldsymbol{\alpha}')]\exp\left(-\frac{V(\phi)}{kT}\right) d\phi$$

$$= \frac{Z}{15}[3\langle\text{trace}(\boldsymbol{\alpha}\boldsymbol{\alpha}')\rangle - \langle\text{trace}(\boldsymbol{\alpha})\rangle\langle\text{trace}(\boldsymbol{\alpha}')\rangle] \tag{6.7.23}$$

By substituting Eqs (6.7.22) and (6.7.23) into Eq. (6.7.10), one finally obtains

$$\langle\Delta\alpha\rangle_F = \frac{F^2}{30}\frac{Z}{Z_F}\left\{\frac{1}{k^2T^2}[3\langle\boldsymbol{\mu}^T\boldsymbol{\alpha}\boldsymbol{\mu}\rangle - \langle\mu^2 \text{trace}(\boldsymbol{\alpha})\rangle]\right.$$

$$\left. + \frac{1}{kT}[3\langle\text{trace}(\boldsymbol{\alpha}\boldsymbol{\alpha}')\rangle - \langle\text{trace}(\boldsymbol{\alpha})\rangle\langle\text{trace}(\boldsymbol{\alpha}')\rangle]\right\} \tag{6.7.24}$$

The ratio Z/Z_F is usually taken to be equal to unity as Eq. (6.7.9) suggests if $F \ll kT$, an assumption that was already considered in the determination of Eq. (6.7.24). This equation can be further simplified by defining the anisotropic parts of the polarizability tensors as

$$\hat{\boldsymbol{\alpha}} = \boldsymbol{\alpha} - \tfrac{1}{3}\text{trace}(\boldsymbol{\alpha})\mathbf{E}_3$$
$$\hat{\boldsymbol{\alpha}}' = \boldsymbol{\alpha}' - \tfrac{1}{3}\text{trace}(\boldsymbol{\alpha}')\mathbf{E}_3$$

$$\tag{6.7.25}$$

where $\mathbf{E}_3$ is 3×3 identity matrix. Let us make $\tfrac{1}{3}\text{trace}(\boldsymbol{\alpha}) = a$, and $\tfrac{1}{3}$ trace $(\boldsymbol{\alpha}') = a'$. Then

$$\langle\boldsymbol{\mu}^T\boldsymbol{\alpha}\boldsymbol{\mu}\rangle = \langle\boldsymbol{\mu}^T(\hat{\boldsymbol{\alpha}} + a\mathbf{E}_3)\boldsymbol{\mu}\rangle = \langle\boldsymbol{\mu}^T\boldsymbol{\alpha}\boldsymbol{\mu}\rangle + \tfrac{1}{3}\langle\mu^2 \text{trace}(\boldsymbol{\alpha})\rangle \tag{6.7.26}$$

where it is considered that $\boldsymbol{\mu}^T\mathbf{E}_3\boldsymbol{\mu} = \mu^2$. Moreover

$$\langle\text{trace}(\boldsymbol{\alpha}\boldsymbol{\alpha}')\rangle = \langle\text{trace}(\hat{\boldsymbol{\alpha}} + a\mathbf{E}_3)(\hat{\boldsymbol{\alpha}}' + a'\mathbf{E}_3)\rangle$$

$$= \langle\text{trace}(\hat{\boldsymbol{\alpha}}\hat{\boldsymbol{\alpha}}')\rangle + a'\langle\text{trace}(\hat{\boldsymbol{\alpha}}\mathbf{E}_3)\rangle$$

$$+ a\langle\text{trace}(\hat{\boldsymbol{\alpha}}'\mathbf{E}_3)\rangle + aa'\langle\text{trace}(\mathbf{E}_3)\rangle$$

$$= \langle\text{trace}(\hat{\boldsymbol{\alpha}}\hat{\boldsymbol{\alpha}}')\rangle + \tfrac{1}{3}\langle\text{trace}(\boldsymbol{\alpha})\rangle\langle\text{trace}(\boldsymbol{\alpha}')\rangle \tag{6.7.27}$$

because trace($\hat{\boldsymbol{\alpha}}$) = trace($\hat{\boldsymbol{\alpha}}'$) = 0. From Eqs (6.7.24), (6.7.26), (6.7.27), and (6.6.1) the following final expression for the average molar Kerr constant is obtained

$$\langle_m K\rangle = \frac{2\pi N_A}{15kT}\left[\frac{1}{kT}\langle\boldsymbol{\mu}^T\hat{\boldsymbol{\alpha}}\boldsymbol{\mu}\rangle + \langle\text{trace}(\hat{\boldsymbol{\alpha}}\hat{\boldsymbol{\alpha}}')\rangle\right] \tag{6.7.28}$$

where N_A is Avogadro's number. This equation suggests that the average Kerr constant can be determined using the conformational statistics of polymer chains in the absence of an electric field despite the fact that the field perturbs the energy of the conformations. By assuming that the atomic contribution amounts to about 10% of the electronic polarizability, the value $\alpha' = 1.10\alpha$ is usually substituted into Eq. (6.7.28). Accordingly, the molar Kerr constant is customarily written as

$$\langle_m K\rangle = \frac{2\pi N_A}{15kT}\left[\frac{1}{kT}\langle\boldsymbol{\mu}^T\hat{\boldsymbol{\alpha}}\boldsymbol{\mu}\rangle + 1.10\langle\text{trace}(\hat{\boldsymbol{\alpha}}\hat{\boldsymbol{\alpha}})\rangle\right] \tag{6.7.29}$$

Usually, in the literature, dipole moments are usually given in debyes ($=3.338 \times 10^{-30}$ C m in SI units) and the polarizability in $\mathring{A}^3 (=10^{-33}$ m^3 in SI units).

6.8. VALENCE OPTICAL SCHEME

In evaluation of molecular dimensions by means of the rotational isomeric states (RIS) model, additivity of bond lengths is assumed.[13] However, simple additivity of the contributions of the dipoles associated with each skeletal bond cannot be used in the computation of the mean-square dipole moments of molecular chains, because the polarity of a bond is affected by the polarities of its neighboring bonds. Because of this inductive effect, it is not possible to assign dipole moments to $C=O$ and $C-O$ bonds that simultaneously reproduce the dipole moments of ketones, esters, and ethers.[16,17] Therefore, the polarity of molecules whose structure resembles that of the groups of bonds of the chains is used in the computation of the dipole moments of molecular chains. For example, the dipole moment of methyl benzoate is assigned to the ester group of poly(diethylene glycol terephthalate).[18] In the same way, the dipole moments of aliphatic ester molecules such as methyl propionate and methyl isobutyrate are taken as the contribution of the bond $RCH-CH_2/RC(CH_3)-CH_2$ of the repeating unit of polymethyl acrylate/polymethyl methacrylate to the polarity of the chains.[19] The contribution of the $CH_2-CHR/CH_2C(CH_3)R$ is considered to be nil.

Another important difference between dipole moments and unperturbed dimensions is that the asymmetry of the groups precludes an unambiguous assignation of the direction of the dipole moment.[20,21] Therefore, the direction must be assigned through analysis of the experimental dipole moments of low molecular weight molecules or determination of the residual charge distribution by quantum mechanical procedures.

The same arguments as those discussed above for dipole moments are valid for polarizabilities. In this case, inductive effects also affect the tensor polarizability of the bonds of polymer chains, and therefore it is advisable to use the polarizability of molecules whose structure resembles that of the bond groups of the polymer chains. Unfortunately, quantum mechanics methods for the computation of the polarizability tensor of model molecules are not well developed.[21] An alternative method can be used that requires the determination of at least two microscopic properties.[22] Both magnetic and electric birefringence, together with the mean-squared optical anisotropy, $\langle \gamma^2 \rangle$, defined as

$$\langle \gamma^2 \rangle = \tfrac{3}{2} \langle \text{trace}(\hat{\alpha}\hat{\alpha}) \rangle \tag{6.8.1}$$

are often used for the determination of the $\hat{\alpha}$ of small molecules. The conformation-dependent property $\langle \gamma^2 \rangle$ can be determined from measurements of depolarized Rayleigh scattering (DRS).

By writing the polarizability tensor α in its main coordinate system, XYZ, so that it becomes diagonal, the anistropic part of the tensor $\hat{\alpha}$ (see Eq. 6.7.25) is expressed by

$$\hat{\alpha} = \text{diag}\left[\left(\alpha_X - \frac{\alpha_X + \alpha_Y + \alpha_Z}{3} \right), \left(\alpha_Y - \frac{\alpha_X + \alpha_Y + \alpha_Z}{3} \right), \right.$$
$$\left. \left(\alpha_Z - \frac{\alpha_X + \alpha_Y + \alpha_Z}{3} \right) \right] \tag{6.8.2}$$

This equation can be expressed in terms of the difference between the polarizability on the longitudinal axis (i.e., the X axis) and the average of the two transverse axes, represented by $\Delta\alpha$, and the difference between the two transverse directions, represented by $\Delta\alpha^+$.

The so-called longitudinal and transversal anisotropies of $\hat{\alpha}$ are given by

$$\Delta\alpha = \alpha_X - \left(\frac{\alpha_Y + \alpha_Z}{2} \right), \qquad \Delta\alpha^+ = \alpha_Y - \alpha_Z \tag{6.8.3}$$

Obviously, for spherical symmetry ($\alpha_X = \alpha_Y = \alpha_Z = 0$), $\Delta\alpha = 0$, whereas for cylindrical symmetry ($\alpha_Y = \alpha_Z$), $\Delta\alpha^+ = 0$. The tensor $\hat{\alpha}$ written in terms of $\Delta\alpha$ and $\Delta\alpha^+$ adopts the following form

$$\hat{\alpha} = \Delta\alpha \, \text{diag}\left(\tfrac{2}{3}, -\tfrac{1}{3}, -\tfrac{1}{3}\right) + \Delta\alpha^+ \text{diag}\left(0, \tfrac{1}{2}, -\tfrac{1}{2}\right)$$
$$= \Delta\alpha \mathbf{K} + \Delta\alpha^+ \mathbf{L} \tag{6.8.4}$$

where $\mathbf{K} = \mathrm{diag}(\frac{2}{3}, -\frac{1}{3}, -\frac{1}{3})$ and $\mathbf{L} = \mathrm{diag}(0, \frac{1}{2}, -\frac{1}{2})$. In some cases, however, $\hat{\boldsymbol{\alpha}}$ needs an additional term in Eq. (6.8.4). For example, in methyl acetate, symmetry suggests that one of the main axes should be perpendicular to the plane of the molecule whereas the other two should be contained in that plane. However, one of the axes makes an angle φ with the C–C bond and, as a result, a polarizability $\alpha_{XY} = \alpha_{YX}$ in the XY plane appears. Hence

$$\hat{\boldsymbol{\alpha}} = \Delta\alpha\,\mathrm{diag}\big(\tfrac{2}{3}, -\tfrac{1}{3}, -\tfrac{1}{3}\big) + \Delta\alpha^{+}\mathrm{diag}\big(0, \tfrac{1}{2}, -\tfrac{1}{2}\big)$$

$$+ \begin{pmatrix} 0 & \alpha_{XY} & 0 \\ \alpha_{YX} & 0 & 0 \\ 0 & 0 & 0 \end{pmatrix} \tag{6.8.5}$$

A thorough discussion on the use of the valence optical scheme to compute the molar Kerr effect of macromolecules can be found in reference [2].

6.9. COMPUTATION OF $_mK$ BY THE RIS MODEL

Polar polymers present a relatively large molar Kerr constant whereas this conformational property is rather small in the case of polymers without permanent dipole moments in their structure. In the first case, the dominant contribution to $_mK$ is $\langle \boldsymbol{\mu}^T\hat{\boldsymbol{\alpha}}\boldsymbol{\mu} \rangle$, whereas in cases with weak polarity but strong anisotropies, as occurs in polystyrene, the most important contribution is $\langle \mathrm{trace}(\hat{\boldsymbol{\alpha}}\hat{\boldsymbol{\alpha}}) \rangle$. For polymers with vanishing dipole moments and with low polarizability groups, the molar Kerr constant is small as occurs in poly(propylene). Obviously, if hydrogen atoms of the backbone are substituted for groups with high polarity or high polarizability, the sign of the molar Kerr constant can be negative. Matricial methods explained in detail elsewhere[13,14] allow the evaluation of $\boldsymbol{\mu}^T\hat{\boldsymbol{\alpha}}\boldsymbol{\mu}$ and $\mathrm{trace}(\hat{\boldsymbol{\alpha}}\hat{\boldsymbol{\alpha}})$ for a given conformation as

$$\boldsymbol{\mu}^T\hat{\boldsymbol{\alpha}}\boldsymbol{\mu} = \mathbf{Q}_1\left[\prod_{i=2}^{N-1}\mathbf{Q}_i\right]\mathbf{Q}_n \tag{6.9.1}$$

and

$$\mathrm{trace}(\hat{\boldsymbol{\alpha}}\hat{\boldsymbol{\alpha}}) = \mathbf{P}_1\left[\prod_{i=2}^{N-1}\mathbf{P}_i\right]\mathbf{P}_N \tag{6.9.2}$$

where N is the number of skeletal bonds of the chains. In these equations, $\mathbf{Q}_i$ and $\mathbf{P}_i$ are generator matrices defined as

$$\mathbf{Q}_i = \begin{bmatrix} 1 & 2\boldsymbol{\mu}^T\mathbf{T} & \hat{\boldsymbol{\alpha}}^R(\mathbf{T}\otimes\mathbf{T}) & (\boldsymbol{\mu}^T\otimes\boldsymbol{\mu}^T)(\mathbf{T}\otimes\mathbf{T}) & 2\hat{\boldsymbol{\alpha}}^R(\boldsymbol{\mu}\otimes\mathbf{T}) & \hat{\boldsymbol{\alpha}}^R(\boldsymbol{\mu}\otimes\boldsymbol{\mu}) \\ 0 & \mathbf{T} & 0 & (\mathbf{E}_3\otimes\boldsymbol{\mu}^T)(\mathbf{T}\otimes\mathbf{T}) & \hat{\boldsymbol{\alpha}}\mathbf{T} & \hat{\boldsymbol{\alpha}}\boldsymbol{\mu} \\ 0 & 0 & \mathbf{T}\otimes\mathbf{T} & 0 & 2\boldsymbol{\mu}\otimes\mathbf{T} & \boldsymbol{\mu}\otimes\boldsymbol{\mu} \\ 0 & 0 & 0 & \mathbf{T}\otimes\mathbf{T} & 0 & \hat{\boldsymbol{\alpha}}^C \\ 0 & 0 & 0 & 0 & \mathbf{T} & \boldsymbol{\mu} \\ 0 & 0 & 0 & 0 & 0 & 1 \end{bmatrix}_i$$

$$(6.9.3)$$

and

$$\mathbf{P}_i = \begin{bmatrix} 1 & 2\hat{\boldsymbol{\alpha}}^R(\mathbf{T}\otimes\mathbf{T}) & \hat{\boldsymbol{\alpha}}^R\hat{\boldsymbol{\alpha}}^C \\ 0 & \mathbf{T}\otimes\mathbf{T} & \hat{\boldsymbol{\alpha}}^C \\ 0 & 0 & 1 \end{bmatrix}_i \qquad (6.9.4)$$

where $\mathbf{T}$ is the transformation matrix defined by Eq. (5.6.2) of Chapter 5 which projects bond $i + 1$ into bond i, and $\boldsymbol{\mu}$ is the dipole moment associated with skeletal bond i. In these equations, the symbol $\otimes$ denotes the direct product of matrices, $\mathbf{E}_3$ is a 3×3 identity matrix, $\hat{\boldsymbol{\alpha}}^C$ and $\hat{\boldsymbol{\alpha}}^R$ are respectively column (9×1) and row (1×9) vectors containing the components of the polarizability tensor in the order $\hat{\alpha}_{xx}, \hat{\alpha}_{xy}, \hat{\alpha}_{xz}, \hat{\alpha}_{yx}, \hat{\alpha}_{yy}, \ldots$ The bold zeros in Eqs (6.9.3) and (6.9.4) are null submatrices required to complete $\mathbf{Q}_i$ and $\mathbf{P}_i$ as 26×26 and 11×11 matrices. $\mathbf{P}_1$ and $\mathbf{Q}_1$ represent the 1×11 and the 1×26 first rows of matrices $\mathbf{P}$ and $\mathbf{Q}$ that are associated with the first bond of the chain. Moreover, $\mathbf{P}_N$ and $\mathbf{Q}_N$ are the 26×1 and the 11×1 last columns of matrices $\mathbf{P}$ and $\mathbf{Q}$ that are associated with the last bond of the chain.

In order to obtain $\langle \boldsymbol{\mu}^T\hat{\boldsymbol{\alpha}}\boldsymbol{\mu} \rangle$ and $\langle \text{Trace}(\hat{\boldsymbol{\alpha}}\hat{\boldsymbol{\alpha}}) \rangle$, the generator matrix associated with each skeletal bond ($\mathbf{Q}_i$ or $\mathbf{P}_i$) must be combined with the corresponding statistical weight matrix ($\mathbf{U}_i$) (see Chapter 5). Thus, the following supermatrices are obtained

$$\mathbb{Q}_i = (\mathbf{U}_i \otimes \mathbf{E}_{27})\|\mathbf{Q}_i\|$$
$$\mathbb{P}_i = (\mathbf{U}_i \otimes \mathbf{E}_{11})\|\mathbf{P}_i\|$$

$$(6.9.5)$$

Here, $\mathbf{E}_{27}$ and $\mathbf{E}_{11}$ are respectively the 27×27 and 11×11 identity matrices, and $\|\|$ denotes a pseudodiagonal matrix with the terms of the generator matrices ($\mathbf{P}_i$ or $\mathbf{Q}_i$) associated with each rotational bond. For example

$$\|\mathbf{Q}_i\| = \text{diag}[\mathbf{Q}_i(\beta), \mathbf{Q}_i(\gamma), \ldots, \mathbf{Q}_i(\nu)] \qquad (6.9.6)$$

where $\beta, \gamma, \ldots, \nu$ are respectively the rotational angles of skeletal bond i. The averages are thus written as

$$\langle \mu^T \hat{\alpha} \mu \rangle = Z^{-1} \mathbb{Q}_1 \left[\prod_{i=2}^{N-1} \mathbb{Q}_i \right] \mathbb{Q}_N$$

$$\langle \text{trace}(\hat{\alpha}\hat{\alpha}) = Z^{-1} \mathbb{P}_1 \left[\prod_{i=2}^{N-1} \mathbb{P}_i \right] \mathbb{P}_N$$

$$(6.9.7)$$

where Z is the rotational partition function (see Eq. (5.8.3)). The supermatrices corresponding to the first and last bonds of the chains are given by

$$\mathbb{Q}_1 = \mathbf{U}_1 \otimes \mathbf{Q}_1, \qquad \mathbb{Q}_N = \mathbf{U}_n \otimes \mathbf{Q}_N$$
$$\mathbb{P}_1 = \mathbf{U}_1 \otimes \mathbf{P}_1, \qquad \mathbb{P}_N = \mathbf{U}_n \otimes \mathbf{P}_N$$

$$(6.9.8)$$

where $\mathbf{U}_1$ and $\mathbf{U}_n$ are defined in Eq. (5.8.4) of Chapter 5.

Let us consider, as an example, the assignment of polarizability tensors and dipole moments to the skeletal bonds of the repeating unit of poly(diethylene glycol terephthalate), the repeating unit of which is shown in Fig. 6.3. The anisotropic polarizability tensor associated with the terephthaloyl residue (bonds 1 to 4) can be written in a first approximation as (see Fig. 6.4)[18]

$$\hat{\alpha}_{DT} = \hat{\alpha}_B + 2\hat{\alpha}_E + 2\hat{\alpha}_{(CH_4)} \tag{6.9.9}$$

where subscripts B and E correspond respectively to benzene and methyl acetate. For symmetry, $\hat{\alpha}_{(CH_4)} = \mathbf{0}$, so that

$$\hat{\alpha}_{DT} = \hat{\alpha}_B + 2\hat{\alpha}_E \tag{6.9.10}$$

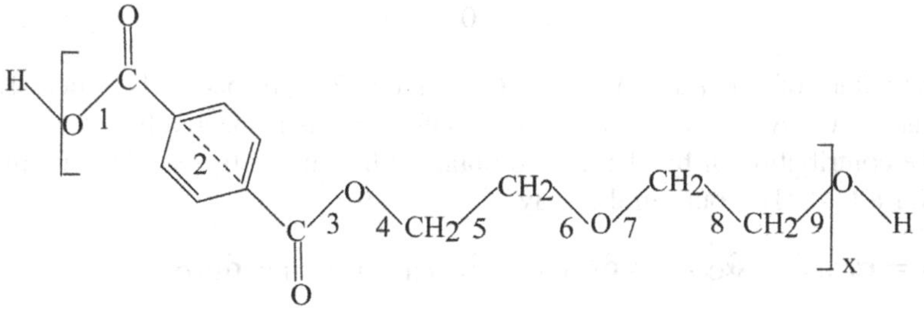

FIG. 6.3

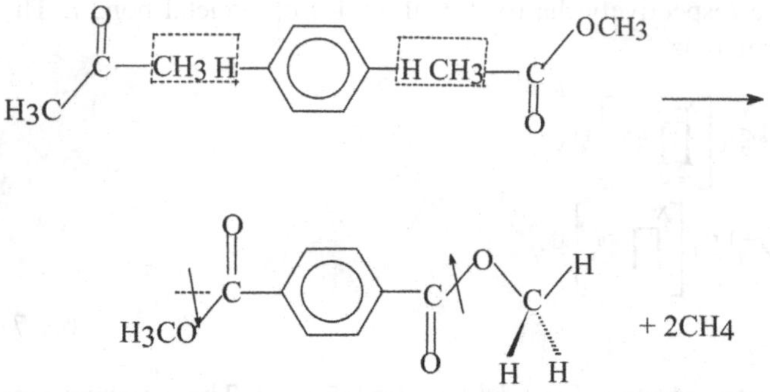

FIG. 6.4 Schematic representation of the formation of dimethyl terephthalate from terephthalic acid and methyl acetate. Dipoles are represented by arrows

The contribution of substituted benzene referred to a coordinate system having both the X and Y axes in the plane of the ring is

$$\alpha_B = \Delta\alpha_B \mathbf{K} + \Delta\alpha_B^+ \mathbf{L} \tag{6.9.11}$$

with $\Delta\alpha_B = 3.85 \text{ Å}^3$ and $\Delta\alpha_B^+ = 3.0 \text{ Å}^3$; $\mathbf{K}$ and $\mathbf{L}$ are defined in Eq. (6.8.4). The contribution of the methyl acetate molecule may be written as

$$\hat{\alpha}_E = \Delta\alpha_E \mathbf{K} + \Delta\alpha_E^+ \mathbf{L} + \begin{pmatrix} 0 & -0.1 & 0 \\ -0.1 & 0 & 0 \\ 0 & 0 & 0 \end{pmatrix} \text{Å}^3 \tag{6.9.12}$$

In this expression, $\Delta\alpha_E = 1.545 \text{ Å}^3$ and $\Delta\alpha_E^+ = 1.21 \text{ Å}^3$ in a coordinate system having the X axis in the direction of the $C^{ar}-C^*$ bond and the Y axis in the plane of the molecule. Therefore, the contribution of the DT molecule written in the coordinate system affixed to the virtual bond of the repeat unit (bond 2) will be

$$\hat{\alpha}_{DT} = \Delta\alpha_{DT} \mathbf{K} + \Delta\alpha_{DT}^+ \mathbf{L} + \begin{pmatrix} 0 & \alpha_{xy} & 0 \\ \alpha_{xy} & 0 & 0 \\ 0 & 0 & 0 \end{pmatrix} \tag{6.9.13}$$

where the values of $\Delta\alpha_{DT}$ and $\Delta\alpha_{DT}^+$ are 6.94 and 5.42 Å^3 respectively, whereas α_{xy} is 0 and -0.2 Å^3 for the *cis* and *trans* conformations respectively.

The contribution of bond 5 can be obtained by substituting a CH bond in bond 4 for a $C-CH_3$ group. In this way

$$\hat{\alpha}_5 = \hat{\alpha}_{C-CH_3} - \hat{\alpha}_{C-CH} = \hat{\alpha}_{C-CH_3} - \hat{\alpha}_{C-CH} + \hat{\alpha}_{C-CH} - \hat{\alpha}_{C-CH}$$

$$= \hat{\alpha}_{C-C} - 2\hat{\alpha}_{C-H} \tag{6.9.14}$$

where it has been considered that

$$\hat{\alpha}_{C-CH_3} + \hat{\alpha}_{C-H} = \hat{\alpha}_{C-C} + \hat{\alpha}_{CH_4} = \hat{\alpha}_{C-C} \qquad (6.9.15)$$

Owing to the cylindrical symmetry of C–C, the contributions of bonds 5 and 8, $\hat{\alpha}_5$, can be written as

$$\hat{\alpha}_5 = \hat{\alpha}_8 = (\Delta\alpha_{C-C} - 2\Delta\alpha_{C-H})\mathbf{K} \qquad (6.9.16)$$

where $\Delta\alpha_{C-C} = 0.95 \text{ Å}^3$ and $\Delta\alpha_{C-H} = 0.21 \text{ Å}^3$. Using the same arguments for bonds 6 and 7

$$\hat{\alpha}_6 = \hat{\alpha}_7 = \hat{\alpha}_{C-O} - \hat{\alpha}_{C-H} = (\Delta\alpha_{C-O} - \Delta\alpha_{C-H})\mathbf{K} \qquad (6.9.17)$$

The contribution of bond 9 as well the contributions of bonds 1, 3, and 4 are already included in $\hat{\alpha}_{DT}$. Therefore

$$\alpha_1 = \alpha_3 = \alpha_4 = \alpha_9 = 0 \qquad (6.9.18)$$

and

$$\hat{\alpha}_2 = \hat{\alpha}_{DT} \qquad (6.9.19)$$

Finally, the dipole moment associated with the ester groups has a value of 1.89 D, and its direction makes an angle of 57° with the C^{ar}–CO(O) bond. Moreover, the values of the dipole moments associated with O–CH$_2$, CH$_2$–O, and CH$_2$–CH$_2$ skeletal bonds are 1.07, −1.07 and 0.00 D respectively.

The molar Kerr constant is a molecular property whose range of values for low molecular weight substances may vary over four orders of magnitude and can even be positive or negative. Therefore, this technique may be very sensitive to the stereochemical and chemical distributions in which at least one of the comonomers is polar. Computations of this kind were carried out on polypropylene–polyvinyl chloride and polystyrene–poly(p-chlorostyrene).[23] This behavior is reflected in Table 6.1, where some results for the molar Kerr constants of polypropylene and poly(vinyl chloride) are shown.

The sensitivity of $_mK$ to the stereochemical compositions of copolymers was confirmed by in the analysis of the Kerr constants of poly(p-chlorostyrene) and poly(p-bromostyrene) performed by Flory and coworkers.[4]

A shortcoming of the Kerr contant is that the permanent and induced dipolar contributions to $_mK$ may have opposite sign and as a result the molar Kerr constant may in some cases be less sensitive to structure than other alternative conformational properties. This is reflected in the analysis of the bond scission occurring in the two CO bonds of 1,3-dioxolane. As a result of the polymerization mechanisms, CH$_2$OCH$_2$CH$_2$O and CH$_2$CH$_2$OCH$_2$O units may appear in the structure of the resulting polymer. By changing the fraction, w, of one of these

TABLE 6.1   Values of the Molar Kerr Constant as a Function of the Steroregularity for Polypropylene (PP), Poly(vinyl chloride) (PVC), and Copolymers of these Monomers[23]

Polymer	Stereoregularity*	Sequence distribution	$_mK^D$	$_mK^P$	$_mK^†$
PP	S		0	49	49
PP	I		0	45	45
PP	A		0	39	39
PVC	S		−13,500	74	−13,462
PVC	I		287	19	306
PVC	A		45	26	71
PVC–PP	S	Reg. alt.	−611	44	−567
PVC–PP	S	50:50 random	−691	45	−646
PVC–PP	I	Reg. alt.	81	28	109
PVC–PP	I	50:50 random	79	29	108
PVC–PP	A	Reg. alt	−416	33	−383
PVC–PP	A	10:90 random	−12	38	26
PVC–PP	A	50:50 random	−40	32	−8
PVC–PP	A	90:120 random	16	27	43

*S = syndiotactic, I = isotactic, A = atactic; superscript D denotes dipolar contribution and superscript P denotes polarizability contribution to $_mK$.
†Expressed per mole of backbone bonds in $(cm^5\,statV^{-2}) \times 10^{12}$.

units from 0 to 50%, the molar Kerr constant only increases by about 5% as a consequence of the cancellation of the two contributions to $_mK$. This increase is about 29% in $\langle \gamma^2 \rangle$, where the only contribution is $\langle trace(\alpha\alpha) \rangle$, and 60% in $\langle \mu^2 \rangle$.[24] However, the change in the mean-square end-to-end distance is only 0.3%.

The usefulness and shortcomings of the molar Kerr constant in the analysis of the conformational properties of molecular chains is discussed in detail in reference.[2]

PROBLEMS
Problem 1

Determine $_mK$ for a solution at infinite dilution.

Solution

According to Eqs (6.3.3) and (6.4.11)

$$_mK = \varphi\left(\frac{M_2}{\rho_2\phi_2}\Delta B + B_1\frac{M_2}{\rho_2} \right)$$

(P.6.1.1)

where subscripts 1 and 2 refer to the solvent and solute, respectively. Since $\rho_2\phi_2/M_2 = C_2$ (mol/m^3), Eq. (P.6.1.1) becomes

$$_mK = \frac{54\lambda n_1}{(n_1^2 + 2)^2(\varepsilon_1^2 + 2)^2}\left[\lim_{C_2 \to 0}\left(\frac{\Delta B}{C_2} + v_2B_1\right)\right] \tag{P.6.1.2}$$

When Eq. (P.6.1.2) is applied to polymers, C_2 is taken as the concentration of repeating units. Remember that $M_2 = xM_0$, where x and M_0 are respectively the degree of polymerization and the molecular weight of the repeating unit. The result thus obtained is $_mK/x$, or the molar Kerr constant per repeat unit of the chain.

Problem 2

Calculate the average of $\sin^2\psi$.

Solution

Let us consider that the orientation of the rod with respect to the z axis is defined by θ as indicated in Fig. 6.2. An infinitesimal area of the sphere is given by $(r\sin\theta\,d\psi)(r\,d\theta)$. Therefore

$$\langle\sin^2\psi\rangle = \frac{\int_0^{2\pi}\sin^2\psi\,d\psi\int_0^{\pi}r\sin\theta r\,d\theta}{\int_0^{2\pi}\int_0^{\pi}r\sin\theta\,d\psi r\,d\theta} = \frac{\int_0^{2\pi}\sin^2\psi\,d\psi}{2\pi} \tag{P.6.2.1}$$

Since

$$\sin^2\psi = \frac{1 - \cos 2\psi}{2} \tag{P.6.2.2}$$

Eq. (P.6.2.1) becomes

$$\langle\sin^2\psi\rangle = \tfrac{1}{2}$$

In the same way

$$\langle\cos^2\psi\rangle = \tfrac{1}{2}$$

Problem 3

Let us assume that the main axis of the polarizability tensor of a molecule coincides with the Z' axis of a coordinate system $OX'Y'Z'$. Let us assume further that the dipole moment of the molecule forms an angle γ with the OZ axis of the coordinate reference frame $OXYZ$. If the angle between the dipole moment and the main axis of the molecules is δ, calculate the averages $\langle\cos^2\theta\cos^2\gamma\rangle$ and $\langle\cos^2\theta\sin^2\gamma\rangle$.

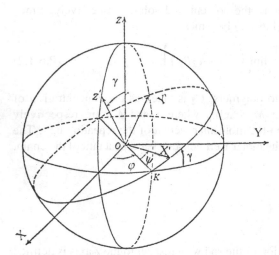

FIG. 6.5   Scheme showing the Euler angles

Solution

The three Euler angles defining the rotatory $OX'Y'Z'$ coordinate axes with respect to the fixed coordinate system $OXYZ$ is defined by the three Euler angles $XOK = \varphi$; $X'OK = \psi$; $Z'OZ = \gamma$.

As can be seen in Fig. 6.5, superimposition of the XYZ coordinate systems upon $OX'Y'Z'$ can be accomplished by:

1. Rotation by an angle φ about the OZ axis.
2. Rotation by an angle γ about the OK axis. This rotation leads OZ to coincide with OZ'.

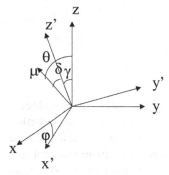

FIG. 6.6   Angles of the dipole moment μ with the Z and Z' axes

3. Rotation by an angle ψ about the OZ' axis which will result in OX and
    OY coinciding with OX' and OY' respectively.

The matrix that transforms a vector in the reference frame $OXYZ$ to $OX'Y'Z'$
is given by

$$
\mathbf{M} = \begin{pmatrix} \cos\psi & \sin\psi & 0 \\ -\sin\psi & \cos\psi & 0 \\ 0 & 0 & 1 \end{pmatrix} \begin{pmatrix} 1 & 0 & 0 \\ 0 & \cos\gamma & \sin\gamma \\ 0 & -\sin\gamma & \cos\gamma \end{pmatrix} \begin{pmatrix} \cos\varphi & \sin\varphi & 0 \\ -\sin\varphi & \cos\varphi & 0 \\ 0 & 0 & 1 \end{pmatrix}
$$

$$
= \begin{pmatrix} \cos\psi\cos\varphi - \cos\gamma\sin\psi\sin\varphi \\ -\sin\psi\cos\varphi - \cos\gamma\cos\psi\sin\varphi \\ \sin\gamma\sin\varphi \end{pmatrix}
$$

$$
\begin{matrix} \cos\psi\sin\varphi + \cos\gamma\sin\psi\cos\varphi & \sin\gamma\sin\psi \\ -\sin\psi\sin\varphi + \cos\gamma\cos\psi\cos\varphi & \sin\gamma\cos\psi \\ -\sin\gamma\cos\varphi & \cos\gamma \end{matrix} \Bigg) \qquad (P.6.3.1)
$$

Accordingly, the unit vector $\mathbf{u}' = (0,\ \sin\delta,\ \cos\delta)$ in the reference frame
$OX'Y'Z'$ is related to $\mathbf{u} = (0,\ \sin\theta,\ \cos\theta)$ in the coordinates system $OXYZ$ by

$$\mathbf{u}' = \mathbf{M}\mathbf{u} \qquad (P.6.3.2)$$

Therefore

$$\mathbf{u} = \mathbf{M}^{-1}\mathbf{u}' \qquad (P.6.3.3)$$

Hence

$$
\begin{pmatrix} 0 \\ \sin\theta \\ \cos\theta \end{pmatrix} = \begin{pmatrix} \cos\psi\cos\varphi - \cos\gamma\sin\psi\sin\varphi \\ -\sin\psi\sin\varphi + \cos\gamma\sin\psi\cos\varphi \\ \sin\psi\sin\gamma \end{pmatrix}
$$

$$
\begin{matrix} -\sin\psi\cos\varphi - \cos\gamma\cos\psi\sin\varphi & \sin\gamma\sin\varphi \\ -\sin\psi\sin\varphi + \cos\gamma\cos\psi\cos\varphi & -\sin\gamma\cos\varphi \\ \sin\gamma\cos\psi & \cos\gamma \end{matrix} \Bigg) \begin{pmatrix} 0 \\ \sin\delta \\ \cos\delta \end{pmatrix}
$$

and

$$\cos\theta = \cos\psi\sin\gamma\sin\delta + \cos\gamma\cos\delta \qquad (P.6.3.4)$$

The averages of $\cos^2 \gamma \cos^2 \theta$ and $\sin^2 \gamma \cos^2 \theta$ will be given by

$$
\begin{aligned}
\langle \cos^2 \gamma \cos^2 \theta \rangle &= \frac{\int_0^{2\pi} d\varphi \int_0^{2\pi} d\psi \int_0^{\pi} \cos^2 \theta \cos^2 \gamma \sin \gamma \, d\gamma}{8\pi^2} \\
&= \frac{1}{4\pi} \int_0^{2\pi} \int_0^{\pi} \cos^2 \gamma (\cos \psi \sin \gamma \sin \delta + \cos \gamma \cos \delta) \sin \gamma \, d\gamma \, d\psi \\
&= \frac{1}{15} (2 \cos^2 \delta + 1) \qquad\qquad\qquad\qquad\qquad\qquad\qquad (P.6.3.5)
\end{aligned}
$$

$$
\begin{aligned}
\langle \sin^2 \gamma \cos^2 \theta \rangle &= \frac{\int_0^{2\pi} d\varphi \int_0^{2\pi} d\psi \int_0^{\pi} \cos^2 \theta \sin^3 \gamma \, d\gamma}{8\pi^2} \\
&= \frac{1}{4\pi} \int_0^{2\pi} \int_0^{\pi} \sin^2 \gamma (\cos \psi \sin \gamma \sin \delta + \cos \gamma \cos \delta) \sin \gamma \, d\gamma \, d\psi \\
&= \frac{1}{15} (2 - \cos^2 \delta) \qquad\qquad\qquad\qquad\qquad\qquad\qquad (P.6.3.6)
\end{aligned}
$$

It should be noted that, since the precession motion defined by φ does not intervene in the average, the matrix $\mathbf{M}$ could be simplified by assuming that $\varphi = 0$ in Eq. (P.6.3.1).

Problem 4

Assign the polarizability tensors to the repeat unit of poly(methyl acrylate).

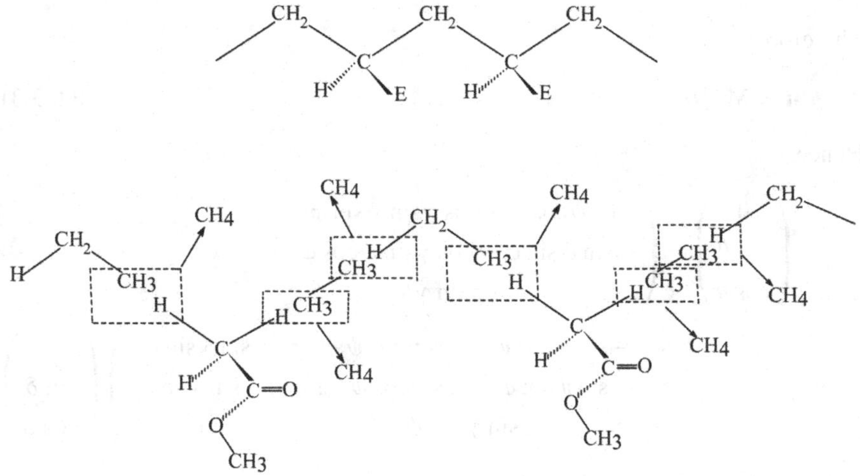

FIG. 6.7 Schematic formation of poly(methyl methacrylate) from ethane and methyl acetate)

Solution[25]

The polarizability associated with the bonds of the repeat unit of poly(methyl acrylate) can be built from that of ethylene and methyl acetate as indicated in Fig. 6.7. The polarizability of ethylene can be written as

$$\hat{\alpha}_i = \hat{\alpha}_{CH_3-CH_3} = \hat{\alpha}_{C-C} - 2\hat{\alpha}_{C-H} \qquad (P.6.4.1)$$

The anisotropic part of the polarizability tensor associated with the ester residue can be obtained from the schematic reaction of ethane and methyl acetate, giving

$$CH_3-CH_2-H + CH_3-COOCH_3 \longrightarrow CH_3-COOCH_4 + CH_4 \qquad (P.6.4.2)$$

Therefore

$$\hat{\alpha}_{i+1} = \hat{\alpha}_{CH_3-CH_3} + \hat{\alpha}_E$$

where $\hat{\alpha}_E$ is given by Eq. (6.9.12).

REFERENCES

1. Freed, K.F. *Renormalization Group Theory of Macromolecules*; Wiley-Interscience: New York, 1987; 25 pp.
2. Riande, E.; Saiz, E. *Dipole Moments and Birefringence of Polymers*; Prentice-Hall: Englewood Cliffs, NJ, 1992.
3. Kerr, J. Phil. Mag. **1875**, *50*, 337; **1875**, *50*, 416; **1879**, *8*, 85; **1879**, *8*, 229; **1880**, *9*, 157; **1882**, *13*, 153; **1882**, *13*, 248; **1894**, *37*, 380; **1894**, *38*, 144.
4. Saiz, E.; Suter, U.W.; Flory, P.J. J. Chem. Soc., Far. Trans. 2 **1977**, *73*, 1538.
5. LeFevre, C.G.; LeFevre, R.J.W. Rev. Pure Appl. Chem. **1955**, *5*, 265.
6. LeFevre, C.G.; LeFevre, R.J.W. Rev. Pure Appl. Chem. **1970**, *20*, 57.
7. LeFevre, C.G.; LeFevre, R.J.W. J. Chem. Soc. **1953**, 4041.
8. O'Konsky, C.T.; Yoshioka, K.; Orttung, W.H. J. Phys. Chem. **1959**, *63*, 1558.
9. Watanabe, H.; Morita, A. Adv. Chem. Phys. **1984**, *56*, 255.
10. Shah, M.J. J. Phys. Chem. **1963**, *67*, 2215.
11. Stuart, H.A.; Peterlin, A. J. Polym. Sci. **1950**, *5*, 551.
12. Nagai, K.; Ishikawa, T. J. Chem. Phys. **1965**, *43*, 4508.
13. Flory, P.J. *Statistical Mechanics of Chain Molecules*; Wiley-Interscience: New York, 1969.
14. Flory, P.J. Macromolecules **1974**, *7*, 381.
15. Flory, P.J.; Jernigan, R.L. J. Chem. Phys. **1968**, *48*, 3823.
16. Saiz, E.; Hummel, J.P.; Flory, P.J.; Plavsic, M. J. Phys. Chem. **1981**, *85*, 3211.
17. Rodrigo, M.M.; Tarazona, M.P.; Saiz, E. J. Phys. Chem. **1986**, *90*, 3236, 5565.
18. Riande, E.; Guzmán, J.; Tarazona, M.P.; Saiz, E. J. Polym. Sci.: Phys. Ed. **1984**, *22*, 917.
19. Ojalvo, E.A.; Saiz, E.; Masegosa, R.M.; Hernández-Fuentes, I. Macromolecules **1979**, *12*, 865.

20. Andre, J.M.; Barbier, C.; Bodart, V.; Delhalle, J. In *Nonlinear Optical Properties of Organic Molecules and Crystals*; Chemula, D.S., Zyss, J., Eds.; Academic Press: Orlando, FL, 1987.
21. Waite, J.; Papadopoulos, M.G. J. Phys. Chem. **1990**, *94*, 1755.
22. Suter, U.W.; Flory, P.J. J. Chem. Soc., Far. Trans. 2 **1977**, *73*, 1521.
23. Tonelli, A.E. Macromolecules **1977**, *10*, 153.
24. Riande, E.; Saiz, E.; Mark, J.E. Macromolecules **1980**, *13*, 448.
25. Saiz, E.; Riande, E.; Mark, J.E. Macromolecules **1984**, *17*, 899.

7

Molecular Dynamics Simulations of Equilibrium and Dynamic Dielectric Properties

7.1. INTRODUCTION

The response of liquids to perturbation force fields depends on the molecular mobility, which in turn is conditioned by both the molecular interactions and the intramolecular flexibility. Intermolecular rather than intramolecular interactions play a leading role in the dynamics of low molecular weight liquids, whereas the opposite may be true for liquids made up of highly flexible molecules. Complex motions are involved in the response of liquids to external force fields, and a single relaxation time is not enough to describe the relaxation behavior of even the simplest low molecular weight substances. The complexity of the response grows anomalously with increase in the internal degrees of freedom of the molecular entities, so that the relaxation times associated with the response may spread from picoseconds, associated with local motions, to seconds or even longer times, corresponding to flow processes.

The molecular motions associated with the response to perturbation fields are reflected in the relaxation spectra of liquids. The isotherm corresponding to a loss property in the frequency domain displays a prominent and slow absorption, called α-relaxation, and a rather weak and fast relaxation, called β-relaxation. These relaxations at temperatures in the vicinity of the glass transition temperature appear at low and high frequencies, respectively, for supercooled liquids. As the temperature increases, both relaxations are shifted to higher frequencies, the displacement being higher in the α-absorption owing to its comparatively high activation energy. As a result, both relaxations overlap until, deep in the liquid state, they coalesce into a single relaxation, called the $\alpha\beta$-process. Polymers with dipoles having a component parallel to the chain contour exhibit the normal mode process, in addition to the α- and β-relaxations. Whereas both the α- and β-relaxations are rather insensitive to molecular weight, the normal mode process shows a strong dependence on chain length. Dielectric relaxation processes in flexible polymers are discussed in Chapter 9. The normal mode process and secondary relaxations are discussed in Chapters 3, 5, and 9.

The normalized response in the time domain to a very small electric perturbation field can be expressed in molecular terms by[1–3]

$$g(t) = \frac{\langle \boldsymbol{\mu}_i(t) \cdot \boldsymbol{\mu}_i(0) \rangle + \langle \sum_{i \neq j} \boldsymbol{\mu}_i(0) \cdot \boldsymbol{\mu}_j(t) \rangle}{\langle \mu^2 \rangle + \langle \sum_{i \neq j} \boldsymbol{\mu}_i(0) \cdot \boldsymbol{\mu}_j(0) \rangle} \tag{7.1.1}$$

where $\boldsymbol{\mu}_i$ is the dipole moment of the molecule i at time t, and $\langle \mu^2 \rangle$ is the mean-square dipole moment of the relaxing species. According to the linear phenomenological theory of dielectrics, the Fourier transform of the time–dipole correlation function gives the complex dielectric permittivity. This function can be written as [Eq. (2.12.7)]

$$\frac{\varepsilon^*(\omega) - \varepsilon_\infty}{\varepsilon_r - \varepsilon_\infty} = 1 - i\omega \int_0^\infty g(t) \exp(-i\omega t) \, dt \tag{7.1.2}$$

where ω is the angular frequency and the subscripts r and infinity refer to the relaxed and unrelaxed dielectric permittivity respectively. Therefore, the relaxation behavior of polymers in the frequency domain can in principle be predicted as a function of the chemical structure if t is known.

7.2. BASIC PRINCIPLES OF MOLECULAR DYNAMICS

7.2.1. Force Fields

Molecular Dynamics tries to mimic all the forces acting in a given molecule and to calculate further the trajectory of each atom as it moves by the effect of those forces.[4–11] Internal and external forces condition the trajectories in the conformational space. Internal forces arise from interactions of each atom with all the other atoms. These interactions depend on both the kind of atoms and their geometric disposition which governs all the interatomic distances. The standard procedure employed for reproducing the internal forces involves the use of an empirical function that allows the evaluation of the energy produced by those interactions for all the atoms contained in the sample as a function of the interatomic distances. The atoms are placed in a force field that can be defined as a potential energy surface represented by a closed set of analytical potential energy functions.[12]

The total potential energy is usually written as a sum of terms, each one of them representing a different kind of interaction

$$
E_P = \sum_{\text{all bonds}} \frac{k_{ij}}{2} (d_{ij} - d_{ij}^0)^2 + \sum_{\text{all angles}} \frac{k'_{ijk}}{2} (\theta_{ijk} - \theta_{ijk}^0)^2
$$
$$
+ \sum_{\text{all rotations}} \frac{k_\varphi}{2} \left[1 + \frac{|s|}{s} \cos(|s|\varphi) \right]
$$
$$
+ \sum_{ij} \frac{B_{ij}}{r_{ij}^{12}} - \frac{A_{ij}}{r_{ij}^6} + \sum_{ij} \frac{q_i q_j}{\varepsilon r_{ij}} + \sum_i k_i d_i^2 \tag{7.2.1.1}
$$

where the terms on the right-hand side of the equation represent respectively the energies due to bond stretching, angle bending, intrinsic rotational barriers, van der Waals, Coulombic, and out-of-plane bending interactions. Some force fields employ cross-terms (e.g., bond stretch-angle bending) for fine tune-up.

The parameters k_{ij} and k'_{ijk} in Eq. (7.2.1.1) represent force constants, whereas d_{ij}^0 and θ_{ijk}^0 are respectively the unstressed bond length and bond angles. The parameter k_φ represents the barrier energy of the rotation φ defined by three consecutive bonds joining four atoms, for example, the sequence of $i-j-k-l$ bonds. The absolute value of s represents the symmetry of the rotation, i.e., the term in brackets passes through $|s|$ maxima and the same number of minima when φ varies from 0 to 360°. The sign of s indicates the sign convention employed to define the barrier. Thus, when s is negative, the term in cosine subtracts and the torsional energy becomes zero for $\varphi = 0$; for $|s|\varphi = 180°$, or any odd integer of this value, the energy is equal to k_φ. Positive values of s render this term equal to k_φ for $\varphi = 0$; for any odd value of $|s|\varphi = 180°$, the torsional energy contribution is nil.

The sum in the van der Waals contributions expands over all pair of atoms i, j whose separation depends on at least one rotation. Here, A_{ij} and B_{ij} represent respectively the strengths of the attractive and repulsive interactions between the i, j atoms. The attractive term coefficient A_{ij} can be determined by applying the Slater–Kirkwood equation in a slightly modified form, as suggested by Scheraga (see Eqs (5.8.1) and (5.8.2) of Chapter 5), while B_{ij} is obtained from the minimum of the van der Waals interactions at $r_{ij} = r_0$, where the parameter r_0 is approximately the sum of the van der Waals radii of atoms ij.

The Coulombic contribution extends to all pairs of atoms that, having residual charges ($q_i, q_j \neq 0$), are neither directly bonded nor bonded to a common third atom. The parameter ε represents the dielectric permittivity of the medium in which the atoms i, j are supposedly to be immersed, and values of 3 or 4 are often used for ε. Partial charges over each atom are computed by quantum chemistry procedures such as AMPC.[13] Finally, the last contribution, the plane bending energy, is used for some force fields to push into planarity some groups whose structure is known to be planar. Here, k_i indicates the strength of the contribution and d_i represents the distance from atom i to the plane defined by its three attached jkl atoms.

Information on the theory and development of force fields for the conformational analysis of molecules can be found elsewhere.[12,14−19] Some of the most popular force fields are MM2,[15] Amber/OPLS,[20] Amber,[21−23] ECEPP,[24] CHARMM,[25,26] CFF/VFF,[27−31] YETI,[32] TRIPOS, [33,34] etc. All of them employ harmonic functions for bond stretching and angle bending energies. The largest discrepancy between the different force fields lies in the parameters used for the functions (Lennard Jones, Buckingham potentials, etc.) that account for the van der Waals interactions. The weakest point for any force field is its applicability to molecules other than those for which it was designed.

7.2.2. Integration Algorithms and Trajectories

Atom k of mass m_k of a molecule is under the action of a force $\mathbf{f}_k$ given by

$$\mathbf{f}_k = m_k \frac{d^2 \mathbf{r}_k}{dt^2} = -\nabla_k E_P \tag{7.2.2.1}$$

where E_P is the force field and $\mathbf{r}_k$ is the vector defining the position of the atom k with respect to an arbitrary reference frame. Since the integration of Eq. (7.2.2.1) is too complicated to allow its analytical solution, the equation of motion is usually solved by means of a series expansion. Actually, if the location of atom k at time t has the value $\mathbf{r}_k(t)$, the location of the atom at a neighboring time $t + \delta$ and $t - \delta$ can be written as

$$\mathbf{r}_k(t + \delta) = \mathbf{r}_k(t) + \left[\frac{\partial \mathbf{r}_k(t)}{\partial t}\right]\delta + \frac{1}{2}\left[\frac{\partial^2 \mathbf{r}_k(t)}{\partial t^2}\right]\delta^2 + \cdots$$

$$\mathbf{r}_k(t - \delta) = \mathbf{r}_k(t) + \left[\frac{\partial \mathbf{r}_k(t)}{\partial t}\right](-\delta) + \frac{1}{2}\left[\frac{\partial^2 \mathbf{r}_k(t)}{\partial t^2}\right](-\delta)^2 + \cdots \tag{7.2.2.2}$$

where the second and third terms between brackets on the right-hand side of Eq. (7.2.2.2) are respectively the velocity and acceleration of atom k. Therefore, if the velocities and acceleration at time t are known, the approximate positions of atom k at time $t + \delta$ and $t - \delta$ are

$$\mathbf{r}_k(t - \delta) = \mathbf{r}_k(t) - \mathbf{v}_k(t)\delta + \tfrac{1}{2}\mathbf{a}_k(t)\delta^2 + \cdots$$

$$\mathbf{r}_k(t + \delta) = \mathbf{r}_k(t) + \mathbf{v}_k(t)\delta + \tfrac{1}{2}\mathbf{a}_k(t)\delta^2 + \cdots$$

(7.2.2.3)

These two equations lead to

$$\mathbf{r}_k(t + \delta) = 2\mathbf{r}_k(t) - \mathbf{r}_k(t - \delta) + \mathbf{a}_k(t)\delta^2 \qquad (7.2.2.4)$$

also known as the Verlet algorithm. The application of this algorithm requires the knowledge of $\mathbf{r}(t - \delta)$ which can be gained using the following scheme

$$\left\{ \begin{array}{l} \mathbf{r}_k(t) \longrightarrow \mathbf{f}_k(t) \longrightarrow \mathbf{a}_k(t) \\ \mathbf{v}_k(t) \end{array} \right\} \longrightarrow \mathbf{r}_k(t - \delta) \qquad (7.2.2.5)$$

The initial position $\mathbf{r}_k(t)$ is employed to compute $\mathbf{f}_k(t)$ from $\nabla_k E_P$ and then $\mathbf{a}_k(t) = \mathbf{f}_k/m_k$. This latter parameter, together with the initial velocity of the atom k, $\mathbf{v}_k(t)$, allows the determination of $\mathbf{r}_k(t - \delta)$. After this, the main integration loop starts, so that for step i one obtains

$$\mathbf{r}_k[t + (i + 1)\delta] = 2\mathbf{r}_k(t + i\delta) - \mathbf{r}_k[t + (i - 1)\delta] + \mathbf{a}_k(t + i\delta) \qquad (7.2.2.6)$$

This step is repeated for $i = 0$ to $i = n - 1$ to obtain the coordinates of the atom k as a function of time. The time increment on the integration loop must be very small to obtain the trajectories of the atoms with enough accuracy.

A shortcoming of the Verlet algorithm is that it does not control the velocities of the atoms and therefore has no command over the temperature of the system. This thermodynamic parameter is related to the velocities of the atoms of the system through the following equation

$$E_{\text{kin}} = \frac{1}{2}\sum_{k=1}^{N} m_k v_k^2 = \frac{k_\text{B} T}{2}(3N - N_{\text{fd}} - 6) \qquad (7.2.2.7)$$

In this expression, $3N$ and N_{fd} represent respectively, the number of degrees of freedom of the molecule and the number of distances between atoms that are kept fixed. Molecular dynamics focuses on relative motions of some atoms that give rise to changes in the geometry (i.e., conformational transitions) or in the distribution of atoms or molecules within the sample (a typical example is the diffusion of small molecules in a polymer matrix). The numerical factor 6 that appears in Eq. (7.2.2.7) comes from the global translations (three degrees of freedom) and rotations (three degrees of freedom assuming that the system is not a single linear molecule that only contains two degrees of rotation). Removal of the global motions of the molecules can be accomplished by subtracting both linear ($\mathbf{v}$) and angular ($\boldsymbol{\omega}$) constant velocities from the velocities of all the atoms of the sample. By representing the velocity and positions of atom k by $\mathbf{v}_k$ and $\mathbf{r}_k$, the removal of the global motions can be attained by means of the following transformation[15]

$$\mathbf{v}_k \longrightarrow \mathbf{v}_k - \mathbf{v} - \boldsymbol{\omega} \times \mathbf{r}_k \qquad (7.2.2.8)$$

The values of $\mathbf{v}$ and $\boldsymbol{\omega}$ are chosen in such a way that the total linear $\mathbf{P}$ and angular $\mathbf{J}$ momenta of the whole system will become zero after the subtraction. Therefore, $\mathbf{v}$ and $\boldsymbol{\omega}$ are computed by solving the following equations

$$\mathbf{P} = \sum_{k=1}^{N} m_k(\mathbf{v}_k - \mathbf{v} - \boldsymbol{\omega} \times \mathbf{r}_k) = 0$$

$$\qquad (7.2.2.9)$$

$$\mathbf{J} = \sum_{k=1}^{N} m_k[\mathbf{r}_k \times (\mathbf{v}_k - \mathbf{v} - \boldsymbol{\omega} \times \mathbf{r}_k)] = 0$$

Owing to the importance of the control of temperature during molecular dynamics (MD) simulations, an alternative method to the Verlet algorithm is employed in the integration of the equation of motion. Let us consider that the actual acceleration of an atom k changes with time, as indicated by the solid line in Fig. 7.1. The interval of integration δ is divided into halves and the value of the acceleration, for example at time t, is assumed to be constant from $t - \delta/2$ to $t + \delta/2$ (see Fig. 7.2). Accordingly, $v(t + \delta/2) = v(t - \delta/2) + a(t)\delta$. It is

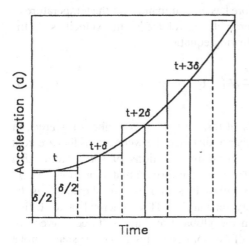

FIG. 7.1 Scheme showing the replacement of the actual acceleration by a stair function

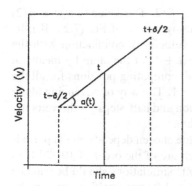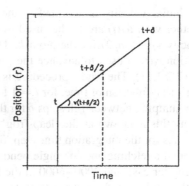

FIG. 7.2 (Left) The acceleration remains constant in the interval $t - \delta/2$ and $t + \delta/2$, and the velocity changes linearly with slope $a(t)$; (right) the velocity at $t + \delta/2$ is taken as the mean velocity in the interval t, $t + \delta$, and the location of the particle changes linearly with time

considered further that $v(t + \delta/2)$ is the average velocity between t and $t + \delta$ (see Fig. 7.2b) so that $r(t + \delta) = r(t) + v(t + \delta/2)\delta$. By generalizing these expressions, the so-called leap-frog algorithm is obtained which can be written as

$$\mathbf{v}_k\left[t + (i + \tfrac{1}{2})\delta\right] = \mathbf{v}_k\left[t + (i - \tfrac{1}{2})\delta\right] + \mathbf{a}_k(t + i\delta)\delta \qquad (7.2.2.10)$$

and

$$\mathbf{r}_k\left[t + (i+1)\delta\right] = \mathbf{r}_k(t + i\delta) + \mathbf{v}_k\left[t + (i + \tfrac{1}{2})\delta\right]\delta \qquad (7.2.2.11)$$

In the first step ($i = 0$), the initial position of each atom ($\mathbf{r}_k(t)$) is used to obtain the force $\mathbf{f}_k$ acting on each atom from which the acceleration $\mathbf{a}_k$ is computed. In this initial step, the atomic velocities are randomly assigned in such a way that they reproduce the macroscopic temperature [Eq. (7.2.2.7)] and the zero values of linear and angular momenta [Eq. (7.2.2.9)]. A Boltzmann distri-bution is generally utilized for this purpose. The easiest procedure for this assignment is to use a routine for the generation of random numbers and employ it to produce a Gaussian distribution of $3N$ numbers (one for each Cartesian coordinate of each atom within the sample), having zero mean and unity variance. Denoting by g_j^k the number associated with the jth coordinate of atom k, its initial velocity will be

$$v_j^k = g_j^k\left[\frac{k_B T}{3Nm_k}(3N_{at} - N_{fd} - 6)\right]^{1/2} \qquad (7.2.2.12)$$

The initial velocity is assigned to the time $t - \delta/2$. The velocity $\mathbf{v}_k(t - \delta/2)$ together with $\mathbf{a}_k(t)$ are further used to compute by means of Eq. (7.2.2.10) the velocity $v_k(t + \delta/2)$ for the atom k. The new velocity in conjunction with the position $\mathbf{r}_k(t)$ is used to advance the position $\mathbf{r}_k(t + \delta)$ of the atom by means of Eq. (7.2.2.11). The whole procedure is repeated, computing positions for all the atoms as a function of time, for $i = 1$ to $i = n - 1$. The way of advancing along time, jumping between full steps δ for the position and half-steps for the velocity, is why this is known as the 'leap-frog' algorithm.

As for the integration time step, δ, the value chosen depends on the periods for bond stretching and for angle bending which are of the order of 10–20 fs in the former case and 2000–6000 in the latter. MD simulations should be capable of distinguishing the atomic trajectories among different positions within the vibration of bonds. Therefore, the time step δ should be substantially smaller than the value of the lowest vibration period, since the positions of the atoms would be blurred otherwise. A good rule of thumb is that the integration time step should not be larger than one-tenth of the smallest value of the vibration period. This is the reason for employing $\delta = 1$ fs in most simulations.

Under configurations of high potential energy, the interactions between the atoms will be strong producing huge forces that will accelerate the movement of the atoms, thus warming up the system. Conversely, the system will cool down in the vicinity of stable conformations where the interatomic interactions are small and therefore produce weak forces. The most intuitive way of controlling the temperature of the system in a canonical ensemble (fixed values of the total number of atoms N, total volume V, and temperature T_0) is to compute its temperature T by means of Eq. (7.2.2.7) after each integration cycle. For this task, the set of values $\{v_i\}$ are rescaled by means of a scaling factor that would render T in accordance with T_0. This can be accomplished by means of the scaling factor S defined as

$$S = \left(\frac{T_0}{T}\right)^{1/2} \tag{7.2.2.13}$$

Then, the rescaled velocities are given by

$$\{\mathbf{v}_k\}_{\text{corr}} = S\{\mathbf{v}_k\} \tag{7.2.2.14}$$

However, when T and T_0 are noticeably different, this strategy can produce sharp variations in the temperature of the system that may render unstable the integration process. To avoid this, it is frequent to use a damping factor that forces T toward T_0 in a more gentle way without producing sharp changes.

The scaling factor often used is

$$S = \left[1 + \left(\frac{\delta}{2d} \right) \left(\frac{T}{T_0} - 1 \right) \right]^{1/2} \qquad \{ \mathbf{v}_k \}_{\text{corr}} = S \{ \mathbf{v}_k \} \qquad (7.2.2.15)$$

where δ is the time step used for the integration cycle and d is the amplitude of damping. For $d = \delta/2$, Eq. (7.2.2.15) becomes Eq. (7.2.2.13), so no damping occurs. On the other hand, large values of d force T to approach T_0 too slowly.

The condition of constant volume is achieved by introducing periodic boundary conditions. It is assumed that the N atoms are contained in a cubic box of length L centered at the origin of coordinates so that it spans from $-L/2$ to $L/2$ on the three coordinate axes. Each time that any of the three coordinates of an atom is larger than $L/2$ or lower than $-L/2$, the atom is forced inside the box by subtracting or adding the value of L to that coordinate.

The easiest method to keep the pressure constant assumes that the pressure at time t, $p(t)$, arises from both the thermal motions of the atoms [i.e., $(Nk_BT)/V$] and the interatomic interactions (i.e., the virial term).[4] Thus

$$p(t) = \frac{Nk_BT}{V} + \frac{1}{3V} \sum_{k=1}^{N} \mathbf{r}_k(t) \cdot \mathbf{f}_k(t) \qquad (7.2.2.16)$$

where N is the number of atoms of the sample, V is the volume of the box, T is the instantaneous temperature obtained from the set of velocities by means of Eq. (7.2.2.7), and $\mathbf{r}_k$ and $\mathbf{f}_k$ are respectively the position occupied by atom k and the force acting on it, both of them computed at time t. After each integration cycle, the pressure is computed by means of Eq. (7.2.2.16) and compared with the target constant pressure P_0 that the system should have. A scaling factor λ is defined as

$$\lambda = 1 - \beta_T \frac{\delta}{d_p} [p(t) - p_0] \qquad (7.2.2.17)$$

where δ represents the time step used for the integration cycle, β_T is the isothermal compressibility, and d_p is a damping factor. The atomic position $\mathbf{r}_k$ and the volume of the box are scaled by a factor λ after each integration cycle. Thus, the components of the position and the dimensions of the box in the coordinate axes are scaled as

$$(r_k^x)_{\text{corr}} = \lambda^{1/3} r_k^x; (L_x)_{\text{corr}} = \lambda^{1/3} L_x$$

$$(r_k^y)_{\text{corr}} = \lambda^{1/3} r_k^y; (L_y)_{\text{corr}} = \lambda^{1/3} L_y \qquad (7.2.2.18)$$

$$(r_k^z)_{\text{corr}} = \lambda^{1/3} r_k^z; (L_z)_{\text{corr}} = \lambda^{1/3} L_z$$

In this way, the pressure of the system is gently pushed towards the constant pressure p_0.

7.2.3. Computation Time Savings

A great deal of time is spent computing the van der Waals and Coulombic contributions to the potential. The time involved scales with the square of atoms contained in the sample. In addition to a correct and efficient programming of the algorithms, substantial computing time savings can be obtained by assuming that the interactions between atoms i and j and the forces they produce are negligible when the separation between those atoms is larger than a predetermined value r_c. These contributions are replaced by

$$E_{ij}(r_{ij}) = \begin{cases} E_{ij}(r_{ij}), & r_{ij} < r_c \\ 0, & r_{ij} > r_c \end{cases} \qquad (7.2.3.1)$$

In order to avoid discontinuities in both the energy and the force at $r = r_c$, shifted potentials are used that gently push both quantities towards zero, avoiding the discontinuity. For example

$$E_{ij}(r_{ij}) = E_{ij}(r_{ij}) - E_{ij}(r_c) - \left[\frac{\partial E_{ij}(r_{ij})}{\partial r_{ij}}\right]_{r_{ij}=r_c} (r_{ij} - r_c) \qquad (7.2.3.2)$$

for $r_{ij} < r_c$, and

$$E_{ij}(r_{ij}) = 0 \qquad (7.2.3.3)$$

for $r_{ij} > r_c$.

Saving time can also be accomplished by using multiple step methods. For example, when the distance between the ij atoms is very small, large accelerations are produced, and therefore very small time steps δ are required. However, as the distance increases, coming close to r_c, larger time steps δ' can be used. Another method of reducing computing time is to keep constant the distance between any selected pair of atoms i, j. This can be used to eliminate fast vibrational motions (e.g., C–H bond stretching) in order to allow the use of larger time step integration times. There are several force fields available nowadays, from the early ones developed for hydrocarbons to those produced by commercial software companies such as Tripos or Byosym that are supposed to be applicable to any kind of molecules.

7.3. TRAJECTORIES OF MOLECULES IN PHASE SPACE AND COMPUTING TIME

A shortcoming of molecular dynamics is the prohibitively large computing time that molecules require to visit all the phase conformational space. For illustrative purposes, let us consider the trajectory of 2-chlorocyclohexyl acetate (Fig. 7.3) in the phase conformational space[35]. Figures 7.4 and 7.5 represent respectively the equatorial (i.e., Cl and ester substituents in eq–eq positions) and axial (i.e., axial–axial positions) conformations of the molecule. The rotation angle over the ClCH–CHO bond, ϕ, governs the relative orientation of the C_yH–Cl and C_yH–O bonds. The evolution of this angle with time can be used to monitor the chair-to-inverse chair conformational transition on the cyclohexane ring, since the C_y–C_y bond is very close to the *trans* state (i.e., $\phi \approx 180°$) for axial conformation and near *gauche* (i.e., $\approx 300°$) in the equatorial orientation. The rotational angle χ (i.e., rotation over the bond HC_y–OC^*) indicating the orientation of the ester group relative to the cyclohexane ring was taken to be 180° for the *trans* conformation.

The evaluation of the trajectories of ϕ, χ, and the dipole moment μ was performed by computing the partial charges with the Sybyl molecular modeling package, employing the MOPAC program and the AM1 procedure. The variation in the dipole of the molecules with time can be obtained from the location of the centers of gravity of the positive and negative residual charges, given by

$$\mathbf{r}^+(t) = \frac{\sum_{k=1} \mathbf{r}_k(t)q_k^+}{\sum_k q_k^+}, \qquad \mathbf{r}^-(t) = \frac{\sum_{k=1} \mathbf{r}_k(t)q_k^-}{\sum_k q_k^-} \qquad (7.3.1)$$

where from the electroneutrality principle

$$\sum_k q_k^+ = \sum_k q_k^- \qquad (7.3.2)$$

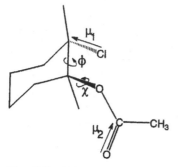

FIG. 7.3 Scheme of 2-chlorocyclohexyl acetate with the halogen atom in the equatorial position

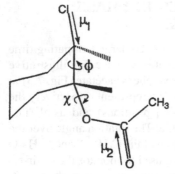

FIG. 7.4   Scheme of 2-chlorocyclohexyl acetate with the halogen atom in the equatorial position

Therefore, the dipole moment of the molecules can be written as

$$\boldsymbol{\mu}(t) = [\mathbf{r}^+(t) - \mathbf{r}^-(t)] \sum_k q_k^+ \qquad (7.3.3)$$

Figures 7.5 and 7.6 represent the variation with time in the dipole moment and rotational angles ϕ and χ obtained during the MD simulation at 300 K. The simulation was carried out using the MD software package Sybyl. Data shown in Fig. 7.6 were obtained employing the equatorial conformation of the molecule as the starting point for the simulation, while Fig. 7.7 presents the results obtained when the MD trajectory starts in the axial conformation. The relative orientations of the ester group and cyclohexane, governed by the rotational angle χ, are almost identical in both cases with values of χ in the range 52–55°. As for the rotational angle ϕ, it remains within the range 297–300° when the simulation is started in the equatorial conformation, and within 174–177° in the case of the axial conformation.

```
 0.139              ┌0.120 (Cl)
   │        0.154   │
   │         │     -0.133
-0.266────-0.263       \0.187
        -0.290───0.021──0.186
-0.260    0.156  -0.287      -0.388
   │             0.352
 0.142
                   │
                 0.120
```

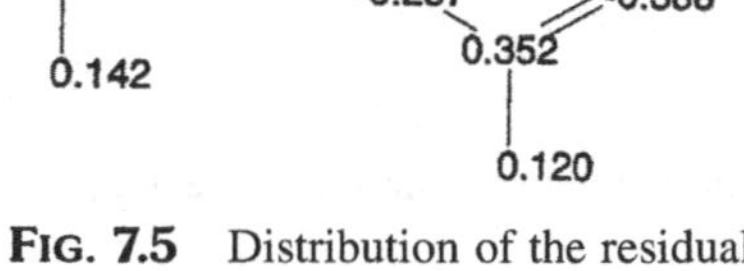

FIG. 7.5   Distribution of the residual charges in 2-chlorocyclohexyl acetate

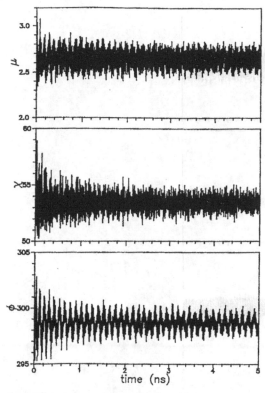

FIG. 7.6 Trajectories of the dipole moment μ and the angles ϕ and χ for 2-chlorocyclohexyl acetate at 300 K. The angles ϕ and χ define respectively the positions of the chlorine atoms and the ester residue. The trajectories started with the chlorine atom in the equatorial position[35]

An important feature exhibited by Figs 7.6 and 7.7 is the absence of chair-to-inverse chair conformational transitions, since the molecule remains during all the simulation in the conformation used as the starting point for the MD trajectory. These results suggest that either much higher times or much higher temperatures (or a combination of both effects) should be used in order to observe interconversion between the two allowed conformations of the substituted cyclohexane moiety. Trajectories obtained at 1000 K for ϕ, χ and μ are shown in Fig. 7.8. It can be seen that at this temperature the C_yH-O bond is nearly freely rotating, whereas the trajectory of ϕ suggests that there is a reasonably fast interconversion between the two allowed conformations of the cyclohexane ring. At 1000 K the dipole moment oscillates approximately in the range 2–6 D.

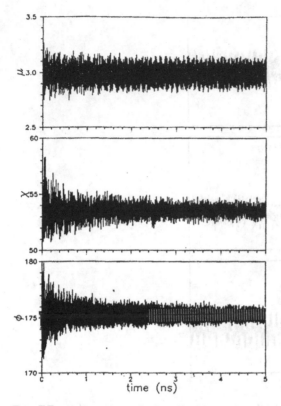

FIG. 7.7 Trajectories of the dipole moment μ and the angles ϕ and χ for 2-chlorocyclohexyl acetate at 300 K. The simulations started with the chlorine atom in the axial conformation[35]

The probability distributions for ϕ at 1000 and 1500 K are shown in Fig. 7.9. The probability curve was obtained by computing the time spent by the molecule at the value of ϕ of interest, with a tolerance of $\pm\,0.5°$, and dividing it by the total time. The areas of the peaks of each curve give the relative incidence of the two conformations allowed to the substituted cyclohexane molecule, thus permitting the evaluation of the fraction of the conformations F_{ax} and F_{eq} in which the chlorine atom is in axial and equatorial positions respectively. Semi-logarithmic plots of $F_{\text{ax}}/F_{\text{eq}}$ against $1/T$ in the temperature interval 1000–1600 K give reasonably straight lines that, extrapolated to $T = 300$ K, lead to $F_{\text{eq}} = 0.987$ and $F_{\text{ax}} = 0.013$. Therefore, the averaged dipole moment of the molecule at room temperature should be similar to that obtained for the equatorial conformation. The value simulated at this temperature is 2.65 D, in good agreement with the experimental result.

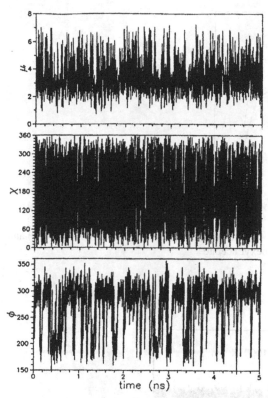

FIG. 7.8 Results of the MD simulations for 2-chlorocyclohexyl acetate at 1000 K[35]

The simulated results show that at 1000 K the chlorine atom spends a total of 1.15 ns in the axial conformation (i.e., ϕ is within the range 140–230°) distributed over 101 visits, and therefore the averaged lifetime τ of this conformation is 0.0114 ns. The chlorine atom spends 3.75 ns in the equatorial position (i.e., $250° \leq \phi \leq 360°$) with 113 visits and $\tau = 0.0332$ s. From the Arrhenius plots of τ, the barrier energies for the equatorial → axial and axial → equatorial conformational transitions are 8.15 and 5.72 kcal mol^{-1} respectively. By extrapolating the Arrhenius plots at room temperature, it is found that the average lifetimes spent by chlorine atoms in equatorial and axial conformations at 300 K are 463 and 9.6 ns respectively.

The barrier energies determine the computing time involved in the simulations, as the trajectory of ϕ for cyclohexane (see Fig. 7.10), in comparison with that of substitute cyclohexane, shows. It can be seen that no chair-to-inverse chair conformational transitions are observed at 1000 K in the

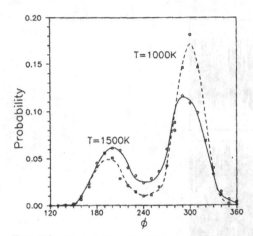

FIG. 7.9   Probability distribution of the rotational angle ϕ obtained at 1000 K (······) and 1500 K (———) for 2-chlorocyclohexyl acetate

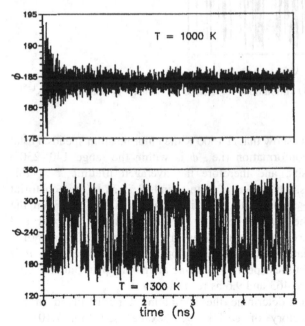

FIG. 7.10   Time dependence of the rotational angle ϕ for cyclohexane obtained at 1000 and 1300 K[35]

interval of 5 ns, and therefore these transitions are slower in cyclohexane than in 2-chlorocyclohexyl acetate. The value of the barrier energy obtained for the chair-to-inverse chair conformational transition is $10.5\, \text{kcal mol}^{-1}$, in good agreement with the value obtained for this quantity from NMR results. These results seem to suggest that some substituted cyclohexanes have lower barrier energies than the unsubstituted ones.

7.4. DETERMINATION OF THE TIME–DIPOLE CORRELATION COEFFICIENT

From the computing results, the time–dipole correlation function can be obtained by means of the following equation[36,37]

$$g(n\Delta) = \frac{1}{N-n}$$
$$\times \sum_{k=1}^{N-n}\left[\frac{\langle \boldsymbol{\mu}_i(k\Delta)\cdot\boldsymbol{\mu}_i[(k+n)\Delta]\rangle + \langle \sum_{i\neq j}\boldsymbol{\mu}_i(k\Delta)\cdot\boldsymbol{\mu}_j[(k+n)\Delta]\rangle}{\langle \boldsymbol{\mu}_i(k\Delta)\cdot\boldsymbol{\mu}_i(k\Delta)\rangle + \langle \sum_{i\neq j}\boldsymbol{\mu}_i(k\Delta)\cdot\boldsymbol{\mu}_j(k\Delta)\rangle}\right]$$

(7.4.1)

where the angular brackets denote averages. Time steps of the order of 1 fs are often used in the integration of the equation of motion, and therefore several millions of integration cycles are necessary to reach times on the order of ns. As a consequence, the conformations obtained every 100–500 integration steps are written down, and Δ in Eq. (7.4.1) is the time gap between these consecutive conformations. Equation (7.4.1) represents an average over the N-n of g obtained with all pairs of conformations that are separated by a time equal to Δn.

Obviously, the time–dipole correlation coefficient arises from a weighted sum of auto- and cross-correlation terms for dipole motions. Although the cross-correlation terms in Eq. (7.4.1) may have an appreciable magnitude and their signs may be positive or negative, it has been observed that these contributions hardly affect the time–dipole correlation function. Therefore, the normalized response in the time domain can be expressed in terms of the time–dipole autocorrelation coefficient given by

$$g(n\Delta) = \frac{1}{(N-n)<\mu^2>}\sum_{k=1}^{N-n}\langle \boldsymbol{\mu}_i(k\Delta)\cdot\boldsymbol{\mu}_i[(k+n)\Delta]\rangle \qquad (7.4.2)$$

Illustrative plots representing the simulated normalized response for 2-(acetoloxy)ethyl-2-(naphthyl) acetate in the time domain are shown in Fig. 7.11. It can be seen that the curves corresponding to Eqs (7.4.1) and (7.4.2)

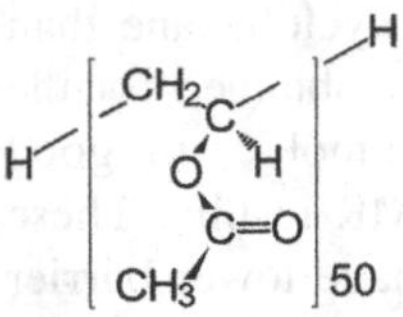

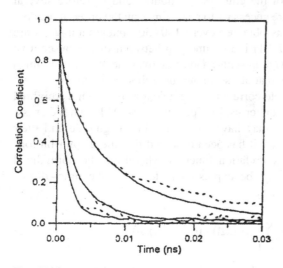

FIG. 7.11   Repeating unit of poly(vinyl acetate)

are similar, bearing out that cross-correlation terms in Eq. (7.4.1) apparently do not affect the normalized response in the time domain.

In what follows, the simulation of the relaxation behavior of an isolated chain of polyvinyl acetate (PVAc), PVAc in bulk, and solutions of PVAc in toluene will be described. The basic molecule employed in the simulations was an H terminated oligomer containing 50 repeat units of PVAc with a 40% contents of meso units and random placement of meso and racemic centers. This molecule will be referred to as the oligomer, and its structure is schematically represented in Fig. 7.12.

Solutions[38] of PVAc in toluene with different weight percentage contents of polymer were simulated by means of cubic lattices having periodic boundary conditions. The lengths of the box sides were of the order of 30 Å and were adjusted so as to reproduce a density of 1 g cm^{-3}. All the systems were initially generated within much larger boxes, with side lengths roughly twice the desired final values, in

FIG. 7.12   Time–dipole correlation coefficients (including cross-correlation terms) (- - - -) and time–dipole autocorrelation function (———) for 2-(acetyloxy)ethyl-2-(2-naphthyl) acetate obtained from top to bottom at 500, 750, and 1000 K[37]

order to avoid interpenetration among the molecules which would produce unrealistic and extremely high values of conformational energies, thus rendering difficult and very inaccurate any strategy for energy minimization. An MD trajectory at high temperature (i.e., $T = 500$ K) was then applied to these initial systems, decreasing the box sides at regular intervals and allowing an equilibration time of 500 fs at each new length. Once the final box size was achieved, the systems were cooled down to 50 K with temperature increments of 50 K and equilibration times of 500 fs at each new temperature. Then the energy was minimized with respect to all internal coordinates and the systems were warmed up, under NpT conditions, to the working temperature, again with increments of 50 K and equilibration times of 500 fs at each new temperature. The data collection stage of the MD simulation was then performed on each lattice. It consisted of 10^6 integration steps (i.e., a time span of 1 ns) performed under NpT conditions, computing and recording the dipole moment of one molecular chain at intervals of 10 fs.

To obtain the normalized response in a reasonable computing time, the simulations were performed at high temperature, namely, 400, 500, 550, 600, and 700 K. For each of these temperatures, the dipole moments of an individual PVAc molecule in the ensemble was calculated as a function of time for the N conformations recorded during the MD simulation. The time–dipole autocorrelation function was obtained by means of Eq. (7.4.2) using time gaps between two consecutive conformations of $\Delta = 10$ fs.

Illustrative plots of the time–dipole correlation function in the time domain for a single PVAc molecule, for PVAc in bulk, and for solutions of PVAc in toluene in which the weight fraction of polymers was 0.20, 0.30, and 0.50 are shown in Figs 7.13 and 7.14. The shape of this curve fitted the empirical Kohlsrausch–Williams–Watts (KWW) equation [given by Eq. (2.12.8)]

$$g(t) = \exp\left[-\left(\frac{t}{\tau^*}\right)^{\bar{\beta}}\right] \tag{7.4.3}$$

where $0 < \bar{\beta} \le 1$ and τ^* is a characteristic relaxation time. The mean-relaxation time $\langle \tau \rangle$ for each system at each temperature of interest was obtained from $g(t)$ by means of the expression

$$\langle \tau \rangle = \int_0^\infty g(t)\,dt \tag{7.4.4}$$

The mean relaxation time $\langle \tau \rangle$ would be identical to the τ^* parameter on the KWW expression for a Debye relaxation, i.e., when $\bar{\beta} = 1$. Values of the complex dielectric permittivity were obtained from $g(t)$ by Eq. (7.1.2).

Values at 550 K of the real and loss components of the simulated complex dielectric permittivity for isolated PVAc chains, PVAc in bulk, and solutions of PVAc in toluene, are plotted as a function of the angular frequency in Figs 7.15–7.20.

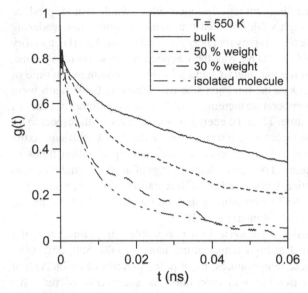

FIG. 7.13   Simulated time–dipole correlation functions for the systems indicated in the inset[38]

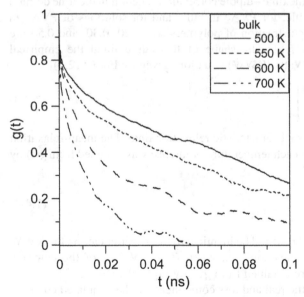

FIG. 7.14   Simulated time–dipole correlation functions for polyvinyl acetate in bulk at the temperatures indicated in the inset[38]

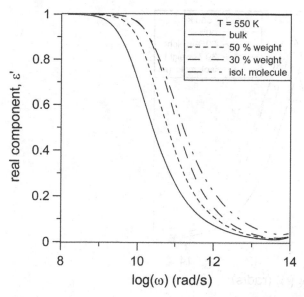

FIG. 7.15 Simulated values at 550 K of the real component of the complex dielectric permittivity in the frequency domain for polyvinyl acetate and its toluene solutions

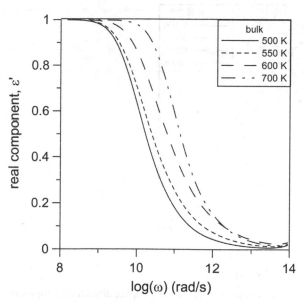

FIG. 7.16 Simulated values of the real component of the complex dielectric permittivity in the frequency domain for polyvinyl acetate at the temperatures indicated in the inset

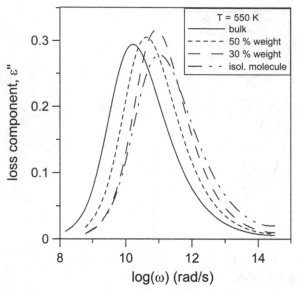

FIG. 7.17 Simulated values at 550 K of the loss component of the complex dielectric permittivity in the frequency domain for polyvinyl acetate and its toluene solutions

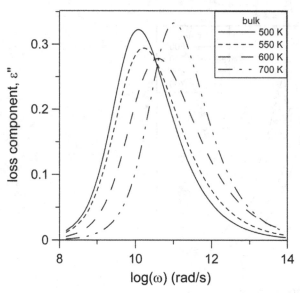

FIG. 7.18 Simulated values of the loss component of the complex dielectric permittivity in the frequency domain for polyvinyl acetate in bulk at the temperatures indicated in the inset

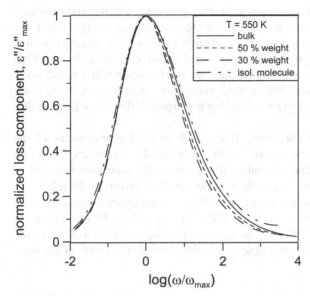

FIG. 7.19 Simulated values of the normalized dielectric loss for poly(vinyl acetate) and its toluene solutions in the normalized frequency domain

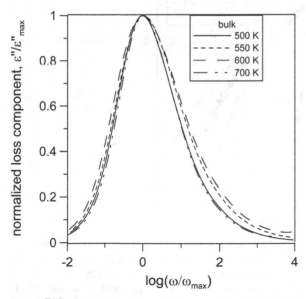

FIG. 7.20 Simulated values of the normalized dielectric loss for poly(vinyl acetate) in bulk plotted in the normalized frequency domain at different temperatures

Arrhenius plots for the loss maximum frequency f_m, computed from the simulated values of the mean relaxation times of these systems $\langle \tau \rangle$ as $f_m = 1/(2\pi \langle \tau \rangle)$, are shown in Fig. 7.21 together with the experimental values of this quantity obtained in the low-temperature region of the supercooled liquid state. It can be seen that, in comparison with the maximum frequencies in the supercooled liquid state, the values computed at very high temperatures show only a weak dependence on temperature. Similar behavior has been observed for the $\alpha\beta$-relaxation process.[1,2]

Owing to the high computing time involved in the simulations of the trajectories of the relaxing species, realistic molecular dynamics is not, in principle, a useful technique to obtain information on the relaxation behavior of materials in the supercooled liquid state. However, the analysis of the relaxation behavior of systems carried out in this study shows an unexpected fact: the widths of the absorption simulated for isolated PVAc chains and for PVAc in bulk are quite similar. The values of $\bar{\beta}$ for the temperature range 400–600 K lie in the interval 0.54–0.51 in the former case and in the interval 0.51–0.60 in the latter

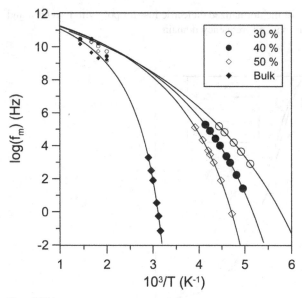

FIG. 7.21 Arrhenius plots of the maximum of the loss frequency f_m computed from the mean relaxation times $\langle \tau \rangle$ of the systems studied [Eq. (6) and Table 3] as $f_m = 1/(2\pi \langle \tau \rangle)$, together with the experimental values obtained in the low-temperature region of the supercooled liquid state. The (——) represent values computed with the VFTH equation [Eq. (1)], while the symbols indicate experimental values ($\diamond \blacklozenge \bullet \bigcirc$) or simulated values ($\diamond \blacklozenge \bullet \circ$)

case. The simulations carried out in the solutions do not show a definite temperature dependence, but on the other hand the values of $\bar{\beta}$ lie in the range of values obtained for PVAc. The scattering in the values of $\bar{\beta}$ obviously arises from the difficulty of fitting the simulated time–dipole correlation function to the KWW equation. However, the simulations suggest that a Debye behavior of the relaxation of liquids at high temperatures is quite unlikely. This conclusion is supported by the plot of the experimental values of $\bar{\beta}$ for PVAc in bulk and its toluene solutions against $1/T$. Extrapolations of these results to infinite temperature give values of $\bar{\beta}$ of the order of 0.57 ± 0.05.

The results obtained for stretched exponential $\bar{\beta}$ of PVC using realistic molecular dynamics simulations are in rather good agreement with those obtained for polyethylene (0.59 ± 0.04), using the same technique and considering each methylene group as a single spherical particle.[37] In the same way, MD simulations of the relaxation of mixtures of soft spheres present an unambiguous example of KWW relaxation with a stretched exponent $\bar{\beta} = 0.62$.[38]

Molecular dynamics simulations of poly(oxyethylene) melts have been performed at temperatures not far away from room temperature. Simulations of this kind were performed for an ensemble of 32 chains of 12 repeat units each.[39] The normalized time–dipole correlation functions for the melt, obtained at several temperatures employing time integration steps of 1 fs, are shown in Fig. 7.22. Transformation of the time–dipole correlation function from the time to the frequency domain gives the reduced complex dielectric permittivity shown at several temperatures in Fig. 7.23a. The peaks are asymmetric, which is

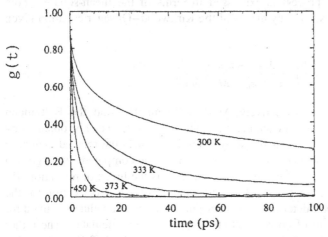

FIG. 7.22 Time–dipole corelation coefficient for polyethylene oxide at different temperatures [39]

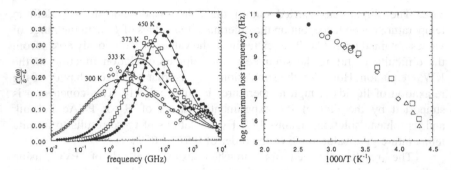

FIG. 7.23 Arrhenius plots for the relaxation time of the dielectric loss of polyethylene. ○ and ● represent computed and experimental results respectively[39]

consistent with a molecular dipole moment following KWW-like behavior. The stretched exponent increases with temperature, reaching at high temperature a value of 0.63, a value that is in the vicinity of that found in the simulations of the dynamics of other systems. Figure 7.23b shows the frequency maximum loss as a function of temperature. Results from experiments on high molecular weight poly(oxyethylene) melts are also shown. It can be seen that the results are in rather good agreement with experiments and show strongly non-Arrhenius behavior.

Simulations allow determination of the reduced complex dielectric permittivity associated with the α-process in the chains with relatively low barrier energies, or with the αβ-process in molecules with high barrier energies. The strength of the process is expressed in terms of the mean-square dipole moment of the chains, $\langle \mu^2 \rangle$, by means of the Kirkwood–Onsager equation given by [Eq. (1.16.8)]

$$\varepsilon_r - \varepsilon_u = \left(\frac{\varepsilon_u + 2}{3} \right)^2 \frac{3\varepsilon_r}{2\varepsilon_r + \varepsilon_u} \frac{4\pi\rho N_A}{M} \frac{\langle \mu^2 \rangle}{3k_B T} \tag{7.4.5}$$

where N_A and k_B are respectively Avogadro's number and the Boltzmann constant, and T is the absolute temperature. The value of $\langle \mu^2 \rangle$ can be obtained by simulation methods, from which and by using Eq. (7.4.1) the relaxed dielectric permittivity can be obtained. The value of ε_u cannot be computed, although in a first approximation $\varepsilon_u = n^2$ can be taken, where n is the index of refraction. By using a value of 2.62 for ε_u at 333 K, and making a volume correction for the value of this parameter determined at other temperatures, the values obtained for the relaxation strength of poly(oxyethylene) melts were calculated. The rather good agreement between experiment and simulation (see Table 7.1) suggests that simulations may successfully predict the α-relaxation strength of polymers as a

TABLE 7.1 Dielectric relaxation parameters of poly(oxyethene) melts

Parameter	300 K, simulation	333 K, simulation	339 K, experiment	373 K, simulation	450 K, simulation
$\langle \mu^2 \rangle^*$, chains	15.1	15.1		16.3	17.1
ε_u	2.69	2.62	2.62	2.54	2.41
ε_r	8.94	7.87	7.82	7.82	5.99
$\varepsilon_r - \varepsilon_u$	6.25	5.25	5.20	5.20	3.58

*Debye.

function of chemical structure. It should be pointed out that the dipole autocorrelation coefficient obtained, $\langle \mu^2 \rangle / 2xm_{C-O}^2 = 0.55$, is in rather good agreement with the experimental value of 0.53 obtained for isolated poly(oxyethylene chains). In the evaluation of the dipole autocorrelation coefficient, the dipole moment associated with CH_2-O bonds was assumed to have a value of 1.07 D. This means that intermolecular interactions play a minor role in the value of the mean-square dipole moments of polymers.

REFERENCES

1. Cook, M.; Watts, D.C.; Williams, G. Trans. Faraday Soc. **1970**, *66*, 250.
2. Williams, G. Chem. Rev. **1972**, *72*, 55.
3. Williams, G. Dielectric relaxation spectroscopy of amorphous polymer systems: the modern approaches In *Keynote Lectures in Selected Topics of Polymer Science*; Riande, E., Ed.; CSIC: Madrid, 1995; 1–39.
4. Berendsen, H.J.C.; Gunsteren, W.F. Molecular dynamics simulations: techniques and approaches. In *Molecular Liquids, Dynamics and Interactions*; Barnes, A.J., Orville-Thomas, W.J., Yarwood, J., Eds.; NATO ASI Series C135; Reidel: New York, 1984; 475–500.
5. Heermann, D.W. *Introduction to Computer Simulation Methods in Theoretical Physics*; Springer: Berlin, 1986.
6. Cicotti, G.; Hoover, W. Eds. *Molecular Dynamics Simulations of Statistical Mechanical Systems*; North Holland: Amsterdam, 1986.
7. Hoover, W.G. *Molecular Dynamics*; Springer: Berlin, 1987.
8. Allen, M.P.; Tildesley, D.J. *Computer Simulation of Liquids*; Clarendon Press: Oxford, 1987.
9. Hockney, R.W.; Eastwood, J.W. *Computer Simulation Using Particles*; Adam Hilger: Bristol, 1988.
10. Frenkel, D.; Smit, B. *Understanding Molecular Simulation*; Academic Press: London, 1996.
11. Riande, E.; Saiz, E. Curr. Trends Polym. Sci. **1997**, *2*, 1.

12. Welsh, W.J. Force-field techniques and their use in estimating conformational stability of polymers. In *Computational Modeling of Polymers*; Bicerano, J., Ed.; Marcel Dekker: New York, 1992; Chap.2.
13. AMPAC. *Quantum Chemistry Program Exchange*; Department of Chemistry, Indiana University, Bloomington, IN.
14. Williams, J.E.; Stang, P.J.; Schleyer, P.V.R. Ann. Rev. Phys. Chem. **1968**, *19*, 531.
15. Burket, U.; Allinger, N.L. *Molecular Mechanics*; ACS Monograph, No. 177; American Chemical Society: Washington, D.C., 1982.
16. Clark, T. *A Handbook of Computational Chemistry*; Wiley-Interscience: New York, 1985.
17. Hirst, D.M. *A Computational Approach to Chemistry*; Blackwell Scientific: London, 1990.
18. Allinger, N.L.; Yuh, Y.H.; Lii, J.H. J. Am. Chem. Soc. **1989**, *111*, 8551.
19. Lii, J.H.; Allinger, N.L. J.Am. Chem. Soc. **1989**, *111*, 8566, 8576.
20. Jorgensen, W.L.; Tirado-Rives, J. J. Am. Chem. Soc. **1988**, *110*, 1657.
21. Weiner, S.J.; Kollman, P.A.; Case, D.A.; Singh, U.C.; Ghio, C.; Alagona, G.; Profeta, S.; Weiner, P. J. Comp. Chem. **1981**, *2*, 257; J. Am. Chem. Soc. **1988**, *106*, 765.
22. Weiner, S.J.; Kollman, P.A.; Nguyen, D.T.; Case, D.A. J. Comp. Chem. **1986**, *7*, 230; Biochemistry **1990**, *29*, 9110.
23. Homans, S.W. Biochemistry **1990**, *29*, 9110.
24. Momany, F.A.; McGuire, R.F.; Burgess, A.W.; Scheraga, H.A. J. Phys. Chem. **1975**, *79*, 2361.
25. Gelin, B.R.; Kerplus, M. Biochemistry **1979**, *18*, 1256.
26. Brooks, B.R.; Brucoleri, R.E.; Olafson, B.D.; States, D.J.; Swaminathan, D.; Karplus, M. J. Comp. Chem. **1983**, *4*, 187.
27. Hagler, A.T.; Huler, E.; Lifson, S. J. Am. Chem. Soc. **1974**, *96*, 5319.
28. Lifson, S.; Hagler, A.T.; Dauber, P. J. Am. Chem. Soc. **1979**, *101*, 5111.
29. Waldman, M.; Hagler, A.T. J. Comp. Chem. **1993**, *14*, 1077.
30. Marple, J.R.; Hwang, M.J.; Stockfisch, T.P.; Dinur, U.; Waldman, M.; Ewing, C.S.; Hagler, A.T. J. Comp. Chem. **1994**, *15*, 162.
31. Hill, J.R.; Sauer, J. J. Phys. Chem. **1994**, *98*, 1238.
32. Vedani, A.J. Comp. Chem. **1988**, *9*, 269.
33. Tripos Associates Inc.: St. Louis, MO.
34. Clark, M.; Cramer III, R.D. III; Opdenbosch, N.V. J. Comp. Chem. **1988**, *10*, 982.
35. Saiz, E.; Riande, E. J. Chem. Phys. **1995**, *103*, 3832.
36. Saiz, E.; Riande, E.; Diaz-Calleja, R. J. Phys. Chem. A **1997**, *101*, 7234.
37. Saez-Torres, P.; Saiz, E.; Díaz-Calleja, R.; Guzmán, J.; Ríande, E. J. Phys. Chem. A **1998**, *102*, 5763.
38. Adachi, A.; Saiz, E.; Riande, E. Phys. Chem. Phys. **2002**, *4*, 635.
39. Smith, G.D.; Yoon, D.Y.; Wade, Ch.G.; O'Leary, D.; Chen, A.; Jaffe, R.L. J. Chem. Phys. **1997**, *106*, 3798.

8

Dielectric Relaxation Processes at Temperatures Above T_g Molecular Chains Dynamics

8.1. INTRODUCTION

Condensed matter has a structural fading memory reflected in the velocity with which a perturbed system forgets the molecular configuration that it had in the past. In ordinary liquids, molecular reorganization occurs very rapidly and structural memory at the molecular level is very short, so that the relaxation time, roughly defined as the time necessary for the system to forget the configuration it had prior to the perturbation, is very small. At the other extreme are solids characterized for having a very large structural memory at the molecular level, reflected in large relaxation times. From a strict point of view, the distinction between solids and liquids cannot be made in absolute terms because it is based on a subjective comparison of the relaxation time of the system and the time of observation. The solid or liquid character of the condensed matter is expressed in terms of the Deborah number defined as

$$N_D = \frac{\tau}{\tau_{exp}} \tag{8.1.1}$$

where τ can be taken as the time required for a relaxation process to approach completion and τ_{exp} is the timescale of the experiment. For ideal liquids $\tau \to 0$ and $N_D = 0$, while for ideal solids $\tau \to \infty$ and $N_D \to \infty$. For the so-called viscoelastic systems τ and τ_{exp} are comparable and their Deborah's number is of the order of unity. Polymers are the most important viscoleastic systems.

8.2. PHENOMENOLOGICAL DIELECTRIC RESPONSE IN THE TIME DOMAIN

Let us consider the response of an isotropic polar system to a perturbation electric field defined as

$$
\begin{aligned}
E(t) &= 0, & t < 0 \\
E(t) &= E_0, & t \geqslant 0
\end{aligned}
\tag{8.2.1}
$$

It is assumed that the electric field is small enough to render the dielectric displacement a linear function of the electric field. In this case the orientation of the dipoles by the effect of the field is reflected in a continuous increase in the dielectric displacement with time until eventually a constant value is reached. The time dependence of the dielectric displacement can be written as

$$
D(t) = \varepsilon(t)E_0
\tag{8.2.2}
$$

where the dielectric permittivity ε depends on time according to the following expression

$$
\varepsilon(t) = \varepsilon_u + (\varepsilon_r - \varepsilon_u)\psi(t)
\tag{8.2.3}
$$

In this expression, ε_u ($\omega = \infty$) is the unrelaxed dielectric permittivity arising from the distortion of both the electronic cloud and the positions of the nuclei of the atoms by the effect of the electric field and therefore it is time independent, and $\psi(t)$ is a monotonous increasing function of time, the extreme values of which are $\psi(0) = 0$ and $\psi(\infty) = 1$. The function $\psi(t)$ is of entropic nature and reflects the molecular motions that take place in the system to accommodate the orientation of the dipoles to the perturbation field. Finally, ε_r ($\omega = 0$) and $\varepsilon_r - \varepsilon_u$ are respectively the relaxed dielectric activity and the dielectric relaxation strength of the system. A schematic representation of the dielectric permittivity in the time domain is shown in Fig. 8.1.

Owing to the entropic nature of the response at $t > 0$, the dielectric displacement does not vanish when the perturbation field ceases and, as a result, $D(t)$ not only depends on the actual perturbation field but also on the electric history undergone by the material in the past. Under a linear behavior regime, the response to perturbation fields is governed by the Boltzmann superposition

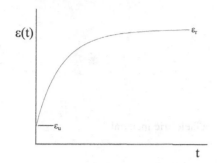

FIG. 8.1 Schematic representation of the dielectric permittivity in the time domain. The parameters ε_r and ε_u are respectively the relaxed and unrelaxed dielectric permittivities

principle. The dielectric displacement can then be written as

$$D(t) = \sum_{\theta_i=-\infty} \varepsilon(t - \theta_i) \, \Delta E(\theta_i) \tag{8.2.4}$$

where $\Delta E(\theta_i)$ is the electric field applied at time θ_i. If the electric field varies with time in very small steps in such a way that it can be assumed to be a continuous function of time, Eq. (8.2.4) adopts the following form [see Eq. (2.12.2)]

$$D(t) = \int_{-\infty}^{t} \varepsilon(t - \theta) \frac{dE(\theta)}{d\theta} \, d\theta \tag{8.2.5}$$

This equation represents the Boltzmann superposition principle for dielectric experiments in continuous form.

Let us consider a history in which the electric field is a continuous function of time in the range $-\infty < t < \theta_1$, it has the constant value $\Delta E(\theta_1)$ in the interval $\theta_1 \leqslant t \leqslant \theta_2$, and again it is a continuous function of time for $t > \theta_2$, as shown in Fig. 8.2. The response for this history can be written as

$$D(t) = \int_{-\infty}^{\theta_1} \varepsilon(t - \theta) \frac{dE(\theta)}{d\theta} \, d\theta + \varepsilon(t - \theta_1) \, \Delta E(\theta_1)$$

$$+ \int_{\theta_2}^{t} \varepsilon(t - \theta) \frac{dE(\theta)}{d\theta} \, d\theta \tag{8.2.6}$$

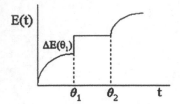

Fig. 8.2   Electric perturbation and response of a dielectric material

Integration of Eq. (8.2.5) by parts leads to

$$D(t) = \varepsilon_u E(t) + \int_0^\infty E(t-u)\frac{d\varepsilon(u)}{du}\,du$$

$$= \varepsilon_u E(t) + (\varepsilon_r - \varepsilon_u)\int_0^\infty E(t-u)\frac{d\psi(u)}{du}\,du \qquad (8.2.7)$$

where the change in variables $t - \theta = u$ was made and $E(-\infty)$ was considered to be zero. As $\varepsilon(t)$ is a continuous function of time, Eq. (8.2.7) is more suitable than Eq. (8.2.5) for expressing the Boltzmann superposition principle. Equation (8.2.7) can alternatively be written as

$$D(t) = \varepsilon_u E(t) + (\varepsilon_r - \varepsilon_u)\int_{-\infty}^t E(\theta)\frac{d\psi(t-\theta)}{d(t-\theta)}\,d\theta \qquad (8.2.8)$$

The function $\psi(t)$, schematically represented in Fig. 8.3, modulates the entropic response to the electric field. Since $\psi(t - \theta)$ is a decreasing function of time, $d\psi(t - \theta)/d(t - \theta)$ is an increasing function of time that increases considerably as θ approaches t. Consequently, $\psi(t)$ behaves as a memory function that accounts

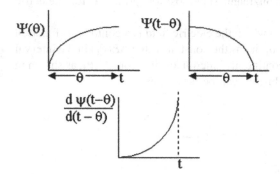

Fig. 8.3   Continuous and discontinuous electric perturbation of material

for the history of the material in such a way that the perturbation effects are larger the closer the time θ, at which the perturbation has occurred, to time t.

8.3. DIELECTRIC RESPONSE IN THE FREQUENCY DOMAIN

If a material is under a sinusoidal electric field $E(t) = E_0 \sin \omega t$, where E_0 is the amplitude of the electric field and ω is the angular frequency, the dielectric displacement lags an angle δ from the electric field (see Fig. 8.4). Accordingly

$$D(t) = D_0 \sin(\omega t - \delta) = D_0 \sin \omega t \cos \delta - D_0 \cos \omega t \sin \delta \qquad (8.3.1)$$

where D_0 is the amplitude of the response. This equation indicates that the dielectric displacement is a complex quantity with a component in phase with the perturbation field ($D_0 \cos \delta$) and another $90°$ out of phase ($D_0 \sin \delta$). The dielectric permittivity is a complex quantity ($\varepsilon^* = \varepsilon' - i\varepsilon''$) whose components are

$$\varepsilon'(\omega) = \frac{D_0}{E_0} \cos \delta(\omega)$$

$$\varepsilon''(\omega) = \frac{D_0}{E_0} \sin \delta(\omega) \qquad (8.3.2)$$

Therefore, Eq. (8.3.1) can be rewritten as [see Eq. (2.12.11)]

$$D(t) = E_0[\varepsilon'(\omega) \sin \omega t - \varepsilon''(\omega) \cos \omega t] \qquad (8.3.3)$$

The angle δ lagging the dielectric displacement from the electric field is given by

$$\delta(\omega) = \tan^{-1} \frac{\varepsilon''(\omega)}{\varepsilon'(\omega)} \qquad (8.3.4)$$

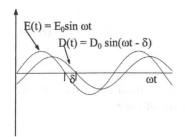

$E(t) = E_0 \sin \omega t$

$D(t) = D_0 \sin(\omega t - \delta)$

ωt

FIG. 8.4 Sinusoidal electric perturbation and response of a material

To relate the dielectric permittivity in the frequency and time domains, let us express Eq. (8.2.5) as

$$D(t) = \int_{-\infty}^{t} \{[\varepsilon_u + (\varepsilon_r - \varepsilon_u)\psi(\infty)] - (\varepsilon_r - \varepsilon_u)[\psi(\infty) - \psi(t - \theta)]\} \frac{dE(\theta)}{d\theta} \, d\theta$$

$$= E(t)\varepsilon_r - \int_{-\infty}^{t} (\varepsilon_r - \varepsilon_u)[\psi(\infty) - \psi(t - \theta)] \frac{dE(\theta)}{d\theta} \, d\theta \qquad (8.3.5)$$

where it has been considered that $\varepsilon(t) = \varepsilon_u + (\varepsilon_r - \varepsilon_u)\psi(t)$ and $\varepsilon_u + (\varepsilon_r - \varepsilon_u)\psi(\infty) = \varepsilon_r$ [see Eq. (8.2.3)] and $E(-\infty) = 0$. For a sinusoidal electric field, for example $E = E_0 \sin \omega t$, Eq. (8.3.5) becomes

$$D(t) = \varepsilon_r E_0 \sin \omega t - E_0(\varepsilon_r - \varepsilon_u)\omega \int_0^{\infty} [\psi(\infty) - \psi(u)] \cos \omega(t - u) \, du$$

$$= E_0[\varepsilon'(\omega) \sin \omega t - \varepsilon''(\omega) \cos \omega t] \qquad (8.3.6)$$

where the change in variables $u = t - \theta$ was made. Hence, the real component $\varepsilon'(\omega)$ and loss component $\varepsilon''(\omega)$ of the complex dielectric permittivity are related to the dielectric permittivity in the time domain by

$$\varepsilon'(\omega) = \varepsilon_r - \omega \int_0^{\infty} [\varepsilon_r - \varepsilon(t)] \sin \omega t \, dt \qquad (8.3.7)$$

and

$$\varepsilon''(\omega) = \omega \int_0^{\infty} [\varepsilon_r - \varepsilon(t)] \cos \omega t \, dt \qquad (8.3.8)$$

where the change $u = t$ was made. Note that Eqs. (8.3.7) and (8.3.8) are alternative forms of writing Eqs. (2.12.2). These expressions can also be written in terms of sine and cosine Fourier transforms as

$$\frac{\varepsilon'(\omega) - \varepsilon_r}{\omega} = -F_s[\varepsilon_r - \varepsilon(t)]$$

$$\frac{\varepsilon''(\omega)}{\omega} = F_c[\varepsilon_r - \varepsilon(t)] \qquad (8.3.9)$$

The permittivity in the time domain can be obtained from the inverse of the Fourier transform of Eq. (8.3.9). The pertinent equations are

$$\varepsilon(t) = \varepsilon_r + \frac{2}{\pi} \int_0^{\infty} \frac{\varepsilon'(\omega) - \varepsilon_r}{\omega} \sin \omega t \, d\omega \qquad (8.3.10)$$

and

$$\varepsilon(t) = \varepsilon_r - \frac{2}{\pi}\int_0^\infty \frac{\varepsilon''(\omega)}{\omega}\cos\omega t\,d\omega \tag{8.3.11}$$

Equation (8.3.10) can be written as

$$\varepsilon(t) = \varepsilon_u + \varepsilon_r - \varepsilon_u + \frac{2}{\pi}\int_0^\infty \frac{\varepsilon'(\omega) - \varepsilon_u + \varepsilon_u - \varepsilon_r}{\omega}\sin\omega t\,d\omega \tag{8.3.12}$$

The fact that according to the complex variable theory

$$\int_0^\infty \frac{\sin\omega t}{\omega}\,d\omega = \frac{\pi}{2} \tag{8.3.13}$$

means that the dielectric permittivity in the frequency domain can alternatively be written as

$$\varepsilon(t) = \varepsilon_u + \frac{2}{\pi}\int_0^\infty \frac{\varepsilon'(\omega) - \varepsilon(u)}{\omega}\sin\omega t\,d\omega \tag{8.3.14}$$

When $t \to 0$, Eq. (8.3.11) becomes

$$\varepsilon_r = \varepsilon_u + \frac{2}{\pi}\int_0^\infty \frac{\varepsilon''(\omega)}{\omega}\,d\omega \tag{8.3.15}$$

This expression allows the determination of the relaxation strength $\varepsilon_r - \varepsilon_u$ from loss measurements. The value of ε_r given by Eq. (8.3.15) substituted into Eq. (8.3.11) gives the relationship

$$\varepsilon(t) = \varepsilon_u + \frac{2}{\pi}\int_0^\infty \frac{\varepsilon''(\omega)}{\omega}(1 - \cos\omega t)\,d\omega \tag{8.3.16}$$

which is an alternative way of writing Eq. (8.3.11). Illustrative curves depicting the real and loss components in the frequency domain for poly(vinyl chloride) above T_g are shown, at different temperatures, in Fig. 8.5.

8.4. DIELECTRIC RELAXATION MODULUS IN THE TIME AND FREQUENCY DOMAINS

The experimental determination of the relaxed permittivity ε_r may involve some difficulties in cases where the conductivity contribution to the dielectric loss

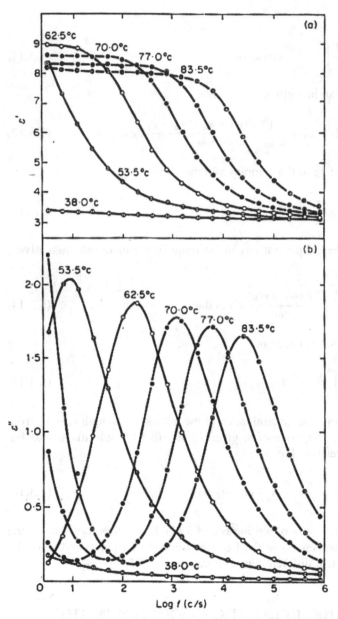

FIG. 8.5 Real and loss components of the complex dielectric permittivity in the frequency domain, at several temperatures, for poly(vinyl acetate) in the α-region [1a]

overlaps the dipolar one. In this situation it is preferable to analyze the dielectric results in terms of the dielectric modulus M.

Let us consider the following electric history

$$D(t) = 0, \quad t < 0$$
$$D(t) = D_0, \quad t \geqslant 0 \tag{8.4.1}$$

If D_0 is very small, the linear dielectric phenomenological theory predicts that

$$E(t) = D_0 M(t) \tag{8.4.2}$$

The dielectric relaxation modulus, $M(t)$, is given by

$$M(t) = M(0)\phi(t) \tag{8.4.3}$$

where $\phi(t)$ is a monotonous decreasing function of time, the extreme values of which are $\phi(0) = 1$ and $\phi(\infty) = 0$. For low dielectric displacements, the Boltzmann superposition principle holds

$$E(t) = \sum_{\theta_i = -\infty}^{t} M(t - \theta_i)\,\Delta D(\theta_i) \tag{8.4.4}$$

or in continuous form

$$E(t) = \int_{-\infty}^{t} M(t - \theta)\frac{\mathrm{d}D(u)}{\mathrm{d}\theta}\,\mathrm{d}\theta$$
$$= M(0)D(t) + \int_{0}^{\infty} D(u)\frac{\mathrm{d}M(t - u)}{\mathrm{d}u}\,\mathrm{d}u \tag{8.4.5}$$

If a sinusoidal dielectric displacement $D(t) = D_0 \sin \omega t$ is imposed on the material, the electric field can be written as

$$E(t) = E_0 \sin(\omega t + \delta) = E_0 \sin \omega t \cos \delta + E_0 \cos \omega t \sin \delta$$
$$= D_0(M' \sin \omega t + M'' \cos \omega t) \tag{8.4.6}$$

where E_0 is the amplitude of the response. Accordingly, the electric field is a complex quantity with a component in phase with the perturbation field ($E_0 \cos \delta$) and another 90° out of phase ($E_0 \sin \delta$). The real component M' and loss component M'' of the complex dielectric modulus $M^* = M' + iM''$ are given by

$$M'(\omega) = \frac{E_0}{D_0} \cos \delta$$

$$M''(\omega) = \frac{E_0}{D_0} \sin \delta$$

(8.4.7)

whereas δ is the out-of-phase angle given by

$$\delta(\omega) = \tan^{-1} \frac{M''(\omega)}{M'(\omega)}$$

(8.4.8)

By comparing Eqs (8.3.2) and (8.4.8) it is established that

$$\varepsilon^*(\omega) = \frac{1}{M^*(\omega)}$$

(8.4.9)

as indicated in Eq. (4.1.11).

In order to relate the dielectric permittivity in the frequency and time domains, a methodology similar to that described above can be used, leading to

$$M'(\omega) = \omega \int_0^\infty M(t) \sin \omega t \, dt$$

$$M''(\omega) = \omega \int_0^\infty M(t) \cos \omega t \, dt$$

(8.4.10)

8.5. KRÖNIG–KRAMERS RELATIONSHIPS

Each component of the dielectric permittivity can be obtained from the other by means of the Krönig–Kramers relations (see also section 2.14). Thus, by substituting Eq. (8.3.15) into Eq. (8.3.7), the following expression is obtained

$$\varepsilon'(\omega) - \varepsilon_r = -\frac{2}{\pi} \omega \int_0^\infty \frac{\varepsilon''(x)}{x} \, dx \times \lim_{R \to \infty} \int_0^R \sin \omega t \cos xt \, dt$$

(8.5.1)

By taking into account that

$$\sin \omega t \cos xt = \frac{\sin (\omega - x)t + \sin (\omega + x)t}{2}$$

(8.5.2)

Eq. (8.5.1) becomes

$$\varepsilon'(\omega) - \varepsilon_r = -\frac{2}{\pi}\omega \int_0^\infty \frac{\varepsilon''(x)}{x} \, dx \left\{ \frac{\omega}{\omega^2 - x^2} - \frac{1}{2} \right.$$

$$\left. \times \lim_{R\to\infty} \left[\frac{\cos(\omega - x)R}{\omega - x} + \frac{\cos(\omega + x)R}{\omega + x} \right] \right\} dx$$

$$= \frac{2}{\pi} \int_{-\infty}^\infty \varepsilon''(x) \frac{\omega^2}{x^2 - \omega^2} \, d\ln x \qquad (8.5.3)$$

It should be noted that the integral of cosines in this equation is zero, as a simple integration by parts shows. By combining Eqs (8.5.3) and (8.3.15), the following expression, which makes it possible to determine the real component of the complex dielectric permittivity from the dielectric loss, is obtained

$$\varepsilon'(\omega) = \varepsilon_u + \frac{2}{\pi} \int_{-\infty}^\infty \frac{\varepsilon''(x)x^2}{x^2 - \omega^2} \, d\ln x \qquad (8.5.4)$$

Note that, in the limit $\omega \to 0$, Eq. (8.5.4) gives the dielectric strength as

$$\varepsilon_r - \varepsilon_u = \frac{2}{\pi} \int_{-\infty}^\infty \varepsilon''(x) \, d\ln x \qquad (8.5.5)$$

which coincides with Eq. (8.3.15). In the same way, the dielectric loss can be obtained from the real component of the complex dielectric permittivity. Actually, substitution of Eq. (8.3.10) into Eq. (8.3.8) gives

$$\varepsilon''(\omega) = -\frac{2}{\pi}\omega \int_0^\infty \frac{\varepsilon'(x) - \varepsilon_r}{x} \, dx \times \lim_{R\to\infty} \int_0^R \sin xt \cos \omega t \, dt \qquad (8.5.6)$$

By writing

$$\sin xt \cos \omega t = \frac{\sin(\omega + x)t - \sin(\omega - x)t}{2} \qquad (8.5.7)$$

Eq. (8.5.6) becomes

$$\varepsilon''(\omega) = \frac{2}{\pi}\omega \int_0^\infty \frac{\varepsilon'(x) - \varepsilon_r}{x} \left\{ \frac{x}{\omega^2 - x^2} \, dx \right.$$

$$\left. \times \lim_{R\to\infty} \left[\frac{\cos(\omega - x)R}{\omega - x} - \frac{\cos(\omega + x)R}{\omega + x} \right] \right\}$$

$$= \frac{2}{\pi} \int_{-\infty}^\infty \frac{[\varepsilon'(x) - \varepsilon_r]\omega x}{\omega^2 - x^2} \, d\ln x \qquad (8.5.8)$$

A more rigorous deduction of the Krönig–Kramers relationships can be done by using a formulation in the complex plane, thus avoiding the singularities appearing in Eqs (8.5.4) and (8.5.7).

The same methodology as that utilized to obtain the relationship between the real and loss components of the complex dielectric permittivity leads to the Krönig–Kramers relations for the dielectric relaxation moduli

$$M'(\omega) = \frac{2}{\pi} \int_{-\infty}^{\infty} M''(x) \frac{\omega^2}{\omega^2 - x^2} \, \mathrm{d} \ln x$$

$$M''(\omega) = \frac{2}{\pi} \int_{-\infty}^{\infty} M'(x) \frac{\omega x}{x^2 - \omega^2} \, \mathrm{d} \ln x$$

(8.5.9)

8.6. ANALYTICAL EXPRESSIONS FOR THE DIELECTRIC PERMITTIVITY AND DIELECTRIC RELAXATION MODULUS IN THE TIME AND FREQUENCY DOMAINS (SEE ALSO SECTION 4.4.1)

The dielectric permittivity in the time domain is a continuous increasing function of time governed by the buildup function $\psi(t)$ whose extremes are $\psi(0) = 0$ and $\psi(\infty) = 1$. The buildup function can be expressed as the identity

$$\psi(t) = \psi(\infty) - [\psi(\infty) - \psi(t)]$$

(8.6.1)

The term $\psi(\infty) - \psi(t)$ is a continuous decreasing function of time that may be viewed as the Laplace transform of an unknown function $N'(s)$, that is

$$\psi(\infty) - \psi(t) = \int_{0}^{\infty} N(s) \exp(-st) \, \mathrm{d}s$$

(8.6.2)

For $t = 0$, Eq. (8.6.2) becomes

$$\psi(\infty) = \int_{0}^{\infty} N(s) \, \mathrm{d}s$$

(8.6.3)

Hence, the buildup function can be written as

$$\psi(t) = \int_{0}^{\infty} N(s)[1 - \exp(-st)] \, \mathrm{d}s$$

(8.6.4)

Since st in this expression is dimensionless, s must be the reciprocal of a time τ ($=1/s$), called the retardation time, because it is associated with motions of mechanisms of response that are delayed with respect to the perturbation of the electric field. By combining Eqs (8.2.3) and (8.6.4), the following expression is obtained

$$\varepsilon(t) = \varepsilon_u + (\varepsilon_r - \varepsilon_u) \int_0^\infty \frac{N(1/\tau)}{\tau} \left[\left(1 - \exp\left(-\frac{t}{\tau}\right) \right) \right] d\ln \tau$$

$$= \varepsilon_u + \int_0^\infty L_\varepsilon(\tau)(1 - e^{-t/\tau}) \, d\ln \tau \tag{8.6.5}$$

where

$$L_\varepsilon(\tau) = \left[(\varepsilon_r - \varepsilon_u) \frac{N(1/\tau)}{\tau} \right] \tag{8.6.6}$$

is the so-called retardation spectrum and represents the contribution to the permittivity of the mechanisms whose logarithms of retardation times lie in the range ln τ to ln $\tau +$ d ln τ.

Let us now proceed with the development of expressions for the components of the complex dielectric permittivity. Equation (8.3.7) can be written as

$$\varepsilon'(\omega) = \varepsilon_r - \omega(\varepsilon_r - \varepsilon_u) \int_0^\infty [\psi(\infty) - \psi(t)] \sin \omega t \, dt$$

$$= \varepsilon_r - \omega(\varepsilon_r - \varepsilon_u) \int_0^\infty N(s) \, ds \int_0^\infty \exp(-st) \sin \omega t \, dt$$

$$= \varepsilon_r - (\varepsilon_r - \varepsilon_u) \int_0^\infty N(s) \frac{\omega^2}{\omega^2 + s^2} \, ds \tag{8.6.7}$$

In the limit $\omega \to \infty$, Eq. (8.6.7) becomes

$$\varepsilon_r = \varepsilon_u + (\varepsilon_r - \varepsilon_u) \int_0^\infty N(s) \, ds \tag{8.6.8}$$

Combining Eqs. (8.6.7) and (8.6.8) gives

$$\varepsilon'(\omega) = \varepsilon_u + (\varepsilon_r - \varepsilon_u) \int_0^\infty N(s) \frac{s^2}{\omega^2 + s^2} \, ds \tag{8.6.9}$$

By substituting $s = 1/\tau$ into this equation, the real component of the complex permittivity is expressed in terms of the retardation times by

$$\varepsilon'(\omega) = \varepsilon_u + \int_{-\infty}^\infty L_\varepsilon(\tau) \frac{1}{1 + \omega^2 \tau^2} \, d\ln \tau \tag{8.6.10}$$

where $L_\varepsilon(\tau) = (\varepsilon_r - \varepsilon_u)N(1/\tau)/\tau$.

To obtain the frequency dependence of the dielectric loss, use must be made of Eq. (8.3.8). Accordingly

$$\varepsilon''(\omega) = \omega(\varepsilon_r - \varepsilon_u) \int_0^\infty [\psi(\infty) - \psi(t)] \cos \omega t \, dt$$

$$= \omega(\varepsilon_r - \varepsilon_u) \int_0^\infty N(s) \, ds \int_0^\infty e^{-st} \cos \omega t \, dt$$

$$= (\varepsilon_r - \varepsilon_u) \int_0^\infty N(s) \frac{\omega s}{\omega^2 + s^2} \, ds \qquad (8.6.11)$$

By replacing s with $1/\tau$, Eq. (8.6.11) becomes

$$\varepsilon''(\omega) = \int_{-\infty}^\infty L(\tau) \frac{\omega \tau}{1 + \omega^2 \tau^2} \, d\ln \tau \qquad (8.6.12)$$

Note that, if conductivity contributions are present, this expression should be written as

$$\varepsilon''(\omega) = \int_{-\infty}^\infty L(\tau) \frac{\omega \tau}{1 + \omega^2 \tau^2} \, d\ln \tau + \frac{\sigma}{e_0 \omega} \qquad (8.6.13)$$

It should be pointed out that the kernels of Eqs (8.6.5), (8.6.10), and (8.6.12) lead the retardation spectrum function to the dielectric permittivity in the time and frequency domains. Note that the kernels of Eqs (8.6.12) and (8.6.13) reach a maximum for $\omega \tau = 1$. Therefore, the mean relaxation time of a loss process in the frequency domain is taken as $\tau = 1/\omega_{max}$, where ω_{max} is the frequency at the peak maximum.

The dielectric relaxation modulus is a monotonic continuous decreasing function of time that can be expressed as the Laplace transform of an unknown function $N'(s)$

$$M(t) = \int_0^\infty N'(s) \exp(-st) \, dt \qquad (8.6.14)$$

Since st is a dimensionless quantity, $s = 1/\tau$, where τ is the so-called relaxation time. Therefore, Eq. (8.6.14) can be written as

$$M(t) = \int_{-\infty}^\infty H_M(\tau) \exp\left(\frac{-t}{\tau}\right) d\ln \tau \qquad (8.6.15)$$

where $H_M(\tau) = N'(1/\tau)/\tau$ is the relaxation time spectrum. Following the same methodology as that utilized for the dielectric permittivity, the storage and loss components of M^* are given in terms of the relaxation times by

$$M'(\omega) = \int_{-\infty}^{\infty} H_M(\tau) \frac{\omega^2 \tau^2}{1 + \omega^2 \tau^2} \, d\ln \tau$$

$$M''(\omega) = \int_{-\infty}^{\infty} H_M(\tau) \frac{\omega \tau}{1 + \omega^2 \tau^2} \, d\ln \tau \tag{8.6.16}$$

Note that the distribution of retardation times is the same for the dielectric permittivity in both the time and frequency domains.

8.7. LOCAL AND COOPERATIVE DYNAMICS: BASIC CONCEPTS

Relaxation response functions of high and low molecular weight liquids to weak external perturbations v (mechanical, dielectric, etc.) may provide information on the actual structural kinetics resulting from interbasin transitions in the configurational hyperspace describing the system.[1b] The normalized response function to an electric field can be expressed in molecular terms by[2]

$$g(t) = \frac{\langle \mathbf{M}(0) \cdot \mathbf{M}(t) \rangle}{\langle \mathbf{M}(0)^2 \rangle} \tag{8.7.1}$$

where

$$\mathbf{M}(t) = \sum_{i=1}^{N} \mathbf{m}_i(t) \tag{8.7.2}$$

In this expression, N is the number of relaxing species and $\mathbf{m}_i(t)$ is the dipole moment of the species i at time t. The area of the normalized relaxation curve $g_v(t)$ vs. t, at temperature T, defines the mean relaxation time $\tau_v(T)$. The value of this quantity in the liquid state undergoes a rapid rise with declining temperature in the vicinity of the glass transition temperature, which can be fitted to a Vogel–Tamman–Fulher–Hesse (VFTH) equation[3–5]

$$\tau(T) = \tau_0 \exp\left(\frac{DT_V}{T - T_V}\right) \tag{8.7.3}$$

where τ_0 is a prefactor of the order of picoseconds and D is the so-called strength parameter, the value of which is lower/higher than 10 for weak/strong liquids.

The importance of the dynamics becomes apparent if a melt is cooled at a cooling rate high enough to avoid its crystallization. The fast cooling process leads the system to a supercooled liquid and then to a glass. The glass transition temperature at which the supercooled liquid forms a glass is a dynamic property. It is often defined as the temperature at which the relaxation time is 200 s, a reasonable maximum relaxation time for dynamic experiments, but obviously T_g increases/ decreases as the timescale of the experiment decreases/increases.[6] The relaxation time that reflects the time involved in the structural rearrangement of the molecules changes by many orders of magnitude from the liquid to the glass. The glass transition is a universal property of condensed amorphous matter, irrespective of its molecular weight.

Equation (8.7.3) suggests a divergence in τ at the Vogel temperature T_V that cannot experimentally be observed in real experiments. For low temperatures the relaxation time becomes longer than the timescale of the experiment and the sample falls out of equilibrium. Below T_g the dynamic response depends on the thermal history, and aging phenomena may occur.

Owing to the latent heat of melting, a supercooled liquid has substantially higher entropy than the crystal. However, the former loses entropy faster than the latter below the melting temperature T_m. The difference between the entropies of the liquid and the crystal becomes rather small just below the glass transition. Extrapolation of the entropy of the glass to low temperatures indicates the existence of a temperature T_K, called the Kauzmann temperature, at which the crystal and the liquid attain the same entropy.[1] The fact that the vibrational entropy is nearly the same for the two phases, and the inherent structural entropy of the ordered crystal vanishes, leads to the conclusion that the fully relaxed glass at T_K (and therefore the extrapolated liquid) must also have vanishing inherent structural entropy. The experimental observation that the Kauzmann temperature is approximately equal to the Vogel temperature at which divergence in the mean relaxation time occurs has generated the concept of an "ideal glass state" that could only be attained if sufficiently slow cooling rates were available.[1]

Many theories follow this notion of a low-temperature phase transition underlying the glass transition. Among the most prominent theoretical approaches, the Adam–Gibbs[7] and the free volume[8] theories stand out. The former theory is the first to consider the glass transition as a cooperative process, whereas in the free volume approach T_V in Eq. (8.7.3) is considered to be the temperature at which the free volume would be zero were it is not for the formation of the glassy state. In the kinetic[7] and fluctuation models[9,10] the volume of the cooperatively rearranging region is defined as the smallest volume that can relax to a new configuration independently. The cooperativity continues well above T_g and increases with decreasing temperature. Experimental testing of the theories has traditionally been carried out in the low-temperature region using slow dynamics.

Another approach to the study of the glass transition is to determine the response of liquids deep in the liquid state. These studies have been stimulated by the mode coupling theory (MCT)[11,12], (see Section 2.31), which explains the glass transition in terms of a dynamic phase transition occurring at a critical temperature T_c significantly above the glass transition temperature. A coupling scheme has been proposed by Ngai and coworkers[12a,12b] in which the first-time derivative of the KWW equation is taken to be the master equation in the time domain for application to isothermal and nonisothermal experiments. A shortcoming of the theories is that they do not present a molecular basis for application to molecular scattering and relaxation phenomena.

The experimental determination of the relaxation response of liquids requires the combination of different experimental methods that include mechanical and dielectric spectroscopy as well as light and neutron scattering techniques. In what follows, the response of amorphous monomers and polymers to electric perturbation fields at temperatures above the glass transition temperature is addressed.

8.8. RESPONSES OF GLASS FORMERS ABOVE T_g TO PERTURBATION FIELDS IN A WIDE INTERVAL OF FREQUENCIES

As indicated above, the temperature dependence of the mean relaxation time associated with the most prominent contribution to the relaxation spectrum of supercooled liquids obeys Eq. (8.7.3). This absorption, called the α-relaxation process, may display at high temperatures conductive contributions on the low-frequency side that lead to a $1/\omega$ divergence of $\varepsilon''(\omega)$. The α-relaxation displays a Kohlrausch–Williams–Watts (KWW) stretched exponential decay[13,14]

$$g_\alpha(t) = \exp\left[-\left(\frac{t}{\tau_{\mathrm{KWW}}} \right)^{\beta_K} \right] \tag{8.8.1}$$

where $0 < \beta_K \leqslant 1$ [see Section 2.12]. For a Debye-type response, $\beta_K = 1$. The wider the α-relaxation, the lower the value of the stretch exponent. In the vicinity of T_g, the temperature dependence of τ_{KWW} is described by the VFTH equation. The mean relaxation time associated with the glass–rubber absorption is given by

$$\langle \tau_\alpha \rangle = \int_0^\alpha g_\alpha(t)\, dt = \left(\frac{\tau_{\mathrm{KWW}}}{\beta_K} \right) \Gamma\left(\frac{1}{\beta_K} \right) \tag{8.8.2}$$

where Γ is the gamma function.

According to the linear phenomenological theory, the glass–liquid relaxation can be obtained in the frequency domain by means of Eq. (8.2.7), giving[15]

$$\frac{\varepsilon_\alpha^*(\omega) - \varepsilon_{u\alpha}}{\varepsilon_{r\alpha} - \varepsilon_{u\alpha}} = \int_0^\infty \left[-\frac{dg_\alpha(t)}{dt}\right] \exp(-i\omega t)\, dt$$

$$= 1 - i\omega \int_0^\infty g_\alpha(t) \exp(-i\omega t)\, dt \qquad (8.8.3)$$

where it was assumed that $g_\alpha(t) = \phi(t) = 1 - \psi(t)$. For a Debye-type substance $g_\alpha = \exp(-t/\tau)$. Substitution of this correlation function into Eq. (8.8.3) leads to the following expressions for the complex permittivity

$$\varepsilon^*(\omega) = \varepsilon_u + \frac{\varepsilon_r - \varepsilon_u}{1 + i\omega\tau} \qquad (8.8.4)$$

Obviously, Eq. (8.8.4) only holds for a Debye-type process. The α-relaxation in the frequency domain is described by the empirical Havriliak–Negami equation [Eq. (4.4.5.1)][16,17]

$$\varepsilon_\alpha^*(\omega) = \varepsilon_{u\alpha} + \frac{\varepsilon_{r\alpha} - \varepsilon_{u\alpha}}{[1 + (i\omega\tau_{HN})^{\beta_{HN}}]^{\alpha_{HN}}} \qquad (8.8.5)$$

In this expression, τ_{HN} is related to the maximum of the α-peak in the frequency domain, whereas β_{HN} and α_{HN} are fractional shape parameters fulfilling the following conditions: $\beta_{HN} > 0$ and $\beta_{HN}\alpha_{HN} \leqslant 1$. These parameters are related through τ_{HN}^{-1} to the limiting values of the slopes in the low- and high-frequency regions of the double logarithmic plot ε'' vs. ω, by means of the expressions (see Chapter 2)

$$\frac{d\log \varepsilon_\alpha''(\omega)}{d\log \omega} = \beta_{HN} = m, \qquad \omega\tau_{HN} \ll 1$$

$$\frac{d\log \varepsilon_\alpha''(\omega)}{d\log \omega} = -\beta_{HN}\alpha_{HN} = -n, \qquad \omega\tau_{HN} \gg 1$$

$$(8.8.6)$$

Although m and n are uncorrelated parameters, n for polymeric materials is found to lie in the interval $0 < n \leqslant 0.5$. It should be noted that Eq. (8.8.5) is a generalization of the Debye, Cole–Cole,[18] and the Dadidson–Cole[19,20] equations. Thus, if $\alpha_{HN} = \beta_{HN} = 1$, the Havriliak–Negami (HN) equation becomes the Debye equation; for $\alpha_{HN} = 1$, the HN equation is converted into the Cole–Cole equation, whereas, for $\beta_{HN} = 1$, Eq. (8.8.5) becomes the Davidson–Cole function.

A more detailed inspection of the α-relaxation in the frequency domain of some low molecular glass formers shows at frequencies several decades above the peak maximum an excess loss or wing not accounted for by the empirical expressions commonly used to describe the α-process. A schematic view of the dielectric loss in glass-forming materials as observed in extremely broadband spectroscopy is shown in Fig. 8.6.[21] Different techniques that must be used to obtain the spectra are shown in Fig. 8.7. For many systems in which the α-relaxation is described by the Davidson–Cole equation, the excess wing contribution can be reasonably described by the scaling law $\varepsilon'' \sim \omega^{-a}$ with the exponent a lower than the value of the exponent of the Davidson–Cole expression.[22-25] No accepted model exists to describe the macroscopic origin of the excess wing.[21]

Besides the α-relaxation, many glass formers, especially polymers, display at high frequencies one or more relaxations (β, γ, δ, etc.) that are believed to be associated with local intramolecular relaxations in the main chain or in side groups. Another type of relaxation is the so-called Johari–Goldstein β-relaxation[26,27] which seems to be a rather universal property of glass formers. This relaxation even appears in relatively simple systems in which intramolecular motions are absent. In some materials, the Johari–Goldstein β-relaxation may overlap the excess wing of the α-relaxation process.

A minimum in $\varepsilon''(\omega)$ is reached at frequencies in the GHz–THz range which in principle could be considered a crossover between the decrease in the β-relaxation and the increase towards the far infrared absorptions typical of glass former materials. It seems, however, that fast processes contribute to $\varepsilon''(\omega)$ in this region, and therefore the minimum depends on the glass former. Finally, the so-called Boson peak known from neutron and light scattering experiments shows up in the THz region. Its name comes from the temperature dependence of the intensity of the peak that obeys Boson statistics. The Boson peak corresponds to the commonly found excess contribution in specific heat measurements at low temperatures, and it is a universal feature of glass-forming materials.

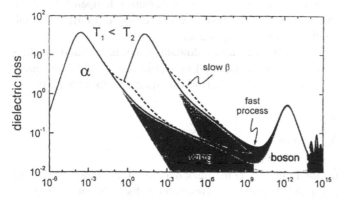

FIG. 8.6 Extreme broadband spectroscopy of a glass-forming liquid[21]

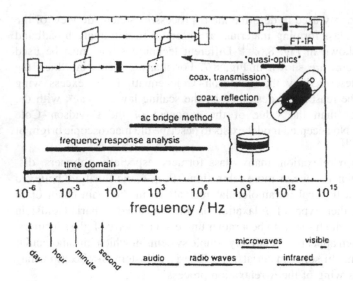

FIG. 8.7 Overview of the experimental techniques used to determine the dielectric loss of glass formers in a wide interval of frequencies[21]

8.9. BROADBAND DIELECTRIC SPECTROSCOPY OF SUPERCOOLED POLYMERS

Some polymers with flexible side groups are characterized for displaying high dielectric activity even in the glassy state. Illustrative curves depicting the dielectric loss in the frequency domain for poly(5-acryloxymethyl-5-ethyl-1,3-dioxacyclohexane) (PAMED) are shown in Fig. 8.8.[28] The repeating unit of the polymer is presented in Fig. 8.9. Below the glass transition temperature, about 20°C, the isotherms present a wide β-relaxation whose maximum shifts, as usual, to higher frequency with increasing temperature.

Usually, the β-peak in the frequency domain is symmetric and displays half-widths of 4–6 decades. The normalized dielectric loss for this process is often described by the empirical Fuoss–Kirkwood equation [Eq. (4.4.3.2)][29]

$$\frac{\varepsilon''_\beta(\omega)}{\varepsilon''_{\beta_{max}}} = \mathrm{sech}\left(m \ln \frac{\omega}{\omega_{max}}\right), \qquad 0 < m \leqslant 1 \tag{8.9.1}$$

where ω_{max} represents the frequency at the peak maximum. An illustrative curve showing the fitting of Eq. (8.9.1) to the experimental results is shown in Fig. 8.10. The parameter m, like β_K in the α-relaxation, accounts for the breadth of the

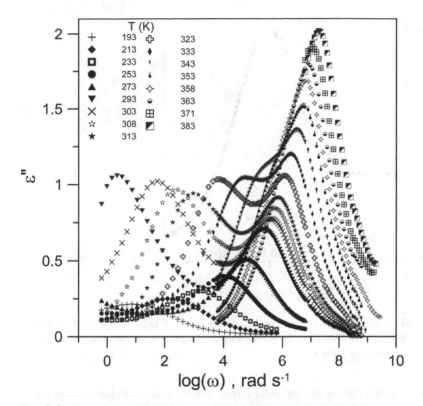

FIG. 8.8 Broadband spectroscopy of poly(5-acryloxy-5-ethyl-1,3-dioxacyclohexane) (PAMED)[28]

β-process in such a way that, the larger the value of m, the narrower is the absorption. Also, the intensity of the β-process increases with increasing temperature. The width of the β-relaxation is often explained in terms of the distribution of both the activation energies and the pre-exponential factor that

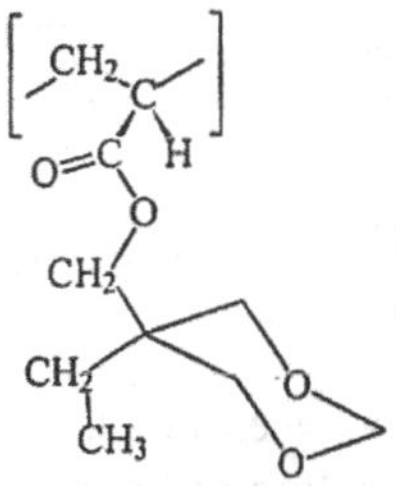

FIG. 8.9 Repeating unit of poly(5-acryloxy-5-ethyl-1,3-dioxacyclohexane) (PAMED)

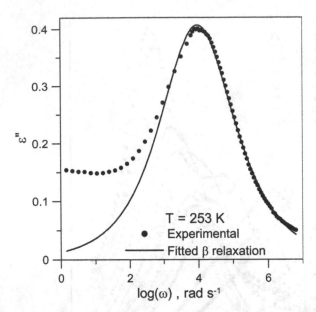

FIG. 8.10   Fitting of the experimental dielectric loss of poly(5-acryloxy-5-ethyl-1,3-dioxacyclohexane) (PAMED) in the glassy state (253 K) to a β-relaxation according to the Fuoss–Kirkwood equation

results from the variety of molecular environments that cause the relaxation.[30] In the supercooled liquid state and in the vicinity of T_g, isotherms display in the lower-frequency range a slow α-relaxation followed by a weaker β-process in the high-frequency range. As the temperature augments, the distance separating the β- from the α-peak decreases as a consequence of the high activation energy of the latter relaxation process. A temperature is reached at which both relaxations coalesce into a single peak, named αβ-relaxation, the intensity of which seems to increase with increasing temperature.

The loss curves at $T > T_g$ can be fitted to a sum of an α- plus a β-relaxation according to the expression

$$\varepsilon''(\omega) = \varepsilon''_\alpha(\omega) + \varepsilon''_\beta(\omega)$$

$$= \varepsilon''_{max,\alpha}\Im\{i\omega F[g_\alpha(t)]\} + \varepsilon''_{max,\beta}\mathrm{sech}\left[m\ln\left(\frac{\omega_{max}}{\omega}\right)\right] \tag{8.9.2}$$

where $\Im$ denotes the imaginary part of the complex function between brackets, and F is the Fourier transform. Examples showing the deconvolution of the loss curves are given in Figs 8.11 and 8.12.

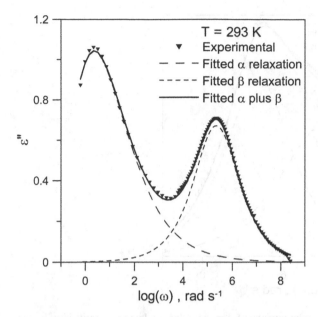

FIG. 8.11 Deconvolution of the experimental values of the dielectric loss at 293 K for poly(5-acryloxy-5-ethyl-1,3-dioxacyclohexane) (PAMED). Individual contributions of α- and β-relaxations are indicated by broken lines and their sum by a solid line

According to Eq. (8.3.15), the relaxation strength is given by

$$\varepsilon_{ri} - \varepsilon_{ui} = \frac{2}{\pi} \int_{-\infty}^{\infty} \varepsilon''(\omega)\, d\ln \omega \qquad (8.9.3)$$

where subscript i refers to the α, β relaxations. It should be noted that the introduction of Eq. (8.9.1) into Eq. (8.9.3) leads to [see Eq. (4.4.3.4)]

$$\varepsilon_{r\beta}'' - \varepsilon_{u\beta}'' = 2\frac{\varepsilon_{max,\beta}''}{m} \qquad (8.9.4)$$

which allows the rapid determination of the relaxation of subglass relaxation processes.

The strength of the α-relaxation decreases with increasing temperature as a consequence of the randomization of the dipolar orientation. The evolution of the strength of the α- and β-processes with temperature for the example of Fig. 8.8 is shown in Fig. 8.13. For example, $\varepsilon_{r\alpha} - \varepsilon_{u\alpha}$ reaches a maximum value at temperatures just slightly above T_g and then undergoes a steep decrease with increasing temperature. Extrapolation to $\Delta\varepsilon_\alpha = 0$ (α-onset) points to an onset

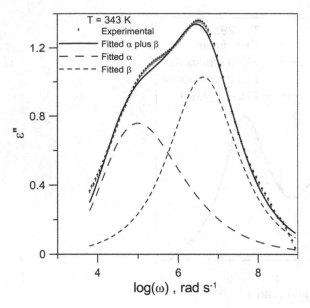

FIG. 8.12 Real and loss components of the complex dielectric permittivity in the frequency domain, at several temperatures, for PAMED in the α-region

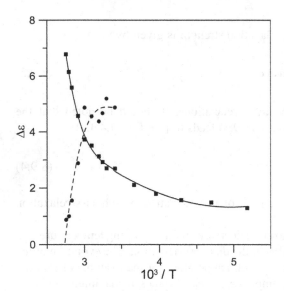

FIG. 8.13 Arrhenius plots for the strength of the α-relaxation (----) and β-relaxation (——) of PAMED[28]

temperature of 97°C in the case analyzed here. Therefore, seen from the high-temperature side, the α-relaxation starts with zero intensity at this temperature and steeply increases with decreasing temperature until it levels off in the vicinity of T_g. The strength of the β-relaxation displays a pattern by which it increases steeply in the temperature range in which the strength of the α-relaxation steeply decreases. However, as pointed out elsewhere,[15] the total strength of the isotherms expressed as $\Delta\varepsilon = \Delta\varepsilon_\alpha + \Delta\varepsilon_\beta$ remains nearly constant in the whole interval of temperatures.

The location of the α-relaxation in the frequency domain depends on the chemical structure but not on molecular weight for high molecular weight polymers. As will be indicated later, this behavior suggests some sort of cooperativity on the motions of the molecular chains, otherwise this relaxation should exhibit a strong dependence on chain length. As indicated above, the relaxation time associated with the α-relaxation shows a stronger dependence on temperature than the secondary relaxation processes. This behavior is shown in Fig. 8.14 where the angular frequencies at the maximum of the peaks for α- and β-relaxations of poly(5-acryloxymethyl-5-ethyl-1,3-dioxacyclohexane) are plotted.

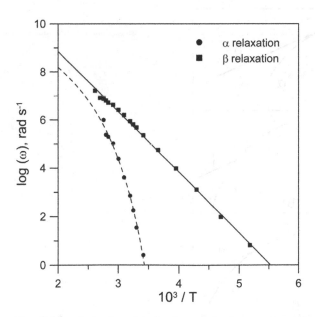

FIG. 8.14 Arrhenius plots for the angular frequencies at the maximum of the peaks for α- and β-relaxations of poly(5-acryloxymethyl-5-ethyl-1,3-dioxacyclohexane) (PAMED)[28]

Based on the analysis of the splitting of the αβ-relaxation of condensed matter, including polymers, it has been speculated that the five scenarios schematically presented in Fig. 8.15 are possible.[31] In the first scenario, also called the conventional case, the β-process continues deep in the liquid state, whereas in the type A scenario the separated α-onset can be characterized by minimum cooperativity for α that cannot be continued to a local, noncooperative process. For the type B scenario there is a locally coordinative β-precursor for the cooperative α-process at high temperatures. The topology of the diagram for the type C scenario is similar to that of the conventional one, but here the α-relaxation curve is above and below the splitting, and α is not the tangent to α. In the type D scenario[32] there are two different but touching α-relaxations, with a sharp crossover between them, and a β-trace aiming towards this common tangential point with a constant of increasing intensity again. Examples of the conventional, A, B, C, and D scenarios are, respectively, solutions of chlorobenzene in *cis*-decaline, poly(*n*-butylmethacrylate), poly(ethylmethacrylate), poly(propylene glycol), and *o*-terphenyl.[31] Although different approaches have been proposed to describe the origin of the observed dielectric response of polymer systems in the merging region,[31–33] it seems that most of the

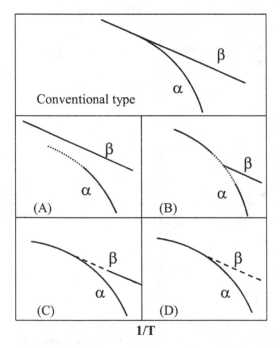

FIG. 8.15   Scenarios suggested for the splitting region in the Arrhenius diagram[31]

experimental cases fall within any of the scenarios described above. Comparison of the results of Fig. 8.14 with the different scenarios suggests that the relaxation behavior of PAMED resembles scenario A schematically depicted in Fig. 8.15.

8.10. TEMPERATURE DEPENDENCE OF THE STRETCH EXPONENT FOR THE α-RELAXATION

There is some dispute concerning the temperature dependence of the stretch exponent in the supercooled and even in the deep liquid states, arising from inconsistent results reported in the literature.[21] Experiments have been cited[34–36] suggesting that β_K increases as the temperature rises, approaching $\beta_K = 1$ at high temperatures. Variation in β_K with temperature was also found in some dielectric experiments.[37–39]

Experimental evidence supporting the temperature independence of β_K mainly arises from light scattering experiments.[21,40–45] For example, the α-absorption in Brillouin spectra of the ionic glass formed by calcium potassium nitrate in the temperature range 120–190°C is fitted by Eq. (8.8.1) with $\beta_K = 0.54$. Moreover, a comparative analysis of the broadband dielectric behavior of propylene carbonate and glycerol,[21] shows a tendency for β_K to level off at a constant value, smaller than unity, at high temperatures. Dis-crepancies observed in the experimental values of the stretch exponent obtained at high temperatures arise from various sources of error. The most important of these is that, when the loss peak shifts towards the high-frequency limit of the available frequency window, there is an increasingly small amount of experimental data on the high-frequency side of the peaks, and as a result the error involved in the determination of β_K is enhanced. As for the stretch exponent associated with the α-relaxation in Fig. 8.8, the value of this quantity lies in the range 0.3–0.5.

Recent studies[46] on the relaxation behavior of poly(vinyl acetate), both in bulk and in solution, show an unexpected fact: the width of the absorption simulated for isolated PVAc chains and for PVAc in bulk are quite similar. The values of β_K for the temperature range 400–600 K lie in the interval 0.54–0.51 in the former case, and in the interval 0.51–0.60 in the latter case. The simulations carried out in the solutions do not show a definite temperature dependence, and, on the other hand, the values of β_K lie in the range of values obtained for PVAc. The scattering observed in the values of β_K obviously arises from the difficulty of fitting the simulated time–dipole correlation function to the KWW equation. However, the simulations suggest that a Debye behavior of the relaxation of liquids at high temperatures is quite unlikely. This conclusion is supported by the analysis of the experimental temperature dependence of β_K for PVAc in bulk and

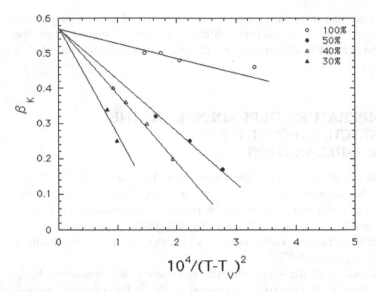

FIG. 8.16 Experimental values of the exponent of the KWW equation against $1/(T - T_V)^2$ for toluene solutions of poly(vinyl acetate). The concentration of polymer is indicated in the inset[46]

its toluene solutions. As can be seen in Fig. 8.16, the extrapolation of these results to infinite temperature gives values of β_K of the order of 0.57 ± 0.05.

8.11. TEMPERATURE DEPENDENCE OF SECONDARY RELAXATIONS

The relaxation rate of secondary relaxations obeys Arrhenius behavior in such a way that the frequency of the peak maximum can be written as

$$\omega_{\text{max},\beta} = \omega_{\infty\beta} \exp\left(-\frac{E_a}{RT}\right) \tag{8.11.1}$$

where $\omega_{\infty\beta}$ is a pre-exponential factor and E_a is the activation energy. The Arrhenius behavior of the β-relaxation spans over a wide interval of temperature from the glassy state to deep in the liquid state, as reflected in Fig. 8.17 where relaxation times obtained from isotherms and isochrones are plotted. The good Arrhenius fitting obtained for the β-relaxation from isotherms obtained in the glassy and liquid states shows that this process detected in the liquid state remains operative below the glass transition temperature.

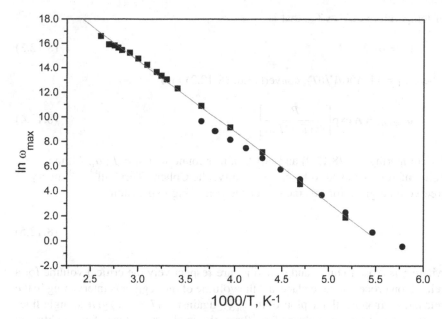

FIG. 8.17 Temperature dependence of the β-relaxation of poly(5-acryloxymethyl-5-ethyl-1,3-dioxacyclohexane) (PAMED). Data from isochrones (●) and from isotherms (■)[281]

8.12. TEMPERATURE DEPENDENCE OF THE α-RELAXATION

Arrhenius plots of the α-relaxation display a curvature as shown in the plot of Fig. 8.14. According to Eq. (8.7.3), the dependence of the peak maximum of the α-relaxation in the frequency domain is given by

$$\ln \omega_{max,\alpha} = \ln \omega_{\infty\alpha} - \frac{m}{T - T_\infty} \qquad (8.12.1)$$

where $T_\infty = T_V$. This equation can be determined from the empirical Doolittle equation[47] which establishes that the relaxation time associated with the α-process depends on the free volume according to the following expression

$$\omega_{max,\alpha} = A \exp\left(-\frac{B}{v_f}\right) \qquad (8.12.2)$$

where v_f is the free volume fraction. According to the Cohen and Turnbull[48] theory, the free volume is zero at T_∞, so the assumption that v_f is a linear function of temperature at $T > T_\infty$, that is

$$v_f = \alpha_f (T - T_\infty) \tag{8.12.3}$$

where $\alpha_f = (1/V)(\partial V/\partial T)$, converts Eq. (8.12.2) into

$$\omega_{max,\alpha} = A \exp\left[\frac{B}{\alpha_f(T - T_\infty)}\right] \tag{8.12.4}$$

By comparing Eqs (8.12.2) and (8.12.4), it is found that $m = B/\alpha_f$. Although the value of B is believed to be close to unity, the Cohen–Turnbull[48,49] theory of free volume gives for this parameter the following expression

$$B = \gamma \frac{v^*}{v_m} \tag{8.12.5}$$

where γ is close to unity and v^* and v_m are respectively the critical volume for a relaxation process to take place and the volume of the segments intervening in the relaxation. In general, the plots of $\ln \omega_{max,\alpha}$ against $1/(T - T_\infty)$ give straight lines from whose slopes the values of α_f/B are obtained. The values of α_f/B obtained for most polymeric systems from these plots lie in the interval $4 \times 10^{-4} K^{-1}$– $6 \times 10^{-4} K^{-1}$, of the order of the values experimentally found for the expansion coefficient. It should be pointed out that T_∞ is usually obtained by fitting the experimental results to Eq. (8.12.1), and its value is about 50 K below T_g.

By comparing Eqs (8.12.2) and (8.12.4), the following expression is found

$$\frac{v_{f,g}}{B} = \frac{T_g - T_\infty}{m} \tag{8.12.6}$$

For many systems investigated, $v_{f,g}/B = 0.0025 \pm 0.005$. If B is assumed to be equal to unity, this would mean that the free volume fraction at T_g would have a universal value lying in the range 0.025 ± 0.005.

If the time–temperature correspondence holds, the time dependence of the shift factor can be written as

$$\ln a_{0T} = \ln \frac{\tau_i(T)}{\tau_i(T_0)} = \ln \frac{\omega_{max}(T_0)}{\omega_{max}(T)} = -\frac{B}{\alpha(T_0 - T_\infty)}\left(\frac{T - T_0}{T - T_\infty}\right) \tag{8.12.7}$$

where T_0 is the reference temperature. By writing

$$C_1 = \frac{B}{\alpha_f(T_0 - T_\infty)}, \qquad C_2 = T_0 - T_\infty \tag{8.12.8}$$

the so-called Williams–Landel–Ferry equation is obtained[50]

$$\ln a_{0T} = -\frac{C_1(T - T_0)}{T + C_2 - T_0} \tag{8.12.9}$$

In this equation, T_0 is a reference temperature and C_1 and C_2 are constants whose values are claimed to be universal for $T_0 = T_g$. However, the values experimentally found are only rough approximations.

8.13. DIELECTRIC STRENGTH AND POLARITY

By using the cavity model as Debye[51] did, Onsager[52] (Section 1.13) developed a theory that relates the strength of the relaxation to the dipoles of the relaxing species. Kirkwood[53] (Section 1.14) and Fröhlich[54] (Section 1.15) included in the theory dipole correlations between the relaxing species. According to Fröhlich, the total relaxation strength can be written as

$$\Delta\varepsilon = \varepsilon_r - \varepsilon_u = F_{\text{Onsager}} \frac{4\pi\langle \mathbf{M}_i \cdot \mathbf{M}_j \rangle}{3kT} = F_{\text{Onsager}} g \frac{4\pi N \mu^2}{3kT} \tag{8.13.1}$$

where $\mathbf{M} = \sum \boldsymbol{\mu}_i$, $\boldsymbol{\mu}_i = \mu$, and N are respectively the dipole moment of the relaxing species and its number, and F_{Onsager} is given by

$$F_{\text{Onsager}} = \frac{\varepsilon_r(\varepsilon_u + 2)^2}{3(2\varepsilon_r + \varepsilon_\infty)} \tag{8.13.2}$$

The value of the dipolar correlation function, g, can be written as

$$g = 1 + \sum_{j \neq i} \langle \cos \gamma_{ij} \rangle^{\text{I}} + \sum_{j \neq i} \langle \cos \gamma_{ij} \rangle^{\text{II}} \tag{8.13.3}$$

where the second term on the right-hand side of this equation refers to the average of the cosine of the angle γ_{ij} made between the dipole associated with the reference unit i and that associated with j within the same chain. The third term is the cosine of the angle between the dipole associated with reference unit i and unit j not belonging to the polymer chain that contains reference unit i. Since the correlation between two dipoles dies away very rapidly when the number of flexible bonds separating them are four or more, it is expected that the contribution of the third term to g is negligible. This is likely to be the case for most flexible polymers, so that g can be written as[55]

$$g = 1 + \sum_{j \neq i} \langle \cos \gamma_{ij} \rangle^{\text{I}} \tag{8.13.4}$$

Therefore, the value of g obtained by using the rotational isomeric state model can, in principle, be used to predict the relaxation strength of polymers provided that ε_∞ is estimated by other methods. In most cases[15]

$$\Delta\varepsilon \cong \Delta\varepsilon_\alpha + \Delta\varepsilon_\beta \tag{8.13.5}$$

and therefore the strengths of the α- and β-relaxations can, in principle, be expressed in terms of the fraction of the dipole of each species that relaxes through each of them. Accordingly

$$\Delta\varepsilon_i = \varepsilon_{ri} - \varepsilon_{ui} = F_{Onsager}g_i \frac{4\pi N\mu_i^2}{3kT} \tag{8.13.6}$$

where subscript i refers to the α- and β-relaxations. Once the strengths of α- and β-relaxations are known, the fraction of dipole relaxed in the two processes can be roughly estimated.

8.14. SEGMENTAL MOTIONS

The glass transition temperature of polymers is related to the molecular weight by the empirical expression

$$T_g = T_{g\infty} - \frac{K}{M_n} \tag{8.14.1}$$

where $T_{g\infty}$ is the glass transition temperature of a polymer of infinite molecular weight, and K is a constant dependent of the concentration of end groups in the system. Accordingly, the glass transition temperature only shows a moderate temperature dependence for molecular weights below the critical value $M_c \cong 2M_e$, where M_e is the molecular weight between entanglements. Since the α-relaxation is related to the glass transition temperature, the average relaxation time associated with this process shows a negligible molecular weight dependence for $M > M_c$. This is reflected in Fig. 8.18 where the peak maximum of the dielectric α-relaxation for different molecular weights of poly(isoprene) is plotted.

The fact that the glass transition is a cooperative phenomenon leads to the conclusion that the α-relaxation in polymers involves cooperative micro-Brownian segmental motions of the chains. Moreover, segmental motions are associated with conformational transitions taking place about the skeletal bonds. The independence of the relaxation α on molecular weight suggests that some sort of cooperativity occurs in the conformational transitions taking place in the intervening segment in order to ensure that the volume swept by the tails of the chains is negligible; otherwise the friction energy, and hence the relaxation times,

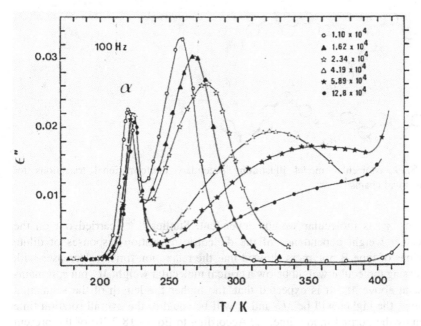

FIG. 8.18 Isochrones at 100 Hz, showing the dependence of the α-relaxation on molecular weight[63a]

would increase with molecular weight. Simulations carried out in simple polymers such as polyethylene show that the conformational transitions are mostly of the following type[56]

$$\ldots g^{\pm}tt \ldots \rightleftarrows \ldots ttg^{\pm} \ldots \qquad (8.14.2a)$$

$$\ldots ttt \ldots \rightleftarrows \ldots g^{\pm}tg^{\mp} \ldots \qquad (8.14.2b)$$

As shown in Fig. 8.19, these transitions produce changes only in the central segments, the extreme segments remaining in positions parallel to the initial ones. Some authors have described the α-process as a damped process of conformational changes of this kind along the chain.[57,58]

Attempts to quantify the number of bonds involved in the segmental motions in relation to the stiffness of the chains have been made. Some authors[59,60] have pointed out the competition between the relaxation through the segmental motions and the overall rotation of the random coil. If the overall rotation is more rapid than the segmental motions, the relaxation of type B dipoles (dipoles perpendicular to the chain contour) proceeds through the overall rotation of the molecule. In this case the time associated with the segmental

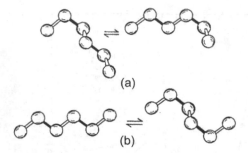

FIG. 8.19 Molecular model illustrating secondary conformational transitions for symmetrical chains[56]

motions, τ_s, is molecular weight dependent. Studies[61,62] carried out on the molecular weight dependence of the dielectric relaxation responses of dilute solutions of type B polymers showed that the relaxation time τ_s decreases with decreasing molecular weight below a critical molecular weight M_c, but τ_s remains constant above M_c. It is expected that, the higher the length of the segmental motions, the higher will be M_c, and τ_s will be equal to the overall rotation time given by the normal mode time, τ_n. According to Eq. (8.18.2.10) of the present chapter, the value of M_c determined as the molecular weight at which $\tau_s = \tau_n$ can be written as

$$M_c = \frac{RT\tau_s}{1.4M\eta_s[\eta]} = \left(\frac{RT\tau_s}{1.4\eta_s K}\right)^{1/(1+a)} \tag{8.14.3}$$

where use was made of the Mark–Houwink equation $[\eta] = KM^a$. The number of segments of the segmental mode, N_s, is strongly dependent on the chemical structure. The value of N_s is less than 10 for poly(ethylene oxide) but of the order of 100 for poly(p-chlorostyrene), poly(p-bromostyrene), and poly(methyl methacrylate). It seems that, as the rigidity of the chains increases, the number of segments involved in the segmental mode also increases. This behavior is reflected in Fig. 8.20, where the values of τ_s for different polymers in dilute solutions are plotted against their glass transition temperatures in bulk.[63] The plot reveals a close correlation between τ_s and the rigidity of the chains expressed in terms of T_g. These results seem to suggest that both the static stiffness and the bulkiness of the side groups govern τ_s in dilute solutions.

8.15. LONG-TIME RELAXATION DYNAMICS

The relaxation behavior of polymer chains at long times (low frequencies) depends on the orientation of the dipoles of bonds, or groups of bonds, relative to

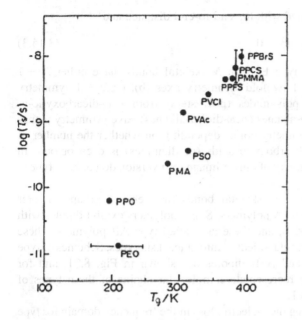

FIG. 8.20 Plot of the logarithm of the segmental mode against the glass transition temperature of poly(ethylene oxide) (PEO), poly(propylelne oxide) (PPO), poly(methyl acrylate) (PMA), poly(styrene oxide) (PSO), poly(vinyl acetate) (PVAc), poly(vinyl chloride) (PVC), poly(p-fluorostyrene) (PPFS), poly(methyl methacrylate) (PMMA), poly(p-chlorostyrene) (PPCS), and poly(p-bromostyrene) (PPBrS)[62]

the chain contour. As mentioned in Chapter 5, Stockmayer[59] classified polymer dipoles into three types: A, B, and C. Dipoles of type A and B are rigidly fixed to the chain backbone in such a way that their orientation in the force field requires motion of the molecular skeleton. However, dipoles of type C are located in flexible side chains, and their mobility is independent of the motions of the molecular skeleton. Dipoles of type A are parallel to the chain contour, and the vector dipole moment associated with a given conformation is proportional to the end-to-end distance vector of that conformation, that is[59]

$$\mu_A \sim r \tag{8.15.1}$$

As a result

$$\langle \mu_A \cdot r \rangle = \text{constant} \times \langle r^2 \rangle \tag{8.15.2}$$

where the angular brackets denote statistical averages, and $\langle r^2 \rangle$ is the mean-square end-to-end distance of the molecular chains. The vector sum of dipoles of type B and C is not correlated with the end-to-end distance and

$$\langle \mu_B \cdot r \rangle = 0, \qquad \langle \mu_C \cdot r \rangle = 0 \tag{8.15.3}$$

Conventional polymers of type B, with N skeletal bonds, have either $N - 1$ symmetry planes (a), $N - 1$ twofold symmetry axes (b), or $N - 1$ symmetry points (c). For example, polyamides (polyesters) from α, ω-dicarboxylic-n-alkanes and α, ω-diamino-n-alkanes (α, ω-diol-n-alkanes) have symmetry planes (and symmetry axes) or symmetry points depending on whether the number of methylene groups in the dicarboxylic acids (or diamines) is even or odd. In general, chains having symmetry of mirror image or inversion do not have type A dipoles.

Polymers with $-X-Y-Z-$ skeletal bonds or chemical groups in their structure are intrinsically type A polymers. Some polymers exhibit dipoles with components of types A and B, and these are called type AB polymers. These latter polymers can be further classified[62] into at least six types. Schemes of type A1, A1B, A2, A2B, A3, and A3B dipoles are shown in Fig. 8.21, and for illustrative purposes, some representative molecular chains of these types of dipole are given in Table 8.1.

The curves representing the dielectric loss in the frequency domain for type A polymers present at low frequencies the normal mode process associated with motions of the whole chain. This relaxation is followed, in increasing order of frequencies, by the α-relaxation, reflecting segmental motions of the chains, and,

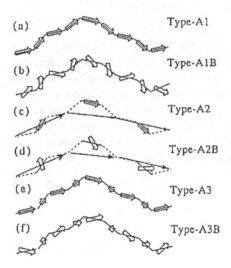

Fig. 8.21   Different type A chains. Arrows indicate dipole vectors, and dotted lines are nonpolar bonds

TABLE 8.1 Types A and AB Dipoles in Polymers[62]

Type	Polymer	Repeat unit		m_A, D
A1	Polyphenylene oxide[a]	$-C_6H_4X_2-O-$	X=Cl	1.56
			X=CH$_3$	
A1B	Polyphenylene oxide[b]	$-C_6H_5X-O-$	X=C$_6$H$_4$Cl	2
			X=C$_6$H$_4$OCH$_3$	3
	Polyacetylene[c]	$-CH=CX-$	X=OC$_6$H$_5$	6.7
	Poliphosphazene[d]	$-PX_2=N-$	X=OCH$_2$CF$_3$	
	Poly(methylene phenylene)[e]	$-(CH_2)_4-C_6H_4X_2-$	X=Cl	1.56
A2B	Polydiene	$-CH_2-CX=CH-CH_2-$	X=Cl	1.03(*cis*)
				0.21(*trans*)
			X=Cl	0.22(*cis*)
			X=CH$_3$	0.2
			X=C$_6$H$_5$	0.82
	Polyester	$-(CH_2)_nO-CO-$	$n=4,5$	
	Polypeptide	$-CHX-CO-NH-$		
	Polylactam	$-(CH_2)_n-CO-NH-$	$n=5$	1.0
A3B	Polyether	$-CH_2-CHX-O-$	X=CH$_3$	0.18
			X=C$_6$H$_5$	0

[a]Poly(2,6-disubstituted-1,4-phenylene oxide)
[b]Poly(2-substituted-1,4-phenylene oxide)
[c]Poly(substituted acetylene)
[d]Poly(organophosphazene)
[e]Poly[tetramethylene(2,6-dichloro-1,4-phenylene)]

finally, by the β-process at very high frequencies, arising from local motions. In what follows we shall deal with the analysis of the normal mode process.

8.16. TIME–DIPOLE CORRELATION FUNCTION FOR POLYMERS OF TYPE A

Let us consider a general case, for example, the polarization of a solution of type AB polymers. The polarization of a sphere of volume V is the sum of all dipoles contained in the sphere

$$\mathbf{M}(t) = \sum \mathbf{\mu}_i(t) + \sum \sum \mathbf{m}_{ij} + \sum \mathbf{m}_k \qquad (8.16.1)$$

where $\mathbf{\mu}_i$ is the permanent dipole moment of individual chain i at time t, $\mathbf{m}_{ij}$ is the induced dipole moment of the repeat unit j of chain i, and $\mathbf{m}_k$ is the dipole moment of molecule k of the solvent. If the solvent has low polarity and the cross-correlation terms between molecules i and j, together with the induced dipole contributions, are neglected, the time–dipole autocorrelation function is given by[1,15]

$$g(t) = \frac{\langle \mathbf{\mu}_i(0) \cdot \mathbf{\mu}_i(t) \rangle}{\langle \mu^2 \rangle} \qquad (8.16.2)$$

For type AB dipoles

$$\mathbf{\mu}_i(t) = \sum_j \mathbf{\mu}_{ij_\parallel}(t) + \mathbf{\mu}_{ij_\perp}(0) \qquad (8.16.3)$$

where the symbols $\perp$ and $\parallel$ in the subscripts denote the components parallel and perpendicular to the chain contour of the dipole moment associated with the repeat unit j of the chain i. The average $\langle \mathbf{\mu}_i(0) \cdot \mathbf{\mu}_i(t) \rangle$ involves four components[63]

$$
\begin{aligned}
\langle \mathbf{\mu}_i(0) \cdot \mathbf{\mu}_i(t) \rangle = &\sum_j \sum_k \langle \mathbf{\mu}_{ij_\parallel}(0) \cdot \mathbf{\mu}_{ik_\parallel}(t) \rangle \\
&+ \sum_j \sum_k \langle \mathbf{\mu}_{ij_\parallel}(0) \cdot \mathbf{\mu}_{ik_\perp}(t) \rangle \\
&+ \sum_j \sum_k \langle \mathbf{\mu}_{ij_\perp}(0) \cdot \mathbf{\mu}_{ik_\parallel}(t) \rangle \\
&+ \sum_j \sum_k \langle \mathbf{\mu}_{ij_\perp}(0) \cdot \mathbf{\mu}_{ik_\perp}(t) \rangle
\end{aligned} \qquad (8.16.4)
$$

where the parallel component is related to the $\mathbf{a}_{ij}$ bond vector of the repeat unit by

$$\mathbf{\mu}_{ij_\parallel}(t) = m_a \mathbf{a}_{ij}(t) \qquad (8.16.5)$$

In this expression, m_a is the parallel dipole moment per unit chain contour. Taking into account that the end-to-end vector is $\mathbf{r}_i(t) = \sum_j a_{ij}(t)$, the time–dipole autocorrelation function can be written as

$$\sum_j \sum_k \langle \boldsymbol{\mu}_{ij_\parallel}(0) \cdot \boldsymbol{\mu}_{ik_\parallel}(t) \rangle = m_a^2 \sum_j \sum_k \langle \mathbf{a}_{ij}(0) \cdot \mathbf{a}_{ik}(t) \rangle$$

$$= m_a^2 \langle \mathbf{r}_i(0) \cdot \mathbf{r}_i(t) \rangle \tag{8.16.6}$$

If cross-correlation terms between parallel and perpendicular components in Eq. (8.16.4) are further ignored, Eq. (8.16.2) can be written as

$$g(t) = \frac{m_a^2 \langle \mathbf{r}_i(0) \cdot \mathbf{r}_i(t) \rangle + \sum_j \sum_k \langle \boldsymbol{\mu}_{ij_\perp}(0) \cdot \boldsymbol{\mu}_{ik_\perp}(t) \rangle}{m_a^2 \langle \mathbf{r}_i(0) \cdot \mathbf{r}_i(0) \rangle + \sum_j \sum_k \langle \boldsymbol{\mu}_{ij_\perp}(0) \cdot \boldsymbol{\mu}_{ik_\perp}(0) \rangle} \tag{8.16.7}$$

The relaxation times associated with the so-called normal mode involve motions of the whole chain and therefore are much larger than the motions corresponding to the perpendicular components of the type AB dipoles. Motions of type B dipoles by the effect of cooperative micro-Brownian motions give rise to the dielectric α-relaxation. The dielectric normal mode is well separated from the dielectric α-relaxation, and the normalized time–dipole relaxation function that accounts for the normal mode process is given by[63]

$$g_n(t) = \frac{\langle \boldsymbol{\mu}_\parallel(0) \cdot \boldsymbol{\mu}_\parallel(t) \rangle}{\langle \mu_\parallel^2 \rangle} = \frac{\langle \mathbf{r}_i(0) \cdot \mathbf{r}_i(t) \rangle}{\langle r^2 \rangle} \tag{8.16.8}$$

where $\mathbf{r}_i(t)$ is the end-to-end distance vector at time t, and $\langle r^2 \rangle$ is the mean-square end-to-end distance of the chains.

8.17. NORMAL RELAXATION TIME FOR POLYMERS HAVING TYPE A AND TYPE AB DIPOLES

In the analysis of long-range dynamics, two cases can be considered:[64,65]

1. The response of the chains to perturbation fields is not affected by interactions with the neighboring molecular chains. This behavior corresponds to both low molecular weight polymers in the melt and very dilute solutions of long chains.
2. Topological constraints arising from intermolecular interactions affect the response of the chains to external perturbations. The response of melts and concentrated solutions of high molecular weight polymers corresponds to this second case.

8.18. MOLECULAR CHAINS DYNAMICS

Spring–bead theories were developed to describe at the molecular level the response of molecular chains to perturbation fields. Spring–bead theories describing the dynamics of molecular chains will be briefly described below.

8.18.1. Rouse Model

A flexible high molecular chain can be represented by a large number of freely jointed segments. The end-to-end distance distribution is of Gaussian type and can be expressed by[65,66]

$$W(r) = \left(\frac{3}{2\pi a^2}\right)^{3/2} \exp\left(-\frac{3}{2}\frac{r^2}{a^2}\right) \tag{8.18.1.1}$$

where a^2 is the mean-square end-to-end distance of the chains. The free energy is given by

$$F(r) = -k_B T \ln W(r) = F(0) + \frac{3k_B T}{2a^2} r^2 \tag{8.18.1.2}$$

Here it has been considered that the internal energy is constant. This is a fundamental formula that gives the spring constant $(3/2)(k_B T/a^2)$ of an ideal chain.

In the free draining approximation used by Rouse to model the dynamics of polymer chains in dilute solutions, hydrodynamic interactions are considered negligible. In this approximation, the velocity of the liquid medium is assumed to be unaffected by the moving polymer molecules. A molecular chain is described as a succession of submolecules large enough to obey Gaussian statistics. Each submolecule is represented by a spring that accounts for its elasticity, while the mass that is responsible for the energy dissipated by the submolecule moving in the viscous medium is represented by a bead (see Fig. 8.22). Therefore, each molecular chain is described by Rouse[67] as a succession of beads $\mathbf{r}_1, \mathbf{r}_2, \ldots, \mathbf{r}_N$ separated by springs along the vectors $\mathbf{a}_1, \mathbf{a}_2, \ldots, \mathbf{a}_N$. By analogy with Eq. (8.18.1.2), the elastic energy for a subchain is

$$F_{n,n+1} = \frac{3k_B T}{2a^2} (\mathbf{r}_{n+1} - \mathbf{r}_n)^2 \tag{8.18.1.3}$$

where it has been assumed that the mean-square end-to-end distance of each submolecule is a^2. The force exerted on the nth submolecule by the two neighbors

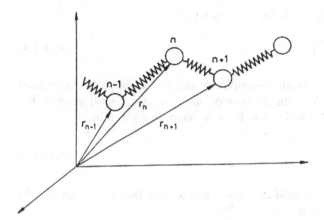

FIG. 8.22 Submolecules of molecular chains represented by springs (elasticity) and beads (viscous dissipation)

is

$$\mathbf{f}_n = \frac{\partial\left(F_{n,n-1} + F_{n,n+1}\right)}{\partial r_n}$$

$$= -\frac{3k_B T}{a^2}\left[(\mathbf{r}_{n+1} - \mathbf{r}_n) - (\mathbf{r}_n - \mathbf{r}_{n-1})\right] \qquad (8.18.1.4)$$

The work done by the driving force $\mathbf{f}_n$ is dissipated by the friction energy of the moving submolecule, represented by a bead, in the viscous medium. Therefore

$$\mathbf{f}_n + \xi_0 \frac{d\mathbf{r}_n}{dt} = 0 \qquad (8.18.1.5)$$

where ξ_0 is the friction coefficient per submolecule. By going to the continuous limit, $\mathbf{r}_{n+1} - \mathbf{r}_n \rightarrow \partial \mathbf{r}_n / \partial n$, the combination of Eqs (8.18.1.4) and (8.18.1.5) leads to

$$\frac{\partial \mathbf{r}_n}{\partial t} = \frac{3k_B T}{\xi_0 a^2} \frac{\partial^2 \mathbf{r}_n}{\partial n^2} \qquad (8.18.1.6)$$

By using the boundary conditions at the ends of the chains

$$\left.\frac{\partial \mathbf{r}_n}{\partial n}\right|_{n=0} = \left.\frac{\partial \mathbf{r}_n}{\partial n}\right|_{n=N} = 0 \qquad (8.18.1.7)$$

the solution of Eq. (8.18.1.6) is the wave function[66]

$$r_{np} = r_{np}^0 \cos\left(\frac{p\pi n}{N}\right) e^{-t/\tau_p} \tag{8.18.1.8}$$

where r_{np}^0 and τ_p are respectively the amplitude and the relaxation time associated with the p mode, and N is the number of submolecules. Substitution of Eq. (8.18.1.8) into Eq. (8.18.1.6) leads to the relaxation time given by

$$\tau_p = \frac{a^2 \xi_0}{3 k_B T p^2 \pi^2} N^2 \tag{8.18.1.9}$$

According to the Rouse model, the time–dipole correlation function for the normal mode process can be written as[67,69]

$$g_n(t) = \frac{\langle \boldsymbol{\mu}_A(t) \cdot \boldsymbol{\mu}_A(0) \rangle}{\langle \mu_A^2 \rangle} = \frac{\langle \mathbf{r}(t) \cdot \mathbf{r}(0) \rangle}{\langle r^2 \rangle_0}$$

$$= \frac{8}{\pi^2} \sum_{p \text{ odd}} p^{-2} \exp\left(-\frac{t}{\tau_p}\right) \tag{8.18.1.10}$$

From Eqs (8.8.3) and (8.18.1.10), transformation of the dielectric permittivity from the time domain to the frequency domain can be accomplished by means of the following expression

$$\frac{\varepsilon^*(\omega) - \varepsilon_u}{\varepsilon_r - \varepsilon_u} = \int_0^\infty \left[-\frac{d}{dt}\left(\frac{\langle \mathbf{r}(0) \cdot \mathbf{r}(t) \rangle}{\langle r^2 \rangle_0}\right) \right] \exp(-i\omega t) \, dt$$

$$= 1 - \int_0^\infty \frac{\langle \mathbf{r}(0) \cdot \mathbf{r}(t) \rangle}{\langle r^2 \rangle_0} \exp(-i\omega t) \, dt \tag{8.18.1.11}$$

where the subscripts r and u refer to the relaxed and unrelaxed dielectric permittivities of the normal mode process. By combining Eqs (8.18.1.10) and (8.18.1.11), one obtains

$$\frac{\varepsilon^*(\omega) - \varepsilon_u}{\varepsilon_r - \varepsilon_u} = \frac{8}{\pi^2} \sum_{p \text{ odd}} \frac{1}{p^2(1 + i\omega\tau_p)} \tag{8.18.1.12}$$

The dielectric response in the normal mode process is mainly governed by the first mode [$p = 1$ in Eq. (8.18.1.9)] given by

$$\tau_1 = \tau_r = \frac{a^2 \xi_0}{3 k_B T \pi^2} N^2 \tag{8.18.1.13}$$

Since $N \sim M$, where M is the molecular weight of the chain, the longest relaxation time in the Rouse model scales as

$$\tau_1 \sim M^2 \tag{8.18.1.14}$$

This scaling law disagrees with that obtained from experiments in ϑ solvents given by[70]

$$\tau_1 \sim M^{3/2} \tag{8.18.1.15}$$

This failure comes from the neglect of hydrodynamic interactions, and, as a consequence, the Rouse model is now regarded as inappropriate as a model in dilute solutions.

8.18.2. Zimm Model

Zimm's theory uses the assumptions of the Rouse theory and, in addition, considers interactions between the moving submolecules and the solvent. The theory[71] also makes use of the method formulated by Kirkwood and Riseman for the evaluation of viscosity solutions. A parameter $h = \nu^{1/2}\xi_0/(12\pi)^{1/2}a^2\eta_s$, where η_s is the viscosity of the solvent, is defined that accounts for the hydrodynamic interaction. This parameter is close to zero ($h \ll 1$) for vanishing interactions (free draining) and much larger than unity when the interactions are dominant (nonfree draining). The mathematical details are beyond the scope of this book. It is convenient to point out, however, that the expression for $\langle \mathbf{r}(t) \cdot \mathbf{r}(0) \rangle$ is similar to Eq. (8.18.1.10). The difference lies in the value of the longest relaxation time that governs the normal mode process.

Scaling arguments[72] could in principle be used to estimate τ_1. Since a low-frequency statistical property of a Gaussian chain does not depend on its local structure, the original Gaussian chain consisting of N segments of length a can be substituted for a new Gaussian chain consisting of N/λ segments. Each of these segments is a Gaussian whose square root of the mean-square end-to-end distance is $a' = a\lambda^\nu$, where $\nu = 1/2$ and 0.6 in ϑ and good solvents, respectively (note that the radius of a coil $\langle r^2 \rangle^{1/2} = R = aN^\nu$, where a is the length of each freely jointed segment). The transformation from the old to the new chain involves the following transformations[72a]

$$N \longrightarrow \frac{N}{\lambda} \quad \text{and} \quad a \longrightarrow a\lambda^\nu \tag{8.18.2.1}$$

The longest relaxation time of polymer chains in very dilute solutions depends on a, the solvent viscosity η_s, N, and the thermal energy $k_B T$. From dimensional analysis one obtains

$$\tau_1 = \frac{a^3 \eta_s}{k_B T} f(N) \tag{8.18.2.2}$$

The transformations indicated in Eq. (8.18.2.1) give

$$\frac{a^3 \eta_s}{k_B T} f(N) = \frac{a^3 \lambda^{3\nu} \eta_s}{k_B T} f\left(\frac{N}{\lambda}\right)$$

(8.18.2.3)

Fulfillment of Eq. (8.18.2.3) for any value of λ requires that

$$f(N) = \text{constant} \times N^{3\nu}$$

(8.18.2.4)

This means that $f(N/\lambda) \sim (N^\nu/\lambda)^3$. Hence, τ_1 can be written as

$$\tau_1 \simeq \frac{\eta_s (aN^\nu)^3}{k_B T} = \frac{\eta_s R^3}{k_B T}$$

(8.18.2.5)

where use is made of the fact that the square root of the mean-square end-to-end distance of a Gaussian chain scales as

$$R = aN^\nu$$

(8.18.2.6)

In a theta solvent $\nu = 1/2$, and consequently $R \sim N^{1/2}$. Therefore, the relaxation time associated with the normal dielectric mode for very highly dilute solutions of polymers with dipoles of either type A or type AB in their structure scales with molecular weight as

$$\tau_1 \sim M^{3/2}$$

(8.18.2.7)

in agreement with the experimental results. To account for excluded volume effects, the longest relaxation time ($p = 1$) associated with the normal mode process for a free draining model obtained rigorously from the Rouse–Zimm theory is expressed by

$$\tau_1 = \frac{12 M \eta_s [\eta]}{\pi^2 RT}$$

(8.18.2.8)

For a nonfree draining model, the expression is

$$\tau_1 = \frac{0.844 M \eta_s [\eta]}{RT}$$

(8.18.2.9)

Note that the intrinsic viscosity $[\eta]$ is included to account for excluded volume effects. The predictions of the Zimm theory have been tested by plotting the normal mode relaxation time against $M \eta_s [\eta]/RT$. The results, shown in Fig. 8.23, fall on a straight line with the slope 1 within experimental error.[62] The best-fit

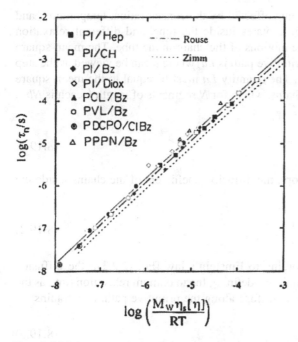

FIG. **8.23** Dependence of the normal relaxation times on $M_w \eta_s[\eta]/RT$ for different dilute solutions. Polymers: PI = *cis*-polyisoprene; PCL = poly(ε-caprolactone); PVL = poly(valerolactone); PDCPO = poly(2,6-dichloro-1,4-phenylene oxide); PPPN = poly-(phenoxy phosphacene). Solvents: Hep = heptane; CH = cyclohexane; Bz = benzene; Diox = dioxane; ClBz = chlorobenzene[62]

curve (solid line) obeys the following equation[73]

$$\tau_1 = \frac{1.4 M \eta_s[\eta]}{RT} \qquad (8.18.2.10)$$

It is worth noting that the front factor 1.4 is somewhat smaller and larger than those of 1.22 and 0.84 predicted, respectively, by the free and nonfree draining models.

8.19. NORMAL MODE RELAXATION TIME FOR MELTS AND CONCENTRATED SOLUTIONS OF POLYMERS HAVING TYPE A AND TYPE AB DIPOLES

Topological constraints in melts and concentrated solutions confine molecular chains in a tubelike region made up of the surrounding chains.[65,74,75] The tube for each chain, schematically depicted in Fig. 8.24, is continuously renewing itself, the time of renewal being affected by the concentration, temperature, and molecular weight. The chain reptates inside the tube, and different relaxation times are associated with the motions of the chain in the tube. The mean-square end-to-end distance of the primitive path is La, where a can be viewed as the step length of the primitive path. The quantity La must be equal to the mean-square end-to-end distance of the chains, which for N segments of length b each is Nb^2. Hence

$$L = \frac{Nb^2}{a} \qquad (8.19.1)$$

According to the Rouse theory, the diffusion coefficient of the chains inside the tube is given by[74]

$$D_t = \frac{k_B T}{\xi_0 N} = \frac{D_1}{N} \sim N^{-1} \qquad (8.19.2)$$

where $D_1 = k_B T/\xi_0$ is, according to Einstein's law [Eq. (2.2.3)], the diffusion coefficient of the monomer unit. By defining the maximum relaxation time as the time necessary for the chain to diffuse along the primitive path, one obtains

$$\tau_1 \cong \frac{L^2}{D_t} \propto \frac{N^3}{D_1} \sim N^3 \qquad (8.19.3)$$

where use was made of Eq. (8.19.2). The scaling law $\tau_1 \sim M^3$ gives a rather weak dependence of the longest relaxation time on molecular weight. The experimental value of the exponent lies in the range 3.2–3.4. Illustrative data showing the molecular weight dependence of the normal mode relaxation time are shown in Fig. 8.25.[76]

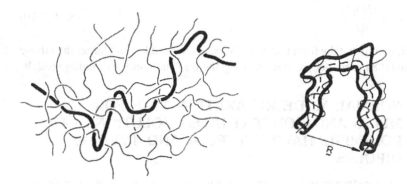

FIG. 8.24 Left: Molecular motions in concentrated polymer solutions. Right: Doi–Edwards tube model

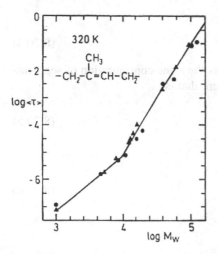

FIG. 8.25 Double logarithmic plot showing the molecular weight dependence of the dielectric normal relaxation time of poly(isoprene)[63b]

8.20. SCALING LAWS FOR THE DIELECTRIC NORMAL MODE OF SEMI-DILUTE SOLUTIONS OF POLYMERS HAVING EITHER TYPE A OR TYPE AB DIPOLES

In the case of semidilute solutions, some conclusions can be drawn from scaling arguments. The relevant parameters to obtain τ_1 are a, N, k_BT, η_s, and the concentration of polymer C. Dimensional analysis suggests that[77,72a]

$$\tau_1 = \frac{\eta_s a^3}{k_B T} f(Ca^3, N) \tag{8.20.1}$$

When λ segments are grouped so that

$$N \longrightarrow \frac{N}{\lambda}, \qquad a \longrightarrow a\lambda^\nu, \qquad C \longrightarrow \frac{C}{\lambda} \tag{8.20.2}$$

Eq. (8.20.1) must be invariant under this transformation, whence[76]

$$a^3 f(Ca^3, N) \longrightarrow a^3 \lambda^{3\nu} f\left(\frac{C}{\lambda} a^3 \lambda^{3\nu}, \frac{N}{\lambda}\right) = a^3 \lambda^{3\nu} f\left(Ca^3 \lambda^{3\nu-1}, \frac{N}{\lambda}\right) \tag{8.20.3}$$

This requires that

$$a^3 f(Ca^3, N) = N^{3\nu} a^3 f(Ca^3 N^{3\nu-1}) \tag{8.20.4}$$

By defining the critical concentration C^* as the same concentration of segments inside the coil of radius R as in the solution, that is

$$C^* \simeq \frac{N}{R^3} = \frac{N}{a^3 N^{3\nu}} = \frac{N^{1-3\nu}}{a^3} \tag{8.20.5}$$

Eq. (8.20.1) becomes

$$\tau_1 = \frac{\eta_s N^{3\nu} a^3}{k_B T} f\left(\frac{C}{C^*}\right) \tag{8.20.6}$$

By assuming that

$$f\left(\frac{C}{C^*}\right) = \text{constant} \times \left(Ca^3 N^{3\nu-1}\right)^x \tag{8.20.7}$$

Eq. (8.20.6) becomes

$$\tau_1 = \frac{\eta_s N^{3\nu} a^3}{k_B T} \left(Ca^3 N^{3\nu-1}\right)^x \tag{8.20.8}$$

The exponent x can be obtained by imposing the condition $\tau_1 \sim N^3$. In this case, $(3\nu - 1)x + 3\nu = 3$ and $x = (3 - 3\nu)/(3\nu - 1)$. For good solvents ($\nu = 0.6$), Eq. (8.20.8) becomes

$$\tau_1 \simeq \frac{\eta_s N^{3\nu}}{k_B T} \left(\frac{C}{C^*}\right)^{(3-3\nu)/(3\nu-1)} \sim c^{3/2} M^3 \tag{8.20.9}$$

where it has been considered that $C \sim c$ (g/ml) and $N \sim M$.

8.21. RELATION BETWEEN MOLECULAR DIMENSIONS AND RELAXATION STRENGTH FOR TYPE A POLYMERS

As indicated in Chapter 5, the mean-square dipole moment of molecular chains can be obtained from Onsager-type equations, the Guggenheim and Smith equation being mostly used. If the electronic contribution to the polarization is negligible in comparison with the orientation polarization, the mean-square dipole moment of the component parallel to the chain contour can be expressed by[78,79]

$$\langle \mu_\parallel^2 \rangle = \frac{3k_B MT}{4\pi\rho N_A F} \frac{\Delta\varepsilon}{c} \tag{8.21.1}$$

in the limit $c \to 0$. Here, the concentration is in wt/vol, and F is the ratio of the internal to the external electric fields, the value of which is close to unity. According to Eq. (8.16.5)

$$\langle \mu_\parallel^2 \rangle = m_a^2 \langle r^2 \rangle \tag{8.21.2}$$

the mean-square end-to-end distance can be obtained as a function of concentration from $\Delta\varepsilon/c$ results if the dipole moment per contour length unit, m_a, is known. Actually

$$\langle r^2 \rangle = \frac{3k_B T}{4\pi N_A m_a^2 F} \frac{\Delta\varepsilon}{c} \tag{8.21.3}$$

Equation (8.21.3) has mainly been used to determine the expansion factor of the molecular dimensions as a function of both the solvent and concentration. This quantity is defined by

$$\alpha_r = \frac{\langle r^2 \rangle}{\langle r^2 \rangle_0} \tag{8.21.4}$$

where $\langle r^2 \rangle_0$ is the unperturbed mean square end-to-end distance.

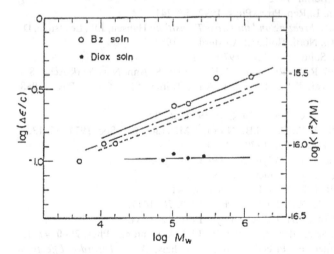

FIG. 8.26 Variation in the relaxation strength and the expansion factor α_r with the molecular weight of *cis*-polyisoprene in a good (benzene) and a theta (dioxane) solvent[62]

Experimental results showing the variation in $\Delta\varepsilon/C$ with molecular weight for solutions of poly(cis-isoprene) in benzene (good solvent) and dioxane (theta solvent) are presented in Fig. 8.26. The values of $\langle r^2 \rangle/M$ were calculated with $\mu_b = 4.8 \times 10^{-12}$ electrostatic units in the cgs system with $F = 1$. The value of $\langle r^2 \rangle/M$ in benzene solution is higher than in dioxane and it increases with M, the slope being 0.20 of the order of that expected from Flory's theory.[79]

REFERENCES

1a. Ishida, Y.; Matsuo, M.; Yamafuji, K. Kolloid Z. **1962**, *180*, 108.

1b. Stillinger, F.H. Science **1995**, *267*, 1935.

2.   Williams, G. Dielectric relaxation spectroscopy of amorphous polymer systems. The modern approach. In *Keynote Lectures in Polymer Science*; Riande, E., Ed.; CSIC: Madrid, 1995; Williams, G. Theory of dielectric properties. In *Dielectric Spectroscopy of Polymeric Materials*; Runt, J.P., Fitzgerald, J.J., Eds.; American Chemical Society: Washington, D.C., 1997.

3.   Vogel, H. Z. Phys. **1921**, *22*, 645.

4.   Fulcher, G.S. J. Am. Ceram. Soc. **1995**, *8*, 339.

5.   Tammann, G.; Hesse, W.Z. Anorg. Allgem. Chem. **1926**, *156*, 245.

6.   Angell, C.A. Science **1995**, *267*, 1924.

7.   Adam, D.; Gibbs, J.H. J. Chem. Phys. **1965**, *43*, 139.

8.   Grest, G.S.; Cohen, M.H. Adv. Chem. Phys. **1961**, *48*, 455.

9.   Donth, E.J. J. Non-Cryst. Solids **1982**, *53*, 325.

10. Donth, E. *Relaxation and Thermodynamics in Polymers: Glass Transition*; Akademic Verlag: Berlin, 1992.

11. Götze, W.; Sjögren, L. Rep. Prog. Phys. **1992**, *55*, 241.

12. Götze, W. In *Liquids, Freezing and the Glass Transition*; Hansen, J.P., Levesque, D., Zinn-Justin, J., Eds.; North-Holland: Amsterdam, 1991.

12a. Ngai, K.L. Comm. Solid State Phys. **1979**, *9*, 127.

12b. Ngai, K.L.; Rendell, R.W.; Rajagopal, A.K.; Teitler, S. Ann. New York Acad. Sci. **1987**, *484*, 150; Ngai, K.L.; Rajagopal, A.K.; Teitler, S. J. Chem. Phys. **1988**, *88*, 5086.

13. Williams, G.; Watts, D.C. Trans. Far. Soc. **1970**, *66*, 80.

14. Williams, G.; Watts, D.C.; Dev, S.B.; North, A.M. Trans. Far. Soc. **1971**, *67*, 1323.

15. Williams, G. Adv. Polym. Sci. **1979**, *33*, 59.

16. Havriliak, S.; Negami, S. J. Polym. Sci. **1966**, *C14*, 99.

17. Havriliak, S.; Negami, S. Polymer **1967**, *8*, 161.

18. Cole, K.S.; Cole, R.H. J. Chem. Phys. **1941**, *9*, 341.

19. Davidson, D.W.; Cole, R.H. J. Chem. Phys. **1950**, *18*, 1417.

20. Davidson, D.W.; Cole, R.H. J. Chem. Phys. **1951**, *19*, 1484.

21. Lunkenheimer, P.; Schneider, U.; Brand, R.; Lidl, A. Contemp. Phys. **2000**, *41*, 15.

22. Hofman, A.; Kremer, F.; Fischer, E.W.; Schönhals, A. In *Disorder Effects in Relaxation Processes*; Richert, R., Blumen, A., Eds.; Springer: Berlin, 1994; 309–331.

23. Lunkenheimer, P.; Pimenov, A.; Schiener, B.; Böhmer, R.; Lidl, A. Europhys. Lett. **1996**, *33*, 611.
24. Leheny, R.L.; Nagel, S.R. Europhys. Lett. **1997**, *39*, 447.
25. Kudlik, A.; Benkhof, S.; Blochowicz, T.; Tschirwitz, C.; Rössler, E. J. Mol. Struct. **1999**, *479*, 201.
26. Johari, G.P.; Goldstein, M. J. Chem. Phys. **1970**, *53*, 2372.
27. Johari, G.P. In *The Glass Transition and Nature of the Glassy State*; Goldstein, M., Simha, R., Eds.; Ann. New York Acad. Sci. **1976**, *279*, 117.
28. Huang, Y.; Saiz, E.; Ezquerra, T.; Guzmán, J.; Riande, E. Macromolecules **2002**, *35*, 2926.
29. Fuoss, R.M.; Kirkwood, J.G. J. Am. Chem. Soc. **1941**, *63*, 385.
30. Wu, L. Phys. Rev. B **1991**, *43*, 9906.
31. Garwe, F.; Schönhals, A.; Lockwenz, H.; Beiner, M.; Schröter, K.; Donth, E. Macromolecules **1996**, *29*, 247.
32. Stickel, F.; Fischer, E.W.; Richert, R.J. Chem. Phys. **1995**, *102*, 1.
33. Bergman, R.; Alvarez, F.; Alegria, A.; Colmenero, J. J. Chem. Phys. **1998**, *109*, 7546.
34. Torell, I.M. J. Chem. Phys. **1982**, *76*, 3467.
35. Angell, C.A.; Torell, I.M. J. Chem. Phys. **1983**, *78*, 937.
36. Cheng, L.T.; Yan, Y.X.; Nelson, K.A. J. Chem. Phys. **1989**, *91*, 6052.
37. Ngai, K.L.; Rajagopal, A.K.; Huang, C.Y. J. Appl. Phys. **1984**, *55*, 1714.
38. Schönhals, A.; Kremer, F.; Hoffmann, F.; Fischer, E.W.; Schlosser, E. Phys. Rev. Lett. **1993**, *70*, 3459.
39. Dixon, P.K.; Menon, N.; Nagel, S.R. Phys. Rev. E **1994**, *50*, 1717.
40. Du, W.M.; Li, G.; Cummins, H.Z.; Fuchs, M.; Toulouse, L.; Knauss, L.A. Phys. Rev. E **1994**, *49*, 2192.
41. Li, G.; Du, W.M.; Sakai, A.; Cummins, H.Z. Phys. Rev. A **1992**, *46*, 3343.
42. Börjesson, L.; Elmroth, M.; Torell, L.M. Chem. Phys. **1990**, *149*, 209.
43. Rössler, E.; Sokolov, A.P.; Kisliuk, A.; Quitmann, D. Phys. Rev. B **1994**, *49*, 14,967.
44. Yang, Y.; Nelson, K.A. J. Chem. Phys. **1996**, *104*, 5429.
45. Cummins, H.Z.; Li, G.; Du, W.M.; Hernandez, J. J. Non-Crystall. Sol. **1994**, *172–174*, 26.
46. Adachi, K.; Saiz, E.; Riande, E. Phys. Chem. Chem. Phys. **2002**, *4*, 635.
47. Doolittle, A.K.; Doolittle, D.B. J. Appl. Phys. **1957**, *28*, 901.
48. Cohen, M.H.; Turnbull, D. J. Chem. Phys. **1959**, *31*, 1164.
49. Turnbull, D.; Cohen, M.H. J. Chem. Phys. **1961**, *34*, 120.
50. Williams, M.L.; Landel, R.F.; Ferry, J.D. J. Am. Chem. Soc. **1955**, *77*, 3701.
51. Debye, P. *Collected Papers*; Wiley Interscience: New York, 1954; 158 pp.
52. Onsager, L. J. Am. Chem. Soc. **1938**, *58*, 1486.
53. Fröhlich, H. *Theory of Dielectrics*; Oxford University Press: London, 1958.
54. Kirkwood, J.G.; Ann. New York Acad. Sci. **1940**, *40*, 315; Trans. Far. Soc. **1946**, *42ᵃ*, 7.
55. McCrum, N.G.; Read, B.E.; Williams, G. *Anelastic and Dielectric Effects in Polymeric Solids*; Wiley: London, 1967; 96 pp.
56. Helfand, H. Science **1984**, *226*, 647.
57. Hall, C.K.; Helfand, H. J. Chem. Phys. **1982**, *77*, 3275.

58. Skolnick, J.; Yaris, R. Macromolecules **1982**, *15*, 1041.
59. Stockmayer, W.H. Pure Appl. Chem. **1967**, *15*, 539.
60. North, A.M. Chem. Soc. Rev. **1972**, *1*, 49.
61. Stockmayer, W.W.; Matsu, K. Macromolecules **1972**, *5*, 766.
62. Adachi, K. In *Dielectric Spectrosocopy of Polymeric Materials*; Runt, J.P., Fitzgerald, J.J., Eds.; J. Am. Chem. Soc.: Washington, 1997.
63. Adachi, K.; Kotaka, T. Macromolecules **1988**, *21*, 157.
63a. Adachi, K.; Kotaka, T. Macromolecules **1984**, *17*, 120.
63b. Adachi, K.; Kotaka, T. Macromolecules **1985**, *18*, 466.
64. Graessley, W.W. In *Conformations and Dynamics of Macromolecules in Condensed Systems*; Nagasawa, M., Ed.; Elsevier: Amsterdam, 1988; 163 pp.
65. Doi, M.; Edwards, S.F. *The Theory of Polymer Dynamics*; Oxford University Press, Clarendon Press: 1986.
66. de Gennes, P.G. *Scaling Concepts in Polymer Physics*, 2nd Ed.; Cornell University Press: Ithaca, 1985.
67. Rouse, P.E. J. Chem. Phys. **1953**, *21*, 1273.
68. Stockmayer, W.H.; Baur, M.E. J. Am. Chem. Soc. **1964**, *43*, 4319.
69. Jones, A.; Stockmayer, W.H.; Molinary, R.J. J. Polym. Sci. Polym. Symp. **1976**, *54*, 227.
70. Doi, M.; Edwards, S.F. *The Theory of Polymer Dynamics*; Oxford University Press, Clarendon Press: 1986; 96 pp.
71. Zimm, B.H. J. Chem. Phys. **1956**, *24*, 269.
72. Doi, M.; Edwards, S.F. *The Theory of Polymer Dynamics*; Oxford University Press, Clarendon Press: 1986; 104 pp.
72a. Riande, E.; Diaz-Calleja, R.; Prolongo, M.G.; Masegosa, R.M.; Salom, C. In *Polymer Viscoelasticity*; Marcel Dekker: New York, 2000; 443 pp.
73. Urakawa, O.; Adachi, K.; Kotaka, T.; Takemoto, Y.; Yasuda, H. Macromolecules **1994**, *27*, 7410.
74. de Gennes, P.G. *Scaling Concepts in Polymer Physics*, 2nd Ed.; Cornell University Press: Ithaca, 1985; 219 pp.
75. Marrucci, G. In *Rheological Fundamentals in Polymer Processing*; Covacs, J.A., et al., Eds.; NATO ASI Ser.; Kluwer Academic Publishers: London, 1994; 37 pp.
76. Boese, D.; Kremer, F.; Fetters, L. Macromolecules **1990**, *23*, 1826.
77. Doi, M.; Edwards, S.F. *The Theory of Polymer Dynamics*; Oxford University Press, Clarendon Press: 1986; 234 pp.
78. Adachi, K.; Kotaka, T. Prog. Polym. Soc. **1993**, *18*, 585.
79. Adachi, K.; Okazaki, H.; Kotaka, T. Macromolecules **1985**, *18*, 1486.

9

Relaxations in the Glassy State. Short-Range Dynamics

9.1. INTRODUCTION

Although molecular mobility is severely restricted below the glass transition temperature, a cascade of subglass relaxation phenomena may occur below T_g that reveal different modes of mobility. These processes, called secondary relaxations, are labeled $\beta, \gamma, \ldots$, etc., in order of decreasing temperatures. The temperature at which a given group of segments undergoes a relaxation usually depends on both the polymer structure and the local molecular environment. It is widely accepted that relaxations in the glassy state basically imply intramolecular processes, and perhaps only the β-relaxation (the highest/lowest temperature/frequency secondary relaxation) is also related to intermolecular effects. In fact, as mentioned in the previous chapter, some authors[1] consider the β-relaxation as the precursor of the glass transitions, and in this sense it would be a "universal feature" of the glass-forming materials, including low molecular weight compounds. As a general picture, it is considered that molecular units associated with a dipole tend to rotate within a cage, cluster, or mobile islands by the effect of the electric field. The motion is restricted by limits of the cluster (island of mobility) made up of other polymer atoms acting as the boundaries of the cage. The motion of the unit is described in terms of the coordinates fixed with respect to the coordinates of the cavity, and an intramolecular potential barrier expressed as a function of these coordinates controls it. Consequently, subglass relaxation processes can be envisaged as a thermally activated motion between two potential wells separated by a potential barrier (Fig. 9.1). The probability of location of molecules at each side of the barrier is determined by the Boltzmann distribution. The application of an external electric field alters the equilibrium distribution by changing the relative depths of the minima, thus causing a redistribution, the rate of which is controlled by the activation energy barrier or activation enthalpy.

9.2. MOLECULAR MODELS ASSOCIATED WITH A SINGLE RELAXATION TIME

As made explicit in Chapter 2, the formulation of a dielectric microscopic theory is difficult because dipoles have a mutual influence on each other. A dipole is not only subject to the external field but also has its own field. In fact, mutual interactions make the response of an assembly of dipoles a cooperative phenomenon. As a result, the local field acting upon a specified dipole is generally higher than the externally applied field. Bearing this in mind, let us consider an ensemble of n dipoles distributed in two equilibrium positions according to the Boltzmann law

$$n_{10} = \frac{1}{1 + \exp(-\Delta U/kT)}$$

$$n_{20} = \frac{1}{1 + \exp(\Delta U/kT)}$$

(9.2.1)

where n_{10} and n_{20} are the respective equilibrium populations and ΔU is the difference in the energies between the two equilibrium states. Obviously, $n_{10} = n_{20}$, if $\Delta U = 0$. Under the influence of an electric field $\mathbf{E}_i$, the new

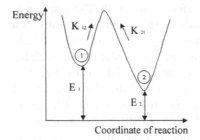

FIG. 9.1 Model of two well potentials to describe a secondary relaxation

populations will be given by

$$n_1 = \frac{1}{1 + \exp[-(\Delta U + \boldsymbol{\mu} \cdot \mathbf{E_i})/kT]}$$

$$n_2 = \frac{1}{1 + \exp[(\Delta U + \boldsymbol{\mu} \cdot \mathbf{E_i})/kT]}$$

(9.2.2)

According to the erghodic hypothesis, n_1 and n_2 represent the populations of the dipolar ensemble at a given time, as well as the interval of residence times for an isolated dipole in each one of the two potential wells. Since $\boldsymbol{\mu} \cdot \mathbf{E_i} \ll kT$, the exponentials in Eqs (9.2.2) can approximately be written as

$$\exp \mp \frac{\Delta U + \boldsymbol{\mu} \cdot \mathbf{E_i}}{kT} \cong \exp \mp \frac{\Delta U}{kT} \left(1 \mp \frac{\boldsymbol{\mu} \cdot \mathbf{E_i}}{kT} \right)$$

(9.2.3)

The number of dipoles moving from one to another potential well will be

$$n_1 - n_{10} \cong \frac{n \boldsymbol{\mu} \cdot \mathbf{E_i}}{4kT} \cosh^{-2} \frac{\Delta U}{2kT}$$

(9.2.4)

Then, the total polarization is given by

$$\mathbf{P} = (n_1 - n_{10}) \boldsymbol{\mu} = \frac{n \mu^2 \mathbf{E_i}}{2kT} \cosh^{-2} \frac{\Delta U}{2kT}$$

(9.2.5)

where the increase in the polarization due to the diminution of the $n_{20} - n_2$ dipoles in the second potential well has been taken into account. Note that $n_{20} - n_2 = n_1 - n_{10}$.

To describe the dynamic process and obtain the relaxation time, the kinetic equations for the jumping process across the barrier energy can be written as

$$\frac{dn_1}{dt} = -n_1 k_{12} + n_2 k_{21}$$

$$\frac{dn_2}{dt} = n_1 k_{21} - n_2 k_{12}$$

(9.2.6)

where k_{12} and k_{21} are the relative transition probabilities from one to another well. Since $n_1 + n_2 = n$, Eq. (9.2.6a) may be written as

$$\frac{dn_1}{dt} = -n_1(k_{12} + k_{21}) + n k_{21}$$

(9.2.7)

At equilibrium, that is, for $t \to \infty$, $dn_1/dt = 0$ and $n_1 = n_{1\infty}$. This leads to

$$n_{1\infty} = n \frac{k_{21}}{k_{12} + k_{21}} \tag{9.2.8}$$

Substitution of Eq. (9.2.8) into Eq. (9.2.7) yields

$$\frac{dn_1}{dt} = -\frac{n_1 - n_{1\infty}}{\tau} \tag{9.2.9}$$

where

$$\tau = (k_{12} + k_{21})^{-1} \tag{9.2.10}$$

Assuming proportionality between the polarization and the number of dipoles, Eq. (9.2.9) may alternatively be written as

$$\tau \frac{d\mathbf{P}}{dt} + \mathbf{P} = \frac{n\mu^2 \mathbf{E}_i}{2kT} \cosh^{-2} \frac{\Delta U}{2kT} \tag{9.2.11}$$

Equation (9.2.11) has the same formal structure as Eq. (2.11.4) which describes the macroscopic polarization of a dielectric specimen characterized by a single relaxation time. By comparing these two equations, the relaxation strength is obtained as

$$\Delta \varepsilon = \varepsilon_0 - \varepsilon_\infty = \frac{2\pi n\mu^2}{kT} \cosh^{-2} \frac{\Delta U}{2kT} \tag{9.2.12}$$

Equation (9.2.12) indicates that the relaxation strength strongly depends on the conformational energy, temperature, dipole moment, and also on the number of dipolar groups participating in the motion. Therefore, the relaxation strength estimated from experimental data, combined with Eq. (9.2.12), may provide an estimation of the number of molecular dipoles participating in the relaxation. Struik[2] has revisited the theory by considering the effect of the matrix en-vironment on the relaxation strength in viscoelastic processes. He concludes that only a small percentage of the potentially active groups intervene in the relaxation.

Since the transition or frequency probability between two wells is given by

$$k_{12} = f_0 \exp\left(-\frac{H + \Delta U}{kT}\right)$$

$$k_{21} = f_0 \exp\left(-\frac{H}{kT}\right) \tag{9.2.13}$$

Eq. (9.2.10) leads to the following expression for the relaxation time

$$\tau = \tau_0 \frac{\exp(H/kT)}{1 + \exp(-\Delta U/kT)} \cong \tau_0 \exp\left(\frac{H}{kT}\right) \tag{9.2.14}$$

or alternatively

$$f = f_0 \exp\left(-\frac{H}{kT}\right) \tag{9.2.15}$$

which is an Arrhenius-type equation where $\tau_0 = f_0^{-1}$ and the inequality $\Delta U \ll kT$ is considered. In fact, varying ΔU from zero to infinity, the relaxation time only changes by a factor of 2, in contrast to the important changes experimentally observed. Equation (9.2.15) offers a physical insight into relaxation times. In this picture, the relaxation time is the inverse of the mean frequency of jump between the two potential wells. In other words, the delay in the polarization is due to the scarcity of such jumps. From Eq. (9.2.15), the shift in frequency of the peak maximum with temperature is given by

$$\ln\frac{f_2}{f_1} = \frac{H}{kT}\left(\frac{1}{T_1} - \frac{1}{T_2}\right) \tag{9.2.16}$$

From Eq. (9.2.16), the activation energy H can be found.

Most secondary maxima obey Eq. (9.2.15) or (9.2.16). In many cases the Arrhenius plots for secondary relaxations of polymers fall in a straight line, giving $\log f_0 \cong 13.5$ (Fig. 9.2). As a consequence, an estimation of the activation energy can be made according to

$$H \cong (0.258 - 0.0083 \ln f)T_m \quad (\text{kJ/mol}) \tag{9.2.17}$$

where T_m is the temperature at which the maximum is observed at a frequency f. For 1 Hz, Eq. (9.2.17) reduces to

$$H \cong 0.258 T_m \quad (\text{kJ/mol}) \tag{9.2.18}$$

It is expected that the activation energy obtained from isochronal curves by

$$H_T = \left.\frac{\partial \ln f}{\partial(1/T)}\right|_{\varepsilon''_{max(T)}} \tag{9.2.19}$$

slightly differs from that obtained from isotherms by using the expression

$$H_f = \left.\frac{\partial \ln f}{\partial(1/T)}\right|_{\varepsilon''_{max(f)}} \tag{9.2.20}$$

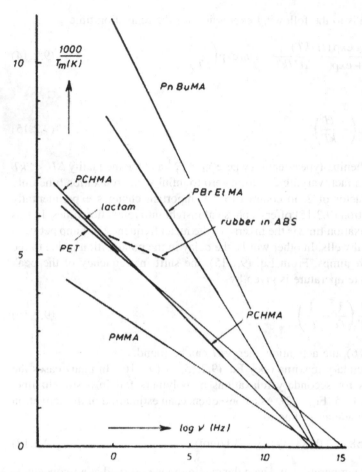

FIG. 9.2 Arrhenius plots of several secondary maxima

A shortcoming in the determination of activation energies from Arrhenius plots is that only the maximum of either the isochrone or isotherm loss curves is usually considered. A better approach is to calculate the activation energy *along* the experimental curves.[3,4] In this case a distribution of activation energy $H = H(f,T)$ revealing the fine structure of the relaxations is obtained. The starting point for such a calculation is the thermodynamic relationship

$$\frac{\partial \ln f}{\partial (1/T)}\bigg|_{\varepsilon'} \frac{\partial (1/T)}{\partial \varepsilon'}\bigg|_{\ln f} \frac{\partial \varepsilon'}{\partial \ln f}\bigg|_{1/T} = -1 \qquad (9.2.21)$$

This equation, in conjunction with

$$H = -kT^2 \frac{\partial \ln f}{\partial T}\bigg|_{\varepsilon'} \cong -kT^2 \frac{\Delta \ln f}{\Delta T} \qquad (9.2.22)$$

and the relation between ε'' and ε'

$$\varepsilon'' \cong -\frac{\pi}{2}\frac{\partial \varepsilon'}{\partial \ln f} \qquad (9.2.23)$$

obtained by means of the Krönig–Kramers relationships {Eq. (2.14.4), see also P6.5.11 from reference[5]} lead to

$$H \cong \frac{\pi kT}{2\varepsilon''}\frac{d\varepsilon'}{dT} \qquad (9.2.24)$$

Thus, an activation energy *map* in terms of temperature and frequency is obtained. In Fig. 9.3, values of the activation energy of the γ-relaxation of polycyclohexyl acrylate at 213 K are represented as a function of frequency. The occurrence of a maximum of the activation energy in this figure is due to the narrow shape and to the isolated character of this relaxation with respect to the relaxation associated with the glass transition. However, a continuous increase in the activation energy, overlapping several relaxation zones, is observed in most cases.[4]

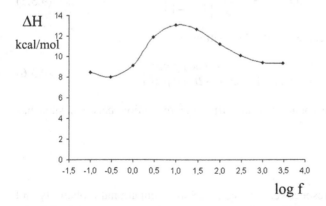

FIG. 9.3 Distribution of the activation energy of the dielectric secondary maximum for polycyclohexyl acrylate (PCHA) at 213 K

9.3. DYNAMICS OF SECONDARY DIELECTRIC RELAXATION IN TWO-SITE MODELS

Equations (9.2.6) may be written in terms of the probabilities of the transitions $(p_1(t), p_2(t))$ instead of the number of the relaxing units. Then, in the case of equal transition probability, we have

$$\frac{d\mathbf{p}}{dt} = \mathbf{T}\mathbf{p}(t) \tag{9.3.1}$$

where $\mathbf{p}(t) = (p_1(t), p_2(t))$ and

$$\mathbf{T} = \begin{vmatrix} -k & k \\ k & -k \end{vmatrix}$$

The solution of Eq. (9.3.1) is

$$\mathbf{p}(t) = \mathbf{p}(0)\exp(\mathbf{T}t) \tag{9.3.2}$$

By diagonalizing $\mathbf{T}$ by means of the usual transformation

$$\mathbf{S}^{-1}\mathbf{T}\mathbf{S} = \mathbf{D} \tag{9.3.3}$$

where $\mathbf{S}$ is an orthogonal matrix containing the eigenvectors of $\mathbf{T}$, Eq. (9.3.2) becomes

$$\mathbf{p}(t) = \mathbf{p}(0)\mathbf{S}|\exp(\mathbf{D}t)|\mathbf{S}^{-1} \tag{9.3.4}$$

because, as is known, $\exp(\text{diag}) = \text{diag}(\exp)$. It is easy to show that

$$\mathbf{D} = \begin{vmatrix} 0 & 0 \\ 0 & -2k \end{vmatrix} \quad \text{and} \quad \mathbf{S} = \frac{1}{\sqrt{2}}\begin{vmatrix} 1 & 1 \\ 1 & -1 \end{vmatrix} \tag{9.3.5}$$

Therefore, Eq. (9.3.4) can explicitly be written as

$$\begin{vmatrix} p_1(t) \\ p_2(t) \end{vmatrix} = \begin{vmatrix} 1 + \exp(-2kt) & 1 - \exp(-2kt) \\ 1 - \exp(-2kt) & 1 + \exp(-2kt) \end{vmatrix}\begin{vmatrix} p_1(0) \\ p_2(0) \end{vmatrix} \tag{9.3.6}$$

The average dipole moment in a site after the relaxation between sites has occurred is

$$\langle \boldsymbol{\mu} \rangle = \sum_i {}^0 p_i \boldsymbol{\mu}_i \tag{9.3.7}$$

where ${}^0 p_i$ and $\boldsymbol{\mu}_i$ are, respectively, the equilibrium occupational probability and the dipole moment at site i. Then, the time–dipole correlation function can be written as the average of the decay functions $[\boldsymbol{\mu}(0) \cdot \boldsymbol{\mu}(t)]_i$ for dipoles starting at sites $i = 1,2$ at $t = 0$, that is

$$\langle \boldsymbol{\mu}(0) \cdot \boldsymbol{\mu}(t) \rangle = \sum_i {}^0 p_i \sum_j p_{ji}(t) \boldsymbol{\mu}_i \cdot \boldsymbol{\mu}_j \qquad (9.3.8)$$

where $p_{ji}(t)$ is the conditional probability that the dipole is at site j at t provided that it was at site i at $t = 0$. Thus, p_{j1} is obtained from Eq. (9.3.6) for $p_1(0) = 1, p_2(0) = 0$. Analogously $p_{j2}(t)$ follows when $p_1(0) = 0, p_2(0) = 1$, that is

$$p_{11}(t) = \tfrac{1}{2}[1 + \exp(-2kt)], \qquad p_{12}(t) = \tfrac{1}{2}[1 - \exp(-2kt)]$$
$$p_{21}(t) = \tfrac{1}{2}[1 - \exp(-2kt)], \qquad p_{22}(t) = \tfrac{1}{2}[1 + \exp(-2kt)] \qquad (9.3.9)$$

By taking into account that $\boldsymbol{\mu}_1 \cdot \boldsymbol{\mu}_1 = \boldsymbol{\mu}_2 \cdot \boldsymbol{\mu}_2 = -\boldsymbol{\mu}_1 \cdot \boldsymbol{\mu}_2 = -\boldsymbol{\mu}_2 \cdot \boldsymbol{\mu}_1 = \mu^2$, Eq. (9.3.8) can be written as

$$\langle \boldsymbol{\mu}(0) \cdot \boldsymbol{\mu}(t) \rangle = \mu^2 \exp(-2kt) \qquad (9.3.10)$$

which corresponds to a single Debye process decaying with a relaxation time $\tau = (2k)^{-1}$.

A more complicated case is analyzed in Problem 1 of the present chapter. Williams and Cook[6] have reviewed other examples. We note that, if the total relaxation strength is under consideration, the average over the conformational space includes all the spatial directions. For this reason, when we consider relaxations in the glassy state, for example relaxations of side groups with respect to a fixed main chain, only very specific conformations are to be considered. In these conditions, the dipole moment of a selected specimen need not average to zero. In such cases, only an effective dipole moment given by

$$\mu_{\text{eff}} = [\langle \mu^2 \rangle - \langle \mu \rangle^2]^{1/2} \qquad (9.3.11)$$

determines the dielectric relaxation.

Recently, Smith and Boyd[7] outlined an entirely equivalent method on the basis of the Fröhlich treatment of the permittivity given in Sections 1.15 and 1.16. According to this, the strength of a secondary relaxation process is given by

$$\varepsilon_0 - \varepsilon_\infty = \frac{3\varepsilon_0}{2\varepsilon_0 + \varepsilon_\infty} \left(\frac{\varepsilon_\infty + 2}{3} \right)^2 \frac{4\pi N_1 g \mu_v^2}{9kT} \qquad (9.3.12)$$

where N_1 is the number of dipoles per unit volume, μ_v is the vacuum dipole moment associated with a repeating unit, and g is the dipolar correlation factor defined as

$$g = 1 + \frac{1}{n} \sum_{i=1}^{n} \sum_{j \neq 1}^{n} \langle \cos \theta_{ij} \rangle = \frac{\langle m^2 \rangle}{n \mu_v^2} \tag{9.3.13}$$

where m is the dipole moment of the molecule [see also Eq. (8.13.3)]. Then, the factor $N_1 g \mu_v^2$ in Eq. (9.3.12) can be written as

$$N_1 g \mu_v^2 = N \langle \mathbf{m}_k \cdot \mathbf{M}^* \rangle \tag{9.3.14}$$

where $\mathbf{M}^*$ is, according to the Fröhlich methodology, the dipole moment of a cavity large enough to consider statistical averages. The left-hand side of Eq. (9.3.14) is associated with the dipolar reorientation, and the right-hand side is the average of $\mathbf{m}_k \cdot \mathbf{M}^*$ over all the configurations of $\mathbf{M}^*$, $\mathbf{m}_k$ being the dipole moment of a specific configuration and $\mathbf{M}^*$ the statistical average of $\mathbf{M}$ with $\mathbf{m}_k$ fixed. Then, by decomposing $\mathbf{M}$ in intra- and intermolecular terms, and neglecting intermolecular interactions, Smith and Boyd obtain

$$N_1 g \mu_v^2 = N \sum_i (\langle \mu_i^2 \rangle - \langle \boldsymbol{\mu}_i \rangle^2) \tag{9.3.15}$$

as in Eq. (9.3.11). In this equation, N is the number of molecular chains. The values between brackets in Eq. (9.3.15) can be obtained from rotational isomeric state statistics. By comparing the values of g thus calculated with those obtained from dielectric results, an estimation of the probability of dipoles associated with different rotational states in the glassy state can be made if the intermolecular interactions are assumed to be negligible (see Problem 3).

9.4. COALESCENT αβ-PROCESS

As mentioned in the preceding chapter, mechanical as well as dielectric relaxation spectra in glass-forming liquids exhibit in the frequency domain, at $T > Tg$, a prominent *rapid* α-relaxation followed by a *slow* β-process. The fact that both relaxations tend to overlap at high temperatures, giving rise to the α + β-process, suggests a "conservation" rule for the relaxation strength, provided that the molecular motions causing these relaxation processes were coupled. For well separated α- and β-processes in the time domain, one can write

$$\psi(t) = a_\alpha \psi_\alpha(t) + a_\beta \psi_\beta(t) \tag{9.4.1}$$

where $a_\alpha + a_\beta = 1$ and ψ_α and ψ_β the respective normalized dipole correlation function for the α- and β-processes. Equation (9.4.1) can be understood as the result of considering a model assuming a partial reorientation controlled by a barrier provided by adjacent molecules, followed by a relaxation of the environment, thus completing the dipolar reorientation. In this way, $\mu^2 - \langle\mu\rangle^2$ is relaxed by the barrier and $\langle\mu\rangle^2$ is subsequently relaxed by the coordinate reorientation. The model states that α- and β-processes are coupled and the α-process completes the dipolar reorientation of the β-process, and consequently

$$\psi(t) = \frac{\langle\mu\rangle^2}{\mu^2} + \left(1 - \frac{\langle\mu\rangle^2}{\mu^2}\right)\psi_\beta(t) \tag{9.4.2}$$

which corresponds to a low-frequency α-process, whose frequency dependence is given by $L[-\dot{\psi}(t)]$ and a higher-frequency β-process whose frequency dependence is controlled by $L[\dot{\psi}(t)]$, L being the Laplace transformation.

9.5. RELAXATION OF N SEGMENTS IN A CHAIN: DYNAMIC ROTATIONAL ISOMERIC STATE MODEL (DRIS)

Long-wave length modes only depend on the chain length, and in their description conformational details can be ignored. However, local modes are strongly dependent on the fine structure or conformational characteristics of the chains. Some models assume local motions as produced by some type of crankshaft mechanisms that do not perturb the chain tails adjacent to the mobile segments.[8–11] Other models assume that conformational transitions of a single bond, followed by compensating rearrangements of the neighboring units, are predominantly responsible for local motions.[12–16] In what follows, the mathematical model of Jernigan, which is essentially the application of the rotational isomeric state model to chain dynamics,[17] will be described. The model, used and improved by others,[18–21] can, in principle, be used to investigate local motions in molecular chains. A portion of segments undergoing a relaxation process is presented by continuous lines in Fig. 9.4. A coordinate reference frame, embedded in the sequence, is defined with the x axis affixed to the segment at which the sequence begins, and the y axis is in the plane defined by this segment and the previous one. The end-to-end vector of the sequence is defined by $\mathbf{r}$, and the position of the xyz coordinate axis with respect to the laboratory fixed coordinate system is denoted by $\mathbf{R}$. The orientational relaxation of a vector $\mathbf{m}$ is described by a time correlation function $\Phi(t)$ which can be viewed as the product of the internal relaxation resulting from conformational transitions in the AB bonds of the sequence, formulated with respect to the xyz axis, and the external orientational relaxation of the xyz system with respect to the fixed laboratory reference frame XYZ, as will be discussed below.

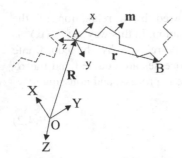

FIG. 9.4 Scheme of a polymeric chain with a unit vector affixed to the central bond of the bond sequence whose end-to-end distance is **r**. Dashed portions represent the chain tails flanking the relaxing sequence. The mobile reference frame $Axyz$ is the embedded chain, and $OXYZ$ is the laboratory fixed coordinate system

9.6. PROBABILITY OF ROTATIONAL STATES AT EQUILIBRIUM

According to the rotational isomeric state model, each skeletal bond has accessibility to a discrete number of rotational states.[22,23] For a sequence of n skeletal bonds, a set of rotational angles $(\phi_1, \phi_2, \ldots, \phi_n)$ specifies a conformation $\{\phi_k\}$, $k = \nu^n$, where ν is the number of rotational states associated with each skeletal bond. The conformational states of each bond are correlated with those of its first neighbors in such a way that, if the rotational states for each skeletal bond are *trans* (t) and *gauche* ($g^{\pm}$), the statistical weights associated with the conformations of a pair of bonds i and $i - 1$ are

$$\mathbf{U}_i = \begin{pmatrix} u_{tt} & u_{tg^+} & u_{tg^-} \\ u_{g^+t} & u_{g^+g^+} & u_{g^+g^-} \\ u_{g^-t} & u_{g^-g^+} & u_{g^-g^-} \end{pmatrix}_i \tag{9.6.1}$$

where $u_{\eta\gamma}$; $i = \exp(-E_{\eta\gamma,i}/kT)$ is the statistical weight or Boltzmann factor for a conformation of energy $E_{\eta\gamma}$ in which the bonds i and $i - 1$ are in the rotational states γ and η respectively. The rotational partition function for a chain of n skeletal bonds is given by

$$Z = \mathbf{U}_1 \left[\prod_{i=2}^{i=n-1} \mathbf{U}_i \right] \mathbf{U}_n \tag{9.6.2}$$

where $\mathbf{U}_1 = \text{row}(100)$ and $\mathbf{U}_n = \text{column}(111)$ convert the matricial product in the scalar Z. The probability of occurrence of the state η in the bond j can be

written as

$$p_{\eta;j} = Z^{-1} U_1 \left(\prod_{i=2}^{j-1} U_i\right) U_{\eta;j} \left(\prod_{l=j+1}^{n-1} U_l\right) U_n = \frac{Z_{\eta;j}}{Z} \tag{9.6.3}$$

where all the statistical weights of the matrix U_j, except those corresponding to the column of the state η, are zero.

Let us now analyze the effect of the local electric field F on the probability of the conformational states. The probability that, under the field, bond i is in rotational state η is given by

$$p_{\eta;i;F} = \frac{Z_{\eta;i;F}}{Z_F} \tag{9.6.4}$$

By analogy with Eq. (5.4.13)

$$Z_{\eta;i;F} = Z_{\eta;i}\left[1 + \frac{1}{6}\left(\frac{F}{kT}\right)^2 \langle \mu_{\eta;i}^2 \rangle + \frac{1}{216}\left(\frac{F}{kT}\right)^4 \langle \mu_\eta^4 \rangle + \cdots \right] \tag{9.6.5}$$

and

$$Z_F = Z\left[1 + \frac{1}{6}\left(\frac{F}{kT}\right)^2 \langle \mu^2 \rangle + \frac{1}{216}\left(\frac{F}{kT}\right)^4 \langle \mu^4 \rangle + \cdots \right] \tag{9.6.6}$$

Hence

$$p_{\eta;i;F} = p_{\eta;i}\left[1 + \left(\frac{1}{6}\right)\left(\langle \mu_{\eta;i}^2 \rangle - \langle \mu^2 \rangle\right)\left(\frac{F}{kT}\right)^2 + \cdots \right] \tag{9.6.7}$$

The effect of the electric field on the probability of rotational states depends on the factor $\langle \mu_{\eta;i}^2 \rangle - \langle \mu^2 \rangle$. If $\langle \mu_\eta^2 \rangle$ is larger than $\langle \mu^2 \rangle$, then $p_{\eta;F}$ will be higher than $p_{\zeta\eta}$, whereas it will be smaller if $\langle \mu_{\zeta\eta}^2 \rangle$ is lower than $\langle \mu^2 \rangle$. In most cases, $F/kT \ll 1$ and $p_{\eta;i;F} \cong p_{\eta;i}$.

9.7. CONFORMATIONAL TRANSITION RATES

An isomer oscillates in a potential well until the thermal energy kT is high enough to overcome the barrier energy separating this isomer from others. The isomer crosses the barrier and a conformational transition occurs. According to the theory of absolute reaction rates, the conformational transition rate of a backbone segment may be considered the product of a front factor that accounts for the frictional drag exerted by the medium and another factor that reflects the barrier energy that must be crossed in the conformational transition. In strict terms, conformational transition rates are given by[24]

$$r = \frac{kT}{h} \frac{F^*}{F} \exp\left(-\frac{E_a}{RT}\right) \tag{9.7.1}$$

where E_a is the barrier energy, F^*/F is the ratio of the partition function of the molecule in the activated state to that in the initial state, and kT/h represents the frequency with which the activated complex crosses the barrier energy at a given temperature. Kramer's equation[25] for the transition rate in a low-friction medium can be written as

$$r = \frac{1}{2\pi} \left(\frac{\gamma}{m}\right)^{1/2} \exp\left(-\frac{E_a}{RT}\right) \tag{9.7.2}$$

whereas in a high-friction medium this expression is given by[14,15]

$$r = (\gamma\gamma^*)^{1/2}(2\pi\xi)^{-1} \exp\left(-\frac{E_a}{RT}\right) \tag{9.7.3}$$

In these expressions, γ and γ^* represent the force constant in the initial and activated states respectively, m is the mass of the segment undergoing the transition, and ξ is the friction coefficient.

9.8. INDEPENDENT CONFORMATIONAL TRANSITIONS

Let us consider that a mobile skeletal bond of a symmetric chain undergoes independent conformational transitions[17]

$$g^- \underset{r_1}{\overset{r_2}{\rightleftarrows}} t \underset{r_2}{\overset{r_1}{\rightleftarrows}} g^+$$

where t, g^+ and g^- represent the *trans*, *gauche positive*, and *gauche negative* rotational states, and r_1 and r_2 are the conformational transition rates. By representing by $\mathbf{p}_j(t) = \{p_{t;j}(t), p_{g^+;j}(t), p_{g^-;j}(t)\}$ the probabilities of the conformations of bond j at time t, the time rate of $\mathbf{p}_j(t)$ obeys the following equation

$$\frac{d\mathbf{p}_j(t)}{dt} = \mathbf{A}_j \mathbf{p}_j(t) \tag{9.8.1}$$

where $\mathbf{A}$ is a 3×3 matrix given by

$$\mathbf{A}_j(t) = \begin{pmatrix} -2r_1 & r_2 & r_2 \\ r_1 & -r_2 & 0 \\ r_1 & 0 & -r_2 \end{pmatrix} \tag{9.8.2}$$

Note that the bond is permitted to turn away from the *trans* state in two directions, and hence the first term of the matrix is $-2r_1$. Each element of A_{kl} of $\mathbf{A}$ describes the momentary rate passage from conformation k to l. The solution of the first-order differential equation is

$$\mathbf{p}_j(t) = \mathbf{p}_j(t = 0) \exp(\mathbf{A}_j t) \tag{9.8.3}$$

where $\mathbf{p}_j(t = 0)$ represents the probability of the rotational states (t, g^+, g^-) of bond j at equilibrium, and its value is given by Eq. (9.6.3). By diagonalizing $\mathbf{A}_j$ using a similarity transformation

$$\mathbf{B}_j^{-1}\mathbf{A}\mathbf{B}_j = \mathbf{\Lambda}_j \tag{9.8.4}$$

where $\mathbf{B}$ is a matrix formed by the eigenvectors of $\mathbf{A}$. Eq. (9.8.13) becomes

$$\mathbf{p}_j(t) = \mathbf{p}_j(t = 0)\mathbf{B}_j \exp(\mathbf{\Lambda}_j t)\mathbf{B}_j^{-1} \tag{9.8.5}$$

The elements of the diagonal of $\mathbf{\Lambda}_j$ are the eigenvalues of $\mathbf{A}$. By assuming that bond rotations about a bond are independent of the rotations of neighboring bonds, the probability vector for a chain of n segments will contain as many terms as conformations, which for segments with n skeletal bonds are 3^{n-2} if each bond is associated with three rotational states. The pertinent expression is

$$\mathbf{p}^{(n)}(t) = \mathbf{p}_2(t) \otimes \mathbf{p}_3(t) \cdots \otimes \mathbf{p}_j(t) \cdots \otimes \mathbf{p}_{n-1}(t) \tag{9.8.6}$$

where the symbol $\otimes$ denotes the direct product. The time derivative of this equation is

$$\begin{aligned}
\frac{d\mathbf{p}^{(n)}(t)}{dt} = {}&\frac{d\mathbf{p}_2(t)}{dt} \otimes \mathbf{p}_3(t) \cdots \otimes \mathbf{p}_j(t) \otimes \cdots \otimes \mathbf{p}_{n-1}(t) + \\
&+ \mathbf{p}_2(t) \otimes \frac{d\mathbf{p}_3(t)}{dt} \cdots \otimes \mathbf{p}_j(t) \otimes \cdots \otimes \mathbf{p}_{n-1}(t) + \cdots \\
&+ \mathbf{p}_2(t) \otimes \mathbf{p}_3(t) \otimes \cdots \otimes \frac{d\mathbf{p}_j(t)}{dt} \cdots \otimes \mathbf{p}_{n-1}(t) + \cdots \\
&+ \mathbf{p}_2(t) \otimes \mathbf{p}_3(t) \otimes \cdots \otimes \mathbf{p}_j(t) \cdots \otimes \frac{d\mathbf{p}_{n-1}(t)}{dt}
\end{aligned} \tag{9.8.7}$$

where it is assumed that all the skeletal bonds undergo conformational transitions at the same time. Since $d\mathbf{p}_j/dt = \mathbf{A}_j\mathbf{p}_i$, Eq. (9.8.1) can be written as

$$
\begin{aligned}
\frac{d\mathbf{p}^{(n)}(t)}{dt} &= \mathbf{A}_2\mathbf{p}_2(t) \otimes \mathbf{p}_3(t) \otimes \cdots \otimes \mathbf{p}_j(t) \cdots \otimes \mathbf{p}_{n-1}(t) \\
&+ \mathbf{p}_2(t) \otimes \mathbf{A}_3\mathbf{p}_3(t) \otimes \cdots \otimes \mathbf{p}_j(t) \otimes \cdots \otimes \mathbf{p}_{n-1}(t) + \cdots \\
&+ \mathbf{p}_2(t) \otimes \mathbf{p}_3(t) \cdots \otimes \mathbf{A}_j\mathbf{p}_j(t) \cdots \otimes \mathbf{p}_{n-1}(t) + \cdots \\
&+ \mathbf{p}_2(t) \otimes \mathbf{p}_3(t) \cdots \otimes \mathbf{p}_j(t) \cdots \otimes \mathbf{A}_{n-1}\mathbf{p}_{n-1}(t)
\end{aligned}
\tag{9.8.8}
$$

By using successively the direct product theorem $(\mathbf{KL}) \otimes (\mathbf{MN}) = (\mathbf{K} \otimes \mathbf{M})(\mathbf{L} \otimes \mathbf{N})$, Eq. (9.8.8) becomes

$$
\begin{aligned}
\frac{d\mathbf{p}^{(n)}(t)}{dt} &= \begin{bmatrix} \mathbf{A}_2 \otimes \mathbf{E}_3 \otimes \ldots \otimes \mathbf{E}_3 \otimes \ldots \otimes \mathbf{E}_3 \otimes \mathbf{E}_3 + \\ \mathbf{E}_3 \otimes \mathbf{A}_3 \otimes \ldots \otimes \mathbf{E}_3 \otimes \ldots \otimes \mathbf{E}_3 \otimes \mathbf{E}_3 + \cdots + \\ \mathbf{E}_3 \otimes \mathbf{E}_3 \otimes \ldots \otimes \mathbf{A}_j \otimes \ldots \otimes \mathbf{E}_3 \otimes \mathbf{E}_3 + \cdots + \\ \mathbf{E}_3 \otimes \mathbf{E}_3 \otimes \ldots \otimes \mathbf{E}_3 \otimes \ldots \otimes \mathbf{A}_{n-1} \end{bmatrix} \mathbf{p}^{(n)}(t) \\
&= \left(\sum_{j=2}^{n-1} \mathbf{E}_3 \otimes \mathbf{E}_3 \otimes \ldots \otimes \mathbf{A}_{j,} \otimes \ldots \otimes \mathbf{E}_3 \otimes \mathbf{E}_3 \right) \mathbf{p}^{(n)}(t) \\
&= \mathcal{A}^{(n)}\mathbf{p}^{(n)}(t)
\end{aligned}
\tag{9.8.9}
$$

where $\mathbf{E}_3$ is the identity matrix of order 3×3 and $\mathcal{A}^{(n)}$ is a matrix of order $3(n-2) \times 3(n-2)$. Integration of this equation gives

$$
\mathbf{p}^{(n)}(t) = \mathbf{B}\exp(\mathbf{\Lambda}t)\mathbf{B}^{-1}\mathbf{p}^{(n)}(t=0) = \mathbf{C}^{(n)}\mathbf{p}^{(n)}(t=0)
\tag{9.8.10}
$$

where $\mathbf{\Lambda}(t)$ is a diagonal matrix whose elements are the eigenvalues of matrix $\mathbf{A}$ defined by Eq. (9.8.9), $\mathbf{B}$ is the matrix with the eigenvectors of $\mathcal{A}^{(n)}$, and $\mathbf{C}^{(n)}$ is the conditional probabilities matrix in which the element C_{kl} denotes the probability of the occurrence of conformation $\{\phi\}_k$ at time t assuming that the conformation was $\{\phi\}_l$ at $t = 0$. The vector $\mathbf{P}^{(n)}(t = 0)$ contains the probabilities of the conformations at equilibrium (or time zero) of a flexible segment of n skeletal bonds. For example, $\mathbf{P}^{(n)}(t = 0) = \{p_{ttt\cdots tt}, \ldots, p_{tt\cdots tg^+}, \cdots p_{tt\cdots tg^-}, \cdots$

$p_{ttt\cdots g^+t}, \cdots p_{g^-g^-\cdots g^-g^-}\}$ has 3^n terms. Practical computations using this approach are restricted to $n \leqslant 7$.

The total joint probability matrix, $\wp^{(n)}$, is given by

$$\wp^{(n)}(t) = \mathbf{C}^{(n)}\mathrm{diag}[\mathbf{p}^{(n)}(t = 0)] \tag{9.8.11}$$

where $\wp^{(n)}$ is a $3n \times 3n$ matrix whose $\wp_{kl}^{(n)}$ element denotes the joint probability of occurrence of the conformation $\{\phi\}_k$ at time t and $\{\phi\}_l$ at $t = 0$.

The normalized time–dipole autocorrelation coefficient is given by

$$\Phi_{\mathrm{int}} = \frac{\mathbf{M}^{\mathrm{T}}(\wp^{(n)} \otimes \mathbf{E}_3)\mathbf{M}}{\langle \mu_0^2 \rangle} \tag{9.8.12}$$

where $\mathbf{M}$ is the dipole moment containing the dipole moments of all the 3^n conformations, specifically, $\mathbf{M} = \{\mathbf{m}_{ttt\cdots tt}, \mathbf{m}_{tt\cdots tg^+}, \mathbf{m}_{tt\cdots tg^-}, \mathbf{m}_{ttt\cdots g^+t}, \cdots \mathbf{m}_{g^-g^-\cdots g^-g^-}\}$, $\mathbf{M}^{\mathrm{T}}$ is its transpose, and $\langle \mu_0^2 \rangle$ is the mean-square dipole moment.

9.9. PAIRWISE DEPENDENT CONFORMATIONAL TRANSITIONS

The treatment for independent rotations discussed above can easily be extended to pairwise conformational transitions. The pertinent scheme is as follows.[18]

In this case, the transition rates about skeletal bond j are given by

$$
\begin{pmatrix}
dp_{tt}(t)/dt \\
dp_{tg^+}(t)/dt \\
dp_{tg^-}(t)/dt \\
dp_{g^+t}(t)/dt \\
dp_{g^+g^+}(t)/dt \\
dp_{g^+g^-}(t)/dt \\
dp_{g^-t}(t)/dt \\
dp_{g^-g^+}(t)/dt \\
dp_{g^-g^-}(t)/dt
\end{pmatrix}_j
=
\begin{pmatrix}
-4r_1 & r_{-1} & r_{-1} & r_{-1} & 0 & 0 & r_{-1} & 0 & 0 \\
r_1 & R & 0 & 0 & r_2 & 0 & 0 & r_{-3} & 0 \\
r_1 & 0 & R & 0 & 0 & r_{-3} & 0 & 0 & r_{-2} \\
r_1 & 0 & 0 & R & r_{-2} & r_{-3} & 0 & 0 & 0 \\
0 & r_2 & 0 & r_2 & -2r_2 & 0 & 0 & 0 & 0 \\
0 & 0 & r_3 & r_3 & 0 & 2r_{-3} & 0 & 0 & 0 \\
r_1 & 0 & 0 & 0 & 0 & 0 & R & r_{-3} & r_{-2} \\
0 & r_3 & 0 & 0 & 0 & 0 & r_3 & 2r_{-3} & 0 \\
0 & 0 & r_2 & 0 & 0 & 0 & r_2 & 0 & -2r_{-2}
\end{pmatrix}_j
\begin{pmatrix}
p_{tt}(0) \\
p_{tg^+}(0) \\
p_{tg^-}(0) \\
p_{g^+t}(0) \\
p_{g^+g^+}(0) \\
p_{g^+g^-}(0) \\
p_{g^-t}(0) \\
p_{g^-g^+}(0) \\
p_{g^-g^-}(0)
\end{pmatrix}_j
\tag{9.9.1}
$$

where $R = -r_2 - r_3 - r_{-1}$. Note that the elements of the 9×9 matrix $\mathbf{A}_{j,j+1}$ denote the transition rates from one pair of isomeric states to others. For example, rows and columns are in the order tt, tg^+, tg^-, g^+t, g^+g^+, g^+g^-, g^-t, g^-g^+, g^-g^-. If simultaneous transitions of bonds j and $j + 1$ are considered, Eq. (9.8.9) becomes

$$
\frac{d\mathbf{p}^{(n)}(t)}{dt} = \left(\sum_{j=1}^{n} \mathbf{E}_3 \otimes \mathbf{E}_3 \otimes \cdots \otimes \mathbf{A}_{j,j+1} \otimes \cdots \otimes \mathbf{E}_3 \otimes \mathbf{E}_3 \right) \mathbf{p}^{(n)}(t) \tag{9.9.2}
$$

This equation holds for sequences of bonds with independent statistics but pairwise dependent dynamics.

Let us consider now a sequence with pairwise cooperative transitions, for example, the following transition

$$
p(\alpha'\beta'\gamma'\delta' \ldots; 0) \to p(\alpha\beta\gamma\delta \ldots; t)
$$

The joint probability is given by[18]

$$p(\alpha\beta\gamma\delta\ldots, t; \alpha'\beta'\gamma'\delta'\ldots, 0)$$
$$= p(\alpha\beta, t; \alpha'\beta', 0)p'(\beta\gamma, t; \beta'\gamma', 0)p'(\gamma\delta, t; \gamma'\delta', 0)\ldots \qquad (9.9.3)$$

where

$$p'(\beta\gamma, t; \beta'\gamma', 0) = \frac{p(\beta\gamma, t; \beta'\gamma', 0)}{\sum_{s'}\sum_{\eta} p(\beta\eta, t; \beta's', 0)} \qquad (9.9.4)$$

It should be pointed out that the summation in Eq. (9.9.4) is performed over all rotational states in such a way that, if the number of rotational states is 3 ($s = t, g^+, g^-; \eta = t, g^+, g^-$) the summation includes nine terms. Therefore, $p'(\beta\gamma, t; \beta'\gamma' = 0)$ represents the conditional probability of occurrence of the joint event ($\beta\gamma, t; \beta'\gamma' = 0$), since the first bond of the pair undergoes the transition from the initial state β' to state β at time t. By applying Eqs (9.9.3) and (9.9.4) to each of the possible $3^n \times 3^n$ transitions for a chain with three rotational states, the time-delayed joint probability matrix $\wp^{(n)}$ is obtained. If conformational transitions occurring in each bond are assumed to be independent of neighboring bonds, Eq. (9.9.3) becomes

$$p(\alpha\beta\gamma\delta\ldots, t; \alpha'\beta'\gamma'\delta'\ldots, 0)$$
$$= p(\alpha, t; \alpha', 0)p(\beta, t; \beta', 0)p(\gamma, t; \gamma', 0)p(\delta, t; \delta', 0)\ldots \qquad (9.9.5)$$

where $p(\alpha, t; \alpha', 0)$, $p(\beta, t; \beta', 0)$, etc., are elements of the matrix $\wp^{(1)}$.

It should be noted that, in the absence of constraints operating on the ends of the sequence, the fluctuation of $\Delta\mathbf{r} = \mathbf{r}(t) - \mathbf{r}(0)$, where $\mathbf{r}(t)$ and $\mathbf{r}(0)$ are the end-to-end vectors of the sequence, will take any value compatible with conformational transitions available to the sequence. However, the motions of the sequence are constrained by the tails of the chains attached to its two ends, and as a result $\Delta\mathbf{r}$ will be confined to within a spherical domain defined by $\Delta\mathbf{r} < \delta$, the value of this quantity varying inversely with the strength of the constraints. Confinement of $\Delta\mathbf{r}$ into a sphere excludes certain conformations. If the transition from conformation $\{\phi\}_k$ at time $t = 0$ to $\{\phi\}_l$ at $t = t$ renders $\Delta\mathbf{r} > \delta$, the element $\wp_{kl}^{(n)}$ of $\wp^{(n)}(t)$ must be replaced by zero. Any other constraint, for example, that the two extreme bonds of the sequence are nearly parallel after the conformational transition, could be introduced.

9.10. TIME–DIPOLE AUTOCORRELATION COEFFICIENT

The orientational relaxation in a sequence involves two independent processes: (a) an internal relaxation due to conformational transitions in the sequence formulated with respect to a referred frame affixed to the first segment of the sequence, and (b) relaxation of this reference frame with respect to the laboratory frame referred to as the external relaxation. Therefore, the time-dependent autocorrelation function can be written as

$$\Phi(t) = \Phi_{int}\Phi_{ext} \tag{9.10.1}$$

where[17]

$$\Phi_{ext} = \exp\left(-\frac{t}{\tau_{rot}}\right) \tag{9.10.2}$$

In this expression, τ_{rot} is the rotational relaxation time of a rigid body with dimensions equivalent to the average dimensions of the relaxing segments, and $\Phi_{int}(t)$ is given by Eq. (9.8.12).

Analysis of the internal relaxation of N-bond sequences in flexible symmetric chains with conformational characteristics similar to polyethylene shows that the relaxation time for internal units decreases with increasing sequence length. Longer sequences have accessibility to a larger number of conformations through which relaxation can more easily occur. On the other hand, environmental resistance due to both friction and size of the relaxation units tends to slow down the motion of the unit under consideration, and this effect becomes increasingly strong as the size of the motional unit increases.

9.11. RELAXATION TIMES

An inspection of Eqs (9.8.3), (9.8.5), and (9.8.12) suggests that the internal time–dipole autocorrelation function can be written as

$$\Phi_{int}(t) = \sum_j k_j \exp(\lambda_j t) \tag{9.11.1}$$

where λ_j is the jth eigenvalue of the master matrix $\mathbf{A}$, and k_j represents the a priori probability of relaxation through internal mode j. The reciprocal of λ_j with negative sign is the relaxation time associated with mode j. It should be stressed that the number of modes is equal to the number of conformations intervening in the relaxation process. The parameter k_j assumes different forms depending on the type of correlation analyzed. By considering $\Phi(t) = \Phi_{int}(t)\Phi_{ext}(t)$, and assuming that Φ_{ext} is described by Eq. (9.10.2), the time–dipole autocorrelation coefficient is given by

$$\Phi(t) = \sum_j k_j \exp[(\lambda_j - 1/\tau_r)t] = \sum_j k_j \exp(\lambda_j' t) \qquad (9.11.2)$$

where $\lambda_j' = \lambda_j - 1/\tau_r$. The mean relaxation time can be obtained from the integral of the normalized time–dipole to yield

$$\langle \tau \rangle = \int_0^\infty \Phi(t)\,dt = \sum_j \int_0^\infty k_j \exp(\lambda_j t) = -\sum_j \frac{k_j}{\lambda_j} \qquad (9.11.3)$$

Here, use was made of the fact that $\Phi(\infty) = 0$ in relaxation processes. The complex dielectric permittivity can be obtained by combining Eqs (9.11.2) and (2.12.7) to give

$$\frac{\varepsilon^*(\omega) - \varepsilon_u}{\varepsilon_r - \varepsilon_u} = L\left[-\frac{d\Phi(t)}{dt}\right] = -\sum_j \int_0^\infty k_j \lambda_j' \exp(\lambda_j' t - i\omega t)\,dt \qquad (9.11.4)$$

where ε_r and ε_u are respectively the relaxed and unrelaxed dielectric permittivities. Solution of Eq. (9.11.4) gives

$$\frac{\varepsilon^*(\omega) - \varepsilon_u}{\varepsilon_r - \varepsilon_u} = \sum_j \frac{k_j \lambda_j'}{\lambda_j' - i\omega} \qquad (9.11.5)$$

From this expression the real and loss components of the complex dielectric permittivity are obtained as

$$\frac{\varepsilon' - \varepsilon_u}{\varepsilon_r - \varepsilon_u} = \sum_j \frac{k_j}{1 + \omega^2 \tau_j^2}$$

$$\varepsilon'' = \sum_j \frac{k_j \tau_j \omega}{1 + \omega^2 \tau_j^2} \qquad (9.11.6)$$

where $\tau_j = -1/\lambda_j'$.

9.12. MOTION OF A SINGLE BOND

Figure 9.5 shows a contour S between points A and B of long flexible chains which undergoes an instantaneous jump from the conformation drawn with a solid line to that depicted by a dashed line. The parts of the chain outside the S are called the tails. Let us consider the orientation of the probe $\mathbf{m}_j$ located at the jth bond. It is assumed that, at a given time, a single rotation takes place, and this

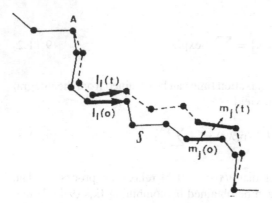

FIG. 9.5   Scheme of the change in the conformation of the contour from the solid line to the dashed line by a rotameric transition in the ith bond. The vectors for the ith bond are $\mathbf{l}_i(0)$ and $\mathbf{l}_r(t)$ at time 0 and t respectively. The same occurs for the vector $\mathbf{m}_j$

transition affects the position and orientation of $\mathbf{m}_j$. The bond $\mathbf{l}_i$ will be both displaced and oriented during the rotational transition, as required for accommodating the transition over the S without appreciably moving the tails. The orientation of $\mathbf{m}_j$ will be the resultant of an internal motion associated with the internal transition of the rotating bond $\mathbf{l}_i$ and the accompanying compensating motions. The extent of the orientation of $\mathbf{m}_j$ resulting from the internal motion depends on the number and states of the bonds between $\mathbf{l}_i$ and $\mathbf{m}_j$.

After rotation, bond $\mathbf{l}_i$ is displaced and reoriented with respect to a laboratory reference frame XYZ. The bond $\mathbf{l}_i$ is assumed to coincide with the x axis of a bond-based coordinate system xyz which moves with $\mathbf{l}_i$. The axes of the molecule fixed coordinate system at time 0 and t are, respectively, $x(0)y(0)z(0)$ and $x(t)y(t)z(t)$. In what follows, subscript j will be omitted in $\mathbf{m}_j$. The probe at time 0 and t may be represented in the laboratory reference frame and the molecule fixed reference frame as[20]

$$\mathbf{m}(0) = \mu_\alpha(0)\mathbf{e}_\alpha(0) = m_i(0)\mathbf{e}_i$$
$$\mathbf{m}(t) = \mu_\beta(t)\mathbf{e}_\beta(t) = m_i(0)\mathbf{e}_i$$

(9.12.1)

where $\mathbf{e}_i$ represent the base vectors of the XYZ coordinate system, and $\mathbf{e}_\alpha$ and $\mathbf{e}_\beta$ are the base vectors along the coordinates $x(0)y(0)z(0)$ and $x(t)y(t)z(t)$ respectively. Moreover, μ_α and m_i are the components of $\mathbf{m}$ with respect to xyz and XYZ respectively. As usual, summation is assumed over repeated indices. The time–dipole autocorrelation coefficient as observed from the laboratory reference frame is defined as

$$\Phi_{ext} = \langle \mathbf{m}(0) \cdot \mathbf{m}(t) \rangle = \langle \mu_\alpha(0)\mu_\beta(t)\mathbf{e}_\alpha(0) \cdot \mathbf{e}_\beta(t) \rangle$$

$$= \langle M_{\alpha\beta}(t)T_{\alpha\beta}(t) \rangle \qquad (9.12.2)$$

where $M_{\alpha\beta}$ and $T_{\alpha\beta}$ are matrices defined as

$$M_{\alpha\beta}(t) = \mu_\alpha(0)\mu_\beta(t); \quad T_{\alpha\beta}(t) = \mathbf{e}_\alpha(0) \cdot \mathbf{e}_\beta(t) \qquad (9.12.3)$$

Here, $M_{\alpha\beta}$ is obtained as the product of components of $\mathbf{m}$ relative to the molecule fixed reference frame, whereas $T_{\alpha\beta}$ represents the transformation of the reference frame from $x(0)y(0)z(0)$ to $x(t)y(t)z(t)$ and hence reflects the effect of the compensating effect on $\Phi(t)$. The isotropic reorientation and displacement of $\mathbf{l}_i$ accompanying its torsion will be more or less random. Therefore, the autocorrelation function can be written as

$$\Phi_{ext} = \langle M_{\alpha\beta}(t) \rangle \langle T_{\alpha\beta}(t) \rangle \qquad (9.12.4)$$

For small rotations, the off-diagonal terms of $\langle T_{\alpha\beta}(t) \rangle$ are negligible with respect to the diagonal elements, which in turn have similar values owing to the isotropic compensation motion. Denoting the latter by $\langle T_{11}(t) \rangle$, Eq. (9.12.4) becomes

$$\Phi_{ext}(t) = \langle T_{11}(t) \rangle \langle M_{11}(t) + M_{22}(t) + M_{33}(t) \rangle = \langle T_{11}(t) \rangle \phi_{int}(t) \qquad (9.12.5)$$

where $\phi_{int}(t)$ is the time–dipole autocorrelation function as observed from the molecule fixed coordinate system; $\langle T_{11}(t) \rangle$ is expected to decay as a single exponential.

The conformational transition from the initial conformation $\{\phi\}_j$ to the final conformation $\{\phi\}_k$ arising from motions about bond i is described by the transition probability matrix.

Transitions between molecular configurations take place through barrier energies separating rotational states (see Fig. 9.6). In fact, the conformational transition rate is given by

$$r_{ij} = A_{ij} = (\gamma\gamma^*)^{1/2}\left(2\pi\xi\sum_{j=i+1}^{n} s_{ij}^2\right)^{-1}\exp\left(-\frac{E_i}{RT}\right) \qquad (9.12.6)$$

where s_{ij} represents the separation of the jth atom in Fig. 9.6 from the axis of rotation that is defined for bond i. Following the usual convention, the bond is the atom connecting bonds $i - 1$ and i. The term $\sum_{j=i+1}^{n} s_{ij}^2$, also called the size effect, is representative of the squared distance swept by the ensemble of atoms accompanying the bond transition about bond i. The computations show that the frequency distribution of internal relaxation modes is not excessively affected by

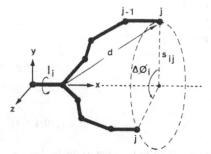

FIG. 9.6 Displacement of the *j*th atom by the effect of a rotation about the *i*th bond[20]

including in the evaluation of transition rates the term $\sum_{j=i+1}^{n} s_{ij}^2$. The main effect of this term is to shift the frequencies/relaxation times to lower/higher values, although a new peak also appears at low frequencies (see Fig. 9.7). Also, it is found that displacements of the ends of the relaxing skeletal bonds are sufficiently small to be easily accommodated by compensating motions of the tails flanking the relaxing segment. Therefore, the restriction of confinement within a small volume to which we referred above is not as severe as conceived.

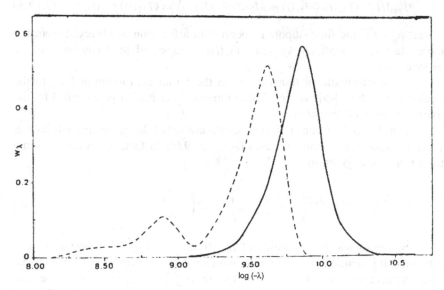

FIG. 9.7 Frequency distribution of the internal rotation modes in a sequence of four mobile bonds in polyethylene at 300 K. The solid and dashed curves are without and with the size effect respectively[20]

9.13. PHENOMENOLOGICAL CLASSIFICATION OF SECONDARY RELAXATION PROCESSES

There are two competing theories concerning the origin of the β-relaxation peak. According to Johari and Goldstein,[26] the appearance of the β-peak does not require any specific molecular motion. The occurrence of the β-peak would be due to the so-called "clusters" or "islands of mobility" present in glass-forming materials. Another school of thought[27] interprets the presence of all the observed secondary losses as the result of some specific molecular motion. We think that these theories are not mutually exclusive; in fact, it is possible to consider specific molecular motions taking place in the above mentioned "islands of mobility". Several cases of molecular motions causing dielectric relaxations are to be considered:

1. In polymers without prominent lateral chains, such as polyvinyl chloride (PVC) or polycarbonates, local main chain motions can give rise to secondary dielectric relaxations.
2. Motions of side groups about the bonds linking them to the main chain, as in the case of poly-*n*-alkyl methacrylates, are probably the best studied examples of the polymers containing dipolar groups in their lateral chains.
3. Internal motions of the side chain groups without cooperation of the main chain are typical of polymers containing flexible units or polar final groups.
4. Another possibility concerns the motions of small molecules, as water, for example, embedded in a polymer matrix.
5. Secondary relaxations in semicrystalline polymers.
6. Secondary relaxations in liquid crystalline polymers.

In any case, the relevant parameters concerning the secondary relaxations are: the relaxation strength, $\Delta\varepsilon$, the frequency of the maximum, and the shape and broadness of the relaxation peak. Unfortunately, in many cases, secondary relaxations mutually overlap or appear as a shoulder of the prominent α-relaxation. The determination of the relaxation strength and the maximum of the peak may be subjected to some degree of uncertainty owing to the errors involved in the deconvolution of the overlapping peaks.

9.13.1. Local Main Chain Motions

Polymers such as polyvinyl chloride (PVC), polycarbonates (PCs), and aromatic polyesters derived from terephthalic acid (PET) and similar polymers or isophthalic (PEIT) acid show secondary peaks which in some cases can be depressed by the effect of additives that increase the modulus and decrease the damping (antiplasticizers). Many authors have reported dielectric relaxation studies on PVC. References[28] and[29] present, respectively, the pioneering work and one of the most recent contributions to the dielectric properties of PVC.

The dielectric behavior of polycarbonates has also been extensively studied, from the 1960s[30,31] to the present day.[32] It is assumed that secondary relaxation in polybisphenol A carbonate is a complex process arising from coupled motions of the phenyl ring and the carbonyl group.[33]

The first detailed study of the dielectric relaxations of polyethylene terephtalate (PET) was carried out by Reddish.[34] More recent results on the dielectric properties of PET have been reported by Coburn and Boyd.[35]

9.13.2. Motions of Side Groups About Their Link to the Main Backbone

In this category are included the β-relaxations of poly(alkyl methacrylates)[36] and poly(itaconates).[37] Experimental data[38] show that the position of the β-relaxation in polyalkyl methacrylates is insensitive to the length of the alkyl group.[39] However, as the temperature of the α-relaxation decreases by the effect of adding successive methylene groups in the lateral chain, the α-relaxation tends to overlap with the β-relaxation (Fig. 9.8), giving rise to the αβ-peak. As mentioned in Chapter 8, different types of scenario have been proposed to explain

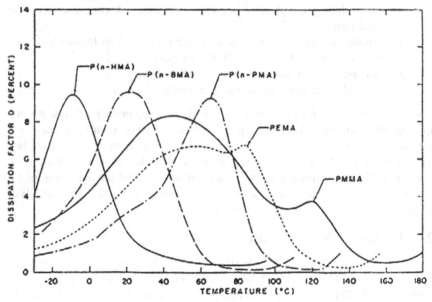

FIG. 9.8 Comparison of the dissipation loss factor at 60 Hz for polyalkyl methacrylates, showing the merging αβ-peak[38]

the $\alpha\beta$-relaxation. Incidentally, the intensity of the β-relaxation in polymethyl methacrylate (PMMA) is higher than that of the α-relaxation, an uncommon fact in dielectric spectroscopy of polymers.

9.13.3. Motions within Side Groups

There are a number of examples corresponding to this category, such as the polymers in which one or more hydrogen atoms of alkyl groups of polyalkyl methacrylates, polyalkylitaconates, etc., are substituted for halogen atoms. This category also includes polymers containing flexible rings as side groups. These polymers may present ostensible β-relaxations.

9.13.4. Motions due to the Presence of Small Molecules in the Polymer Matrix

Low molecular weight compounds not only act as plasticizers depressing the glass transition temperature of polymers but also interact with the motions that cause subglass activity. A typical case is the effect of water on the secondary relaxations of polymers containing hydrophilic groups, such as hydroxylic or amide groups.

9.13.5. Secondary Relaxations in Semicrystalline Polymers

The analysis of the dielectric spectroscopy of semicrystalline polymers is more complicated than that of amorphous polymers.[40,41] The reason is that, in addition to the difficulties concerning the molecular assignation of the observed peaks, the phase in which the relaxations occur must also be elucidated. To start with, the existence of a double α-process in many semicrystalline polymers is well known. The lower-temperature process, called the α_a-peak, is usually related to the cooperative relaxation of the amorphous phase. This fraction is sometimes called "intercrystalline", because it refers to the fringing material existing between lamellar structures. Typical WLF behavior corresponding to an amorphous polymer is expected for this peak. At slightly higher temperatures, another narrower peak appears that is related to some sort of mechanism in which crystalline entities of the material are implied. The dielectric spectrum of PE displays three characteristic relaxational zones conventionally designated as γ, β, and α in order of increasing temperatures.

Semicrystalline halogen-polymers such as polyvinyl fluoride (PVF), polytetrafluorethylene (PTFE), and related polymers also show considerable sub-T_g activity.[42] The low-temperature relaxations in these polymers are understood as "cooperative", local mode, main chain motions.

9.13.6. Dielectric Relaxations in Liquid Crystalline Polymers (LCPs)

The electric properties of liquid crystalline polymers LCP will be analyzed in Chapter 11. They present very specific dielectric behavior. Moreover, in main chain liquid crystalline polymers (MCLCP), the appearance of the Goldstone and soft modes above the glass transition are noticeable facts linked to the presence of ferroelectricity.[43,44] In side chain liquid crystalline polymers (SCLCP) these phenomena do not appear, but the degree of orientation of the lateral nematic or smectic mesophases determines the shape and size of the dielectric loss peaks above the glass transition temperature (see Chapter 11). By using broadband dielectric spectroscopy and by varying the chemical structure of a series of combined main chain, side group, liquid crystalline polymers, it has been possible to assign the observed processes to specific molecular motions.[45]

PROBLEMS

Problem 1

Let us consider a three-site barrier model in which sites 2 and 3 are equivalent but are different in energy from site 1. Calculate the correlation, the decay function, and the dielectric permittivity[48].

Solution

The scheme corresponding to the transition probabilities can be expressed as

$$1 \underset{k_2}{\overset{k_1}{\rightleftharpoons}} 2 \underset{k_3}{\overset{k_3}{\rightleftharpoons}} 3 \underset{k_1}{\overset{k_2}{\rightleftharpoons}} 1$$

The rate equations for the populations of the three sites are

$$\dot{n}_1 = -2k_1 n_1 + k_2 n_2 + k_2 n_3$$
$$\dot{n}_2 = k_1 n_1 - (k_2 + k_3)n_2 + k_3 n_3 \qquad (P.9.1.2)$$
$$\dot{n}_3 = k_1 n_1 + k_3 n_2 - (k_2 + k_3)n_3$$

and, according to this, the transition matrix is given by

$$\mathbf{T} = \begin{vmatrix} -2k_1 & k_2 & k_2 \\ k_1 & -(k_2 + k_3) & k_3 \\ k_1 & k_3 & -(k_2 + k_3) \end{vmatrix} \qquad (P.9.1.3)$$

Then, an orthogonal matrix $\mathbf{Q}$ is found[46,47] on account of the symmetry of the site model under consideration and the assignation of its point group. In this case

$$
\mathbf{Q} = \begin{vmatrix} 1 & 0 & 0 \\ 0 & b & b \\ 0 & b & -b \end{vmatrix}
\tag{P.9.1.4}
$$

where the orthogonality condition requires $2b^2 = 1$. A block matrix $\mathbf{W}$ is obtained according to

$$
\mathbf{Q}^{-1}\mathbf{T}\mathbf{Q} = \mathbf{W} = \begin{vmatrix} -2k_1 & 2bk_2 & 0 \\ 2bk_1 & -k_2 & 0 \\ 0 & 0 & -(k_2 + 2k_3) \end{vmatrix}
\tag{P.9.1.5}
$$

In this way, the group theory reduces a 3×3 matrix problem to 2×2 and 1×1 matrices. The eigenvalues of $\mathbf{W}$ form the matrix $\mathbf{D}$ from the usual expression $\mathbf{U}^{-1}\mathbf{W}\mathbf{U} = \mathbf{D}$, that is

$$
\mathbf{D} = \begin{vmatrix} 0 & 0 & 0 \\ 0 & -(k_2 + 2k_1) & 0 \\ 0 & 0 & -(k_2 + 2k_3) \end{vmatrix}
\tag{P.9.1.6}
$$

and a simple calculation gives

$$
\mathbf{U} = \begin{vmatrix} a & c & 0 \\ \dfrac{ak_1}{bk_2} & -\dfrac{c}{2b} & 0 \\ 0 & 0 & 1 \end{vmatrix}
\tag{P.9.1.7}
$$

where $a^2[1 + 2(k_1/k_2)] = 1$ and $\frac{3}{2}c^2 = 1$, as required by the normalization condition, and

$$
\frac{k_1}{k_2} = \exp{-\frac{E}{kT}}
\tag{P.9.1.8}
$$

where k is the Boltzmann constant and E represents the energy difference between sites 1 and 2 or 1 and 3.

Then, by calculating $\mathbf{Q}\mathbf{U}|\exp \mathbf{D}t|\mathbf{U}^{-1}\mathbf{Q}^{-1}$, and according to the fact that the equilibrium probabilities for sites 1 to 3 are respectively

$$
{}^0p_1 = \frac{1}{1 + 2x}, \qquad {}^0p_2 = \frac{x}{1 + 2x}, \qquad {}^0p_3 = \frac{x}{1 + 2x}
$$

we obtain the matrix for the occupational probabilities

$$\mathbf{p}(t) = \frac{1}{1+2x} \begin{vmatrix} 1 + 2x\psi_1(t) & 1 - \psi_1(t) \\ x[1 - \psi_1(t)] & x + \frac{1}{2}\psi_1(t) + \frac{1}{2}(1 + 2x)\psi_3(t) \\ x[1 - \psi_1(t)] & x + \frac{1}{2}\psi_1(t) - \frac{1}{2}(1 + 2x)\psi_3(t) \end{vmatrix}$$

$$\begin{matrix} 1 - \psi_1(t) \\ x + \frac{1}{2}\psi_1(t) - \frac{1}{2}(1 + 2x)\psi_3(t) \\ x + \frac{1}{2}\psi_1(t) + \frac{1}{2}(1 + 2x)\psi_3(t) \end{matrix}$$

$$(P.9.1.9)$$

where $x = k_1/k_2$, $\psi_1(t) = \exp[-(k_2 + 2k_1)t]$, and $\psi_3(t) = \exp[-(k_2 + 2k_3)t]$.

The time-dependent correlation function can be found from Eq. (9.3.8) on account of the fact that the angle between the dipole direction at sites 1 and 2 or 3 is θ. Hence, the angle between sites 2 and 3 will be $2\pi - 2\theta$, and consequently $\cos(2\pi - 2\theta) = \cos 2\theta$

$$\langle \boldsymbol{\mu}(0) \cdot \boldsymbol{\mu}(t) \rangle$$

$$= \sum_{i=1}^{3} {}^0p_i \sum_{j=1}^{3} p_{ji}\boldsymbol{\mu}_i \cdot \boldsymbol{\mu}_j = \sum_{i=1}^{3} {}^0p_i(p_{1i}\boldsymbol{\mu}_i \cdot \boldsymbol{\mu}_1 + p_{2i}\boldsymbol{\mu}_i \cdot \boldsymbol{\mu}_2 + p_{3i}\boldsymbol{\mu}_i \cdot \boldsymbol{\mu}_3)$$

$$= \frac{\mu^2}{(1+2x)^2} \{[1 + 2x\psi_1(t)] + [x - x\psi_1(t)]\cos\theta + [x - x\psi_1(t)]\cos\theta\}$$

$$+ \frac{\mu^2 x}{(1+2x)^2} \{[1 - \psi_1(t)]\cos\theta + [x + \frac{1}{2}\psi_1(t) + \frac{1}{2}(1 + 2x)\psi_3(t)]$$

$$+ [x + \frac{1}{2}\psi_1(t) - \frac{1}{2}(1 + 2x)\psi_3(t)]\cos 2\theta\} + \frac{\mu^2 x}{(1+2x)^2} \{[1 - \psi_1(t)]\cos\theta$$

$$+ [x + \frac{1}{2}\psi_1(t) - \frac{1}{2}(1 + 2x)\psi_3(t)]\cos 2\theta$$

$$+ [x + \frac{1}{2}\psi_1(t) + \frac{1}{2}(1 + 2x)\psi_3(t)]\}$$

$$= \frac{\mu^2}{(1+2x)^2} [(1 + 2x\cos\theta)^2 + 2x(1 - \cos\theta)^2\psi_1(t)$$

$$+ 2x(1 + 2x)(1 - \cos^2\theta)\psi_3(t)] \tag{P.9.1.10}$$

It should be noted that the correlation does not decay to zero for $t \to \infty$ but to

$$\mu^2 \left(\frac{1 + 2x\cos\theta}{1 + 2x}\right)^2$$

which is equivalent to

$$\langle \mu \rangle^2 = \left[\sum_{i=1}^{3} {}^0 p_i \mu_i \right]^2$$

This part of the dipole moment is in fact not relaxed in secondary processes (see the comment at the end of Section 9.3).

According to the definition given by Eq. (2.12.7), that is

$$\frac{\varepsilon^*(\omega) - \varepsilon_\infty}{\varepsilon_0 - \varepsilon_\infty} = L[-\dot\phi(t)], \qquad \phi(t) = \frac{\langle \mu(0) \cdot \mu(t) \rangle}{\mu^2(0)} \qquad \text{(P.9.1.11)}$$

where the field factor is omitted, the permittivity of the system can be found. The result, omitting the time-independent term, is given by the following equation

$$\frac{\varepsilon^*(\omega) - \varepsilon_\infty}{\varepsilon_0 - \varepsilon_\infty} = \frac{2x(1 - \cos\theta)^2}{(1 + 2x)^2} \frac{1}{1 + i\omega\tau_1} + \frac{2x(1 - \cos^2\theta)}{(1 + 2x)} \frac{1}{1 + i\omega\tau_3}$$

$$\text{(P.9.1.12)}$$

where two relaxation processes, with relaxation times given by $\tau_1 = (2k_1 + k_2)^{-1}$ and $\tau_2 = (k_2 + 2k_3)^{-1}$, are present.

Note that, for $x = 1$, that is, $k_1 = k_2 = k_3$, Eq. (P.9.1.3) reduces to

$$\mathbf{T}' = \begin{vmatrix} -2k & k & k \\ k & -2k & k \\ k & k & -2k \end{vmatrix} \qquad \text{(P.9.1.13)}$$

and in this case $\mathbf{Q} = \mathbf{S}$ and $\mathbf{W} = \mathbf{D}$ to give

$$\mathbf{p}(t) = \frac{1}{3} \begin{vmatrix} 1 + 2\psi(t) & 1 - \psi(t) & 1 - \psi(t) \\ 1 - \psi(t) & 1 + 2\psi(t) & 1 - \psi(t) \\ 1 - \psi(t) & 1 - \psi(t) & 1 + 2\psi(t) \end{vmatrix} \qquad \text{(P.9.1.14)}$$

where $\psi(t) = \exp(-3kt)$.

Problem 2

Let us consider a nonsymmetric two-site barrier model. Calculate the normalized dipolar correlation function according to Eq. (9.4.2).

Solution

According to the Boltzmann distribution, the probabilities for sites 1 and 2 are respectively

$$p_1(x) = \frac{1}{1+x}, \qquad p_2(x) = \frac{x}{1+x} \tag{P.9.2.1}$$

where $x = \exp(-U/kT)$ and

$$\langle \mu \rangle = p_1(x)\mu_1 + p_2(x)\mu_2 \tag{P.9.2.2}$$

From Eqs (P.9.2.1) and (P.9.2.2) one obtains

$$\langle \mu \rangle^2 = \frac{1}{(1+x)^2}\mu_1^2 + \frac{x^2}{(1+x)^2}\mu_2^2 + \frac{2x}{(1+x)^2}\mu_1 \cdot \mu_2 \tag{P.9.2.3}$$

If the angle θ between μ_1 and μ_2 is different from $180°$, then

$$\langle \mu \rangle^2 = \mu^2 [1 - \frac{2x}{(1+x)^2}(1 - \cos \theta)]$$

$$\mu^2 - \langle \mu \rangle^2 = \mu^2 \frac{2x}{(1+x)^2}(1 - \cos \theta)$$

$$\tag{P.9.2.4}$$

and the normalized dipole moment correlation function is given by

$$\psi(t) = 1 - \frac{2x}{(1+x)^2}(1 - \cos \theta)[1 - \psi_\beta(t)] \tag{P.9.2.5}$$

Problem 3

Experimental data concerning the β-relaxation of poly-o-biphenyl acrylate (PBFA)[49] show that at $-60°C$ $\varepsilon_{0\beta} = 2.92$, $\varepsilon_{\infty\beta} = 2.80$, the index of refraction $n = 1.6186$, and density $\rho = 1.22$ g/cm^3.

It is assumed, as usual in methacrylates, that each bond in the main chain is restricted to two rotational states t and g. Each one is split into two others, which corresponds to angles χ around the C^α—C^* bond of the lateral group in which the =C=O bond is in *cis* ($\chi = 0°$) or *trans* ($\chi = 180°$) with the methine bond. In our case, the dipolar moment of the lateral group is 1.72 D, forming an angle of $123°$ with the C^α—C^* bond. The components of the dipolar moment in the reference configurations are

$$\mu_{1x} = 1.018, \qquad \mu_{1y} = -1.387, \qquad \mu_{1z} = -0.024 \qquad \text{for } \chi = 0$$

$$\mu_{2x} = -0.285, \qquad \mu_{2y} = 0.457, \qquad \mu_{2z} = 1.634, \qquad \text{for } \chi = 180°$$

By considering a two-site model to represent the relaxation as in the preceding problem, make an estimation of the percentage of groups intervening in this relaxational process.

Solution

According to Eq. (9.3.12), we can write

$$\varepsilon_0 - \varepsilon_\infty = \frac{3\varepsilon_0}{2\varepsilon_0 + \varepsilon_\infty} \left(\frac{\varepsilon_\infty + 2}{3} \right)^2 \frac{4\pi N_A \rho g \mu_v^2}{9kMT} \tag{P.3.3.1}$$

where M is the molecular weight of the structural unit, ρ is the density, and N_A is the Avogadro number. By taking n^2 (the square of the index of refraction) instead of $\varepsilon_{\infty\beta}$ in Eq. (P.3.3.1), we obtain $g = 0.0883$.

From the data of the given dipolar moment components

$$\cos \theta = \frac{\boldsymbol{\mu}_1 \cdot \boldsymbol{\mu}_2}{|\boldsymbol{\mu}_1| \cdot |\boldsymbol{\mu}_2|}$$

$$= \frac{\mu_{1x}\mu_{2x} + \mu_{1y}\mu_{2y} + \mu_{1z}\mu_{2z}}{(\mu_{1x}^2 + \mu_{1y}^2 + \mu_{1z}^2)^{1/2}(\mu_{2x}^2 + \mu_{2y}^2 + \mu_{2z}^2)^{1/2}} = -0.3256 \tag{P.3.3.2}$$

which corresponds to $\theta = 109°$.

From Eq. (P.9.2.4b) we can write

$$\frac{\mu^2 - \langle \mu \rangle^2}{\mu^2} = \frac{2x}{(1+x)^2}(1 - \cos \theta) \tag{P.3.3.3}$$

If all the dipoles under consideration are equivalent, Eqs (9.3.11) and (9.3.15) lead to

$$\mu_{\text{eff}}^2 = g\mu^2 = \frac{2x}{(1+x)^2}(1 - \cos \theta)\mu^2 \tag{P.3.3.4}$$

from which we can easily obtain $x = 0.033$.

Thus, only about 3% of the lateral groups participate in the motion responsible for the relaxation. Note that, if we take for $\varepsilon_{\infty\beta}$ the experimental value (2.80), the number of groups participating in the relaxation is lower (about 1%).

REFERENCES

1. Johari, G.P. The glass transition and the nature of the glassy state; Goldstein, M., Simha, R., Eds.; Ann. New York Acad. Sci. **1976**, *279*, 117.
2. Struik, L.C.E. In *Molecular Dynamics and Relaxations*; Dorfmuller, Th., Williams, G., Eds.; Lecture Notes in Physics; Springer: New York, 1987.
3. Díaz-Calleja, R.; Riande, E.; San Román, J. J. Polym. Sci. Polymer Phys. Ed. **1992**, *30*, 1239.
4. Steeman, P.A.M.; Van Turnhout, J. Macromolecules **1994**, *27*, 5421.
5. Riande, E. *et al. Polymer Viscoelasticity, Stress and Strain in Practice*; M. Dekker: 2000.
6. Williams, G.; Cook, M. Trans. Far. Soc. **1971**, *67*, 990.
7. Smith, G.D.; Boyd, R. Macromolecules **1991**, *24*, 2731.
8. Verdier, P.H.; Stockmayer, W.H. J. Chem. Phys. **1962**, *36*, 227.
9. Verdier, P.H. J. Chem. Phys. **1966**, *45*, 2118.
10. Monnerie, L.; Gény, F. Chim. Phys. **1969**, *66*, 1708, 1961.
11. Geny, F.; Monnerie, L. J. Polym. Sci. Polym. Phys. Ed. **1979**, *17*, 131, 147.
12. Helfand, E. J. Chem. Phys. **1978**, *69*, 1010.
13. Helfand, E.; Wassermann, Z.; Weber, T.A. Macromolecules **1980**, *13*, 526.
14. Skolnik, J.; Helfand, E. J. Chem. Phys. **1980**, *72*, 5489.
15. Helfand, E.; Skolnik, J. **1982**, *77*, 5714.
16. Helfand, E. Science **1984**, *226*, 647.
17. Jernigan, R.L. Internal relaxation in short chains bearing terminal polar groups. In *Dielectric Properties of Polymers*; Karasz, F., Ed.; Plenum: N. York, 1972; 99 pp.
18. Bahar, I.; Erman, B. Macromolecules **1987**, *20*, 1368.
19. Bahar, I. J. Chem. Phys. **1989**, *91*, 6525.
20. Bahar, I.; Erman, B. Macromolecules **1990**, *23*, 1174.
21. Bahar, I.; Erman, B. Adv. Polym. Sci. **1994**, *116*, 145.
22. Flory, P.J. *Statistical Mechanics of Chain Molecules*; Wiley-Interscience: New York, 1969.
23. Flory, P.J. Macromolecules **1974**, *7*, 381.
24. Eyring, H. J. Chem. Phys.**1935**, *3*, 107.
25. Kramers, H.A. Physica **1940**, *7*, 284.
26. Johari, G.P.; Goldstein, M. J. Chem. Phys. **1971**, *55*, 4245.
27. Heijboer, J. In *Static and Dynamic Properties of the Polymeric Solid State*; Pethrick, R.A., Richards, R.W., Eds.; D. Riedel Pub. Co., 1982; 197–211.
28. Fuoss, R.M. J. Am. Chem. Soc. **1941**, *63*, 369, 378.
29. Steeman, P.A.M.; Gondard, C.; Scherrenberg, R.L. J. Polym. Sci. Polym. Phys. Ed. **1994**, *32*, 119.
30. Krum, F.; Müller, F.H. Kolloid. Z. **1959**, *164*, 81.
31. Matsuoka, S.; Ishida, Y. J. Polym. Sci. Part C **1966**, *14*, 247 see also Am. Chem. Soc. Poly. Preprints **1965**, *6* (2), 795.
32. Pratt, G.J.; Smith, M.J.A. Polymer **1989**, *30*, 1113; Polym. Degradation and Stability **1989**, *25*, 267; Plastics, Rubber and Composites, Processing and Applications **1991**, *16*, 67; Br. Polym. J. **1986**, *18*, 105; Polym. Int. **1996**, *40*, 239; and **1997**, *43*, 137.
33. Pocham, J.M.; Pocham, D.F. Macromolecules **1980**, *13*, 1577; see also Macromolecules **1978**, *11*, 165.

34. Reddish, W. Trans. Far. Soc. **1950**, *46*, 459.
35. Coburn, J.C.; Boyd, R. Macromolecules **1986**, *19*, 2238.
36. McCrum, N.G.; Read, B.E.; Williams, G. *Anelastic and Dielectric Effects in Polymeric Solids*; Wiley: London, 1967; 238 pp and ss.
37. Díaz-Calleja, R.; Gargallo, L.; Radic, D. Macromolecules **1995**, *28*, 6963.
38. Steck, N.S. SPE Trans. **1964**, *34*.
39. Mikhailov, G.P.; Borisova, T. Sov. Phys. Tech. Phys. **1958**, *3*, 120.
40. Boyd, R.H. Polymer **1985**, *26*, 323, 1123.
41. Matsuoka, S.; Roe, R.J.; Cole, H.F. A comparative study of the dielectric behavior of polyethylene and chlorinated polyethylene. In *Dielectric Properties of Polymers*; Karasz, F., Eds.; Plenum: New York, 1972; 255 pp.; Ashcraft, C.R., Boyd, R. J. Polym. Sci, Polym. Phys. Ed. **1976**, *14*, 2153.
42. Sauer, B.B.; Avakian, P.; Starkweather, H.W. Jr. J. Polym. Sci. Polym. Phys. Ed. **1996**, *34*, 517.
43. Moscicki, J.K. *Dielectric Relaxation in Macromolecular Liquid Crystals in Liquid Crystal Polymers: From Structures to Applications*; Collyer, A.A., Ed.; Elsevier: London, 1992; 142 pp.
44. Kremer, F.; Schönfeld, A.; Hofmann, A.; Zentel, R.; Poths, H. Polym. for Advd. Technol **1992**, *3*, 249; **1990**, *2*, 145. See also Liquid Crystals, **1990**, *8*, 719; **1990**, *8*, 889; **1991**, *9*, 145; **1993**, *13*, 403; Ferroelectrics **1990**, *109*, 273; **1991**, *121*, 13; **1991**, *121*, 69.
45. Kremer, F.; Vallerien, S.U.; Zentel, R.; Kapitza, H. Macromolecules **1989**, *22*, 4040.
46. Schonland, D. *Molecular Symmetry*; Van Nostrand: London, 1965.
47. Cotton, F.A. *Chemical Applications of Group Theory*; Wiley: New York, 1963.
48. Cole, R.H. J. Chem. Phys. **1965**, *42*, 637.
49. Diaz-Calleja, R.; Riande, E.; San Romain, J. J. Phys. Chem. **1992**, *96*, 6843.

10

Electric Birefringence Dynamics

10.1. INTRODUCTION

In Chapter 6 we have shown that the equilibrium birefringence of polymer solutions, expressed by the Kerr effect, is an important tool for analyzing the conformational properties of polymers as a function of their chemical structure. This chapter is devoted to the study of the rate at which macromolecules in a solution are oriented under the influence of an applied electric field.

Let us consider a molecular system under the following force field history

$$E(t) = E_0 \, \Delta(t), \quad \Delta(t) = \begin{cases} 0; & t < 0 \\ 1; & t \geqslant 0 \end{cases} \tag{10.1.1}$$

where E_0 is the electric field. The interaction between the field and the dipole moment is the driving force that orients the molecule in an antiparallel direction to the field. The birefringence of the system will increase with time as the molecules rotate to align with the field, eventually reaching a constant value at equilibrium. The time required for a molecule to rotate depends on its size and shape. However, thermal motion of the molecules tends to randomize their orientations so that the birefringence will decrease with increasing temperature. Small molecules in ordinary liquids rotate with lifetimes of 10^{-9}–10^{-10} s. For polymers, these lifetimes are much larger and their measurement by analysis of the time dependence of the birefringence provides a method for studying the size and shape of macromolecules.

The increase in birefringence with time after application of the field is the buildup process. If the electric field is removed after the birefringence reaches equilibrium, for example at $t = t_0$, that is

$$E(t) = E_0\,\Delta(t - t_0), \qquad \Delta(t - t_0) = \begin{cases} 1; & t \leqslant t_0 \\ 0; & t > t_0 \end{cases} \qquad (10.1.2)$$

the birefringence will decrease with time as the aligned molecular system, by the effect of Brownian motion, randomizes the orientations of the molecules until eventually the birefringence vanishes. This is the decay process.

Both the buildup and decay processes can be detected in glasses where the rotatory relaxation times are of the order of seconds or more. However, the relaxation times may be very small for solutions of low molecular weight compounds in such a way that the time dependence of the birefringence in buildup and decay processes cannot be easily measured. Rotatory relaxation times associated with macromolecules in solution are somewhere between these two extremes, and buildup and decay processes can often be measured in these systems.

10.2. ORIENTATION FUNCTION

The probability of finding the symmetry axis of a molecule forming at time t an angle θ with the direction of the applied electric field is given by

$$f(\theta,t) = \frac{\exp\left[-V_F(\theta,t)/k_B T\right]}{2\pi \int \exp\left[-V_F(\theta,t)/k_B T\right] \sin\theta\,d\theta} \qquad (10.2.1)$$

where according to Eq. (6.7.1)

$$V_F(\theta,t) = V(\theta) - F(t)\mu\cos\theta - \tfrac{1}{2}F^2(t)\Lambda'\cos^2\theta \qquad (10.2.2)$$

For a rigid rod molecule, the electrostatic anisotropy is represented by

$$\Lambda' = \alpha'_\parallel - \alpha'_\perp = \alpha'_L - \alpha'_T \qquad (10.2.3)$$

where the subscripts L and T represent the longitudinal and transverse directions. The time dependence of the orientation factor can be written as[1]

$$\Phi(\theta,t) = 2\pi \int_0^\pi f(\theta,t) P_2(\cos\theta)\sin\theta\,d\theta \qquad (10.2.4)$$

where $P_2(\cos\theta) = (3\cos^2\theta - 1)/2$ is the orientation factor (second Legendre polynomial).

The function $f(\theta,t)$ depends on:

(a)   the thermal energy that causes the particles to undergo a random walk or Brownian motion in the orientation sense;

(b)   the friction between the particle and the surrounding medium;

(c)   the orientation torque due to the electric field.

For macromolecules of ordinary size, inertial effects are negligible because rotational motions are quickly dissipated as the molecules expend their inertial energy on viscous drag in the surrounding medium. The dynamic process can be viewed as a Brownian diffusion process biased by the torque, M, produced by the applied field on the dipole moments of the particles. In this contest, the partial differential equation for the diffusion process can be written as [see Eq. (2.8.11)]

$$\frac{\partial f(\theta,t)}{D\,\partial t} = \frac{\partial f(\theta,t^*)}{\partial t^*} = \frac{1}{\sin\theta}\frac{\partial}{\partial\theta}\left\{\sin\theta\left[\frac{\partial f(t^*,\theta)}{\partial\theta} + \frac{1}{k_B T}\frac{\partial V}{\partial\theta}\,f(t^*,\theta)\right]\right\} \quad (10.2.5)$$

where θ is the angle between the symmetry axis of the macromolecule under consideration and the applied field and $t^* = Dt$, D being the rotatory diffusion coefficient for the reorientation of the symmetry axis of the molecule by rotation about the transverse axis. Equation (10.2.5) has been solved for torques produced by static and dynamic electric fields. The first attempts to describe the buildup and decay processes in birefringence using low electric fields were made by Benoit.[2] Morita and Watanabe extended these studies to fields of any strength.[3]

10.3. DECAY FUNCTION

Let us consider that a molecular system under the field history described by Eq. (10.1.1) reaches steady state conditions. If the field is removed, as indicated in Eq. (10.1.2), reorientation of the main axes of the molecules gives rise to a free field decay of the molecular orientation. In this case, by analogy with Eq. (2.9.4), Eq. (10.2.5) becomes

$$\sin\theta\,\frac{\partial f(\theta,t^*)}{\partial t^*} = \frac{\partial}{\partial\theta}\left[\sin\theta\,\frac{\partial f(t^*,\theta)}{\partial\theta}\right] \quad (10.3.1)$$

According to Eq. (10.2.4), the dependence of Φ on t^* can be written as

$$\frac{d\Phi(t)}{dt} = 2\pi\int_0^\pi \frac{\partial(t^*,\theta)}{\partial t^*}P_2(\cos\theta)\sin\theta\,d\theta$$

$$= 2\pi\int_0^\pi P_2(\cos\theta)\frac{\partial}{\partial\theta}\left[\sin\theta\frac{\partial f(t^*,\theta)}{\partial\theta}\right]d\theta \quad (10.3.2)$$

After integration by parts, Eq. (10.3.2) becomes

$$\frac{d\Phi(t^*)}{dt^*} = 2\pi \int_0^\pi f(t^*,\theta) \cos(\theta) \frac{dP_2(\cos\theta)}{d\theta} d\theta$$
$$+ 2\pi \int_0^\pi f(t^*,\theta) \sin(\theta) \frac{d^2P_2(\cos\theta)}{d\theta^2} d\theta \qquad (10.3.3)$$

By taking into account that $dP_2(\cos)/d\theta = -3\cos\theta\sin\theta$ and $d^2P_2(\cos)/d\theta^2 = 3\sin^2\theta - 3\cos^2\theta$, Eq. (10.3.3) adopts the following form

$$\frac{d\Phi(t^*)}{dt^*} = -12\pi \int_0^\pi f(t^*,\theta)P_2(\cos\theta)\sin\theta\, d\theta = -6\Phi(t^*) \qquad (10.3.4)$$

Hence

$$\Phi(t) = \Phi_0 \exp(-6t^*) = \Phi_0 \exp(-6Dt) = \exp\left(-\frac{t}{\tau}\right) \qquad (10.3.5)$$

where $\tau = 1/6D$ is the relaxation time of the decay process. Note that the extreme values of the decay function are Φ_0 and 0 for $t = 0$ and $t = \infty$ respectively. Taking into account the relationships between the correlation factor and anisotropy [Eq. (6.5.3)] and between this quantity and birefringence [Eq. 6.4.5)], the time dependence of $\langle\Delta\alpha\rangle$ and $\langle\Delta n\rangle$ can be written as

$$\langle\Delta\alpha\rangle(t) = \langle\Delta\alpha\rangle_0 \frac{\Phi(t)}{\Phi_0}$$
$$\qquad (10.3.6)$$
$$\Delta n(t) = \Lambda\langle\Delta\alpha\rangle(t) = \Lambda\langle\Delta\alpha\rangle_0 \exp\left(-\frac{t}{\tau}\right) = \Delta n(0) \exp\left(-\frac{t}{\tau}\right)$$

10.4. BUILDUP ORIENTATION FUNCTION

If the field history is described by Eq. (10.1.1), then the buildup orientation factor $\Phi'(t)$ increases from 0, at $t = 0$, up to the limiting value of 1 when $t \to \infty$. By neglecting terms depending on powers higher than E^2 in the case of weak force fields, the following expression is obtained[2] (see Problem 1 for details)

$$\frac{\Phi'(t)}{\Phi'(\infty)} = 1 - \frac{3\Gamma}{2\Gamma+4}\exp(-2Dt) - \frac{4-\Gamma}{2\Gamma+4}\exp(-6Dt) \qquad (10.4.1)$$

where $\Gamma = \beta^2/\gamma = 2\mu^2/k_BT\Lambda'$ represents the relative strength of permanent versus induced dipoles [see Eq. (6.5.6)]. An inspection of Eq. (10.4.1) indicates that the buildup orientation factor is a combination of two contributions, a fast contribution that, like the decay correlation function, has a short retardation time equal to $1/6D$, and a slower contribution with a higher retardation time $\tau = 1/2D$.

Two extreme cases can be analyzed:

(a)   molecules with non-permanent dipole moments ($\mu = 0$),
(b)   molecules with permanent dipole moments ($\mu \neq 0$).

For the first case, $\Gamma = 0$ and Eq. (10.4.1) becomes

$$\frac{\Phi'(t)}{\Phi'(\infty)} = 1 - \exp(-6Dt) = 1 - \exp\left(-\frac{t}{\tau}\right) \tag{10.4.2}$$

In this situation, the time dependence of the birefringence is given by

$$\Delta n(t) = \Delta n(\infty)\left[1 - \exp\left(-\frac{t}{\tau}\right)\right] \tag{10.4.3}$$

For the second case, $\Gamma \neq 0$. If $\Gamma \gg 1$,

$$\frac{\Phi'(t)}{\Phi'(\infty)} = 1 - \frac{3}{2}\exp\left(-\frac{t}{\tau_1}\right) + \frac{1}{2}\exp\left(-\frac{t}{\tau_2}\right) \tag{10.4.4}$$

and

$$\Delta n(t) = \Delta n(\infty)\left[1 - \frac{3}{2}\exp\left(-\frac{t}{\tau_1}\right) + \frac{1}{2}\exp\left(-\frac{t}{\tau_2}\right)\right] \tag{10.4.5}$$

where τ_1 and τ_2 are given by $1/2D$ and $1/6D$ respectively. The illustrative plots in Fig. 10.1 indicate that the increase in the buildup correlation function is slower

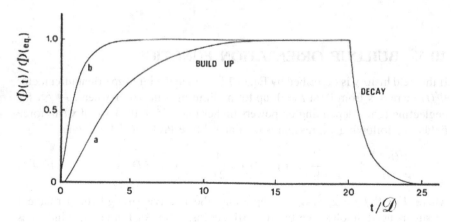

FIG. 10.1   Schematic representation of the normalized orientation factor for buildup and decay processes. (a) polar systems; (b) nonpolar systems

for molecules with larger dipole moments. For molecules with $\Gamma > 0$, the buildup function is somewhat intermediate to that corresponding to the two extreme cases discussed above.

To explain at the qualitative level the differences observed between the buildup functions for rigid molecules with and without permanent dipoles, let us consider first that the polar molecule has a dipole in the direction of the largest axis. The molecule will be asymmetric with respect to a rotation of 180° about the transversal axis in such a way that, if the interaction with the field for an orientation angle θ is V, it will be $-V$ for $\pi - \theta$. By taking as $\theta = 0$ the most favorable orientation (the lowest value of V), $\theta = \pi$ will be the most unfavorable orientation. Although the polarizability tensor is symmetric with respect to a π rotation over the transversal axis, and therefore the anisotropy is equivalent for $\theta = 0$ and $\theta = \pi$, energetic requirements are only fulfilled for $\theta = 0$. Therefore, the molecules with an orientation close to $\theta = \pi$ will start to rotate until $\theta = 0$. At this orientation the permanent dipole moment is antiparallel to the electric field. However, for nonpolar molecules, the induced dipoles will be antiparallel to the electric field for $\theta = 0$ and $\theta = \pi$. In this case the molecules can choose to rotate towards $\theta = 0$ or $\theta = \pi$ and the orientation is faster.

When the electric field is removed, the orientation $\pi - \theta$ is as ordered/disordered as θ, regardless of the polarization of the molecule. Since the polarizability tensor only depends on the degree of order, the decay function is similar for polar and nonpolar rigid molecules.

10.5. PULSED FIELDS

Pulsed field techniques can be used to study the electrooptical properties of aqueous solutions of macromolecules displaying electrical conductivity and electrophoresis. By applying pulses of equal and opposite polarity, the net electrophoretic motion is zero. Moreover, Joule heating effects in solutions can be minimized by adjusting the pulse duration, amplitude, and the period between pulses. Experiments have been performed using electronically generated antisymmetric square waves of the kind illustrated in Fig. 10.2.

In the reverse polarity square wave technique, the electric field $\mathbf{E}$ is suddenly applied and kept constant during a time t. Then the polarity is suddenly reversed, $-\mathbf{E}$ is kept constant during the same time t, and so on. For molecules with cylindrical symmetry under weak electric fields, the time dependence of the orientation function can be written as[4]

$$\frac{\Phi(t)}{\Phi_{eq}} = 1 + \frac{3\Gamma}{\Gamma + 2} [\exp(-6Dt) - \exp(-2Dt)] \qquad (10.5.1)$$

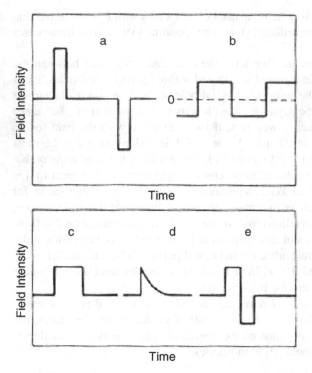

FIG. 10.2 Electric field waveforms in dynamic birefringence experiments. (a) continuously repeated experiments of opposite polarity; (b) continuous square waves; (c) single pulse; (d) exponential pulse; (e) reversing field pulse[1]

where Φ_{eq} is the equilibrium value for a field of constant strength $+E$ or $-E$. For molecules with nonpermanent dipole moments $\Gamma = 2\mu^2/k_B T \Lambda' = 0$ and $\Phi(t)/\Phi_{eq} = 1$, regardless of t. This is due to the fact that the interaction with the induced dipole does not change during the field reversal, because the symmetry of the polarizability makes molecules orientate in the same direction for $+E$ and $-E$ fields.

As a consequence of the asymmetry of permanent dipoles, the orientation in polar molecules with cylindrical symmetry produced by $+E$ and $-E$ differs in $180°$. When the field is reversed from $+E$ to $-E$, the molecules oriented under the field $+E$ start to become disordered giving rise to a decrease in the birefringence until the molecules begin again to be oriented under the new polarity. Therefore, the birefringence must pass through a minimum before reaching equilibrium under the field $-E$. From the minimum, both the rotatory diffusion coefficient and the parameter Γ can be determined. Actually, from the time, t_{min}, for which $\Phi(t)$ passes through a minimum Φ_{min}, the rotatory diffusion coefficient can be obtained as

$$D = -\frac{1}{4t_{min}} \ln \frac{1}{3} \qquad (10.5.2)$$

On the other hand, the parameter Γ is given by

$$\Gamma = \frac{2(1 - \Phi_{min}/\Phi_{eq})}{0.1547 + \Phi_{min}/\Phi_{eq}} \qquad (10.5.3)$$

where Φ_{min}/Φ_{eq} is the minimum value of the normalized orientation function. Therefore, the pulse technique is suitable for determining the rotatory diffusion and the relation between the permanent and induced dipole moments expressed by Γ.

10.6. DISPERSION OF THE BIREFRINGENCE IN SINE WAVE FIELDS

In the evaluation of the birefringence under an alternating force field $E = E_0 \sin \omega t$, the orientation factor becomes both time and frequency dependent and therefore is represented by $\Phi(\omega, t)$. Here, birefringence depends on the strength and frequency of the field and on the optical, polarity, and rotational characteristics of the molecules. The treatment developed by Thurston and Bowling[5] for weak fields and cylindrical particles, which gives a good compromise between relatively simple equations and possibilities of application, will be presented below. A more general treatment has been reported[3].

If $\lim \Phi_{\omega \to 0}(\omega, t) = \Phi_0$, then according to Eq. (6.5.7)

$$\Phi_0 = \frac{\beta_0^2 + \gamma_0}{15} \qquad (10.6.1)$$

where

$$\beta_0 = \frac{\mu F_0}{k_B T} \quad \text{and} \quad \gamma_0 = \frac{\Lambda' F_0^2}{2k_B T} \qquad (10.6.2)$$

so that $\Gamma = \beta_0^2/\gamma_0$. Note that F_0 is the amplitude of the effective electric field.

The orientation factor expressed in terms of Φ/Φ_0 is made up of two contributions, one stationary and the other time dependent given by

$$\frac{\Phi}{\Phi_0} = \Phi_{st} + \Phi_{alt} \cos(2\omega t - \delta) \qquad (10.6.3)$$

where δ is the difference of phase between the response and the applied field. The static component can be written as

$$\Phi_{st} = \frac{1}{1 + \Gamma}\left(1 + \frac{4D^2\Gamma}{\omega^2 + 4D^2}\right) \tag{10.6.4}$$

Obviously, $\Phi_{st} \to 1$ when $\omega \to 0$.

The frequency-dependent component is given by

$$\Phi_{alt} = (\Phi_1^2 + \Phi_2^2)^{1/2} \tag{10.6.5}$$

so that the phase difference is $\delta = \tan^{-1} \Phi_2/\Phi_1$. The parameters Φ_1 and Φ_2 are given by

$$\Phi_1 = \frac{6D^2(6D^2 - \omega^2)\Gamma + 9D^2(4D^2 + \omega^2)}{(\Gamma + 1)(4D^2 + \omega^2)(9D^2 + \omega^2)}$$

$$\Phi_2 = \frac{30D^3\omega\Gamma + 3D\omega(4D^2 + \omega^2)}{(\Gamma + 1)(4D^2 + \omega^2)(9D^2 + \omega^2)} \tag{10.6.6}$$

Let us consider the frequency dependence of Φ/Φ_0 for two extreme cases, nonpolar ($\Gamma = 0$) and very polar molecules ($\Gamma \to \infty$). In the first case ($\Gamma = 0$, $\mu = 0$), Eqs (10.6.6), (10.6.5), and (10.6.3) lead to

$$\frac{\Phi}{\Phi_0} = 1 + \frac{3D}{(9D^2 + \omega^2)^{1/2}}\cos(2\omega t - \delta) \tag{10.6.7}$$

where

$$\tan \delta = \frac{\omega}{3D}$$

$$\Phi_0 = \frac{\Lambda'}{30k_B T}F_0^2 \tag{10.6.8}$$

According to Eq. (10.6.7), the orientation of nonpolar molecules reaches a maximum for $+E_0$ and $-E_0$ because their polarization, by the effect of an electric field, is nearly instantaneous. Moreover, as indicated above when buildup processes were discussed, maximum orientation is accomplished when the field reaches the value of the amplitude ($+E_0$ or $-E_0$). The orientation decreases from $E_0 \to 0$, increases from $0 \to -E_0$, decreases from $-E_0 \to 0$, and increases from $0 \to E_0$. As a result, both Φ/Φ_0 and therefore the birefringence oscillate with twice the angular frequency of the force field, as indicated by Eq. (10.6.7). However, because of the friction between the rotating molecules and the surrounding medium, the orientation of the molecules lags an angle δ with respect to the oscillating electric field. As indicated in Eq. (10.6.8a), the phase angle increases with the angular frequency of the field and decreases as the rotatory diffusion coefficient increases. Obviously, the measurement of δ allows the determination of the rotatory diffusion coefficient.

Two extreme cases appear. At very high frequencies the orientation of the molecules cannot follow the field and, in the limit $\omega \to \infty$, $\Phi = \Phi_0$, the value of which is given by Eq. (10.6.8b). At low frequencies, in the limit $\omega \to 0$, the value of $\Phi(t)$ oscillates between $2\Phi_0$ and 0.

Experimental values showing the variation in the maximum and minimum values of the birefringence dispersion for tobacco mosaic virus is presented in Fig. 10.3. The theory reproduces rather well the evolution of δ^2/E with ω in the low-frequency region. However, it does not treat the relaxation responsible for the anomalous dispersion around $0.1-1$ MHz. Polar molecules can be analyzed in the light of the theory if $\beta_0 \gg \gamma_0$ so that $\Gamma \to 0$. In this situation, Eqs (10.6.3) to (10.6.6) become

$$\frac{\Phi(t)}{\Phi_0} = \frac{4D^2}{\omega^2 + 4D^2} + \frac{6D^2}{[(\omega^2 + 4D^2)(\omega^2 + 9D^2)]^{1/2}} \cos(2\omega t - \delta) \qquad (10.6.9)$$

with

$$\tan \delta = \frac{5\omega D}{6D^2 - \omega^2}$$

$$\Phi_0 = \frac{\mu^2 F_0^2}{15 k_B T^2} \qquad (10.6.10)$$

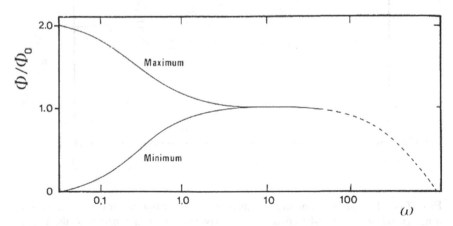

FIG. 10.3 Maximum and minimum values of the orientation factor Φ for a nonpolar polymer under a sine wave electric field of frequency ω

As in the case of nonpolar molecules, molecular friction is responsible for the lag between the response and the oscillating field. According to Eq. (10.6.10), the phase angle increases with frequency, but it is not a linear function of this variable. Actually, the molecules are oriented in the direction of the field twice each cycle and, taking into account that the directions of $+E_0$ and $-E_0$ differ in 180°, the molecules are forced to rotate 180° followed by −180° during each cycle. As the frequency increases, the molecules cannot follow the field and they remain disoriented as a response to the two opposing pulls of the field. When $\omega \to \infty$, both the static and time-dependent components of Φ tend to zero, as Eq. (10.6.9) suggests.

If the frequency of the oscillating field is low enough for the molecules to be able to follow the field, they adopt twice in each cycle the direction in which

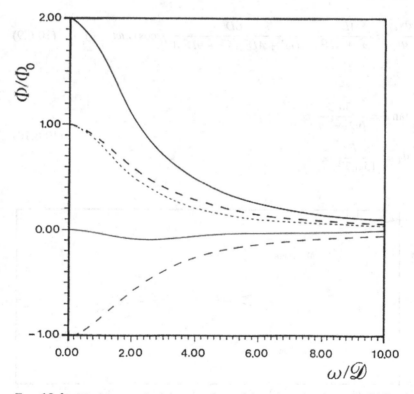

FIG. 10.4 Maximum and minimum values of the orientation factor (solid lines) and its time-dependent (——) and stationary (········) components as a function of the angular frequency of molecules whose orientations are produced only by their permanent dipole moments[6]

they are perpendicular to the field. The birefringence produced in these moments has the opposite sign to that induced by a static field. In the limit $\omega \to 0$, polar molecules behave as nonpolar ones. The static component of Φ tends to unity whereas the time-dependent component oscillates between 0 and $2\Phi_0$ with a mean value of Φ_0 given by Eq. (10.6.10b). Curves representing the variation in stationary and time-dependent components of Φ as well as the total value of this latter quantity, calculated as a function of the frequency of the field, are shown in Fig. 10.4.

PROBLEMS

Problem 1

If $E = 0$ for $t < 0$ and $E = E_0$ for $t \geq 0$, determine the time dependence of the buildup response for a birefringence experiment.

Solution

According to the Schmoluchowski equation [Eq. (10.2.5)]

$$\frac{\partial f(t^*, \theta)}{\partial t^*} = \frac{1}{\sin \theta} \frac{\partial}{\partial \theta} \left[\sin \theta \left(\frac{\partial f}{\partial \theta} + \frac{1}{k_B T} \frac{\partial V}{\partial \theta} f(t^*, \theta) \right) \right] \qquad \text{(P.10.1.1)}$$

where $t^* = Dt$ and the potential can be written as [see Eq. (6.5.5)]

$$\frac{V(\theta)}{k_B T} = -\frac{F\mu \cos \theta}{k_B T} - \frac{1}{2} \frac{\Lambda' \cos^2 \theta}{k_B T} = -\beta \cos \theta - \gamma \cos^2 \theta \qquad \text{(P.10.1.2)}$$

Here

$$\beta = \frac{\mu F}{k_B T}, \qquad \gamma = \frac{\alpha_1 - \alpha_2}{2 k_B T} \qquad \text{(P.10.1.3)}$$

Let us assume that

$$f = \sum_n a_n(t^*) P_n(\cos \theta) \qquad \text{(P.10.1.4)}$$

is a general solution of Eq. (P.10.1.1), where P_n is the nth-order term of Legendre's polynomial. From Eq. (P.10.1.4) and using the identities

$$\frac{\partial}{\partial \theta} = -\sin \theta \frac{\partial}{\partial \cos \theta},$$

$$\frac{\partial^2}{\partial \theta^2} = -\cos \theta \frac{\partial}{\partial \cos \theta} + (1 - \cos^2 \theta) \frac{\partial^2}{\partial (\cos \theta)^2} \qquad \text{(P.10.1.5)}$$

the following expression is obtained

$$
\frac{1}{\sin \theta} \frac{\partial}{\partial \theta} \left(\sin \theta \frac{\partial f}{\partial \theta} \right) = \frac{1}{\sin \theta} \left(\cos \theta \frac{\partial f}{\partial \theta} + \sin \theta \frac{\partial^2 f}{\partial \theta^2} \right)
$$

$$
= \sum_n a_n(t^*) \left[(1 - u^2) \frac{\partial^2 P_n}{\partial u^2} - 2u \frac{\partial P_n}{\partial u} \right] \qquad \text{(P.10.1.6)}
$$

where $\cos \theta = u$. According to Legendre's differential equation

$$
(1 - u^2) \frac{\partial^2 P_n}{\partial u^2} - 2u \frac{\partial P_n}{\partial u} + n(n+1)P_n = 0 \qquad \text{(P.10.1.7)}
$$

Eq. (P.10.1.6) can be written as

$$
-\sum_n a_n(t^*) n(n+1)P_n = \sum_n a_n(t^*) \left[(1 - u^2) \frac{\partial^2 P_n}{\partial u^2} - 2u \frac{\partial P_n}{\partial u} \right] \qquad \text{(P.10.1.8)}
$$

Equation (P.10.1.5), in conjunction with Eqs (P.10.1.2) and (P.10.1.3), leads to

$$
\frac{1}{k_B T} \frac{\partial V}{\partial \theta} = \beta u + 2\gamma u (1 - u^2)^{1/2},
$$

$$
\frac{1}{k_B T} \frac{\partial^2 V}{\partial \theta^2} = \beta u + \gamma(2u^2 - 1) \qquad \text{(P.10.1.9)}
$$

so that

$$
\frac{1}{\sin \theta} \frac{\partial}{\partial \theta} \left(\sin \theta \frac{1}{k_B T} \frac{\partial V}{\partial \theta} f \right)
$$

$$
= \frac{1}{\sin \theta} \left(\cos \theta \frac{1}{k_B T} \frac{\partial V}{\partial \theta} f + \sin \theta \frac{1}{k_B T} \frac{\partial^2 V}{\partial \theta^2} f + \sin \theta \frac{1}{k_B T} \frac{\partial V}{\partial \theta} \frac{\partial f}{\partial \theta} \right)
$$

$$
= \sum_n a_n(t^*) \left\{ [2\beta u + 2\gamma(3u^2 - 1)]P_n - (1 - u^2)(\beta + 2\gamma u) \frac{\partial P_n}{\partial n} \right\}
$$

$$
\text{(P.10.1.10)}
$$

Substitution of Eqs (P.10.1.4), (P.10.1.8), and (P.10.1.10) into Eq. (P.10.1.1) gives

$$
\sum_n \frac{da_n(t^*)}{dt^*} P_n = \sum_n a_n(t^*) \left\{ -n(n+1)P_n - (1 - u^2)(\beta + 2\gamma u) \frac{\partial P_n}{\partial u} \right.
$$

$$
\left. + [2\beta u + 2\gamma(3u^2 - 1)]P_n \right\} \qquad \text{(P.10.1.11)}
$$

In order to eliminate the variable u in Eq. (P.10.1.11), it is necessary to make use of Legendre's polynomial recurrence formulae

$$(2n+1)uP_n = (n+1)P_{n+1} + nP_{n-1} \tag{P.10.1.12}$$

and

$$(1-u^2)\frac{dP_n}{du} = nP_{n-1} - nuP_n \tag{P.10.1.13}$$

Thus, by combining Eqs (P.10.1.13) and (P.10.1.11), the following expression is obtained

$$\sum_n \frac{da_n(t^*)}{dt^*} P_n = \sum_n a_n(t^*)\{\beta(-nP_{n-1} + nuP_n + 2uP_n) + 2\gamma[(3u^2 - 1)$$

$$\times P_n - unP_{n-1} + nu^2 P_n]\} \tag{P.10.1.14}$$

On the other hand, Eq. (13) makes it possible to obtain uP_n, $u^2 P_n$, uP_{n+1} and uP_{n-1} as

$$uP_n = \frac{(n+1)P_{n+1} + nP_{n-1}}{2n+1}$$

$$u^2 P_n = \frac{(n+1)uP_{n+1} + nuP_{n-1}}{2n+1}$$

$$= \frac{1}{2n+1}\left[\frac{(n+1)(n+2)P_{n+2} + (n+1)^2 P_n}{2n+3} + \frac{n^2 P_n + n(n-1)P_{n-2}}{2n+1}\right]$$

$$uP_{n+1} = \frac{(n+2)P_{n+2} + (n+1)P_n}{2n+3}$$

$$uP_{n-1} = \frac{nP_n + (n-1)P_{n-2}}{2n-1}$$

$$\tag{P.10.1.15}$$

These expressions convert Eq. (P.10.1.14) to

$$\sum_n \frac{da_n(t^*)}{dt^*} P_n$$

$$= \sum_n a_n(t^*) \left\{ \begin{array}{l} -n(n+1)P_n - 2\gamma\left[1 + \dfrac{n^2(n-2)}{4n^2-1} - \dfrac{(n+1)^2(n+3)}{(2n+1)(2n+3)}\right]P_n - \\[3mm] \beta\dfrac{n(n-1)}{2n+1}P_{n-1} + \beta\dfrac{(n+1)(n+2)}{2n+1}P_{n+1} + \\[3mm] 2\gamma\dfrac{(n+1)(n+2)(n+3)}{(2n+1)(2n+3)}P_{n+2} - 2\gamma\dfrac{n(n-1)(n-2)}{4n^2-1}P_{n-2} \end{array} \right\}$$

$$\tag{P.10.1.16}$$

By multiplying the left- and right-hand sides of Eq. (P.10.1.16) by $P_m(u)$, and taking into account the orthogonal properties of Legendre's polynomials, we obtain

$$\int_{-1}^{1} P_n(u)P_m(u)\,du = \frac{2}{2m+1}\,\delta_{mn} \tag{P.10.1.17}$$

where δ_{mn} is Kronecker's delta, and, by further integration over u in the interval $[-1, 1]$, the following set of differential equations for $n = 0, 1, 2$ is obtained

$$\frac{da_0}{dt} = 0$$

$$\frac{da_1(t^*)}{dt} = 2\beta a_0 - 2a_1(t^*)\left(1 - \frac{2\gamma}{5}\right) - \frac{2}{5}\beta a_2(t^*) - \frac{12}{35}\gamma a_3(t^*)$$

$$\frac{da_2(t^*)}{dt} = 4\gamma a_0 + 2\beta a_1(t^*) - 2a_2(t^*)\left(3 - \frac{2\gamma}{7}\right) - \frac{6}{7}\beta a_3(t^*) - \frac{16}{21}\gamma a_4(t^*)$$

$$\tag{P.10.1.18}$$

The normalization condition $\int_{-1}^{1} 2\pi f\,du = 1$ imposes that $a_0 = 1/4\pi$. By assuming, for simplicity, that $a_3(t^*) = a_4(t^*) = 0$, Eq. (P.10.1.18) can be written as

$$\frac{da_1(t^*)}{dt^*} = 2\beta a_0 - 2a_1(t^*)\left(1 - \frac{2\gamma}{5}\right) - \frac{2}{5}\beta a_2(t^*)$$

$$\tag{P.10.1.19}$$

$$\frac{da_2(t^*)}{dt^*} = 4\gamma a_0 + 2\beta a_1(t^*) - 2a_2(t^*)\left(3 - \frac{2\gamma}{7}\right)$$

Solution of this system of differential equations can be carried out by Laplace transforms. Let $a_1(0) = a_2(0) = 0$. In this case

$$s\bar{a}_1 = \frac{2\beta}{s} - 2\left(1 - \frac{2\gamma}{5}\right)\bar{a}_1 - \frac{2}{5}\beta\bar{a}_2 \rightarrow \bar{a}_1 = \frac{2\beta((1/s) - (\bar{a}_2/5))}{s + 2(1 - (2\gamma/5))}$$

$$s\bar{a}_2 = \frac{4\gamma}{s} + 2\beta\left\{\frac{2\beta(5 - \bar{a}_2 s)}{5s[s + 2(1 - (2\gamma/5))]}\right\} - 2\left(3 - \frac{2\gamma}{7}\right)\bar{a}_2 \tag{P.10.1.20}$$

where $\bar{a}_1 = 4\pi L[(a, (t^*)]$ and $\bar{a}_0 = 4\pi L[a_2(t^*)]$, the symbol L being the Laplace transform. Hence

$$\bar{a}_2 = \frac{4\gamma(s + 2 - (4\gamma/5)) + 4\beta^2}{s[(s + 6 - (4\gamma/7))(s + 2 - (4\gamma/5)) + 4\beta^2/5]} \tag{P.10.1.21}$$

For very small fields ($\gamma \to 0$, $\beta^2 \to 0$), this equation can approximately be written as

$$\bar{a}_2 \cong \frac{4\gamma}{s(s+6)} + \frac{4\beta^2}{s(s+2)(s+6)} \tag{P.10.1.22}$$

Since

$$\frac{1}{s(s+6)} = \frac{1}{6}\left(\frac{1}{s} - \frac{1}{s+6}\right),$$

$$\frac{1}{s(s+2)(s+6)} = \frac{1/12}{s} - \frac{1/8}{s+2} + \frac{1/24}{s+6} \tag{P.10.1.23}$$

Eq. (P.10.1.22) becomes

$$\bar{a}_2 = \frac{4\gamma}{6}\left(\frac{1}{s} - \frac{1}{s+6}\right) + \beta^2\left[\frac{1}{3s} - \frac{1}{2(s+2)} + \frac{1}{6(s+6)}\right] \tag{P.10.1.24}$$

The inverse of the Laplace transform of this equation gives

$$a_2(t^*) = \frac{1}{12\pi}\left[(\beta^2 + 2\gamma) - \frac{3}{2}\exp(-2t^*) + \left(\frac{\beta^2}{2} - 2\gamma\right)\exp(-6t^*)\right] \tag{P.10.1.25}$$

In the same way

$$a_1(t^*) = \frac{\beta}{4\pi}\left[1 - \exp(-2t^*) + \frac{1}{15}\left(\frac{\beta^2}{4} - \gamma\right)\exp(-6t^*)\right] \tag{P.10.1.26}$$

Once $a_2(t)$ is known, the buildup function $\Phi'(t)$ can be determined. Actually

$$\Phi'(t) = \frac{\int_0^\pi P_2(\cos\theta)f(\theta)\sin\theta\,d\theta}{\int_0^\pi f(\theta)\sin\theta\,d\theta} \tag{P.10.1.27}$$

where $f(\theta)$ can be written as

$$f(\theta) = a_0 P_0(\cos\theta) + a_1(t^*)P_1(\cos\theta) + a_2(t^*)P_2(\cos\theta) \tag{P.10.1.28}$$

The numerator of Eq. (P.10.1.27) is given by

$$\int_0^\pi P_2(\cos\theta)f(\theta)\sin\theta\,d\theta = \int_0^\pi a_2(t^*)[P_2(\cos\theta)]^2 f(\theta)\sin\theta\,d\theta$$

$$= \frac{2}{5}a_2(t^*) \tag{P.10.1.29}$$

whereas the denominator of Eq. (P.10.1.27) is $1/4\pi$. Hence

$$\Phi'(t) = \frac{8\pi}{5} a_2(t^*) = \frac{2}{5} \left[\frac{\beta^2 + 2\gamma}{3} - \frac{\beta^2}{2} \exp(-2Dt) \right.$$
$$\left. + \frac{1}{3} \left(\frac{\beta^2}{2} - 2\gamma \right) \exp(-6Dt) \right] \tag{P.10.1.30}$$

and

$$\Phi'_{\infty} = \frac{2(\beta^2 + 2\gamma)}{15} \tag{P.10.1.31}$$

The normalized buildup function is given by

$$\frac{\Phi'(t)}{\Phi'_{\infty}} = 1 - \frac{3\beta^2}{2(\beta^2 + 2\gamma)} \exp(-2Dt)$$
$$+ \frac{\beta^2 - 4\gamma}{2(\beta^2 + 2\gamma)} \exp(-6Dt) \tag{P.10.1.32}$$

By writing $\Gamma = \beta^2/\gamma$, Eq. (P.10.1.32) becomes

$$\frac{\Phi'(t)}{\Phi'_{\infty}} = 1 - \frac{3\Gamma}{2(\Gamma + 2)} \exp(-2Dt) + \frac{\Gamma - 4}{2(\Gamma + 2)} \exp(-6Dt) \tag{P.10.1.33}$$

REFERENCES

1. Krause, S.; O'Konsky, Ch.T. Electric birefringence dynamics. In *Molecular Electrooptics*; Krause, S., Ed.; Plenum Press: New York, 1981.
2. Benoit, H. Comp. Rend. **1949**, *228*, 1716; Ann. Physik. **1951**, *6*, 561; J. Chim. Phys. **1950**, *47*, 719; **1952**, *49*, 517.
3. Watanabe, H.; Morita, A. Adv. Chem. Phys. **1984**, *56*, 255.
4. Tinoko, I.; Yamaoka, K. J. Phys. Chem. **1959**, *63*, 423.
5. Thurston, G.B.; Bowling, D.I. J. Colloid Interf. Sci. **1969**, *30*, 34.
6. O'Konski, C.T.; Haltner, A.J. J. Am. Chem. Soc. **1957**, *79*, 5634.

11

Dielectric Properties of Liquid Crystals

11.1. INTRODUCTION: LIQUID CRYSTAL GENERALITIES

The crystalline and amorphous states are the two extreme states of condensed matter. In the crystalline state each molecule is constrained to occupy a certain place in the molecular arrangement, keeping also an orientation in a specific way. Thus, the crystalline state is an arrangement of molecules displaying both positional and orientational order. However, molecules in the liquid state neither occupy a specific average position nor remain oriented in a particular way. Therefore, the liquid state is an arrangement of molecules having neither positional nor orientational order.[1]

When most crystalline solids melt, the molecules move and tumble randomly, losing their positional and orientational order. However, there are some substances in which the arrangement of the molecules in the melt may preserve some orientational order though the positional order is lost. A new state of aggregation of the matter thus appears, called the liquid crystalline order, characterized by the molecules being free to move as in a liquid, though as they do so they tend to spend a little more time pointing along the direction of orientation than along some other direction. An arrow, called the director of the liquid crystal, represents the direction along which the molecules tend to align.

The order parameter is computed from the second term of Legendre's polynomial

$$S_2 = \langle P_2(\cos\theta) \rangle = \frac{3\langle\cos^2\theta\rangle - 1}{2} \tag{11.1.1}$$

where the angular brackets denote average and θ is the angle that the molecule makes with the director. The values of $\langle\cos^2\theta\rangle$ for crystalline solids and liquids are 1 and 1/3 respectively, so the values of S_2 for these two extreme cases are 1 and 0. Typical values of the order parameter in liquid crystals lie between 0.3 and 0.9. Obviously the order parameter decreases with increasing temperature.

Elongated molecules tend to have stronger attractive forces when they are aligned parallel to one another, and rigidity in their central part hinders bumping motions when the molecules are all pointing in the same direction. On the other hand, flexible attachments to the ends of rigid molecular structures favor the easier positioning of a molecule among other molecules as they chaotically move about. Therefore, the development of liquid crystalline phases or mesophases requires that the length of molecules is significantly greater than their width and they must also have some rigidity in their central regions. It is also advantageous that molecules have a flexible moiety attached at their ends.

Nematic mesophases or nematic liquid crystals are molecular arrangements in which the molecules are preferentially oriented along the director, as schematically shown in Fig. 11.1a. A representative molecule developing nematic mesophases is shown in Fig. 11.2. If the intermolecular forces favor alignment between molecules at a slight angle to one another, the director is not fixed in space but rotates throughout the sample. The mesophase developed is called chiral nematic (see Fig. 11.1b). The distance needed for the director to rotate a full turn is called the pitch of the liquid crystal. A substance may develop nematic or chiral nematic mesophases, but not both.

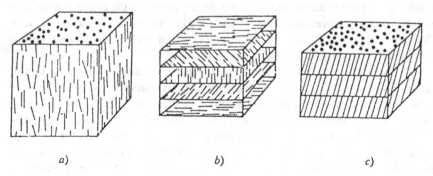

$a)$ $b)$ $c)$

FIG. 11.1   Schemes of a nematic (a), cholesteric (b), and smectic (c) phase. Segments represent molecular rods

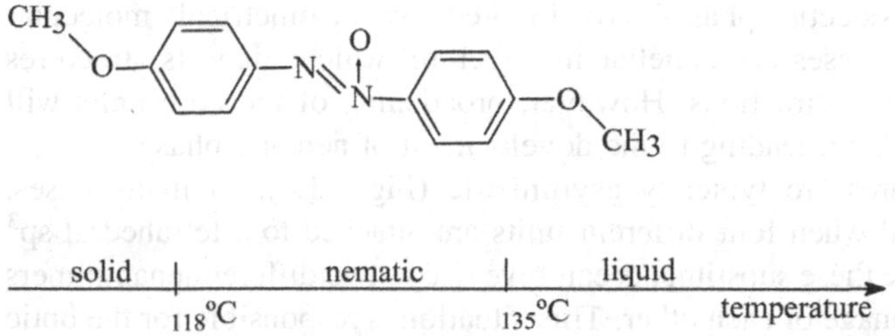

FIG. 11.2 Molecular structure and phase diagram for p-azoxyanisole

Another type of liquid crystal may appear in which there is also a small amount of positional order. In this case molecules are free to bounce quite freely, pointing towards the director, but they also spend slightly more time in regularly spaced arranged layers than in other available positions in-between, giving rise to the so-called smectic mesophase. This third mesophase may be the only liquid crystalline phase that a substance possesses, though in some cases it can also appear at a temperature below the nematic or chiral nematic mesophases of such a substance. Molecules tend randomly to diffuse within each layer in smectic liquid crystals, and therefore the positional order is in one dimension only. The smectic mesophases may be of type A and C which differ in that the director is perpendicular to the layers in the former case and forms an angle other than 90° in the latter, as shown in Fig. 11.1c. A representative molecule developing smectic mesophases of type A and C and nematic mesophases is shown in Fig. 11.3. In some smectic liquid crystals the molecules diffusing within each plane spend more time at special locations than at other locations. The positional order is three-dimensional in this case; several arrangements of the positional positions within each plane are now possible, giving rise to smectic phases A to K. Lateral substituents destabilize the smectic phases, which require more order than

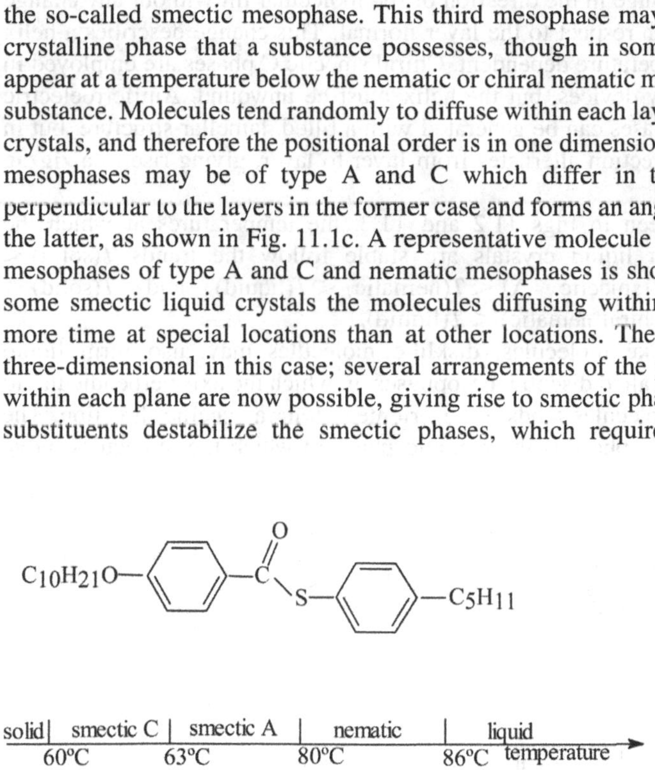

FIG. 11.3 Molecular structure and phase diagram of 4-n-pentylbenzenethio-n-decycloxybenzoate

nematics. Hence, smectic phases are favored by symmetrical molecular structures. Smectic phases are lamellar in structure, which suggests structures with lateral molecular attractions. However, broadening of the core units will disrupt lamellar packing, leading to the development of nematic phases.

Chiral structures are typically asymmetric (Fig. 11.4). In many cases, chirality is produced when four different units are attached to a tetrahedral sp^3 carbon atom because these substituents can give rise to two different enantiomers that are the mirror image of each other. This situation is responsible for the optic activity observed in these materials. Chiral molecules can generate liquid crystalline phases. As indicated above, chiral nematic phases are called cholesteric phases because the first thermotropic liquid crystalline substance found was an ester of cholesteric acid. However, many chiral nematics have nothing in common with cholesterol derivatives. There are many chiral smectic mesophases. In the case of a chiral smectic C phase (S_{C^*}), which is by far the most important, the molecules are also organized in layers, but chirality causes a slight and progressive change in the direction of the molecular tilt without any change in the tilt angle with respect to the layer normal. This change describes a helix whose pitch is temperature dependent. Chiral smectic C phases are employed in ferroelectric display devices, but the helix must be unwound. Antiferroelectric chiral smectic C phases can be generated with a tilted, lamellar structure, but in this case the tilt direction alternates from layer to layer, giving rise to a zigzag structure.

As can be seen in Figs 11.2 and 11.3, the temperatures at which the different phases of liquid crystals are stable follow the trends T(solid) < T(smectic C) < T(smectic A) < T(nematic) < T(liquid) and T(solid) < T(smectic A) < T(chiral nematic) < T(liquid).

Besides rodlike molecules, disklike molecules may also form liquid crystalline phases, called discotic mesophases, in which the axis perpendicular to the plane of the molecules tends to be oriented along a specific direction. The most simple discotic phase is the nematic phase in which the molecules move

$C_{13}H_{27}-$ (molecular structure)

solid	smectic A	chiral nematic	liquid
	71°C	79°C	85°C temperature

FIG. 11.4 Phase diagram and molecular structure for cholesteryl myristate

around randomly, though on average the axis perpendicular to the plane of the molecule spends slightly more time along a special direction called the director. In smectic mesophases most of the molecules tend to position themselves quite randomly in columns arranged in a hexagonal lattice. The positional order in these phases is two-dimensional. Disklike molecules may also display chiral nematic phases in which the director rotates helicoidally throughout the sample, as in the case of rodlike chiral nematic liquid crystals. The general chemical structure of a discotic liquid crystal is given in Fig. 11.5. The central core is a symmetric disk-shaped unit such as benzene or triphenylene. The peripheral dendritic units are identical in order to preserve the discotic structure. As in rodlike systems, the core provides the rigidity, and the flexibility of peripheral units reduces the melting point. Star or macrocyclic structures can also generate mesophases. In many cases the center of the molecule can hold metallic ions, and consequently these systems are potentially useful to transport

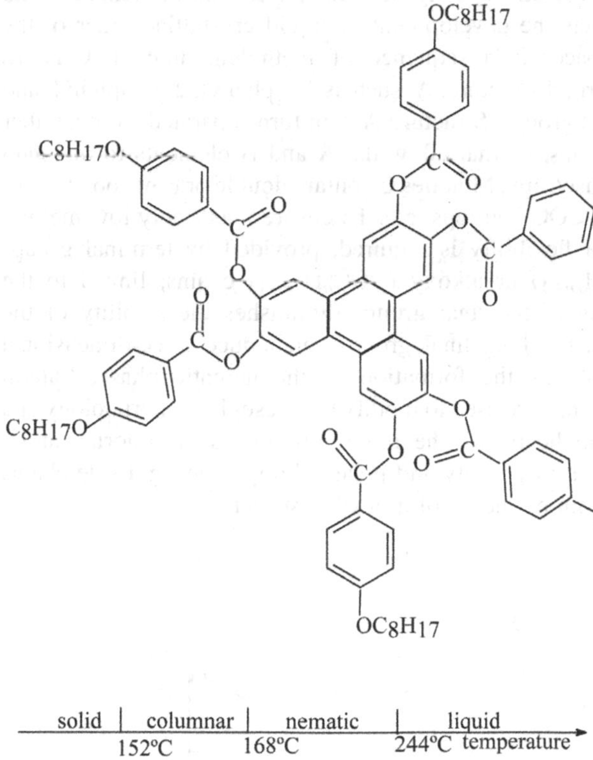

solid	columnar	nematic	liquid
	152°C	168°C	244°C temperature

FIG. 11.5   Scheme and phase diagram of a typical discotic liquid crystal

ions or small molecules (molecular wires or permeable membranes). In these systems, a small perturbation of the discotic structure can eliminate the anisotropy, which is needed to generate mesophases.

Polymers are long-chain molecules or macromolecules in which a chemical motive repeats along the chains. If a rigid, anisotropic, and highly polarizable moiety, such as the central part of a molecular liquid crystal, forms part of the backbone of the repeating unit molecule, the resulting polymer may exhibit a liquid crystal character. The rigid moieties may be separated by flexible segments or spacers whose lengths affect the thermal properties of the polymers, as shown in Fig. 11.6. If the spacers are short, the polymers melt at high temperatures and are often insoluble in common organic solvents. The melting points of liquid crystalline polymers with rigid moieties located in the backbone decrease as both the flexibility and length of the spacer increase.

Side chain polymer liquid crystals may also be obtained by attaching rigid and highly polarizable moieties as side chains (Fig. 11.7) to highly flexible polymers such as polyacrylates and polysiloxanes. The moiety can be separated from the main chain by a relatively long flexible spacer, and the length of the spacer also strongly affects the development of liquid crystalline order of the polymers. The usual spacer is a sequence of methylene units $(-CH_2-)_n$. The spacer is linked to a rigid structure, A, such as 1,4-phenyl, 2,6-naphthyl, and even $trans$-1,4-cyclohexyl groups. Structure A is in turn separated from another rigid moiety, B, by moieties, C, that allow the A and B elements to maintain linearity and dielectric anisotropy. Moieties C contain double or triple bonds (e.g. $-N=N-$ or $-C\equiv C-$), $-COO-$ groups, etc. To ensure reasonably low melting points, a certain level of flexibility is required, provided by terminal groups D, such as alkyl $(-C_nH_{2n+1})$ or alkoxy $(-OC_nH_{2n+1})$ chains, linked to the rigid group B. Branching of the final groups diminishes the stability of the liquid crystalline phases, but long final groups can induce three-dimensional crystallization, which inhibits the formation of the nematic phase. Lateral substituents in the core groups are used to modify the mesophase morphology and enhance properties for applications. The more common is the fluorine atom, which combines high electronegativity and polarizability. The rigid side chains are linked to the main chain by means of a flexible spacer.

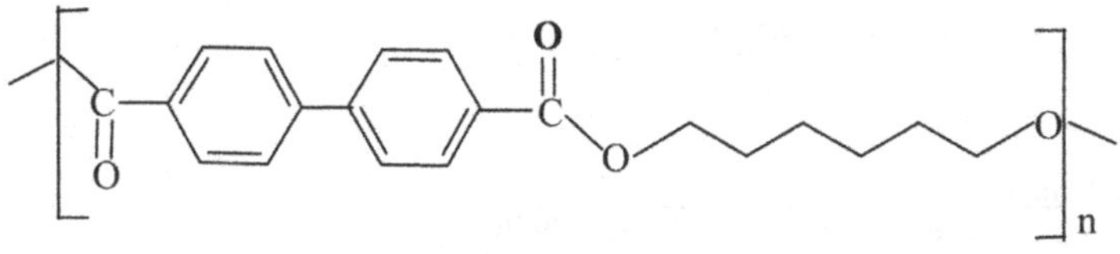

FIG. 11.6   Schemes of liquid crystals with mesogenic groups in the main chain

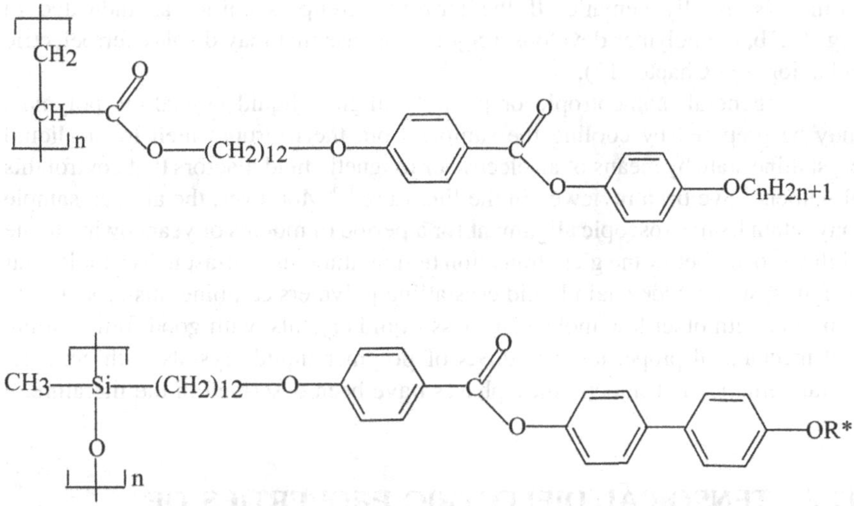

$$CH_2$$
$$CH-C$$
$$\Big]_n$$
$$O-(CH_2)_{12}-O-\bigcirc-C$$
$$O-\bigcirc-OC_nH_{2n+1}$$

$$CH_3-Si-(CH_2)_{12}-O-\bigcirc-C$$
$$O-\bigcirc-\bigcirc-OR^*$$
$$O$$
$$\Big]_n$$

FIG. 11.7 Lateral components of side chain liquid crystalline polymers (see text for details)

The length and flexibility of the spacer affects the thermotropic transitions of liquid crystalline polymers with mesogenic groups in the main chain. The greater the flexibility of the spacer, the lower are the temperatures at which the transitions occur. For liquid crystalline polymers with mesogenic groups in the side chains, the role of the spacer is to decouple the motions of the main backbone from those of the mesogenic groups. For this reason, if the spacer is absent, mesomorphic behavior is not usually generated. However, in practice, the decoupling is only partial. In fact, the spacer length and the flexibility of the backbone are mutually influenced. Thus, siloxane polymers may display a lone nematic phase. However, a transition from smectic to nematic menophases develops as the spacer length increases. In the case of the less flexible acrylic chains with no spacer, mesophases are not developed. For short segments a nematic phase is induced that may give way to smectic phases as the spacer length increases. The spacer length also influences the glass transition temperature through a mechanism of internal plasticization. The odd−even effect of the spacer is displayed in the clearing point or temperature at which the mesophase → isotropic liquid transition occurs. All the possibilities for the shape and size of the mesogenic groups used for low molecular weight compounds can be utilized for polymers. With reference to the terminal groups, high polar groups, such as for example −CN, have a tendency to form smectic phases, whereas in the case of groups with low polarity, such as −OCH$_3$, the phase formed is usually nematic. If the terminal group is chiral, as indicated in Fig. 11.7b, the polymer develops a S_{C^*} mesophase that may display ferroelectric behavior (see Chapter 12).

In general, homeotropic or planarly aligned liquid crystalline polymers may be prepared by cooling the sample from the isotropic melt to the liquid crystalline state by means of an electric or magnetic field. Factors that control this alignment have been reviewed in the literature.[1] Moreover, the aligned sample may retain its macroscopic alignment for a period of months or years owing to the high viscosity below the glass transition temperature, in contrast to low molecular weight systems. Side chain liquid crystalline polymers combine this property, in common with other low molecular mass liquid crystals, with good film-forming and mechanical properties. Hundreds of polymer liquid crystals with nematic, chiral nematic, and smectic mesophases have been described in the literature.

11.2. TENSORIAL DIELECTRIC PROPERTIES OF ANISOTROPIC MATERIALS

For anisotropic systems, the polarization **P** does not lie only in the direction of the applied electric field **E**, and the relation between these magnitudes, in contrast to Eq. (1.8.5), is of tensorial character

$$P_i = \chi_{ij} E_j \tag{11.2.1}$$

Here, χ_{ij} represents the nine coefficients of a second-order tensor, called the dielectric susceptibility tensor. For very low electric fields, the polarization is a linear function of the field. However, to take account of the effects of strong electric fields, as, for example, occurs in nonlinear optics, nonlinear terms must be included in Eq. (11.2.1), that is (see Chapter 13)

$$P_i = \chi_{ij} E_j + \chi_{ijk} E_j E_k + \cdots \tag{11.2.2}$$

where the Einstein summation convention for repeated subscripts has been used. Here, χ_{ijk} is a third-order tensor. This type of tensor also appears in the context of the piezoelectric effects in systems with special crystal symmetry, where the dielectric displacement vector is expressed in terms of the second-order stress tensor (see Chapter 12).

The fact that the second derivative of the free energy of a linear dielectric is independent of the order of differentiation leads to the symmetry of χ_{ij}, and thus only six different coefficients for the susceptibility persist [see Eq. (1.8.10)]. The symmetry of the dielectric susceptibility for a liquid crystal also expresses the intrinsic symmetry of the crystal unit of the system. A set of orthogonal axes can be related to another by means of the three Euler angles. According to Fig. 11.8, the matrix that transforms a vector in a laboratory frame (X, Y, Z) to a new reference frame (x, y, z) is given by

$$
a_{ij} = \begin{vmatrix} a_{Xx} & a_{Xy} & a_{Xz} \\ a_{Yx} & a_{Yy} & a_{Yz} \\ a_{Zx} & a_{Zy} & a_{Zz} \end{vmatrix}
$$

$$
= \begin{vmatrix} \cos\psi & \sin\psi & 0 \\ -\sin\psi & \cos\psi & 0 \\ 0 & 0 & 1 \end{vmatrix} \begin{vmatrix} 1 & 0 & 0 \\ 0 & \cos\theta & \sin\theta \\ 0 & -\sin\theta & \cos\theta \end{vmatrix} \begin{vmatrix} \cos\varphi & \sin\varphi & 0 \\ -\sin\varphi & \cos\varphi & 0 \\ 0 & 0 & 1 \end{vmatrix}
$$

$$
= \begin{vmatrix} \cos\psi\cos\varphi - \cos\theta\sin\psi\sin\varphi & \cos\psi\sin\varphi + \cos\theta\cos\varphi\sin\psi & \sin\psi\sin\theta \\ -\sin\psi\cos\varphi - \cos\theta\cos\psi\sin\varphi & -\sin\psi\sin\varphi + \cos\theta\cos\varphi\cos\psi & \cos\psi\sin\theta \\ \sin\theta\sin\varphi & -\sin\theta\cos\varphi & \cos\theta \end{vmatrix}
$$

$$(11.2.3)$$

It should be noted that, as corresponds to a rotation, the sums of squares of components in columns or rows are unity, while the sums of products of components in adjacent columns or rows are zero. This means that the six different components of the matrix given by Eq. (11.2.3) are not independent. Then, the transformation rule for the dielectric susceptibility in terms of the Euler

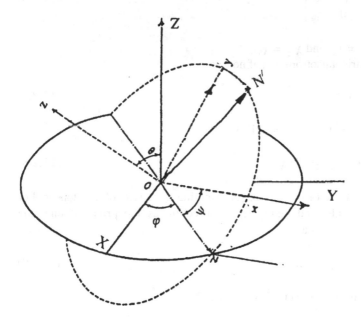

FIG. 11.8 Euler angles that define the three rotations appearing in Eq. (11.2.3)

angles can be expressed as

$$\chi'_{ij} = a_{ik}a_{jm}\chi_{km} \tag{11.2.4}$$

where X' the repeated subscripts indicate summation. Equation (11.2.4) can alternatively be written as

$$\chi' = \mathbf{A}\chi\mathbf{A}^T \tag{11.2.5}$$

A symmetric matrix representing a second-order tensor can be diagonalized in its three principal components. The axes thus defined are called principal axes. In this way, the tensor representing the susceptibility can also be diagonalized by means of the transformation given by Eq. (11.2.5). If the material under study has some type of symmetry, the principal axes are defined accordingly. For uniaxial liquid crystalline phases such as nematic or smectic A, for example, only one symmetry axis can be defined, which is parallel to the director, and there are only two independent components for the diagonalized susceptibility tensor. If the z axis is chosen as the symmetry axis, then

$$\chi_{ij} = \begin{vmatrix} \chi_\perp & 0 & 0 \\ 0 & \chi_\perp & 0 \\ 0 & 0 & \chi_\parallel \end{vmatrix} \tag{11.2.6}$$

where $\chi_{xx} = \chi_{yy} = \chi_\perp$ and $\chi_{zz} = \chi_\parallel$.

The dielectric anisotropy is defined as

$$\Delta\chi = \chi_\parallel - \chi_\perp \tag{11.2.7}$$

and the mean value of the susceptibility as

$$\bar{\chi} = \tfrac{1}{3}(\chi_\parallel + 2\chi_\perp) = \tfrac{1}{3}\mathrm{Tr}\,\chi \tag{11.2.8}$$

where $\mathrm{Tr}\,\chi$ is the trace of the matrix representing the susceptibility tensor. For biaxial liquid crystals, and taking the z direction as the principal one, the anisotropy can be written as

$$\Delta\chi = \chi_{zz} - \tfrac{1}{2}(\chi_{xx} + \chi_{yy}) \tag{11.2.9}$$

The mean value of the susceptibility is given by

$$\bar{\chi} = \tfrac{1}{3}(\chi_{xx} + \chi_{yy} + \chi_{zz}) = \tfrac{1}{3}\mathrm{Tr}\,\chi \tag{11.2.10}$$

11.3. MACROSCOPIC ORDER PARAMETER

The anisotropy of liquid crystals is originated by the orientational order at molecular level. However, macroscopic anisotropy can only be expressed in terms of macroscopic tensorial properties. In the present context, the macroscopic anisotropy can be defined in terms of the anisotropic part of the dielectric susceptibility, that is

$$\chi_{ij}^a = \chi_{ij} - \bar{\chi}\delta_{ij} \tag{11.3.1}$$

where δ_{ij} is the Kronecker symbol. If the maximum value for the principal component (taken along the z axis) is conventionally set equal to one, the macroscopic order parameter Q can be written as

$$Q_{zz\,max} = 1 = C(\chi_{zz} - \bar{\chi}) = \tfrac{2}{3}C[\chi_{zz} - \tfrac{1}{2}(\chi_{xx} + \chi_{yy})] = \tfrac{2}{3}C\,\Delta\chi_{max} \tag{11.3.2}$$

where C is a constant and $\Delta\chi_{max}$ is the anisotropy of a fully aligned sample for which the order parameter is the unit. The following expression is obtained for the macroscopic order parameter

$$Q_{ij} = \tfrac{3}{2}(\Delta\chi_{max})^{-1}(\chi_{ij} - \tfrac{1}{3}\chi_{ll}\delta_{ij}) \tag{11.3.3}$$

where the first subscript in Q_{ij} indicates the macroscopic character of the order parameter and χ_{ll} is the trace of the susceptibility tensor. According to this definition, the traceless macroscopic order parameter Q for a uniaxial liquid crystal in the principal axis is given by

$$Q = \begin{vmatrix} -\tfrac{1}{2}Q & 0 & 0 \\ 0 & -\tfrac{1}{2}Q & 0 \\ 0 & 0 & Q \end{vmatrix} \tag{11.3.4}$$

where

$$Q = (\Delta\chi_{max})^{-1}(\chi_\| - \chi_\perp) \tag{11.3.5}$$

For a system with different susceptibilities along the three principal axes, that is, for a biaxial phase, the corresponding macroscopic order parameter in the principal axis will be given by

$$Q = \begin{vmatrix} -\tfrac{1}{2}(Q - P) & 0 & 0 \\ 0 & -\tfrac{1}{2}(Q + P) & 0 \\ 0 & 0 & Q \end{vmatrix} \tag{11.3.6}$$

where

$$Q = (\Delta\chi_{max})^{-1}[\chi_{zz} - \tfrac{1}{2}(\chi_{xx} + \chi_{yy})] \quad \text{and}$$
$$P = \tfrac{3}{2}(\Delta\chi_{max})^{-1}(\chi_{xx} - \chi_{yy}) \quad\quad\quad\quad\quad (11.3.7)$$

It should be noted that there is no universally accepted way to define the order parameter, and in fact other definitions are possible (see, for example, reference [1]).

11.4. MICROSCOPIC ORDER PARAMETER

As has been seen in the preceding sections, the connection between macroscopic and molecular properties is an issue of the highest interest and a challenge to statistical mechanics. The most important factor to be considered in this regard is to express the dielectric permittivity in terms of the dipole moment of a single molecule, though in many cases the interaction between molecules is also relevant. This is the situation in more complicated cases already discussed in Chapter 2. In the context of liquid crystals, the molecular contribution of the orientation to the macroscopic orientation will be considered additive, and consequently the angular distribution of molecular orientations for a single particle is enough for our purposes. This distribution function gives the probability of a single molecule having a specific orientation with respect to a fixed reference frame. In this respect, the director **n** is defined as the unit vector along the symmetry axis of the orientation distribution. This direction is usually taken along the Z axis in the laboratory reference frame. It is interesting to note that the probability distribution of the principal axis of the molecules can be represented by an ellipsoidal surface with rotational symmetry about the director. Thus, the probability of a molecule having a particular direction is proportional to the distance from the origin of coordinates to which the surface is referred to the surface along this direction. This is in fact a special feature of symmetric second-order tensors.

For axially symmetrical molecules an axial distribution of orientations with respect to the director can be defined. The distribution can also be interpreted as a distribution of local directors. For this purpose it is convenient to define the angular distribution in terms of Legendre polynomials (see Appendix A of Chapter 1), a special type of spherical harmonics. These are a set of orthogonal polynomials whose odd components are not required for molecules for which the local directors $\mathbf{n_l}$ and $-\mathbf{n_l}$ are equivalent, that is, for molecules with inversion symmetry. Hence, for uniaxial molecules in uniaxial phase, the angular distribution can be written as

$$f(\theta) = \tfrac{1}{2}[1 + a_2 P_2(\cos\theta) + a_4 P_4(\cos\theta) + \cdots] \quad\quad\quad (11.4.1)$$

The coefficients of Eq. (11.4.1) can be calculated by multiplying both sides of this equation by $P_{2n}(\cos\theta)$ and further integration over θ. Then, according to Eq. (1.A.7), and taking into account that

$$\int_0^\pi f(\theta)\sin\theta\,d\theta = 1 \qquad (11.4.2)$$

the generalized expression for the coefficients of Eq. (11.4.1) is given by

$$a_{2n} = (4n+1)\frac{\int_0^\pi P_{2n}(\cos\theta)f(\theta)\sin\theta\,d\theta}{\int_0^\pi f(\theta)\sin\theta\,d\theta} \qquad (11.4.3)$$

Since one of the factors is the average value of $P_{2n}(\cos\theta)$ over the distribution function $f(\theta)$, the coefficients adopt the following values

$$a_{2n} = 1, 5, 9, \ldots \qquad \text{for } n = 0, 1, 2, \ldots \qquad (11.4.4)$$

Consequently, the distribution function can be written as

$$f(\theta) = \tfrac{1}{2}[1 + 5\langle P_2(\cos\theta)\rangle P_2(\cos\theta) + 9\langle P_4(\cos\theta)\rangle P_4(\cos\theta) + \cdots] \qquad (11.4.5)$$

Thus, the coefficients in the preceding expression can be interpreted as the order parameters for the distribution function. If the distribution function of a molecule whose orientation is defined by Euler's angles is required, then $f(\theta, \psi, \varphi)$ must be expanded in terms of a set of orthogonal functions, and for this purpose the choice of the Wigner rotation matrices is convenient. Attard and Nordio et al.[2] developed the general procedure described below. In the case of a uniaxial liquid crystalline phase, having uniaxial or biaxial molecules with inversion symmetry, the distribution function can be expressed in terms of two parameters as follows

$$f(\theta,\psi) = \frac{1}{4\pi}(1 + 5S_{ij}l_il_j + 9S_{ijkm}l_il_jl_kl_m + \cdots) \qquad (11.4.6)$$

The factor 2π in the denominator of the prefactor arises from the integration over φ. In this equation, $S_{ij\ldots}$ is a molecular order parameter of tensorial character and $l_{ij\ldots}$ represents the cosine of the angle formed by the macroscopic director with the molecular axis. In particular, the second-order tensor S_{ij} is the so-called Saupe-order parameter matrix[3] whose definition as averages of l_{ij} is given by

$$S_{ij} = \tfrac{1}{2}\langle 3l_il_j - \delta_{ij}\rangle \qquad (11.4.7)$$

The Saupe-order parameter for uniaxial molecules reduces to

$$S_{zz} = \tfrac{1}{2}\langle 3\cos^2\theta - 1\rangle = \langle P_2(\cos\theta)\rangle \qquad \text{and}$$

$$S_{xx} = S_{yy} = -\tfrac{1}{2}S_{zz} \tag{11.4.8}$$

In matricial notation, the Saupe-order parameter takes the same form as Eq. (11.3.4) but now without subscripts.

For biaxial order at molecular level, the matrix must be expressed in terms of a new order parameter defined as

$$D = S_{xx} - S_{yy} \tag{11.4.9}$$

which, taking into account Eq. (11.2.3), can be written in tensorial notation as

$$D_{ij} = \tfrac{3}{2}\langle a_{Zx}a_{Zx} - a_{Zy}a_{Zy}\rangle = -\tfrac{3}{2}\langle \sin^2\theta\cos 2\varphi\rangle \tag{11.4.10}$$

The corresponding Saupe matrix is

$$\mathbf{S} = \begin{vmatrix} -\tfrac{1}{2}(S-D) & 0 & 0 \\ 0 & -\tfrac{1}{2}(S+D) & 0 \\ 0 & 0 & S \end{vmatrix} \tag{11.4.11}$$

For uniaxial molecules in a biaxial phase, for example, tilted smectic phases, an order parameter P must be introduced with the following definition

$$P = \tfrac{3}{2}\langle a_{Xz}a_{Xz} - a_{Yz}a_{Yz}\rangle = -\tfrac{3}{2}\langle \sin^2\theta\cos 2\psi\rangle \tag{11.4.12}$$

In this case, the Saupe matrix is given by

$$\mathbf{S} = \begin{vmatrix} -\tfrac{1}{2}(S-P) & 0 & 0 \\ 0 & -\tfrac{1}{2}(S+P) & 0 \\ 0 & 0 & S \end{vmatrix} \tag{11.4.13}$$

which is similar to the matrix given by Eq. (11.3.6).

The tensorial representation can be used to describe the orientational ordering of biaxial molecules in biaxial phases. In this case, a fourth-order tensor is introduced, namely

$$S_{ij,km} = \tfrac{1}{2}\langle 3 l_{k,i}l_{m,j} - \delta_{ij}\delta_{km}\rangle \tag{11.4.14}$$

In Eq. (11.4.14), $l_{k,i}$ represents the cosine of the angles between the molecular and the laboratory reference frames. By means of a suitable choice of the sets of axes corresponding to these two reference frames, the number of components of $S_{ij,km}$ can be reduced from 81 to 9 when $i = j$ and $k = m$. Then, Eq. (11.4.14) can be

written as

$$S_{ij}^k = \tfrac{1}{2}\langle l_{k,i} l_{k,j} - \delta_{ij}\rangle \tag{11.4.15}$$

which represents three diagonal matrices, one for each axis. In tensorial notation

$$S_{ij}^k = \begin{vmatrix} S_{xx}^X & S_{yy}^X & S_{zz}^X \\ S_{xx}^Y & S_{yy}^Y & S_{zz}^Y \\ S_{xx}^Z & S_{yy}^Z & S_{zz}^Z \end{vmatrix} \tag{11.4.16}$$

The components of this matrix can be interpreted as follows: The long-axis order is described by $S_{zz}^Z = S$. For uniaxial molecules in a biaxial phase, the order is given by $P = S_{zz}^X - S_{zz}^Y$. For biaxial molecules in a uniaxial phase, the order is described by $D = S_{xx}^Z - S_{yy}^Z$. In this case, the biaxial order at molecular level can also be defined with respect to the X and Y axes of the laboratory frame as $D' = S_{xx}^X - S_{yy}^X$ and $D'' = S_{xx}^Y - S_{yy}^Y$. In the case of uniaxial phases, X and Y are equivalent, and $D' = D''$. However, for biaxial phases, the biaxiality can be described in terms of a new order parameter

$$C = D' - D'' = (S_{xx}^X - S_{yy}^X) - (S_{xx}^Y - S_{yy}^Y) \tag{11.4.17}$$

C can be expressed in terms of Euler's angles by using the definition (11.4.15) and the matrix (11.2.3) as follows

$$S_{xx}^X = \tfrac{3}{2}(\cos^2\theta \sin^2\varphi \sin^2\psi + \cos^2\varphi\cos^2\psi - \tfrac{1}{2}\cos\theta\sin 2\varphi\sin 2\psi)$$
$$S_{yy}^X = \tfrac{3}{2}(\cos^2\theta \sin^2\psi \cos^2\varphi + \cos^2\psi\sin^2\varphi + \tfrac{1}{2}\cos\theta\sin 2\varphi\sin 2\psi)$$
$$S_{xx}^Y = \tfrac{3}{2}(\cos^2\theta \sin^2\varphi\cos^2\psi + \cos^2\varphi\sin^2\psi + \tfrac{1}{2}\cos\theta\sin 2\varphi\sin 2\psi)$$
$$S_{yy}^Y = \tfrac{3}{2}(\cos^2\theta \cos^2\varphi\cos^2\psi + \sin^2\varphi\sin^2\psi - \tfrac{1}{2}\cos\theta\sin 2\varphi\sin 2\psi)$$

$$\tag{11.4.18}$$

Substitution of Eq. (11.4.18) into Eq. (11.4.17) gives

$$C = \tfrac{3}{2}\langle(1 + \cos^2\theta)\cos 2\varphi\cos 2\psi - 2\cos\theta\sin 2\varphi\sin 2\psi\rangle \tag{11.4.19}$$

Although some translational order can be present in some type of mesophase, only these orientational order parameters will be taken into account to relate molecular to macroscopic properties as experimentally measured.

It should be noted that, besides the mesophase order parameters S, D, and C, a *director order parameter* S_d should be defined for aligned phases. This new parameter characterizes the degree of alignment of the macroscopic sample. The director order parameter is related to the angle between the director and the field

ϑ by the following expression

$$S_{\mathrm{d}} = \tfrac{1}{2}(3\langle\cos^2\vartheta\rangle - 1) \tag{11.4.20}$$

Thus, in a homeotropically aligned mesophase, the director and the aligning electric field are parallel ($\mathbf{n} \parallel \mathbf{E}$), and the director order parameter S_{d} is unity. On the other hand, for a planarly aligned mesophase, the director and the aligning field are perpendicular ($\mathbf{n} \perp \mathbf{E}$) and the director order parameter is $-1/2$. For uniaxial nematic liquid crystals, a method has been proposed (see Problem 3) for obtaining expressions for the intermediate degree of alignment in terms of the averaged dielectric permittivities and the permittivities parallel and perpendicular to the director $\mathbf{n}$.

11.5. DIELECTRIC SUSCEPTIBILITIES

Let α_{xx}^l, α_{yy}^l, α_{zz}^l be the principal components of the dipolar polarizability in a local reference frame. According to Eq. (11.2.4), the macroscopic susceptibility in the laboratory frame will be

$$\chi_{ij} = N_1 \langle a_{ik} a_{jm} \rangle \alpha_{km}^l \tag{11.5.1}$$

where N_1 is the number of molecules per unit volume, and the angular brackets denote the average of the molecular property over all orientations of these molecules. The anisotropic part of the macroscopic susceptibility χ_{ij}^{a}, as given by Eq. (11.3.1), can be expressed in terms of the principal components of the polarizability in the local reference frame according to

$$\chi_{ij}^a = \tfrac{1}{3} N_1 [\langle 3a_{iz}a_{jz} - \delta_{ij}\rangle [\alpha_{zz}^l - \tfrac{1}{2}(\alpha_{xx}^l + \alpha_{yy}^l)] \\ + \tfrac{1}{2}(3a_{ix}a_{jx} - 3a_{iy}a_{jy})(\alpha_{xx}^l - \alpha_{yy}^l)] \tag{11.5.2}$$

If the phase is uniaxial but formed by biaxial molecules

$$\chi_{\parallel}^a = \tfrac{2}{3} N_1 [S(\alpha_{zz}^l - \tfrac{1}{2}(\alpha_{xx}^l + \alpha_{yy}^l)) + \tfrac{1}{2}D(\alpha_{xx}^l - \alpha_{yy}^l)] \tag{11.5.3}$$

where S and D are respectively given by Eqs (11.4.8) and (11.4.10). From $\chi_{\perp}^a = -\tfrac{1}{2}\chi_{\parallel}^a$, the corresponding value for the transversal susceptibility is obtained, namely

$$\chi_{\perp}^a = -\frac{N_1}{3}\left[S\left(\alpha_{zz}^l - \frac{1}{2}(\alpha_{xx}^l + \alpha_{yy}^l)\right) + \frac{1}{2}D(\alpha_{xx}^l - \alpha_{yy}^l) \right] \tag{11.5.4}$$

If this result is compared with those obtained in Eq. (11.3.1) for the macroscopic order parameter, it is found that

$$\Delta\chi_{\text{max}} = N_1\left[\alpha^1_{zz} - \tfrac{1}{2}(\alpha^1_{xx} + \alpha^1_{yy})\right] \tag{11.5.5}$$

Accordingly, $\Delta\chi_{\text{max}}$ is N times the molecular anisotropy.

It is worth noting that measurement of the macroscopic anisotropy does not allow splitting of the contributions of parameters S and D. For a biaxial phase, the diagonal elements of the anisotropic tensor are independent. In this case, the three parameters to describe the anisotropy are

$$\chi^a_{\parallel} = \frac{2N_1}{3}\left\{S\left[\alpha^1_{zz} - \frac{1}{2}(\alpha^1_{xx} + \alpha^1_{yy})\right] + \frac{1}{2}D(\alpha^1_{xx} - \alpha^1_{yy})\right\}$$

$$\chi^a_{\perp 1} = -\frac{N_1}{3}\left\{\left[(S+P)\left(\alpha^1_{zz} - \frac{1}{2}(\alpha^1_{xx} + \alpha^1_{yy})\right)\right] + \frac{1}{2}(D+C)(\alpha^1_{xx} - \alpha^1_{yy})\right\}$$

$$\chi^a_{\perp 2} = -\frac{N_1}{3}\left\{(S-P)\left[\alpha^1_{zz} - \frac{1}{2}(\alpha^1_{xx} + \alpha^1_{yy})\right] + \frac{1}{2}(D-C)(\alpha^1_{xx} - \alpha^1_{yy})\right\}$$

$$\tag{11.5.6}$$

If $C = 0$ in Eq. (11.5.6), Eqs (11.5.3) and (11.5.4) are recovered.

The present theory assumes that molecular properties contribute additively to the macroscopic experimental data, and, on the other hand, intermolecular interactions are not considered.

11.6. HIGH-FREQUENCY RESPONSE

The optical properties of liquid crystals determine their high-frequency dielectric response. It has been shown in Chapter 1 that, if the Lorentz local field is assumed, Eq. (1.9.15) gives the relationship between the high-frequency polarizability and the refractive index. In anisotropic liquid crystals, an isotropic model for the internal field can be assumed, and the refractive indexes along the principal directions can be written as

$$\frac{n^2_i - 1}{n^2 + 2} = \frac{4\pi}{3}N_1\langle\alpha_{ii}\rangle \tag{11.6.1}$$

where $\langle\alpha_{ii}\rangle$ is the average value along the principal refractive index, and

$$n^2 = \tfrac{1}{3}(n^2_x + n^2_y + n^2_z) \tag{11.6.2}$$

The birefringence Δn is defined as

$$\Delta n = n_z - \tfrac{1}{2}(n_x + n_y) \tag{11.6.3}$$

According to the results of the preceding section, the mean polarizabilities can be expressed in terms of the molecular components and the orientational order parameters. For uniaxial liquid crystals the optical axis lies along the macroscopic director and the polarizabilites parallel and perpendicular to the axis are

$$\begin{aligned}
\langle \alpha_\parallel \rangle &= \bar{\alpha} + \tfrac{2}{3}[S(\alpha_{zz} - \tfrac{1}{2}(\alpha_{xx} + \alpha_{yy}))] + \tfrac{1}{2}D(\alpha_{xx} - \alpha_{yy}) \\
\langle \alpha_\perp \rangle &= \bar{\alpha} - \tfrac{1}{3}[S(\alpha_{zz} - \tfrac{1}{2}(\alpha_{xx} + \alpha_{yy}))] + \tfrac{1}{2}D(\alpha_{xx} - \alpha_{yy})
\end{aligned} \tag{11.6.4}$$

where $\bar{\alpha} = \tfrac{1}{3}\operatorname{trace} \boldsymbol{\alpha}$.

In future calculations, the parameters α_1 and α_2 defined as

$$\alpha_{zz} - \tfrac{1}{2}(\alpha_{xx} + \alpha_{yy}) = \alpha_1, \qquad \alpha_{xx} - \alpha_{yy} = \alpha_2 \tag{11.6.5}$$

will be used. For molecules with axial symmetry ($\alpha_2 = 0$), Eqs (11.6.1) and (11.6.4) lead to

$$\frac{n_\parallel^2 - 1}{n^2 + 2} - \frac{n_\perp^2 - 1}{n^2 + 2} = \frac{4\pi}{3}N_1\langle \alpha_\parallel - \alpha_\perp \rangle = \frac{4\pi}{3}N_1 S(\alpha_{zz} - \alpha_{xx})$$

$$= \frac{4\pi}{3}N_1 S \Delta\alpha \tag{11.6.6}$$

In a similar way

$$\sum_{i=1}^{3} \frac{n_i^2 - 1}{n^2 + 2} = \frac{4\pi}{3}N_1 \sum \langle \alpha_{ii} \rangle \tag{11.6.7}$$

This expression can be written as

$$\frac{n_x^2 + n_y^2 + n_z^2 - 3}{n^2 + 2} = \frac{4\pi}{3}N_1\langle \alpha_\parallel + 2\alpha_\perp \rangle = 4\pi N_1 \bar{\alpha} \tag{11.6.8}$$

where $\bar{\alpha} = \tfrac{1}{3}(\alpha_\parallel + 2\alpha_\perp)$. Equations (11.6.2) and (11.6.8) lead to

$$\frac{n^2 - 1}{n^2 + 2} = \frac{4\pi}{3}N_1 \bar{\alpha} \tag{11.6.9}$$

Finally, from Eqs (11.6.6) and (11.6.9) one obtains

$$\frac{n_{\parallel}^2 - n_{\perp}^2}{n^2 - 1} = S\frac{\Delta\alpha}{\bar{\alpha}} \tag{11.6.10}$$

Thus, from a plot of $(n_{\parallel}^2 - n_{\perp}^2)/(n^2 - 1)$ vs. $\Delta\alpha/\bar{\alpha}$, the order parameter S can be found.

For biaxial liquid crystalline phases, the three refractive indexes can be expressed in terms of three parameters as in Eq. (11.5.6), namely

$$\langle\alpha_{\parallel}\rangle = \bar{\alpha} + \tfrac{2}{3}(S\alpha_1 + \tfrac{1}{2}D\alpha_2)$$

$$\langle\alpha_{\perp 1}\rangle = \bar{\alpha} - \tfrac{1}{3}[(S+P)\alpha_1 + \tfrac{1}{2}(D+C)\alpha_2] \tag{11.6.11}$$

$$\langle\alpha_{\perp 2}\rangle = \bar{\alpha} - \tfrac{1}{3}[(S-P)\alpha_1 + \tfrac{1}{2}(D-C)\alpha_2]$$

11.7. STATIC DIELECTRIC PERMITTIVITIES

The purpose of this section is to obtain an expression similar to the Onsager equation [Eq. (1.13.12)] for anisotropic materials. The Onsager equation is the first step to obtaining the Fröhlich–Kirkwood formulation for the static permittivity [Eqs (1.16.6) or (1.16.8)] valid for condensed matter and, in particular, polymers. A convenient starting point is Eq. (1.13.11), valid for isotropic materials and written in the following form

$$\chi_0\mathbf{E}_0 = N_1\left(\alpha\mathbf{E}_i + \frac{\mu_v^2\mathbf{E}_d}{3kT}\right) \tag{11.7.1}$$

where χ_0 is taken in the low-frequency limit and $\mathbf{E}_i$ and $\mathbf{E}_d$ are respectively given by Eqs (P.1.7.4) and (P.1.7.2). In fact, substitution of these equations into Eq. (11.7.1) leads to

$$\chi_0 = N_1\left\{\alpha + \frac{1}{1 - \alpha/a^3[2(\varepsilon-1)/(2\varepsilon+1)]}\frac{\mu_v^2}{3kT}\right\}$$

$$\times \frac{(3\varepsilon)/(2\varepsilon+1)}{1 - \alpha/a^3[2(\varepsilon-1)/(2\varepsilon+1)]} \tag{11.7.2}$$

or

$$\chi_0 = N_1\left(\alpha + \frac{1}{1 - \alpha f}\frac{\mu_v^2}{3kT}\right)\frac{g}{1 - \alpha f} \tag{11.7.3}$$

where

$$f = \frac{1}{a^3} \frac{2(\varepsilon - 1)}{2\varepsilon + 1} \qquad \text{and} \qquad g = \frac{3\varepsilon}{2\varepsilon + 1} \tag{11.7.4}$$

In Eqs (11.7.1) to (11.7.3), α is the mean polarizability. For this reason, in the case of anisotropic materials, the isotropic part of the dielectric susceptibility can be written as

$$\frac{1}{3}\chi_{0ii} = N_1 g h \left(\bar{\alpha} + \frac{\mu_v^2}{3kT} h \right) \tag{11.7.5}$$

where $h = 1/(1 - \alpha f)$ and $\bar{\alpha} = \frac{1}{3}(\alpha_{xx} + \alpha_{yy} + \alpha_{zz})$. The anisotropic part of the susceptibility, given by Eq. (11.3.1), can be calculated in a similar way by taking the principal components of the microscopic susceptibility along the molecular axis. For a symmetrically axial molecule, an average value of the dipole moment must be taken over the two directions as follows

$$\bar{\mu}_i = \frac{\mu_{i+} \exp -[(u_0 - \mu_i E_{id})/kT] + \mu_{i-} \exp -[(u_0 + \mu_i E_{id})/kT]}{\exp -[(u_0 - \mu_i E_{id})/kT] + \exp -[(u_0 + \mu_i E_{id}/kT]} \tag{11.7.6}$$

where $\mu_{i+} = -\mu_{i-} = \mu_i$ and u_0 is the energy in the absence of an electric field. Accordingly, Eq. (11.7.6) can be written as

$$\bar{\mu}_i = \mu_i \tanh \frac{\mu_i E_{id}}{kT} \tag{11.7.7}$$

For low electric fields, $\mu_i E_{id} \ll kT$ and

$$\tanh \frac{\mu_i E_{id}}{kT} \simeq \frac{\mu_i E_{id}}{kT} \tag{11.7.8}$$

Then, Eq. (11.7.7) can be approximated by

$$\bar{\mu}_i = \frac{\mu_i^2 E_{id}}{kT} \tag{11.7.9}$$

In these conditions, the anisotropic part of the susceptibility along the principal (molecular) axis can be written as

$$\chi_{0ii} = N_1 g h \left(\alpha_{ii} + \frac{\mu_i^2 h}{kT} \right) \tag{11.7.10}$$

The principal components for the susceptibility and the dielectric permittivity can be obtained from Eqs (11.7.10), (11.5.6), and (11.6.11), together with the definitions (11.6.5), as follows

$$\chi_{0\parallel} = \frac{1}{4\pi}(\varepsilon_{\parallel} - 1)$$

$$= N_1 gh \left[\bar{\alpha} + \frac{2}{3}\alpha_1 S + \frac{1}{3}\alpha_2 D + \frac{h}{3kT}\mu_z^2(1+2S) \right.$$

$$\left. + \mu_x^2(1 - S + D) + \mu_y^2(1 - S - D) \right]$$

$$\chi_{0\perp 1} = \frac{1}{4\pi}(\varepsilon_{\perp 1} - 1)$$

$$= N_1 gh \left\{ \bar{\alpha} - \frac{1}{3}\alpha_1(S+P) - \frac{1}{6}\alpha_2(D+C) + \frac{h}{6kT}[2\mu_z^2(1 - S - P) \right.$$

$$\left. + \mu_x^2(2 + S + P - D - C) + \mu_y^2(2 + S + P + D + C)] \right\}$$

$$\chi_{0\perp 2} = \frac{1}{4\pi}(\varepsilon_{\perp 2} - 1)$$

$$= N_1 gh \left\{ \bar{\alpha} - \frac{1}{3}\alpha_1(S-P) - \frac{1}{6}\alpha_2(D-C) + \frac{h}{6kT}[2\mu_z^2(1 - S + P) \right.$$

$$\left. + \mu_x^2(2 + S - P - D + C) + \mu_y^2(2 + S - P + D - C)] \right\}$$

$$(11.7.11)$$

It is easy to show that

$$\frac{1}{4\pi}(\bar{\varepsilon}_0 - 1) = N_1 gh \left(\bar{\alpha} + \frac{\mu^2 h}{3kT} \right) \tag{11.7.12}$$

where

$$\mu^2 = \mu_z^2 + \mu_x^2 + \mu_y^2 \tag{11.7.13}$$

Equation (11.7.12) indicates that the mean value of the permittivity is independent of the orientational order. Experiments are not always in agreement with this prediction. In fact, changes in $\bar\varepsilon$ can be observed in liquid crystalline phase transitions.

For a uniaxial nematic liquid crystalline phase, and defining the longitudinal and transversal components of the dipole moment according to $\mu_z = \mu_1$ and $\mu_t^2 = \mu_x^2 + \mu_y^2$ respectively, Eqs (11.7.11) can be written as

$$\frac{1}{4\pi}(\varepsilon_{0\|} - 1) = N_1 hg\left\{\bar\alpha + \frac{2}{3}\Delta\alpha S + \frac{h}{3kT}[\mu_1^2(1 + 2S) + \mu_t^2(1 - S)]\right\}$$

$$\frac{1}{4\pi}(\varepsilon_{0\perp} - 1) = N_1 hg\left\{\bar\alpha - \frac{1}{3}\Delta\alpha S + \frac{h}{3kT}\left[\mu_1^2(1 - S) + \mu_t^2\left(1 + \frac{1}{2}S\right)\right]\right\}$$

$$(11.7.14)$$

where $\Delta\alpha = \alpha_{zz} - \alpha_{xx}$.

By making

$$N_1 hg\left(\bar\alpha + \frac{2}{3}\Delta\alpha S\right) = \frac{1}{4\pi}(\varepsilon_{\infty\|} - 1)$$

$$N_1 hg\left(\bar\alpha - \frac{1}{3}\Delta\alpha S\right) = \frac{1}{4\pi}(\varepsilon_{\infty\perp} - 1) \qquad (11.7.15)$$

$$G = \frac{4\pi N_1 h^2 g}{3kT}$$

alternative forms for Eqs (11.7.14) are obtained, namely

$$\varepsilon_{0\|} = \varepsilon_{\infty\|} + \frac{G}{3kT}[\mu_1^2(1 + 2S) + \mu_t^2(1 - S)]$$

$$= \varepsilon_{\infty\|} + \frac{G}{3kT}\mu^2[1 + S(3\cos^2\beta - 1)]$$

$$(11.7.16)$$

$$\varepsilon_{0\perp} = \varepsilon_{\infty\perp} + \frac{G}{3kT}[\mu_1^2(1 - S) + \mu_t^2(1 + S/2)]$$

$$= \varepsilon_{\infty\perp} + \frac{G}{3kT}\mu^2\left[1 - \frac{S}{2}(3\cos^2\beta - 1)\right]$$

which are the so-called Maier–Meier equations [4]. In these equations, β is the angle between the dipole and the direction of the principal axis of the molecule.

By considering dipole–dipole interactions, generalizations of Eqs (11.7.11) lead to

$$\chi_{0\|} = \frac{1}{4\pi}(\varepsilon_\| - 1)$$

$$= N_1 g h \left[\bar{\alpha} + \frac{2}{3}\alpha_1 S + \frac{1}{3}\alpha_2 D + \frac{hg_\|}{3kT}\mu_z^2(1 + 2S) \right.$$

$$\left. + \mu_x^2(1 - S + D) + \mu_y^2(1 - S - D) \right]$$

$$\chi_{0\perp 1} = \frac{1}{4\pi}(\varepsilon_{\perp 1} - 1)$$

$$= N_1 g h \left\{ \bar{\alpha} - \frac{1}{3}\alpha_1(S + P) - \frac{1}{6}\alpha_2(D + C) + \frac{hg_{\perp 1}}{6kT}[2\mu_z^2(1 - S - P) \right.$$

$$\left. + \mu_x^2(2 + S + P - D - C) + \mu_y^2(2 + S + P + D + C)] \right\}$$

$$\chi_{0\perp 2} = \frac{1}{4\pi}(\varepsilon_{\perp 2} - 1)$$

$$= N_1 g h \left\{ \bar{\alpha} - \frac{1}{3}\alpha_1(S - P) - \frac{1}{6}\alpha_2(D - C) + \frac{hg_{\perp 2}}{6kT}[2\mu_z^2(1 - S + P) \right.$$

$$\left. + \mu_x^2(2 + S - P - D + C) + \mu_y^2(2 + S - P + D - C)] \right\}$$

$$(11.7.17)$$

where $g_\|$, $g_{\perp 1}$ and $g_{\perp 2}$ are the corresponding anisotropic Kirkwood correlation factors, the isotropic counterpart of which was defined in Eq. (1.14.14).

While correlation factors have been introduced in partially ordered phases, the problem of the internal field in a macroscopic anisotropic material still remains without solution. For this reason, isotropic internal field factors are assumed.

11.8. SMECTIC PHASES

Changes in the components of the dielectric permittivity have been detected as a consequence of smectic $\rightarrow$ nematic transitions. Differences in the orientational order may affect the parameters appearing in Eqs (11.7.17). On the other hand, translational order only affects the dielectric permittivity through the macroscopic internal field or short-range dipolar interactions. In principle, the effect of different local interactions in a change of phase could be measured through the anisotropic Kirkwood correlation factor. However, this requires detailed information about the molecular structure of the liquid crystal. A very simple approach is to assume total orientational order ($S = 1$) in such a way that the dipole is constrained to be oriented parallel or antiparallel to the normal to layer l, that is, the director or z axis. This situation is similar to the two-well problem. In this case the dipole correlation factor can be written as

$$g_\parallel = 1 + n\frac{\langle \mu_{1z}\mu_{2z}\rangle}{\langle \mu_{1z}^2\rangle} \tag{11.8.1}$$

where n is the number of neighboring dipoles. The probability of a dipole jumping from one well to another is determined by writing the probabilities in terms of the dipole–dipole interaction energy [see Eq. (P.1.3.4)] to give

$$g_\parallel = 1 + n\frac{\langle \exp -(u_\mathrm{p}/kT) - \exp -(u_\mathrm{ap}/kT)\rangle}{\langle \exp -(u_\mathrm{p}/kT) + \exp -(u_\mathrm{ap}/kT)\rangle} \tag{11.8.2}$$

where subscripts p and ap denote parallel and antiparallel respectively. Expanding the exponentials and assuming that $u \ll kT$, Eq. (11.8.2) becomes

$$g_\parallel = 1 + n\frac{\langle u_\mathrm{p} - u_\mathrm{ap}\rangle}{kT} \tag{11.8.3}$$

Setting $\gamma = 0$ and $\theta_1 = \theta_2$ in Eq. (P.1.3.4) (see Fig. 11.9), the correlation factor can be evaluated to give

$$g_\parallel = 1 - \frac{n\mu^2\langle 3\cos^2\theta - 1\rangle}{r^3 kT} \tag{11.8.4}$$

where $\cos\theta = r_z/r$.

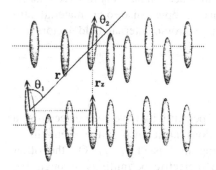

FIG. 11.9 Dipole correlation in smectic phases

In a smectic phase, the average separation between layers is r_z times greater than in-plane separation and $g_{\parallel} < 1$, whereas $g_{\perp}$ is unity because the normal component of the dipolar moment is zero. This model can be extended to molecules where a dipole forms an angle β with the molecular long axis. In such a case

$$g_{\parallel} = 1 - \frac{n\mu^2 \cos^2 \beta \langle 3(r_z/r)^2 - 1 \rangle}{r^3 kT}$$

$$g_{\perp} = 1 - \frac{n\mu^2 \sin^2 \beta \langle 3(r_x/r)^2 - 1 \rangle}{2r^3 kT}$$

(11.8.5)

If $\langle r_x^2 \rangle < \frac{1}{3}\langle r^2 \rangle$, the normal dipole correlation factor is greater than unity, and the material prefers the parallel alignment of dipoles in the smectic layer. In general, the dielectric study of smectic phases is complicated owing to their biaxiality.

The dielectric properties of ferroelectric chiral smectic phases have been recently studied[5,6] on account of their growing applications. In this case, adequate field factors need to be chosen to give account of the tilted layered structure of the material. Moreover, from the experimental point of view, the problem is to choose suitable alignments that allow independent measurements of the principal components of the permittivity. A first approach is to measure the permittivity of a homeotropically aligned sample in which the measurement direction is normal to the smectic layer. A second permittivity can be measured for the planar orientation for which a chevron structure is presumed with a tilt angle δ[5] (see Fig. 11.10). The corresponding permittivities are given by the following equations (Appendix A)

$$\varepsilon_{\text{hom}} = \varepsilon_z \cos^2 \theta + \varepsilon_x \sin^2 \theta = \varepsilon_x + (\varepsilon_z - \varepsilon_x) \cos^2 \theta$$

$$\varepsilon_{\text{planar}} \sin^2 \theta = \varepsilon_y \sin^2 \theta - (\varepsilon_y - \varepsilon_x) \sin^2 \delta$$

(11.8.6)

where $\varepsilon_z - \varepsilon_x$ and $\varepsilon_y - \varepsilon_x$ are the anisotropy and the dielectric biaxiality respectively.

On the other hand, according to Eq. (11.2.10), the mean value of the permittivity is given by

$$\bar{\varepsilon} = \frac{1}{3}(\varepsilon_x + \varepsilon_y + \varepsilon_z)$$

(11.8.7)

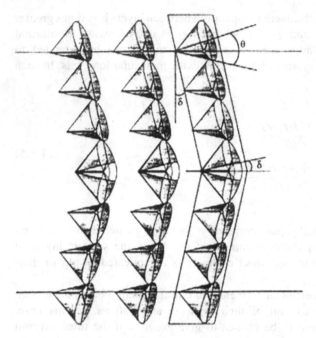

FIG. 11.10 Chevron structure of a smectic phase

The three principal components of the permittivity obtained from the system defined by Eqs (11.8.6) and (11.8.7) are given by

$$\varepsilon_x = \frac{(3\bar{\varepsilon}\cos^2\theta - \varepsilon_{\text{hom}})(\sin^2\theta - \sin^2\delta) - \varepsilon_{planar}\sin^2\theta\cos^2\theta}{\cos 2\theta(\sin^2\theta - \sin^2\delta) - \sin^2\delta\cos^2\theta}$$

$$\varepsilon_y = \frac{-(3\bar{\varepsilon}\cos^2\theta - \varepsilon_{\text{hom}})\sin^2\delta + \varepsilon_{planar}\sin^2\theta\cos^2\theta}{\cos 2\theta(\sin^2\theta - \sin^2\delta) - \sin^2\delta\cos^2\theta}$$

$$\varepsilon_z = \frac{(\varepsilon_{\text{hom}} - 3\bar{\varepsilon}\sin^2\theta)(\sin^2\theta - \sin^2\delta) - \varepsilon_{\text{hom}}\sin^2\delta + \varepsilon_{planar}\sin^4\theta}{\cos 2\theta(\sin^2\theta - \sin^2\delta) - \sin^2\theta\cos^2\theta}$$

$$(11.8.8)$$

Tilted chiral structures such as smectics C* lack a center of symmetry and therefore they can be ferroelectric. A helical structure can be developed in chiral tilted liquid crystalline smectic phases having planar surface alignment. This can be used to measure the principal components of the dielectric permittivity tensor by means of three measurements[6] (see Fig. 11.11): in the homeotropic state, as above, in the helical state, and in the unwounded state, by applying a field in order

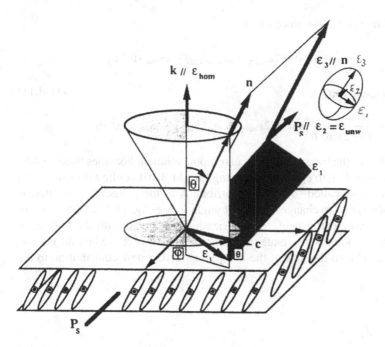

FIG. 11.11   Relative orientation of the dielectric tensor components ε_x, ε_y, ε_z with respect to the measured dielectric permittivities $\varepsilon_{\mathrm{hom}}$, $\varepsilon_{\mathrm{unwound}}$ (see Eq. 11.8.9 and details in reference [7])

to unwind the helical structure (Appendix B). The pertinent results are

$$\varepsilon_{\mathrm{helix}} = \tfrac{1}{2}\cos^2\delta(\varepsilon_x\cos^2\theta + \varepsilon_z\sin^2\theta + \varepsilon_y) + \sin^2\delta(\varepsilon_x\sin^2\theta + \varepsilon_z\cos^2\theta)$$
$$\varepsilon_{\mathrm{unwound}} = \varepsilon_y\cos^2\delta + \sin^2\delta(\varepsilon_x\sin^2\theta + \varepsilon_z\cos^2\theta)$$

$$(11.8.9)$$

The three principal components of the permittivity obtained solving the system formed by Eqs (11.8.6a) and (11.8.9) are

$$\varepsilon_x = \frac{1}{1 - 2\sin^2\theta}\left[\frac{\cos^2\theta}{\cos^2\delta}(2\varepsilon_{\mathrm{helix}} - \varepsilon_{\mathrm{unwound}} - \varepsilon_{\mathrm{hom}}\sin^2\delta) - \varepsilon_{\mathrm{hom}}\sin^2\theta\right]$$

$$\varepsilon_y = \frac{1}{\cos^2\delta}(\varepsilon_{\mathrm{unwound}} - \varepsilon_{\mathrm{hom}}\sin^2\delta)$$

$$\varepsilon_z = -\frac{1}{1 - 2\sin^2\theta}\left[\frac{\sin^2\theta}{\cos^2\delta}(2\varepsilon_{\mathrm{helix}} - \varepsilon_{\mathrm{unwound}} - \varepsilon_{\mathrm{hom}}\sin^2\delta) - \varepsilon_{\mathrm{hom}}\cos^2\theta\right]$$

$$(11.8.10)$$

For $\delta = 0$, Eqs (11.8.10) reduce to

$$\varepsilon_x = \frac{1}{1 - 2\sin^2\theta}\left[\cos^2\theta(2\varepsilon_{\text{helix}} - \varepsilon_{\text{unwound}}) - \varepsilon_{\text{hom}}\sin^2\theta\right]$$

$$\varepsilon_y = \varepsilon_{\text{unwound}} \qquad\qquad\qquad\qquad (11.8.11)$$

$$\varepsilon_z = -\frac{1}{1 - 2\sin^2\theta}\left[\sin^2\theta(2\varepsilon_{\text{helix}} - \varepsilon_{\text{unwound}}) - \varepsilon_{\text{hom}}\cos^2\theta\right]$$

The orientation of the transverse molecular dipole moment becomes biased along the axis corresponding to ε_2 of the preceding Eq. (11.8.10) in tilted smectics. This gives rise to the so-called soft mode contribution to the dielectric permittivity. The soft mode is due to changes in the tilt angle θ (see Fig. 11.12). Moreover, in the absence of an electric field, the polarization associated with the transverse dipole moment components rotates with the helix. However, when an electric field perpendicular to the axis of the helix is applied, a new contribution to the

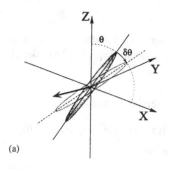

(a)

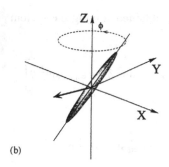

(b)

FIG. 11.12   Contributions to (a) soft mode and (b) Goldstone mode

permittivity, known as the Goldstone mode, appears. This mode is due to the polarization produced by changes in the azimuthal angle (see Fig. 11.13) and is only present in helicoidal structures.

Although the theory of the dielectric properties of chiral smectic liquid crystalline phases is far from complete, simple expressions for the susceptibilities of the soft and Goldstone modes are given by Landau[7] as

$$\chi_S = \frac{\varepsilon_{\infty y}^2 \mu_p^2}{kq_0^2 + 2a(T_c - T)}$$

$$(11.8.12)$$

$$\chi_G = \frac{\varepsilon_{\infty y}^2 \mu_p^2}{2kq_0^2}$$

where μ_p is the piezoelectric coefficient in the Landau free energy, $\varepsilon_{\infty y}$ is the high-frequency permittivity component along the y axis, k is an elastic constant for the helicoidal distortion, $q_0 = 2\pi/p$ is the wave vector of the helical pitch p, and a is a coefficient in the Landau free energy.

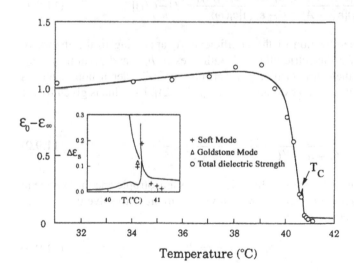

FIG. 11.13 Soft and Goldstone mode contributions to the dielectric permittivity for a main chain liquid crystalline polymer. Full lines correspond to the Landau theory

11.9. DYNAMIC DIELECTRIC PERMITTIVITY. INTERNAL FIELD FACTORS

The frequency dependence of the permittivity, showing dispersion as well as absorption, has established in Chapter 2 for isotropic materials. However, the situation in the case of liquid crystals is more complicated. The symmetry of the static permittivity can be proved on the basis of thermodynamic (strictly speaking, thermostatic) arguments. However in a variable field, out of thermodynamic equilibrium, the same proof is no longer valid. The symmetry of the dynamic permittivity tensor must be based on the principle of the symmetry of kinetic coefficients according to the methodology of the classic thermodynamics of irreversible processes.

Since a liquid crystalline phase is constitutively anisotropic, the rotational motion of the dipoles is also nonisotropic. The dipole moment does not lie along the molecular axis, and the dipolar motion that gives rise to the frequency dielectric response is not directly coupled with the motion of the molecule. The problem is further complicated if short-range dipole–dipole interactions and the macroscopic internal field factors arising from the anisotropy of the mesogenic phases are considered. Böttcher and Bordewijk[8] considered this last effect for the simplest case in which the applied external field lies along the principal direction of the permittivity. Then, the normalized time derivative of the correlation function is obtained as

$$\frac{[\varepsilon_i(\omega) - \varepsilon_{i\infty}]\{\varepsilon_i(\omega) - A_i^*[\varepsilon_i(\omega) - \varepsilon_{i\infty}]\}\varepsilon_{i0}}{(\varepsilon_{i0} - \varepsilon_{i\infty})[\varepsilon_{i0} - A_i^*(\varepsilon_{i0} - \varepsilon_{i\infty})]\varepsilon_i(\omega)} = L[-\dot{C}(t)] \tag{11.9.1}$$

where A_i^* is a generalization of the coefficients A_i appearing in the theory of the polarization of a dielectric ellipsoid with axes a, b, c and permittivity ε_1, embedded into a dielectric medium of permittivity ε_2 under a homogeneous unidirectional (on the z axis) external field [Eq. (1.C.2)]. Its value is given by the elliptic integral

$$A_c^* = \frac{a'b'c'}{2} \int_0^\infty \frac{ds}{(s + a'^2)^{1/2}(s + b'^2)^{1/2}(s + c'^2)^{3/2}} \tag{11.9.2}$$

In the present situation the surrounding dielectric is anisotropic and the change to the coordinate system indicated below is advisable in order to solve the Laplace equation in the outer dielectric

$$x' = x\varepsilon_{21}^{-1/2}, \qquad y' = y\varepsilon_{22}^{-1/2}, \qquad z' = z\varepsilon_{23}^{-1/2} \tag{11.9.3}$$

Here, the first subscript refers to the outer medium and the second to the principal values of the permittivity.

For a transformation such as those given by Eqs (11.9.3), a spherical surface changes into an ellipsoidal one, and in the case of ellipsoids the magnitudes of their axes also change. If one assumes that the axes of the ellipsoid coincide with the principal ones of ε_1 and ε_2, then

$$a' = a\varepsilon_{21}^{-1/2}, \qquad b' = b\varepsilon_{22}^{-1/2}, \qquad c' = c\varepsilon_{23}^{-1/2} \tag{11.9.4}$$

Substitution of Eq. (11.9.4) into Eq. (11.9.2) yields

$$A_c^* = \frac{abc}{2}\varepsilon_{23} \int_0^\infty \frac{ds}{(s\varepsilon_{21} + a^2)^{1/2}(s\varepsilon_{22} + b^2)^{1/2}(s\varepsilon_{23} + c^2)^{3/2}} \tag{11.9.5}$$

A_c^* depends on both the shape of the ellipsoid and the anisotropy of ε_2. For a spherical specimen, Eq. (11.9.5) reduces to

$$A_c^* = \frac{1}{2}\varepsilon_{21} \int_0^\infty \frac{ds}{(s\varepsilon_{21} + 1)^{1/2}(s\varepsilon_{22} + 1)^{1/2}(s\varepsilon_{23} + 1)^{3/2}} \tag{11.9.6}$$

which corresponds to the value to be used in Eq. (11.9.1).

The field in the ellipsoid under a homogeneous external field along the z axis is given by (see reference [7] p. 41 or reference [8] p. 430)

$$\mathbf{E}_1 = \frac{\varepsilon_{23}}{\varepsilon_{23} + A_c^*(\varepsilon_{13} - \varepsilon_{23})}\mathbf{E}_0 \tag{11.9.7}$$

Similar equations for the other axes are obtained.

For arbitrary directions of the external field and principal axes of the permittivities ε_1 and ε_2, $\mathbf{A}^*$ is a tensor with principal axes coinciding with those of ε_2 and eigenvalues given by equations similar to (11.9.2). Note that Eqs (11.9.1) and (11.9.7) are, respectively, generalizations of Eqs (2.23.13) and (1.C.10a) for the case of anisotropic systems.

11.10. DIELECTRIC RELAXATIONS IN NEMATIC UNIAXIAL MESOPHASES

In the case of rigid rodlike molecules in a uniaxial phase, as in nematic liquid crystals, two rotational diffusion constants, parallel and perpendicular to the inertial axis, are needed. For this reason, different dipolar moment contributions to the principal components of the permittivity are expected. For rodlike molecules, the frequency dependence of the orientational polarization can be calculated from the Debye model for Brownian rotational diffusion by modifying the classical potential to include a nematic potential. This problem has been addressed by Martin et al. [9] The starting point is the Smoluchowski equation (2.8.11) where the dependence on the φ angle can be ignored for the purpose of analyzing the polarization along the director, that is

$$\dot{f} = \frac{D}{kT \sin \theta} \frac{\partial}{\partial \theta} \left[\sin \theta \left(kT \frac{\partial f}{\partial \theta} + f \frac{\partial U}{\partial \theta} \right) \right] \tag{11.10.1}$$

where

$$U = U_1 + U_2 = -\mu E \cos \theta + q \sin^2 \theta \tag{11.10.2}$$

is the addition of the electric and the nematic potentials, and the remaining parameters are defined in Chapter 2. The expression for the nematic potential is commonly used in long-range order theories.[3] At equilibrium, $\dot{f} = 0$ and $U_1 = 0$, and consequently Eq. (11.10.1) becomes

$$\frac{\partial f_0}{\partial \theta} = -\frac{f_0}{kT} \frac{\partial U_2}{\partial \theta} \tag{11.10.3}$$

from which an equation similar to Eq. (2.8.12) but for the nematic potential is obtained.

Equation (11.11.1) can be rewritten using a new function $F(\theta, t)$ defined as the deviation of equilibrium as

$$F(\theta, t) = \frac{f(\theta, t)}{f_0(\theta)} \tag{11.10.4}$$

Taking into account Eqs (11.10.4), (11.10.2), and (11.10.3), as well as their derivatives, equation (11.10.1) can be written as

$$D^{-1}\dot{F} = \left[\frac{\partial^2 F}{\partial \theta^2} + \left(\frac{\cos \theta}{\sin \theta} - \frac{2q}{kT} \sin \theta \cos \theta \right) \frac{\partial F}{\partial \theta} \right]$$
$$+ \frac{\mu E}{kT} \left[2 \left(\cos \theta - \frac{q}{kT} \sin^2 \theta \cos \theta \right) F + \sin \theta \frac{\partial F}{\partial \theta} \right] \tag{11.10.5}$$

where $D^{-1} = 2\tau_D$. Note that, for $q = 0$ in Eq. (11.10.5), the Debye equation (2.9.2) is recovered.

A similar calculation can be performed taking into account the dependency of the function F on the angle φ. For the situation in which the field is perpendicular to the director, the result is

$$D^{-1}F = \left[\frac{\partial^2 F}{\partial\theta^2} + \left(\frac{\cos\theta}{\sin\theta} - \frac{2q}{kT}\sin\theta\cos\theta\right)\frac{\partial F}{\partial\theta}\right]$$
$$+ \frac{\mu E}{kT}\left[-\cos\theta\cos\varphi\frac{\partial F}{\partial\theta} + \frac{\sin\varphi}{\sin\theta}\frac{\partial F}{\partial\varphi} + 2\sin\theta\cos\varphi\left(1 + \frac{q}{kT}\cos^2\theta\right)F\right]$$
$$+ \frac{1}{\sin^2\theta}\frac{\partial^2 F}{\partial\varphi^2} \tag{11.10.6}$$

Both Eqs (11.10.5) and (11.10.6) can be solved by choosing a convenient perturbation function F. For example, by assuming that

$$F(\theta, t) = h(t)g(x) \tag{11.10.7}$$

where $x = \cos\theta$, Eq. (11.10.5) gives the following expresssion for $F(\theta, t)$

$$F(\theta, t) = 1 + \frac{\mu E}{kT}\sum_{n=1,3,5} b_n g_n(\cos\theta)\exp-\frac{c_n t}{2\tau_D} \tag{11.10.8}$$

where c_n are eigenvalues for which the series converges at $\theta = 0$, and $2/c_n$ can be interpreted as a retardation factor that gives rise to a larger relaxation time than the ordinary τ_D. Similar equations are obtained for the other cases considered in the theory.[9]

On the other hand, the solution for the rotational diffusion equation, including a more general single-particle potential of the symmetry corresponding to rodlike molecules, has been given by Nordio et al. in reference [2] and, in fact, they obtain more complicated expressions. It should be pointed out that the formulation of the problem for the longitudinal relaxation of nematic liquid crystals is mathematically identical to that of the magnetic relaxation for single-domain ferromagnetic particles. This has been fully described by Coffey et al. else-where.[10] An exact solution for the transverse component of the polarizability and the corresponding relaxation time has been proposed by Coffey et al. [11] The same results can also be obtained without resorting to the Debye–Smoluchowski equation by direct averaging of the noninertial Langevin equation for the rotational Brownian motion of linear molecules in a nematic field potential[12] (Appendix C).

According to Nordio et al.,[2] the dipole correlation function to be related to the frequency-dependent permittivity is the sum of a set of exponential terms, each characterized by a single relaxation time. By setting the angular correlation terms between different molecules to zero, the time-dipole correlation function in many cases can be approximated by the first term of the set. This approach gives rise to four relaxation times corresponding to four dipole contributions when the dipole moment is split into its longitudinal and transverse components (see Fig. 11.14).

$$C_{\parallel}(t) = \frac{1}{3\mu_{\parallel\text{eff}}^2}[\mu_l^2(1+2S)\varphi_{00} + \mu_t^2(1-S)\varphi_{01}]$$

$$(11.10.9)$$

$$C_{\perp}(t) = \frac{1}{3\mu_{\perp\text{eff}}^2}\left[\mu_l^2(1-S)\varphi_{10} + \mu_t^2\left(1+\frac{S}{2}\right)\varphi_{11}\right]$$

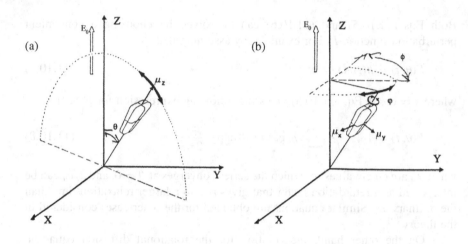

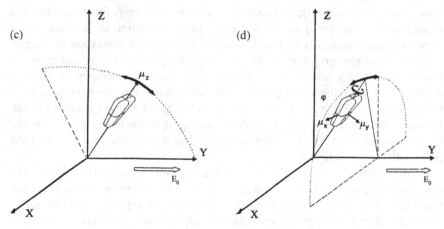

FIG. 11.14 Schematic representation of the fundamental molecular motions relaxing the square dipole moment: (a) τ_{00}^1; (b) τ_{10}^1; (c) τ_{01}^1; (d) τ_{11}^1

where $\mu_{\|eff}^2 = \frac{1}{3}[\mu_l^2(1+2S) + \mu_t^2(1-S)]$ and $\mu_{\perp eff}^2 = \frac{1}{3}[\mu_l^2(1-S) + \mu_t^2(1+S/2)]$.

In these equations, $\varphi_{ij}(t)$ represents the decay functions for the four relaxational modes corresponding to the molecular dipole in an anisotropic environment. As a consequence, the principal complex permittivities for the uniaxial material can be written as

$$\frac{\varepsilon_{\|}^*(\omega) - \varepsilon_{\|\infty}}{\varepsilon_{\|0} - \varepsilon_{\|\infty}} = \frac{1}{3\mu_{\|eff}^2}\{\mu_l^2(1+2S)L[-\dot\varphi_{00}(t)] + \mu_t^2(1-S)L[-\dot\varphi_{01}(t)]\}$$

$$= \frac{1}{3\mu_{\|eff}^2}[\mu_l^2(1+2S)F_\|^l + \mu_t^2(1-S)F_\|^t]$$

$$\frac{\varepsilon_{\perp}^*(\omega) - \varepsilon_{\perp\infty}}{\varepsilon_{\perp0} - \varepsilon_{\perp\infty}} = \frac{1}{3\mu_{\perp eff}^2}\left\{\mu_l^2(1-S)L[-\dot\varphi_{10}(t)] + \mu_t^2\left(1+\frac{S}{2}\right)L[-\dot\varphi_{11}(t)]\right\}$$

$$= \frac{1}{3\mu_{\perp eff}^2}\left[\mu_l^2(1-S)F_\perp^l + \mu_t^2\left(1+\frac{S}{2}\right)F_\perp^t\right]$$

(11.10.10)

where L indicates the Laplace transformation. The field factors are neglected and

$$F_\|^l = 1 - i\omega L[\varphi_{00}(t)]$$

$$F_\|^t = 1 - i\omega L[\varphi_{10}(t)]$$

$$F_\perp^l = 1 - i\omega L[\varphi_{01}(t)]$$

$$F_\perp^t = 1 - i\omega L[\varphi_{11}(t)]$$

(11.10.11)

As above, S in these equations represents the local microscopic order parameter that indicates the degree of alignment of the mesogen with respect to the local director. Equations (11.10.10) are a generalization of the Maier–Meier equations [11.7.16] for the dynamic case, and they are generally independent of the model assumed for reorientation. In the small-step rotational diffusion model, each decay function can be written in terms of the nematic potential and the rotational diffusion constants $D_\|$ and $D_\perp$ which are related to the molecular shape and local viscosity as indicated in references [2] and [9]. The simplest motion is that of the 00 mode corresponding to the end-over-end rotation at the lowest frequency and can be accurately represented by a single exponential, that is

$$\varphi_{00}(t) = \langle \cos\theta(t) \cdot \cos\theta(0)\rangle = \exp\left(-\frac{t}{\tau_{00}}\right)$$

(11.10.12)

where θ is the angle between the z axis of the molecule and the Z axis of the reference frame. If $D_\| = D_\perp$, then $\varphi_{01} = \varphi_{10}$. The rotational modes are represented in Fig. 11.14. In this case, the effect of the order parameter upon the relaxation frequencies, τ^{-1} is shown in Fig. 11.14. The results indicate that the relaxation frequency of the end-over-end rotation decreases as the order parameter decreases. On the other hand, the converse is true for the frequency corresponding to the rotation about the long axis.

The relative magnitudes of the frequencies at which the loss maxima are present in dynamic dielectric measurements carried out in mesogenic phases of nematic side chain liquid crystalline polymers (SCLCPs) are

$$\tau_{00}^{-1} < \tau_{01}^{-1} \leqslant \tau_{11}^{-1} \leqslant \tau_{10}^{-1} \tag{11.10.13}$$

in good agreement with the predictions of the theory described by Nordio *et al.* [2] Equations (11.10.10) show that $\varepsilon_{\parallel}$ measured with the electric field parallel to the director giving a nearly homeotropically aligned sample has two clearly separated relaxations. However, $\varepsilon_{\perp}$ measured with the electric field normal to the director only shows a broad peak resulting from the superposition of modes 10 and 11. For this reason, in samples with intermediate alignment, two relaxation peaks are usually present. The lower-frequency peak is relatively narrow, corresponds to mode 00, and is governed by τ_{00}. The other peak on the high-frequency side of the spectrum is a superposition of the remaining modes (01,10,11) and is broader in shape [2b,10]. These facts are in good agreement with the experimental evidence. For example, Attard and Williams[13–15] showed that a liquid crystalline siloxane polymer can be homeotropically aligned using an a.c. electric field. In this case, the broad loss peak observed in the unaligned material is replaced by a narrower (close to a single relaxation time process) peak on homeotropic alignment. The dominant low-frequency process (δ-process) is due to the 00 mode, whereas the broad higher-frequency process (α-process) results from the overlapping of the 01, 10, and 11 modes in the unaligned specimen. The broadness of the δ-peak in some cases[16] suggests that it may be the result of coupled motions of both the mesogenic and the ester groups.

The determination of the permittivity parallel and perpendicular to the applied field requires averaging over the orientations of the directors with respect to the electric field. This implies introduction of the director order parameter S_d as defined in Eq. (11.4.20). As mentioned there, the procedure is outlined in Problem 3, and the final result is expressed by Eqs (P.11.3.10). The pertinent results are[15]

$$\frac{\varepsilon_Z^*(\omega) - \varepsilon_{Z\infty}}{\varepsilon_{Z0} - \varepsilon_{Z\infty}} = \frac{1}{3\mu_{Z\text{eff}}^2} \left\{ (1 + 2S_d)[\mu_l^2(1 + 2S)F_{\parallel}^l + \mu_t^2(1-S)F_{\parallel}^t] + 2(1 - S_d) \right.$$
$$\left. \times \left[\mu_l^2(1 - S)F_{\perp}^l + \mu_t^2\left(1 + \frac{S}{2}\right)F_{\perp}^t \right] \right\}$$

$$\frac{\varepsilon_I^*(\omega) - \varepsilon_{I\infty}}{\varepsilon_{I0} - \varepsilon_{I\infty}} = \frac{1}{3\mu_{Ieff}^2}\left\{(1 - S_d)[\mu_l^2(1 + 2S)F_\parallel^l + \mu_t^2(1 - S)F_\parallel^t] + 2\left(1 + \frac{S_d}{2}\right)\right.$$

$$\left. \times \left[\mu_l^2(1 - S)F_\perp^l + \mu_t^2\left(1 + \frac{S}{2}\right)F_\perp^t\right]\right\}$$

$$(11.10.14)$$

where $I = X, Y$ and

$$\mu_{Zeff}^2 = \frac{1}{3}\left\{(1 + 2S_d)[\mu_l^2(1 + 2S) + \mu_t^2(1 - S)] + 2(1 - S_d)\right.$$

$$\left. \times \left[\mu_l^2(1 - S) + \mu_t^2\left(1 + \frac{S}{2}\right)\right]\right\}$$

$$\mu_{Ieff}^2 = \frac{1}{3}\left\{(1 - S_d)[\mu_l^2(1 + 2S) + \mu_t^2(1 - S)] + 2\left(1 + \frac{S}{2}\right)\right.$$

$$\left. \times \left[\mu_l^2(1 - S) + \mu_t^2\left(1 + \frac{S}{2}\right)\right]\right\}$$

Accordingly, the strengths of the four relaxations can be written as

$$\Delta\varepsilon_{00} = c(1 + 2S_d)(1 + 2S)\mu_l^2$$

$$\Delta\varepsilon_{10} = c(1 + 2S_d)(1 - S)\mu_t^2$$

$$\Delta\varepsilon_{01} = 2c(1 - S_d)(1 - S)\mu_l^2 \qquad (11.10.15)$$

$$\Delta\varepsilon_{11} = 2c(1 - S_d)\left(1 + \frac{S}{2}\right)\mu_t^2$$

where c is a factor involving the dipoles density. The total relaxation strength, $\Delta\varepsilon$, and those corresponding to the δ-process, $\Delta\varepsilon_\delta$, and α-process, $\Delta\varepsilon_\alpha$, are

$$\Delta\varepsilon = 3c[\mu_l^2(1 + 2SS_d) + \mu_t^2(1 - SS_d)]$$

$$\Delta\varepsilon_\delta = c(1 + 2S)(1 + 2S_d)\mu_l^2 \qquad (11.10.16)$$

$$\Delta\varepsilon_\alpha = c[2\mu_l^2(1 - S)(1 - S_d) + 3\mu_t^2(1 - SS_d)]$$

Let us consider three different cases:

   (a)   homeotropic alignment ($S_d = 1$),
   (b)   an unaligned sample ($S_d = 0$),
   (c)   planar alignment ($S_d = -1/2$).

For case (a), the total strength of the homeotropic oriented sample and the strengths for the δ- and α-relaxations are given by

$$\Delta\varepsilon_h = \Delta\varepsilon_{00h} + \Delta\varepsilon_{10h} = 3c[(\mu_l^2 + \mu_t^2) + S(2\mu_l^2 - \mu_t^2)]$$

$$\Delta\varepsilon_{00h} = \Delta\varepsilon_{h\delta} = 3c(1 + 2S)\mu_l^2 \qquad\qquad\qquad (11.10.17)$$

$$\Delta\varepsilon_{10h} = \Delta\varepsilon_{h\alpha} = 3c(1 - S)\mu_t^2$$

For a sample in which the mesogenic groups are not oriented ($S_d = 0$), Eqs (11.10.16) become

$$\Delta\varepsilon_{00u} = c(1 + 2S)\mu_l^2$$

$$\Delta\varepsilon_{10u} = c(1 - S)\mu_t^2$$

$$\Delta\varepsilon_{01u} = 2c(1 - S)\mu_l^2 \qquad\qquad\qquad (11.10.18)$$

$$\Delta\varepsilon_{11u} = 2c\left(1 + \frac{S}{2}\right)\mu_t^2$$

Hence

$$\Delta\varepsilon_u = 3c(\mu_l^2 + \mu_t^2)$$

$$\Delta\varepsilon_{u\delta} = c(1 + 2S)\mu_l^2 \qquad\qquad\qquad (11.10.19)$$

$$\Delta\varepsilon_{u\alpha} = c[2(1 - S)\mu_l^2 + 3\mu_t^2]$$

In this case, only a part of the dipolar moment $\mu_\parallel^2$ is relaxed by the δ-process. For planar alignment ($S_d = -0.5$), the pertinent expressions are

$$\Delta\varepsilon_p = \Delta\varepsilon_{p\alpha} = 3c\left[(\mu_l^2 + \mu_t^2) + S\left(\frac{\mu_t^2}{2} - \mu_l^2\right)\right]$$

$$\Delta\varepsilon_{01p} = 3c(1 - S)\mu_l^2 \qquad\qquad\qquad (11.10.20)$$

$$\Delta\varepsilon_{11p} = 3c\left(1 + \frac{S}{2}\right)\mu_t^2$$

It should be noted that the relaxation strength parameters can be estimated from the experimental curves by deconvolution of the fitted data to an appropriate empirical model. Assuming symmetric shape for δ- and α-relaxations, the Fuoss–Kirkwood equation is a good choice. Hence, Eqs (11.10.15) to (11.10.20) could supply interesting relationships for estimating the microscopic order parameter from the values of the relaxation strength of the observed peaks. For example, Eqs (11.10.17b) and (11.10.19b) lead to

$$\frac{\Delta\varepsilon_{h\delta}}{\Delta\varepsilon_{u\delta}} = 3 \tag{11.10.21}$$

This indicates that the strength of the δ-relaxation in a homeotropically aligned material is 3 times larger than that corresponding to the unaligned sample. Other relations of this type can be found in references [15] and [16].

Unfortunately, in many cases it is difficult to obtain a fully homeotropic (or planar) specimen, but even in this case useful relations can be obtained from the preceding strategy. For example, dividing the equation corresponding to the strength of the δ-process in a partially aligned sample [Eq. (11.10.16b)] by the corresponding equation for a sample as received (unaligned) [Eq. (11.10.18b)], the following expression holds[15]

$$\frac{\Delta\varepsilon_{pa\delta}}{\Delta\varepsilon_{u\delta}} = 1 + 2S_d \tag{11.10.22}$$

from which the macroscopic order parameter S_d can be estimated. The microscopic order parameter could be estimated from data of unaligned samples by dividing Eqs (11.10.19c) and (11.10.19b) to give

$$\frac{\Delta\varepsilon_{u\alpha}}{\Delta\varepsilon_{u\delta}} = \frac{2(1-S) + 3y}{1 + 2S} \tag{11.10.23}$$

where $y = \mu_t^2/\mu_l^2$. However, in this case, information on the longitudinal and transversal components of the dipolar moment is needed. On the other hand, if a fully homeotropic material has been obtained, then Eqs (11.10.17b) and (11.10.17c) lead to the following expression for the ratio of the δ- to the α-process

$$\frac{\Delta\varepsilon_{h\alpha}}{\Delta\varepsilon_{h\delta}} = \frac{1-S}{1+2S}y \tag{11.10.24}$$

From Eqs (11.10.23) and (11.10.24), tentative values of S and y could in principle be estimated.

Recently[17] a method to predict the loss curves with different degrees of alignment has been proposed for samples where the permittivity is measured parallel to the electric field. The method is based on the knowledge of the loss curves corresponding to an unaligned sample and those of a sample with S_d known (from Eq. 11.10.22, for example).

Let us write Eq. (P.11.1.10) in terms of the imaginary parts of the permittivity as follows

$$\varepsilon_{pa}''(\omega) = \frac{1 + 2S_d}{3} \varepsilon_h''(\omega) + \frac{2(1 - S_d)}{3} \varepsilon_p''(\omega) \tag{11.10.25}$$

where subscript pa denotes partially aligned. For an unoriented sample, the dielectric loss can be written as

$$\varepsilon_u''(\omega) = \tfrac{1}{3} \varepsilon_h''(\omega) + \tfrac{2}{3} \varepsilon_p''(\omega) \tag{11.10.26}$$

For a sample with an arbitrary (but known) alignment, an expression similar to Eq. (11.10.25) holds

$$\varepsilon_i''(\omega) = \frac{1 + 2S_{di}}{3} \varepsilon_h''(\omega) + \frac{2(1 - S_{di})}{3} \varepsilon_p''(\omega) \tag{11.10.27}$$

Eliminating ε_h'' and ε_p'' from Eqs (11.10.25) to (11.10.27), we obtain

$$\varepsilon_{pa}''(\omega) = \frac{S_d}{S_{di}} \varepsilon_i''(\omega) + (1 - \frac{S_d}{S_{di}})\varepsilon_u''(\omega) \tag{11.10.28}$$

Equation (11.10.28) suggests that a family of loss curves can be calculated with S_d as a parameter from two experimental loss curves, one of them corresponding to the unaligned sample. An example is given in Fig. 11.15 where the higher curve corresponds to the homeotropic alignment, and the lower curve to the planar alignment. These curves can be deconvoluted in terms of two Fuoss–Kirkwood equations corresponding, to δ- and α-relaxations respectively. The curves in Fig. 11.15 intercept at an isobestic point, as expected.

Equation (11.10.28) can be rearranged to give

$$S_d = S_{di} \frac{\varepsilon_{pa}''(\omega) - \varepsilon_u''(\omega)}{\varepsilon_i''(\omega) - \varepsilon_u''(\omega)} \tag{11.10.29}$$

Equations (11.10.28) and (11.10.29) can be easily generalized to predict the loss permittivity of a partially aligned sample with the macroscopic order parameter given by S_d in terms of two loss curves corresponding to macroscopic order parameters S_{d1} and S_{d2}. The pertinent equations are

$$\varepsilon_{pa}'' = \frac{S_d - S_{d2}}{S_{d1} - S_{d2}} \varepsilon_1''(\omega) + \frac{S_d - S_{d1}}{S_{d2} - S_{d1}} \varepsilon_2''(\omega) \tag{11.10.30}$$

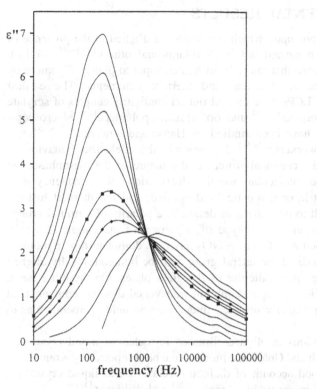

FIG. 11.15 Dielectric loss permittivity of the LC95 at different alignments. Calculated curves are continuous lines; ◆ relates to unoriented sample, and ■ to the partially aligned sample ($S_d = 0.18$)

and

$$S_d = \frac{S_{d1}[\varepsilon''_{pa}(\omega) - \varepsilon''_2(\omega)] - S_{d2}[\varepsilon''_{pa}(\omega) - \varepsilon''_1(\omega)]}{\varepsilon''_1(\omega) - \varepsilon''_2(\omega)} \qquad (11.10.31)$$

These equations may be useful in cases where the knowledge of the experimental data corresponding to homeotropic and planar alignments are precluded for experimental reasons. For low molecular weight compounds the second relaxational peak appears at high frequencies (>50 MHz), and its experimental measurement may present some difficulties.

11.11. EXPERIMENTAL RESULTS

A study of the conditions upon which polymers are aligned in the presence of electric fields has been carried out by Williams and others.[13,14,18] In fact, polymers can be produced that range from homeotropic to planar alignment by the appropriate choice of electrical and thermal treatments. The earliest dielectric studies of SCLCPs were carried out on unaligned samples of acrylate and methacrylate polymers.[19-21] and on siloxane polymers.[22] Macroscopically aligned SCLCPs have been studied by Haase and coworkers.[23-26] and by Williams and coworkers.[27-31] In general, the dielectric behavior of SCLCPs depends on the chemical structure, the nature of the mesophase, the molecular weight and molecular weight distribution, the frequency and amplitude of the electric or magnetic field applied, and the thermal history. Therefore, it is difficult to describe it in detail. The polarity of terminal groups is important in order to ensure the type of alignment of the sample. Thus,[32] highly polar groups such as $-CN$ or $-NO_2$ having a dominant dipole moment component along the axis of the lateral group will be homeotropically aligned on applying the voltage perpendicular to the glass plates. On the other hand, groups such as $-OCH_3$, having a dominant transversal dipolar moment, tend to be homogeneous or planarly oriented under electric fields perpendicular to the plates.

Dielectric relaxations in aligned films of ferroelectric liquid crystalline polymers exhibiting soft and Goldstone modes have been reported by Kremer and coworkers.[33-35] A good account of dielectric relaxation in liquid crystalline polymers can be found in papers by Moscicki[36] and Williams.[37]

APPENDIX A

For materials with a phase transition sequence given by $N^* - S_A - S_C^*$, two cell geometries based on the homeotropic and planar alignments are available. In the first, the smectic layers are parallel to the cell plane and the director precesses about the layer normal. On the other hand, planar alignment with a low tilt angle δ leads to a chevron layer structure. Two director profiles based on the chevron structure are possible: the first is uniform within each domain, but a triangular profile is also possible for a surface stabilized state (Fig. 11.16). In what follows we are concerned with the first case. For a triangular profile, see reference [5].

Let us consider the geometry shown in Fig. 11.17 where the local director is referred to a frame (x,y,z or $1,2,3$) and the laboratory frame is denoted by (X,Y,Z). The director $\mathbf{n}$ is tilted from the layer normal $\mathbf{q}$ by the angle θ, the position of the director around the cone is described by the angle φ, and the layer normal is tilted with respect to the laboratory Z axis system by an angle δ. The properties in the

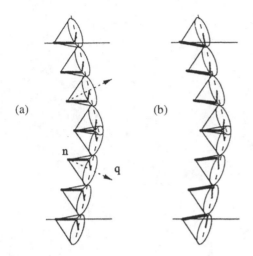

(a)   (b)

FIG. 11.16 Uniform and triangular director profiles based on the chevron layer structure

laboratory system are related to the director system by the rotation product

$$\mathbf{R} = \mathbf{R}_\delta \mathbf{R}_{-\varphi} \mathbf{R}_{-\theta} \tag{11.A.1}$$

which represents successive rotations about the 2, 3 and 1 directions in conventionally specified senses. The matrix $\mathbf{R}$ has the following form

$$
\mathbf{R} =
\begin{vmatrix}
1 & 0 & 0 \\
0 & \cos\delta & -\sin\delta \\
0 & \sin\delta & \cos\delta
\end{vmatrix}
\begin{vmatrix}
\cos\varphi & \sin\varphi & 0 \\
-\sin\varphi & \cos\varphi & 0 \\
0 & 0 & 1
\end{vmatrix}
\begin{vmatrix}
\cos\theta & 0 & \sin\theta \\
0 & 1 & 0 \\
-\sin\theta & 0 & \cos\theta
\end{vmatrix}
$$

$$
=
\begin{vmatrix}
\cos\varphi\cos\theta & -\sin\varphi & \cos\varphi\sin\theta \\
-\cos\delta\sin\varphi\sin\theta + \sin\delta\sin\theta & \cos\delta\cos\varphi & -\sin\varphi\cos\delta\sin\theta - \sin\delta\cos\theta \\
-\sin\varphi\sin\delta\cos\theta - \cos\delta\sin\theta & \sin\delta\cos\varphi & -\sin\delta\sin\varphi\sin\theta + \cos\delta\cos\theta
\end{vmatrix}
$$

$$\tag{11.A.2}$$

In the laboratory frame the director has the following components

$$(n_X, n_Y, n_Z) = (\cos\varphi\cos\theta, \ -\sin\varphi\cos\delta\sin\theta - \sin\delta\cos\theta,$$
$$-\sin\delta\sin\varphi\sin\theta + \cos\delta\cos\theta) \tag{11.A.3}$$

In a similar way, the permittivity in the laboratory frame $\boldsymbol{\varepsilon}'$ is related to the permittivity in the director frame $\boldsymbol{\varepsilon}$ by means of the following transformation

$$\boldsymbol{\varepsilon}' = \mathbf{R}\boldsymbol{\varepsilon}\mathbf{R}^\mathrm{T} \tag{11.A.4}$$

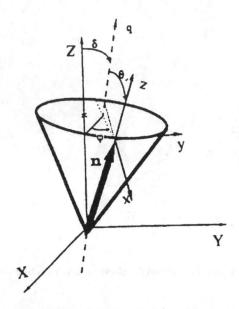

FIG. 11.17 Relationship between the director **n** and the laboratory frames of reference

where

$$\varepsilon = \begin{vmatrix} \varepsilon_x & 0 & 0 \\ 0 & \varepsilon_y & 0 \\ 0 & 0 & \varepsilon_z \end{vmatrix} \tag{11.A.5}$$

is the permittivity tensor in the local or director frame.

According to Eqs (11.A.2), (11.A.4), and (11.A.5), the components of the permittivity in the laboratory frame are

$$\varepsilon'_{XX} = \varepsilon_x \cos^2 \varphi \cos^2 \theta + \varepsilon_y \sin^2 \varphi + \varepsilon_z \cos^2 \varphi \sin^2 \theta$$

$$\varepsilon'_{XY} = \varepsilon'_{YX} = \varepsilon_x \cos \varphi \cos \theta (-\cos \delta \sin \varphi \cos \theta + \sin \delta \sin \theta)$$
$$+ \varepsilon_y \cos \delta \cos \varphi \sin \varphi$$
$$- \varepsilon_z \cos \varphi \sin \theta (\cos \delta \sin \varphi \sin \theta + \sin \delta \cos \theta)$$

$$\varepsilon'_{XZ} = \varepsilon'_{ZX} = -\varepsilon_x \cos \varphi \cos \theta (\sin \delta \sin \varphi \cos \theta + \cos \delta \sin \theta)$$
$$+ \varepsilon_y \sin \delta \sin \varphi \cos \varphi$$
$$+ \varepsilon_z \cos \varphi \sin \theta (-\sin \delta \sin \varphi \sin \theta + \cos \delta \cos \theta)$$

$$\varepsilon'_{YY} = \varepsilon_x (-\cos \delta \sin \varphi \cos \theta + \sin \delta \sin \theta)^2 + \varepsilon_y \cos^2 \delta \cos^2 \varphi$$
$$+ \varepsilon_z (\cos \delta \sin \varphi \sin \theta + \sin \delta \cos \theta)^2$$

$$\varepsilon'_{YZ} = \varepsilon'_{ZY} = -\varepsilon_x(-\cos\delta\sin\varphi\cos\theta$$
$$+ \sin\delta\sin\theta)(\sin\delta\sin\varphi\cos\theta + \cos\delta\sin\theta)$$
$$+ \varepsilon_y\sin\delta\cos\delta\cos^2\varphi - \varepsilon_z(\cos\delta\sin\varphi\sin\theta + \sin\delta\cos\theta)$$
$$\times (-\sin\delta\sin\varphi\sin\theta + \cos\delta\cos\theta)$$
$$\varepsilon'_{ZZ} = \varepsilon_x(\sin\delta\sin\varphi\cos\theta + \cos\delta\sin\theta)^2 + \varepsilon_y\sin^2\delta\cos^2\varphi$$
$$+ \varepsilon_z(-\sin\delta\sin\varphi\sin\theta + \cos\delta\cos\theta)^2$$

$$(11.A.6)$$

In order to proceed further, it is convenient to represent the director orientation by the degree of tilt out of the cell plane ζ and the in-plane twist β away from the preferred alignment director but remaining parallel to the cell walls (Fig. 11.18). The words "tilt" and "twist" are only used to denote the director orientation. For a uniform profile director, β is the angle between the optical axis and the preferred alignment direction. From the present scheme, the in-plane twist angle is defined as the angle between the Z axis of the laboratory frame and the component of the director and the XZ plane, given according to Eq. (11.A.3) by

$$\tan\beta = \frac{\cos\varphi\sin\theta}{-\sin\delta\sin\varphi\sin\theta + \cos\delta\cos\theta} \qquad (11.A.7)$$

In a similar way, the out-of-cell tilt angle is

$$n_Y = \sin\zeta = -\sin\varphi\cos\delta\sin\theta - \sin\delta\cos\theta \qquad (11.A.8)$$

When the electric field is applied parallel to the Y axis, the measured permittivity is given by ε'_{YY} which, on account of Eq. (11.A.8), can alternatively

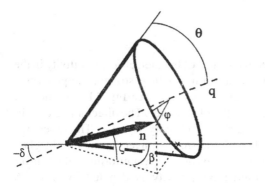

FIG. 11.18 Definition of the tilt ζ and twist β angle in the laboratory reference frame

be written as

$$\varepsilon'_{YY} = \varepsilon_x(-\cos\delta\sin\varphi\cos\theta + \sin\delta\sin\theta)^2 + (\varepsilon_y - \varepsilon_x)\cos^2\delta\cos^2\varphi$$
$$+ \varepsilon_x\cos^2\delta\cos^2\varphi + (\varepsilon_z - \varepsilon_x)(\cos\delta\sin\varphi\sin\theta - \sin\delta\cos\theta)^2$$
$$+ \varepsilon_x(-\cos\delta\sin\varphi\sin\theta - \sin\delta\cos\theta)^2$$
$$= \varepsilon_x + (\varepsilon_z - \varepsilon_x)\sin^2\zeta + (\varepsilon_y - \varepsilon_x)\cos^2\delta\cos^2\varphi \qquad (11.A.9)$$

In homeotropic geometry, $\delta = \pi/2$ and Eq. (11.A.9) reduces to

$$\varepsilon_{\mathrm{hom}} = \varepsilon_x + (\varepsilon_z - \varepsilon_x)\cos^2\theta \qquad (11.A.10)$$

which agrees with Eq. (11.8.6a) and with $\varepsilon'_{ZZ}(\delta = 0)$.

In the case of a planarly aligned cell with symmetric chevron layers, the director is confined to the XZ plane, that is, $\zeta = 0$, and from Eq. (11.A.8) we obtain

$$\sin\varphi_{\mathrm{i}} = -\frac{\tan\delta}{\tan\theta} \qquad (11.A.11)$$

where subscript i indicates that the azimuthal angle is taken at the interface. Then, on account of Eq. (11.A.11), the following result holds

$$\varepsilon'_{YY} = \varepsilon_{\mathrm{p}} = \varepsilon_x + (\varepsilon_y - \varepsilon_x)\cos^2\delta\cos^2\varphi = \varepsilon_y + (\varepsilon_y - \varepsilon_x)(\cos^2\delta\cos^2\varphi - 1)$$
$$= \varepsilon_y + (\varepsilon_y - \varepsilon_x)[\cos^2\delta(1 - \sin^2\varphi) - 1]$$
$$= \varepsilon_y - (\varepsilon_y - \varepsilon_x)\left(\sin^2\delta + \frac{\sin^2\delta}{\tan^2\theta}\right) = \varepsilon_y - (\varepsilon_y - \varepsilon_x)\frac{\sin^2\delta}{\sin^2\theta} \qquad (11.A.12)$$

which agrees with Eq. (11.8.6b).

Typically, the layer tilt angle is nearly equal to the cone angle, and the largest contribution to the permittivity is made by ε_x.

APPENDIX B

The determination of the three principal values of the dielectric permittivity in the case of a chiral smectic C liquid crystal with a helical structure requires two samples and three different measurement geometries (see Fig. 11.19). The first sample has the smectic layers parallel to the glass plates. The director makes a tilt angle θ with the field, and the dielectric permittivity is $\varepsilon_{\mathrm{hom}}$. The other two measurements are made on a sample with the smectic layers normal to the glass plates, allowing bookshelf geometry, chevron, or a uniformly tilted layer structure in the sample. The measured permittivity is denoted by $\varepsilon_{\mathrm{helix}}$. A nonhelical (unwounded) state can be obtained by applying a biased field, and the

(a)

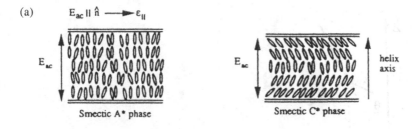

(b)

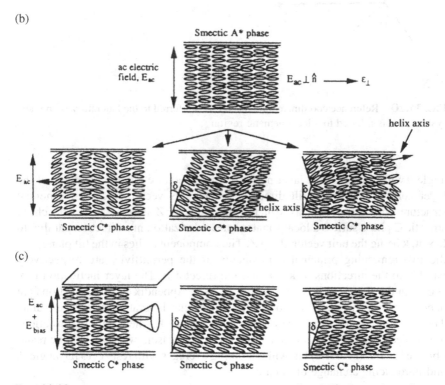

FIG. 11.19 Measurement geometries in the smectic $-A^*$ and C^* phases (see reference [7] for details)

permittivity measured is $\varepsilon_{unwound}$. In order to visualize the problem, the relative orientation of the local tensor ellipsoid and the laboratory frame (X, Y, Z) are shown in Fig. 11.20, where two of the axes along which the field is applied to obtain ε_{hom} and $\varepsilon_{unwound}$ are also shown. In the figure, the director **n** is tilted an

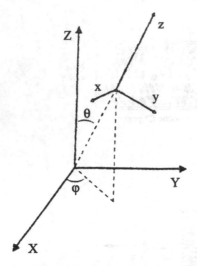

FIG. 11.20 Reference coordinate system, X,Y,Z is referred to the laboratory frame, and system x,y,z is referred to a local nematic region

angle θ with respect to the layer normal. A unit vector **c** lying in the plane of the layer corresponds to the tilt direction. When this vector precesses, a helical structure with the helical axis in the direction of Z is generated. The chiral smectic C phase suffers a local spontaneous polarization in a direction parallel to $\mathbf{k} \times \mathbf{n}$, **k** being the unit vector along Z. The component ε_1 lies in the tilt plane, and the two remaining principal components of the permittivity are respectively parallel to the directions of $\mathbf{k} \times \mathbf{n}$ and the director **n**. The layer inclination (not shown in the figure) will be denoted by δ. As in Appendix A, three rotations are needed in order to carry the local ellipsoid to the laboratory reference frame. The procedure generalizes that outlined in Problem 1, where only two rotations are needed for a nematic uniaxial system. For this reason, we start with the result obtained as Eq. (P.11.4.8) but without taking averages with respect to the angle θ, and introducing a change of notation

$$\begin{vmatrix} \frac{1}{2}(\varepsilon_x \cos^2\theta + \varepsilon_y + \varepsilon_z \sin^2\theta) & 0 & 0 \\ 0 & \frac{1}{2}(\varepsilon_x \cos^2\theta + \varepsilon_y + \varepsilon_z \sin^2\theta) & 0 \\ 0 & 0 & \varepsilon_x \sin^2\theta \\ & & + \varepsilon_z \cos^2\theta \end{vmatrix}$$

$$(11.B.1)$$

This tensor, which is diagonal in the laboratory frame, describes a uniaxial crystal with the symmetry axis parallel to the helix axis. The third rotation, with angle δ, is represented by the matrix

$$
\begin{vmatrix}
1 & 0 & 0 \\
0 & \cos \delta & -\sin \delta \\
0 & \sin \delta & \cos \delta
\end{vmatrix}
\tag{11.B.2}
$$

The transformed permittivity tensor will be given by

$$
\boldsymbol{\varepsilon}' = \mathbf{R}_\delta \boldsymbol{\varepsilon} \mathbf{R}_\delta^{\mathrm{T}} =
\begin{vmatrix}
1 & 0 & 0 \\
0 & \cos \delta & -\sin \delta \\
0 & \sin \delta & \cos \delta
\end{vmatrix}
$$

$$
\times
\begin{vmatrix}
\frac{1}{2}(\varepsilon_x \cos^2 \theta + \varepsilon_y + \varepsilon_z \sin^2 \theta) & 0 & 0 \\
0 & \frac{1}{2}(\varepsilon_x \cos^2 \theta + \varepsilon_y + \varepsilon_z \sin^2 \theta) & 0 \\
0 & 0 & \varepsilon_x \sin^2 \theta + \varepsilon_z \cos^2 \theta
\end{vmatrix}
$$

$$
\times
\begin{vmatrix}
1 & 0 & 0 \\
0 & \cos \delta & \sin \delta \\
0 & -\sin \delta & \cos \delta
\end{vmatrix}
\tag{11.B.3}
$$

In the homeotropic orientation, the electric field is applied along Z and

$$
\varepsilon'_{ZZ} = \varepsilon_{\mathrm{hom}} = \tfrac{1}{2}(\varepsilon_x \cos^2 \theta + \varepsilon_y + \varepsilon_z \sin^2 \theta) \sin^2 \delta
$$
$$
+ (\varepsilon_x \sin^2 \theta + \varepsilon_z \cos^2 \theta) \cos^2 \delta
\tag{11.B.4}
$$

Equation (11.B.4) applies when the smectic layers are tilted. However, if the smectic layers are kept parallel to the glass plates, $\delta = 0$, and then Eq. (11.8.6a) is recovered, that is

$$
\varepsilon_{\mathrm{hom}} = \varepsilon_x + (\varepsilon_z - \varepsilon_x) \cos^2 \theta
\tag{11.B.5}
$$

When the samples are measured in planar orientation, the field is applied parallel to Y and the pertinent dielectric value of the tensor ε'_{22}, also denoted by $\varepsilon_{\mathrm{helix}}$, is given by

$$
\varepsilon'_{YY} = \varepsilon_{\mathrm{helix}} = \tfrac{1}{2}(\varepsilon_x \cos^2 \theta + \varepsilon_z \sin^2 \theta) \cos^2 \delta
$$
$$
+ (\varepsilon_x \sin^2 \theta + \varepsilon_z \cos^2 \theta) \sin^2 \delta
\tag{11.B.6}
$$

To obtain the expression for the dielectric permittivity in the unwound helical state, the component ε'_{YY} must be chosen as above, but disregarding now the rotation angle φ. This means that the rotation to be carried out can be written as

$$\varepsilon' = \mathbf{R}_\delta \mathbf{R}_\theta \varepsilon \mathbf{R}_\theta^T \mathbf{R}_\varphi^T \tag{11.B.7}$$

The corresponding $\varepsilon'_{YY} = \varepsilon_{\text{unwound}}$ element is given by

$$\varepsilon_{\text{unwound}} = \varepsilon_y \cos^2 \delta + (\varepsilon_x \sin^2 \theta + \varepsilon_z \cos^2 \theta) \sin^2 \delta \tag{11.B.8}$$

Equations (11.B.6) and (11.B.8) are those appearing in Eqs (11.8.9). Note that, if the smectic layers are perpendicular to the glass plates, $\delta = 0$, and in this case

$$\varepsilon_{\text{helix}} = \tfrac{1}{2}(\varepsilon_x \cos^2 \theta + \varepsilon_y + \varepsilon_z \sin^2 \theta)$$

$$\varepsilon_{\text{unwound}} = \varepsilon_y \tag{11.B.9}$$

APPENDIX C

The theory of the dielectric relaxation of uniaxial nematic liquid crystals can be developed, without resorting to the Debye-Smoluchowski equation, by averaging the noninertial Langevin equation (Chapter 2, Section 2.17) for the rotational Brownian motion of a molecule in a nematic force field. The average of the equation for the dipole moments is linearized, and the Laplace transform of the equations is closed in a way equivalent to that of the effective eigenvalue method. The main lines followed here are taken from the work by Coffey and coworkers.[12,38–40]

The Langevin equation in the noninertial version for the Brownian motion of a molecular dipole is

$$\xi\dot{\boldsymbol{\theta}} = \boldsymbol{\Lambda}(t) + \boldsymbol{\mu} \times \mathbf{E}(t) \tag{11.C.1}$$

where the white noise driving torque satisfies

$$\overline{\Lambda_i(t)} = 0$$

$$\overline{\Lambda_i(t)\Lambda_j(t')} = 2kT\delta_{ij}\delta(t - t') \tag{11.C.2}$$

where $i = x, y, z$ and the overbar denotes the statistical average. The total field $\mathbf{E}(t)$ includes the nematic field $\mathbf{E}_0$ and the externally applied field $\mathbf{E}(t)$.

Equation (11.C.1), in combination with the kinematic equation (2.17.4), written here as

$$\dot{\boldsymbol{\mu}} = \dot{\boldsymbol{\theta}} \times \boldsymbol{\mu} \tag{11.C.3}$$

leads to

$$\xi\dot{\boldsymbol{\mu}} = \Lambda(t) + (\boldsymbol{\mu} \times \mathbf{E}) \times \boldsymbol{\mu} = \Lambda(t) + \mu^2\mathbf{E} - \boldsymbol{\mu}(\boldsymbol{\mu} \cdot \mathbf{E}) \tag{11.C.4}$$

If the uniaxial nematic potential as well as the external field is taken along the z axis, and the field is suddenly applied at $t = 0$, the three-component form of Eq. (11.C.4) can be written as

$$\dot{\mu}_x = \frac{1}{\xi}(\Lambda_y\mu_z - \Lambda_z\mu_y) - \frac{\mu_x\mu_z}{\xi}[E_0 + E_1U(t)]$$

$$\dot{\mu}_y = \frac{1}{\xi}(\Lambda_z\mu_x - \Lambda_x\mu_z) - \frac{\mu_y\mu_z}{\xi}[E_0 + E_1U(t)] \tag{11.C.5}$$

$$\dot{\mu}_z = \frac{1}{\xi}(\Lambda_x\mu_y - \Lambda_y\mu_y) + \frac{(\mu^2 - \mu_z^2)}{\xi}[E_0 + E_1U(t)]$$

where $U(t)$ is the unit step function.

Multiplicative noise terms appear on the right-hand side of Eqs (11.C.5) which can be evaluated by means of a procedure outlined by Risken[43] to give $-2kT\mu_i/\xi$. Then, averaging Eqs (11.C.5) over a density function given by $\exp -(\boldsymbol{\mu} \cdot \mathbf{E}(t)/kT)$, the longitudinal and transversal cases, respectively, are obtained

$$\frac{d}{dt}\langle\mu_z\rangle + \frac{2kT}{\xi}\langle\mu_z\rangle = \frac{\mu^2[E_0 + E_1(t)]}{\xi}(1 - \langle\cos^2\theta\rangle)$$

$$\frac{d}{dt}\langle\mu_x\rangle + \frac{2kT}{\xi}\langle\mu_x\rangle = \frac{\mu^2E_1(t)}{\xi}(1 - \langle\cos^2\theta\rangle) - \frac{\langle\mu_x\mu_zE_0\rangle}{\xi} \tag{11.C.6}$$

Note that the electric field in the transversal case is applied parallel to the x axis.

The calculation of the relaxation times can be carried out by assuming that the perturbation field $E_1(t)$ is small. In this way, the value of the averages of the linearized equations for $\mu_z = \mu\cos\theta$ and $\cos^2\theta$ can be expressed as follows

$$\langle\mu_z\rangle = \langle\mu_z\rangle_0 + \langle\mu_z\rangle_1 \rightarrow \langle\cos\theta\rangle = \langle\cos\theta\rangle_0 + \langle\cos\theta\rangle_1$$

$$\langle\cos^2\theta\rangle = \langle\cos^2\theta\rangle_0 + \langle\cos^2\theta\rangle_1 \tag{11.C.7}$$

where subscript 0 denotes the ensemble average in the absence of an external electric field, that is, the average over the density function $\exp -(\boldsymbol{\mu} \cdot \mathbf{E}_0/kT)$, where $\mathbf{E}_0$ is the nematic potential. On the other hand, subscript 1 denotes the part of the ensemble average that is linear in $E_1(t)$.

Mean equilibrium values appearing in Eqs (11.C.7) are evaluated (Problem 10) to give

$$\langle \cos \theta \rangle_0 = \coth \zeta - \zeta^{-1} = L(\zeta)$$

$$\langle \cos^2 \theta \rangle_0 = 1 - \frac{2}{\zeta} \coth \zeta + \frac{2}{\zeta^2} = 1 - \frac{2}{\zeta} L(\zeta) \tag{11.C.8}$$

where $\zeta = \mu E_0 / kT$, and L is the Langevin function. Note that the first expression in (11.C.8) agrees with Eq. (1.9.9) and also coincides with the equilibrium average of the first Legendre polynomial.

According to the linearization procedure, the nonequilibrium average can be expressed as

$$L(\zeta + \zeta_1) \cong L(\zeta) + \zeta_1 L'(\zeta) \tag{11.C.9}$$

where $\zeta_1 = \mu E_1 / kT$. Equation (11.C.9) can also be interpreted as the long-time $(t \to \infty)$ behavior of the Langevin function. For future purposes it is convenient to calculate $L'(\zeta)$. This can be done by differentiating the integral form of the Langevin function (Problem 10) to obtain

$$L'(\zeta) = 1 - \frac{2}{\zeta} L(\zeta) - L^2(\zeta) \tag{11.C.10}$$

According to Eqs (11.C.8) and (11.C.9), the nonequilibrium contributions to Eqs (11.C.7) are

$$\langle \cos \theta \rangle_1 = \zeta_1 L'(\zeta)$$

$$\langle \cos^2 \theta \rangle_1 = \left[1 - \frac{2}{\zeta + \zeta_1} L(\zeta + \zeta_1) \right] - \left[1 - \frac{2}{\zeta} L(\zeta) \right]$$

$$= \frac{2 \zeta_1}{\zeta^2} (\zeta L'(\zeta) - L(\zeta)) \tag{11.C.11}$$

In a similar way, Eq. (11.C.6a) can be written as

$$\frac{d}{dt} \langle \mu_z \rangle_1 + \frac{1}{\tau_D} \langle \mu_z \rangle_1 = \frac{\mu^2 E_0}{\xi} (1 - \langle \cos^2 \theta \rangle_1) + \frac{\mu^2 E_1}{\xi} (1 - \langle \cos^2 \theta \rangle_0) \tag{11.C.12}$$

where $\tau_D = \xi / 2kT$ and, as above, only linear terms have been retained. Equation (11.C.12) represents a three-term recurrence equation driven by a forced term containing E_1. In order to determine the eigenvalues, τ_D^{-1}, the unforced equation is considered, namely

$$\frac{d}{dt}\langle\mu_z\rangle_1 + \frac{1}{\tau_D}\langle\mu_z\rangle_1 = \frac{\mu^2 E_0}{\xi}(1 - \langle\cos^2\theta\rangle_1) \tag{11.C.13}$$

The Laplace transform of Eq. (11.C.13) with initial condition $\langle\mu_z(0)\rangle_1 = 0$ is given by

$$L(\langle\mu_z\rangle_1)\left[s + \tau_D^{-1} - \xi^{-1}\mu E_0 \frac{L\langle1 - \cos^2\theta\rangle_1}{L\langle\cos\theta\rangle_1}\right] = 0 \tag{11.C.14}$$

Then, following Morita's approach,[44] one can write

$$\frac{L\langle1 - \cos^2\theta\rangle_1}{L\langle\cos\theta\rangle_1} = \frac{L\langle1 - \cos^2\theta\rangle - L\langle1 - \cos^2\theta\rangle_0}{L\langle\cos\theta\rangle - L\langle\cos\theta\rangle_0} \tag{11.C.15}$$

The former statement corresponds to the effective eigenvalue method described by San Miguel and coworkers[45] to obtain the precise details of the initial decay of the polarization.

According to the final (equilibrium) theorem

$$\lim_{s\to0}\frac{sL\langle1 - \cos^2\theta\rangle_1}{sL\langle\cos\theta\rangle_1} = \lim_{t\to\infty}\frac{\langle1 - \cos^2\theta\rangle_1}{\langle\cos\theta\rangle_1} \tag{11.C.16}$$

Taking into account Eqs (11.C.11) and (11.C.14) to (11.C.16), the lowest eigenvalue obtained with this approach is given by

$$\lambda_\parallel = \tau_\parallel^{-1} = \tau_D^{-1} - \xi^{-1}\frac{2\mu E_0}{\zeta}\left[1 - \frac{L(\zeta)}{\zeta L'(\zeta)}\right] \tag{11.C.17}$$

Then, from the definitions of ξ and ζ, the following expression is obtained for the lowest relaxation time $\tau_\parallel^{-1}$

$$\tau_\parallel = \tau_D\left[\frac{\zeta L'(\zeta)}{L(\zeta)}\right] \tag{11.C.18}$$

which, with the aid of Eq. (11.C.10), can be transformed into

$$\tau_\parallel = \tau_D\frac{1 - (2/\zeta)L(\zeta) - L^2(\zeta)}{L(\zeta)/\zeta} \tag{11.C.19}$$

By neglecting the third term in the numerator of Eq. (11.C.19), this equation can be written in terms of the order parameter $S = \langle P_2 \rangle_0 = \frac{3}{2} \langle \cos^2 \theta \rangle_0 - \frac{1}{2}$ as follows

$$\tau_\parallel = \tau_D \frac{2S + 1}{1 - S} \tag{11.C.20}$$

in agreement with the results of Meier who proposes $g_\parallel = (2S + 1)/(1 - S)$ for the retardation factor.[46]

In a similar way,[12] it is possible to obtain

$$\tau_\perp = \tau_D \frac{2(1 - S)}{2 + S} \tag{11.C.21}$$

From Eqs (11.C.20) and (11.C.21), the following relationships between the retardation factors can be obtained

$$g_\parallel = \frac{2 - g_\perp}{g_\perp}, \qquad g_\perp = \frac{2}{1 + g_\parallel} \tag{11.C.22}$$

PROBLEMS

Problem 1

Determine the electric field produced by a dipole in a homogeneous anisotropic medium whose principal permittivity values are $\varepsilon_x = \varepsilon_y = \varepsilon_\perp$ and $\varepsilon_z = \varepsilon_\parallel$. Study the case where the dipole is along the principal axis z.

Solution

Let us consider first a point charge at the origin. The Maxwell equation for the electric displacement gives

$$\mathrm{div}\mathbf{D} = 4\pi q \delta(\mathbf{r}) \tag{P.11.1.1}$$

where δ is the Dirac function. For anisotropic media

$$D_i = \varepsilon_{ij} E_j = -\varepsilon_{ij} \frac{\partial \varphi}{\partial x_j} \tag{P.11.1.2}$$

where φ is the potential of the field. By choosing the coordinate axis along the principal values of the permittivity tensor, Eqs (P.11.1.1) and (P.11.1.2) lead to

$$\varepsilon_1 \frac{\partial^2 \varphi}{\partial x^2} + \varepsilon_2 \frac{\partial^2 \varphi}{\partial y^2} + \varepsilon_3 \frac{\partial^2 \varphi}{\partial z^2} = -4\pi q \delta(x)\delta(y)\delta(z) = -4\pi q \delta(\mathbf{r}) \tag{P.11.1.3}$$

A change of variables

$$x' = x\varepsilon_1^{-1/2}, \qquad y' = y\varepsilon_2^{-1/2}, \qquad z' = z\varepsilon_3^{-1/2} \tag{P.11.1.4}$$

together with the properties of the δ function

$$\delta(\mathbf{r}') = (\varepsilon_1\varepsilon_2\varepsilon_3)^{1/2}\delta(\mathbf{r}) \tag{P.11.1.5}$$

converts Eq. (P.11.1.3) to

$$\frac{\partial^2\varphi}{\partial x'^2} + \frac{\partial^2\varphi}{\partial y'^2} + \frac{\partial^2\varphi}{\partial z'^2} = \Delta'\varphi = -\frac{q}{(\varepsilon_1\varepsilon_2\varepsilon_3)^{1/2}}\delta(\mathbf{r}') \tag{P.11.1.6}$$

The solution of this equation is given by

$$\varphi = -\frac{q'}{r'} = -\frac{q}{(\varepsilon_1\varepsilon_2\varepsilon_3)^{1/2}(x^2/\varepsilon_1 + y^2/\varepsilon_2 + z^2/\varepsilon_3)^{1/2}} \tag{P.11.1.7}$$

The electric field can be written as

$$\mathbf{E} = -\mathrm{grad}\varphi = \frac{q(x/\varepsilon_1\mathbf{i} + y/\varepsilon_2\mathbf{j} + z/\varepsilon_3\mathbf{k})}{(\varepsilon_1\varepsilon_2\varepsilon_3)^{1/2}(x^2/\varepsilon_1 + y^2/\varepsilon_2 + z^2/\varepsilon_3)^{3/2}} \tag{P.11.1.8}$$

For a uniaxial phase, taking the z axis as the optical axis, the potential is given by

$$\varphi = -\frac{q}{\varepsilon_\perp\varepsilon_\parallel^{1/2}(x^2/\varepsilon_\perp + y^2/\varepsilon_\perp + z^2/\varepsilon_\parallel)^{1/2}}$$

$$= -\frac{q}{(\varepsilon_\perp\varepsilon_\parallel)^{1/2}[x^2 + y^2 + (\varepsilon_\perp/\varepsilon_\parallel)z^2]^{1/2}} \tag{P.11.1.9}$$

and the corresponding field is

$$\mathbf{E} = -\mathrm{grad}\varphi = \frac{q(x\mathbf{i} + y\mathbf{j} + (\varepsilon_\perp/\varepsilon_\parallel)z\mathbf{k})}{(\varepsilon_\perp\varepsilon_\parallel)^{1/2}(x^2 + y^2 + (\varepsilon_\perp/\varepsilon_\parallel)z^2)^{3/2}} \tag{P.11.1.10}$$

Assuming now that the charges $+q$ and $-q$ are at the points $(0, 0, c)$ and $(0, 0, -c)$, the potential of the corresponding dipole can be expressed as

$$\varphi = -\frac{q}{\varepsilon_\perp\varepsilon_\parallel^{1/2}}\left\{\left[\frac{x^2 + y^2}{\varepsilon_\perp} + \frac{(z-c)^2}{\varepsilon_\parallel}\right]^{-1/2} - \left[\frac{x^2 + y^2}{\varepsilon_\perp} + \frac{(z+c)^2}{\varepsilon_\parallel}\right]^{-1/2}\right\} \tag{P.11.1.11}$$

The components of the electric field are

$$
E_x = \frac{qx}{\varepsilon_\perp^2 \varepsilon_\parallel^{1/2}} \left\{ \left[\frac{x^2 + y^2}{\varepsilon_\perp} + \frac{(z+c)^2}{\varepsilon_\parallel} \right]^{-3/2} - \left[\frac{x^2 + y^2}{\varepsilon_\perp} + \frac{(z-c)^2}{\varepsilon_\parallel} \right]^{-3/2} \right\}
$$

$$
E_y = \frac{qy}{\varepsilon_\perp^2 \varepsilon_\parallel^{1/2}} \left\{ \left[\frac{x^2 + y^2}{\varepsilon_\perp} + \frac{(z+c)^2}{\varepsilon_\parallel} \right]^{-3/2} - \left[\frac{x^2 + y^2}{\varepsilon_\perp} + \frac{(z-c)^2}{\varepsilon_\parallel} \right]^{-3/2} \right\}
$$

$$
E_z = \frac{q}{\varepsilon_\perp \varepsilon_\parallel^{3/2}} \left\{ (z+c) \left[\frac{x^2 + y^2}{\varepsilon_\perp} + \frac{(z+c)^2}{\varepsilon_\parallel} \right]^{-3/2} \right.
$$

$$
\left. - (z-c) \left[\frac{x^2 + y^2}{\varepsilon_\perp} + \frac{(z-c)^2}{\varepsilon_\parallel} \right]^{-3/2} \right\}
$$

$$(\text{P.11.1.12})$$

By rewriting Eq. (P.11.1.11) as

$$
\varphi = -\frac{q}{\varepsilon_\perp \varepsilon_\parallel^{1/2}} \left(\frac{\varepsilon_\perp}{x^2 + y^2} \right)^{1/2}
$$

$$
\times \frac{[1 + (\varepsilon_\perp/\varepsilon_\parallel)(z^2/(x^2 + y^2))(1 + (c^2/z^2))]^{1/2}}{[1 + (\varepsilon_\perp/\varepsilon_\parallel)(z^2/(x^2 + y^2))(1 + (c^2/z^2))]^{1/2}} \quad (\text{P.11.1.13})
$$

$$
\frac{-[1 + (\varepsilon_\perp/\varepsilon_\parallel)(z^2/(x^2 + y^2))(1 - (c^2/z^2))]^{1/2}]}{[1 + (\varepsilon_\perp/\varepsilon_\parallel)(z^2/(x^2 + y^2))(1 - (c^2/z^2))]^{1/2}}
$$

the use of the binomial expansion leads to the following approximate expression for the potential

$$
\varphi \cong \frac{\mu \varepsilon_\perp^{1/2}}{\varepsilon_\parallel^{3/2}} \frac{z}{(x^2 + y^2)^{1/2}} \frac{1}{x^2 + y^2 + (\varepsilon_\perp/\varepsilon_\parallel)(z^2 + c^2)}
$$

$$
\simeq \frac{\mu \varepsilon_\perp^{1/2}}{\varepsilon_\parallel^{3/2}} \frac{z}{(x^2 + y^2)^{3/2}} \quad (\text{P.11.1.14})
$$

where $\mu = 2qc$ is the dipole moment of the system. Note that the potential in the plane xy is zero.

Following the same procedure, the components of the electric field are given by

$$
\begin{aligned}
E_x &\cong -\frac{3\mu\varepsilon_\perp^{1/2}}{\varepsilon_\parallel^{3/2}} \frac{xz}{(x^2+y^2)^{3/2}} \frac{1}{x^2+y^2+3(\varepsilon_\perp/\varepsilon_\parallel)z^2(1+(c^2/z^2))} \\
&\simeq -\frac{3\mu\varepsilon_\perp^{1/2}}{\varepsilon_\parallel^{3/2}} \frac{xz}{(x^2+y^2)^{5/2}} \\[4pt]
E_y &\cong -\frac{3\mu\varepsilon_\perp^{1/2}}{\varepsilon_\parallel^{3/2}} \frac{yz}{(x^2+y^2)^{3/2}} \frac{1}{x^2+y^2+3(\varepsilon_\perp/\varepsilon_\parallel)z^2(1+(c^2/z^2))} \\
&\simeq -\frac{3\mu\varepsilon_\perp^{1/2}}{\varepsilon_\parallel^{3/2}} \frac{yz}{(x^2+y^2)^{5/2}} \\[4pt]
E_z &\cong \frac{\mu\varepsilon_\perp^{1/2}}{\varepsilon_\parallel^{3/2}} \frac{1}{(x^2+y^2)^{3/2}} \frac{1-\frac{3}{2}(\varepsilon_\perp/\varepsilon_\parallel)(z^2/(x^2+y^2))(1-(c^2/z^2))}{1+3(\varepsilon_\perp/\varepsilon_\parallel)(z^2/(x^2+y^2))(1+(c^2/z^2))} \\
&\simeq \frac{\mu\varepsilon_\perp^{1/2}}{\varepsilon_\parallel^{3/2}} \frac{1}{(x^2+y^2)^{3/2}}
\end{aligned}
$$

$$(P.11.1.15)$$

For points such that $z \gg (x^2+y^2)^{1/2}$, that is, close to the z axis, the potential will be approximately given by

$$
\begin{aligned}
\varphi &\cong -\frac{\mu}{\varepsilon_\perp} \frac{1-\frac{1}{2}(\varepsilon_\perp/\varepsilon_\parallel)(x^2+y^2)/z^2}{z^2[1-(c^2/z^2)+(x^2+y^2)/z^2(\varepsilon_\perp/\varepsilon_\parallel)(1+(c^2/z^2))/(1-(c^2/z^2))]} \\
&\simeq -\frac{\mu}{\varepsilon_\perp z^2}
\end{aligned}
\qquad (P.11.1.16)
$$

The pertinent equations for the field can also be obtained. Other specific cases can be analyzed by using the same methodology.

Problem 2

An anisotropic medium contains a spherical cavity with radius a. Determine the field in the cavity as a function of the external field $\mathbf{E}_0$.

Solution

The transformation given by Eqs (P.11.1.4) reduces the problem to the isotropic case. The cavity is transformed into an ellipsoid with semiaxes given by

$$
d' = a\varepsilon_{21}^{-1/2}, \qquad b' = a\varepsilon_{22}^{-1/2}, \qquad c' = a\varepsilon_{23}^{-1/2}
\qquad (P.11.2.1)
$$

where the first subscript indicates that the permittivities are referred to the external medium.

According to Eq. (11.9.7), the components of the electric field along the three principal axes for $\varepsilon_{11} = \varepsilon$ and $\varepsilon_{12} = \varepsilon_{13} = 1$ are

$$E_1 = \frac{\varepsilon_{21}}{\varepsilon_{21} - A_1(\varepsilon_{21} - 1)} E_{10}$$

$$E_2 = \frac{\varepsilon_{22}}{\varepsilon_{22} - A_2(\varepsilon_{22} - 1)} E_{20} \qquad \text{(P.11.2.2)}$$

$$E_3 = \frac{\varepsilon_{23}}{\varepsilon_{23} - A_3(\varepsilon_{23} - 1)} E_{30}$$

where the depolarizing factors can be taken from Eq. (1.C.2) and the semiaxes are now given by Eq. (P.11.2.1).

For an oblate spheroid, $a' = b' \gg c'$, with eccentricity given by

$$e = \sqrt{\left(\frac{a'}{c'}\right)^2 - 1} = \sqrt{\frac{\varepsilon_\parallel}{\varepsilon_\perp} - 1} \qquad \text{(P.11.2.3)}$$

and immersed in an infinite medium with principal permittivities $\varepsilon_{21} = \varepsilon_{22} = \varepsilon_\perp$ and $\varepsilon_{23} = \varepsilon_\parallel$, that is, uniaxial, the components of the electric field according to Eq. (1.C.7a) are

$$E_\parallel = \frac{3\varepsilon_\parallel}{2\varepsilon_\parallel + 1 - \frac{2}{5}(\varepsilon_\parallel - 1)((\varepsilon_\parallel/\varepsilon_\perp) - 1)} E_{\parallel 0}$$

$$E_\perp = \frac{3\varepsilon_\parallel}{2\varepsilon_\parallel + 1 + \frac{1}{5}(\varepsilon_\parallel - 1)((\varepsilon_\parallel/\varepsilon_\perp) - 1)} E_{\perp 0} \qquad \text{(P.11.2.4)}$$

Note that the field in the empty ellipsoid in an infinite medium of principal permittivities $\varepsilon_{21}, \varepsilon_{22}, \varepsilon_{23}$ is equivalent to the field of a dielectric ellipsoid with principal permittivities $\varepsilon_{11}^{-1}, \varepsilon_{12}^{-1}, \varepsilon_{13}^{-1}$ in vacuum.

Problem 3

Let a sample of unixial nematic liquid crystal be characterized by a local director axis $\mathbf{n}$ and a dielectric diagonalized permittivity tensor $\varepsilon^l = \{\varepsilon_\perp, \varepsilon_\perp, \varepsilon_\parallel\}$, where $\varepsilon_\parallel$ and $\varepsilon_\perp$ are respectively the permittivities parallel and normal to the director axis. An electric field $\mathbf{E}$ is applied to the sample, in such a way that a uniform electric field results in the sample. Deduce expressions for the average dielectric permittivity for intermediate alignments between the homeotropic alignment ($\mathbf{n} \parallel \mathbf{E}$) and planar alignment ($\mathbf{n} \perp \mathbf{E}$)[47].

Solution

Let us consider a sample aligned by an electric field along the Z axis in the laboratory reference frame. A particular molecule (or region) has the local orientation given by the axis system (x, y, z) (see Fig. 11.20 taking into account that the angles θ and φ are respectively the angles ϑ and ϕ in the problem). If a local region is under an electric field with components (E_x, E_y, E_z) in the laboratory frame, then the permittivity in the laboratory frame (L) in terms of the permittivity in the molecular frame (l) will be given by

$$\varepsilon^L = \mathbf{T}\varepsilon^l \mathbf{T}^{\mathrm{T}} \tag{P.11.3.1}$$

where $\mathbf{T}$ is a matrix that represents the product of two rotations: the first with angle ϑ around the Y axis

$$\begin{vmatrix} \cos\vartheta & 0 & -\sin\vartheta \\ 0 & 1 & 0 \\ \sin\vartheta & 0 & \cos\vartheta \end{vmatrix} \tag{P.11.3.2}$$

and the second with angle ϕ around the new Z' axis

$$\begin{vmatrix} \cos\phi & \sin\phi & 0 \\ -\sin\phi & \cos\phi & 0 \\ 0 & 0 & 1 \end{vmatrix} \tag{P.11.3.3}$$

The product of these matrices is given by

$$\begin{bmatrix} \cos\vartheta\cos\phi & -\mathrm{sen}\phi & -\mathrm{sen}\vartheta\cos\phi \\ \cos\vartheta\,\mathrm{sen}\phi & \cos\phi & -\mathrm{sen}\vartheta\,\mathrm{sen}\phi \\ \mathrm{sen}\vartheta & 0 & \cos\vartheta \end{bmatrix} \tag{P.11.3.4}$$

a matrix that transforms a vector in the laboratory frame into a vector in the molecular frame.

Carrying Eq. (P.11.3.4) into Eq. (P.11.3.1), we obtain

$$\begin{bmatrix} \begin{pmatrix} \varepsilon_\perp \cos^2\vartheta\cos^2\phi \\ +\varepsilon_\perp \mathrm{sen}^2\phi \\ +\varepsilon_\parallel \mathrm{sen}^2\vartheta\cos^2\phi \end{pmatrix} & \begin{pmatrix} \varepsilon_\perp \cos^2\vartheta\cos\phi\,\mathrm{sen}\phi \\ -\varepsilon_\perp \mathrm{sen}\phi\cos\phi \\ +\varepsilon_\parallel \mathrm{sen}^2\vartheta\,\mathrm{sen}\phi\cos\phi \end{pmatrix} & \begin{pmatrix} (\varepsilon_\parallel - \varepsilon_\perp)\mathrm{sen}\vartheta \\ \times \cos\vartheta\cos\phi \end{pmatrix} \\[2em] \begin{pmatrix} \varepsilon_\perp \cos^2\vartheta\,\mathrm{sen}\phi\cos\phi \\ -\varepsilon_\perp \mathrm{sen}\phi\cos\phi \\ +\varepsilon_\parallel \mathrm{sen}^2\vartheta\,\mathrm{sen}\phi\cos\phi \end{pmatrix} & \begin{pmatrix} \varepsilon_\perp \cos^2\vartheta\,\mathrm{sen}^2\phi \\ +\varepsilon_\perp \cos^2\phi \\ +\varepsilon_\parallel \mathrm{sen}^2\vartheta\,\mathrm{sen}^2\phi \end{pmatrix} & \begin{pmatrix} (\varepsilon_\parallel - \varepsilon_\perp)\mathrm{sen}\vartheta \\ \times \cos\vartheta\cos\phi \end{pmatrix} \\[2em] \begin{pmatrix} (\varepsilon_\parallel - \varepsilon_\perp)\mathrm{sen}\vartheta \\ \times \cos\vartheta\,\mathrm{sen}\phi \end{pmatrix} & \begin{pmatrix} (\varepsilon_\parallel - \varepsilon_\perp)\mathrm{sen}\vartheta \\ \times \cos\vartheta\,\mathrm{sen}\phi \end{pmatrix} & \begin{pmatrix} \varepsilon_\perp \cos^2\vartheta \\ +\varepsilon_\parallel \mathrm{sen}^2\vartheta \end{pmatrix} \end{bmatrix} \tag{P.11.3.5}$$

where $\varepsilon_\perp = \varepsilon_x = \varepsilon_y$ and $\varepsilon_\parallel = \varepsilon_z$.

For a macroscopic specimen, the permittivity is the average over the distribution function $f(\phi, \vartheta)$. Accordingly, $f(\phi, \vartheta) \sin \vartheta \, d\vartheta \, d\phi$ is the probability of obtaining a region oriented inside an element of solid angle $d\Omega = \sin \vartheta \, d\vartheta \, d\phi$. However, since the sample has axial symmetry with respect to the Z axis, and noting that ϕ varies between 0 and 2π, we can write

$$f(\vartheta, \phi) = 2\pi F(\vartheta) \tag{P.11.3.6}$$

The relevant mean values are

$$\langle \sin^2 \phi \rangle = \int_0^{2\pi} \sin^2 \phi \, d\phi = \tfrac{1}{2}$$

$$\langle \cos^2 \phi \rangle = \int_0^{2\pi} \cos^2 \phi \, d\phi = \tfrac{1}{2} \tag{P.11.3.7}$$

$$\langle \sin \phi \cos \phi \rangle = \langle \sin \phi \rangle = \langle \cos \phi \rangle = 0$$

After substitution into Eq. (P.11.3.5), we obtain

$$e = \begin{bmatrix} \tfrac{1}{2}\varepsilon_\perp \langle \cos^2 \vartheta \rangle + \tfrac{1}{2}\varepsilon_\perp \\ + \tfrac{1}{2}\varepsilon_\parallel \langle \operatorname{sen}^2 \vartheta \rangle & 0 & 0 \\ 0 & \tfrac{1}{2}\varepsilon_\perp \langle \cos^2 \vartheta \rangle + \tfrac{1}{2}\varepsilon_\perp \\ & + \tfrac{1}{2}\varepsilon_\parallel \langle \operatorname{sen}^2 \vartheta \rangle & 0 \\ 0 & 0 & \varepsilon_\parallel \langle \cos^2 \vartheta \rangle \\ & & + \varepsilon_\perp \langle \operatorname{sen}^2 \vartheta \rangle \end{bmatrix} \tag{P.11.3.8}$$

The order parameter is given by Eq. (11.4.8)

$$S_d = \tfrac{1}{2}(3 \langle \cos^2 \vartheta \rangle - 1) \tag{P.11.3.9}$$

and consequently the following expressions are obtained

$$\langle \varepsilon_Z \rangle = \tfrac{1}{3}\varepsilon_\parallel (1 + 2S_d) + \tfrac{2}{3}\varepsilon_\perp (1 - S_d)$$
$$\langle \varepsilon_X \rangle = \langle \varepsilon_Y \rangle = \tfrac{1}{3}\varepsilon_\parallel (1 - S_d) + \tfrac{2}{3}\varepsilon_\perp \left(1 + \tfrac{1}{2}S_d\right) \tag{P.11.3.10}$$

Since the trace is invariant, the following equation holds

$$\langle \varepsilon_X \rangle = \langle \varepsilon_Y \rangle = \tfrac{1}{2}(\varepsilon_\parallel + 2\varepsilon_\perp - \langle \varepsilon_Z \rangle) \tag{P.11.3.11}$$

The matrix given by (P.11.3.7) can also be written as

$$e = \begin{bmatrix} \varepsilon_\perp & 0 & 0 \\ 0 & \varepsilon_\perp & 0 \\ 0 & 0 & \varepsilon_\perp \end{bmatrix}$$

$$+ \begin{bmatrix} \frac{1}{2}(\varepsilon_\parallel - \varepsilon_\perp)\langle \operatorname{sen}^2 \vartheta \rangle & 0 & 0 \\ 0 & \frac{1}{2}(\varepsilon_\parallel - \varepsilon_\perp)\langle \operatorname{sen}^2 \vartheta \rangle & 0 \\ 0 & 0 & (\varepsilon_\parallel - \varepsilon_\perp)\langle \cos^2 \vartheta \rangle \end{bmatrix} \tag{P.11.3.12}$$

which in indicial notation is given by

$$\varepsilon_{ij} = \varepsilon_\perp \delta_{ij} + (\varepsilon_\parallel - \varepsilon_\perp)\langle n_i n_j \rangle \tag{P.11.3.13}$$

where n_i are the components of the director with respect to the laboratory frame. Taking into account the definition (1.8.7), and disregarding averages, the relationship between the electric displacement and the electric field is obtained

$$D_i = \varepsilon_\perp E_i + (\varepsilon_\parallel - \varepsilon_\perp) n_i n_j E_j \tag{P.11.3.14}$$

or equivalently

$$\mathbf{D} = \varepsilon_\perp \mathbf{E} + (\varepsilon_\parallel - \varepsilon_\perp)(\mathbf{n} \cdot \mathbf{E})\mathbf{n} \tag{P.11.3.15}$$

The corresponding relationships between the polarizability and the electric field are

$$P_i = \chi_\perp E_i + (\chi_\parallel - \chi_\perp) n_i n_j E_j \tag{P.11.3.16}$$

and

$$\mathbf{P} = \chi_\perp \mathbf{E} + (\chi_\parallel - \chi_\perp)(\mathbf{n} \cdot \mathbf{E})\mathbf{n} \tag{P.11.3.17}$$

Note that, if $\mathbf{n}$ is the director, then the components of the electric field along and normal to it are respectively given by $E_\parallel = \mathbf{E} \cdot \mathbf{n}$ and $E_\perp = |\mathbf{E} - (\mathbf{n} \cdot \mathbf{E})\mathbf{n}|$.

Problem 4

Deduce the free energy and the torque for unit volume and constant voltage for the nematic liquid crystal of the preceding example.

Solution

From the definition of dielectric free energy (see reference [7], p. 48) and Eqs (1.8.7) and (P.11.3.14), we have, after integration, the following expression

$$F = -\int \mathbf{E}\, d\mathbf{P} = F_0 - \frac{1}{2}\chi_\perp E^2 - \frac{1}{2}(\chi_\parallel - \chi_\perp)(\mathbf{n}\cdot\mathbf{E})^2 \qquad (P.11.4.1)$$

Note that the last term on the right-hand side is the only term dependent on orientation. If the dielectric anisotropy $(\varepsilon_\parallel - \varepsilon_\perp)$ is positive, homeotropic alignment is favored under electric fields. On the other hand, if the molecule has a dipole moment predominantly normal to the long axis, then the anisotropy is negative and the electric field favors planar alignment.

The torque is given by

$$\boldsymbol{\Gamma} = \mathbf{P} \times \mathbf{E} = (\chi_\parallel - \chi_\perp)(\mathbf{n}\cdot\mathbf{E})(\mathbf{n}\times\mathbf{E}) \qquad (P.11.4.2)$$

Problem 5

Taking into account Maxwell's equations [Eqs 4.3.1b]: (a) show that in an anisotropic nonmagnetic medium

$$\mathbf{D} = n^2\mathbf{E} - (\mathbf{n}\cdot\mathbf{E})\mathbf{n}$$

where $\mathbf{n} = \mathbf{k}/\omega$ and $\mathbf{k}$ is the wave vector; (b) deduce the compatibility condition for the preceding equation (Fresnel's equation).

Solution

(a) For a nonmagnetic medium $\mathbf{B} = \mathbf{H}$, and the Laplace transformation of Eqs (4.3.1b) gives

$$\begin{aligned} \mathbf{k} \times \mathbf{H} &= -\omega \mathbf{D} \\ \mathbf{k} \times \mathbf{E} &= \omega \mathbf{H} \end{aligned} \qquad (P.11.5.1)$$

From Eqs (P.11.5.1) one obtains

$$\mathbf{k} \times (\mathbf{k} \times \mathbf{E}) = -\omega^2\mathbf{D} \qquad (P.11.5.2)$$

From the definition given for $\mathbf{n}$ the required result holds

$$\mathbf{D} = \mathbf{n} \times (\mathbf{E} \times \mathbf{n}) = n^2\mathbf{E} - (\mathbf{n} \cdot \mathbf{E})\mathbf{n} \tag{P.11.5.3}$$

(b) According to Eq. (1.8.7), a system of three linear homogeneous equations for the three components of the electric field is obtained. The compatibility condition requires that the determinant of the coefficients of this system vanishes, that is

$$\det |n^2\delta_{ij} - n_in_j - \varepsilon_{ij}| = 0 \tag{P.11.5.4}$$

In order to evaluate this determinant, it is advisable to choose as axes the principal axes of the dielectric tensor. If the principal values of the dielectric tensor are ε_1, ε_2, ε_3, the condition given by Eq. (P.11.5.4) can be written as

$$\begin{vmatrix} n^2 - n_1^2 - \varepsilon_1 & -n_1n_2 & -n_1n_3 \\ -n_2n_1 & n^2 - n_2^2 - \varepsilon_2 & -n_2n_3 \\ -n_3n_1 & -n_3n_2 & n^2 - n_3^2 - \varepsilon_3 \end{vmatrix} = 0 \tag{P.11.5.5}$$

After algebraic calculations, we obtain

$$n^2(n_1^2\varepsilon_1 + n_2^2\varepsilon_2 + n_3^2\varepsilon_3) - [n_1^2\varepsilon_1(\varepsilon_2+\varepsilon_3)$$
$$+ n_2^2\varepsilon_2(\varepsilon_1 + \varepsilon_3) + n_3^2\varepsilon_3(\varepsilon_1 + \varepsilon_2)] + \varepsilon_1\varepsilon_2\varepsilon_3 = 0 \tag{P.11.5.6}$$

which is the required condition.

Note that the lack of sixth-order terms in Eq. (P.11.5.6) is due to the fact that the wave has two, instead of three, independent directions of polarization. For a uniaxial material where $\varepsilon_1 = \varepsilon_2 = \varepsilon_\perp$ and $\varepsilon_3 = \varepsilon_\parallel$, Eq. (P.11.5.6) reduces to

$$(n^2 - \varepsilon_\perp)[n_3^2\varepsilon_\parallel + (n_1^2 + n_2^2)\varepsilon_\perp - \varepsilon_\parallel\varepsilon_\perp] = 0 \tag{P.11.5.7}$$

This means that the fourth-order equation degenerates into two second-order equations: the first is a sphere whose equation is given by

$$n^2 = \varepsilon_\perp \tag{P.11.5.8}$$

and the second is an ellipsoid with the following equation

$$\frac{n_3^2}{\varepsilon_\perp} + \frac{n_1^2 + n_2^2}{\varepsilon_\parallel} = 1 \tag{P.11.5.9}$$

This ellipsoid is (externally or internally) tangential to the abovementioned sphere.

With respect to the wave corresponding to Eq. (P.11.5.8), the crystal behaves as an isotropic material with refractive index $\varepsilon_\perp^{1/2}$, and this is called an *ordinary* wave. The wave corresponding to Eq. (P.11.5.9) is called *extraordinary* and the magnitude of the wave vector depends on the angle θ that it makes with the optical axis. By choosing the plane zx as the plane in which the ray vector is coplanar with the wave vector and the optical axis, we have

$$n_3 = n\cos\theta, \qquad n_1 = n\sin\theta \tag{P.11.5.10}$$

Then Eq. (P.11.5.9) gives

$$\frac{\sin^2\theta}{\varepsilon_\parallel} + \frac{\cos^2\theta}{\varepsilon_\perp} = \frac{1}{n^2} \tag{P.11.5.11}$$

Problem 6

Calculate the mean dipole moment along the two principal directions for a nematic liquid crystal averaging over the Euler angles[41].

Solution

The mean dipole moments will be obtained from

$$\bar{\mu}_z = \frac{\int_0^\pi \int_0^{2\pi} \int_0^{2\pi} \mu_z \exp(-U/kT) \sin\theta \, d\theta \, d\varphi \, d\psi}{\int_0^\pi \int_0^{2\pi} \int_0^{2\pi} \exp(-U/kT) \sin\theta \, d\theta \, d\varphi \, d\psi}$$

$$\bar{\mu}_x = \frac{\int_0^\pi \int_0^{2\pi} \int_0^{2\pi} \mu_x \exp(-U/kT) \sin\theta \, d\theta \, d\varphi \, d\psi}{\int_0^\pi \int_0^{2\pi} \int_0^{2\pi} \exp(-U/kT) \sin\theta \, d\theta \, d\varphi \, d\psi} \tag{P.11.6.1}$$

where $U = -g\mu\mathbf{E}$.

Expanding the exponential to the second term gives

$$\bar{\mu}_z = \frac{\int_0^\pi \int_0^{2\pi} \int_0^{2\pi} \mu_z(1 + ghE_1\mu_z/kT) \sin\theta \, d\theta \, d\psi \, d\varphi}{\int_0^\pi \int_0^{2\pi} \int_0^{2\pi} (1 + ghE_1\mu_z/kT) \sin\theta \, d\theta \, d\psi \, d\varphi}$$

$$\bar{\mu}_x = \frac{\int_0^\pi \int_0^{2\pi} \int_0^{2\pi} \mu_x(1 + ghE_2\mu_x/kT) \sin\theta \, d\theta \, d\psi \, d\varphi}{\int_0^\pi \int_0^{2\pi} \int_0^{2\pi} (1 + ghE_2\mu_x/kT) \sin\theta \, d\theta \, d\psi \, d\varphi} \tag{P.11.6.2}$$

where E_1 and E_2 are the directing fields and

$$\mu_z = \mu^*(a_{zz'}\cos\beta + a_{xz'}\sin\beta)$$
$$= \mu^*(\cos\beta\cos\theta + \sin\beta\sin\psi\sin\theta)$$
$$\mu_x = \mu^*(a_{zx'}\cos\beta + a_{xx'}\sin\beta)$$
$$= \mu^*[\cos\beta\sin\theta\sin\varphi + \sin\beta(\cos\psi\cos\varphi - \cos\theta\sin\psi\sin\varphi)]$$

$$(P.11.6.3)$$

Here β is the angle defining the longitudinal and transversal components of the dipole moment, and $\mu^* = h\mu$. After integration we obtain

$$\bar{\mu}_z = \mu^2\frac{gh^2E_1^2}{kT}\left(\cos^2\beta\,\overline{\cos^2\theta} + \frac{1}{2}\sin^2\beta\,\overline{\sin^2\theta}\right)$$

$$= \frac{gh^2E_1^2}{kT}\left(\mu_l^2\,\overline{\cos^2\theta} + \frac{1}{2}\mu_t^2\,\overline{\sin^2\theta}\right)$$

$$(P.11.6.4)$$

$$\bar{\mu}_x = \mu^2\frac{gh^2E_2^2}{2kT}\left[\cos^2\beta\,\overline{\sin^2\theta} + \sin^2\beta\left(1 - \frac{1}{2}\overline{\sin^2\theta}\right)\right]$$

$$= \frac{gh^2E_2^2}{2kT}\left[\mu_l^2\,\overline{\sin^2\theta} + \mu_t^2\left(1 - \frac{1}{2}\overline{\sin^2\theta}\right)\right]$$

where an overbar denotes the mean value.

According to the definition of the order parameter, Eqs (P.11.6.4) can be written as follows

$$\bar{\mu}_z = \frac{gh^2E_1}{3kT}[\mu^2(1-S) + 3\mu_l^2 S]$$

$$\bar{\mu}_x = \frac{gh^2E_2}{3kT}\left[\mu^2(1-S) + \frac{3}{2}\mu_t^2 S\right]$$

$$(P.11.6.5)$$

Problem 7

Evaluate the mean order parameter on the basis of a nematic potential given by[42].

$$U = q\sin^2\theta$$

Solution

The mean order parameter in terms of the nematic potential is defined as

$$\langle P_2(\cos\theta)\rangle = \frac{\int_0^\pi \sin\theta\left(\frac{3}{2}\cos^2\theta - \frac{1}{2}\right)\exp[-(q\sin^2\theta)/kT]\,d\theta}{\int_0^\pi \sin\theta\exp[-(q\sin^2\theta)/kT]\,d\theta}$$

$$= \frac{3}{2}\frac{\int_0^\pi \sin\theta\cos^2\theta\exp[-(q\sin^2\theta)/kT]\,d\theta}{\int_0^\pi \sin\theta\exp[-(q\sin^2\theta)/kT]\,d\theta} - \frac{1}{2} \qquad \text{(P.11.7.1)}$$

The integral in the numerator of the second equality in Eq. (P.11.7.1) can also be written as

$$\int_0^\pi \sin\theta\cos^2\theta\exp\left(-\frac{q\sin^2\theta}{kT}\right)d\theta = \int_0^\pi \sin\theta(1-\sin^2\theta)\exp$$

$$-\left(\frac{q\sin^2\theta}{kT}\right)d\theta = I_1 - I_2 \qquad \text{(P.11.7.2)}$$

Accordingly, one can write

$$\langle P_2(\cos\theta)\rangle = \frac{3}{2}\left(1 - \frac{I_2}{I_1}\right) - \frac{1}{2} = 1 - \frac{3}{2}\frac{I_2}{I_1} \qquad \text{(P.11.7.3)}$$

Let us evaluate I_1 by parts

$$I_1 = \int_0^\pi \sin\theta\exp\left(-\frac{q\sin^2\theta}{kT}\right)d\theta = 2 - \frac{2q}{kT}(I_1 - I_2) \qquad \text{(P.11.7.4)}$$

from which one obtains

$$\frac{I_2}{I_1} = 1 + \frac{kT}{2q} - \frac{kT}{qI_1} \qquad \text{(P.11.7.5)}$$

By substituting Eq. (P.11.7.5) into Eq. (P.11.7.3), the following expression is obtained

$$\langle P_2(\cos\theta)\rangle = \frac{3kT}{4q}\left(1 + \frac{2}{I_1}\right) - \frac{1}{2} \qquad \text{(P.11.7.6)}$$

Finally, by making the substitution $\sin^2 \theta = 1 - x^2$, I_1 can be evaluated as

$$I_1 = \int_{-1}^{1} \exp\left[-\frac{q(1-x^2)}{kT}\right] dx = 2 \int_{0}^{1} \exp\left[-\frac{q(1-x^2)}{kT}\right] dx$$

$$= 2\exp\left(-\frac{q}{kT}\right)\left(\frac{q}{kT}\right)^{-1/2} \int_{0}^{q/kT} \exp y^2 \, dy \qquad (P.11.7.7)$$

By carrying Eq. (P.11.7.7) into Eq. (P.11.7.6), the second-order Legendre polynomial is obtained as

$$\langle P_2(\cos\theta)\rangle = \frac{3kT}{4q}\left[\frac{(q/kT)^{1/2}\exp(q/kT)}{\int_{0}^{q/kT}\exp y^2 \, dy} - 1\right] - \frac{1}{2}$$

$$= \frac{3kT}{4q}\left[\frac{\exp(q/kT)}{\int_{0}^{1}\exp[q/(kT)u^2]\,du} - 1\right] - \frac{1}{2} \qquad (P.11.7.8)$$

The integral in the denominator must be evaluated numerically (see reference [38]).

Problem 8

Meier and Saupe[42] proposed the following differential equation to take account of the homeotropic polarization of a nematic liquid crystal

$$\frac{dP}{dt} = -\frac{1}{\tau_D}\frac{f_0(\pi/2)}{\langle|\cos\theta|\rangle}P$$

where $\langle|\cos\theta|\rangle$ is the expectation value of the absolute value of $\cos\theta$ in the undisturbed distribution f_0, and here $f_0 = \exp(-q/kT)\sin^2\theta$. Show that the retardation factor $g = \tau/\tau_D = \langle|\cos\theta|\rangle/f_0(\pi/2)$ is close to $(q/kT)^{-1}\exp(q/kT)$.

Solution

$$\langle|\cos\theta|\rangle = 2\int_{0}^{\pi/2} \sin\theta\cos\theta\exp-\frac{q\sin^2\theta}{kT}\,d\theta$$

$$= \int_{0}^{\pi/2} \sin 2\theta\exp-\frac{q\sin^2\theta}{kT}\,d\theta \qquad (P.11.8.1)$$

Changing the variable $\cos 2\theta = t$, the following result is obtained

$$\langle|\cos\theta|\rangle = \frac{kT}{q}\left(1 - \exp-\frac{q}{kT}\right) \qquad (P.11.8.2)$$

On the other hand

$$f_0\left(\frac{\pi}{2}\right) = \exp{-\frac{q}{kT}} \qquad\qquad (P.11.8.3)$$

from which

$$g = \frac{\tau}{\tau_D} = \frac{kT}{q}\left(\exp{\frac{q}{kT}} - 1\right) \cong \frac{kT}{q}\exp{\frac{q}{kT}} \qquad\qquad (P.11.8.4)$$

Problem 9

Let us consider a liquid crystal consisting of molecules having a dipole making an angle β with the principal molecular axis. This material is biaxial, but we will assume that the biaxial order parameter D averages to zero. Calculate the corresponding permittivity anisotropy.

Solution

We set $P = D = C = 0$ in Eqs (5.7.10). Then, $\varepsilon_{0\perp1} = \varepsilon_{0\perp2}$. The permittivity anisotropy is given by

$$\Delta\varepsilon = \varepsilon_{0\parallel} - \varepsilon_{0\perp} \qquad\qquad (P.11.9.1)$$

Then, from Eq. (5.7.10), we obtain

$$\Delta\varepsilon_0 = 4\pi N_1 g h S\left\{\Delta\alpha + \frac{h}{kT}\left[\mu_z^2 - \frac{1}{2}(\mu_x^2 + \mu_y^2)\right]\right\} \qquad\qquad (P.11.9.2)$$

If the molecules are assumed to be uniaxial, $\mu_x = \mu_y$, then

$$\Delta\varepsilon_0 = 4\pi N_1 g h S\left[\Delta\alpha + \frac{h}{kT}(\mu_z^2 - \mu_x^2)\right] \qquad\qquad (P.11.9.3)$$

Moreover, $\mu^2 = \mu_z^2 + 2\mu_x^2$, $\mu_z = \mu\cos\beta$, and $\mu_x = \mu\sin\beta$. Substitution of these equations into Eq. (P.11.9.3) leads to

$$\Delta\varepsilon_0 = 4\pi N_1 g h S\left[\Delta\alpha + \frac{h\mu^2}{2kT}(3\cos^2\beta - 1)\right] \qquad\qquad (P.11.9.4)$$

For values of $\beta \leqslant 57.4°$, the dipolar term is positive. The limit value of β for a positive anisotropy at a given temperature is

$$\beta = \cos^{-1}\left(\frac{h\mu^2 - 2kT\Delta\alpha}{3h\mu^2}\right)^{1/2} \qquad\qquad (P.11.9.5)$$

Note that, for specific values of the molecular properties, dipolar terms and polarizability anisotropy mutually cancel out at a temperature given by

$$T = \frac{h\mu^2}{2k\Delta\alpha}(1 - 3\cos^2\beta) \tag{P.11.9.6}$$

At this temperature the permittivity anisotropy changes from negative to positive.

Problem 10

Evaluate $\langle\cos^2\theta\rangle_0$, $\langle\cos^3\theta\rangle_0$, and the derivative of the Langevin function $L(\zeta)$ with respect to ζ.

Solution

By the definition in Eq. (1.9.8)

$$\langle\cos^2\theta\rangle_0 = \frac{\int_0^\pi \cos^2\theta \exp(\mu E_0 \cos\theta/kT) \sin\theta\, d\theta}{\int_0^\pi \exp(\mu E_0 \cos\theta/kT) \sin\theta\, d\theta} \tag{P.11.10.1}$$

With the change in variable $x = \cos\theta$, and taking $\zeta = \mu E_0/kT$, we obtain

$$\langle\cos^2\theta\rangle_0 = \frac{\int_{-1}^1 x^2 \exp(\zeta x)\, dx}{\int_{-1}^1 \exp(\zeta x)\, dx} \tag{P.11.10.2}$$

After integration by parts, the following expression holds

$$\langle\cos^2\theta\rangle_0 = 1 - \frac{2}{\zeta}\coth\zeta + \frac{2}{\zeta^2} \tag{P.11.10.3}$$

According to Eq. (1.9.9)

$$\langle\cos\theta\rangle_0 = L(\zeta) = \coth\zeta - \frac{1}{\zeta} \tag{P.11.10.4}$$

and Eq. (P.11.10.3) may be written as

$$\langle\cos^2\theta\rangle_0 = 1 - \frac{2}{\zeta}L(\zeta) \tag{P.11.10.5}$$

Following the same procedure, we obtain

$$\langle \cos^3 \theta \rangle_0 = L(\zeta)\left(1 + \frac{6}{\zeta^2}\right) - \frac{2}{\zeta}$$ (P.11.10.6)

Finally

$$L'(\zeta) = \frac{d}{d\zeta}\left[\frac{\int_{-1}^{1} x \exp \zeta x \, dx}{\int_{-1}^{1} \exp \zeta x \, dx}\right] = \frac{d}{d\zeta}\left(\frac{I_2}{I_1}\right)$$

$$= \frac{I_1(dI_2/d\zeta) - I_2(dI_1/d\zeta)}{I_1^2}$$ (P.11.10.7)

Moreover

$$\frac{dI_1}{d\zeta} = I_1 L(\zeta), \qquad \frac{dI_2}{d\zeta} = I_1\left[1 - \frac{2}{\zeta}L(\zeta)\right]$$ (P.11.10.8)

and from Eqs (P.11.10.7) and (P.11.10.8) we obtain

$$L'(\zeta) = 1 - \frac{2}{\zeta}L(\zeta) - L^2(\zeta)$$ (P.11.10.9)

REFERENCES

1. Collings, P.J. *Liquid Crystals*; Zadam Hilger: Bristol, 1990; de Gennes, P.G.; Prost, J. *The Physics of Liquid Crystals*; Clarendon Press: Oxford, 1995.
2. Nordio, P.L.; Rigatti, G.; Segre, U. Mol. Phys. **1973**, *25*, 129; Attard, G.S. Mol. Phys. **1986**, *58*, 1087.
3. Maier, W.; Saupe, A. Naturforschung **1958**, *13a*, 564.
4. Maier, W.; Meier, G. Z. Naturforschung A **1961**, *16*, 262.
5. Jones, C.; Raynes, E.P. Liq. Cryst. **1992**, *11*, 199.
6. Gouda, F.; Kuczinsky, W.; Lagerwall, S.T.; Matuszczyk, M.; Sharp, K. Phys. Rev. A **1992**, *46*, 951.
7. Landau, L.D.; Lifshitz, E.M.; Pitaevskii, L.P. *Electrodynamics of Continuous Media*, 2nd Ed.; Butterworth-Heinemann: Oxford, 1993.
8. Böttcher, C.J.F.; Bordewijk, P. *Theory of Electric Polarization*, 2nd Ed.; Elsevier: Amsterdam, 1978; Vol. 2, 443 pp.
9. Martin, A.; Meier, G.; Saupe, A. Far. Soc. Symp. **1971**, (5), 119.
10. Coffey, W.T.; Crothers, D.S.F.; Kalmykov, Y.P.; Massawe, E.S.; Waldron, J.T. Phys. Rev. E **1994**, *49*, 1869.
11. Coffey, W.T.; Crothers, D.S.F.; Kalmykov, Y.P.; Waldron, J.T. Physica A **1995**, *213*, 551.
12. Coffey, W.T.; Kalmykov, Y.P. Liquid Crystals **1993**, *14*, 1227.

13. Attard, G.S.; Williams, G. Polymer **1986**, *27*, 2.
14. Attard, G.S.; Williams, G. Polymer **1986**, *27*, 66.
15. Attard, G.S.; Williams, G. Liq. Crystals **1986**, *1*, 253.
16. Díaz Calleja, R.; Sanchis, M.J.; Riande, E.; Pérez, E.; Pinto, M. J. Mol. Struct. **1999**, *479*, 135.
17. García-Bernabé, A.; Díaz-Calleja, R. Polym. Int. **2001**, *50*, 165.
18. Nazemi, A.; Williams, G.; Attard, G.S.; Karasz, F.E. Polym. for Advd Technol. **1992**, *3*, 157.
19. Kresse, H.; Talroze, R.V. Makromol. Chem. Rapid Comm. **1981**, *2*, 369.
20. Kresse, H.; Kostromin, S.; Shibaev, V.P. Makromol. Chem. Rapid Comm. **1982**, *3*, 509.
21. Zentel, R.; Strobl, G.; Rinsdorf, H. Macromolecules **1985**, *18*, 960.
22. Attard, G.S.; Williams, G.; Gray, G.W.; Lacey, D.; Gemmel, P.A. Polymer **1986**, *27*, 185.
23. Pranoto, H.; Bormuth, F.J.; Haase, W.; Kiechle, U.; Finkelmann, H. Makromol. **1986**, *187*, 2453.
24. Bormuth, F.J.; Haase, W. Liq. Crystals **1988**, *3*, 881.
25. Bormuth, F.J.; Haase, W. Mol. Crystals Liq. Crystals **1987**, *153*, 207.
26. Bormuth, F.J.; Haase, W.; Zentel, R. Mol. Crystals Liq. Crystals **1987**, *148*, 1.
27. Attard, G.S.; Araki, K.; Moura-Ramos, J.J.; Williams, G. Liq. Crystals **1988**, *3*, 861.
28. Attard, G.S.; Williams, G. J. Mol. Electronics **1986**, *2*, 107.
29. Attard, G.S.; Araki, K.; Williams, G. J. Mol. Electronics **1987**, *3*, 1.
30. Attard, G.S.; Moura-Ramos, J.J.; Williams, G. J. Polym. Sci. Polym. Phys. Ed. **1987**, *25*, 1099.
31. Kozak, A.; Simom, G.P.; Williams, G. Polym. Comm. **1989**, *30*, 102.
32. Haase, W. Field-induced effects in side chain liquid crystal polymers. In *Side Chain Liquid Crystal Polymers*; McArdle, C.B., Ed.; Blakie: Glasgow, 1989; Chap. 11, 309 pp.
33. Vallerien, S.U.; Kremer, F.; Kapitza, H.; Zentel, R.; Frank, W. Phys. Lett. **1989**, *138*, 219.
34. Vallerien, S.U.; Zentel, R.; Kremer, F.; Kapitza, H.; Fisher, E.W. Makromol. Chem. Rapid Commun. **1989**, *10*, 33.
35. Vallerien, S.U.; Kremer, F.; Kapitza, H.; Zentel, R.; Fisher, E.W. Ferroelectrics **1990**, *109*, 273.
36. Moscicki, J.K. Dielectric relaxations in macromolecular liquid crystals. In *Liquid Crystal Polymers*; Collyer, A.A., Ed.; Elsevier: Cambridge, 1992; 143 pp.
37. Williams, G. Dielectric relaxation behavior of liquid crystals. In *The Molecular Dynamics of Liquid Crystals*; Kluwer: Amsterdam, 1994; 431 pp.
38. Coffey, W.T.; Kalmykov, Y.P.; Quinn, K.P. J. Chem. Phys. **1992**, *96*, 5471.
39. Coffey, W.T.; Cregg, P.J.; Kalmykov, Y.P. Adv. Chem. Phys.; Prigogine and Rice, Eds.; Wiley-Interscience: 1993; Vol. 88, 263 pp.
40. Coffey, W.T.; Kalmykov, Y.P.; Waldron, J.T. *The Langevin Equation, World Scientific Series in Contemporary Chemical Physics*; World Scientific: Singapore, 1996; Vol. 10, Chap. 7.
41. Meier, G.; Saupe, A. Molecular Crystals **1966**, *1*, 515.
42. Maier, W.; Saupe, A. Naturforschung **1959**, *14a*, 882.

43.   Risken, H. *The Fokker-Plank Equation*, 2nd Ed.; Springer-Verlag: Berlin, 1989.
44.   Morita, A. J. Phys. D **1978**, *11*, 1357.
45.   SanMiguel, M.; Pescara, L.; Rodriguez, M.A.; Hernández-Machado, A. Phys. Rev. **1987**, *A35*, 208.
46.   Meier, G. Freiburg i. B, 1960; Thesis.
47.   Attard, G.S.; Araki, K.; Williams, G. British Polym. J. **1987**, *19*, 119.

12

Piezoelectric and Pyroelectric Materials

12.1. INTRODUCTION

Certain crystals under mechanical stress become electrically polarized. P. and J. Curie discovered this phenomenon, known as the piezoelectric effect, in 1880. They found that the polarization of the crystal is proportional to the applied stress. The most important piezoelectric materials are some inorganic crystals such as quartz, barium titanate ($BaTiO_3$), ammonium dihydrogen phosphate ($NH_4H_2PO_4$), potassium sodium tartrate tetrahydrate ($NaKC_4H_4O_6 \cdot 4H_2O$), and zincblende (ZnS). An applied stress between two opposite surfaces of a piezoelectric crystal generates an electric potential. Also, an electric potential between two opposite surfaces of the crystal generates a piezoelectric strain, a phenomenon known as the converse piezoelectric effect.

At the molecular level, piezoelectricity appears when elastic deformations in crystals are accompanied with unequal displacement of the centers of gravity of positive and negative charges, giving rise to a net polarization of the material. This can also be accomplished in some cases by a difference in temperature, and this effect is called pyroelectricity. Barium titanate, cane sugar, and tourmaline are among the best-known pyroelectric crystals.

Ferroelectric crystals, like the pyroelectric crystals, exhibit spontaneous polarization. However, ferroelectric crystals present the additional property of reversing their polarization by applying a sufficiently large electric field. This

497

latter property involves a small relative shift of the atoms in the crystal, turning the crystal in its electric twin. A more symmetrical nonpolar configuration can be accomplished if only half of the atoms make this shift. That configuration is usually obtained by a change in temperature and, as a consequence, most ferroelectrics have a transition temperature, called the Curie point, above which they are normal and nonpolar. The phase transition at the Curie temperature is related to the change in the lattice symmetry of the crystal. The best-known ferroelectric crystals are inorganic monocrystals such as $BaTIO_3$, KH_2PO_4, and $NaKC_4H_4O_6 \cdot 4H_2O$. A crystal in its ferroelectric state, with spontaneous polarization, has lower symmetry than the same crystal in its nonpolar state. Ferroelectrics must belong to one of the classes listed as pyroelectric. Ferroelectric materials below the Curie temperature are also piezoelectric, because the polarized material has no center of symmetry.

Although piezoelectricity and pyroelectricity were first studied in inorganic crystals, these effects were also observed, as early as 1941, in wool and hair. Somewhat later, Fukada reported direct and converse shear piezoelectricity in wood. Subsequent studies carried out by this author revealed that many kinds of biological macromolecules such as polysaccharides, proteins, etc., exhibit piezoelectric properties.[1,2] Among polymers, polyvinylidene fluoride (PVDF) is one of the most important ferroelectric materials. Some types of liquid crystalline polymers with mesogenic side groups containing chiral moieties also display ferroelectric behavior. Fukada's work led to the finding of strong piezoelectric and pyroelectric effects in uniaxially drawn and poled polyvinylidene fluoride (PVDF) about 30 years ago.[3,4] Earlier studies on the origin of piezoelectric and pyroelectric activities in PVDF revealed that these properties primarily arise from ferroelectric dipole orientation rather than from trapped space charges.[5] Polymer films, especially of poly(vinylidene fluoride), have been recognized as thermoelectric and electromechanical transducers and have found their way into new polymer transducer technology, for example, in devices to convert electric signals to mechanical signals, or mechanical signals to electric signals, such as microphones, loudspeakers, ultrasound generators, quartz clocks, etc.[6] The pyroelectric effect has been used for electromagnetic radiation detection.

12.2. BASIC CONCEPTS

By assuming linear conditions, the polarization of a piezoelectric crystal under stress can be written as

$$P_i = d_{ijk}\sigma_{jk} \tag{12.2.1}$$

This phenomenon is known as the direct piezoelectric effect. In Eq. (12.2.1), d_{ijk} is a third-rank tensor with 27 components. However, the symmetry of the stress tensor ($\sigma_{jk} = \sigma_{kj}$) leads to $d_{ijk} = d_{ikj}$, and the number of independent components of the piezoelectric tensor decreases to 18. Accordingly, Eq. (12.2.1) can be written in explicit form as

$$P_1 = d_{111}\sigma_{11} + d_{112}\sigma_{12} + d_{113}\sigma_{13} + d_{122}\sigma_{22} + d_{123}\sigma_{23} + d_{133}\sigma_{33}$$
$$P_2 = d_{211}\sigma_{11} + d_{212}\sigma_{12} + d_{213}\sigma_{13} + d_{222}\sigma_{22} + d_{223}\sigma_{23} + d_{233}\sigma_{33}$$
$$P_3 = d_{311}\sigma_{11} + d_{312}\sigma_{12} + d_{313}\sigma_{13} + d_{322}\sigma_{22} + d_{323}\sigma_{23} + d_{333}\sigma_{33}$$

$$(12.2.2)$$

In SI units, d is given in C/N.

Under an electric field, the shape of a piezoelectric crystal changes slightly. This is known as the converse dielectric effect, the existence of which is a thermodynamic consequence of the direct effect. Experiments show that there is a linear relation between the components of the electric field vector within the crystal and the components of the strain tensor γ_{ij} that describe the change in shape. The converse effect is written as

$$\gamma_{jk} = d_{ijk}E_i \qquad\qquad (12.2.3)$$

The symmetry of the strain tensor ($\gamma_{jk} = \gamma_{kj}$) also leads to the symmetric relationship $d_{ijk} = d_{ikj}$ of the piezoelectric tensor. As will be shown later, the components of the coefficient connecting strain and electric field are the same, in certain conditions, as those connecting polarity and stress.

The direct piezoelectric effect can be expressed in terms of the electric displacement, **D**. Actually

$$dD_i = \varepsilon_{ij}\, dE_j \qquad\qquad (12.2.4)$$

where ε_{ij} is the permittivity tensor. The electric displacement is related to the polarization in SI units by (see Eqs 1.8.5a and 1.8.7a)

$$D_i = e_0 E_i + P_i \qquad\qquad (12.2.5)$$

where e_0 is the permittivity in vacuum. Therefore, if the electric field in the crystal is held constant, $dD_i = dP_i$ and Eq. (12.2.1) can be written as

$$dD_i = d_{ijk}\, d\sigma_{jk} \qquad\qquad (12.2.6)$$

Either Eq. (12.2.1) or Eq. (12.2.6) can be used to express the direct piezoelectric effect.

The polarization arising from a difference in temperature is given by

$$dP_i = p_i\, dT \qquad\qquad (12.2.7)$$

where p_i ($i = 1,2,3$) are the components of the pyroelectric vector **p**. In SI units, p is given in C/m^2 K. If the temperature in the crystal is changed while keeping the electric field in the crystal constant, Eq. (12.2.7) can be expressed in terms of the electric displacement by

$$dD_i = p_i \, dT \qquad\qquad (12.2.8)$$

The converse effect, or the so-called electrocalorimetric effect, is caused by the application of an electric field in pyroelectrics. Therefore

$$dT = p_i \, dE_i \qquad\qquad (12.2.9)$$

Thermodynamic arguments discussed later show that p_i is the same for the direct and converse pyroelectric effects.

The lack of a crystallographic center of symmetry is a prerequisite for the development of both piezoelectricity and pyroelectricity in a crystal. However, pyroelectricity also requires a polar axis in the crystal, thus reducing the number of possible crystal classes in comparison with piezoelectrics. A polar direction is any of which the two ends are not related by any symmetry element of the point group.[7] Out of the 32 possible crystal classes, 20 classes fulfill the requirements for piezoelectricity, whereas pyroelectric effects can be observed in only 10 classes. As a consequence of the presence of a polar axis, pyroelectric materials have permanent electric polarization compensated by free charges on the surface of the material. A change in temperature causes electric polarization, and, conversely, a change in polarity gives rise to a change in temperature. A crystal in its ferroelectric state exhibits spontaneous polarization and has lower symmetry than the same crystal in its nonpolar state.

12.3. THERMODYNAMICS

Entropy, S, electric displacement, D_i, and strain, γ_{ij}, are respectively the direct responses of a crystal to the thermal field, T, electric field, E_i, and mechanical field, σ_{ij}, at equilibrium. Perturbation and responses are represented in the three outer corners of Fig. 12.1.[8] In a reversible change, an increase in temperature causes a change in entropy per unit volume, given by

$$dS = \frac{C_V}{T} \, dT \qquad\qquad (12.3.1)$$

where C_V is the specific heat at constant volume per unit volume, and T is the absolute temperature. Also, a change in electric displacement arises from a small variation in the electric field, as Eq. (12.2.4) indicates. Finally, a small change in

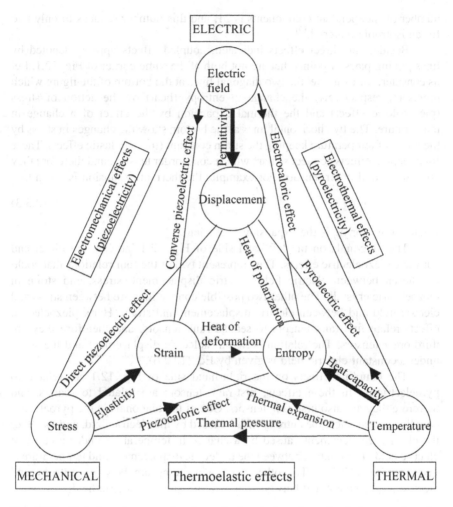

FIG. 12.1 Relations between the thermal, electrical, and mechanical properties of a crystal[7]

stress gives rise to a small variation in the strain according to the expression

$$d\gamma_{ij} = s_{ijkl}\, d\sigma_{kl} \tag{12.3.2}$$

where s_{ijkl} is the elastic compliance. The number of components of s_{ijkl} is 81, which reduces to 36 if the symmetry of stress ($\sigma_{ij} = \sigma_{ji}$) and strain ($\gamma_{kl} = \gamma_{lk}$) is considered. The number of independent components decreases as the symmetry of the material increases. Thus, for a system without elements of symmetry, the number of independent components is 21, and this number reduces to only two for an isotropic material.[9]

Besides the direct effects indicated, coupled effects appear, denoted by lines joining pairs of points that are not both at the same corner of Fig. 12.1. Let us consider, for example, the two diagonal lines at the bottom of the figure which represent, respectively, the change in entropy (heat) by the action of stress (piezocaloric effect) and the thermal expansion by the effect of a change in temperature. The two horizontal lines at the bottom show the changes in stress by the effect of temperature keeping the strain constant (thermoelastic effect). These four coupled effects connect scalars with second-order tensors, and therefore they are also second-order tensors. For example, the thermal expansion is given by

$$d\gamma_{ij} = \alpha_{ij}\, dT \tag{12.3.3}$$

where the tensor α_{ij} is the expansion coefficient.

The diagonals on the left-hand side of Fig. 12.1 indicate the direct and converse piezoelectric effects. They represent two of the four relations that could be chosen between electric field, electric displacement, stress, and strain in piezoelectric crystals. The other two possible connections are between stress and electric field and between electric displacement and strain. Here, piezoelectric effects relate first-rank tensors to second-rank tensors, and therefore they are third-order tensors. The relation between dielectric displacement and the stress under a constant electric field is given by Eq. (12.2.6).

The coupling effects on the right-hand side of Fig. 12.1 are related to pyroelectricity. In these effects, first-rank tensors are related to scalars, and therefore they are also first-rank tensors. One of the diagonals in the pyroelectric coupling effects show the entropy (heat) caused by the electric field, and the other the electric displacement caused by a change in temperature under a constant electric field. The relation between the dielectric displacement and temperature is given by Eq. (12.2.8). The other two connections are between entropy and electric displacement and between temperature and the electric field.

12.4. PIEZOELECTRICITY

Piezoelectricity is a cross-coupling effect between the elastic variables (stress σ, strain γ) and the electric variables (electric displacement D, electric field E). By combining these variables, four tensional parameters d, e, g, and h are defined, as indicated below. According to the first and second principles of thermodynamics, the variation in the internal energy per unit volume of a piezoelectric system under a stress σ_{ik} is given by

$$dU = T\, dS + dW = T\, dS + E_i\, dD_i + \sigma_{ik}\, d\gamma_{ik} \tag{12.4.1}$$

where γ_{ik} is the strain, $dQ = T\,dS$ and dW is the work done on the system arising from the mechanical ($\sigma_{ik}\,d\gamma_{ik}$) and induced electric work ($E_i\,dD_i$). Hence, the variation in Helmholtz free energy A can be written as

$$dA = dU - d(TS) = E_i\,dD_i + \sigma_{ik}\,d\gamma_{ik} - S\,dT \qquad (12.4.2)$$

Obviously, A is a function of ($\mathbf{D}$, γ,T), and its differential form may be written as

$$dA = \left(\frac{\partial A}{\partial D_i}\right)_{\gamma,T} dD_i + \left(\frac{\partial A}{\partial \gamma_{ik}}\right)_{D,T} d\gamma_{ik} + \left(\frac{\partial A}{\partial T}\right)_{\gamma,D} dT \qquad (12.4.3)$$

From the comparison of the coefficients of Eqs (12.4.2) and (12.4.3) one obtains

$$\left(\frac{\partial A}{\partial D_i}\right)_{\gamma,T} = E_i, \qquad \left(\frac{\partial A}{\partial \gamma_{ik}}\right)_{D,T} = \sigma_{ik}, \qquad \left(\frac{\partial A}{\partial T}\right)_{\gamma,D} = -S \qquad (12.4.4)$$

Differentiating the first of these equations with respect to γ_{ik} and the second with respect to D_j after changing the suffixes from i to j gives

$$\left(\frac{\partial^2 A}{\partial \gamma_{jk}\partial D_i}\right)_T = \left(\frac{\partial E_i}{\partial \gamma_{jk}}\right)_{T,D}, \qquad \left(\frac{\partial^2 A}{\partial \gamma_{jk}\partial D_i}\right)_T = \left(\frac{\partial \sigma_{jk}}{\partial D_i}\right)_{T,\gamma} \qquad (12.4.5)$$

Hence, the following Maxwell relation is obtained

$$h = \left(\frac{\partial E_i}{\partial \gamma_{jk}}\right)_{T,D} = \left(\frac{\partial \sigma_{jk}}{\partial D_i}\right)_{T,\gamma} \qquad (12.4.6)$$

By using Legendre transformations, the three following potentials can be defined

$$A_1 = A - E_i D_i \qquad (12.4.7)$$
$$A_2 = A - \sigma_{ik}\gamma_{ik} \qquad (12.4.8)$$
$$A_3 = A - \sigma_{ik}\gamma_{ik} - E_i D_i \qquad (12.4.9)$$

Equations (12.4.7) and (12.4.2) give

$$dA_1 = \sigma_{ik}\,d\gamma_{ik} - D_i\,dE_i - S\,dT \qquad (12.4.10)$$

which leads to the Maxwell relationship

$$e = \left(\frac{\partial D_j}{\partial \gamma_{ik}}\right)_{T,E} = -\left(\frac{\partial \sigma_{ik}}{\partial E_j}\right)_{T,\gamma} \qquad (12.4.11)$$

In the same way, Eqs (12.4.8) and (12.4.2) give

$$dA_2 = E_i\, dD_i - \gamma_{ik}\, d\sigma_{ik} - S\, dT \tag{12.4.12}$$

thus obtaining the Maxwell relationship

$$g = \left(\frac{\partial E_j}{\partial \sigma_{ik}}\right)_{T,D} = -\left(\frac{\partial \gamma_{ik}}{\partial D_j}\right)_{T,\sigma} \tag{12.4.13}$$

Finally, Eqs (12.4.9) and (12.4.2) lead to

$$dA_3 = -\gamma_{ik}\, d\sigma_{ik} - D_i\, dE_i - S\, dT \tag{12.4.14}$$

from which the following Maxwell relation is obtained

$$d = \left(\frac{\partial D_j}{\partial \sigma_{ik}}\right)_{T,E} = \left(\frac{\partial \gamma_{ik}}{\partial E_j}\right)_{T,\sigma} \tag{12.4.15}$$

From Eqs (12.2.3), (12.2.6), and (12.4.15) it is concluded that d_{ijk} is the same for the direct and converse piezoelectric effects. However, this is only true if the sample dimensions do not change.[10,11] Although changes in dimensions may be small in hard and brittle ferroelectric materials, they may be important in soft substances such as polymers. Let us first write Eq. (12.4.15) in terms of $Q_i = P_j A_j$, where Q_j is the free charge in the area A_j. Since $dQ_j = P_j\, dA_j + A_j\, dP_j$, Eq. (12.4.15) is given by

$$\left(\frac{\partial D_i}{\partial \sigma_{jk}}\right)_{E,T} = \frac{1}{A_i}\left(\frac{\partial Q_i}{\partial \sigma_{jk}}\right)_{E,T} = \left(\frac{\partial P_i}{\partial \sigma_{jk}}\right)_{E,T} + \frac{P_i}{A_i}\left(\frac{\partial A_i}{\partial \sigma_{jk}}\right)_{E,T} = \left(\frac{\partial \gamma_{jk}}{\partial E_i}\right)_{\sigma,T} \tag{12.4.16}$$

Taking into account that $dP_i = dD_i$ if E is kept constant, Eq. (12.4.16) can alternatively be written as

$$\left(\frac{\partial D_i}{\partial \sigma_{jk}}\right)_{E,T} + \frac{P_i}{A_i}\left(\frac{\partial A_i}{\partial \sigma_{jk}}\right)_{E,T} = \left(\frac{\partial \gamma_{jk}}{\partial E_i}\right)_{\sigma,T} \tag{12.4.17}$$

This expression suggests that d_{ijk} may differ for direct and converse piezoelectric effects. The relations between the direct and converse piezoelectric effects through parameters d, e, g, and h are schematically shown in Fig. 12.2.

The dielectric permittivity of piezoelectric materials depends on the nature of the mechanical constraints. The free permittivity ($d\sigma = 0$) is larger

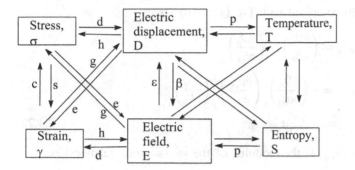

FIG. 12.2 Definition of piezoelectric (d, e, g, and h) and pyroelectric (p) constants

than the clamped permittivity ($d\gamma = 0$) because in the first case the piezoelectric produces a polarity resulting from the combination of the converse and direct effects. The difference between them can be obtained by using thermodynamic arguments. Let us consider a piezoelectric material under an electric and mechanical force field. In isothermal conditions, the shear deformation and the electric displacement are both functions of the stress and the electric field. Therefore

$$d\gamma_{ij} = \left(\frac{\partial\gamma_{ij}}{\partial E_k}\right)_\sigma dE_k + \left(\frac{\partial\gamma_{ij}}{\partial\sigma_{lm}}\right)_E d\sigma_{lm} \tag{12.4.18}$$

and

$$dD_n = \left(\frac{\partial D_n}{\partial E_k}\right)_\sigma dE_k + \left(\frac{\partial D_n}{\partial\sigma_{pq}}\right)_E d\sigma_{pq} \tag{12.4.19}$$

Writing down $d\gamma_{ij} = 0$ in Eq. (12.4.18), and multiplying the resulting equation by $(\partial\sigma_{pq}/\partial\gamma_{ij})_E$, leads to

$$0 = \left(\frac{\partial\gamma_{ij}}{\partial E_k}\right)_\sigma\left(\frac{\partial\sigma_{pq}}{\partial\gamma_{ij}}\right)_E dE_k + \left(\frac{\partial\gamma_{ij}}{\partial\sigma_{lm}}\right)_E\left(\frac{\partial\sigma_{pq}}{\partial\gamma_{ij}}\right)_E d\sigma_{lm} \tag{12.4.20}$$

Since the second term on the right-hand side is

$$\delta_{pl}\delta_{qm} d\sigma_{lm} = d\sigma_{pq} \tag{12.4.21}$$

Equation (12.4.19) combined with Eq. (12.4.20) yields

$$
\begin{aligned}
(\varepsilon_{nk})_\gamma - (\varepsilon_{nk})_\sigma &= \left(\frac{\partial D_n}{\partial E_k}\right)_\gamma - \left(\frac{\partial D_n}{\partial E_k}\right)_\sigma \\
&= -\left(\frac{\partial D_n}{\partial \sigma_{pq}}\right)_E \left(\frac{\partial \sigma_{pq}}{\partial \gamma_{ij}}\right)_E \left(\frac{\partial \gamma_{ij}}{\partial E_k}\right)_\sigma \\
&= -d_{npq}d_{ijk}c_{pqij}
\end{aligned}
\tag{12.4.22}
$$

In this expression, c_{pqij} is the modulus of the stress–strain relationship

$$
\sigma_{pq} = c_{pqij}\gamma_{ij} \tag{12.4.23}
$$

where c_{pqij} has 21 independent components for a system without any element of symmetry.[9] Equation (12.4.22) allows the determination of the difference between the dielectric permittivity at strain constant and stress constant, once both the piezoelectric and the elastic relaxation tensors are known.

12.5. PYROELECTRICITY

From Eq. (12.4.14) it follows that

$$
\left(\frac{\partial D_i}{\partial T}\right)_{\sigma,E} = \left(\frac{\partial S}{\partial E_i}\right)_{T,\sigma} = p_i \tag{12.5.1}
$$

This expression suggests the similarity of the pyroelectric coefficients for the direct and converse effects. However, if changes in the sample dimensions occur by the effect of temperature, the data must be corrected. For the reasons outlined above, [Eq. (12.4.16)]

$$
\frac{1}{A_j}\left(\frac{\partial Q_j}{\partial T}\right)_{\sigma,E} = \frac{P_j}{A_j}\left(\frac{\partial A_j}{\partial T}\right)_{\sigma,E} + \left(\frac{\partial P_j}{\partial T}\right)_{\sigma,E} \tag{12.5.2}
$$

Therefore, Eq. (12.5.1) is given by

$$
\frac{P_i}{A_i}\left(\frac{\partial A_i}{\partial T}\right)_{\sigma,E} + \left(\frac{\partial P_i}{\partial T}\right)_{\sigma,E} = \left(\frac{\partial S}{\partial E_i}\right)_{T,\sigma} = p_i \tag{12.5.3}
$$

Primary and secondary pyroelectricity can be defined, respectively, as the change in polarization with temperature when the dimensions of the sample are kept constant and when they are permitted to relax to their equilibrium values. Similar arguments as for dielectric permittitivity lead to

$$d\gamma_{ij} = \left(\frac{\partial\gamma_{ij}}{\partial T}\right)_\sigma dT + \left(\frac{\partial\gamma_{ij}}{\partial\sigma_{lm}}\right)_T d\sigma_{lm} \tag{12.5.4}$$

and

$$dD_n = \left(\frac{\partial D_n}{\partial T}\right)_\sigma dT + \left(\frac{\partial D_n}{\partial\sigma_{pq}}\right)_T d\sigma_{lm} \tag{12.5.5}$$

Writing $d\gamma_{ij} = 0$ in Eq. (12.5.4), and multiplying the resulting expression by $(\partial\sigma_{pq}/\partial\gamma_{ij})_T$, gives

$$0 = \left(\frac{\partial\gamma_{ij}}{\partial T}\right)_\sigma \left(\frac{\partial\sigma_{pq}}{\partial\gamma_{ij}}\right)_T dT + \left(\frac{\partial\gamma_{ij}}{\partial\sigma_{lm}}\right)_T \left(\frac{\partial\sigma_{pq}}{\partial\gamma_{ij}}\right)_T d\sigma_{lm} \tag{12.5.6}$$

Since the second term on the right-hand side is

$$\delta_{pl}\delta_{qm}\,d\sigma_{lm} = d\sigma_{pq} \tag{12.5.7}$$

Equation (12.5.5) becomes

$$
\begin{aligned}
(p_n)_\gamma - (p_n)_\sigma &= \left(\frac{\partial D_n}{\partial T}\right)_\gamma - \left(\frac{\partial D_n}{\partial T}\right)_\sigma \\
&= -\left(\frac{\partial D_n}{\partial\sigma_{pq}}\right)_T \left(\frac{\partial\sigma_{pq}}{\partial\gamma_{ij}}\right)_T \left(\frac{\partial\gamma_{ij}}{\partial T}\right)_\sigma \\
&= -d_{npq}\alpha_{ij}c_{pqij} \tag{12.5.8}
\end{aligned}
$$

where α_{ij} is the expansion coefficient. This equation allows the determination of the difference between the primary and secondary pyroelectric coefficients once the piezoelectric, expansion, and relaxation modulus coefficients are known.

12.6. PIEZOELECTRIC AND PYROELECTRIC POLYMERS

Semicrystalline polymers are usually isotropic materials where electric responses to mechanical or thermal perturbations are absent. To achieve these responses requires the polymers to have certain directionality in their structure, which in some cases is accomplished by subjecting semicrystalline polymer films to strong electric fields and/or mechanical drawing. Permanent anisotropic polarization is obtained by some structural rearrangement involving either crystal packing or dipole alignment of dipoles. Anisotropic polarization under a strong electric field, called the poling process, creates an orientation of dipoles in polar polymers. The oriented polymers exhibit $C_{v\infty}$ symmetry (class 1). In this case, the orientation of molecular dipoles is characterized by the first-order Legendre polynomial P_1 defined by

$$P_1 = \langle \cos \theta \rangle \qquad (12.6.1)$$

where θ is the angle between the dipole and poling axis. Mechanical drawing causes a P_2 orientation of chain molecules given by the second-order Legendre polynomial

$$P_2 = \frac{3\langle \cos^2 \theta \rangle - 1}{2} \qquad (12.6.2)$$

where θ is the angle between the z axis of the crystalline entities of the polymer and the z drawing axis. A uniaxial drawing film of a chiral (optically active) polymer exhibits piezoelectricity associated with D_∞ symmetry (class 2). The P_1 orientation of dipoles and the P_2 orientation of chiral molecules are schematically shown in Fig. 12.3.[12]

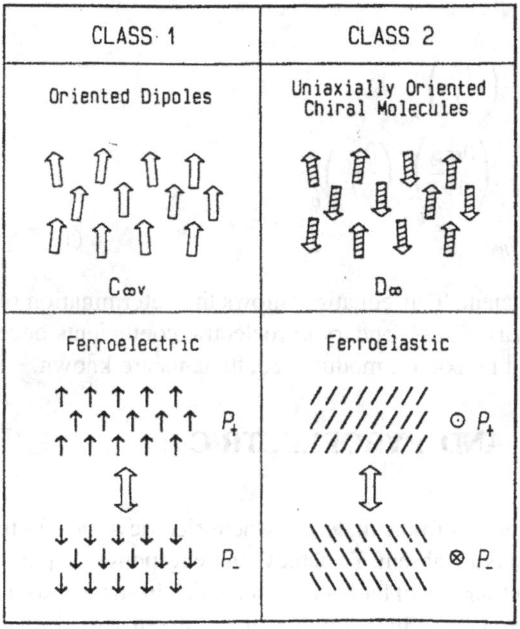

CLASS 1	CLASS 2
Oriented Dipoles	Uniaxially Oriented Chiral Molecules
$C_{\infty v}$	D_∞
Ferroelectric	Ferroelastic

FIG. 12.3 Schematic representations of the dipolar and molecular orientations associated with piezoelectricity and pyroelectricity of class 1 and 2 polymers

Poly(vinylidene fluoride) (PVDF) is one of the most important polymers corresponding to class 1. The electric dipoles of PVDF can reorient reversibly following inversion of an externally applied electric field without any change in the structure before and after the reorientation. Therefore, PVDF is a ferroelectric material. PVDF can take at least three types of molecular conformation:[13,14] TGTḠ, TTTT, and TTTGTTTḠ. The packing mode of the chains in the crystals gives rise to four structures: α, β, γ, and δ (see Fig. 12.4). The α-structure formed by cooling the melt at a normal cooling rate of 20 K/min or higher, is nonpolar, whereas the other three structures are ferroelectric. Slow cooling of the melt gives the γ-structure. Although spherulitic β-structures are also obtained, they grow 4–5 times more slowly than the γ-structure. A scheme showing the interconversion between different crystal modifications is shown in Fig. 12.5.[14]

The preparation of ferroelectric PVDF films involves stretching the film and then poling. During the uniaxial or biaxial stretching processes, usually carried out in the range 60–100°C, crystallites in the α-form convert to the β-form. As can be seen in Fig. 12.6, the extended chain structure of the β-form places the dipoles in parallel position and therefore results in a high-polarity configuration. Moreover, there are two chains in the unit cell aligned in the same direction. There is evidence that, the higher the piezoelectric and pyroelectric properties, the higher are both the crystallite number in the β-phase and the facility with which they can align during the orientation process. The γ- and δ-structures have lower polarity per unit cell than the β-structure. Orientation probably increases the crystallinity of this already

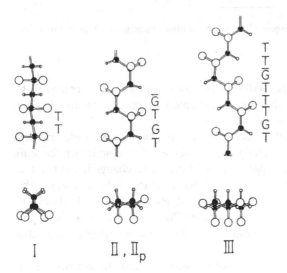

FIG. 12.4 Molecular structures of four crystalline modifications of poly(vinylidene fluoride)[14]

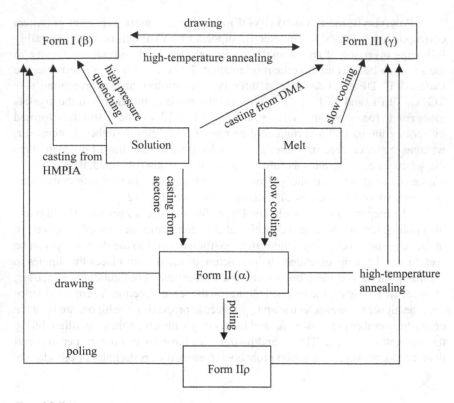

FIG. 12.5 Interconversion between different crystalline structures of poly(vinylidene fluoride)[14]

highly crystalline material. The ability of residues of amorphous regions to be incorporated in the crystalline phases and to interconvert their orientation in grain boundaries increases the polarity of the films.

In the second stage, PVDF films do not exhibit a large external electric field owing to the fact that surface charges are injected that counteract the bulk polarization. However small the application of a stress or a change in temperature is, a change in bulk polarization is induced that produces a field change. This property provides the mechanism for the use of PVDF devices. It should be stressed that PVDF retains the polarization until 70°C, and the decay is slow at temperatures up to 130°C. The resulting material may have a large piezoelectric coefficient (≈ 6.7 pC N^{-1}).

PVDF must be drawn prior to poling with the aim of converting the nonpolar α-structure to the ferroelectric β-structure. Electrodes are evaporated onto the oriented film, which is subsequently poled. The poling process consists

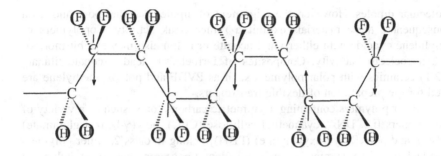

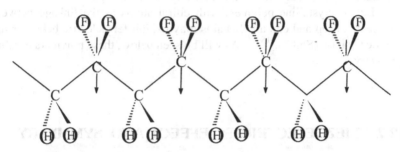

FIG. 12.6   Dipole entities in the α- and β-configurations of poly(vinylidene fluoride). Dipoles are indicated by arrows

in the application of an electric field E_p at a temperature T_p for a time t_p. The poling conditions depend on the system, and the use of a stronger field, a higher temperature, and a longer poling time does not always yield better orientation of the dipoles. A field of 100–300 kV/cm for about 20 min at 80–90°C is commonly used. The poled polymer under the field is cooled to room temperature and then the field is removed.

Crystallization of copolymers of vinylidene fluoride (VDF) with trifluoroethylene (TrFE) or tetrafluoroethylene (TFE) from the melt or solution directly leads to a ferroelectric phase analogous to the β-PVDF and therefore does not require mechanical treatment of the PVDF films. To accomplish the alignment of dipoles, E_p should be greater than the critical field, called the coercive field. These copolymers exhibit clear indications of the ferroelectric → paraelectric properties missing in PVDF.[12] Here, nearly perfect alignment of dipoles can be accomplished by applying sufficiently high poling fields.

In principle, thermal poling of polar polymers such as poly(vinyl chloride) (PVC) and poly(vinyl acetate) (PVAc) could induce frozen-in orientation of molecular dipoles. However, the alignment of dipoles is rather poor and, as a consequence, these materials exhibit a rather weak activity. Copolymers of vinylidene cyanide with either vinyl acetate or nylon also may exhibit more or less ferroelectric activity. Composites of ferroelectric lead zirconate titanate (PZT) ceramics with polar polymers such as PVDF and polyoxymethylene are used for the preparation of flexible transducers.

Polar polymers containing asymmetric carbon atoms, such as a variety of biopolymers [i.e. DNA, cyanoethyl cellulose (CEC), poly(γ-benzyl glutamate) (PBG), and poly(β-hydroxybutyrate) (PBH)] belong to class 2. Other polymers also lacking mirror symmetry, such as isotactic polypropylene oxide, belong to this category. As a consequence of the lack of spontaneous polarity, these polymers are not pyroelectric, but, as will be indicated below, the axial orientation P_2 is enough for piezoelectricity to appear.

Liquid crystalline polymers with chiral atoms as the linkage between a mesogenic group and end alkyl chains may exhibit ferroelectric behavior in the smectic C phase (SmC*).[16–20] As will be seen below, these polymers exhibit C_2 symmetry.

12.7. PIEZOELECTRICITY EFFECT AND SYMMETRY

The piezoelectric effect depends on the symmetry of the material in such a way that, the greater the degree of symmetry of the piezoelectric, the lower is the number of independent terms of d_{ijk}. Let us first consider the reduction in the terms of d_{ijk} for symmetry C_∞. In a coordinate system in which axis 3, x_3, is the axis of orientation (Fig. 12.7), three components of polarization, P_1, P_2, and P_3, are connected with the six components of the stress, σ_{11}, σ_{12}, σ_{13}, σ_{22}, σ_{23}, and σ_{33}. In order to write Eq. (12.2.1) in matricial form, the reductions $11 \to 1$, $22 \to 2$, $33 \to 3$, $23 \to 4$, $13 \to 5$ and $12 \to 6$ are customarily made in the jk subscripts of both the stress and d_{ijk} tensors. Accordingly

$$
\begin{pmatrix} P_1 \\ P_2 \\ P_3 \end{pmatrix} = \begin{pmatrix} d_{11} & d_{12} & d_{13} & d_{14} & d_{15} & d_{16} \\ d_{21} & d_{22} & d_{23} & d_{24} & d_{25} & d_{26} \\ d_{31} & d_{32} & d_{33} & d_{34} & d_{35} & d_{36} \end{pmatrix} \begin{pmatrix} \sigma_1 \\ \sigma_2 \\ \sigma_3 \\ \sigma_4 \\ \sigma_5 \\ \sigma_6 \end{pmatrix}
\tag{12.7.1}
$$

Let us now consider a reference frame x_1, x_2, x_3 and a coordinate system x_1', x_2', x_3', the x_3 and x_3' axes coinciding in both coordinate systems. The cosine directors, after the x_1', x_2', x_3' coordinate system (see Fig. 12.8) has rotated an

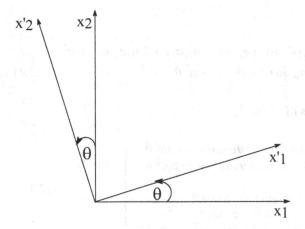

FIG. 12.7 Angle forming the $OX'Y'Z'$ coordinate axes with respect to the $OXYZ$ reference frame. In both cases the OZ (OZ') axis is perpendicular to the plane

angle θ about the x_3 (x'_3) axis, are

$$a_{ij} = \begin{pmatrix} \cos\theta & \sin\theta & 0 \\ -\sin\theta & \cos\theta & 0 \\ 0 & 0 & 1 \end{pmatrix} \qquad (12.7.2)$$

The components of the stress tensor in the coordinate system (x'_1, x'_2, x'_3) are related to the components of the same tensor in the reference frame (x_1, x_2, x_3) by

$$\sigma'_{ij} = a_{ik}a_{jl}\sigma_{kl} \qquad (12.7.3)$$

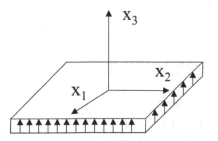

FIG. 12.8 Coordinate system assigned to a poled sample in a perpendicular direction to the surface of the film, and coordinate system assigned to a uniaxially drawn film

For example

$$\sigma'_{11} = a_{1k}a_{1l}\sigma_{kl} = a_{11}a_{1l}\sigma_{1l} + a_{12}a_{1l} = a_{11}^2\sigma_{11} + 2a_{11}a_{12}\sigma_{12} + a_{12}^2\sigma_{22}$$
$$= \sigma_1 \cos^2\theta + 2\sigma_6 \sin\theta\cos\theta + \sigma_2 \sin^2\theta \qquad (12.7.4)$$

Therefore, the components of σ'_i are

$$[\sigma'_i] = \begin{pmatrix} \sigma_1 \cos^2\theta + 2\sigma_6 \sin\theta\cos\theta + \sigma_2 \sin^2\theta \\ \sigma_1 \sin^2\theta - 2\sigma_6 \sin\theta\cos\theta + \sigma_2 \cos^2\theta \\ \sigma_3 \\ -\sigma_5 \sin\theta + \sigma_4 \cos\theta \\ \sigma_5 \cos\theta + \sigma_4 \sin\theta \\ (\sigma_2 - \sigma_1)\sin\theta\cos\theta + \sigma_6(\cos^2\theta - \sin^2\theta) \end{pmatrix} \qquad (12.7.5)$$

On the other hand, the components of the polarization vector in the coordinate axes (x'_1, x'_2, x'_3) and (x_1, x_2, x_3) are related by

$$\begin{pmatrix} P'_1 \\ P'_2 \\ P'_3 \end{pmatrix} = \begin{pmatrix} \cos\theta & \sin\theta & 0 \\ -\sin\theta & \cos\theta & 0 \\ 0 & 0 & 1 \end{pmatrix} \begin{pmatrix} P_1 \\ P_2 \\ P_3 \end{pmatrix} \qquad (12.7.6)$$

Since the symmetry of the system is C_∞, the symmetry operation is invariant for any angle, say $\theta = 90°$. As a result

$$[\sigma'_i] = \begin{pmatrix} \sigma_2 \\ \sigma_1 \\ \sigma \\ -\sigma_5 \\ \sigma_4 \\ -\sigma_6 \end{pmatrix}, \qquad [P'_i] = \begin{pmatrix} P_2 \\ -P_1 \\ P_3 \end{pmatrix} \qquad (12.7.7)$$

In the (x'_1, x'_2, x'_3) reference frame, Eq. (12.2.1) can be written as

$$P_1 = -d_{21}\sigma_2 - d_{22}\sigma_1 - d_{23}\sigma_3 + d_{24}\sigma_5 - d_{25}\sigma_4 + d_{26}\sigma_6$$
$$P_2 = d_{11}\sigma_2 + d_{12}\sigma_1 + d_{13}\sigma_3 - d_{14}\sigma_5 + d_{15}\sigma_4 - d_{16}\sigma_6$$
$$P_3 = d_{31}\sigma_2 + d_{32}\sigma_1 + d_{33}\sigma_3 - d_{34}\sigma_5 + d_{15}\sigma_4 - d_{36}\sigma_6$$

$$(12.7.8)$$

By comparing Eqs (12.7.8) and (12.7.1), one obtains

$$
\begin{aligned}
d_{11} &= d_{22} = -d_{22} = 0, & d_{12} &= d_{21} = -d_{21} = 0, \\
d_{13} &= d_{23} = -d_{23} = 0, & d_{16} &= d_{26} = -d_{26} = 0, \\
d_{35} &= d_{34} = -d_{34} = 0, & d_{14} &= -d_{25}, \\
d_{15} &= d_{24} & d_{31} &= d_{32}
\end{aligned}
\tag{12.7.9}
$$

Hence, the tensor d_{ij} can be written as

$$
[d_{ij}] = \begin{pmatrix}
0 & 0 & 0 & d_{14} & d_{15} & 0 \\
0 & 0 & 0 & d_{15} & -d_{14} & 0 \\
d_{31} & d_{31} & d_{33} & 0 & 0 & 0
\end{pmatrix}
\tag{12.7.10}
$$

This means that, in a system with C_∞ symmetry, the number of independent components of d_{ij} decreases from 18 to 4. These components without subscripts reduction are d_{123}, d_{113}, d_{311}, and d_{333}. The symmetry of C_∞ is found in certain structures of living systems such as bone and tendon. The components of the polarization vector are given by

$$
\begin{aligned}
P_1 &= d_{14}\sigma_4 + d_{15}\sigma_5 \\
P_2 &= d_{15}\sigma_4 - d_{14}\sigma_5 \\
P_3 &= d_{31}\sigma_1 + d_{31}\sigma_2 + d_{33}\sigma_3
\end{aligned}
\tag{12.7.11}
$$

Piezoelectric polymers of class 1 have symmetry $C_{\infty v}$. In this case there is an additional plane of symmetry perpendicular to either axis x_1 or axis x_2. The cosine directors in the first case are

$$
[a_{ij}] = \begin{pmatrix}
-1 & 0 & 0 \\
0 & 1 & 0 \\
0 & 0 & 1
\end{pmatrix}
\tag{12.7.12}
$$

Following the methods outlined above, one finds

$$
\begin{aligned}
\sigma_1' &= \sigma_1, & \sigma_2' &= \sigma_2, & \sigma_3' &= \sigma_3, & \sigma_4' &= \sigma_4, \\
\sigma_5' &= -\sigma_5, & \sigma_6' &= -\sigma_6
\end{aligned}
\tag{12.7.13}
$$

and

$$
P_1' = -P_1, \qquad P_2' = P_2, \qquad P_3' = P_3
\tag{12.7.14}
$$

Therefore

$$P_1 = -d_{14}\sigma_4 + d_{15}\sigma_5$$
$$P_2 = d_{15}\sigma_4 + d_{14}\sigma_5$$
$$P_3 = d_{31}\sigma_1 + d_{31}\sigma_1 + d_{33}\sigma_3$$

$$(12.7.15)$$

By comparing Eqs (12.7.15) and (12.7.11), it follows that $d_{14} = -d_{14}$ and so $d_{14} = 0$. Therefore, for piezoelectric polymers of class 1, the piezoelectric tensor can be written as

$$[d_{ij}] = \begin{pmatrix} 0 & 0 & 0 & 0 & d_{15} & 0 \\ 0 & 0 & 0 & d_{15} & 0 & 0 \\ d_{31} & d_{31} & d_{33} & 0 & 0 & 0 \end{pmatrix} \qquad (12.7.16)$$

Hence, the independent components are reduced to d_{15}, d_{31} and d_{33} or, without subscript reduction, d_{113}, d_{311}, and d_{333}. The symmetry of $C_{\infty v}$ holds for poled composite films consisting of piezoelectric ceramic powders and polymers. When elongated films of polar polymer are poled, the symmetry of C_{2v} appears. The x_3 axes for both symmetries are in the poling direction.

Let us now consider C_{2v} symmetry. The cosine directors for C_2 are

$$a_{ij} = \begin{pmatrix} -1 & 0 & 0 \\ 0 & -1 & 0 \\ 0 & 0 & 1 \end{pmatrix} \qquad (12.7.17)$$

In this case

$$\sigma_1' = \sigma_1, \qquad \sigma_2' = \sigma_2, \qquad \sigma_3' = \sigma_3, \qquad \sigma_4' = -\sigma_4,$$
$$\sigma_5' = -\sigma_5, \qquad \sigma_6' = \sigma_6 \qquad (12.7.18)$$

and

$$P_1' = -P_1, \qquad P_2' = -P_2, \qquad P_3' = P_3 \qquad (12.7.19)$$

In the reference frame (x_1', x_2', x_3'), one obtains

$$P_1 = -d_{11}\sigma_1 - d_{12}\sigma_2 - d_{13}\sigma_3 + d_{14}\sigma_4 + d_{15}\sigma_5 - d_{16}\sigma_6$$
$$P_2 = -d_{21}\sigma_1 - d_{22}\sigma_2 - d_{23}\sigma_3 + d_{24}\sigma_4 + d_{25}\sigma_5 - d_{26}\sigma_6 \qquad (12.7.20)$$
$$P_3 = d_{31}\sigma_1 + d_{32}\sigma_2 + d_{33}\sigma_3 - d_{34}\sigma_4 + d_{35}\sigma_5 + d_{36}\sigma_6$$

By comparing Eqs (12.7.20) and (12.7.1), the components of d_{ij} are

$$[d_{ij}] = \begin{pmatrix} 0 & 0 & 0 & d_{14} & d_{15} & 0 \\ 0 & 0 & 0 & d_{24} & d_{25} & 0 \\ d_{31} & d_{32} & d_{33} & 0 & 0 & d_{36} \end{pmatrix} \qquad (12.7.21)$$

Hence

$$P_1 = d_{14}\sigma_4 + d_{15}\sigma_5$$
$$P_2 = d_{24}\sigma_4 + d_{25}\sigma_5$$
$$P_3 = d_{31}\sigma_1 + d_{32}\sigma_2 + d_{33}\sigma_3 + d_{36}\sigma_6$$

$$(12.7.22)$$

For C_{2v} symmetry there is an additional vertical plane of symmetry, for example perpendicular to x_1. This leads to $d_{14} = d_{25} = d_{36} = 0$. Therefore, the components of d_{ij} for C_{2v} symmetry are

$$[d_{ij}] = \begin{pmatrix} 0 & 0 & 0 & 0 & d_{15} & 0 \\ 0 & 0 & 0 & d_{24} & 0 & 0 \\ d_{31} & d_{32} & d_{33} & 0 & 0 & 0 \end{pmatrix} \qquad (12.7.23)$$

Polymers of class 2 can be approximated by D_∞ symmetry (Fig. 12.9). Here there is a twofold axis, perpendicular to the C_∞ axis (x_3). The cosine directors are

$$a_{ij} = \begin{pmatrix} -1 & 0 & 0 \\ 0 & 1 & 0 \\ 0 & 0 & -1 \end{pmatrix} \qquad (12.7.24)$$

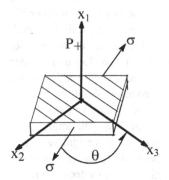

FIG. 12.9   Shear piezoelectricity in uniaxially oriented crystals

Accordingly

$$\sigma'_1 = \sigma_1, \qquad \sigma'_2 = \sigma_2, \qquad \sigma'_3 = \sigma_3, \qquad \sigma'_4 = -\sigma_4,$$
$$\sigma'_5 = \sigma_5, \qquad \sigma'_6 = -\sigma_6 \tag{12.7.25}$$

Moreover

$$P'_1 = -P_1, \qquad P'_2 = P_2, \qquad P'_3 = -P_3 \tag{12.7.26}$$

Hence

$$P_1 = d_{14}\sigma_4 - d_{15}\sigma_5$$
$$P_2 = -d_{15}\sigma_4 - d_{14}\sigma_5 \tag{12.7.27}$$
$$P_3 = -d_{31}\sigma_1 - d_{31}\sigma_1 - d_{33}\sigma_1$$

From the comparison of Eqs (12.7.27) and (12.7.11) it follows that

$$d_{15} = -d_{15} = 0, \qquad d_{31} = -d_{31} = 0, \qquad d_{33} = -d_{33} = 0 \tag{12.7.28}$$

Therefore, there is only one independent term in d_{ij}, so that this tensor can be written as

$$[d_{ij}] = \begin{pmatrix} 0 & 0 & 0 & d_{14} & 0 & 0 \\ 0 & 0 & 0 & 0 & -d_{14} & 0 \\ 0 & 0 & 0 & 0 & 0 & 0 \end{pmatrix} \tag{12.7.29}$$

Accordingly, the shear stress in the plane including the orientation axis (x_3) causes electric polarization in the direction perpendicular to the plane of shear (on the x_1 axis). This behavior is also explained by the illustrations of Fig. 12.10. Actually, for polycrystalline materials with D_∞ symmetry in which crystallites with no polarity are uniaxially oriented, the polarity induced by the shear stress in the crystallites of stretched piezoelectric materials do not cancel each other.[21] Let us consider a sample that consists of a large number of crystals in which $d_{14} = -d_{25} < 0$. The x_1, x_2, x_3 axes, parallel to the X_1, X_2, X_3 axes of the sample, define the crystals. In this case, axis x_3 refers to the drawn axis. Let us consider further that the x_3 axis of crystal (a) in Fig. 12.10 is oriented in the direction of X_3. Crystal (b) in Fig. 12.10 is obtained from (a) by 180° rotation about the x_1 axis, whereas crystals (c) and (d) are obtained from (a) by 180° rotation about the x_3 and x_2 axes respectively. Under a shear stress in the plane X_2X_3, the angle x_3ox_2 for crystals (a) and (b) increases so that the sign of σ_{23} becomes negative. Since $d_{14} < 0$, $P_1 = d_{14}\sigma_{23} > 0$, that is, the polarization occurs along the x_1 axis and it is positive. As for crystals (c) and (d), the shear stress gives rise to a decrease in the x_3ox_2 angle, and this results in P_1 becoming negative. Then the piezoelectric

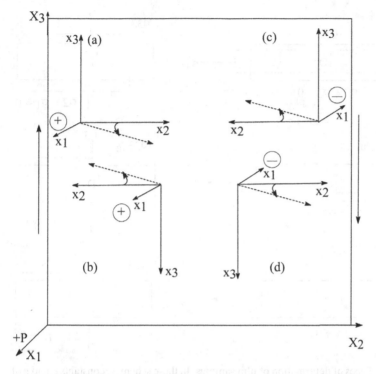

FIG. 12.10 Scheme illustrating the shear piezoelectricity in uniaxially oriented crystals (a), (b), (c), and (d). Note that each crystal is oriented in either the positive or negative direction of the Z axis of the sample. The shear stress σ applied to the crystal induces the $+P$ polarization

polarization of every crystal a, b, c, and d results in the positive polarization P on axis X of the sample.

Most piezoelectric and pyroelectric polymers are used in the form of thin films. For polymers of class 1, the unknown components of d_{ij} are d_{15}, d_{31}, and d_{33}. Here, the x_3 axis is perpendicular to the surface of the film (see Fig. 12.8). It is not easy to measure d_{15} because of the difficulty involved in determining the dielectric permittivity in the direction of the surface of thin film samples. This is not the case for d_{31} (transverse effect) and d_{33} (longitudinal effect) because determination of the dielectric permittivity in the direction perpendicular to the surface of the film is required. A schematic representation of different deformation and stress conditions is shown in Fig. 12.11. The experimental techniques are described in Section 12.12. The direct effect is suitable to determine d_{31}. It is nearly impossible to measure $e_{31} = (\partial D_3/\partial \gamma_1)$ owing to the

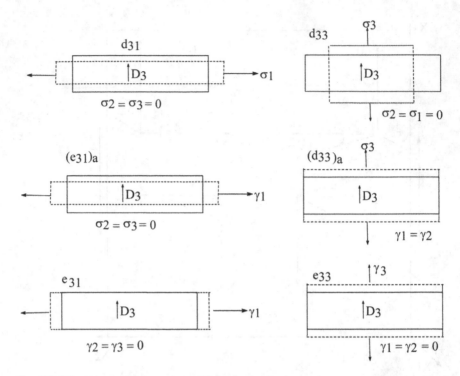

FIG. 12.11   Types of deformation of film samples. In these schemes, constants d and e of the transversal and longitudinal effects are defined

difficulty involved in keeping $\gamma_2 = \gamma_3 = 0$. Since

$$d_{3j} = \left(\frac{\partial D_3}{\partial \sigma_j}\right) = \left(\frac{\partial D_3}{\partial \gamma_k}\right)\left(\frac{\partial \gamma_k}{\partial \sigma_j}\right) = e_{3k}s_{kj} \tag{12.7.30}$$

where s_{ij} is the elastic compliance, we obtain

$$d_{3j} = e_{31}s_{1j} + e_{32}s_{2j} + e_{33}s_{3j} \tag{12.7.31}$$

An apparent value of e_{31} could be defined from d_{31} by means of the expression

$$e_{31}^{a} = \frac{d_{31}}{s_{11}} \tag{12.7.32}$$

where the superscript a is used to distinguish it from a true e_{31}.

The converse effect for a mechanical free sample, $(\partial\gamma_3/\partial E_3)$, gives d_{33}. The resonance method gives the true e_{33} because the longitudinal resonance frequency is much higher than the transverse frequency. Since

$$e_{3j} = \left(\frac{\partial D_3}{\partial\gamma_j}\right) = \left(\frac{\partial D_3}{\partial\sigma_k}\right)_\gamma \left(\frac{\partial\sigma_k}{\partial\gamma_k}\right)_D = d_{3k}c_{kj} \qquad (12.7.33)$$

where c_{kj} is the elastic relaxation modulus, it is possible to define an apparent d_{33} given by

$$d_{33}^a = \frac{e_{33}}{c_{33}} \qquad (12.7.34)$$

where the natural constraints $x_1 = x_2 = 0$ hold. It should be stressed that the resonance method gives the true value of e_{33} because the resonance frequency in the transversal direction is much higher than in the longitudinal one.

The constants g and h of Eqs (12.4.13) and (12.4.6) can be obtained from the permittivity in the x_3 direction, ε_3. The pertinent expressions are

$$d_{3j} = \left(\frac{\partial D_3}{\partial\gamma_j}\right) = \left(\frac{\partial D_3}{\partial E_3}\right)_\gamma \left(\frac{\partial E_3}{\partial\gamma_j}\right)_D = -\varepsilon_3 g_{3j} \qquad (12.7.35)$$

and

$$e_{3j} = \left(\frac{\partial D_3}{\partial\sigma_j}\right) = \left(\frac{\partial D_3}{\partial E_3}\right)_\sigma \left(\frac{\partial E_3}{\partial\sigma_j}\right)_D = -\varepsilon_3 h_{3j} \qquad (12.7.36)$$

The pyroelectric effect is directly obtained from the charge response to a thermal excitation. The excitation may involve a heater controlled temperature variation, laser beam heating, and dielectric heating.[22] Illustrative values of the piezo-electric and pyroelectric properties of PVDF, copolymers, and composites are given in Table 12.1.

12.8. PIEZOELECTRIC AND PYROELECTRIC MECHANISMS

Let us consider a film of thickness l and electrode area A. The polarization remaining after the material has been poled and E_p has been removed is called the remnant polarization P_r. Applying an electric field in the opposite direction can reverse the direction of P_r. This process is called ferroelectric switching. The poling conditions of ferroelectric polymers are strongly related to the switching characteristics.

TABLE 12.1   Values of piezoelectric, pyroelectric, and permittivity coefficients for some polymers[12]

Class 1: Poled polymers		d_{31}, pC N^{-1}	ε_{33}	p_3, μC m^{-2} K^{-1}
A. Ferroelectric polymers	PVDF[a]	30	0.2	35
	VDF/TrFE[b]	30	0.3	50
	VDF/TFE[c]	30	0.25	35
B. Polar polymers	VDCN/VAc[d]	8	0.25	20
	Nylon 11	4		5
	PVC[e]	0.2		1
C. Composite	PZT7PVDF[f]	−25	0.1	100
	PZT[g]	−180	0.6	270
Class 2: Drawn polymers		d_{14}, pC N^{-1}		
A. Chiral polymers	PBG[h]	1.7		
	CEC[i]	3.0		
	PHB[j]	1.5		

[a]Poly(vinylidene fluoride); [b]Vinylidene fluoride/trifluoroethylene copolymer; [c]Vinylidene fluoride/tetrafluoroethylene copolymer; [d]Vinylidene cyanide/vinyl acetate copolymer; [e]Poly(vinyl chloride); [f]Lead zirconate titanate; [h]Poly(γ-benzyl glutamate); [i]Cyanoethyl cellulose; [j]Poly(β-hydroxybutyrate)

In the pole sample, schematically represented in Fig. 12.12, there are N dipoles of moment μ located in crystalline regions, which in turn are dispersed in the amorphous phase. The dipoles within each crystalline entity are oriented with respect to the poling field, $\mathbf{E}_p$, and the average orientation is P_1 or $\langle \cos \theta \rangle$. The remnant polarization is given by

$$P_r = \frac{N\mu\langle \cos \theta \rangle}{Al} \qquad\qquad (12.8.1)$$

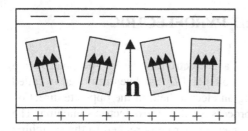

FIG. 12.12   Model containing N dipoles of moment μ in a film of thickness l. The director is indicated by **n**

The charge Q in the electrodes under a voltage V is given by

$$Q = \frac{N\mu\langle\cos\theta\rangle}{Al} + CV \tag{12.8.2}$$

where C is the capacitance. The sample receives the force

$$F = \int_0^Q E \, dQ = \int_0^Q \frac{V}{l} \, dQ = \frac{Q^2}{2lC} \tag{12.8.3}$$

where it has been considered that $V = Q/C$. The piezoelectric constant e obtained with this configuration is given by Eq. (12.4.11). By taking into account that $Q = DA$, $F = \sigma A$, $\gamma = \Delta l/l$, and $E = V/l$, Eq. (12.4.11) can be written as

$$e = \frac{l}{A}\left(\frac{\partial Q}{\partial l}\right)_{V=0} = -\frac{l}{A}\left(\frac{\partial F}{\partial V}\right)_{\Delta l=0} \tag{12.8.4}$$

The mechanisms governing the piezoelectric effect can be analyzed in terms of changes in the parameters of Eqs (12.8.2) and (12.8.3) that arise from variations in the thickness and volume of the poled films. Since the dipoles in polymers are rigid, and are maintained fixed during orientation, the piezoelectric response arises from changes in the thickness of the film. This is the so-called dimensional effect.[23,24] Therefore, among the deformations in three directions, only the change in thickness along the polar axis produces piezoelectricity. Taking into account that parameters other than l are fixed, it is straightforward to obtain

$$e_{33} = -P_{\mathrm{r}}; \; e_{32} = e_{32} = 0 \tag{12.8.5}$$

The dimensional mechanism is important for polymers for which dipoles are associated with strong chemical bonding, as occurs with most polar molecules. In the case of space charge polarization in a homogeneous dielectric, positive and negative free charges comprising a dipole move affine to the macroscopic deformation, and as a result no piezoelectric response appears.

Structural studies show that molecular chains in the ferroelectric phase of PVDF and VDF/TrVF copolymers are all in *trans*-conformation, packed in a parallel manner. The ferroelectric $\rightarrow$ paraelectric transition taking place in the copolymers (this transition occurs above the melting temperature in PVDF and therefore cannot be observed in this polymer) gives rise to a mixture of TT, TG, and TḠ conformations.[25,26] The remnant polarization is therefore lost as a result of this intramolecular disorder. Attempts have been made at calculations of piezoelectric constants, taking as a basis the structural changes occurring in the conformations of the crystal by the effect of deformations.

Changes in crystallinity induced by the effect of strain have also been proposed as the origin of the piezoelectric effect.[27,28] This mechanism involves conformational changes in the surface of the crystal that are analogous to those occurring within the crystal to which we have referred above. Therefore, both mechanisms are essentially the same. The only difference is that one is the event occurring inside the crystal and the other at its surface.

While the dimensional effect is paramount in piezoelectric responses, it plays a minor role in the pyroelectric response. Actually, the pyroelectric constant, according to Eq. (12.5.1), is given by

$$p = \frac{1}{A}\left(\frac{\partial Q}{\partial T}\right) \tag{12.8.6}$$

In accordance with the dimensional piezoelectric effect, the dimensional pyroelectric effect can be written as

$$p_3^a = \alpha_{33}P_r \tag{12.8.7}$$

where α_{33} is the expansion coefficient along the thickness direction. The fact that $\alpha_3 \cong 10^{-4}\ K^{-1}$ leads to a negligible value for p_3^a. The pyroelectric response arises from the effect of temperature on the orientation of dipoles, expressed by $\langle\cos\theta\rangle$. The remnant polarization can be viewed as the result of dipolar alignment by the effect of cooperative forces and dipolar randomization by the effect of entropic forces. The latter effect increases very rapidly as the temperature of the system approaches the Curie temperature. As can be seen in Fig. 12.13, the remnant polarization of fluorovinylidene/trifluoroethylene copolymers slightly decreases as temperature increases until it drops sharply in the vicinity of the Curie point. Therefore, the strong pyroelectric properties of fluorovinylidene copolymers should be attributed to intrinsic pyroelectricity arising from temperature-dependent conformational disorder.

12.9. PIEZOELECTRICITY AND PYROELECTRICITY IN POLAR POLYMERS

Piezoelectric materials can also be prepared from amorphous polymers, such as poly(vinyl chloride), poly(vinyl acetate), etc., containing permanent dipoles in their structure. Applying an electric field at temperatures well above the glass transition temperature results in dipole orientation. By the effect of the field, dipole orientation is accomplished, and then the temperature is lowered below the glass transition temperature where frozen-in dipole orientation is accomplished. The remnant polarization of the films in SI units is given by

$$P_r = e_0\Delta\varepsilon E_p \tag{12.9.1}$$

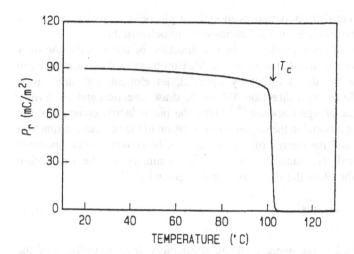

FIG. 12.13 Scheme of an S_C^* mesophase with the mesogenic groups represented as rods. The angle between the director and the normal to the layer is indicated by angle θ

where $e_0 (= 8.854 \text{ pF/m})$ is the permittivity in vacuum, $\Delta\varepsilon$ is the relative relaxation strength, and E_p is the polarization electric field. The only way of getting high remnant polarizations is to use as high as possible electric fields and polymers with large dielectric relaxation strengths. However, E_p is limited to about 100 MV/m, and $\Delta\varepsilon$ for PVCl and PVAc lies in the range 5–10. These values yield remnant polarizations of only 4–8 mC/m². High dielectric strengths are obtained in copolymers in which one of the comonomers has high polarity. For example, copolymers of vinyl acetate and vinylidene cyanide may reach dielectric relaxation strengths[29,30] of 125 at their $T_g \approx 180°C$. The remnant polarization of this copolymer poled under an electric field of 50 MV/m is 55 mC/m², comparable with that of PVDF.[31] It should be pointed out that the dimensional effect dominates the piezoelectricity arising from frozen-in dipoles, as occurs in ferroelectrics.

12.10. UNIAXIALLY ORIENTED, OPTICALLY ACTIVE POLYMERS

Piezoelectricity in class 2 polymers arises from the uniaxial orientation of chiral molecules. Polymers of class 2 include a variety of biopolymers such as poly(γ-benzyl-L-glutamate) (PBG),[23] cyanoethyl cellulose, and DNA. These polymers contain asymmetric carbon atoms and lack mirror symmetry. This class also includes isotactic poly(propylene oxide)[29] and poly(β-hydroxybutyrate).[32]

The P_2 orientation of these polymers is enough for piezoelectricity to appear, but the lack of polarity precludes the appearance of pyroelectricity.

In polymers of class 2 oriented in the direction of the x_3 axis, the only unknown component of the d_{jk} tensor is d_{14}. Measurement of this parameter can be obtained from the direct effect by applying an elongational stress to a rectangular sample cut in a direction $45°$ to the draw direction and by further measurement of the charge response.[33] Here, the piezoelectric constant of the sample, d_{14}, is proportional to the degree of orientation of the orientation function P_2 [Eq. (12.6.2)] and the degree of crystallinity ϕ. Moreover, the components other than d_{14} and d_{25} vanish owing to the symmetry of the orientation distribution, and therefore the d_{14} component is given by[34]

$$d_{14} = \frac{\phi P_2(d_{14}^c - d_{25}^c)}{2} \tag{12.10.1}$$

where $(d_{14}^c - d_{25}^c)/2$ is the average of the shear piezoelectric constants of the crystal. This equation suggests that a necessary condition for these oriented polymers to show piezoelectricity is $d_{14}^c - d_{24}^c \neq 0$. Of 20 piezoelectric constant classes, only nine satisfy this condition. They are C_1, C_2, D_2, C_3, D_3, C_4, D_4, C_6, and D_6. These classes lack not only a center of symmetry but also a mirror image, and, as a consequence, they are optically active.

12.11. FERROELECTRIC LIQUID CRYSTALLINE POLYMERS

Liquid crystals are made up of rodlike or disklike molecules that give rise to a fourth state of matter. At their melting points, they lose some crystalline order, generating a fluid but ordered state known as mesophases. The rods of mesogenic groups can approximately be oriented along a director, forming the so-called nematic phases, or be stacked with different kinds of order in layers, forming the so-called smectic phases. Disklike molecules mostly form columnar phases.

In eight (S_C, S_I, S_F, S_G, S_H, S_J, and S_K) of the about 20 smectic phases known hitherto, the long axes of the rods are tilted with respect to the layer normal. If these mesophases consist of chiral molecules, then they have the requirements for intrinsic polarization. This property has been detected in six (S_C^*, S_I^*, S_F^*, S_G^*, S_J^*, and S_M^*) of the tilted chiral phases. The S_C^* mesophase has the highest fluidity because it is the less ordered and therefore responds very rapidly to the perturbation of electric fields. As a consequence, S_C^* mesophases are prominently used in electrooptical devices. To explain the cause of the appearance of spontaneous polarization in S_C^* mesophases, let us consider first a single S_C layer consisting of achiral molecules. As indicated in Fig. 12.14, a single smectic layer possesses C_2 symmetry (a twofold axis of rotation) and mirror symmetry in the $x-z$ plane. If we assume an arbitrary polarization $\mathbf{P}_s$ and apply the C_2 symmetry operation that involves 180° rotation around the y axis, we have

$$\mathbf{P}_s = \begin{pmatrix} P_x \\ P_y \\ P_z \end{pmatrix} \rightarrow \begin{pmatrix} -P_x \\ P_y \\ -P \end{pmatrix} \rightarrow \mathbf{P}_s = \begin{pmatrix} 0 \\ P_y \\ 0 \end{pmatrix} \qquad (12.11.1)$$

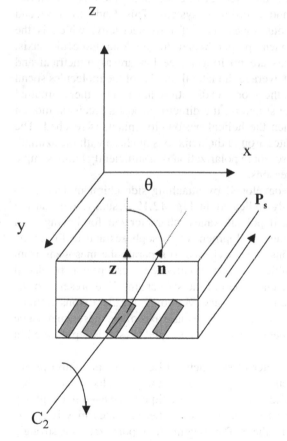

FIG. 12.14 Variation in the remnant polarization with temperature for a VDF(65)/TrFE(35) copolymer poled at 20°C

Accordingly, $\mathbf{P}_s$ only has a component in the y direction. The mirror symmetry leads to

$$\mathbf{P}_s = \begin{pmatrix} 0 \\ P_y \\ 0 \end{pmatrix} \rightarrow \begin{pmatrix} 0 \\ -P_y \\ 0 \end{pmatrix} \rightarrow \mathbf{P}_s = \begin{pmatrix} 0 \\ 0 \\ 0 \end{pmatrix} \tag{12.11.2}$$

Since the symmetry operations lead to $\mathbf{P}_s = 0$, S_C smectic mesophases cannot display ferroelectric behavior. However, if chiral molecules are present, the mirror image is lost while preserving the C_2 symmetry within each smectic layer. In this case, the C_2 axis is the polar axis that admits a spontaneous polarization along it. Within a smectic layer the director is located in the plane zx so that $\mathbf{P}_s = P_0(\mathbf{z} \times \mathbf{n}) = P_0 \sin \theta \mathbf{y}$, and the direction of $\mathbf{P}_s$ lies along the y axis, as shown in Fig. 12.14 for a right-hand coordinate system. This leads to a second prerequisite for the spontaneous appearance of ferroelectricity, which is the existence of a dipole component perpendicular to the long molecular axis. Rotations around the long axis are no longer free but are asymmetrical and hindered in such a way that on average the lateral dipoles of the molecules spend a longer time in pointing along the y (or $-y$) direction than in any other azimuthal direction. Owing to the helical structure, the director makes a precision motion that completes a full turn when the helical periodicity (pitch) is reached. The polarization vector $\mathbf{P}_s$ also rotates around the helix axis in phase with the azimuth angle of the director. The macroscopic polarization of a sufficiently bulky sample averages zero for symmetry reasons.

Mesophases S_C^* can be developed by attaching side chiral molecules to the backbone, as schematically indicated in Fig. 12.15. Usually, the starting polymers are polyacrylates and polysiloxanes, characterized for having high flexibility. In order to favor the development of mesophases and to lower the temperature of phase transitions, it is advisable to separate the mesogens from the backbone by means of flexible spacers. Optically active groups may be placed in the backbone, spacer, or the end group of the side chains. The presence of an alkyl moiety as an end side chain is necessary. Moreover, for the reasons outlined above, a dipole component perpendicular to the main axis of the mesogen must be present. An example of a ferroelectric side chain liquid crystal is represented in Fig. 12.16.

Owing to the lack of a center of symmetry, FLC polymers exhibit piezo-electric and pyroelectric effects. For C_2 symmetry, the piezoelectric tensor has the form indicated by Eq. (12.7.21). Since FLCs are liquid-like even in the glassy state, and therefore cannot resist mechanical stress because they may relax via flow, it is preferable to crosslink them. The magnitude of polarization is strongly dependent on the molecular structure of the dielectric, and its magnitude may vary between 2 and 500 nC/m^2.[35] Note that the spontaneous polarization in S_C^*

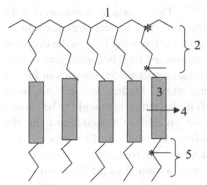

1: backbone; 2 : spacer; 3 : messogen
4: lateral dipole moment; 5 : terminal alkyl chain
asterisk: possible chiral centers

FIG. 12.15 Scheme indicating the requirements that a side chain liquid crystalline polymer must fulfill to develop ferroelectric properties

of FLC polymers is not the result of dipole–dipole interaction, as occurs, for example, in PVDF, but is due to the characteristic effect of intermolecular interactions of the optically active molecules. This is a characteristic of improper ferroelectrics. Values of p of about 13 nC/cm^2 have been reported for a diluted S_C^* polysiloxane.[36] A thorough review of FLC polymers and their applications can be found in reference [37].

In general, the dielectric relaxation spectra of polymers with ferroelectric moieties in the side groups display in the frequency range below 1 MHz a very strong absorption, called the Goldstone mode [see Section 11.8], which is

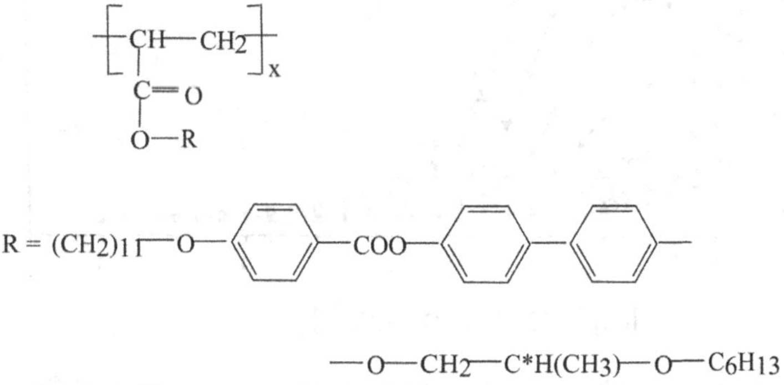

$$\left[CH-CH_2 \right]_x$$
$$\underset{\displaystyle O-R}{\overset{\displaystyle C=O}{|}}$$

R = (CH$_2$)$_{11}$—O—⬡—COO—⬡—⬡—

—O—CH$_2$—C*H(CH$_3$)—O—C$_6$H$_{13}$

FIG. 12.16 Scheme of a liquid crystalline polymer displaying ferroelectricity

restricted to the smectic S_C^* mesophases.[39] This mode is associated with fluctuations of the phase of helical superstructure. The helical structure can be unwound by superimposing a d.c. bias field on the alternating field as the decrease in Goldstone mode with increasing d.c. superimposed field becomes greater (see Fig. 12.17). With the suppression of this mode, a soft mode emerges at higher frequencies, associated with the unwound state. Studies carried out with increasing d.c. field bias indicate that the soft mode, which is thought to be caused by the tilt, increases in the SmA phase, reaching a maximum in the SmA $\rightarrow$ SmC* transition, and decreasing in the SmC* phase (see Fig. 12.18).[38] Even at higher frequencies, in the frequency regime above 1 MHz, one β-relaxation appears that has been assigned to the libration of the mesogen around its long molecular axis. At very low frequencies, conductive processes are dominant and the dielectric loss obeys the equation [see Section 4.3.3]

$$\varepsilon'' = \frac{\sigma}{\varepsilon_0 \omega} \qquad\qquad (12.11.3)$$

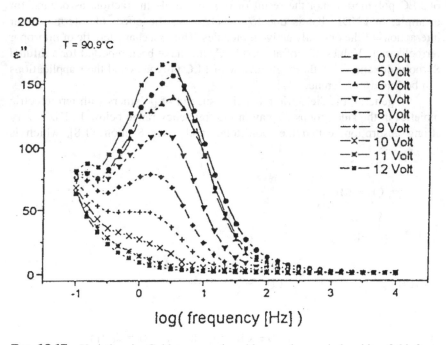

FIG. 12.17 Variation in Goldstone mode with superimposed d.c. bias field for a ferroelectric polymer[38]

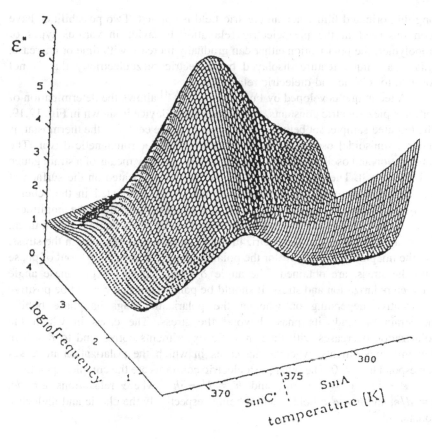

FIG. 12.18 Variation in Goldstone mode with frequency and temperature for a low molecular weight sample[39]

12.12. MEASUREMENTS OF PIEZOELECTRIC CONSTANTS

The simplest way of measuring direct piezoelectric constants is to evaporate metal electrodes on the surface of long thin films of the oriented sample. It is important that the metallic layer is thin enough not to contribute to the mechanical properties of the sample. The charge displaced from one electrode to the other by the effect of a small weight attached to the end of the film can be measured with an electrometer as a function of time. The corresponding converse piezoelectric constants can be obtained by measuring the change in length of a long thin oriented film when an electric field is applied. Two possibilities have been observed in the piezoelectric relaxation behavior in various types of biopolymer: the polarization either can gradually increase with time or decrease. This is a unique feature displayed by dielectric piezoelectricity that is not possible for elastic and dielectric relaxations.

A technique developed by Fukada *et al.* [40,41] allows the determination of complex piezoelectric constants. A scheme of this device is shown in Fig. 12.19. The oriented sample, set between metal clamps at the center of the thermostat, is under a sinusoidal oscillation by the action of an electromagnetic driver. The deformation and oscillation are detected, respectively, by means of a strain gauge and a load cell. The electric charge on the electrodes located on the surface of the sample is detected with a charge amplifier. As indicated in the scheme of Fig. 12.19, operational circuits can be used to obtain the ratio of the amplitudes of the electric charge to either stress or strain and the phase angle between them. The real part of d^*, d', or the polarization per unit stress in phase with the stress, and the imaginary part and d'', or the polarization per unit stress 90° out of phase with the stress, are obtained. The angle $\delta = \tan^{-1}(d''/d')$ is the phase angle between polarization and stress. It should be pointed out that δ could be positive or negative depending on whether the polarization lags its phase behind the strain or leads its phase beyond the stress. The cases in which the polarization decreases with time in static experiments correspond to $\delta < 0$ in dynamic experiments, whereas the cases in which the polarization increases correspond to $\delta > 0$. The other piezoelectric constants are the complex quantities $e^* = e' - ie''$, $g^* = g' - ig''$, and $h^* = h' - ih''$, where relaxations $e = dc$, $g = d/\varepsilon$, and $h = e/\varepsilon$ hold, c and ε being respectively the elastic and dielectric constants.[26]

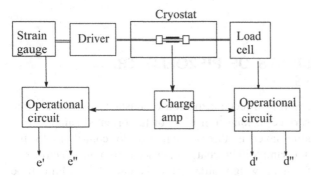

FIG. 12.19 Scheme of the experimental device for measurement of the dynamic piezoelectric constants $d^* = d' - id''$ and $e^* = e' - ie''$

For a system having a single relaxation time, the parameters c^*, ε^*, and d^* are given by

$$c^* = c_\infty + \frac{c_r - c_\infty}{1 + i\omega\tau}, \qquad c_r - c_\infty < 0 \tag{12.12.1}$$

$$\varepsilon^* = \varepsilon_\infty + \frac{\varepsilon_r - \varepsilon_\infty}{1 + i\omega\tau}, \qquad \varepsilon_r - \varepsilon_\infty > 0 \tag{12.12.2}$$

$$d^* = d_\infty + \frac{d_r - d_\infty}{1 + i\omega\tau}, \qquad d_r - d_\infty \lessgtr 0 \tag{12.12.3}$$

where the subscripts r and ∞ refer to the values of the complex quantities at zero and infinite frequency respectively, $\tau = \tau_0 \exp(\Delta E/kT)$ is the relaxation time, τ_0 is the frequency factor, ΔE is the activation energy, k is the Boltzmann constant, and T is the absolute temperature.

Biopolymers including amylose, chitin, wood cellulose, bone and collagen, keratin, various proteins, dexoxyribonucleic acid, poly-γ-benzyl-L-glutamate, etc., exhibit piezoelectric properties. Optically active polymers, such as poly-D-propylene oxide, poly-β-hydroxybutyrate, and its copolymers produced by microorganisms, also present piezoelectric activity. Illustrative curves showing the piezoelectric constants d and e for oriented films of oriented poly(γ-benzyl-L-glutamate) are shown in Fig. 12.20. If the curves are shifted horizontally, and small vertical shifts are used, then master curves are obtained.[20]

12.13. RELATION BETWEEN THE REMNANT POLARIZATION AND THE PIEZOELECTRIC CONSTANTS IN FERROELECTRIC POLYMERS

Illustrative semilogarithmic plots representing the evolution of the dielectric displacement D and its derivative $dD/d \log t$ with the time elapsed after applying step fields are shown in Fig. 12.21.[12] Before starting each measurement, the samples were polarized in the opposite direction to achieve a negative remnant polarization from which D is calculated. The maximum of the peak corresponding to the maximum of $dD/d \log t$ indicates the occurrence of the switching process. The value of the switching time decreases as the value of the electric field increases, being of the order of nanoseconds for electric fields of the order of 400 MV/m. The fact that the amount of reversed polarization is independent of the electric field indicates that complete reversal polarization from $+P_r$ to $-P_r$ takes place. On the other hand, P_r is small for a copolymer in which the amount of VDF comonomer is lower than 40%, but it rises very rapidly as the fraction of VDF increases, reaching a maximum in the vicinity of 80%. Since piezoelectricity arises in the crystalline regions, the increase in this effect is

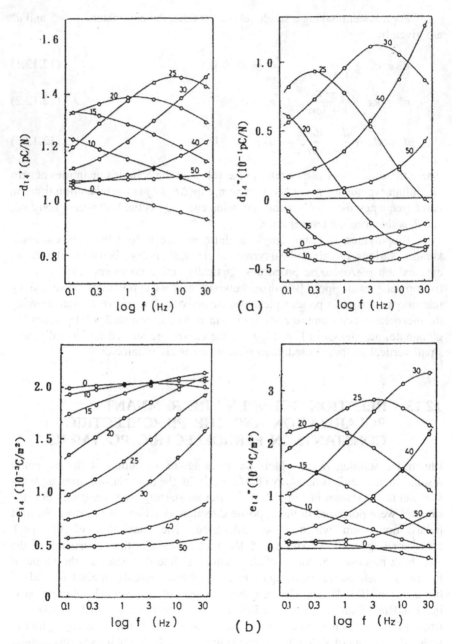

FIG. 12.20 Frequency dependence of the real and loss components of d_{14}^* and e_{14}^* for poly(γ-benzyl-L-glutamate)[33]

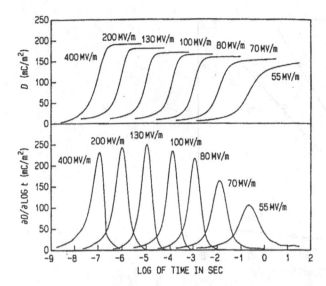

FIG. 12.21 Variation in the dielectric displacement and dD/dt for a VDF(65)/TrFE(35) copolymer at 20°C under various fields (TrFE = trifluoroethylene)[12]

a consequence of the increase in crystallinity that accompanies an increase in the amount of the latter comonomer in the copolymer. Experiments carried out in uniaxially drawn PVDF show that the piezoelectric constants d_{31}, d_{32}, and d_{33} are linear functions of the remanent polarity. This is not the case for copolymers.[12] However, the dependence of both e_{33} and e_{31}^a on P_r is linear (Fig. 12.22).

The variation in the pyroelectric constant p_3 with P_r for PVDF and copolymers, displayed in Fig. 12.23, is linear. Here, however, contrary to what occurs with the piezoelectric constants, drawing has a minor effect on pyroelectric activity. The ferroelectric phase transition has not been observed below the β-phase melting point of PVDF, 175–180°C. However, it has been found that the ferroelectric transition of vinylidene fluoride–tetrafluorethylene copolymers varies linearly with vinylidene fluoride content.[42] By extrapolating the data to 100% vinylidene fluoride, the Curie temperature was estimated to be 195–197°C.

12.14. FERROELECTRIC COMPOSITES

Flexible composites obtained by dispersing an inorganic piezoelectric ceramic into a polymer host allow the development of ferroelectric materials that suit particular properties, such as mechanical, electric, thermal properties, and/or

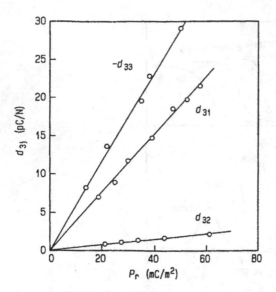

FIG. 12.22 Dependence of d_{31}, d_{32}, and d_{33} of PVDF on the remanent polarization P_r.[12]

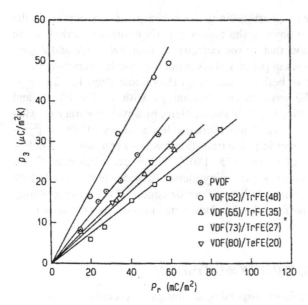

FIG. 12.23 Dependence of the pyroelectric constant p_3 on the remanent polarization P_r for the ferroelectric polymers indicated[12]

coupling between these properties. To obtain ferroelectric composites with good piezoelectric and pyroelectric properties requires the use of ceramics with good piezoelectric and pyroelectric activity. The properties also depend on the ceramic polymer mixture and on the processing employed in the manufacture of the composites. Another important factor is related to the connectivity of the phases. It has been reported that composites consisting of embedded rods of piezoelectric material in polymers forming a two-dimensional parallel network exhibit better properties than the simpler powder dispersed composites. Other ceramic–polymer configurations forming a three-dimensional connected ceramic phase have been investigated. However, cheap piezoelectric and pyroelectric materials that can impart in a reliable way special properties to various structures by coating processes have been developed. A scheme of different connectivities in a two phases system is shown in Fig. 12.24. A general review on inorganic ceramic polymer ferroelectric composites, dealing with models, connectivity, etc., can be found in reference [43] and the references therein.

Ceramic composites may be prepared by heating the polymer in a hot rolling machine to a temperature between the glass transition and the melting temperature of the polymer. Then, the powder ceramic is gradually added while mixing until a blend is obtained. An alternative way is to dissolve the polymer in a suitable solvent and then to add the ceramic to the fluid. The solvent is removed

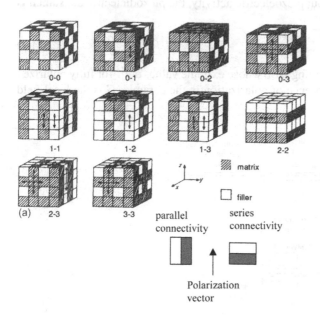

FIG. 12.24   Connectivity patterns in a two phases composite system[43]

by evaporation and the polymer–ceramic mixture is further processed to obtain polymer films with the desired thickness. Orientation of the spontaneous polarization in the ceramic phase of the composite is carried out by means of an external electric field of such magnitude that the electric field across each of the ceramic grains is greater than the so-called coercitive field of the ceramic. The poling can be performed in two ways: by applying an electric field to the film composite sandwiched between two electrode plates, or by the corona method where a one-side electrode sample is charged on the nonelectrode face by means of a corona current controlled by a grid placed in-between the corona plate and the sample. The electrode face of the film rests on a grounded plate. Schemes of the processing and poling of the films are shown in Fig. 12.25.

In the spontaneous polarization of the ceramic towards an applied electric field, important discontinuity of the electric displacement at the polymer–ceramic interface occurs which impedes the dielectric poling of the composite. However, this difficulty can be overcome by poling the composite at high temperature. Then, as shown in Fig. 12.26, space charges in the polymer phase are accumulated near the interface, increasing the internal field in the ceramic phase and removing the discontinuity of the displacement at the interface.

Studies on the piezoelectric and pyroelectric properties of binary systems were carried out by Yamada et al.,[44] assuming the system to be made up of ellipsoidal particles dispersed in a continuous medium. For a continuous medium or polymer matrix without piezoelectric activity, the piezodielectric constant d is given by

$$d = \alpha \phi^c G d^c \tag{12.14.1}$$

where α is the ceramic poling ratio whose extreme values are 1 for fully polarized ceramic and 0 for the absence of polarization in the ceramic, G is the local field

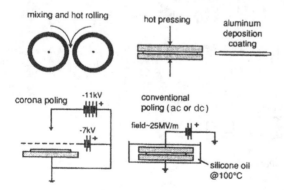

FIG. 12.25 Different methods of films preparation and further polarization

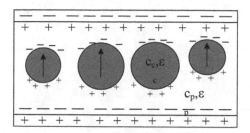

FIG. 12.26 Scheme of a PTZ/polymer composite where the polarization of the filler is fully compensated by space changes

constant, and ϕ^c and d^c are respectively the volume fraction and the piezoelectric constant of the ceramic. The local field using the shape parameter η is written as [see Eq. (1.C.1)]

$$G = \frac{\eta\varepsilon}{\eta\varepsilon + (\varepsilon^c - \varepsilon)} E \qquad (12.14.2)$$

where ε^c is the permittivity of the ceramic and ε is the permittivity of the composite, given by

$$\varepsilon = \varepsilon^p \left[1 + \frac{\eta\phi^c(\varepsilon^c - \varepsilon^p)}{\eta\varepsilon^p + (\varepsilon^c - \varepsilon^p)(1 - \phi^c)} \right] \qquad (12.14.3)$$

where ε^p is the dielectric permittivity of the polymer. For the same configuration, the model gives the following expression for the pyroelectric constant

$$p = \alpha\phi^c G p^c \qquad (12.14.4)$$

where p and p^c are respectively the pyroelectric constant for the composite and the ceramic. Equations (12.14.1) and (12.14.4) give a good account of the experimental results found for the experimental values of d and p in composites made of PZT powder embedded in a PVDF matrix.

A configuration (Fig. 12.27) consisting of spherical inclusions of ceramic embedded in a polymer matrix that in turn is covered with a homogeneous medium whose properties average approximately the average composite properties was studied by Furukawa et al. [45,46] The permittivity and elastic constant can be obtained from the pertinent values for the ceramic and polymer by

$$\varepsilon = \frac{2\varepsilon^p + \varepsilon^c - 2\phi^c(\varepsilon^p - \varepsilon^c)}{2\varepsilon^p + \varepsilon^c + \phi^c(\varepsilon^p - \varepsilon^c)} \varepsilon^p$$

$$\qquad (12.14.5)$$

$$c = \frac{3c^p + 2c^c - 3\phi^c(c^p - c^c)}{3c^p + 2c^c + 2\phi^c(c^p - c^c)} c^p$$

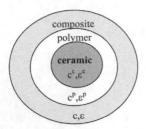

FIG. 12.27 Scheme of a ceramic inclusion embedded in a polymer which in turn is surrounded by a homogeneous medium whose properties approximate the average composite properties

where the superscripts c and p refer to the ceramic and polymer respectively, and the elastic constant c is the Young's modulus.

By definition, the piezoelectric constant d is given by

$$d = \left[\frac{D}{\sigma}\right]_E \tag{12.14.6}$$

where D and σ are respectively the dielectric displacement and the stress. A stress σ applied to the composite produces the local stress

$$\sigma^c = L_\sigma \sigma \tag{12.14.7}$$

where L_σ is the local field. On the other hand, according to Eq. (12.14.7)

$$D^c = d^c \sigma^c = d^c L_\sigma \sigma$$

Hence, the elastic displacement in the composite that is due to the local D^c is given by

$$D = d\sigma = \phi^c L_E D^c = \phi^c L_E L_\sigma d^c \sigma \tag{12.14.8}$$

where L_E is the local electric field coefficient defined as the ratio of the field applied to the composite to the local field produced in the inclusion. From Eq. (12.14.8) it follows that

$$d = \phi^c L_E L_\sigma d^c \tag{12.14.9}$$

Following the same strategy, the values of the constants e, g, and h are given by

$$\begin{aligned}
e &= \phi^c L_\gamma L_E e^c \\
g &= \phi^c L_D L_\sigma g^c \\
h &= \phi^c L_\gamma L_D h^c
\end{aligned} \tag{12.14.10}$$

where the local fields L_σ, $L\gamma$, L_E, and L_D were found to be

$$L_\sigma = \frac{5c^c}{3(1 - \phi^c)c^p + (2 + 3\phi^c)c^c}$$

$$L_\gamma = \frac{5c^p}{(3 + 2\phi^c)c^p + 2(1 - \phi^c)c^c}$$

$$L_E = \frac{3\varepsilon^c}{(2 + \phi^c)\varepsilon^p + 2(1 - \phi^c)\varepsilon^c}$$

$$L_D = \frac{3\varepsilon^c}{2(1 - \phi^c)\varepsilon^p + (1 + 2\phi^c)\varepsilon^c}$$

(12.14.11)

Ceramic/polymer composites of PZT/epoxy, PZT/PVDF, PZT/polyethylene, and PZT/polyvinyl alcohol were analyzed for volumes of ceramic lower than 21% using the method indicated above. The highest piezoelectric activity was found in the PZT/PVDF system, the piezoelectric activity of which is reasonably predicted for volume fractions of ceramic lower than 10%.

REFERENCES

1. Fukada, E. J. Phys. Soc. Japan **1955**, *10*, 149.
2. Fukada, E. Adv. Biophys. **1974**, *6*, 121.
3. Kawai, H. Jap. J. Appl. Phys. **1969**, *8*, 975.
4. Bergman, J.G.; McFee, J.H.; Crane, G.R. Appl. Phys. Lett. **1971**, *18*, 203.
5. Wada, Y. Jap. J. Appl. Phys. **1976**, *15*, 2041.
6. Wong, C.P. Ed. *Polymers for Electronic and Photonic Applications*; Academic Press: San Diego, 1993.
7. Nye, J.F. *Physical Properties of Crystals*; Clarendon Press: Oxford, 1987; 79 pp.
8. Nye, J.F. *Physical Properties of Crystals*; Clarendon Press: Oxford, 1987; 170 pp.
9. Riande, E.; Díaz-Calleja, R.; Prolongo, M.G.; Masegosa, R.; Salom, C. *Polymer Viscoelasticity*; Marcell Dekker: New York, 2000; Ch. 4.
10. Anderson, R.A.; Kepler, R.G. Ferroelectrics **1981**, *32*, 13.
11. Kepler, R.G.; Anderson, R.A. J. Appl. Phys. **1974**, *45*, 3360; Kepler, R.G.; Anderson, R.A. J. Appl. Phys. **1978**, *49*, 4490.
12. Furukawa, T. IEEE Trans. Elect. Ins. **1989**, *24*, 375.
13. Tashiro, K.; Tadokoro, H.; Kobayashi, M. Ferroelectrics **1981**, *32*, 167.
14. Tashiro, K. Crystal structure and phase transition of PVDF and related copolymers. In *Ferroelectric Polymers*; Nalwa, H.S., Ed.; Marcel Dekker: New York, 1995.
15. Tashiro, K.; Tadokoro, H.; Kobayashi, M. Ferroelectrics **1981**, *32*, 167.
16. Meyer, R.B.; Liebert, L.; Strzelecki, L.; Keller, P. J. Phys. (Paris) **1975**, *36*, 169.
17. Cahn, R.W.; Davis, E.A.; Ward, I.M.; Donald, A.M.; Windle, A.H. *Liquid Crystalline Polymers*; Cambridge University Press: London, 1992.

18. Scherowsky, G. Ferroelectric liquid crystal (FLC) polymers. In *Ferroelectric Polymers*; Nalwa, H.S., Ed.; Marcel Dekker: New York, 1995.
19. Le Barny, P.; Dubois, J.C. The chiral smectic liquid crystal side chain polymers. In *Side Chain Liquid Crystal Polymers*; McArdle, C.B., Ed.; Blackie, Chapman and Hall: New York, 1989; 130 pp.
20. Kozlowsky, M.V.; Beresnew, L.A. Phase Transition **1992**, *40*, 129.
21. Fukada, E. Q. Rev. Biophys. **1983**, *16*, 59–87.
22. Furukawa, T.; Wang, T. *Measurements and Properties of Ferroelectric Materials*; Wang, T.T., Herbert, J.M., Glass, A.M., Eds.; Blackie: UK, 1987; 66–117.
23. Wada, Y. Jap. J. Appl. Phys. **1976**, *15*, 2041.
24. Broadhurst, M.G.; Davis, G.T.; McKinney, J.E.; Collins, R.E. J. Appl. Phys. **1978**, *49*, 4992.
25. Lovinger, A.J.; Furukawa, T.; Davis, G.T.; Broadhurst, M. Polymer **1984**, *24*, 1225.
26. Tashiro, K.; Takano, K.; Kobayashi, M.; Chatani, Y.; Tadokoro, H. Ferroelectrics **1984**, *57*, 197.
27. Kepler, R.G.; Andersom, R.A.; Lagasse, R.R. Ferroelectrics **1984**, *57*, 51.
28. Dvey-Aharon, H.; Slucking, T.J.; Taylor, P.L. Ferroelectrics **1981**, *32*, 25.
29. Furukawa, T. Nature **1969**, *225*, 1235.
30. Nakajima, K.; Seo, I. Jap. J. Appl. Phys. **1986**, *25*, 1178.
31. Tasaka, S.; Miyasato, K.; Yoshikawa, M.; Miyata, S.; Ko, M. Ferroelectrics **1984**, *57*, 267.
32. Fukada, E.; Ando, Y. Int. J. Macromol. **1986**, *8*, 361.
33. Furukawa, T.; Fukada, E. J. Polym. Sci.: Polym. Phys. Ed. **1976**, *14*, 1979.
34. Fukada, E. Q. Rev. Biophys. **1983**, *16*, 59.
35. Kiefer, R. Applications of ferroelectric liquid crystalline polymers. In *Ferroelectric Polymers*; Nalwa, H.S., Ed.; Marcel Dekker, 1995; Ch. 19.
36. Sharp, K.; Anderson, G.; Dahlgren, A.; Porth, H.; Zentel, R. Proc. 14th Int. Liq. Cryst. Conf., Pisa, Italy, 1992; Vol. II, 905 pp.
37. Scherowsky, G. Ferroelectric liquid crystal (FLC) polymers. In *Ferroelectric Polymers*; Nalwa, H.S., Ed.; Marcel Dekker, 1995; Ch. 10.
38. Schönfeld, A.; Kremer, F.; Poths, H.; Zentel R. Mol. Cryst. Liq. Cryst. **1994**, *49*, 254.
39. Kremer, F. Broadband dielectric spectroscopy on collective and molecular dynamics in ferroelectric liquid crystals. In *Dielectric Spectroscopy of Polymeric Materials*; Runt, J.P.; Fitzgerald, J.J., Eds.; American Chemical Society, Washington, DC, 1997.
40. Fukada, E.; Date, M.; Hara, K. Jap. J. Appl. Phys. **1969**, *8*, 151.
41. Fukada, E.; Date, M.; Emura, T. J. Soc. Mat. Sci. Jap. **1968**, *17*, 335.
42. Lovinger, A.J.; Davis, D.D.; Casi, R.E.; Kometani, J.M. Macromolecules **1986**, *19*, 1491.
43. Dias, C.J.; Das-Gupta, D.G. IEEE Trans. Diel. Electr. Insul. **1996**, *3*, 706.
44. Yamada, T.; Ueda, T.; Kitayama, T. J. Phys. Phys. **1982**, *53*, 4328.
45. Furukawa, T.; Fujino, K.; Fukada, E. Jap. J. Appl. Phys. **1976**, *15*, 2119.
46. Furukawa, T.; Ishida, K.; Fukada, E. J. Appl. Phys. **1979**, *50*, 4904.

13

Nonlinear Optical Polymers

13.1. INTRODUCTION

Trading electrons for high-speed photons as vehicles of information transfer is a revolutionary process in telecommunications systems that began in the last decade. High-speed optical roadways have been designed that are made of optical fibers with optical on-ramps, off-ramps, and interchanges to direct the photons to their destinies. Special devices are necessary to carry out these processes as well as to generate and decipher the flashes of light that carry the messages. Important components of these devices are nonlinear optical (NLO) materials characterized by undergoing intriguing changes when light passes through them, which in turn alter the light as it moves through.[1] For example, an infrared laser beam passing through an inorganic crystal may come out as green. In the same way, the NLO crystal can act as a switcher, jumping the beam from one optical fiber to another.

NLO materials are made up of molecules whose electronic charges are easily polarizable under the influence of a passing wave of energy such as the original beam of light itself, an additional control beam, and even an applied electric voltage. Negatively charged electrons and positively charged ions of the material vibrate under the action of the electric field of the incoming wave. If that vibration is strong enough, it can alter the frequency of the output beam, change the speed of the light in the material (the refractive index), and change the material transparency.

The NLO materials mostly used in the last decades have been inorganic crystals such as lithium niobate or quartz. However, the use of these materials in light-based communications systems has several drawbacks: they are expensive, fragile, and difficult to integrate with the materials that nowadays make up communications systems, such as optical fibers and semiconductors. Polymers have proven to be suitable alternatives to classical NLO materials and have begun to challenge the supremacy of the inorganics. In this chapter we study the characteristics and structures of NLO polymeric materials.

13.2. BASIC PRINCIPLES OF HARMONIC GENERATION IN CRYSTALS

Before the laser era, the conventional strength of light sources was of the order of 10^5 V/m, much smaller than the field strengths of atomic and interatomic fields, which are about $10^9 - 10^{12}$ V/m. However, MW pulses lasting a few tenths of a nanosecond can nowadays be generated using moderately powerful lasers. Owing to the coherence of lasers, the light beam can be focused onto the surface of a crystal of area $A \approx \lambda^2$, where λ is the wavelength of the light. If λ is about 1 μm, then $A = 10^{-12}$ m^2 and the energy of such a laser is $W \approx 10^{20}$ MW/m, corresponding to an electric field peak of $E \approx 3 \times 10^{10}$ V/m which lies in the range of atomic electric fields. In this situation, the relationship between the electric polarization and the electric field is no longer linear, and, as a result, interesting nonlinear effects arise. The experiment carried out by Franken and coworkers,[2] who observed that ultraviolet light of a ruby laser passing through a quartz crystal emerged from the crystal with a frequency twice that of the incident light, marks the beginning of nonlinear optics.

Let us consider the polarization of a dielectric noncentrosymmetric crystal by the action of an electric field. As stated elsewhere (Chapter 1), the dipole moment per unit volume, $\mathbf{P}$, of a polarized dielectric is

$$\mathbf{P} = \sum_i \mathbf{p}_i \tag{13.2.1}$$

where the summation extends over all the dipole moments of the molecules in the unit volume. Under the effect of an external field, $\mathbf{E}$, the dipoles of the molecules tend to be oriented in the opposite direction to the field, and $\mathbf{P}$ in SI units is given by

$$\mathbf{P} = e_0 \chi \mathbf{E} \tag{13.2.2}$$

where e_0 is the dielectric permittivity in vacuum and χ is the electric susceptibility tensor. Under electric fields of high intensity, the linear relationship between polarization and electric field no longer holds, and Eq. (13.2.2) has to be generalized to

$$P = e_0(\chi^{(1)}E + \chi^{(2)}E^2 + \chi^{(3)}E^3 + \cdots) \tag{13.2.3}$$

where $\chi^{(1)}$ is the same as χ in Eq. (13.2.2) and $\chi^{(2)}$, $\chi^{(3)}$, ... are known as nonlinear susceptibilities. The parameter $\chi^{(1)}$ is dimensionless, whereas the units of $\chi^{(2)}$ and $\chi^{(3)}$ are m/V and m²/V² respectively. Note that, if the field is low, only the first term in Eq. (13.2.3) is retained, and Eq. (13.2.2) is recovered.

Let us assume that the nonlinear terms are not negligible and the incident electric field has the form

$$E = E_0 \cos \omega t \tag{13.2.4}$$

Substituting E into Eq. (13.2.3) gives

$$
\begin{aligned}
P &= e_0(\chi^{(1)}E_0 \cos \omega t + \chi^{(2)}E_0^2 \cos^2 \omega t + \chi^{(3)}E_0^3 \cos^3 \omega t + \cdots) \\
&= \tfrac{1}{2}e_0\chi^{(2)}E_0^2 + e_0(\chi^{(1)} + \tfrac{3}{4}\chi^{(3)}E_0^2)E_0 \cos \omega t \\
&\quad + \tfrac{1}{2}e_0\chi^{(2)}E_0^2 \cos 2\omega t + \tfrac{1}{4}e_0\chi^{(3)}E_0^3 \cos 3\omega t \ldots
\end{aligned}
\tag{13.2.5}
$$

where the following trigonometric relationships were used

$$\cos^2 \omega t = \frac{1 + \cos 2\omega t}{2}, \qquad \cos^3 \omega t = \frac{\cos 3\omega t + 3\cos \omega t}{4} \tag{13.2.6}$$

The term $\tfrac{1}{2}e_0\chi^{(2)}E_0$ is constant. It gives rise to a d.c. field across the medium and it has no practical importance. The terms $\cos \omega t$, $\cos 2\omega t$ and $\cos 3\omega t$ are called, respectively, the first (or fundamental), second, and third harmonics of polarization.

13.3. SECOND HARMONIC GENERATION AND COHERENCE LENGTH

According to Eq. (13.2.5), an electromagnetic radiation at frequency ω traveling through a crystal produces a polarization oscillating at frequency 2ω which radiates an electromagnetic wave of the same frequency that propagates with the same velocity, monochromaticity, and direction as the incident wave.[3] This phenomenon is known as the second harmonic generation (SHG). A simple scheme showing the generation of a second harmonic is presented in Fig. 13.1. When the light beam travels through the crystal, the nonlinear polarizability depends on the direction of propagation, the polarization of the electric field, and the orientation of the optical axis. As a result, $\chi^{(2)}$ has a tensorial character and the polarization $\mathbf{P}^{(2)}$ is related to the electric field by

$$P_i^{(2)} = e_0\chi_{ikl}^{(2)}E_kE_l \tag{13.3.1}$$

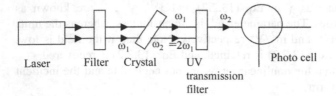

FIG. 13.1 Scheme showing the frequency splitting of a beam light by the effect of a crystal

where Einstein's convection has been used, that is, the sum extends over all repeated indices. Note that the generation of a second-order harmonic cannot take place either in an isotropic medium, such as liquids or gases, or in a crystal symmetrical about a point. Actually, the susceptibility χ_{ikl} is independent of the direction in an isotropic medium, so that, by reversing the direction of the axes $(x_1 \rightarrow -x_1,\ x_2 \rightarrow -x_2,\ x_3 \rightarrow -x_3)$, leaving the electric field and dipole unchanged in direction, the following expression holds

$$-P_i^{(2)} = e_0 \chi_{ikl}^{(2)}(-E_k)(-E_l) = P_i^{(2)} \tag{13.3.2}$$

Hence, $P_i^2 = -P_i^2 = 0$ and $\chi_{ikl}^{(2)} = 0$. For noncentrosymmetric crystals, the polarization can be written as

$$P = \varepsilon_0 \chi^{(1)} E + e_0 \chi^{(2)} E^2 + \varepsilon_0 \chi^{(3)} E^3 \tag{13.3.3}$$

In most cases, $\chi^{(3)}$ is much lower than $\chi^{(2)}$, and the third term in Eq. (13.2.3) can be ignored.

The relation of phase between the fundamental and second harmonics generated as they propagate in an ordered medium depends not only on the intensity of the exciting radiation but also on the direction in which they propagate in the crystals having optical dispersion. Let us consider a plane wave at frequency ω and the second harmonic wave at frequency 2ω propagating in the x direction through a crystal of length L, as shown in Fig. 13.2. To find the second

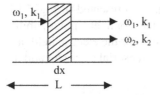

FIG. 13.2 Scheme for the mathematical description of second harmonic generation

harmonic of frequency 2ω at the exit of the crystal, we should consider that the production of the second harmonic in a slab of thickness dx in Fig. 13.2 is proportional to the second harmonic dipole moment per unit volume induced at frequency 2ω, i.e. $P_x^{(2)}$, which in turn is proportional to the square of the electric field.[3]

$$dE_x^{(2)} \propto P_x^{(2)} \, dx \propto \exp[2i(k_1 x - \omega t)] \, dx \tag{13.3.4}$$

where $E \propto \exp[i(k_1 x - \omega t)]$ and $k(= 2\pi n/\lambda)$ is the propagation wave number, n and λ being, respectively, the index of refraction of the material and the wavelength of the radiation. The second harmonic produced by the slab at the exit surface of the crystal in Fig. 13.2 is given by

$$dE_L^{(2)} = dE_x^{(2)} \exp[ik_2(L - x)] \tag{13.3.5}$$

where k_2 is the wave propagation number of the second harmonic. By substituting Eq. (13.3.4) into Eq. (13.3.5) and further integrating the resulting expression, it is established that

$$
\begin{aligned}
E_L^{(2)} &\propto \int_0^L \exp[i(2k_1 - k_2)x] \exp[i(k_2 L - 2\omega t)] \, dx \\
&= \frac{\exp[i(k_2 L - 2\omega t)] \exp[i(2k_1 - k_2)L - 1]}{i(2k_1 - k_2)}
\end{aligned}
\tag{13.3.6}
$$

By taking into account that $e^{ix} = \cos x + i \sin x$ and $\sin a - \sin b = 2 \sin[(a - b)/2] \cos[(a + b)/2]$, Eq. (13.3.6) becomes

$$E_L^{(2)} \propto \frac{\sin[(2k_1 - k_2)/2]L}{(2k_1 - k_2)/2} \tag{13.3.7}$$

Equation (13.3.7) suggests that the field of the second harmonic will reach a maximum when

$$\left(\frac{2k_1 - k_2}{2}\right)L = \frac{\pi}{2} \tag{13.3.8}$$

or

$$L_c = \frac{\pi}{2k_1 - k_2} = \frac{\lambda}{4(n_\omega - n_{2\omega})} \tag{13.3.9}$$

where $k_1 = 2\pi n_\omega/\lambda$ and $k_2 = 4\pi n_{2\omega}/\lambda$. The value of L_c given by Eq. (13.3.9) is called the coherence length of the second harmonic radiation. Increasing the value of L beyond the coherence length will not increase the value of L_c.[3]

The intensity, given by

$$I^{(2)} \propto \frac{\sin^2 \left[(2k_1 - k_2)/2 \right] L}{\left[(2k_1 - k_2)/2 \right]^2} \tag{13.3.10}$$

is sharply peaked at $(2k_1 - k_2)L/2 = 0$ or $k_2 = 2k_1$. This requirement is known as the phase matching criterion. Second harmonic generation allows the effective conversion of infrared radiation into visible.

13.4. NONLINEAR POLARIZATION AND FREQUENCY MIXING

In the development of Eq. (13.2.3), a single-frequency coherent light perturbing the system was assumed. Let us consider now that two coherent beams of unequal frequencies ω_1 and ω_2 are transversing the crystal in such a way that the perturbing field is given by

$$E = E_1 \cos \omega_1 t + E_2 \cos \omega_2 t \tag{13.4.1}$$

By substituting Eq. (13.4.1) into Eq. (13.2.3), the second term becomes

$$\begin{aligned} P^{(2)} &= e_0 \chi^{(2)} (E_1 \cos \omega_1 t + E_2 \cos \omega_2 t)^2 \\ &= e_0 \chi^{(2)} (E_1^2 \cos^2 \omega_1 t + E_2^2 \cos^2 \omega_2 t) \\ &\quad + e_0 \chi^{(2)} [E_1 E_2 \cos (\omega_1 + \omega_2)t + \cos (\omega_1 - \omega_2)t] \end{aligned} \tag{13.4.2}$$

where the trigonometric relationship $2 \cos \alpha \cos \beta = \cos(\alpha + \beta) + \cos(\alpha - \beta)$ was used. This means that the nonlinear polarization emitting radiation contains frequencies $\omega_1 + \omega_2$ and $\omega_1 - \omega_2$. The energy conversion between the beams can only take place over significant distances if the beams travel in the same direction and at the same velocity. It should be noted that the first term of Eq. (13.4.2) includes both the d.c. and the frequency doubling terms.

The phase matching condition is even more stringent in frequency mixing than in second harmonic generation because of the number of frequencies involved. As indicated above, k_1 must be equal to k_2 for second harmonic generation. For the sum of difference of frequencies, three waves must be matched. If $\omega_3 = \omega_1 \pm \omega_2$, the condition $k_3 = k_1 \pm k_2$ must be satisfied.

13.5. NONLINEAR POLARIZATION IN POLYMERS

It became clear in the middle of the 1980s that polar polymers with chromophore groups in their structure could display nonlinear optical behavior.[4−8] However, the second-order NLO effect cannot be observed in these materials unless the isotropy of the systems is destroyed. As will be shown later, this can be accomplished by inducing a polar order in the polymer by an external means such as an electric field poling process.

The origin of the optical nonlinearity in materials depends on the polarization of molecular units. For an isolated molecule, the polarization is given by[9,10]

$$p_i(\omega) = e_0[\alpha_{ij}(\omega)F_j(\omega) + \beta_{ijk}(-\omega; \omega_1, \omega_2)F_j(\omega_1)F_k(\omega_2)]$$
$$+ e_0[\gamma_{ijkl}(-\omega; \omega_1', \omega_2', \omega_3')F_j(\omega_1')F_k(\omega_2')F_l(\omega_3')] + \cdots \quad (13.5.1)$$

where the subscripts i, j, k refer to a molecule-based reference frame. Moreover, F_k is the local field along the axis k, $\alpha_{ij}(\omega)$ is the linear polarizability, and $\beta_{ijk}(-\omega; \omega_1, \omega_2)$ and $\gamma_{ijkl}(-\omega; \omega_1', \omega_2', \omega_3')$ are respectively the frequency-dependent first and second hyperpolarizabilities. Note that Eq. (13.5.1) assumes that the local fields along the various molecular axes are monochromatic, just as the induced polarization is. The SI units of α_{ij} and β_{ijk} are m^3/molecule and m^4/V molecule, respectively.

The second harmonic generation for a molecule exposed to local field of single frequency $F_j(\omega) = F_0$; $(\omega) \cos(\omega t)$ is given by[7]

$$p_i(2\omega) = (1/2)e_0 [\beta_{ijk}(-2\omega; \omega, \omega)F_{0j}^2(\omega) \cos 2\omega t] \quad (13.5.2)$$

For frequency up-conversion processes, two frequency-dependent local fields are present, i.e. $F_j(\omega_1)$ and $F_k(\omega_2)$, so that

$$p_i(2\omega) = Ke_0\beta_{ijk}(-\omega; \omega_1, \omega_2)F_{0j}(\omega_1)F_{0k}(\omega_2) \cos(\omega = \omega_1 + \omega_2)t \quad (13.5.3)$$

where the values of K for different second-order processes are given in Table 13.1.[7] It should be remarked that the factor $\frac{1}{2}$ is associated with the second harmonic generation but not with the up-conversion process.

TABLE 13.1 Values of K in Eq. (13.5.3) for Different Second Order Phenomena[6]

ω	ω_1	ω_2	K	Process
$\omega_1 + \omega_2$	ω_1	ω_2	1	Up-conversion
$\omega_1 - \omega_2$	ω_1	$-\omega_2$	1	Down-conversion
0	ω	$-\omega$	1/2	Optical rectification
2ω	ω	ω	1/2	Frequency doubling
ω	0	ω	2	de Pockels effect

At the macroscopic level, and by analogy to Eq. (13.5.1), the polarization of the material can be written as

$$P_I(\omega) = e_0[\chi_{IJ}^{(1)}(\omega)E_J(\omega) + \chi_{IJK}^{(2)}(-\omega; \omega_1, \omega_2)E_J(\omega_1)E_K(\omega_2)]$$
$$+ \varepsilon_0[\chi_{IJKL}^{(3)}(-\omega; \omega_1', \omega_2', \omega_3')E_J(\omega_1')E_K(\omega_2')E_L(\omega_3')] + \cdots \quad (13.5.4)$$

where the subscripts I, J, K refer to the reference frame of the material in the bulk, $P_I(\omega)$ is the component in direction I of the induced polarization at frequency ω, E_J is the applied electric field in direction J, and $\chi_{IJ}^{(1)}(\omega)$, $\chi_{IJK}^{(2)}(-\omega; \omega_1, \omega_2)$, and $\chi_{IJKL}^{(3)}(-\omega; \omega_1', \omega_2', \omega_3')$ are respectively components of the first-order, second-order and third-order susceptibility tensor in the coordinate axes of the material. Note that the induced polarization acts as a source of radiation, whereas the susceptibility describes the type and magnitude of the response. On the other hand, the frequency arguments in Eqs (13.5.1) and (13.5.4) indicate the type of nonlinear optical effect and, the frequency dependence of the polarizabilities and susceptibilities, and refer to energy conservation requiring that $\omega = \omega_1 + \omega_2$ and $\omega = \omega_1' + \omega_2' + \omega_3'$.

As indicated in Eq. (Chapter 1), the local field F_I and the applied field E_I are related by

$$F_I(\omega) = f_I(\omega)E_I(\omega) \quad (13.5.5)$$

where f_I is the field factor. The Lorentz–Lorenz field correction for a spherical cavity is given by Eq. (1.9.3), that is

$$f = \frac{\varepsilon_\infty + 3}{2} = \frac{n^2 + 3}{2} \quad (13.5.6)$$

where the subindex ∞ denotes that the factor f is only valid at optical frequencies. Note that $\varepsilon_\infty = n^2$, where n is the index of refraction. If the system contains a concentration of chromophores, N, Eqs (13.5.1), (13.5.4), and (13.5.6) lead to

$$\chi_{IJK}^{(2)}(-\omega; \omega_1, \omega_2) = Nf_I(\omega)f_J(\omega_1)f_K(\omega_2)\langle\beta_{ijk}(-\omega; \omega_1, \omega_2)\rangle_{IJK} \quad (13.5.7)$$

which relates macroscopic and microscopic nonlinear magnitudes. As usual, the angular brackets denote a weighted sum over all the orientations of the chromophore in the polymer.[11–14] It is important to note that field factors alone do not account for all the effects of the dielectric medium surrounding the chromophore. The shift in molecular energy levels of the chromophores by the effect of the solvent polarity brings the dependence of β on the solvent. Therefore, hyperpolarizabilities measured in solution may sometimes not be appropriate for polymers.[15]

For second harmonic generation processes, the analogy with Eq. (13.5.2) makes it possible to define a coefficient d_{IJK} given by

$$P_I(2\omega) = (\tfrac{1}{2})e_0\chi^{(2)}_{IJK}(-2\omega; \omega, \omega)E_J(\omega)E_K(\omega)$$
$$= e_0 d_{IJK}(-2\omega; \omega, \omega)E_J(\omega)E_K(\omega) \tag{13.5.8}$$

Most of the SHG data are given in terms of d_{ijk} components, which, according to Eq. (13.5.8), are related to $\chi^{(2)}$ by $d_{ijk} = \chi^{(2)}_{ijk}/2$.[14] Using this terminology, Eq. (13.5.8) can explicitly be written as

$$
\begin{pmatrix} P_1 \\ P_2 \\ P_3 \end{pmatrix} = e_0 \begin{pmatrix} d_{11} & d_{12} & d_{13} & d_{14} & d_{15} & d_{16} \\ d_{21} & d_{22} & d_{23} & d_{24} & d_{25} & d_{26} \\ d_{31} & d_{32} & d_{33} & d_{34} & d_{35} & d_{36} \end{pmatrix} \begin{pmatrix} E_1^2 \\ E_2^2 \\ E_3^2 \\ 2E_2 E_3 \\ 2E_1 E_3 \\ 2E_1 E_2 \end{pmatrix} \tag{13.5.9}
$$

Note that, as in Section 12.7 of Chapter 12, the following reductions of indices were made: $11 \to 1$; $22 \to 2$, $33 \to 3$, $23 \to 4$, $13 \to 5$, and $12 \to 6$. Poled polymers have $C_{\infty v}$ symmetry and, for the reasons outlined in Section 12.7 of Chapter 12, the only non-null components are d_{15}, $d_{24} = d_{15}$, $d_{31} = d_{32}$, and d_{33}, or, expressed without index reduction, d_{113}, $d_{223} = d_{113}$, $d_{311} = d_{322}$, and d_{333}.

The linear change in the index of refraction of a system under the action of an external electric field is known as the de Pockels effect. The IJ component of the change in the electric permittivity can be written as[9,11]

$$\Delta\left(\frac{1}{\varepsilon}\right)_{IJ} = \frac{1}{\varepsilon_{IJ}} - \frac{1}{\varepsilon_{IJ}(0)} \cong -\frac{\Delta\varepsilon_{IJ}}{\varepsilon_{IJ}^2} = r_{IJ,K}E_K \tag{13.5.10}$$

where $\varepsilon_{IJ}(0)$ is the IJ component of the dielectric permittivity in the absence of an electric field. The $r_{IJ,K}$ tensor, called the electrooptic coefficient, is related to the susceptibility tensor (in mks units) by[11]

$$\chi^{(2)}_{IJK}(-\omega; \omega, 0) = -(1/2)\varepsilon_{II}(\omega)\varepsilon_{JJ}(\omega)r_{IJ,K}(-\omega; \omega, 0) \tag{13.5.11}$$

For $C_{\infty v}$ symmetry, the nonzero elements are r_{13}, r_{33}, and r_{15}. By assuming that the values of the index of refraction of the material are n and $n + \Delta n_{Ij}$ in the absence and in the presence of an electric field respectively, Eq. (13.5.10) can be written as

$$\Delta\left(\frac{1}{\varepsilon}\right)_{IJ} = \frac{1}{(n + \Delta n_{Ij})^2} - \frac{1}{n^2} = r_{IJ,K}E_K \tag{13.5.12}$$

By taking into account that Δn is a small quantity, Eq. (13.5.12) is given by

$$\Delta n_{IJ} \cong -\frac{n^3}{2} r_{IJ,K} E_K \tag{13.5.13}$$

Note that the SI units of r_{ijk} are m/V.

13.6. POLING PROCESS

As indicated above, second harmonic generation requires the absence of a center of symmetry in the whole system. This can be accomplished by poling the system in a d.c. electric field at a temperature such that the dipoles of the chromophores have enough mobility (Fig. 13.3a). By spin coating the polymer containing NLO chromophores on a conducting substrate such as indium–thin oxide (ITO) coated glass, films with thicknesses of the order of a micron can

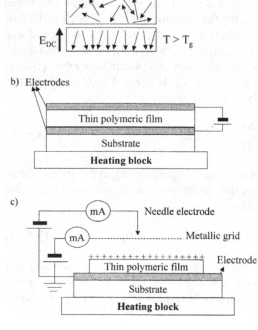

FIG. 13.3   (a) Orientation of dipoles in amorphous polymers before (up) and after the orientation process; (b) schematic poling configuration under a static electric field; (c) a typical poling corona set-up; sometimes a metallic grid is used to control the poling current

be prepared. Then, the system is sandwiched between two electric plates (Fig. 13.3b).[7] In a subsequent step, the polymer is heated at a temperature slightly above its T_g, where it becomes rubbery, and then a strong electric field is applied to align the NLO groups in the matrix, thus destroying the centrosymmetry of the system. In these conditions, chromophore groups of the polymer chains rotate, yielding a poled polymer with the dipoles oriented in the direction of the electric field. Then, under the electric field, the film is cooled below T_g to freeze the molecular motions that would lead to the random reorientation of the chromophores once the field is removed. It is desirable to use the highest possible electric field to reach the highest degree of order. However, charge injection can occur, and pinholes can substantially reduce the magnitude of the field that can be applied.

An alternative way for electric field poling of polymers is the corona procedure.[7] A scheme of a typical corona poling system is shown in Fig. 13.3c. It basically consists of a sharp needle, wire, or grid, which is charged to several kilovolts until breakdown of the surrounding atmosphere occurs. As a consequence of the extremely high electric field close to the corona needle, electrons in the surrounding molecules are accelerated to energies high enough to ionize molecules and atoms in the gas. Either positive or negative ions can be deposited on the surface of the film, depending on the polarity, positive or negative, of the corona needle. In this way, fields exceeding 4 MV/cm can be accomplished across the polymer film. Negative coronas are less stable than positive ones, and, moreover, they are very dependent on the chemical nature of the gas in which discharge occurs.

13.7. RELATION BETWEEN HYPERPOLARIZABILITY AND SECOND-ORDER SUSCEPTIBILITY IN UNIAXIALLY POLED ORIENTED POLYMERS

The relation between the second-order susceptibility and the hyperpolarizability β may become rather complicated even for macroscopic systems presenting uniaxial symmetry. However, this relation can be established if only one component of β, for example β_{zzz}, where the z direction is along the direction of the chromophore permanent dipole moment, needs to be considered. Let us assume that the orientation of chromophore groups is accomplished with an external electric field applied in direction 3 of the space. From Eq. (13.5.7), the relation between $\chi^{(2)}_{333}(-\omega; \omega_1, \omega_2)$ and $\chi^{(2)}_{113}(-\omega; \omega_1, \omega_2)$ with β_{zzz} is given by

$$\chi^{(2)}_{333}(-\omega; \omega_1, \omega_2) = Nf_Z f_{Z_1} f_{Z_2} \int \beta_{zzz} \cos^3(z, 3) G(\Omega) \, d\Omega$$

$$= N\beta^0_{zzz} \langle \cos^3 \theta \rangle \qquad (13.7.1)$$

where $G(\Omega)$ is the normalized distribution function for the chromophores over the solid angle Ω and β_{zzz}^0 incorporates the factor field in the hyperpolarizability. Note that, for an isotropic distribution, $G(\Omega) = 1/8\pi^2$. In the same way

$$\chi_{113}^2(-\omega; \omega_1, \omega_2) = Nf_Z f_{Z_1} f_{Z_2} \int \beta_{zzz} \cos^2(z, 1) \cos(z, 3) G(\Omega) \, d\Omega$$

$$= Nf_Z f_{Z_1} f_{Z_2} \int \beta_{zzz} [1 - \cos^2(z, 3)] \cos(z, 3) G(\Omega) \, d\Omega$$

$$= N\beta_{zzz}^0 \tfrac{1}{2} [\langle \cos \theta \rangle - \langle \cos^3 \theta \rangle] \tag{13.7.2}$$

where $\cos(z, 3) = \cos \theta$.

The evaluation of $\langle \cos^n \theta \rangle$ requires knowledge of $G(\Omega)$ which is difficult to calculate unless both microscopic and macroscopic systems have axial symmetry. In these important cases one can write

$$\langle \cos^n \theta \rangle = \int_0^\pi \cos^n \theta \, G(\theta) \sin \theta \, d\theta \tag{13.7.3}$$

where $G(\theta)$ is related to the Helmholtz free energy by

$$G(\theta) = \frac{\exp[-A(\theta)/kT]}{\int_0^\pi \exp[-A(\theta)/kT] \sin \theta \, d\theta} \tag{13.7.4}$$

where $A = U - TS$. Integration over the other two Euler angles has been cancelled out in the numerator and denominator. Moreover, the contribution of the entropy S of the system is usually neglected. On the other hand, $U(\theta) = U_c - \mathbf{\mu} \cdot \mathbf{F} - (1/2)\mathbf{F}\alpha\mathbf{F}$, where $U_c(\theta)$ is the internal energy which for an isotropic system is independent of the direction, α is the polarizability tensor, and $\mathbf{F}$ is the internal field. For these systems, $U(\theta) \cong -\mathbf{\mu} \cdot \mathbf{F} = -\mu F \cos \theta$, and $\langle \cos^n \theta \rangle$ can be written as[17]

$$L_n(\mu F/kT) = \langle \cos^n \theta \rangle = \frac{\int_0^\pi \exp[\mu F \cos \theta/kT] \cos^n \theta \sin \theta \, d\theta}{\int_0^\pi \exp[\mu F \cos \theta/kT] \sin \theta \, d\theta} \tag{13.7.5}$$

where $L_n(\mu F/kT) = L_n(u)$ are the set of Langevin functions related to the averaged values of the Legendre polynomials. The Langevin functions for $n = 1$, 2, and 3 can be expressed in terms of the values of the thermal average Legendre polynomials by

$$L_1(u) = \coth u - \frac{1}{u} = \langle P_1(\cos \theta) \rangle$$

$$L_2(u) = 1 + \frac{2}{u^2} - \frac{2}{u} \coth u = \frac{1}{3} [2 \langle P_2(\cos \theta) \rangle + 1]$$

$$L_3(u) = \left(1 + \frac{6}{u^2}\right) \coth u - \frac{3}{u} \left(1 + \frac{2}{u^2}\right) = \frac{1}{5} [\langle 2P_3(\cos \theta) \rangle + 3 \langle P_1(\cos \theta) \rangle]$$

$$(13.7.6)$$

where $u = \mu F / kT$. Hence, Eqs (13.7.1) and (13.7.2) can be written as

$$\chi^{(2)}_{333}(-\omega; \omega_1, \omega_2) = N\beta^0_{zzz} \langle \cos^3 \theta \rangle$$

$$= N\beta^0_{zzz} \left[\frac{3 \langle P_1(\cos \theta) \rangle + 2 \langle P_3(\cos \theta) \rangle}{5} \right] \qquad (13.7.7)$$

and

$$\chi^2_{113}(-\omega; \omega_1, \omega_2) = N\beta^0_{zzz} \frac{1}{2} [\langle \cos \theta \rangle - \langle \cos^3 \theta \rangle]$$

$$= N\beta^0_{zzz} \frac{1}{5} [\langle P_1(\cos) \rangle - \langle P_3(\cos \theta) \rangle] \qquad (13.7.8)$$

According to Eqs (13.7.2) and (13.7.6), $\chi^{(2)}_{113}$ can be written in terms of the Langevin equations as

$$\chi^{(2)}_{113}(-\omega; \omega_1, \omega_2) = N\beta^0_{zzz} \left[\frac{L_1(u) - L_3(u)}{2} \right] \qquad (13.7.9)$$

On the other hand, Eqs (13.7.1) and (13.7.6) lead to

$$\chi^{(2)}_{113}(-\omega; \omega_1, \omega_2) = N\beta^0_{zzz} L_3(u) \qquad (13.7.10)$$

For values of $u < 1$ and even for $u = 1$, the following expressions hold[15-17]

$$L_1(u) \approx u/3$$
$$L_3(u) \approx u/5$$

$$(13.7.11)$$

Hence

$$\frac{\chi_{113}}{\chi_{333}} \approx \frac{1}{3} \qquad (13.7.12)$$

In many cases this approximation is used in the analysis of experimental results.[11]

For modest degrees of poling, Eqs (13.7.10) and (13.7.11b) lead to

$$\chi_{113}^{(2)}(-\omega; \omega_1, \omega_2) \approx N\beta_{zzz}^0(-\omega; \omega_1, \omega_2)\frac{\mu F}{5kT} \tag{13.7.13}$$

This expression suggests that the second-order susceptibility is a linear function of the concentration of chromophore groups. This linear relationship has been observed to hold in a variety of polymeric systems.[18-20]

13.8. POLING DECAY

The order parameter, Φ, may be obtained from measurements of the index of refraction parallel, $n_\parallel = n_0 + \Delta n_3(\omega)$, and perpendicular, $n_\perp = n_0 + \Delta n_1(\omega)$, to the poling field where n_0 is the isotropic index of refraction before poling. The pertinent expression is[7]

$$\Phi = \frac{\Delta n_3(\omega) - \Delta n_1(\omega)}{\Delta n_3(\omega) + 2\Delta n_1(\omega)} = \frac{n_\parallel - n_\perp}{n_\parallel + 2n_\perp - 3n_0} \tag{13.8.1}$$

Above the glass transition temperature, micro-Brownian molecular motions promote the relaxation of the oriented chromophore groups to the disordered state. From Chapter 2 it follows that $\Delta n_3(\omega)$, $\Delta n_2(\omega)$, and $\Delta n_1(\omega)$ are given by

$$\Delta n_3(\omega) = \frac{2\pi N}{n}(\alpha_\parallel - \alpha_\perp)\left(\langle \cos^2 \theta \rangle - \frac{1}{3}\right)$$

$$\Delta n_1(\omega) = \Delta n_2(\omega) = -\frac{\Delta n_3(\omega)}{2}$$

$$\tag{13.8.2}$$

Note that $1/3$ is the value of $\langle \cos^2 \theta \rangle$ in a totally disordered system.

The time dependence of $\Delta n_3(\omega)$ and $\chi_{333}(\omega)$ can be obtained from the Schmolukowski equation for rotational diffusion. Taking into account that only the relaxation of the angle formed by the direction of the dipole moment of the chromophore and the poling direction must be considered, the Smoluchowski equation can be written as (see Section 2.9)

$$\frac{1}{D_r}\frac{\partial f(t,\theta)}{\partial t} = \nabla^2 f(t,\theta) \tag{13.8.3}$$

If the left-hand side of this equation is made equal to zero, the solutions of $\nabla^2 f(t,\theta) = 0$ are the Legendre polynomials. Owing to the fact that these functions form a complete set, the time-dependent distribution function can be expanded as

$$f(t, \theta) = \sum_{n=0} \frac{2n+1}{2} C_n a_n(t) P_n(\cos \theta) \qquad (13.8.4)$$

where the set of coefficients C_n describe the initial orientation at $t = 0$, and therefore they are given by[11]

$$C_n = \langle P_n(\cos \theta) \rangle_0 \qquad (13.8.5)$$

By substituting Eq. (13.8.4) into Eq. (13.8.3), and using the well-known relationship

$$\nabla^2 P_n(\cos \theta) = -n(n+1) P_n(\cos \theta) \qquad (13.8.6)$$

it follows that

$$\frac{\partial a_n(t)}{\partial t} = -a_n D_r n(n+1) \qquad (13.8.7)$$

This expression leads to

$$a_n(t) = \exp[-n(n+1) D_r t] \qquad (13.8.8)$$

From Eqs (13.8.4) and (13.8.8) we obtain

$$\langle P_n(\cos \theta) \rangle_t = C_n a_n(t) = C_n \exp(-D_n t) \qquad (13.8.9)$$

where $D_n = n(n+1) D_r$.

Equations (13.8.2a), (13.7.6b), and (13.8.9) predict the decay of the birefringence with time as

$$\Delta n_3(\omega|t) = \frac{2\pi N}{n} (\alpha_\parallel - \alpha_\perp) \left(\langle P_2(\cos \theta) \rangle - \frac{1}{3} \right)$$

$$= \frac{2\pi N}{n} (\alpha_\parallel - \alpha_\perp) \left(\frac{2}{3} \right) C_2 \exp(-D_2 t) \qquad (13.8.10)$$

Equations (13.7.7) and (13.8.9) give the following expression for the decay of the second-order nonlinearity

$$\chi_{333}^{(2)}(-\omega; \omega_1, \omega_2|t) = N\beta_{zzz}^0 \left[\frac{\langle P_1(\cos \theta) \rangle_t - \langle P_3(\cos \theta) \rangle_t}{5} \right]$$

$$= N\beta_{zzz}^0 (1/5)[3C_1 \exp(-D_1 t) + 2C_3 \exp(-D_3 t)] \qquad (13.8.11)$$

Comparison of the averages of the Lagrange polynomials in terms of the Langevin functions it is found that Δn_3 decays faster than $\chi_{333}^{(2)}$.

Rather than a single decay or relaxation time, dipoles exhibit a continuous range of relaxation times in such a way that the time dependence of the order function at temperatures not far above T_g is described by the KWW equation[21,22]

$$\Phi(t) = \exp\left[-\left(\frac{t}{\tau^*}\right)^\beta\right]$$ (13.8.12)

where $0 < \beta < 1$ and τ^* is a magnitude close to the mean relaxation time that shows a Vogel–Fulcher–Tamman–Hesse dependence in the vicinity of the glass transition temperature, T_g, that is

$$\tau^*(T) = A' \exp\left[-\frac{B'}{T - T_\infty}\right]$$ (13.8.13)

It should be stressed that the time dependence of the mean relaxation time of the relaxation process at temperatures below T_g follows Arrhenius behavior. Moreover, A' and B' are positive constants that seem to be independent of the nature of both the polymer and the chromophore group either attached to the polymer or embedded in the polymer matrix.

The evolution of the SHG decay with time has shown to be a suitable tool for studying polymer physics dynamics over very long timescales. For example, relaxation dynamics studies have been reported with chromophore groups located in the backbone and in side chains of polymers.[23–28]

13.9. DETERMINATION OF $\chi_{ijk}^{(2)}$ SUSCEPTIBILITIES

A typical experimental set-up to evaluate $\chi_{ijk}^{(2)}$ is shown in Fig. 13.4. For this purpose, the Maker fringe method is commonly used. According to Eq. (13.3.10), the intensity of the light beam associated with second harmonic generation is proportional to $\sin^2(\pi L/2L_c)$. For poled polymers with $C_{\infty v}$ symmetry it can be shown that[28–31]

$$I_{2Z}^S \propto d_{31}^2 \sin^2 \phi_{2\omega}' I_\omega^2 \sin^2\left[\pi\left(\frac{L}{2L_c}\right)\right]$$ (13.9.1)

and

$$I_{2\omega}^P \propto d_{\text{eff}}^2 I_\omega^2 \sin^2\left[\pi\left(\frac{L}{2L_c}\right)\right]$$ (13.9.2)

where the superscripts P and S refer, respectively, to the SHG intensity of the fundamental beam polarized perpendicular and parallel to the poling direction of

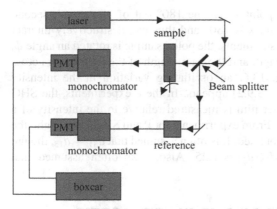

FIG. 13.4   Experimental set-up for the determination of second-order susceptibilities

the poled polymer film (see Fig. 13.5). The effective SHG coefficient is given by

$$d_{\text{eff}} = d_{33} \sin^2 \phi'_\omega \sin \phi'_{2\omega} + d_{31} \cos^2 \phi'_\omega \sin \phi'_{2\omega}$$
$$+ 2d_{15} \sin \phi'_\omega \cos \phi'_\omega \cos \phi'_{2\omega} \qquad (13.9.3)$$

where the angles are defined in Fig. 13.5. The coherence length L_c can be obtained in terms of the refraction index n_ω at ω and $n_{2\omega}$ at 2ω as[7] [see Eq. (13.3.9)]

$$L_c = \frac{\lambda}{4(n_\omega \cos \phi'_\omega - n_{2\omega} \cos \phi'_{2\omega})} \qquad (13.9.4)$$

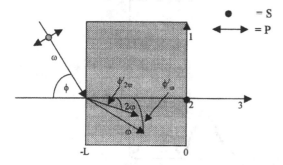

FIG. 13.5   Maker fringe method for determining second-order susceptibilities in a poled polymer. Incident coherent light at frequency ω at an angle ϕ to a poled film of thickness L poled along axis 3. The angles of the fundamental- and second-harmonic beams inside the polymers are ϕ_ω and $\phi''_{2\omega}$[7a]

The second harmonic light at point x will be 180° out of phase with second harmonic light generated at point x + 180° and thus can destructively interact with it. In the Maker fringe measurements, the poled sample is rotated an angle ϕ, thus changing the coherence length and the path length of the light beam across the sample. As shown in Fig. 13.6, an oscillating variation in the intensity described by Eqs (13.9.1) and (13.9.2) appears. In these experiments, the SHG intensity from the poled polymer film is measured relative to the intensity of a reference material, often quartz. From experiments for P and S polarized light, the values of parameters d can be obtained. It is often assumed that $d_{31} = d_{15}$, though some authors have reported $d_{15}/d_{31} = 1.25$. Also, it is often assumed that $d_{31} = \frac{1}{3} d_{33}$.

13.10. MEASUREMENT OF THE ELECTROOPTIC EFFECT

Equation (13.5.13) suggests that measurement of the index of refraction of an NLO material under an electric field is a suitable method for determining electrooptic coefficients. By using a term that is second-order in the applied field, Eq. (13.5.13) adopts the following form

$$\Delta n_{IJ} = -\left(\frac{n^3}{2}\right)(r_{IJK} + s_{IJKL}E_L)E_K \tag{13.10.1}$$

where s_{IJKL} is a fourth-order tensor responsible for the quadratic electrooptic or Kerr effect.[32] Note that a contraction notation has been used for subscripts of Eq. (13.10.1). The quadratic electrooptic contribution to Δn_{IJ} can arise from electrode

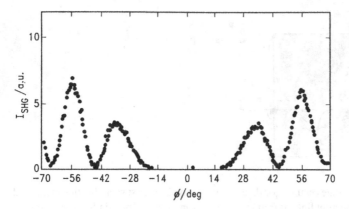

FIG. 13.6 Typical Maker fringe curves versus the incident beam angle as defined in Fig. 13.5 [7a]

attraction, electronic and orientational effects, motions of trapped charges themselves, electrostriction, and heating.[33] However, this contribution is in most cases negligible unless the d.c. voltage is quite large. It is worth noting that internal charges can set up an internal field opposing the applied electric field, thus affecting the measured values of r_{IJk}.

The determination of electrooptic coefficients is often carried out using the Mach–Zehnder interferometric technique schematically shown in Fig. 13.7.[34,35] A coherent light beam splits into two paths of intensities I_s and I_r traveling, respectively, through the sample and a wedge-shaped reference material. The signal of the two beams are recombined at the detector, and its intensity $S(\Phi)$ is given by

$$S(\Phi) = \left(\tfrac{1}{2}\right)(I_r + I_s) + (I_r I_s)^{1/2} \cos(\Phi - \phi) \qquad (13.10.2)$$

where Φ and ϕ are respectively the reference controlled phase and the phase in the sample arm. This latter quantity is given by

$$\phi = 2\pi(n - 1)\frac{t}{\lambda} \qquad (13.10.3)$$

where t is the sample thickness and λ is the laser wavelength. If a modulated voltage is applied to a sample, a change in the refraction index will take place, given by Eq. (13.10.1). Note that the modulation frequencies of the voltage must be small compared with optical frequencies. Changes in sample thickness, Δt, may occur, arising from piezoelectric changes and affecting the phase ϕ. The value of Δt can be written as

$$\Delta t = t d^P E \qquad (13.10.4)$$

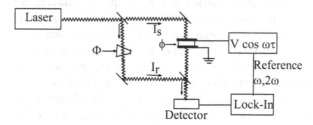

FIG. 13.7 Scheme of Mach–Zehnder interferometer for measuring electrooptic coefficients, r. Continuous and wrinkled lines represent, respectively, electric connections and the light path

where d^P is a piezoelectric parameter that controls the piezoelectric thickness. The value of S under a modulated voltage at frequency ω is given by

$$S_\omega(\Phi) = \frac{2\pi}{\lambda} V(I_r I_s)^{1/2} [\tfrac{1}{2} n^3 r_{13} + (n-1)d^P] \sin(\Phi - \phi) \qquad (13.10.5)$$

where V is the amplitude of the modulated voltage. By measuring the modulated signal as a function of the reference phase Φ, and further comparing it with the signal for $V = 0$, the value of the quantity in brackets in Eq. (13.10.5) can be obtained. If the laser plane is parallel to the poling direction, then only r_{13} can be measured assuming that the piezoelectric contribution is negligible.

13.11. GUIDELINES FOR DESIGNING NLO POLYMER SYSTEMS

The large second-order NLO properties of certain polymers combined with their relatively high laser damage threshold and moderate processing conditions convert these materials into promising candidates for nonlinear optical applications such as frequency doublers, optical storage devices, electrooptic switches, and modulators.[4-8,36] The importance of NLO polymers arises from having good film-forming properties to develop waveguide structures, which are compatible with existing semiconductor technologies.

Frequency doublers based on NLO materials can be used for tight packing of information in optical disks, increasing their storage capacity. A typical frequency-doubling waveguide configuration is shown in Fig. 13.8. The poled polymer can reach the quasi-phase-matched condition ($n_\omega = n_{2\omega}$) by periodically poling in the positive x direction in some regions and in the negative x direction in others. These poled regions are shown in the figure. It has been shown that the value of $\chi^{(2)}$ required in a periodically poled device to accomplish 10% doubling efficiency is of the order of 60 pm/V.

The ability of NLO materials to change their index of refraction under the action of an electric field forms the basis for electrooptic modulators, which can transform a single beam of light into a series of discrete pulses, suitable for transmitting information. A typical modulator, analogous to the Mach–Zehnder interferometer shown in Fig. 13.7 is represented in Fig. 13.9. Note that the beam light enters a parallel channel carved in the polymeric NLO material, and an electric field can be applied in one of the two channels. In the absence of an electric field, the NLO material in both channels has the same refractive index, and the peaks and troughs of the waves traveling through the channels are completely synchronized. The two waves recombine in a single channel at the end of the device and the original beam is restored. However, if an electric field is applied across one of the channels, an electric wave traveling alongside the photons is produced. The traveling electric waves may boost the index of

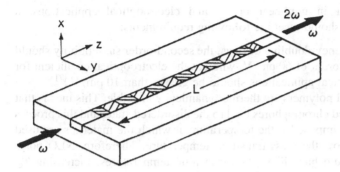

FIG. 13.8 Scheme of frequency-doubling wave configuration. The NLO material forms a channel waveguide. The material has been periodically poled in the $+x$ and $-x$ directions[7a]

refraction of the NLO material, thus slowing the light in such a way that the peaks now line up with the troughs of the unaffected beams. Thus a pi-shift is produced. When these out-of-phase waves meet at the single channel, they cancel each other and the pulse is stopped. Then, by varying the electric field, a single beam of laser light can be chopped into separate pulses.

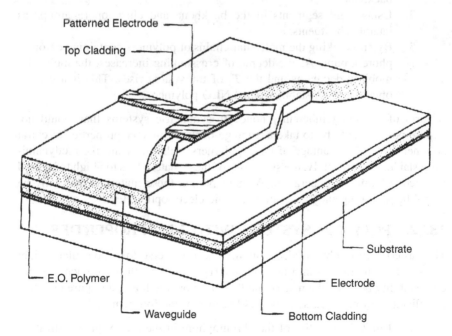

FIG. 13.9 Scheme of a Mach–Zehnder modulator[7a]

To be useful in nonlinear optics and electrooptical applications, a polymeric material should meet the following requirements:

1. For frequency-doubling devices, the second-order susceptiblity should be of the order of 60 pm/V, whereas the electrooptical coefficient for electrooptical applications should be greater than 10 pm/V.[7]
2. The poled polymers are thermodynamically unstable. This means that the aligned chromophores tend to the disordered state, and this process is mostly impeded if the temperature at which the materials are used is far below the glass transition temperature. Therefore, NLO polymers should have high glass transition temperatures. Increasing T_g, increases the poled stability.[37]
3. During device processing, the temperatures of NLO polymers of high T_g can be up to 300°C. Moreover, these materials may endure temperatures of the order of 100°C during operation. Therefore, NLO polymers should exhibit good mechanical and thermal properties.[38]

NLO polymers of high thermal and mechanical stability can be obtained in different ways. For example:

1. Using chromopore groups combined with rigid segments in the backbone.
2. Using rigid segments in the backbone and chromopore groups as lateral substituents.
3. By crosslinking the molecular chains of polymers containing chromophore groups. As the degree of crosslinking increases, the molecular mobility decreases and the T_g of the system rises. This is a way of obtaining high-stability poled NLO polymers.

A form of obtaining inherently noncentrosymmetric systems that would not require poling would be to take advantage of the symmetry properties of chiral polymers.[39] The advantage of these polymers is that they are thermodynamically stable. Unfortunately, the symmetry of these materials is too high to be used for second harmonic generation. Without poling, these materials may only be useful for sum frequency generation and the electrooptic effect.

13.12. POLYMER SYSTEMS WITH NLO PROPERTIES

The easiest way to obtain NLO polymers is to disperse low molecular weight chromophore molecules in a polymer matrix. However, this approach requires compatibility between the matrix and the chromophore hosts, a requirement that is difficult to meet in glassy polymers because of the following:[38]

1. Limited solubility of the chromophore in the glassy matrix (in most cases not higher than 10% before phase separation), which in turn limits the NLO response.

2. The chromophore can act as a plasticizer, thus lowering the glass transition temperature of the polymer and therefore increasing the instability of the poled chromophores.
3. Finally, the chromophore can sublime out of the matrix.

As an example, guest–host systems have been prepared by dissolving 2,4,5-triarylimidazoles in the isotropic polyetherimide Ultem.[40] The chromophores plasticize the Ultem, lowering its T_g from 200 to 170°C at a 20% loading charge. The value of d_{33} for a poled sample of this concentration was 12 pm/V, measured at 1.06 μm.

Phase separation and sublimation processes observed in NLO chromophores dissolved in glassy polymers can be avoided by linking nonlinear chromophores to the polymer backbone, either as a side group or forming part of the polymer backbone itself. Polymers with chromophore groups in the side chain have been obtained by functionalizing polymers with relatively high glass transition temperatures (about 100°C), such as polystyrene or polymethacrylate. Another approach is to copolymerize styrene or methyl methacrylate with monomers appropriately functionalized.[41–45] Some representative results obtained from copolymers of polymethacrylates functionalized with different nonlinear chromophores, shown in Fig. 13.10, are given in Table 13.2. The azo dye exhibits better SHG performance than the stilbene dye, as the comparison of the results for P6 and P7 show. Moreover, all P6 and P7 copolymers show $d_{33}/d_{31} = 3$. These polymers present good poled order stability since the d_{33} parameters only slightly decrease over a period of 4 months at room temperature.

The chromophores can be incorporated in the polymer backbone itself. The disadvantage of these polymers compared with those with the chromophore moiety in the side chains is that large segmental motions of the polymer backbone are required for poling. The advantage is that the relaxation of the oriented chromophore is more difficult. Moreover, subglass relaxations are largely inhibited. Polyesters have been obtained in which azo dye chromophores in the main chain are separated by polymethylene spacers (see Fig. 13.11).[46] Owing to the flexibility of the spacers, the glass transition temperature is relatively low, about 50°C, and as a result a rapid decay of the SHG effect in the vicinity of room temperature occurs. Even at short times the parameter d_{33} only amounts to about 0.57, presumably as a consequence of the fact that $\mu\beta_{zzz}$ [see Eq. (13.7.13)] is rather low for the chromophore incorporated in the backbone. A representative polymer with a rigid backbone is shown in Fig. 13.12. Thin polymer films of this polyurea poled at 130°C exhibit d_{33} values of the order of 5–8.3 pm/V at 1.06 μm.

Linear and branched polyimides containing dye nonlinear groups are in principle promising candidates for the development of NLO polymers. Examples of branched polyimides are shown in Figs 13.13 and 13.14.[47–50] As a consequence of the rigidity of the backbone, the glass transition temperature of

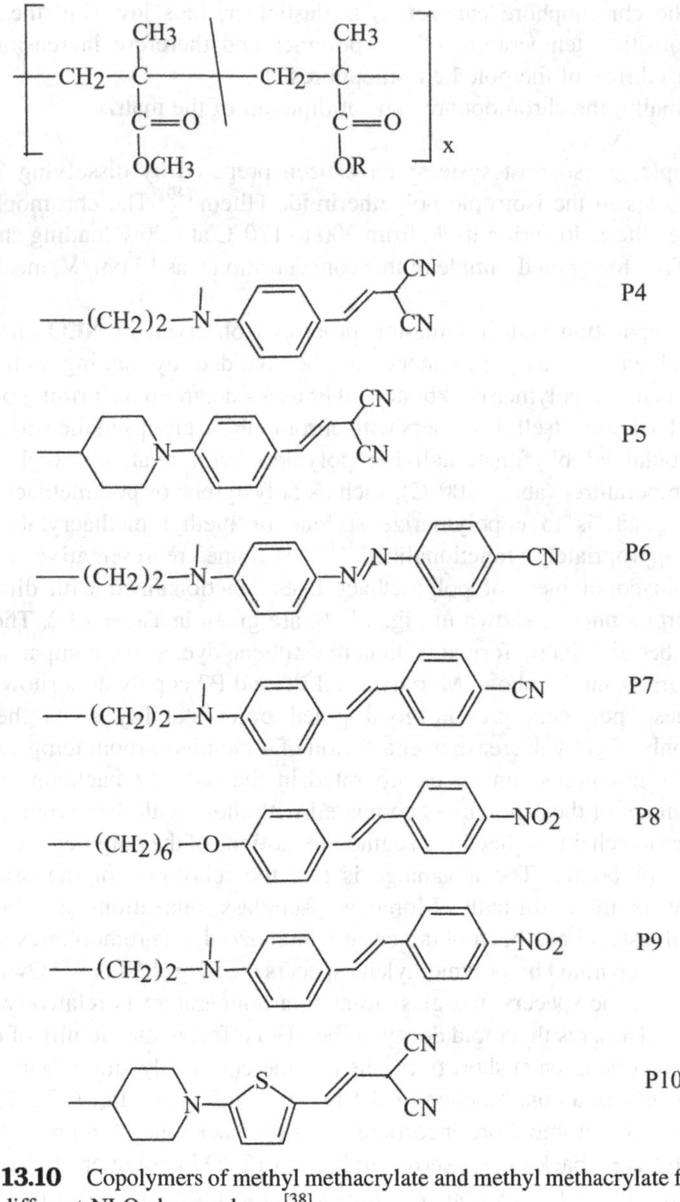

FIG. 13.10 Copolymers of methyl methacrylate and methyl methacrylate functionalized with different NLO chromophores[38]

TABLE 13.2 Glass Transition Temperatures and SHG Results for Polymethacrylate Copolymers in Fig. 13.8[38]

Polymer	$\omega_{dye\ monomer}$	T_g, °C	d_{33}, pm/V
P4-10	0.088	119	12.1
P4-20	0.18	118	21.8
P4-35	0.31	109	26.4
P4-100	1.00	120	65.8
P5-20	0.17	135	45.7
P5-40	0.41	142	20.1
P5-60	0.61	131	9.0
P6-10	0.119	129	5.2
P6-20	0.164	128	21
P6-35	0.315	134	68
P6-40	0.440	128	45
P6-60	0.490	125	32
P6-80	0.717	124	31
P6-100	1.000	181	26
P7-10	0.103	107	8.8
P7-20	0.164	107	11.5
P7-30	0.225	75	12
P7-50	0.430	75	12
P8-30	0.295	98	23.6
P9-20	0.190	120	31.4
P10-20	0.148	135	41.5

the polymers is rather high, 230, 250, and 240°C for the polymers Pa (X = NO₂) and Pb (X = SO₂CH₃), both of Fig. 13.13 and Pc (X = SO₂CH₃) of Fig. 13.14. The values of d_{33} for the fundamental wavelength of 1.06 μm were 27 pm/V for Pa, 10 pm/V for Pb, 18 pm/V for Pc, and 16 pm/V for Pd. As a consequence of the high glass transition temperature of these polymers, their SHG performance was rather good even at relatively high temperatures. Thus, the polyimide Pa did not show any decay at room temperature or even 90°C. The polyimide exhibited

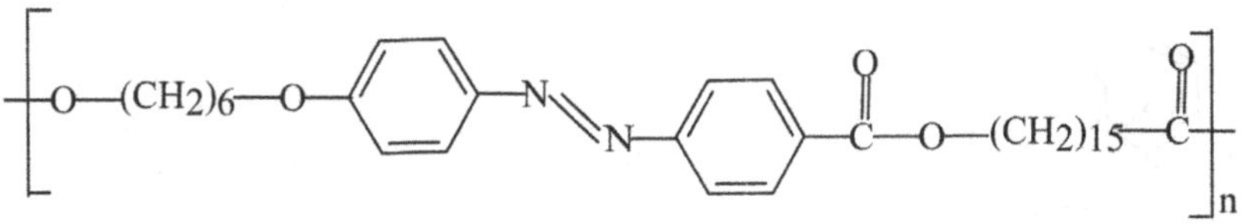

FIG. 13.11 Polyester with NLO chromophores in the backbone, separated by flexible spacers

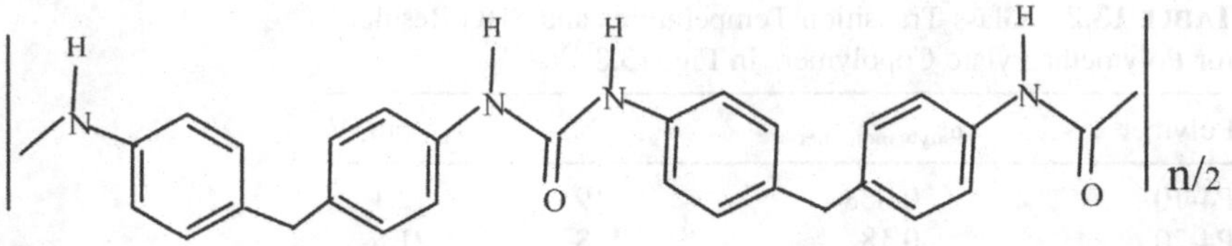

FIG. **13.12**   Scheme of the repeating unit of a polyurea

an initial decay of 15% at 150°C, and then the signal stabilized at this level. Similar stability was observed for Pb. The polyimide Pc showed 85% of the initial nonlinearity at 150°C, and the poled polymer nearly retained 60% after being kept at 170°C for almost 150 h.

Reducing chain mobility via the formation of chemical crosslinks can also enhance the glass transition temperature. Polymers with a high density of crosslinked points are called thermosets. The system must be poled before the curing process is completed. Epoxy resins containing chromophore groups in their structures are examples of NLO thermoset polymers. In most cases, bis-epoxide compounds react with NLO diamines, as indicated in Fig. 13.15.[51] Melting the reactives and heating the mixture at 120°C produces soluble prepolymers with T_g of about 5°C. The prepolymers were coated with diglyme and cured at 120°C under a corona for 16 h. The values of d_{33} and d_{31} obtained from Maker fringe studies were 50 and 16 pm/V respectively at 1.06 μm. The relatively high values of d reflect the high concentration of chromophore nonlinear groups in the films. No decay in the second harmonic was detected at room temperature, and at 80°C, after a small initial loss (~10–15%), no further decay in the values of d was observed.

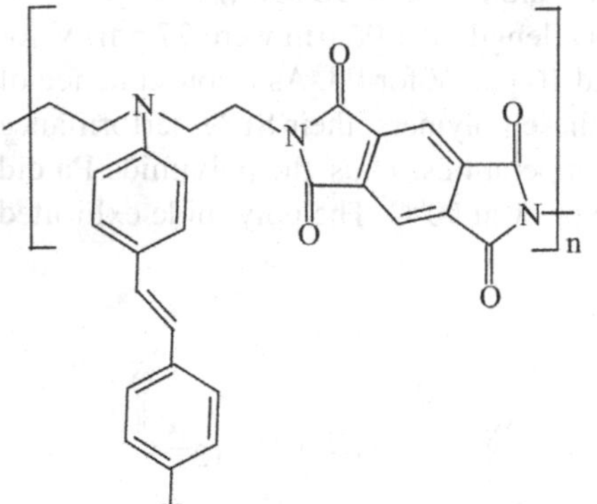

FIG. **13.13**   Polyimide with chromophore groups in the side chains. Polymers have been synthesized with either X = NO$_2$ (Pa) or SO$_2$CH$_3$ (Pb)

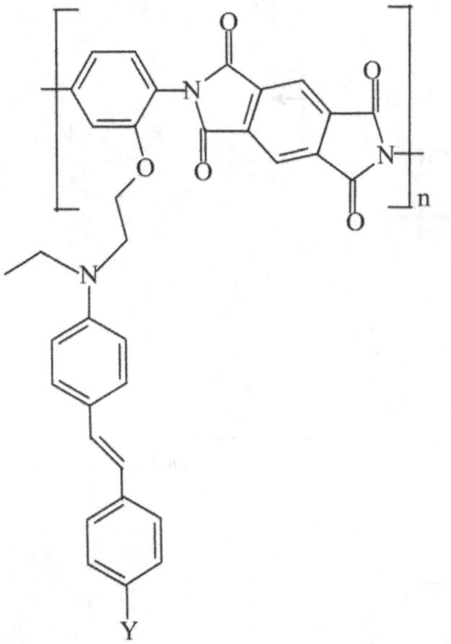

FIG. 13.14 Polyimide with chromophore groups separated from the main chain by flexible segments. Polymers have been studied with $= SO_2CH_3$

Crosslinking agents such as isocyanates can also be used to prepare thermosets containing NLO groups. An example is depicted in Fig. 13.16. Here, glycol containing a highly nonlinear chromophore reacts with 4,4′-diisocyanato-3,3′-dimethoxy biphenyl in stoichiometric conditions such that isocyanate terminated chains are obtained.[52,53] Then, a precise amount of trifunctional crosslinking agent is added to the purified polymer solutions and polymer films are readily prepared by spin coating. The films are poled in a corona discharge

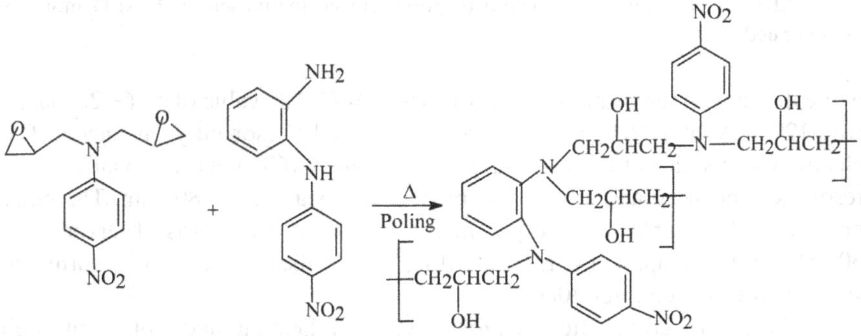

FIG. 13.15 Scheme of the formation of a poled NLO incorporated epoxy

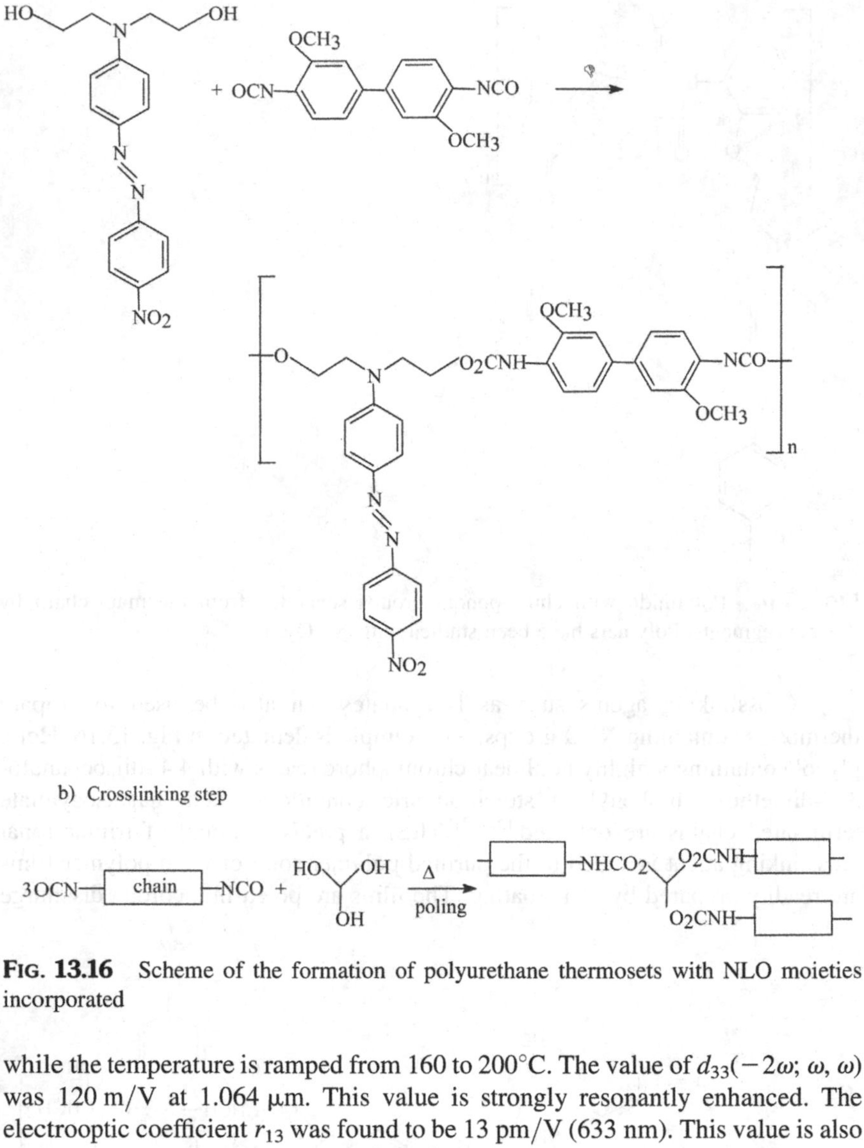

FIG. 13.16 Scheme of the formation of polyurethane thermosets with NLO moieties incorporated

while the temperature is ramped from 160 to 200°C. The value of $d_{33}(-2\omega; \omega, \omega)$ was 120 m/V at 1.064 μm. This value is strongly resonantly enhanced. The electrooptic coefficient r_{13} was found to be 13 pm/V (633 nm). This value is also resonance enhanced and drops to 5 pm/V when measured at 800 nm. The polar order is highly stable in these polymers, so that no decay was observed over 3000 h at room temperature. Even at 90°C, 70% of the original second harmonic signal was observed after 3000 h.

Chemical curing often requires extended heating and poling at high temperatures, a procedure that may lead to degradation of the NLO chromophore.

Recently it has been shown that, in certain cases, photochemical crosslinking procedures during poling can be performed at relatively low temperatures.[54,55]

Dendrimers are a relatively new class of macromolecules known for having interesting properties arising from their nanoscopic dimensions and their regular, well-defined and highly branched three-dimensional architecture. Their treelike monodispersed structure confers upon these substances peculiar characteristics such as globular, void-containing shapes and unusual physical properties.[56-59] Globular shaped and void-containing dendrimers may exhibit large electrooptic coefficients. As shown in Fig. 13.17, multiple NLO chromophore building blocks can be placed into a dendrimer to build a precise molecular architecture with a predetermined chemical composition. As usual, the sequential hardening/crosslinking reactions should take place under the high-temperature electric field poling process. Very large electrooptical coefficients ($r_{33} = 60$ pm/V at 1.55 μm) have been reported for poled dendrimers with the star configuration shown in Fig. 13.18. These dendrimers may retain more than 90% of their original r_{33} value at 85°C for more than 1000 h.[60]

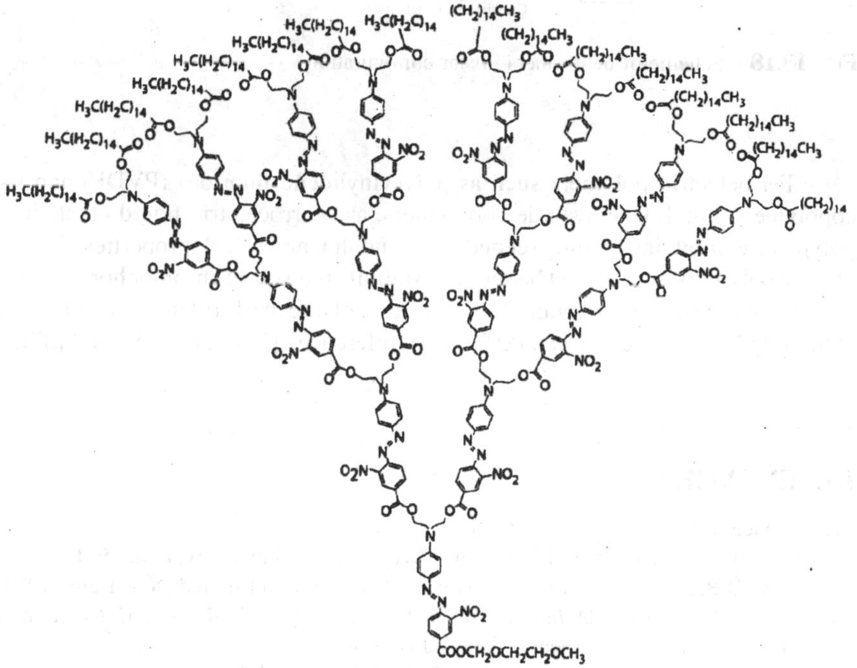

FIG. 13.17   Scheme of a dendrimer molecular architecture[36]

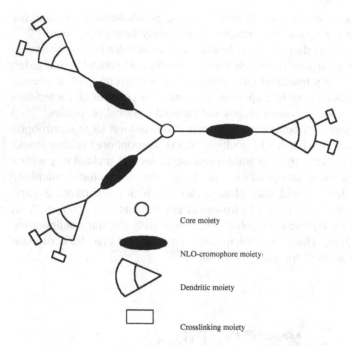

FIG. 13.18   Scheme of dendrimers in star configuration

Ferroelectric polymers such as poly(vinylidene fluoride) (PVDF) and its copolymers, vinylidene cyanide copolymers, and ferroelectric liquid crystalline polymers exhibit interesting second-order nonlinear optical properties. Ferroelectric polymers may exhibit better SHG stability than NLO chromophore grafted conventional polymers. Detailed descriptions of the performance of polymers exhibiting NLO properties can be found in references [5] to [7], [35], and [37].

REFERENCES

1.   Service, R.F. Science **1995**, *267*, 1918.
2.   Franken, P.A.; Hill, A.E.; Peters, C.W.; Weinreich, G. Phys. Rev. Lett. **1961**, *7*, 118.
3.   Laud, B.B. *Lasers and Non-linear Optics*; Wiley Eastern Limited: New Delhi, 1985.
4.   Prasad, P.; Williams, D. *Introduction to Nonlinear Optical Effects in Molecules and Polymers*; Wiley-Interscience: New York, 1992.
5.   Marder, R.S.; Sohn, J.E.; Stucky, G.D., Eds. *Materials for Nonlinear Optics: Chemical Perspectives*; ACS Symposium Series 455; American Chemical Society: Washington, 1991.

6. Wijekoon, W.M.K.P.; Prasad, P.N. Nonlinear optical properties of polymers. In *Physical Properties of Polymers Handbook*; Mark, J.E., Ed.; AIP: Woodbury, NY, 1996; Ch. 38.

7. Burland, D.M.; Miller, R.D.; Welsh, C.A. Chem. Rev. **1994**, *94*, 31.

7a. Eich, M.; Bjorklund, G.; Yoon, D. Poly. Adv. Technol. **1990**, *1*, 189.

8. Watanabe, T.; Miyata, S.; Nalwa, H.S. Nonlinear optical properties of ferroelectric polymers. In *Ferroelectric Polymers*; Nalwa, H.S., Ed.; Marcel Dekker: New York, 1995.

9. Kaminow, I. *An Introduction to Electro-optic Devices*; Academic Press: New York, 1974; 56.

10. Shen, Y.R. *Principles of Nonlinear Optics*; Wiley-Interscience: New York, 1988.

11. Singer, K.; Kuzyk, M.; Sohn, J. J. Opt. Soc. Am. B **1987**, *4*, 968.

12. Kielich, S. IEEE J. Quantum Electron. **1969**, *QE-5*, 562.

13. Dick, B. Chem. Phys. **1985**, *96*, 199.

14. Grange, J.L.; Kuzyk, M.; Singer, K. J. Mol. Struct. **1987**, *150b*, 567.

15. Stähelin, M.; Burland, D.; Rice, J. Chem. Phys. Lett. **1992**, *191*, 245.

16. Böttcher, C. *Theory of Electric Polarization*; Elsevier: Amsterdam, 1973; Vol. I, 161–165.

17. Page, R.; Jurich, M.; Reck, B.; Sen, A.S.; Twieg, R.; Swalen, J.; Bjorklund, G.; Wilson, C. J. Opt. Soc. Am. B **1990**, *7*, 1239.

18. Wijekoon, W.; Zhang, Y.; Karna, S.; Prasad, P.; Giffin, A.; Bhatti, A. J. Opt. Soc. Am. B **1992**, *9*, 1832.

19. Ye, C.; Minami, N.; Marks, T.; Yang, J.; Wong, G. Macromolecules **1998**, *21*, 2899.

20. Rikken, G.; Seppen, C.; Nijhuis, S.; Meijer, E. Appl. Phys. Lett. **1998**, *58*, 435.

21. Jones, P.; Jones, W.; Williams, G. J. Chem. Soc. Far. Trans. **1992**, *86*, 1013.

22. Cicerone, M.; Ediger, M. J. Chem. Phys. **1992**, *97*, 2156.

23. Köhler, W.; Robello, D.R.; Dao, P.T.; Willand, C.S.; Williams, D.J. J. Chem. Phys. **1990**, *93*, 9157.

24. Köhler, W.; Robello, D.R.; Willand, C.S.; Williams, D.J. Macromolecules **1991**, *24*, 4589.

25. Heldmann, C.; Schulze, M.; Wegner, G. Macromolecules **1996**, *29*, 4686.

26. Heldmann, C.; Neher, D.; Winkelhahn, H.J.; Wegner, G. Macromolecules **1996**, *29*, 4697.

27. Firestone, M.A.; Ratner, M.A.; Marks, T.J. Macromolecules **1995**, *28*, 6296.

28. Kaatz, P.; Prêtre, P.; Meier, U.; Stalder, U.; Bosshard, C.; Günter, P.; Zysset, B.; Stähelin, M.; Ahlheim, M. Macromolecules **1996**, *29*, 1666.

29. Maker, P.D.; Terhune, R.W.; Nisenoff, M.; Savage, C.M. Phys. Rev. Lett. **1962**, *8*, 21.

30. Jerphagnon, J.; Kurtz, S.K. J. Appl. Phys. **1970**, *41*, 1667.

31. Kurtz, S. In *Quantum Electronics: A Treatise*; Rabin, H., Tang, C.L., Eds.; Academic Press: New York, 1975; Vol. I, 209–281.

32. Yariv, A. *Introduction to Optical Electronics*; Holt, Rinehardt, Winston: New York, 1978; 361.

33. Kuzyk, M.; Sohn, J.; Dirk, C. J. Opt. Soc. Am. B **1990**, *7*, 842.

34. Singer, K.; Kuzyk, M.; Sohn, J. J. Opt. Soc. Am. B **1987**, *4*, 968.

35. Sigelle, M.; Hierle, R. J. Appl. Phys. **1981**, *53*, 4199.
36. Kajzar, F.; Lex, K.-S.; Jen, A.K.-Y. Adv. Polym. Sci. **2002**, *161*, 1 (an interesting review on NLO polymers with many recent references).
37. Walsh, C.; Burland, D.; Lee, V.; Miller, R.; Smith, B.; Twieg, R.; Volksen, W. Macromolecules **1993**, *26*, 3720.
38. Samyn, C.; Verbiest, T.; Perssons, A. Macromol. Rapid. Commun. **2000**, *21*, 1.
39. Beljonne, D.; Shuai, Z.; Brédas, J.L.; Kauranen, M.; Verbiest, T.; Persoons, A. J. Chem. Phys. **1998**, *108*, 1301.
40. Stähelin, M.; Walsh, C.; Burland, D.; Miller, R.; Twieg, R.; Volksen, W. J. Appl. Phys. **1993**, *73*, 8471.
41. Rondou, P.; Van Beylen, M.; Samyn, C.; S'heeren, G.; Persoons, A. Macromol. Chem. **1992**, *193*, 3045.
42. S'heeren, G.; Persoons, A.; Rondou, P.; Wiersma, J.; Van Beylen, M.; Samyn, C. Macromol. Chem. **1993**, *194*, 1733.
43. S'heeren, G.; Persoons, A.; Rondou, P.; Wiersma, J.; Van Beylen, M.; Samyn, C. Eur. Polym. J. **1993**, *29*, 975.
44. S'heeren, G.; Persoons, A.; Rondou, P.; Wiersma, J.; Van Beylen, M.; Samyn, C. Eur. Polym. J. **1994**, *30*, 7975.
45. Samyn, C.; Claes, G.; Van Beylen, M.; De Wachter, A.; Persoons, A. Macromol. Symp. **1996**, *102*, 145.
46. S'heeren, G.; Persoons, A.; Bolink, H.; Heylen, M.; Van Beylen, M.; Samyn, C. Eur. Polym. J. **1993**, *29*, 981.
47. Peng, Z.; Yu, L. Macromolecules **1994**, *27*, 2638.
48. Yang, S.; Peng, Z.; Yu, L. Macromolecules **1994**, *27*, 5858.
49. Yu, D.; Yu, L. Macromolecules **1994**, *27*, 6718.
50. Yu, D.; Gharavi, A.; Yu, L. Macromolecules **1996**, *27*, 6139.
51. Jungbauer, D.; Reck, B.; Twieg, R.; Yoon, D.; Willson, C.; Swallen, J. Appl. Phys. Lett. **1990**, *56*, 2610.
52. Chen, M.; Dalton, L.R.; Xu, L.P.; Shi, X.Q.; Steier, W.H. Macromolecules **1992**, *25*, 4032.
53. Shi, Y.; Steier, W.H.; Chen, M.; Yu, L.; Dalton, L.R. Appl. Phys. Lett. **1992**, *60*, 2577.
54. Mandal, B.K.; Jeng, J.; Kumar, J.; Tripathy, S.K. Makromol. Chem. Rapid Commun. **1991**, *12*, 607.
55. Zhu, X.; Chen, Y.M.; Li, L.; Jeng, R.J.; Mandal, B.K.; Kumar, J.; Tripathy, S.K. Opt. Commun. **1992**, *88*, 77.
56. Hecht, S.; Frechet, J.M. Angew. Chem. Int. Ed. **2001**, *40*, 74.
57. Vögtl, F.; Gestermann, S.; Hesse, R.; Schwierz, H.; Windisch, B. Prog. Polym. Sci. **2000**, *25*, 987.
58. Bosman, A.W.; Janssen, H.M.; Meijer, E.W. Chem. Rev. **1999**, *99*, 1665.
59. Newkome, G.R.; He, E.; Moorefield, C.N. Chem. Rev. **1999**, *99*, 1689.
60. Ma, H.; Chen, B.Q.; Sassa, T.; Dalton, L.R.; Jen, A.K.-Y. J. Am. Chem. Soc. **2001**, *123*, 986.

14

Conducting Polymers

14.1. INTRODUCTION

According to the classical band theory, a solid for which a certain number of bands of energy are completely filled, the other bands being completely empty, is an electric insulator.[1] The effective number of free electrons is different from zero in a solid containing an incomplete energy band, and in this case the solid has a metallic character. This situation is schematically depicted in Fig. 14.1 where the electron distribution at 0 K is shown. At this temperature, the solid is in the lowest state of energy, but, as the temperature increases, some electrons of the upper filled band are excited into the next empty band and electronic conduction becomes possible. When the energy gap between the upper filled band and the empty one is of the order of several electron volts, the material remains an insulator. For example, for a gap with energy of 4 eV, the fraction of electrons excited across the gap at room temperature is roughly of the order of $e^{-E_g/2k_BT} \approx 10^{-35}$. However, if this gap is small, below 1 eV, the number of thermally excited electrons may become appreciable, and in this case the material is an intrinsic semiconductor. For example, for an energy gap of only 0.25 eV, the fraction of carriers is $e^{-E_g/2k_BT} \approx 10^{-2}$, and observable conduction will occur. If some electrons are thermally excited from the filled band into the conduction band, some states normally near the top of the filled band become unoccupied, and holes in the filled band are formed. At 0 K all intrinsic semiconductors are insulators, but at high temperatures all the insulators may be considered semiconductors. The resistivity of semiconductors at room temperature lies in the interval 10^{-3}–10^9 ohm cm, whereas the values of this parameter for insulators and metals are of the order of 10^{22} ohm cm and 10^{-6} ohm cm respectively.

575

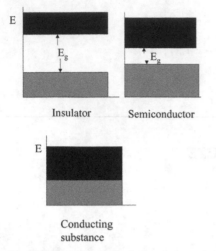

FIG. 14.1   Energy gap in insulators, semiconductors, and conducting materials

The fraction of electrons thermally excited to the conduction band varies exponentially with the reciprocal of the absolute temperature. As a result, the electric conductivity of semiconductors increases with temperature, in contrast to metals in which the conductivity declines with increasing temperature.

14.2.   CHEMICAL STRUCTURE AND CONDUCTING CHARACTER

Although most semiconductors are of inorganic nature, it has long been known that conjugated organic molecules may exhibit semiconductor behavior. These structures can be present in polymers, and, as a consequence of the specific properties of these materials, such as high flexibility, high impact resistance, and low density, which make them specially attractive, efforts have been devoted to the preparation of conducting polymers.

In polydienes, such as *trans*-polyacetylene, each carbon is sp^2 hybridized (Fig. 14.2, upper) and as a consequence this polymer can be treated as a one-dimensional analogue of graphite. However, while in the graphite layers the C–C bond lengths are equivalent, in polyacetylene the backbone bond lengths are alternately slightly longer and slightly shorter (Fig. 14.2, middle). This is due to the so-called Peierls distortion. The bond non-equivalence has important effects on the electronic properties of poly(acetylene) because it opens a gap between the HOMO level of the fully occupied π-band (valence band) and the LUMO level corresponding to the empty π^*-band (conduction band) (see Fig. 14.2, lower). In

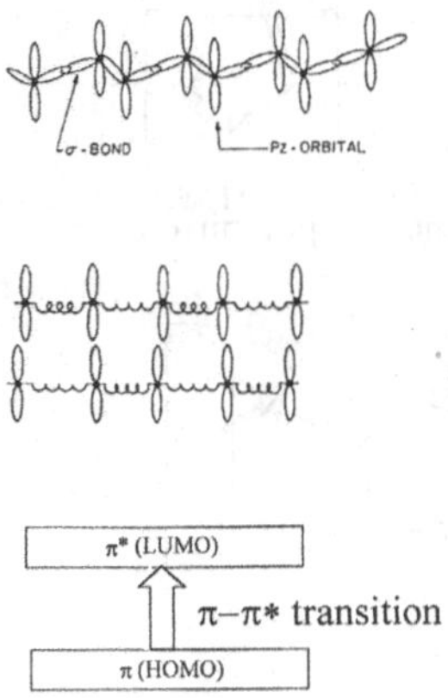

FIG. 14.2   Schematic representation of σ-bonds and p_z orbitals in polydienes (upper); Peierls distortion chain for two ground states of polyacetylene (middle); HOMO–LUMO transitions (lower)

principle, polyacetylene can be considered as a semiconductor with an energy gap of 1.5 eV.

With the explosion of research in the field of organic semiconductors, several polyene-type and aromatic conjugated polymers were synthesized and characterized. The most extensively studied are shown in Fig. 14.3. All the polymers shown in the figure are colored as a result of the strong absorption in the visible range of the spectrum, usually attributed to $\pi \rightarrow \pi^*$ optical transition (Fig. 14.2, lower). Since polymers with a π-electron backbone can be oxidized and reduced more easily and more reversibly than conventional polymers, they are usually transformed from semiconductors to organic metals by doping with p-type (electron-acceptor) or n-type (electron-donor) dopants. In contrast to the conventional ion implantation used in three-dimensional semiconductors, doping processes may be performed in some cases by exposing films of conjugated polymers to vapors of solutions of dopants.

In this chapter we shall survey the routes of synthesis of conducting polymers, their conductivity performance, and applications. The electronic characteristics of these polymers and applications are also discussed.

14.3.   ROUTES OF SYNTHESIS OF CONJUGATED POLYMERS

Polymers are long-chain molecules in which a unique or various chemical motives repeat along the chains. If the chemical structure unit is unique, we have poly(trans-acetylene)   poly(cis-acetylene)   polithiophene   polypirrole

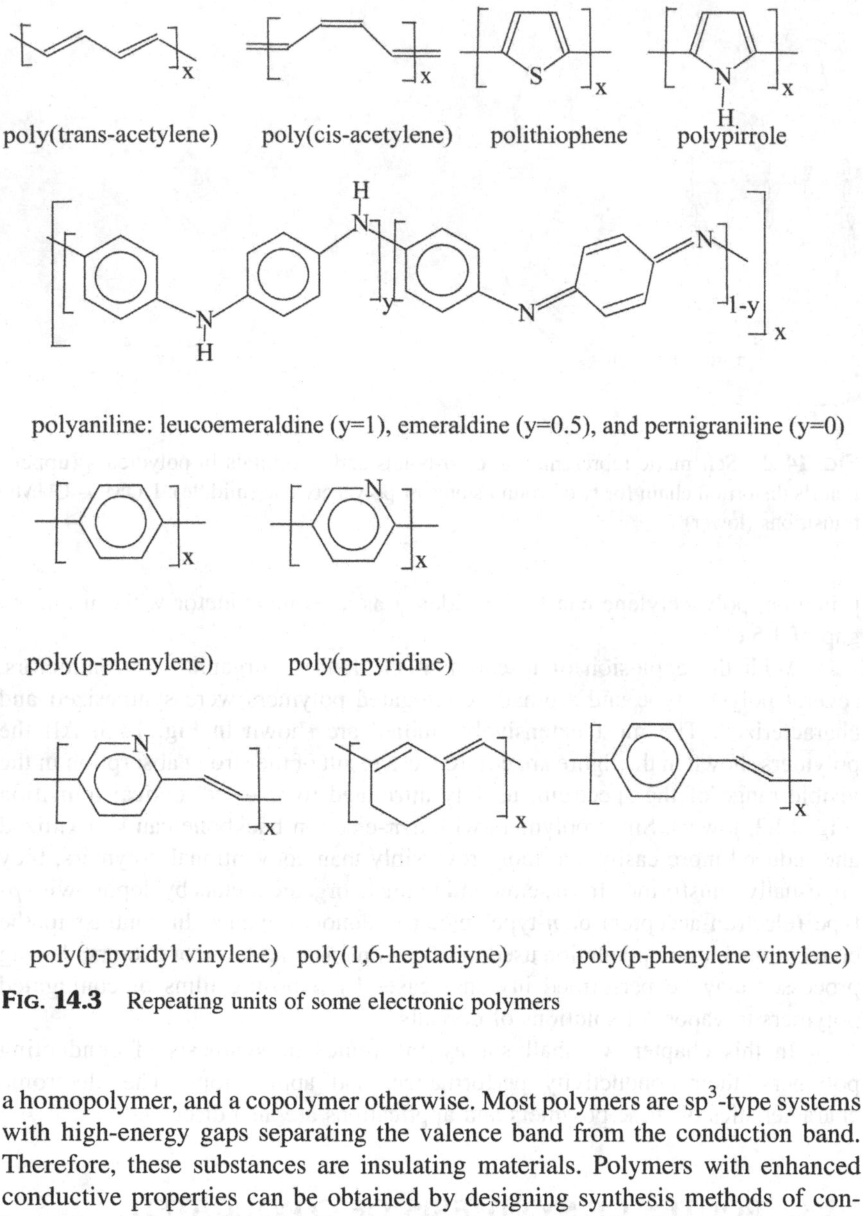

polyaniline: leucoemeraldine (y=1), emeraldine (y=0.5), and pernigraniline (y=0)

poly(p-phenylene)   poly(p-pyridine)

poly(p-pyridyl vinylene) poly(1,6-heptadiyne)   poly(p-phenylene vinylene)

FIG. 14.3   Repeating units of some electronic polymers

a homopolymer, and a copolymer otherwise. Most polymers are sp^3-type systems with high-energy gaps separating the valence band from the conduction band. Therefore, these substances are insulating materials. Polymers with enhanced conductive properties can be obtained by designing synthesis methods of conjugated structures. Chemists have synthesized over 100 conducting polymers with a wide range of specific electrical conductivities.[2] The pristine forms of several families of polymers, including *trans* and *cis* polyacetylene, polypyrrole (PPy), poly(*p*-phenylene) (PPP), polythiophene (PT), polypyridine (Ppyr), poly(*p*-phenylene vinylene) (PPV), poly(*p*-pyridyl vinylene) (PpyV), poly(1,6-heptadiyne), and the leucoemeraldine base (LEB), emeraldine base (EB), and pernigraniline bases of polyaniline (PAN), are shown schematically in Fig. 14.3. Here, the electronic ground states of these polymers are varied.[3] A thorough review on the synthesis of conjugated polymers can be found elsewhere.[4,5]

The synthesis of pristine polymers that, conveniently doped, can exhibit a metallic conducting character can be carried out by chemical methods, involving addition polymerization and a step reaction, and electrochemical polymerization of suitable low molecular weight compounds or monomers. In most cases, addition polymerization involves the repeated addition of an unsaturated monomer to an active center such as a free radical, an ion, or a coordination complex. In this way, long chains are obtained in which branching can appear as a consequence of the formation of radicals in chains already formed that may serve as an initiation center for other growing chains. Stereochemical defects can arise from the addition head-to-head, head-to-tail, and tail-to-tail additions in asymmetric monomers.

The simplest conjugated structure $-CH=CH-CH=CH-$ can be obtained by polymerization addition of gas acetylene exposed to high concentrations of the Ziegler–Natta catalyst $Ti(OBu)_4/AlEt_3$ in a nonstirred reactor.[6–8] The possible structures of poly(acetylene) are shown in Fig. 14.4. It should be pointed out that the *trans–cisoid* configuration has not been detected. The polymerization apparently proceeds at low temperature ($-78°C$) via the repeated *cis*-insertion of acetylene into a Ti–C bond, and the resulting polymer, polyacetylene, is crystalline, with a predominantly *cis* configuration, as shown by ^{13}NMR spectroscopy.[9] The polymer thus obtained is a poor insulator with a conductivity of 10^{-10} S cm^{-1}.

The synthesis conditions and thermal treatment may affect the conductivity of doped polyacetylene. For example, isomerization processes that lead to the transition from the metastable *cis* configuration to the thermodynamically more stable *trans* form can be achieved by postpolymerizing at high temperatures

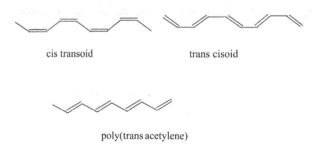

cis transoid   trans cisoid

poly(trans acetylene)

FIG. 14.4 Geometric isomers of polyacetylene chains

the predominantly *cis* polyacetylene.[4] It is believed that the isomerization process involves the formation of residual defects in the chains, such as unpaired electrons, which lead to a significant increase in the conductivity of the polymer in the *trans* form.

Different catalyst systems have been described for the polymerization of acetylene. A common characteristic is that in all cases the polymer obtained exhibits the properties of a poor insulator. However, polyacetylene with rather high conductivities can be obtained by doping. For example, the polymerization of acetylene in a highly viscous and complexing solvent[10,11] such as silicone oil, using catalysts derived from titanium alkoxides or titanium tetrabenzyl, gave a film that was claimed to have far fewer sp^3 defects. This polymer could be oriented to a draw ratio[12] of 6.5. It seems that the solvent controls the morphology of polyacetylene. This material can be doped with iodine to a conductivity of 10^4 S cm^{-1}, one of the highest conductivities reported to date and comparable with the highest metallic range.[13]

The morphology of the films of polyacetylene can be fibrillar or globular depending on the catalytic system used. Since polyacetylene is neither solution nor melt processable, it is important to look for alternative synthetic routes that will not only provide the desired morphology for the polymer but also somewhat enhance its processability. This can be accomplished by using a three-step procedure[14] described in Scheme 14.1. The first step involves the reaction of cyclooctatetraene (COT) with hexafluorobutyne to give the monomer.[4] The polymerization of the monomer with $WCl_6/Sn(C_6H_5)_4$ metathesis catalysts yields the precursor polymer.[5] Casting from acetone or chloroform solutions can produce thin films of polymer. Further heating of the precursor leads to polyacetylene, the *cis/trans* ratio of which depends on the temperature of the thermal transformation. In the presence of air, polyacetylene undergoes a slow oxidation giving rise to the formation of carbonyl groups that break the conjugation of the polymer backbone, thus losing the good semiconductor properties associated with a polyconjugated π-bonding system.[15]

Steric hindrances in mono- and disubstituted polyacetylenes limit the extent of π-overlapping because not only do the substituents separate the chains in the solid, thus reducing intermolecular π-overlapping but also the electron-

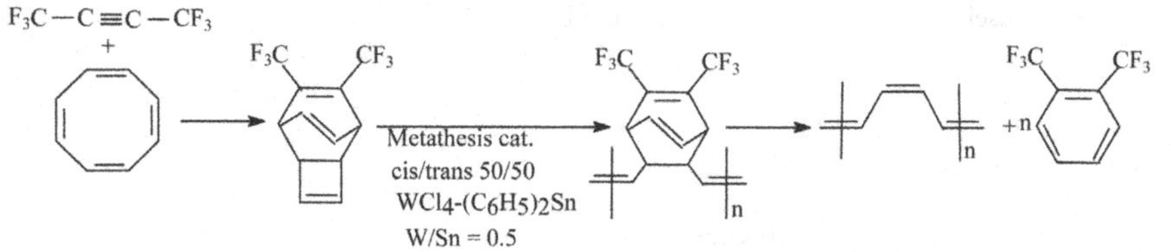

SCHEME 14.1

donating or electron-accepting substituents increase the nonequivalence of the atoms, thus forcing higher bond alternation and thus a higher band gap.

An alternative route to the synthesis of polymers is the step-polymerization that takes place in molecules having two reactive groups with elimination of a small molecule. Step polymerization is an equilibrium process, while addition polymerization is only reversible at temperatures above the ceiling temperature. Polymers of industrial importance such as polyesters and polyamides are obtained by this method.

Step reactions can be used to obtain poly(p-phenylene) from dihaloben-zenes via the decomposition of the corresponding Grignard reagent according to the following scheme[16,17]

$$nBr-C_6H_4-Br + nMg \longrightarrow nBr-C_6H_4-BrMg$$
$$\longrightarrow (-C_6H_4-)_n + nMgBr$$

The reaction is carried out in the presence of catalysts such as Fe(III), Co(II), and Ni(II), and the polymers thus obtained are low-molecular (10–12 rings), insoluble, and infusible powders with an apparently high degree of p-coupling.[18] Poly(p-phenylene) can directly be obtained by oxidation of benzene using an appropriate oxidant to remove the hydrogen liberated in the coupling process. As the oxidant, Cu(II) can be utilized.[19,20] The pertinent reaction is

$$nC_6H_5 + 2nCuCl_2 \longrightarrow (-C_6H_4-)_n + 2nCuCl + 2nHCl$$

Here, AlCl$_3$ with traces of water is used as the catalyst. The polymers obtained by this procedure also have a low molecular weight and probably more structural irregularities than those obtained from dihalobenzenes.

Whereas the infusibility and insolubility of most conducting polymers impede their physical and physicochemical characterization, poly(phenylene sulfide) is soluble in organic solvents such as chlorobenzene and diphenyl ether at high temperatures, and its glass transition and melting temperatures are 85 and 280°C respectively.[21] This polymer, commercially available from Philips Petroleum Company under the trademark Rikton, is obtained by polycondensa-tion of 1,4-dichlorobenzene and sodium sulfide[22] as shown below

$$nCl-C_6H_4-Cl + Na_2S \longrightarrow [-C_6H_4-S-] + 2nNaCl$$

This reaction is carried out at 200–300°C in a polar organic solvent such as N-methylpyrrolidone. It should be pointed out that the true nature of the reaction mechanisms is not well known since long chains are already obtained at very low conversions and unreacted monomers are found at high conversions. Other not so successful routes of synthesis have been devised. For example, low molecular weight poly(phenyl sulfide) can be obtained by the self-condensation of alkali metal salts of 4-halothiophenols[23], according to the following scheme

$$nXC_6H_4SMe \quad \text{at } 250°C \longrightarrow [-C_6H_4-S] + nMX$$

Poly(thiophene), soluble with a hydrogen in position 3 of the 2,5-thienylene ring replaced by an alkyl group longer than the propyl group, is a soluble, conjugated material.[24,25] The substituent does not seem to perturb the conjugated system too much, as indicated by the fact that the optical absorption with a maximum at ca 440 nm corresponding to the $\pi \to \pi^*$ transition only slightly depends on the size of the alkyl substituent. The polymer in its neutral form exhibits solvatochromic[26] and thermochromic effects.[27] As a result of solvatochromism, the absorption ascribed to $\pi \to \pi^*$ transition is red shifted by comparison with the solution one. Owing to the lack of centrosymmetry of the route of synthesis indicated in Scheme 14.2 of alkylthiophenes, three different ways of coupling of thiophene units, head-to-head, head-to-tail, and tail-to-head, may appear along the chains. Poly(alkylthiophenes) with head-to-tail coupling of nearly 99% are obtained by the procedure represented in Scheme 14.3.[28–30]

Alkylthiophene is obtained from 3-bromothiophene by Grignard coupling using Ni(dppp)Cl₂ as a catalyst. The compound is brominated in position 2 by using N-bromosuccinimide (NBS), and then lithium is introduced in position 5 by using lithium diisopropylamide (LDA). After transforming the lithium-containing compound with the corresponding Grignard reagent, the resulting product condenses regiospecifically to regioregular poly(alkylthiophene). Regioregular poly(alkylthiophene) can alternatively be obtained from 2,5-dibromo-3-alkylthiophene as depicted in Scheme 14.4. This reacts with active zinc to give a mixture of isomers that in the presence of Ni(dppp)Cl₂ leads to the regioregular polymer.[31,32]

Poly(p-phenylene vinylene) can be synthesized by several direct routes.[33] An alternative and easy way is to use the precursor route. As shown in Scheme

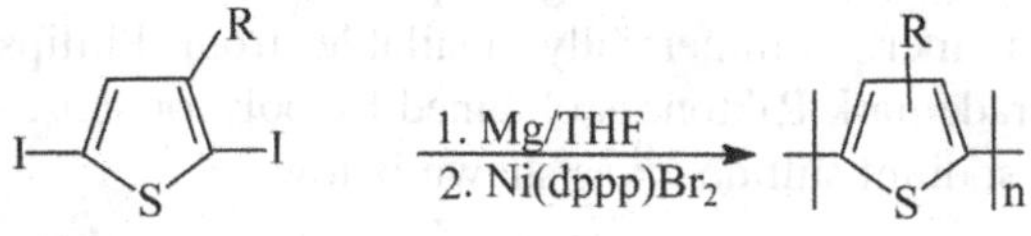

SCHEME 14.2

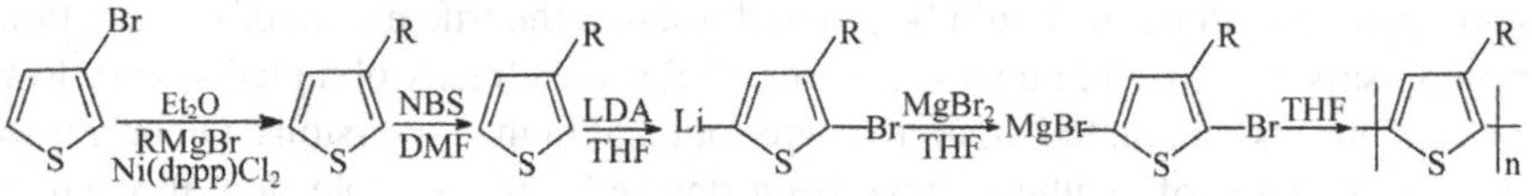

SCHEME 14.3

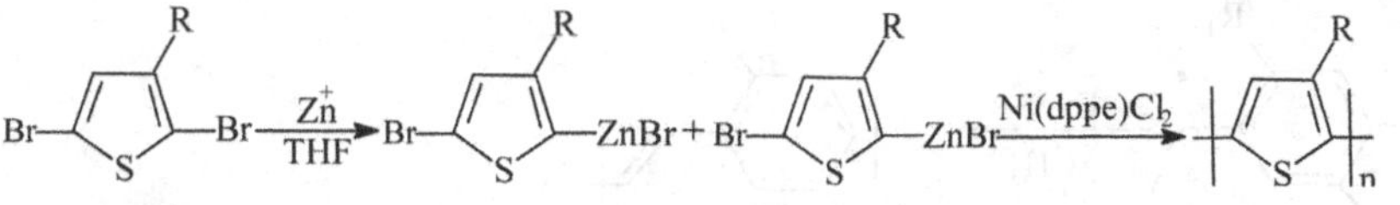

SCHEME 14.4

14.5, *p*-xylylene sulfonium[34] with a base yields a precursor soluble poly-electrolyte from which, with previous purification by dialysis, films can be obtained by casting. Thermal treatment of the films leads to poly(*p*-phenylene vinylene). By ring opening methathesis polymerization (ROMP) of bicyclo-octadiene derivatives (see Scheme 14.6), a precursor soluble in organic solvents is obtained.[35] Further thermal treatment of the films leads to polyphenylene vinylene. Enhancement of the polymer solubility can be achieved by the introduction of alkyl or alkoxy groups into the phenylene ring. A soluble polymer can be prepared by reaction of dihalo-*p*-xylene and a base such as potassium *tert*-butoxide (see Scheme 14.7).[36–38]

Another interesting kind of polymer that, conveniently doped, may show metallic conductivity comprises the aromatic poly(azomethines), represented in Scheme 14.8. These polymers present special characteristics in the sense that they exhibit redox-type chemistry and acid–base chemistry both in the Brönsted as well as in the Lewis sense. The most representative member of this family is polyaniline which, as shown in Fig. 14.3, can exist in a variety of forms that differ in their

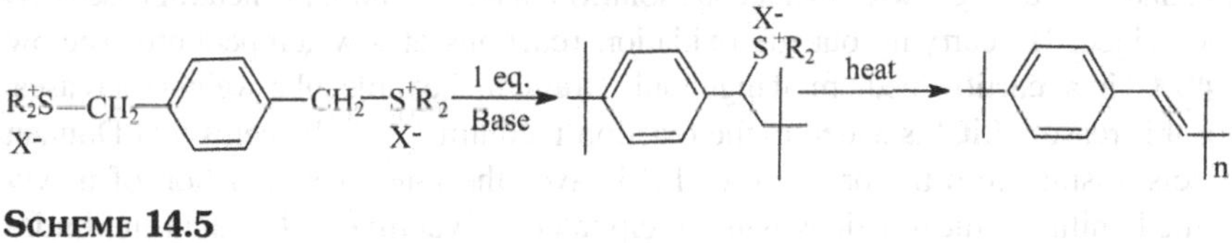

SCHEME 14.5

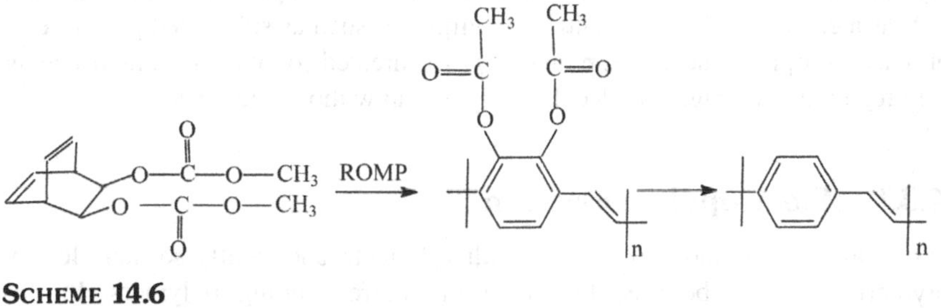

SCHEME 14.6

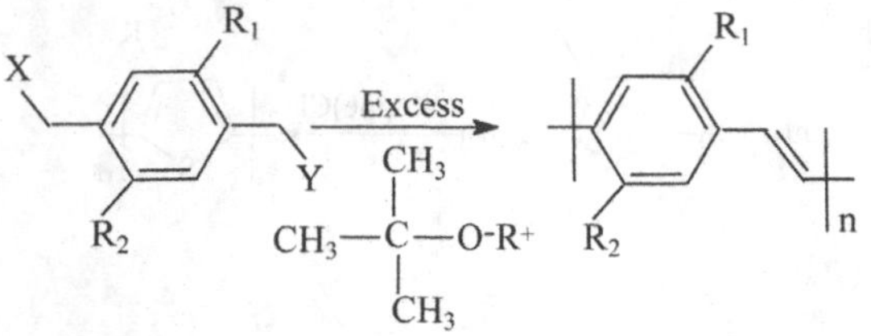

SCHEME 14.7

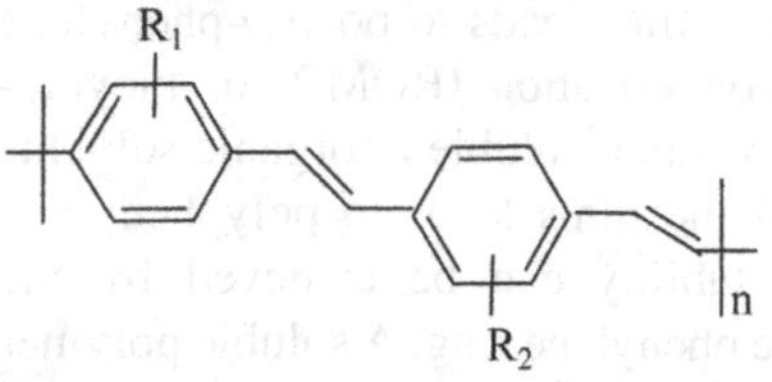

SCHEME 14.8

oxidation states. From a practical point of view, the best way of obtaining the different forms of polyaniline is the synthesis of the emeraldine base (EB), which can be oxidized to a pernigraniline base (PNB) or reduced to a leucoemeraldine base (LEB).[39] EB can be obtained by oxidation of aniline in acid aqueous solutions using ammonium persulfate, potassium chlorate, potassium dichromate, etc., as oxidants.[40] Obviously, a deficit of oxidant with respect to aniline must be utilized. This reaction leads to emeraldine in protonated form, also known as emeraldine salt, and, by treating it with an aqueous solution of ammonium, the neutral base (EB) is obtained. By carrying out the oxidation reactions at low temperatures (below $-40°C$) in a reaction medium of high ionic strength, the molecular weight increases. For this reason, LiCl is added to the reaction medium.[41–44] It seems that Donnan effects arising from the presence of LiCl favor the interphase reaction of newly formed aniline radical cations with precipitated polyaniline rather than the initiation reaction resulting in the growth of a new chain. More recently, enzymatic polymerization of aniline was performed using horseradish peroxidase together with H_2O_2 as a catalyst.[45,46] The addition of templates such as sulfonated polystyrene, polyvinyl phosphonic acid, etc., promotes *para*-directed coupling leading to a better chain regularity and higher molecular weight than without a template.

14.3.1. Electropolymerization

Most conjugated monomers can be oxidized electrochemically so that electropolymerization can be used to obtain the corresponding polymers. In this technique, an active species (anion, cation, or radical) is formed on the electrodes, in which the polymerization starts. Electrochemical polymerization is carried out in a reaction medium of suitable conductivity, and the polymer usually precipitates as a film on the electrode surface and may be removable from it. Usually, the oxidation potential of the conducting polymer is lower than that of the monomer, and as a result the polymer is simultaneously oxidized to the conducting state and kept electrically neutral by incorporation of the electrolyte anion as a counterion. The conducting character of the polymer avoids the passivation of the electrodes.

For the polymerization of pyrrole, a surface electrode, usually platinum, is immersed in a solution containing the monomer, an aprotic solvent (typically acrylonitrile), and a salt such as tetraethylammonium tetrafluoroborate.[47,48] The presence of some reducible species together with traces of water and oxygen favors the reaction in the cathode at sensible potentials; silver salts have been commonly used. Electrochemical polymerization seems to involve the oxidation of the monomer to yield a radical cation. Two radical cations will couple to give a dimeric dication that, after eliminating two protons and rearomatizing, will yield the pyrrole dimer. The oxidation potential of the dimer is lower than that of the pyrrole, and as a result the polymerization proceeds by oxidative coupling of the monomer units to the growing chain, as shown in Fig. 14.5.[4] As the molecular weight of the growing chains increases, the polymer becomes insoluble and precipitates on the surface electrode where growth apparently continues to give a high molecular weight polymer. The electropolymerization conditions, specifically the nature of both solvent and

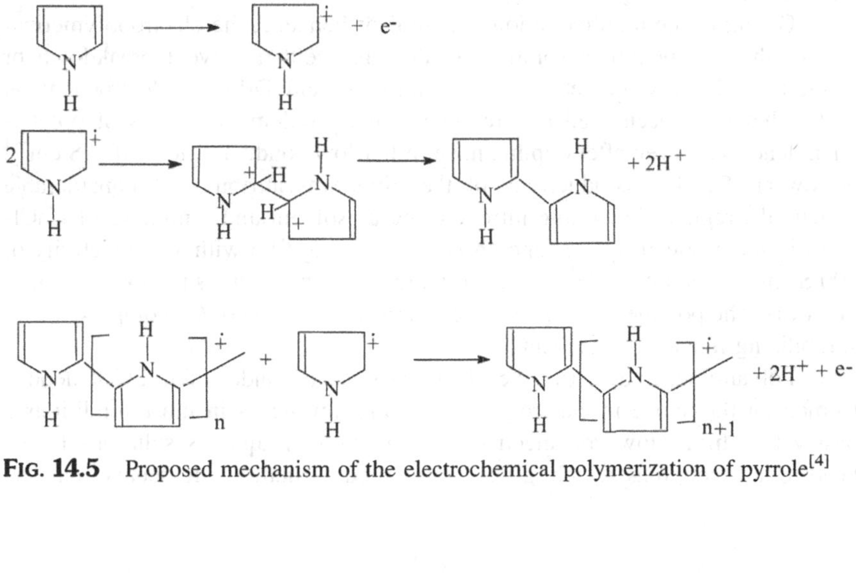

FIG. 14.5 Proposed mechanism of the electrochemical polymerization of pyrrole[4]

counterions, affect the morphology of the polymer which in turn depends on its stereoregularity.

The conductivity of polypyrrole strongly depends on the nature of the counterion. Pyrrole can be polymerized in aqueous solutions of suitable electrolytes[49] in which N-alkyl sulfates, sulphonates, phosphates, or phosphonates[50,51] may act as counterions. The large counterions impose some local structural ordering, thus affecting the packing of the polypyrrole chains.

N-substituted pyrroles have also been electropolymerized. In general, substituted polypyrroles may have a conductivity 5 or 6 orders of magnitude lower than the unsubstituted polymer, presumably as a consequence of the fact that substitution prevents coplanarity of the rings and disrupts effective conjugation.[52,53] Substitution in positions 3 and 3,4 only reduces the conductivity by a factor of 10.[54]

Polythiophene can schematically be obtained from polypyrrole by replacing the nitrogen atom in the rings for a sulfur atom. In practice, the polymer is obtained by electropolymerization of thiophene, a monomer whose oxidation potential is significantly higher than that of pyrrole. The polymerization conditions are rather similar to those of pyrrole.[55–57] The relatively high reactivity of the cations enhances structural irregularities in the polymer, and as a result polythiophene is amorphous. There is some experimental evidence indicating the presence of a significant fraction of units linked through position 3.[58] The electropolymerization of 3-substituted thiophenes gives rise to a regular polymer whose conductivity is ca 2 orders of magnitude higher than that of the unsubstituted polymer having the same counterion. Although the solubility of thiophene in water is very low, it is still high enough to permit electrochemical polymerization from aqueous solutions containing an appropriate electrolyte.[59]

Owing to the high oxidation potential of benzene, the electropolymerization of this compound in solution would cause either solvent breakdown or reactions of the very energetic cations with the solvent. Different electrochemical routes that have been used to circumvent this problem give films of poly(p-phenylene) that, even after doping, have rather low conductivities (10^{-3} S cm^{-1} or lower). The low conductivity of the films was attributed to considerable structural irregularities. Using nitrobenzene as solvent and a mixture of $CuCl_2$ and $LiAsF_6$ as electrolyte, a smooth nonpassivating film with a conductivity of 100 S cm^{-1} was obtained.[60,61] A shortcoming of this route is the low solubility of $CuCl_2$. The polymerization, however, is improved if Ni or Co complexes with solubilizing ligands replace $CuCl_2$.[62]

Polyaniline, known for over 100 years as the undesirable black deposit formed on the anode in electrolysis involving aniline, is in most conditions a passivating film. However, electropolymerization in aqueous solutions favors head-to-tail coupling which gives conjugated structures. It seems that the polymer thus formed can be regarded as a copolymer of reduced (amine) and oxidized (imine) units

$$[(-C_6H_4-NH-C_6H_4-NH-)_y(-C_6H_4-N\!\!=\!\!C_6H_4\!\!=\!\!N-)_{1-y}]_x$$

When $y = 1$ (fully reduced form), $y = 0$ (fully oxidized form), and $y = 0.5$ (50% oxidized structure), the products are called leucoemeraldine, pernigraniline, and emeraldine. The protonation of leucoemeraldine and further oxidation of the resulting product gives $(\!\!=\!\!C_6H_4\!\!=\!\!NH-C_6H_4-N\!\!=\!\!)_x^+$, a structure with metallic characteristics.

Electrochemical synthesis can be carried out by constant current,[63-65] constant potential,[66,67] or by cyclic voltammetry polymerization. The cyclic voltammetry method (Fig. 14.6) seems to lead to the most regular chain of polyaniline. In this procedure, film is deposited layer by layer by potential scanning between the potential characteristic of the most reduced state of polyaniline, leucoemeraldine, and the potential of the onset of the peak of aniline polymerization.[68,69] The oxidation state of the polymer depends on the potential of the working electrode. The typical cyclic voltammogram registered in acidified aqueous solutions shows two oxidation peaks (See Fig. 14.6). The first corresponds to the oxidation of leucoemeraldine to emeraldine, whereas the second, which appears at higher potential, is attributed to the oxidation of emeraldine to pernigraniline. At the potential of emeraldine stability, the polymer is protonated.[70]

Polyaniline is redox active and, as indicated above, can exist in at least three oxidation states of which only one is metallic. However, the macroscopic form of this polymer does not change during reversible oxidation or reduction, and, as a consequence, it can act as a redox catalyst. Two of the three states of

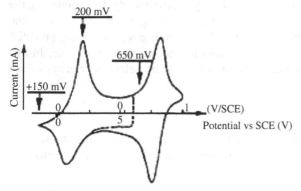

FIG. 14.6 Cyclic voltammetry of electrochemically deposited polyaniline in aqueous acid solution[36]

oxidation are stable under *normal* conditions, from ambient temperature up to 200°C in the presence of air. Moreover, polyaniline is electro- and chemochromic, i.e., it changes its transparent color upon oxidation or reduction (the stable metal form is green, the stable oxidized or neutralized form is blue, the reduced form—readily reoxidized to the green or blue form—is colorless[71]).

Polyaniline is totally unmoldable and insoluble. Polymerization processes have been developed that make it possible to obtain excellent dispersions that can be used for surface coating. In this way, it may be possible to coat glasses and, with a special design, to change the optical and IR transparency properties electrochemically.

14.4. ENERGY GAPS IN CONDUCTING POLYMERS

Two important properties conditioning the conductive characteristics of π-conjugated structures are the ionization potential and the band gap. These properties can be described by using adequate quantum chemistry techniques that take into account the fine details of the whole chemical structure. Thus, these techniques make it possible to predict the effects of chemical substitutions and dielectric medium on the geometric structure and electronic properties. In what follows, we shall give a short overview of some of the quantum mechanical methods often applied to conjugated polymers.

In strict terms, the Schrödinger equation applied to a given conformation of a macromolecular system can be written as

$$H(N)\Psi(N) = E(N)\Psi(N) \tag{14.4.1}$$

where $H(N)$ is the N-electron Hamiltonian describing the system, and $\Psi(N)$ (the eigenfunction) and E (the eigenvalue) are, respectively, the N-electron wave function and the total energy of the system. Since $H(N)$ is nonseparable and $\Psi(N)$ is nonspecific, Eq. (14.4.1) can be solved, only approximately, by introducing the concept of molecular orbitals.[72] According to this approach, each one of the N electrons moves throughout the molecule in a path governed by the potential generated by all the other all electrons and the nuclei of the molecule. In this context, Eq. (14.4.1) is replaced by a set of N one-electron Hartree–Fock equations given by[72,73]

$$H_{HF}(i)\phi_i = \varepsilon_i\phi_i \tag{14.4.2}$$

where $H_{HF}(i)$, ϕ_i, and ε_i are respectively the one-electron Hamiltonian, the wave function, and the energy associated with the ith molecular orbital. Therefore

$$E(N) = \sum_{i=1}^{N} \varepsilon_i \tag{14.4.2a}$$

and

$$\Psi(N) = \prod_{i=1}^{N} \phi_i \qquad (14.4.2b)$$

The Hamiltonian contains four terms

$$H_{\mathrm{HF}}(i) = -\frac{1}{2}\nabla^2(i) - \sum_j^{\mathrm{atoms}} \frac{Z_j}{r_{ji}} + \sum_j^{\mathrm{occ.MOs}} J_j(i) - \sum_j^{\mathrm{occ.MOs}} K_j(i) \qquad (14.4.3)$$

The first term on the right-hand side of this equation represents the kinetic energy of the electron, whereas the nuclear attraction between nuclei j of charge Z_j and electron i is included in the second term. The third term represents the Coulomb repulsion of electron i with all the other electrons. Finally, the last term represents the nonlocal exchange of electron i with all electrons with parallel spin, and J_j and K_j are respectively the Coulomb and the Pauli exchange operators. It is possible to write J_j explicitly

$$J_j(1) = \int \phi_j^*(2)\left(\frac{1}{r_{12}}\right)\phi_j(2)\, d\tau(2) \qquad (14.4.3a)$$

The exchange integral is written explicitly in conjunction with the function in which it is operating

$$K_j(1)\phi_i(1) = \int \phi_j^*(2)\left(\frac{1}{r_{12}}\right)\phi_j(2)\, d\tau(2)\phi_j(1) \qquad (14.4.3b)$$

Let us make the assumption that each molecular orbital (MO) can be written as a linear combination of atomic orbitals (LCAO)

$$\phi_i = \sum_{j=1}^{N} C_{ij}\chi_j \qquad (14.4.4)$$

where χ_j represents the atomic orbital j, and C_{ij} is the weighting coefficient. Since the sum extends over all N electrons, Eq. (14.4.2) can be written as

$$H_{\mathrm{HF}}(i) \sum_{i=1}^{N} C_{ij}\chi_j = \varepsilon_i \sum_{j=1}^{N} C_{ij}\chi_j \qquad (14.4.5)$$

The solution of Eq. (14.4.5) subject to the normalization constraint

$$\int_\tau \phi^*\phi\, d\tau = 1 \qquad (14.4.6)$$

(where τ is the volume) requires premultiplication of each side of Eq. (14.4.5) by the appropriate expression for ϕ_i, leading to

$$\int_\tau \sum_{k=1}^N C_{ik}\chi_j H_{HF}(i) \sum_{j=1}^N C_{ij}\chi_j = \varepsilon_i \int_\tau \sum_{k=1}^N C_{ik}\chi_j \sum_{j=1}^N C_{ij}\chi_j \tag{14.4.7}$$

This expression can be written as

$$\sum_{k=1}^N \sum_{j=1}^N C_{ik} C_{ij} \int_\tau \chi_k H_{HF}(i)\chi_j \, d\tau = \varepsilon_i \sum_{k=1}^N \sum_{j=1}^N C_{ik} C_{ij} \int_\tau \chi_k \chi_j \, d\tau \tag{14.4.8}$$

By defining

$$H_{kj}^i = \int_\tau \chi_k H_{HF}(i)\chi_j \, d\tau \tag{14.4.9}$$

and

$$S_{kj}^i = \int_\tau \chi_k \chi_j \, d\tau \tag{14.4.10}$$

Equation (14.4.8) becomes

$$\varepsilon = \frac{\sum_{k=1}^N \sum_{j=1}^N C_k C_j H_{kj}}{\sum_{k=1}^N \sum_{j=1}^N C_k C_j S_{kj}} \tag{14.4.11}$$

where subscript i is omitted from ε_i because the LCAO sum extends over all N electrons, and therefore molecular orbital N will be obtained from Eq. (14.4.11).

The set of functions $\chi_1, \chi_2, \ldots, \chi_N$ is called the basis set for the calculation. The variational principle states that, for any wave function satisfying the boundary conditions of the problem, the expected value of the energy calculated from this function will always be higher than the true energy of the ground state. Accordingly, the coefficients C_i are to be determined by minimizing ε in Eq. (14.4.11) with respect to C_l, that is

$$\left(\frac{\partial \varepsilon}{\partial C_l}\right)_{C_k} = 0, \qquad k = 1, 2, \ldots, N = l \tag{14.4.12}$$

In this way, we arrive at the so-called secular equation

$$\sum_{i=1}^N (H_{ij} - S_{ij}\varepsilon) C_i = 0, \qquad j = 1, 2, \ldots, N \tag{14.4.13}$$

The condition that must be met by the coefficients of a set of linear homogeneous equations in order that nontrivial solutions exist is that their determinants vanish. Therefore

$$\left| H_{ij} - S_{ij}\varepsilon \right| = 0, \qquad j = 1, 2, \ldots, N \tag{14.4.14}$$

This determinant is called the secular determinant. Expansion of this determinant gives a single equation containing the unknown values of ε. Each value of ε is associated with a nontrivial set of coefficients C_i. The lowest value of ε substituted into the secular equation [Eq. (14.4.13)] gives n simultaneous equations for the N coefficients (see Problem 1).

Two of the four terms of a matrix element of the Hartree–Fock Hamiltonian (kinetic energy and nuclear attraction) are one-electron integrals because they depend on the coordinates of a single electron, whereas the other two correspond to calculation of two-electron integrals in the Hartree–Fock Hamiltonian. Since the computation of the two-electron integrals is highly time consuming, methods have been developed for the approximate calculation of H_{ij} that drastically reduce the computation time. The methods used to evaluate H_{ij} and S_{ij} distinguish the various one-electron schemes from one another. One of the methods widely used in conjugated molecules and polymers is the Hückel approximation. This technique only treats π-electrons and only one π-atomic orbital per site is considered. This is a semiempirical method because it involves some parametrization usually performed on the basis of empirical data. It is assumed that $S_{ij} = 1$ if $i = j$, and 0 otherwise.

In the extended Hückel method, all valence electrons σ and π are taken into account. In this method, each H_{ii}, called the Coulomb integral, is taken as the ionization potential for the appropriate electronic state in the appropriate isolated atom. The off-diagonal elements H_{ij} ($i \neq j$), called the exchange or resonance integrals, are approximated by

$$H_{ij} = \tfrac{1}{2}\left[k_1 \left(H_{ii} + H_{jj} \right) S_{ij} \right] \tag{14.4.15}$$

or

$$H_{ij} = -k_2 \left(H_{ii} H_{jj} \right)^{1/2} S_{ij} \tag{14.4.16}$$

where k_1 and k_2 are adjustable parameters that reproduce some experimental value (for example, the barrier energy of ethane). When atoms i and j are not directly bonded to one another, $H_{ij} = 0$.

The elements of the overlap matrix are obtained from Eq. (14.4.5) by using the Slater wave functions, which for several different electronic states are given in Table 14.1.[74] In these wave functions, M is a normalization constant which makes the probability of χ^2 over all space equal to unity, and $c = Z - S$ is the

TABLE 14.1   Slater Wave Functions for Different States*

n	l	State	χ
1	0	1s	$M \exp(-cr)$
2	0	2s	$Mr \exp(-cr/2)$
2	1	$2p_x$	$Mx \exp(-cr/2)$
3	0	3s	$Mr^2 \exp(-cr/3)$
3	1	$3p_x$	$Mxr \exp(-cr/3)$

*x and r in the fourth column are respectively a variable distance and the internuclear distance.

effective nuclear charge which is computed by modifying the atomic number with a shielding constant S. The value of this parameter is zero for all electrons outside the principal quantum number being considered. For each electron with the same quantum number, $S = 0.35$. If the electron being considered is s or p, then $S = 0.85$ from each electron in the next inner cell, and $S = 1.00$ for each electron further in.

Surveys of models for quantum mechanical approaches to the electronic structure of conducting and conjugated polymers can be found elsewhere.[75–77] The nonempirical pseudopotential valence effective hamiltonian (VEH) technique has been widely used for conjugated polymers.[78,79] In this technique, an effective Hamiltonian is defined as the sum of the kinetic term in atomic units and of the various atomic potentials in the molecule

$$H_{\text{eff}} = -\frac{1}{2}\nabla^2 + \sum_A V_A \qquad (14.4.17)$$

where V_A is the effective potential of atom A. The atomic potentials are expressed as a linear combination of normalized χ_{lm} Gaussian functions with exponent α_i

$$V_A = \sum_l \sum_{ij} C^A_{l,ij} |\chi_{il}\rangle\langle\chi_{jl}| \qquad (14.4.18)$$

where l denotes the Gaussian function angular dependence (it corresponds to the second quantum number); the sums over i and j define the complexity of the potential and are carried out up to 2. The χ_{il} functions have the form

$$\chi_{il} = N_i r^l \ e^{-\alpha_i r^2} Y_l(\theta,\varphi) \qquad (14.4.19)$$

where N_i is a normalization coefficient and Y_l represents the usual spherical harmonics. The parameters of the potential (the linear coefficients $C_{l,ij}$ and the nonlinear exponent α_i) are optimized for each atomic potential type on model molecules so as to reproduce the one-electron energy levels obtained on these molecules from high-quality Hartree–Fock *ab initio* split valence basis set calculations. Since only products of one-electron integrals have to be evaluated, the calculations are very fast, and the computing time is roughly of the same order as that for semiempirical AM1-like calculations.

As indicated above, no experimental information is used in deriving the VEH technique. The only information needed is the molecular geometry which, if unknown, can be obtained by semiempirical techniques such as MNDO. With the VEH technique, ionic potentials and photoelectron have been calculated, and remarkably good agreement has been found between experimental and theoretical results.[80–82] It has also been used to predict band gaps.[83] Since the calculations are carried out in single chains, the values obtained must be corrected in order to account for the polarization energy of the lattice. Calculated and experimental values of the ionization potential (IP) are in rather good agreement after subtracting 1.9 eV from the potentials obtained in the gas phase to correct for polarization energy of the lattice.[84] The values of the redox potentials of conducting polymers, calculated by the VEH method, are also in good agreement with the experimental values. Therefore, the VEH technique is an important tool in the design of new conducting polymers as well as in the prediction of how substituents will affect the conductivity of materials of this kind containing carbon, hydrogen, sulfur, nitrogen, and oxygen atoms.

14.5. DOPING PROCESSES

As indicated above, a conjugated π-electron backbone confers low ionization potential, high electron affinity, and low energy optical transitions to polymers. Most polymers containing these structures behave, in their pristine forms, like insulators with conductivities of about 10^{-12} S cm^{-1} or lower. The conductivity, σ, depends on the carrier concentration, n, and their mobility, μ, in such a way that $\sigma = n\mu e$, where e is the electronic charge unit. Therefore, the insulating character displayed by double-bond conjugated polymers, in spite of the favorable conditions they present for high carrier mobility, must be attributed to the low concentrations of carriers in these materials.

An increase in carriers in conjugated polymers can be achieved by doping. This process involves the oxidation or reduction of the conjugated polymers using electron acceptors or electron donors respectively. In the first case, one electron is removed, creating a cation or hole on the chains, whereas in the second an electron is added, creating an anion. If, as a result of the thermal energy, the hole overcomes the Coulomb binding energy to the counterion acceptor (anion), it will travel through the polymer, contributing to the conductivity. Whereas the dopant in inorganic semiconductors is in the range of parts per million, the amount of dopant in organic semiconductors may be up to 50% of the total weight. The conductivity of polyacetylene can be changed by more than 14 orders of magnitude from 10^{-12} S/cm (undoped polymer) to 10^5 S/cm (highly doped polymer). Another significant difference between conventional semi-conductors and conducting polymers is that, in the latter case, the atomic or molecular dopant ions are located interstitially between chains forming new three-dimensional structures. There is a wealth of these structures, with a wide range of local order, associated with different dopant levels. Therefore, these systems are conducting charge-transfer complexes rather than doped semiconductors.

The doping process involves the exposure of polymer films or powders to vapors or solutions of the dopant, or the use of electrochemical means.[85] In some cases the polymer and the dopant are dissolved in the same solvent before forming the film or powder. Many electron aceptors (I_2, AsF_5, SbF_5, $AlCl_3$, $FeCl_3$, Br_2, IF_5, etc.) have been used in the oxidation of conjugated polymers. However, clusters of different sizes formed by association of inorganic ions make it difficult to identify the counterion resulting from the doping process. For example, counterions of stoichiometries I^-, I_3^-, and I_5^- have been identified by Mössbauer spectroscopy in I_2-doped polyacetylene. The clusters formed in this process may correspond to I_2-solvated I . Accordingly, the chemical reaction describing the oxidative p-type doping of poly(acetylene) with halogens (Br and I_2) can be written as[36]

$$(-CH_2-)_x + (3/2)yxX_2 \longrightarrow [(-C_2H_2-)^{y+}(X_3^-)_y]_x$$

The polyhalogen anions are inserted into the polymeric matrix, worsening its structural order. Note that oxidation with bromine may not stop at the stage of polycarbonium cation. If mild conditions are not used, the reaction can proceed by addition of the halogen to the double bond, transforming segments of polyacetylene into poly(dibromovinylidene), which is an insulator.

The doping process of poly(acetylene) with AsF_5 involves a redox reaction

$$(-CH_2-)_x - yxe^- \longrightarrow [(-C_2H_2-)^{y+}]_x$$
$$(3/2)yxAsF_5 + yxe^- \longrightarrow (1/2)yxASF_3 + yxAsF_6^-$$

followed by an acid–base reaction

$$[(-C_2H_2-)^{y+}]_x + yxAsF_6^- \rightarrow [(-C_2H_2-)^{y+}(AsF_6^-)_y]$$

The Mössbauer spectroscopy identifies SbF_6^-, SbF_5, and SbF_3 in SbF_5-doped polyacetylene.

In doping processes with nitronium salts of antimonates, arsenate, or hexafluorophosphates ($NO_2^+XF_6^-$, where $X = P$, As, Sb) or nitrosonium salts

$(NO^+XF_6^-$, where $X = P$, As, Sb), the redox-type reaction is carried out in solution

$$yxNO_2^+XF_6^- + (-C_2H_2-) \longrightarrow [(-C_2H_2-)^{y+}(XF_6^-)]_x + yxNO_2$$

Other doping agents for polyacetylene are transition metal halides such as $FeCl_3$ which is incorporated in the polymer matrix as $FeCl_4^-$.[86,87] The systems described above can be used to dope all the families of conjugated polymers.

It is usual for n-type doping of polyacetylene to be performed with radical sodium naphthalide (Na^+Naph^-) in tetrahydrofuran solution

$$yxNa^+Naph^{-*} + (-CH_2-)_x \longrightarrow [Na_y^+(-C_2H_2-)^{y-}]_x$$

Here, the radical anion acts as the reducing agent. It is worth noting that n-doped conjugated polymers are much more reactive than the p-doped ones. The conductivity of the n-doped polymers drops very rapidly after short exposures at ambient atmosphere.

In the creation of holes by electrochemical methods, the polymer should be dissolved in a solution containing an electrolyte. A typical arrangement could be: Pt (electrode)|solution [polymer, electrolyte (for example, $Li^+AsF_6^-$), solvent (for example, propylene carbonate)]| Li (counterelectrode).

In the doping process, voltages of 3.5–5.0 V vs. Li are applied so that the polymer is at a sufficiently positive potential for the oxidation to occur. The electrochemical reactions are

$$Polymer \longrightarrow Polymer^+ + e$$
$$Li^+ + 1e \longrightarrow Li$$

Obviously, as the polymer is oxidized, the counterion (AsF_6^-) diffuses to the polymer to maintain charge balance. The same configuration could be used for electrochemical donor doping, though in this case the polymer is positioned at a negative potential relative to that of the neutral polymer. In this situation

$$Li \longrightarrow Li^+ + 1e$$
$$Polymer + 1e \longrightarrow Polymer^-$$

In this process, the polymer incorporates Li^+ from the electrolyte as a counterion.

There exist some polymers in which doping is achieved by protonation, a process in which electrons are not added or removed. For example, the emeraldine base (see Fig. 14.7) is an insulator that for protonation gives the emeraldine salt, which is a conductor. The emeraldine salt can also be obtained by oxidation of the leucoemeraldine base. The effect of doping on some polymeric semiconductors is shown in Fig. 14.8.

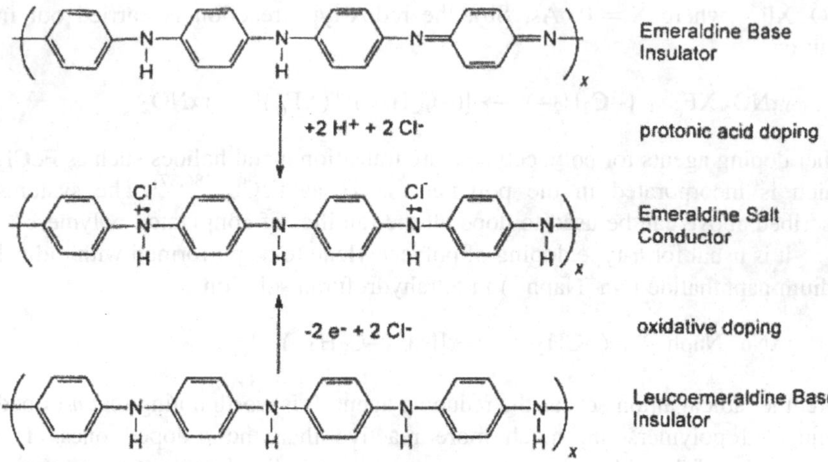

FIG. 14.7   Scheme showing that the protonic acid doping of the emeraldine base and the oxidative doping of the leucoemeraldine base lead to the conducting emeraldine salt

14.6. CHARGE TRANSPORT

The concentration of dopant governs the conductivity of polymers having extended structures of conjugated double bonds. The negative charges initially supplied to the polymer chain by electron donors begin to fill the lowest occupied molecular orbital (or LUMO) of the rigid conduction band. However, the positive charges provided to the chains by electron acceptors fill the highest occupied molecular orbital (or HOMO level) of the valence band. Both mechanisms cause metallic behavior in the systems. Distortions of the bond lengths in the vicinity of the doped charges occur as a consequence of the strong coupling between electrons and phonons (vibrations).[88,89]

Trans-polyacetylene presents an unusual ground state geometry that gives rise to two degenerate states for infinite chain length or for cyclic model chains. The two energetically equivalent representations for the polymer are

$$-CH=CH-CH=CH-CH= \quad \text{and} \quad =CH-CH=CH-CH=CH-$$

As shown in Fig. 14.9, the charge added to the backbone can be stored as a radical, cation, or anion defect, the motion of which in the chain, owing to the degenerate ground states of the polymer, can be described by the motion of a solitary wave, called a soliton in field theory notation. Photoexcitation also leads to the generation of neutral solitons in degenerate polymers and excitons in nondegenerate systems.[90,91] The radical defect is called a neutral soliton, and both the cation and anion defects are referred to as charged solitons. These latter carriers are postulated

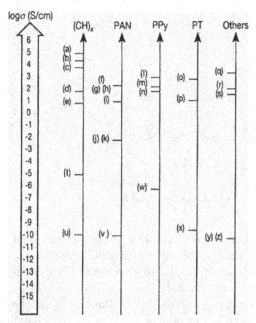

FIG. 14.8 Conductivity of some conducting polymers at room temperature: (a) stretched [CH(I₃)]ₓ; (b) stretched [CH(I₃)]ₓ; (c) [CH(I₃)]ₓ; (d) [CH(I₃)]ₓ; (e) [CH(I₃)]ₓ; (f) stretched PAN–HCl; (g) PAN–CSA (camphor sulfonic acid) from *m*-cresol, (h) PAN–CSA from *m*-cresol; (i) PAN derivative: POT–CSA fiber from *m*-cresol (POT = poly-*o*-toluidine); (j) POT–HCl; (k) 40% sulfonated PAN; (l) stretched PPy (PF₆); (m) PPY(PF₆); (n) PPy (TSO); (o) iodine-doped poly(dodecylthiophene); (p) FeCl₄-doped PT; (q) PPV (H₂SO₄); (r) PPP (AsF₅); (s) ^{84}Kr-implanted poly(phenylenebenzobisoazole); (t) undoped *trans*-(CH)ₓ; (u) undoped *cis*-(CH)ₓ; (v) undoped PAN (EB); (w) undoped PPy; (x) undoped PT; (y) undoped PPV, (z) undoped PPP[105]

to explain the spinless transport observed in polymers with degenerate ground states, because they carry charge but not spin. The energy of solitons and the distortion of the sigma-bond framework have been estimated by quantum chemical treatments using Hückel-type descriptions of π-electrons.[92] High doping in *trans*-polyacetylene causes the soliton energy levels essentially to overlap the filled valence and empty conduction bands, leading to a conducting polymer.[93,94]

Charged solitons are postulated to explain the spinless transport observed in polymers with degenerate ground states, because they carry charge but not spin. Spinless transport also occurs in polymers with nondegenerate ground states, such as poly(*p*-phenylene),[95] although evidently this polymer does not have two energetically equivalent structures, a requirement theoretically

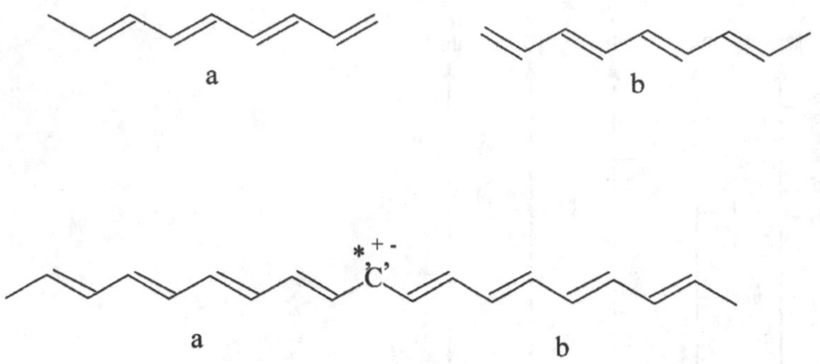

Fig. 14.9 Degenerate states of polydienes. A radical cation or anion defect on the polyacetylene backbone divide the polymer into sections a and b

necessary for mobile solitons. Motions of the cation defect that create stable benzonoid structures give rise to an energy gain due to the stabilizing effects of the aromatic rings. An energy loss occurs if the motion takes place in the direction in which quinoidal structures are formed.

For nondegenerate systems (i.e., polypyrrole, polythiophene, poly(p-phenylene, etc.) the charges introduced at low doping levels or photoexcitation are stored as charged polarons or bipolarons. A polaron is a radical cation or a radical anion plus a lattice distortion around the charge.[92,96] Therefore, these species have spin and charge. An inspection of the positive polaron in poly(p-phenylene) (see Fig. 14.10) shows that the radical and the cation are bound

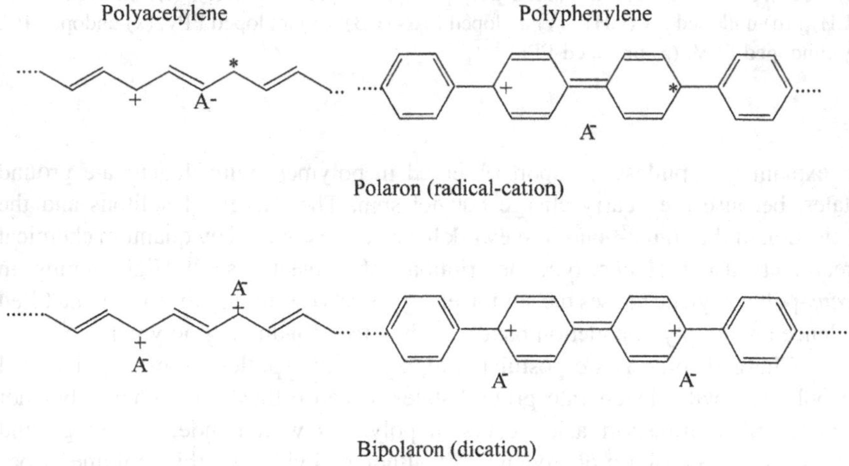

Polaron (radical-cation)

Bipolaron (dication)

Fig. 14.10 Charge defects in polyacetylene and poly(p-phenylene)

together, since an increase in the separation of these two defects would require the creation of additional high-energy quinoid units. The motion of the polaron in polyacetylene involves motions of the neutral soliton and the charged soliton making the radical cation. Model calculations[92,96] suggest that both motions are strongly correlated. For nondegenerate polymers, high doping results in the polarons interacting to form a *polaron lattice* or electrically conducting, partially filled band. According to theoretical models,[92,96] from the exothermic reaction of two polarons on the same chain, a dication or dianion, called a bipolaron, is formed. Some models also suggest equilibrium between polarons and bipolarons.[97] Bipolarons have been detected spectroscopically in several doped polymers, though it is not clear whether these species arise from the combination of two radical ions or a second ionization of one polaron.

Charged solitons can travel in chains of infinite length having degenerate ground states. In principle, soliton transport in polymers would not be possible because hopping of the charged soliton to an adjacent defect free chain should overcome high-barrier energies associated with the necessary geometrical reorganization. However, this barrier will be rather low if hopping of two charged solitons to an adjacent chain occurs. This mechanism would explain the spinless transport observed in polymers with nondegenerate ground states. This transport mechanism is illustrated for polyacetylene in Fig. 14.11.

Summing up, charge doped into the polymer is stored in novel states such as solitons, polarons, and bipolarons, which involves distortion of the lattice surrounding it.

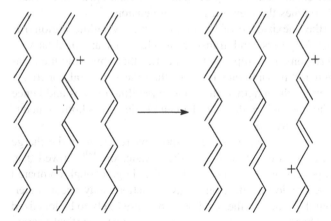

FIG. 14.11 Interchain transport of bipolarons in polyacetylene

14.7. METALLIC CONDUCTIVITY

Our understanding of the metallic state of conducting polymers is somewhat limited by the disorder that these systems present. Thus, properties of metallic character, such as finite d.c. conductivity as T approaches 0 K, a linear dependence of thermoelectric power on temperature, metallic refractivity in the infrared, and negative dielectric constants, were not observed earlier in conducting polymers. The presence of strong disorder in these materials masked their metallic behavior. However, the improvement in the methods of synthesis and processing has led more recently to the preparation of some more ordered materials, which display typical metallic features. For example, conductivity of the order of 10^5 S/cm has been reported for doped polyacetylene,[98] and also a weaker temperature dependence of this material by comparison with that reported for polyacetylene of lower conductivity. Conductivity studies carried out in materials with improved conductivity have made it possible to obtain a better understanding of how disorder affects the insulator-metal transition in doped conducting polymers.

The conductivity of conducting polymers is strongly dependent on the structural disorder of these materials, arising from sample quality, doping procedure, and aging. The results at hand suggest that the carriers become more delocalized as the structural order is improved.[99]

An important and still unsolved issue is to elucidate the effect of the disorder and the one-dimensionality of the polymer on the nature of the metallic state and the insulator–metal transition (IMT). The one-dimensionality of polymers leads to the localization of the electron wave functions. For this reason, although the experimental evidence suggests that the metallic states are three-dimensional, the transport properties are highly anisotropic. Macroscopic transport in conducting materials having well ordered polymer chains is not possible unless the carriers are able to hop to an adjacent chain prior to resonant backscattering which confines the electron in one dimension.

Since disorder (the coexistence of amorphous and crystalline regions) is an intrinsic quality of polymers, and transport mechanisms are qualitatively different in these two regions, it is important to quantify them by using the localization length, L_{lc}. When this parameter is larger than the structural coherence length (which characterizes the length scale of the crystalline regions and hence the length scale for inhomogeneity), the disorder can be viewed as homogeneous because *the system sees an average.*

In highly disordered homogeneous systems, the wave functions of the charge carriers may become localized to a few atomic sites. Anderson[100] showed that, when the random component of the disorder potential is large enough compared with the bandwidth, localization of electronic wave functions may occur. Later, Mott[101,102] pointed out that states in the band tails are more likely to be localized than in the center of the band and, as a consequence, there exists a critical energy,

E_c, called the mobility edge, that separates the localized states from the nonlocalized or extended states in the center of the band. If the Fermi level E_F (the energy below which the one-electron levels are occupied and above which they are unoccupied in the ground state) lies in the region of extended (nonlocalized) states, σ_{dc} is finite as $T \to 0$. In this case, the slope of the double logarithmic plot σ_{dc} vs. T ($W \equiv$ d $\ln\sigma_{dc}$/d $\ln T$) is positive at very low temperatures.[103] If the disorder is strong enough to cause E_F to be in the region of the localized states, the conductivity will decrease very rapidly at low temperatures. In this situation the W plot will have a negative slope, characteristic of hopping systems.[104]

If the localization length L_{lc} is comparable or smaller than the crystalline coherence lengths, there are large-scale inhomogeneities, as in granular metals. In this case, the disorder is considered inhomogeneous. The extent of disorder can be controlled to some degree by details of sample preparation and processing. Tensile drawing leads to chain extension, chain orientation, and interchain order. Inhomogeneous systems can be viewed as composites comprising metallic ordered regions and disordered regions.[105] In the metallic regions, charge carriers are delocalized. The one-dimensional electronic structure of polymer chains in the disordered regions causes electrons to be localized, and therefore hopping along and between polymer chains realizes carrier transport in these regions. For systems comprising disordered regions in which the chains are tightly coiled, the in-chain localized length is short and the coupling between metallic regions is poor. As a result, the free electrons remain confined in the metallic regions. In this case, the temperature-dependent transport depends on the polymer conducting morphology. Hopping, phonon-induced delocalization in the disordered regions, and even tunneling between metallic islands may occur. When the in-chain localization length is larger than the separation between the metallic islands, as occurs with polymers with high persistence lengths (rigid polymers), the carriers are able to diffuse macroscopically among the metallic regions. Thus, a fraction of the carriers will percolate through these ordered paths. A crossover appears in the plots for W(d ln σ/dT) as percolation occurs. According to this model, the value of σ_{dc} strongly depends on the number of well-coupled metallic regions across the sample. Not only will a fraction of the carriers show a free carrier response on the metallic side of the IMT but also phonon-induced delocalization in the disordered region will induce the percolation of carriers in the disordered region at room temperature, even for samples on the insulating side of the IMT.

The temperature dependence of conductivity of Mott's model for a strongly disordered system such that ΔV (disordered energy) is much larger than the bandwidth is given by[101,102]

$$\sigma(T) = \sigma_0 \exp\left[-\left(\frac{T_0}{T}\right)^{1/d+1}\right] \tag{14.7.1}$$

where d is the dimensionality. For three-dimensional systems, $T_0 = c/k_B N(E_F)L_{1c}^3$, where c is the proportionality constant and k_B is the Boltzmann constant. If E_F is such that the electronic states are extended, then finite conductivity at 0 K is expected. Therefore, as the IMT is crossed, the electronic localization length L_{1c} diverges and the system becomes more metallic, displaying higher σ_{dc} values with a lower temperature dependence.

By taking into account that interactions between localized electrons and holes play an important role in hopping transport, it is expected that the variation in the conductivity with temperature will obey the equation[106,107]

$$\sigma(T) = \sigma_0 \exp\left[-\left(\frac{T_0'}{T}\right)^{1/2}\right]$$

(14.7.2)

In this equation $T_0' = e^2/\varepsilon L_{1c}$, where e is the electron charge and ε is the dielectric permittivity.

14.8. MICROWAVE DIELECTRIC PERMITTIVITY

The charge delocalization in conducting materials can be obtained from measurements of microwave frequency dielectric permittivity, ε_{mw}. For a series of protonated esmeraldine samples, the low-temperature ε_{mw} is proportional to ξ^2, where ξ is the crystalline coherence length, independent of the orientation of the sample with respect to the microwave function electric field.[108,109] This behavior suggests that the charge is delocalized within the crystalline regions of the sample. The dielectric permittivity for a simple metallic box model is given by[99]

$$\varepsilon = \varepsilon_\infty + \frac{2^{9/2}}{\pi^3} e^2 N(E_F)L^2$$

(14.8.1)

where $N(E_F)$ is the density of states and L is the coherence length of the material obtained by x-rays. By using this approach, the value of $N(E_F)$ is found to be 1.23 states/eV 2 rings) for PAN–HCl.

14.9. OPTICAL DIELECTRIC PERMITTIVITY AND CONDUCTIVITY

Metals with free electrons are highly reflecting in such a way that the reflectance approaches unity at low frequencies and remains high up to high frequencies. From reflectance data and by using Krönig–Kramers analysis, the optical parameters of interest, including dielectric permittivities and conductivity, can be obtained. The method for this is described in reference [110].

The electric transport in conventional metals is described by the Drude model[110–113] which assumes electrons to be free particles in a gas with a single scattering time τ. According to the model the real part, ε', and imaginary part, ε'', of the dielectric permittivity are given by

$$\varepsilon' = \varepsilon_b - \frac{\omega_p^2 \tau^2}{1 + \omega^2 \tau^2}$$

$$\varepsilon'' = \frac{\omega_p^2 \tau}{\omega(1 + \omega^2 \tau^2)}$$

(14.9.1)

where ω is the external frequency, ε_b is the background dielectric permittivity, and ω_p is the plasma frequency of the electrons

$$\omega_p = \left(\frac{4\pi n e^2}{m^*} \right)^{1/2}$$

(14.9.2)

In this equation n is the density of carriers and m^* is the carriers effective mass. For low frequencies, the components of the complex dielectric permittivity become

$$\varepsilon' \simeq -\omega_p^2 \tau^2$$

$$\varepsilon'' \simeq \frac{\omega_p^2 \tau}{\omega}$$

(14.9.3)

In spite of the simplified assumption, the Drüde model explains that ε' is negative below the screened plasma function ω_p. According to the model, when $\omega < \omega_p$, no radiation can propagate; however, for $\omega > \omega_p$, radiation can propagate and the metal should become transparent.

The curves in the frequency domain corresponding to the real component of the dielectric permittivity of doped conducting polymers present zero, two, three, or one zero crosses as the frequency decreases.[3,104] The number of zero crosses depends on the conductivity of the materials. The real permittivity of the least conducting material is positive in the entire interval of optical frequencies, reaching values of several hundreds at microwave frequencies. However, the permittivity of the higher conducting materials crosses zero between 1 and 3 eV, becomes positive again below 0.1 eV, and reaches a value exceeding 10^4 for microwave frequencies. The permittivity of polymers with d.c. conductivity of the order of 400 S/cm, such as doped polypyrroles and polyanilines, presents, in the frequency domain, the previous two zero crossings and an additional crossing to negative values which occurs at a *delocalized electron plasma frequency* at several hundredths of eV.

By taking into account that in the model $\varepsilon^*(\omega) = 1 + (4\pi\sigma i/\omega)$, from Eq. (14.9.1b) the Drüde conductivity is obtained as

$$\sigma_{\text{Drüde}}(\omega) = \frac{\omega_p^2 \tau}{4\pi(1 + \omega^2\tau^2)} \tag{14.9.4}$$

The Drüde response, which involves a decrease in optical conductivity with increasing frequency, is only observed above a critical frequency $\omega_c \sim D/L_{\text{lc}}^2$, where D and L_{lc} are respectively the diffusion coefficient and the localization length for the electron. When $\omega < \omega_c$, conductivity suppression occurs because the carriers would diffuse a distance greater than the localization length within the period of the a.c. wave. For three-dimensional materials, localization corrections to the frequency-dependent conductivity gives[114-116]

$$\sigma(\omega) = \sigma_{\text{Drüde}}\left[1 - \frac{C}{(k_F v_F \tau)^2} + \frac{C(3\omega)^{1/2}}{(k_F v_F)^2 \tau^{3/2}}\right] \tag{14.9.5}$$

where C is an undetermined universal constant, k_F is the Fermi wave vector, v_F is the Fermi velocity, and τ is the scattering time. Since $k_F v_F \tau$ is a large quantity for more ordered, higher-conductivity materials, the three-dimensional conductor should obey the Drüde formula [Eq. (14.9.4)].

14.10. APPLICATIONS OF SEMICONDUCTOR POLYMERS

As indicated above, undoped conjugated polymers are intrinsic semiconductors whose energy gap is strongly affected by both chemical structure and functionalization of substituents. These materials may emit light under electric perturbations, a phenomenon known as electroluminescence. Electroluminescent diodes can be prepared from a one-layer two-electrode device which in its simplest form consists of a single layer of electroluminescent polymer and two electrodes, one of them, usually the anode, transparent to the light created during the eloctroluminescence effect. The anode or hole-injecting electrode may consist of a thin layer of indium–tin oxide (ITO) with a high work function Φ, whereas the cathode must be a metal of low work function such as calcium, magnesium, or aluminum. Upon application of an external electric field, holes and electrons are injected, respectively, in the highest profile of the valence (π) band (HOMO level) and the lowest profile of the conduction (π^*) band (LUMO level). In most conjugated polymers the barrier for electron injection from the metal electrode ΔE_e is higher than that for hole injection from the ITO electrode, ΔE_h, and holes are the dominant carriers (see Fig. 14.12). Injected charge carriers of different sign drift in opposite directions in the conjugated polymer matrix to form excited species, namely, singlet or triple polaron-excitons. The radioactive decay of singlet excitons gives out light whose wavelength depends on the band gap of the polymer and the relaxation process taking place in the excited state. Note that the name electroluminescent diode given to the configuration of Fig. 14.13 comes from the fact that the relation between intensity I and voltage V is typical of diodes. Actually, no current flow occurs until an onset voltage is reached, above which the current increases very rapidly with increasing voltage. For quantum efficiency constant, electroluminescence–voltage characteristics follow I–V characteristics. It is advisable to move the area of exciton formation from the vicinity of the cathode in order to decrease the probability of nonradioactive recombination. This can be accomplished by fabricating a two-layer device in which the cathode and the electroluminescent polymer are separated by an electron transport layer.[117,118] By varying the $\pi-\pi^*$ optical gap via appropriate functionalization of polymers, it is possible to tune red, green, and blue polymeric light emitting diodes (PLEDs) with a relatively narrow emission peak. For example, regioregular head-to-tail coupled poly(3-decylthiophene) is an electroluminescent polymer with a good color emitting red. Its isomer poly(4,4'-didecyl-2,2'-bithiophene), which shows head-to-head tail-to-tail coupling sequences, emits in green. Detailed information on the luminescent properties of conjugated polymers can be found elsewhere.[119–123]

Undoped conjugated polymers can also be used to construct polymer photovoltaic cells. Actually, as a consequence of the electron-donor nature of conjugated polymers, photoinduced charge separation may occur owing to photoinduced electron transfer if an electron-accepting molecule is in the vicinity

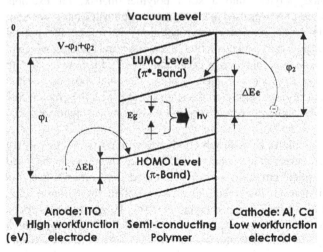

FIG. 14.12   Band diagram for a single-layer polymer light-emitting diode

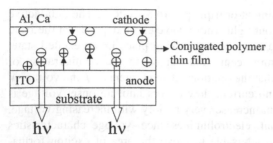

FIG. **14.13**   Carrier transport and electroluminescence in a single-layer polymer light-emitting diode

of a conjugated chain. In this way, a positive polaron, which is highly delocalized and mobile, appears in the conjugated backbone. Fullerenes (C_{60}) are suitable acceptors, and sandwiching a fullerene layer between the anode and the polymer layer[124] could form a photovoltaic structure. The power conversion efficiency of the conjugated polymer-based photovoltaic cell increases considerably if a composite layer consisting of C_{60} dispersed in the conjugated polymer matrix[125] replaces the bilayer heterostructure.

Conjugated polymers can also be used for the preparation of polymer-based photopumped lasers.[126-128] As is known, lasers are devices emitting a spectrally narrow radiation, spatially coherent and strongly polarized. The experience at hand indicates that it is paramount to remove interchain interactions in order for conjugated polymers to exhibit stimulated emission. This can be accomplished by dispersing a conjugated polymer into a solid polymer matrix, for example polystyrene. In most cases, polymers used in lasers have a conjugated backbone with grafted σ-bonded side groups to favor their solution. Therefore, the optical properties of conjugated polymers depend on their microstructural features, which in turn are affected by the solution processing conditions.[129-135] It is worth noting that, as a consequence of the four-level electronic systems that most conjugated polymers comprise, the stimulated emission does not overlap with the ground state absorption spectrum. The shift between absorption and emission minimizes self-absorption.

Besides optoelectronic devices in which electronic energy is transformed into radiation, or vice versa, conjugated polymers may in principle be used for other electronic applications such as polymer-based field-effect transistors (FETs). In this configuration, two metal electrodes called the "source" and "drain" are deposited on a semiconductor layer (Fig. 14.14). The two electrodes are separated from a third electrode, called the "gate," by a thin layer of a dielectric. By applying a voltage between the source and gate electrodes, carriers are created that flow between the source and the drain. It is important to note that

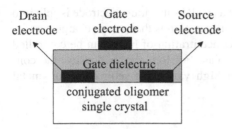

FIG. 14.14 Scheme of a conjugated oligomer based FET device

the gate voltage controls the carrier concentration. For practical applications, the ratio of the conductivities with the gate voltage switched on and off, also called the ON/OFF ratio, should exceed 10^6. Therefore, it is important to use purified conjugated polymers without traces of impurities or doping, which enhance their conductivity when the voltage is switched off. High ON/OFF ratios require high carrier mobility, and this is the weakest point of polymer-based FETs. Carrier charge mobility in highly ordered layers of conjugated oligomers easily exceed $1.0 \, \text{cm}^2 \, \text{V}^{-1} \, \text{s}^{-1}$.[136,137] These mobilities in the best polymer-based FET approach are only $0.1-0.2 \, \text{cm}^2 \, \text{V}^{-1} \, \text{s}^{-1}$.[138,139] Polymers with highly ordered structures are the best candidates for FETs because charge mobility is much larger in these regions than in the amorphous regions. Therefore, it is important to use conjugated polymers with a regular chain structure because these systems may easily form supramolecular ordered structures. If the ensemble of highly ordered zones percolates in the polymer layer, then high charge carrier mobility can be measured for macrosocopic layer. Polymers from the poly(p-phenylene vinylene) family and poly(pyrrole) have relatively high carrier mobility but low ON/OFF ratios, mostly owing to incomplete undoping.[140]

It has been found that single crystals of polysulfide nitride obtained via solid-state polymerization of S_2N_2 become superconductors at extremely low temperature ($T_c = 0.26$ K). The development of superconductivity in conjugated doped polymers requires regular chemical structures, narrow molecular weight distributions, and special processing conditions from solutions to obtain highly ordered microstructures. Finally, the polymer must be doped, but, unfortunately, this process induces disorder in the system. Crystallinity in doped polymers rarely exceeds 40%.[141] Recently it has been demonstrated that substituted poly(thiophenes), such as poly(3-alkyl thiophenes), with high regular chemical structures along the chains, may form highly ordered macrostructures when deposited in the form of thin films on an appropriate substrate.[142] Since chemical doping worsens crystallinity, charge carriers have been introduced electronically using FET configuration.[143] The configuration of the FET used in the studies of poly(3-hexylthiophene) superconductivity differs from the classical configuration in that the source electrode as well as the drain electrode is split into two electrodes. The advantage of this configuration is that p-type charge carriers can be injected electronically, and the concentration of holes can be controlled over a wide range by the applied gate bias. For low carrier density, the conductivity is thermally activated, but for high values the semiconductor–metal transition occurs.

14.11. CONDUCTING POLYMERS

Doped conjugated polymers can be used as organic conductors in the preparation of conductive layers, antistatic coatings, fibers, and transparent electrodes. The combination of high electrical conductivity with the excellent mechanical properties of plastics increases the applications of conjugated doped polymers. The insulating character of polymers may be a drawback in some applications that require electrical conductivity in these materials. Synthetic fibers and textiles coated with conducting layers deposited by electrodeless deposition frequently suffer from poor metal–polymer adhesion. Metallic fibers can be prepared by deposition of doped conjugated polymers on the surface of insulating fibers, in such a way that a surface conductivity of $0.2 \, S \, cm^{-1}$ is easily reached.[144] This conductivity, though low, is much more than that necessary for charge dissipation. Methods have been developed that involve the polymerization *in situ* of suitable monomers such as pyrrole or aniline that allow the deposition of doping conjugated polymer on the fiber surface in the form of an adhering layer.[145,146]

Coated fibers as well as blends of doped conjugated polymers with industrial polymers have many technological applications, for example in the fabrication of carpets, filtration devices, conveyor belts, etc. These blends may even be used as resistive heaters provided that rather lower power per m^2 is required. It is worth noting that doped conjugated polymers display an appropriate level of conductivity and relatively flat attenuation over a wide range of frequencies, which makes them appropriate as components of radio-absorbing materials.

By doping conjugated polymers with large anions of the Keggin type such as $[XM_{12}O_{40}]^{n-}$, organic–inorganic hybrids rather than doped polymers are obtained.[147] These substances are excellent polymer-supported catalysts in several important reactions of industrial importance, and their advantage over the classical catalysts is the molecular dispersion of the catalytic active species via the doping reaction as well as their chemical bonding to the support which prevents desorption. Several surface-doped conjugated polymers have been tested as heterogeneous catalysts, among them polyacetylene, polypyrrole, poly(aniline), and poly(azomethine).[148–151]

In the application of conjugated polymers, the different properties of the doped and undoped states are often exploited. Doping-associated optical changes may be promoted electrochemically, leading to the so-called electrochromic effect. Electrochromic windows can be prepared by sandwiching a thin layer of a conducting polymer electrode, a suitable electrolyte and a transparent counter-electrode. Electrochromic effects are displayed in polyaniline by changing the pH.[152–155]

Conducting polymers can also be used in electromagnetic interference (EMI) shielding, principally in the radio and microwave frequency ranges in computer and telecommunication technologies.[105] These materials are of interest in these applications because of their slight weights, flexibility, recycling potential, etc., compared with standard metals. Although the conducting properties and therefore the shielding efficiency of doped polymers depend on the preparation and the resulting structural order, the shielding of EMI increases as the thickness of the films increases. The shielding capabilities are in the range of utility for many commercial ($\sim$40 dB) and military ($\sim$80–100 dB) applications.

Corrosion is an important problem in the use of metals, principally steel. In order to avoid corrosion in steel and other metals, which causes huge economic losses worldwide, many methods have been developed for the protection of the surface of the metal. These methods involve reduction or oxidation of the surface of the metal (cathodic or anodic protections respectively). An alternative way is to seal off the surface of the metal by deposition of a surface layer of another material, commonly a metal oxide. Conducting polymers can also be used as anticorrosion coatings. Thus, polyaniline exhibits corrosion-protecting properties in its neutral form[156,157] and doped form.[158] The protection of this conducting polymer in steel depends on the thickness of the iron oxide layer at the polymer–metal interface. The best results are obtained when the oxides are removed from the surface of the metal prior to the polyaniline deposition. Owing to the fact that the redox potential of polyaniline is close to that of silver, this material behaves like a noble metal. It ennobles the surface of the metal to be protected by transforming it into a thin but dense metal oxide layer. In other words, polyaniline passivates metals according to the reactions shown in Scheme 14.9.[159] At difference of the mixture of several iron oxides and hydroxides with salt inclusions, called rust, the Fe_2O_3 formed, according to Scheme 9, does not ever build new surfaces for corrosion attack nor offers iron cations able to autocatalytically enhance corrosion velocity. Accordingly, polyaniline withdraws electrons from the surface of the steel so that the protection is of the anodic type.

It is assumed that a similar passivation process occurs on other metals such as copper, aluminum and zinc, all of them ennobled by polyaniline.

Conductive polymers can also be used in the welding of thermoplastics and thermosets. For example, a blend of the intrinsically conducting polymer and powder of the thermoplastic or thermoset is placed between the two pieces to be

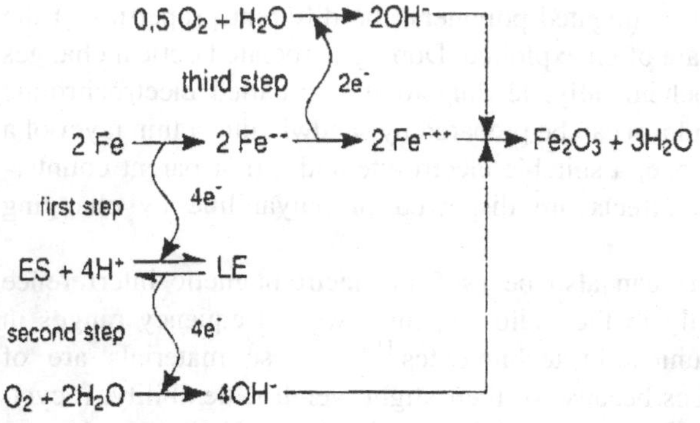

$$0,5\ O_2 + H_2O \longrightarrow 2OH^-$$

third step $\quad 2e^-$

$$2\ Fe \longrightarrow 2\ Fe^{++} \longrightarrow 2\ Fe^{+++} \longrightarrow Fe_2O_3 + 3H_2O$$

first step $\quad 4e^-$

$$ES + 4H^+ \rightleftharpoons LE$$

second step $\quad 4e^-$

$$O_2 + 2H_2O \longrightarrow 4OH^-$$

SCHEME 14.9

joined. Absorption of microwave radiation produces heating of the joint and subsequent welding. The resulting joint may be permanent or not, depending on whether or not the conducting polymer loses its conducting character during the welding process.

Finally, the doping process is exploited in applications not related to electrical conductivity, such as the permselectivity of membranes to gases. It has been found that membranes prepared from polyaniline,[160] poly(dimethoxy-p-phenylene vinylene),[161] poly(N-methylpyrrole),[162] poly(pyrrole),[163] and poly[3-(2-acetoxyethyl)thiophene][164] exhibit favorable permselective properties. It is possible to cast dense polyaniline membranes from solutions with permselectivity coefficients of oxygen with respect to nitrogen close to 10. Protonation of polyaniline in an aqueous solution of HCl decreases both permeability and selectivity because void spaces between the chains are filled with hydrated chloride ions which hinder gas diffusion.[165] Permeation measurements carried out in cast membranes of poly[3-(2-acetoxyethyl)thiophene] give selectivities of 5.1 and 18.5 for O_2/N_2 and CO_2/CH_4.[164] By hydrolysis of the ester group of the latter polymer, membranes are formed that show a decrease in permeability accompanied with a dramatic increase in selectivity for both base treatment ($O_2/N_2 = 11.7$, $CO_2/CH_4 = 20.0$) and acid treatment ($O_2/N_2 = 11.7$, $CO_2/CH_4 = 45.0$).

14.12. POLYMERS FOR RECHARGEABLE BATTERIES

As a consequence of the redox reactions that can take place in conducting polymers, these materials can be used as electrodes in rechargeable batteries. Some battery applications of electroactive polymers are discussed elsewhere.[166] Owing to both the facile processability of polymers to form films (in comparison with the high-temperature processing of metals) and their low density, polymer electrodes satisfy the requirements for portable devices and vehicles. In principle, the specific charge of polymers is not as high as that of metal electrodes, though it is comparable with that of polymer oxides. In the design of a conventional battery, it is an advantage to use conducting polymers as positive electrodes in combination with negative electrodes such as Li, Na, Mg, Zn, and MeH$_x$.[167] In the configuration shown in Fig. 14.15a, the cathode is a positively charged polymer having a negative counterion A$^-$. Here, arrows indicate the chemical

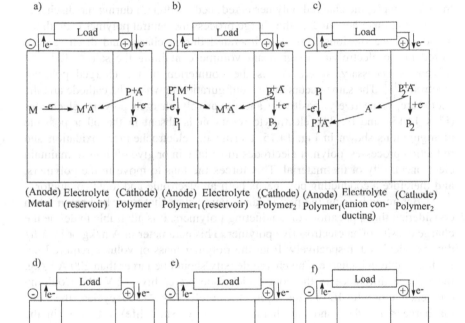

FIG. 14.15 Polymers as electroactive materials in cell assemblies. All batteries are in their discharge configuration[167]

reactions taking place in the electrodes. During the discharge process

$$Li \longrightarrow Li^+ + 1e$$
$$P^+ + 1e \longrightarrow P$$

In the charge process the opposite reactions occur

$$Li^+ + 1e \longrightarrow Li$$
$$P \longrightarrow P^+ + 1e$$

In this example, the charged polymer is reduced (undoped) during the discharge process. On the other hand, in the charge process the neutral polymer is oxidized (doped). Since the ions formed or liberated during charge and discharge are stored in the electrolyte, a significant volume containing the salt $A^- Li^+$ and solvent is necessary, where A^- is the counterion of the charged polymer electrode.[167] The same occurs in the configuration in which the cathode and the anode are, respectively, positively and negatively charged polymer electrodes (Figs 14.15a and b). The electrolyte reservoir is absent in the other possible configurations shown in Fig. 14.15. During the electrochemical oxidation and reduction processes, polymer electrodes must take in or give off ions to maintain electroneutrality of the material. This forces the ions to move in the polymers, and therefore a conducting polymer is both an electronic and ionic conductor.

Before addressing the performance parameters that must be known when considering the applications of conducting polymers, it is advisable to define the charge density of an electroactive polymer. This parameter in A h/kg or in A h/dm^3 is calculated, respectively, from the polymer mass or volume required per exchangeable unit charge. The charge density should be larger than 200 A h/kg, the voltage difference to the counterelectrode preferably >2 V, the Coulomb efficiency close to 100%, the cycle life preferably >500 cycles, the self-discharge $<1\%$/day, and the chemical stability (shelf life) >1 year (in the charged or, more commonly, uncharged state).

Whereas the redox potential of metal electrodes is more or less constant, the potential of polymer electrodes is strongly dependent on their state of charge or discharge.[167] Therefore, the specific energy of a cell having polymer electrodes should be expressed as the product of the specific charge and the mean value of the differences in potential at the start and end of discharge. The highest specific energies are obtained using as the negative electrode Li, which has a high specific charge. In order to obtain batteries of high efficiency, electrodes of Li or other metals, such as Na, Mg, and Zn, acting as anodes, are combined with a cation-exchanging polymer of low equivalent weight which acts as a cation insertion electrode, as shown in configuration (e) in Fig. 14.15. However, this configuration has not been explored extensively until now, and most of the work found in the literature uses configuration (a). Configuration (f) has often been used and it is even marketed. Attempts have been made to use polymer electrodes properly doped as anodes with the aim of obtaining a completely organic battery. All the configurations that are possible in principle are shown in Fig. 14.15.[167]

Among the many conducting polymers reported in the literature, only those prepared from monomers (possibly substituted), such as acetylene, aniline, pyrrole, thiophene, and benzene, have been investigated in depth for rechargeable batteries. A deep and careful review on the preparation of electrochemically active polymers for rechargeable batteries can be found in reference [167].

14.13. OTHER APPLICATIONS

A so-called "semiconducting layer" is placed between the metallic conductor and the insulating layer in the technology of high-voltage coaxial cables. This intermediate layer suppresses local increase in electric field caused by imperfections and defects of various types at the conductor–insulator interface that may produce an electric breakdown. The intermediate layer generally used consists of a mixture of insulating polymer and carbon black, the quantity of the latter material being such that the percolation threshold is exceeded. However, carbon can be replaced with undoped polyaniline,[168] a "self-adapting" material whose conductivity actually depends on the electric field. For low electric fields, polyaniline behaves like an insulator, whereas under electric fields exceeding $1 \, kV \, mm^{-1}$ the intensity is no longer a linear function of the voltage, leading to a significant increase in the polymer conductivity.[168] In those areas in which the electric field locally increases owing to imperfections of the material, polyaniline becomes conductive and the accumulation of charge is dissipated.

PROBLEMS

Problem

Calculate the energy of the orbitals of butadiene using the Hückel theory.[169]

Solution

Polybutadiene may adopt the *cis* and *trans* configurations indicated below. By denoting the components of the diagonal and off-diagonal of the Hamiltonian H_{ij}

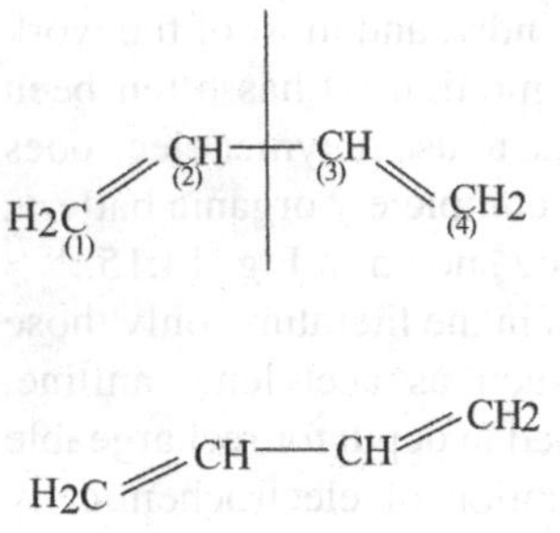

$$
\begin{array}{c}
\text{H}_2\text{C} \underset{(1)}{\overset{\displaystyle{\underset{(2)}{\text{CH}}} \!\!-\!\! \underset{(3)}{\text{CH}}}{\diagup}} \!\!\diagdown\!\! \underset{(4)}{\text{CH}_2}
\end{array}
$$

$$
\text{H}_2\text{C} \overset{\diagup}{\underset{}{}} \text{CH} \!-\! \text{CH} \overset{\diagup \text{CH}_2}{}
$$

FIG. 14.16

by α and β respectively, Eq. (14.4.13) can be written as

$$
\begin{array}{ll}
\text{Electron (1):} & (\alpha - \varepsilon)C_1 + \beta C_2 = 0 \\
\text{Electron (2):} & \beta C_1 + (\alpha - \varepsilon)C_2 + \beta C_3 = 0 \\
\text{Electron (3):} & \beta C_2 + (\alpha - \varepsilon)C_3 + \beta C_4 = 0 \\
\text{Electron (4):} & \beta C_3 + (\alpha - \varepsilon)C_4 = 0
\end{array}
\tag{P.14.1.1}
$$

where $\alpha = H_{ii}$ and $\beta = H_{ij}$

Let us first consider the *cis* configuration. According to Eq. (14.4.4), the wave function describing this configuration is given by

$$
\phi = C_1\chi_1 + C_2\chi_2 + C_3\chi_3 + C_4\chi_4
\tag{P.14.1.2}
$$

By reflection in the plane indicated in the scheme, electrons 1 and 4 as well as 2 and 3 are interchangeable. Accordingly, $C_1 = C_4$ and $C_2 = C_3$. Therefore

$$
\phi = C_1\chi_1 + C_2\chi_2 + C_2\chi_3 + C_1\chi_4
\tag{P.14.1.3}
$$

By exchanging orbitals ($\chi_1 = \chi_4$, $\chi_2 = \chi_3$)

$$
\phi' = C_1\chi_4 + C_2\chi_3 + C_2\chi_2 + C_1\chi_1 = \phi
\tag{P.14.1.4}
$$

The fact that $\phi = \phi'$ indicates that the wave function is symmetric. Obviously, system of Eqs (P.14.1.1) only display two independent variables, namely C_1 and C_2 becoming

$$
\begin{array}{l}
wC_1 + C_2 = 0 \\
C_1 + C_2(w + 1) = 0
\end{array}
\tag{P.14.1.5}
$$

where $w = (\alpha - \varepsilon)/\beta$. The secular determinant for the *cis* configuration is

$$\begin{vmatrix} w & 1 \\ 1 & w+1 \end{vmatrix} = 0 \qquad (P.14.1.6)$$

Solution of this determinant gives $w = -1.618$ and $w = 0.618$, so that $\varepsilon_1 = \alpha + 1.618\beta$ and $\varepsilon_3 = \alpha + 0.618\beta$. The values of C_i can be obtained using the normalization conditions $\sum_i C_i^2 = 1$. Thus, the values obtained are $C_1 = C_4 = 0.3717$ and $C_2 = C_3 = 0.6015$ for $w = -1.618$. For $w = 0.618$, $C_1 = C_4 = 0.6015$ and $C_2 = C_3 = -0.3717$.

Let us now consider the *trans* configuration. Rotation of $180°$ about an axis that passes through the middle of the central C_2–C_3 bond exchanges the positions of atoms 1 and 4 and 2 and 3. In this situation $C_1 = -C_4$ and $C_2 = -C_3$. The molecular orbital has the general form

$$\phi = C_1\chi_1 + C_2\chi_2 - C_2\chi_3 - C_1\chi_4 \qquad (P.14.1.7)$$

Upon interchange, atomic orbitals

$$\phi' = C_1\chi_4 + C_2\chi_3 - C_2\chi_2 - C_1\chi_1 = -\phi \qquad (P.14.1.8)$$

The molecular orbital changes its sign and ϕ' is called the antisymmetric wave function. The secular determinant for the *trans* configuration can be written as

$$\begin{vmatrix} w & 1 \\ 1 & w-1 \end{vmatrix} = 0 \qquad (P.14.1.9)$$

Hence, $w = -0.618$ and $w = 1.618$, so that $\varepsilon_2 = \alpha + 0.618\beta$ and $\varepsilon_4 = \alpha - 1.618\beta$. The fact that β is a negative quantity indicates that the energy increases from ε_1 to ε_4. The symmetry, energy, and molecular orbitals of butadiene are as follows:

Symmetry	Energy, e_i	Molecular orbital, ϕ_i
S	$\alpha + 1.618\beta$	$\phi_1 = 0.3717\chi_1 + 0.6015\chi_2 + 0.6015\chi_3 + 0.3717\chi_4$
A	$\alpha + 0.618\beta$	$\phi_2 = 0.6015\chi_1 + 0.3717\chi_2 - 0.3717\chi_3 - 0.6015\chi_4$
S	$\alpha - 0.618\beta$	$\phi_3 = 0.6015\chi_1 - 0.3717\chi_2 - 0.3717\chi_3 + 0.6015\chi_4$
A	$\alpha - 1.618\beta$	$\phi_4 = 0.3717\chi_1 - 0.6015\chi_2 + 0.6015\chi_3 - 0.3717\chi_4$

It is worth noting that the separation of energy levels becomes smaller with increasing n, tending to a continuum as n tends to infinity.

REFERENCES

1. Ashcroft, N.W.; Mermin, N.D. *Solid State Physics*; CBS Publishing Asia Ltd: Saunders College Publishing, International Edition, 1988.
2. Unsworth, J.; Costa, C.; Zheshi, J.; Kaynak, A.; Ediriweera, R.; Innis, P.; Booth, N. J. Intell. Mater. Syst. Struct. **1994**, *5*, 595.
3. Kohlman, R.S.; Joo, J.; Epstein, A.J. Conducting polymers: electrical conductivity. In *Physical Properties of Polymers Handbook*; Mark, J.E., Ed.; American Institute of Physics: Woodbury, NY, 1996.
4. Billingham, N.C.; Calvert, P.D. Adv. Polym. Sci. **1989**, *90*, 1.
5. Skotheim, T.A.; Elsenbaumer, R.L.; Reynolds, J.R., Eds. *Handbook of Conducting Polymers*; Marcel Dekker: New York, 1998; 197–409.
6. Shirikawa, H.; Ikeda, S. Polym. J. **1971**, *2*, 231.
7. Shirikawa, H.; Ikeda, S. Polym. J. **1974**, *12*, 929.
8. Shirikawa, H.; Louis, E.J.; MacDiarmid, A.G.; Chiang, C.K.; Heeger, A.J. J. Chem. Soc.: Chem. Commun. **1977**, 578.
9. Clarke, T.C.; Yannoni, C.S.; Katz, T.J. J. Am. Chem. Soc. **1983**, *105*, 7787.
10. Theophilou, N.; Aznar, R.; Munardi, A.; Sledz, J.; Schue, F.; Naarmann, H. J. Macromol. Sci. **1987**, *A24*, 797.
11. Munardi, A.; Theophilou, N.; Aznar, R.; Sledz, J.; Schue, F.; Naarmann, H. Makromol. Chem. **1987**, *188*, 395.
12. Naarmann, H.; Theophilou, N. Synth. Met. **1987**, *22*, 1.
13. Basescu, N.; Liu, Z.X.; Moses, D.; Heeger, A.J.; Naarmann, H.; Theophilou, N. Nature **1987**, *327*, 403.
14. Edwards, J.H.; Feast, W.J. Polym. Commun. **1980**, *21*, 595.
15. Chien, J.C.W. *Polyacetylene Chemistry, Physics and Material Science*; Academic Press: Orlando, 1984.
16. Speight, J.G.; Kovacik, P.; Koch, F.W. J. Macromol. Sci.: Rev. **1971**, *C5*, 295.
17. Kovacik, P.; Jones, M.B. Chem. Rev. **1987**, *87*, 357.
18. Yamamoto, T.; Hayashi, Y.; Yamamoto, A. Bull. Chem. Soc. Jap. **1978**, *51*, 2091.
19. Kovacik Kyriakis, P. J. Am. Chem. Soc. **1963**, *85*, 454.
20. Tiecke, B.; Bubeck, C.; Lieser, G. Makromol. Chem.: Rapid Commun. **1982**, *3*, 261.
21. Hill, H.W. *Hirk–Othmer Encyclopedia of Chemical Technology*; Wiley: New York, 1982; 793.
22. Cleary, J.W. Polym. Sci. Technol. **1986**, *31*, 173.
23. Lenz, R.W.; Handlovits, C.E.; Smiths, H.A. J. Polym. Sci. **1962**, *58*, 351.
24. Jen, K.Y.; Miller, G.G.; Elsenbaumer, R.L. J. Chem. Soc.: Chem. Commun. **1987**, *17*, 1346.
25. Elsenbaumer, R.L.; Jen, K.Y.; Oboodi, R. Synth. Met. **1986**, *15*, 169.
26. Rughooputh, S.; Hotta, S.; Heeger, A.J.; Wudl, F. J. Polym. Sci. Pol. Phys. **1987**, *25*, 1071.
27. Iganas, O.; Salaneck, W.R.; Österholm, J.E.; Laakso, J. Synth. Met. **1988**, *22*, 395.

28. Mcullough, R.D.; Lowe, R.D. J. Chem. Soc.: Chem. Comm. **1992**, *1*, 70.
29. Mcullough, R.D.; Tristam-Nagle, S.; Williams, S.P.; Lowe, R.D.; Jayaraman, M. J. Am. Chem. Soc. **1993**, *115*, 4910.
30. Mcullough, R.D.; Lowe, R.D.; Jayaraman, M.; Anderson, D.L. J. Org. Chem. **1993**, *58*, 904.
31. Chen, T.A.; Rieke, R.D. J. Am. Chem. Soc. **1992**, *114*, 10087.
32. Chen, T.A.; Wu, X.; Rieke, R.D. J. Am. Chem. Soc. **1995**, *117*, 233.
33. Moratti, S.C. The chemistry and use of polyphenylene vinylenes. In *Handbook of Conducting Polymers*; Skotheim, T.A., Elsenbaumer, R.L., Reynolds, J.R., Eds.; Marcel Dekker: New York, 1998; 343–361.
34. Wessling, R.A. J. Polym. Sci. Phys. Ed. **1985**, *72*, 55.
35. Conticello, V.P.; Gin, D.L.; Grubbs, R.H. J. Am. Chem. Soc. **1992**, *114*, 9708.
36. Pron, A.; Rannou, P.; Prog. Polym. Sci. **2002**, *27*, 135.
37. Wudl, F.; Allemand, P.M.; Srdanov, G.; Ni, Z.; McBranch, D. ACS Symp. Ser. **1991**, *455*, 683.
38. Louwet, F.; Vanderzande, D.; Gelan, J. Synth. Met. **1995**, *69*, 509.
39. Wei, Y.; Hsueh, K.F.; Jang, G.-W. Macromolecules **1994**, *27*, 518.
40. Pron, A.; Genoud, F.; Menardo, C.; Nechtschein, M. Synth. Met. **1998**, *24*, 193.
41. Matosso, L.H.C.; Faria, R.; Bulhoes, L.O.S.; MacDiarmid, A.G. J. Polym. Sci. Pol. Chem. **1994**, *32*, 2147.
42. Adams, P.N.; Laughlin, P.J.; Monkman, A.P. Synth. Met. **1996**, *76*, 157.
43. Adams, P.N.; Laughlin, P.J.; Monkman, A.P.; Kenwright, A.M. Polymer **1996**, *37*, 3411.
44. Beadle, P.M.; Nicolau, Y.F.; Banka, E.; Rannou, P.; Djurado, D. Synth. Met. **1998**, *95*, 29.
45. Liu, W.; Kumar, J.; Tripathy, S.K.; Senecal, K.J.; Samuelson, L. J. Am. Chem. Soc. **1999**, *121*, 71.
46. Nagarajan, R.; Tripathy, S.; Kumar, J.; Bruno, F.F.; Samuelson, L. Macromolecules **2000**, *33*, 9542.
47. Diaz, A.F. Chem. Scripta **1981**, *17*, 145.
48. Clarke, T.C.; Clarke, J.C.; Street, G.B. J. Res. Dev. **1983**, *27*, 313.
49. Mengoli, G.; Musiani, M.; Fleischmann, M.; Pletcher, D. J. Appl. Electrochcm. **1984**, *14*, 285.
50. Wernet, W.; Monkenbusch, M.; Wegner, G. Makromol. Chem.: Rapid Commun. **1984**, *5*, 157.
51. Wernet, W.; Monkenbusch, M.; Wegner, G. Mol. Cryst. Liq. Cryst. **1985**, *18*, 193.
52. Wernet, W.; Monkenbusch, M.; Wegner, G. Makromol. Chem.: Rapid Commun. **1984**, *5*, 157.
53. Wernet, W.; Monkenbusch, M.; Wegner, G. Mol. Cryst. Liq. Cryst. **1985**, *18*, 193.
54. Street, G.B.; Clarke, T.C.; Geiss, R.H.; Lee, Y.V.; Nazzal, A.; Pfluger, P.; Scott, J.C. J. Phys.: Colloq. **1983**, *C3*, 599.
55. Tourillon, G.; Garnier, F. J. Electroanal. Chem. **1982**, *135*, 173; J. Phys. Chem. **1984**, *87*, 2289; J. Electroanal. Chem. **1984**, *161*, 407.
56. Kaneto, K.; Kohno, Y.; Yoshino, K.; Inuishi, Y. J. Chem. Soc.: Chem. Commun. **1983**, 382.

57. Kaneto, K.; Yoshino, K.; Inuishi, Y. Jap J. Appl. Phys. **1982**, *L21*, 567; **1983**, *L22*, 412.

58. Roncalli, J.; Lemaire, M.; Garreau, R.; Garnier, F. Synth. Met. **1987**, *18*, 139.

59. Czerwinski, A.; Zimmer, H.; Pham, C.V.; Mark Jr., H.B. J. Electrochem. Soc. **1985**, *132*, 2669.

60. Satoh, M.; Kaneto, K.; Yoshino, K. J. Chem. Soc.: Chem. Commun. **1984**, 1199.

61. Satoh, M.; Tabat, M.; Kaneto, K.; Yoshino, K. Polym. Commun. **1986**, *27*, 247.

62. Oshawa, T.; Nishihara, H.; Aramaki, K.; Takeda, S.; Yoshino, K. Polym. Commun. **1987**, *28*, 140.

63. Kobayashi, T.; Yoneyama, H.; Tamura, H. J. Electroanal. Chem. **1984**, *177*, 293.

64. Stilwell, D.E.; Park, S.M. J. Electrochem. Soc. **1988**, *135*, 2497.

65. Watanabe, A.; Mori, K.; Iwasaki, Y.; Nakamura, Y.; Niizuma, S. Macromolecules **1987**, *20*, 1793.

66. Katani, A.; Izumi, J.; Yang, J.; Hiromoto, Y.; Sasaki, K. Bull. Chem. Soc. Jap. **1984**, *57*, 2254.

67. Li, Y.; Yan, B.; Cao, Y.; Qian, B. Synth. Met. **1988**, *25*, 79.

68. Genies, E.M.; Tsintavis, C. J. Electroanal. Chem. **1985**, *195*, 109.

69. Genies, E.M.; Lapkowski, M. J. Electroanal. Chem. **1987**, *236*, 189.

70. Quillard, S.; Berrada, K.; Louarn, G.; Lefrant, S.; Lapkowski, M.; Pron, A. New J. Chem. **1999**, *19*, 365.

71. Wessling, B. Synth. Met. **1997**, *85*, 1313.

72. Loewe, J.P. *Quantum Chemistry*; Academic Press: New York, 1978.

73. Brédas, J.L. Synth. Met. **1997**, *84*, 3.

74. Hopfinger, A.J. *Conformational Properties of Macromolecules*; Academic Press: New York, 1973; 97.

75. Sutherland, B. Synth. Met. **1997**, *84*, 11.

76. Brédas, J.L. Synth. Met. **1997**, *84*, 3.

77. Förner, W. Ind. J. Chem. **1997**, *36 A*, 355.

78. Nicolas, G.; Durand, P. J. Chem. Phys. **1979**, *70*, 2020; **1980**, *72*, 453.

79. André, J.M.; Burke, L.A.; Delhalle, J.; Nicolas, G.; Durand, P. Int. J. Quantum. Chem. Symp. **1979**, *13*, 283.

80. Niwa, O.; Tamamura, T. J. Chem. Soc.: Chem. Commun. **1984**, 817.

81. Brédas, J.L.; Elsenbaumer, R.L.; Chance, R.R.; Silbey, R. J. Chem. Phys. **1983**, *78*, 5656.

82. Brédas, J.L.; Chance, R.R.; Silbey, R.J.; Nicolas, G.; Durand, P. J. Chem. Phys. **1982**, *77*, 371.

83. Brédas, J.L.; Chance, R.R.; Bauham, R.H.; Silbey, R.J. J. Chem. Phys. **1982**, *76*, 3673.

84. Frommer, J.E.; Chance, R.R. *Encyclopedia of Polymers*; Wiley-Interscience: New York, 1986; Vol. 5, 462.

85. Chiang, C.K.; Druy, M.A.; Gau, S.C.; Heeger, A.J.; Louis, E.; MacDiarmid, A.C.; Park, Y.W.; Shirikawa, S. J. Am. Chem. Soc. **1978**, *100*, 1013.

86. Pron, A.; Kulszewick, I.; Brillaud, D.; Przyluski, J. J. Chem. Soc.: Chem. Commun. **1981**, *15*, 783.

87. Pron, A.; Zagorska, M.; Kucharski, Z.; Lukasiak, M.; Suwalski, J. Mat. Res. Bull. **1982**, *17*, 1505.

88. Haeger, A.J.; Kivelson, S.A.; Schrieffer, J.R.; Su, W.P. Rev. Mod. Phys. **1988**, *60*, 781.
89. Ginder, J.M.; Epstein, A.J. Phys. Rev. B **1990**, *41*, 10674.
90. Mizes, H.A.; Conwell, E.M. Phys. Rev. B **1994**, *50*, 11243.
91. Yan, M.; Rothberg, L.J.; Papadimitrakopoulos, F.; Galvin, M.E.; Miller, T.M. Phys. Rev. Lett. **1994**, *72*, 156.
92. Bredas, J.L.; Chance, R.R.; Silbey, R. Phys. Rev. B **1982**, *26*, 58431; Mol. Cryst. Liq. Cryst. **1982**, *77*, 319.
93. Mizes, H.A.; Conwell, E.M. Phys. Rev. B **1994**, *50*, 11243.
94. Stafström, S. Phys. Rev. B **1991**, *43*, 12437.
95. Peo, M.; Roth, S.; Dransfeld, K.; Tieke, B.; Hocker, J.; Gross, H.; Grupp, A.; Sixl, H. Solid State Commun. **1980**, *35*, 119.
96. Brédas, J.L.; Scott, J.C.; Yakushi, K.; Street, G.B. Phys. Rev. B **1984**, *30*, 1023.
97. Genoud, F.; Guglielmi, M.; Nechtschein, M.; Genies, E.; Samon, M. Phys. Rev. Lett. **1985**, *55*, 18.
98. Tsukamoto, J. Adv. Phys. **1992**, *41*, 509, and references therein.
99. Joo, J.; Oblakowski, Z.; Du, G.; Pouget, J.P.; Oh, E.J.; Weisinger, J.M.; Min, Y.; MacDiarmid, A.G.; Epstein, A.J. Phys. Rev. B **1994**, *69*, 2977.
100. Anderson, P.W. Phys. Rev. B **1958**, *109*, 1492.
101. Mott, N.F.; Davis, E.A. *Electronic Process in Noncrystalline Materials*; Oxford University Press: Oxford, 1979.
102. Mott, N.F. *Metal–Insulator Transition*, 2nd Ed.; Taylor & Francis: London, 1990.
103. Zabrodskii, G.; Zeninova, K.N. Zh. Eksp. Teor. Fiz. **1984**, *86*, 727.
104. Kohlman, R.S.; Epstein, A.J. Insulator–metal transition and inhomogeneous metallic state in conducting polymers. In *Handbook of Conducting Polymers*, 2nd Ed.; Skothein, T.A., Elsenbaumer, R.L., Reynolds, J.R., Eds.; Marcel Dekker: New York, 1998; 90.
105. Epstein, A.J. MRS Bull. **1997**, *22*, 16.
106. Efros, A.L.; Shkolowski, B.I. J. Phys. **1975**, *C8*, L49.
107. Efros, A.L.; Shkolowski, B.I. In *Electronic Properties of Doped Semiconductors*; Springer-Verlag: Heidelberg, 1984.
108. Wang, Z.H.; Ray, A.; MacDiarmid, A.G.; Epstein, A.J. Phys. Rev. B **1991**, *43*, 4373.
109. Joo, J.; Prigodin, V.N.; Min, Y.G.; MacDiarmid, A.G.; Epstein, A.J. Phys. Rev. B **1994**, *50*, 12226.
110. Ashcroft, N.W.; Mermin, N.D. *Solid State Physics*; CBS Publishing Asia Ltd.: Saunders College Publishing, International Edition, 1976; 778 pp.
111. Drüde, P. Ann. Phys. **1900**, *1*, 566; **1900**, *3*, 669.
112. Burns, G. *Solid State Physics*; Academic Press: New York, 1985; 187.
113. Ashcroft, N.W.; Mermin, N.D. *Solid State Physics*; CBS Publishing Asia Ltd.: Saunders College, PA, 1976; 1–28.
114. Mott, N.F.; Davis, E.A. *Electronic Process in Noncrystalline Materials*; Oxford University Press: Oxford, 1979; 6.
115. Lee, K.; Haeger, A.J.; Cao, Y. Phys. Rev. B **1993**, *48*, 14884.
116. Mott, N.F.; Kaveh, M. Adv. Phys. **1985**, *34*, 329.

117. Yang, Y.; Pei, Q. J. Appl. Phys. **1995**, *77*, 4807.
118. Kido, J.; Hongawa, K.; Okuyama, K.; Nagal, K. Appl. Phys. Lett. **1994**, *64*, 815.
119. Friend, R.H.; Greenham, N.C. Electroluminescence in conjugated polymers. In *Handbook of Conducting Polymers*, 2nd Ed.; Skotheim, T.A., Elsenbaumer, R.L., Reynolds, J.R., Eds.; Marcel Dekker: New York, 1998; 823–845.
120. Leising, G.; Tasch, S.; Graupner, W. Fundamentals of electroluminescence in paraphenylene-type conjugate polymers and oligomers. In *Handbook of Conducting Polymers*, 2nd Ed.; Skotheim, T.A., Elsenbaumer, R.L., Reynolds, J.R., Eds.; Marcel Dekker: New York, 1998; 847–880.
121. Kraft, A.; Grimsdale, A.C.; Holmes, A.B. Angew. Chem. Int. Engl. Ed. **1998**, *7*, 403.
122. Friend, R.H.; Gymer, R.W.; Holmes, A.B.; Burroughes, J.H.; Marks, R.N.; Taliani, C.; Bradley, D.D.C.; Dos Santos, D.A.; Brédas, J.L.; Löglund, M.; Salaneck, W.R. Nature **1999**, *397*, 121.
123. Mitschke, U.; Baüerle, P. J. Mater. Chem. **2000**, *10*, 1471.
124. Sariciftci, N.S.; Braun, D.; Zhang, C.; Srdanov, V.; Heeger, A.J.; Stucky, G.; Wudl, F. Appl. Phys. Lett. **1993**, *62*, 585.
125. Yu, G.; Gao, J.; Hummelen, J.C.; Wudl, F.; Heeger, A.J. Science **1995**, *270*, 1789.
126. Hide, F.; García, M.A.; Schwartz, B.J.; Heeger, A.J. Acc. Chem. Res. **1997**, *30*, 430.
127. Tressler, N. Adv. Mater. **1999**, *11*, 363.
128. McGehee, M.D.; Heeger, A.J. Adv. Matyer. **2000**, *12*, 1655.
129. Hide, F.; Díaz-García, M.A.; Schwartz, B.; Andersson, M.; Pei, Q.; Heeger, A.J. Science **1996**, *273*, 1833.
130. Andersson, M.R.; Yu, G.; Heeger, A.J. Synth. Met. **1997**, *85*, 1275.
131. Nguyen, T.; Martini, I.B.; Schwartz, B.J. J. Phys. Chem. B **1997**, *104*, 237.
132. Harrison, N.T.; Hayes, G.R.; Philips, R.T.; Friend, R.H. Phys. Rev. Lett. **1996**, *77*, 1881.
133. Denton, G.J.; Tessler, N.; Harrison, N.T.; Friend, R.H. Phys. Rev. Lett. **1997**, *78*, 733.
134. Nguyen, T.; Doan, V.; Schwartz, B.J. J. Chem. Phys. **1999**, *110*, 4068.
135. Hsieh, B.R.; Yu, Y.; Forysthe, E.W.; Schaaf, G.M.; Feld, W.A. J. Am. Chem. Soc. **1998**, *120*, 231.
136. Würthner, F. Angew. Chem. Int. Engl. Ed. **2001**, *40*, 1037.
137. Kraft, A. Chem. Phys. Chem. **2001**, *2*, 163.
138. Fuchigami, H.; Tsumura, A.; Koezuka, H. Appl. Phys. Lett. **1993**, *63*, 1372.
139. Sirringhaus, H.; Tessler, N.; Friend, R.H. Science **1998**, *280*, 1741.
140. Kou, C.T.; Liou, T.R. Synth. Met. **1996**, *82*, 167.
141. Luzny, W.; Trznadel, M.; Pron, A. Synth. Met. **1996**, *81*, 71.
142. Mena-Osteritz, E.; Meyer, A.; Longeveld-Voss, B.M.V.; Janssen, R.A.J.; Meijer, E.W.; Braüerle, P. Angew. Chem. Int. Ed. **2000**, *39*, 2680.
143. Schön, J.H.; Dodabalpur, A.; Bao, Z.; Kloc, Ch.; Schenker, G.; Batlogg, B. Nature **2001**, *410*, 189.
144. Thieblemont, J.C.; Planche, M.F.; Petrescu, C.; Bouvier, J.M.; Bidan, G. Synth. Met. **1993**, *59*, 81.
145. Gregory, R.V.; Kimbrell, W.; Kuhn, H.H. Synth. Met. **1989**, *8*, C823.

146. Jun, H.H.; Child, A.D. Electrically conducting textiles. In *Handbook of Conducting Polymers*, 2nd Ed.; Skothein, T.A., Elsenbaumer, R.L., Reynolds, J.R., Eds.; Marcel Dekker: New York, 1998; 993–1113.

147. Gómez-Romero, P. Adv. Mat. **2001**, *13*, 163.

148. Pozniczek, J.; Kulszewicz-Bajer, I.; Zagorska, M.; Kruczala, K.; Dyrek, K.; Bielanski, A.; Pron, A.; J. Catal. **1991**, *132*, 311.

149. Hasik, M.; Turek, W.; Stochmal, E.; Lapkowski, M.; Pron, A. Mater. Res. Bull. **1995**, *30*, 1571.

150. Turek, W.; Stochmal-Pomarzanska, E.; Pron, A.; Haber, J. J. Catal. **2000**, *189*, 297.

151. Pozniczek, J.; Kulszewicz-Bajer, I.; Zagorska, M.; Kruczala, K.; Dyrek, K.; Bielanski, A.; Pron, A. J. Mol. Catal. **1991**, *69*, 223.

152. Pringsheim, E.; Terpetsschnig, E.; Wolhbeis, O.S. Anal. Chim. Acta **1997**, *357*, 257.

153. Grummt, U.-W.; Pron, A.; Zagorska, M.; Lefrant, S. Anal. Chim. Acta **1997**, *357*, 253.

154. Bossi, A.; Piletsky, S.A.; Piletska, E.V.; Righetti, P.G.; Turner, A.P.F. Anal. Chem. **2000**, *72*, 4296.

155. Pringsheim, E.; Zimin, D.; Wolfbeiss, O.S. Adv. Mater. **2001**, *13*, 819.

156. Jasty, S.; Epstein, A.J. Polym. Mat. Sci. Eng. **1995**, *72*, 565.

157. Jasty, S.; Epstein, A.J. Polym. Mat. Sci. Eng. **1995**, *72*, 565.

158. Epstein, A.J.; Joo, J.; Wu, C.-Y.; Benatar, A.; Faisst, C.F., Jr.; Zegarski, J.; MacDiarmid, A.G. In *Intrinsically Conducting Polymers: An Emerging Technology*; Aldissi, M., Ed.; Kluwer Academic Publishers: The Netherlands, 1993; 165.

159. Wessling, B. Synth. Met. **1997**, *85*, 1313.

160. Kuwabata, S.; Martin, C.R. J. Memb. Sci. **1994**, *91*, 1.

161. Anderson, M.R.; Mattes, B.R.; Reiss, H.; Kaner, R.B. Science **1991**, *252*, 1412.

162. Liang, W.; Martin, C.R. Chem. Mater. **1991**, *3*, 390.

163. Parthasarathy, R.V.; Menon, V.P.; Martin, C.R. Chem. Mater. **1997**, *9*, 560.

164. Reid, B.D.; Ebron, V.H.M.; Musselaman, I.H.; Ferraris, J.P.; Balkus Jr., K.J. J. Memb. Sci. **2002**, *195*, 181.

165. Conklin, J.A.; Su, T.M.; Huang, S.-C.; Kraner, R. Gas and liquid separation applications of polyaniline membranes. In *Handbook of Conducting Polymers*, 2nd Ed.; Skotheim, T.A., Elsenbaumer, R.L., Reynolds, J.R., Eds.; Marcel Dekker: New York, 1998; 945–961.

166. Sorosati, B., Ed. *Application of Electroactive Polymers*; Chapman and Hall: London, 1993.

167. Müller, K.; Santhanam, K.S.V.; Haas, O. Electrochemically active polymers for rechargeable batteries. Chem. Rev. **1997**, *97*, 207–281, and references therein.

168. Cottevieille, D.; Le Méhauté, A.; Challioui, C. J. Chim. Phys. **1999**, *5*, 1502.

169. Higasi, K.; Baba, H.; Rembaum, A. *Quantum Organic Chemistry*; Interscience: New York, 1965; p. 55.

References Related to Fig. 14.7

(a) Tsukamoto, J. Adv. in Phys. **1992**, *41*, 509.

(b) Naarman, H.; Theophilou, N. Synth. Met. **1987**, *22*, 1.

(c) Shirikawa, H.; Zhang, Y.X.; Okuda, T.; Sakamaki, K.; Akagi, K. Synth. Met. **1994**, *65*, 93.

(d)   Chiang, J.-C.; MacDiarmid, A.G. Synth. Met. **1986**, *13*, 193.

(e)   Epstein, A.J.; Rommelmann, H.; Bigelow, R.; Gibson, H.W.; Hoffmann, D.M.; Tanner, D.B. Phys. Rev. Lett. **1983**, *50*, 1866.

(f)   Adams, P.N.; Laughlin, P.; Monkman, A.P.; Bernhoeft, N. Solid State Commun. **1994**, *91*, 895.

(g)   Cao, Y.; Smith, P.; Heeger, A.J. Synth. Met. **1992**, *48*, 91.

(h)   Joo, J.; Oblakowski, Z.; Du, G.; Pouget, J.P.; Oh, E.J.; Weisinger, J.M.; Min, Y. Phys. Rev. B **1994**, *49*, 2977.

(i)   Wang, Z.H.; Joo, J.; Hsu, C.-H.; Pouget, J.P.; Epstein, A.J. Phys. Rev. B **1994**, *50*, 16,811.

(j)   Wang, Z.H.; Havadi, H.H.S.; Ray, A.; MacDiarmid, A.G.; Epstein, A.J. Phys. Rev. B **1990**, *42*, 5411.

(k)   Yue, J.; Wang, Z.H.; Cromack, K.R.; Epstein, A.J.; MacDiarmid, A.G. J. Am. Chem. Soc. **1991**, *113*, 2655.

(l)   Yamaura, M.; Hagiwara, T.; Iwata, K. Synth. Met. **1988**, *26*, 209.

(m)   Kohlman, R.S.; Joo, J.; Wang, Y.Z.; Pouget, J.P.; Kaneko, H.; Ishiguro, T.; Epstein, A.J. Phys. Rev. Lett. **1995**, *74*, 773; Sato, K.; Yamaura, M.; Hagiwara, T.; Murata, K.; Tokumoto, M. Synth. Met. **1991**, *40*, 35.

(n)   Ibid.

(o)   McCullough, R.D.; Williams, S.P.; Tristan-Nagle, S.; Jayaraman, M.; Ewbank, P.C.; Miller, L. Synth. Met. **1995**, *69*, 279.

(p)   Österholm, J.-E.; Passiniemi, P.; Isotalo, H.; Strubb, H. Synth. Met. **1987**, *18*, 283.

(q)   Ohnishi, T.; Noguchi, T.; Nakano, T.; Hirooka, M.; Murase, I. Synth. Met. **1991**, *41–43*, 309.

(r)   Shacklette, L.W.; Chance, R.R.; Ivory, D.M.; Miller, G.G.; Baughman, R.W. Synth. Met. **1979**, *1*, 307.

(s)   Du, G.; Prigodin, V.N.; Burns, A.; Wang, C.S.; Epstein, A.J., unpublished results.

(t)   Epstein, A.J.; Rommelmann, H.; Abkowtiz, M.; Gibson, H.W. Phys. Rev. Lett. **1981**, *47*, 1549.

(u)   Epstein, J.; Rommelmann, H.; Gibson, H.W. Phys. Rev. B **1985**, *31*, 2502.

(v)   Zuo, F.; Angelopoulos, M.; MacDiarmid, A.G.; Epstein, A.J. Phys. Rev. B **1989**, *39*, 3570.

(w)   Scott, J.C.; Pfluger, P.; Krounbi, M.T.; Street, G.B. Phys. Rev. B **1983**, *28*, 2140.

(x)   Österholm, J.-E.; Passiniemi, P.; Isotalo, H.; Strubb, H. Synth. Met. **1987**, *18*, 213.

(y)   Wnek, G.E.; Chien, J.C.; Karasz, F.E.; Lillya, C.P. Polymer **1979**, *20*, 1441.

(z)   Shacklette, L.W.; Chance, R.R.; Ivory, D.M.; Miller, G.G.; Baugbman, R.H. Synth. Met. **1979**, *1*, 307.

Index